Traffic Engineering

Third Edition

Roger P. Roess
Professor of Transportation Engineering
Polytechnic University

Elena S. Prassas
Associate Professor of Transportation Engineering
Polytechnic University

William R. McShane
Professor of Transportation and System Engineering
Polytechnic University

PEARSON
Prentice
Hall

Upper Saddle River, NJ 07458

Library of Congress Cataloging-in-Publication Data on File

Vice President and Editorial Director, ECS: *Marcia Horton*
Vice President and Director of Production and Manufacturing, ESM: *David W. Riccardi*
Acquisitions Editor: *Laura Fischer*
Editorial Assistant: *Andrea Messineo*
Executive Managing Editor: *Vince O'Brien*
Managing Editor: *David A. George*
Production Editor: *Rose Kernan*
Director of Creative Services: *Paul Belfanti*
Creative Director: *Carole Anson*
Art Director: *Jayne Conte*
Cover Designer: *Bruce Kenselaar*
Art Editor: *Connie Long*
Manufacturing Manager: *Trudy Pisciotti*
Manufacturing Buyer: *Lisa McDowell*
Marketing Manager: *Holly Stark*

© 2004 by Pearson Education, Inc.
Pearson Prentice Hall
Pearson Education, Inc.
Upper Saddle River, New Jersey 07458

The author and publisher of this book have used their best efforts in preparing this book. These efforts include the development, research, and testing of the theories and programs to determine their effectiveness. The author and publisher shall not be liable in any event for incidental or consequential damages with, or arising out of, the furnishing, performance, or use of these programs.

Printed in the United States of America

10 9 8 7 6 5 4 3

ISBN 0-13-142471-8

Pearson Education Ltd., *London*
Pearson Education Australia Pty., Ltd., *Sydney*
Pearson Education Singapore, Pte. Ltd.
Pearson Education North Asia Ltd., *Hong Kong*
Pearson Education Canada, Inc., *Toronto*
Pearson Educación de Mexico, S.A. de C.V.
Pearson Education—Japan, *Tokyo*
Pearson Education Malaysia, Pte, Ltd.
Pearson Education, Inc., *Upper Saddle River, New Jersey*

Contents

Preface

Traffic engineering covers a broad range of engineering applications with a common focus: the nation's system of highways and streets. Often defined as the nation's "lifeblood circulation system," this important part of the national infrastructure supports the vast majority of inter- and intra-city movement of both people and goods. Thus, the system plays a role in every important aspect of our society—including the economy, the environment, assurance of public safety and security, basic mobility for all societal functions, and basic access to the most remote regions of the country.

Traffic engineering involves a variety of engineering and management skills—including planning, management, design, construction, operation, control, maintenance, and system optimization. Because the focus of the traffic engineer's work is a most visible part of the public infrastructure, it is a field that also involves politics at virtually every level of government. Thus, the traffic engineer is called on to exercise a broad range of skills and must be sensitive to a wide range of issues to be effective.

This is the third edition of this textbook. It incorporates new standards and analysis techniques from the *Manual on Uniform Traffic Control Devices* (Millennium Edition), the *Highway Capacity Manual* (Fourth Edition, 2000), the *Policy on Geometric Design of Highways and Streets* (Fourth Edition, 2001), and other current standards. Like the first two editions, the text can be used for a survey course at the undergraduate or graduate level, as well as for a series of more detailed courses. At Polytechnic, the text is used in a two-course undergraduate sequence and a series of four graduate courses.

The text is organized in four major functional parts:

- Part I: Components of the Traffic System and their Characteristics
- Part II: Traffic Studies and Programs
- Part III: Applications to Freeway and Rural Highway Systems
- Part IV: Applications to Urban and Suburban Street Systems

Chapters have been added on Intelligent Transportation Systems; Parking, Signing, and Marking; Analysis of Unsignalized Intersections; and Arterial Planning and Management. Additional material on functional and geometric design and on marking and signing of facilities has also been added.

As in the first two editions, the text contains many sample problems and a wide variety of homework and project assignments that can be used in conjunction with course material. A solutions manual is available. The authors hope that faculty, practicing professionals, and students find this text useful and informative, and they invite comments and/or criticisms that will help them continue to improve the material.

The authors wish to thank the following reviewers for their comments and helpful suggestions: Carroll J. Messer, Texas A&M University; Emily Parentella, California State University, Long Beach; Mark Virkler, University of Missouri—Columbia; and William Sproule, Michigan Technological University.

ROGER P. ROESS
ELENA S. PRASSAS
WILLIAM R. MCSHANE

CHAPTER
1

Introduction to Traffic Engineering

1.1 Traffic Engineering as a Profession

The Institute of Transportation Engineers defines traffic engineering as a subset of transportation engineering as follows [1]:

Transportation engineering is the application of technology and scientific principles to the planning, functional design, operation, and management of facilities for any mode of transportation in order to provide for the safe, rapid, comfortable, convenient, economical, and environmentally compatible movement of people and goods.

and:

Traffic engineering is that phase of transportation engineering which deals with the planning, geometric design and traffic operations of roads, streets, and highways, their networks, terminals, abutting lands, and relationships with other modes of transportation.

These definitions represent a broadening of the profession to include multimodal transportation systems and options, and to include a variety of objectives in addition to the traditional goals of safety and efficiency.

1.1.1 Safety: The Primary Objective

The principal goal of the traffic engineer remains the provision of a safe system for highway traffic. This is no small concern. In recent years, fatalities on U.S. highways have ranged between 40,000 and 43,000 per year. While this is a reduction from the highs experienced in the 1970s, when highway fatalities reached over 55,000 per year, it continues to represent a staggering number. More Americans have been killed on U.S. highways than in all of the wars in which the nation has participated, including the Civil War.

While total highway fatalities per year have remained relatively constant over the past two decades, accident rates based on vehicle-miles traveled have consistently declined. That is because U.S. motorists continue to drive more miles each year. With a stable total number of fatalities, the increasing number of annual vehicle-miles traveled produces a declining fatality rate.

1

Improvements in fatality rates reflect a number of trends, many of which traffic engineers have been instrumental in implementing. Stronger efforts to remove dangerous drivers from the road have yielded significant dividends in safety. Driving under the influence (DUI) and driving while intoxicated (DWI) offenses are more strictly enforced, and licenses are suspended or revoked more easily as a result of DUI/DWI convictions, poor accident record, and/or poor violations record. Vehicle design has greatly improved (encouraged by several acts of Congress requiring certain improvements). Today's vehicles feature padded dashboards, collapsible steering columns, seat belts with shoulder harnesses, air bags (some vehicles now have as many as eight), and antilock braking systems. Highway design has improved through the development and use of advanced barrier systems for medians and roadside areas. Traffic control systems communicate better and faster, and surveillance systems can alert authorities to accidents and breakdowns in the system.

Despite this, however, over 40,000 people per year still die in traffic accidents. The objective of safe travel is always number one and is never finished for the traffic engineer.

1.1.2 Other Objectives

The definitions of transportation and traffic engineering highlight additional objectives:

- Speed
- Comfort
- Convenience
- Economy
- Environmental compatibility

Most of these are self-evident desires of the traveler. Most of us want our trips to be fast, comfortable, convenient, cheap, and in harmony with the environment. All of these objectives are also relative and must be balanced against each other and against the primary objective of safety.

While speed of travel is much to be desired, it is limited by transportation technology, human characteristics, and the need to provide safety. Comfort and convenience are generic terms and mean different things to different people. Comfort involves the physical characteristics of vehicles and roadways, and is influenced by our perception of safety. Convenience relates more to the ease with which trips are made and the ability of transport systems to accommodate all of our travel needs at appropriate times. Economy is also relative. There is little in modern transportation systems that can be termed "cheap." Highway and other transportation systems involve massive construction, maintenance, and operating expenditures, most of which are provided through general and user taxes and fees. Nevertheless, every engineer, regardless of discipline, is called upon to provide the best possible systems for the money.

Harmony with the environment is a complex issue that has become more important over time. All transportation systems have some negative impacts on the environment. All produce air and noise pollution in some forms, and all utilize valuable land resources. In many modern cities, transportation systems utilize as much as 25% of the total land area. "Harmony" is achieved when transportation systems are designed to minimize negative environmental impacts, and where system architecture provides for aesthetically pleasing facilities that "fit in" with their surroundings.

The traffic engineer is tasked with all of these goals and objectives and with making the appropriate trade-offs to optimize both the transportation systems and the use of public funds to build, maintain, and operate them.

1.1.3 Responsibility, Ethics, and Liability in Traffic Engineering

The traffic engineer has a very special relationship with the public at large. Perhaps more than any other type of engineer, the traffic engineer deals with the daily safety of a large segment of the public. Although it can be argued that any engineer who designs a product has this responsibility, few engineers have so many people using their product so routinely and frequently and depending upon it so totally. Therefore, the traffic engineer also has a special obligation to employ the available knowledge and state of the art within existing resources to enhance public safety.

The traffic engineer also functions in a world in which a number of key participants do not understand the traffic and transportation issues or how they truly affect a particular project. These include elected and appointed officials with decision-making power, the general public, and other professionals with whom traffic engineers work on an overall project team effort. Because all of us interface regularly with the transportation system, many overestimate their understanding of transportation and traffic issues. The traffic engineer must deal productively with problems associated with naïve assumptions, plans and designs that are oblivious to transportation and traffic needs, oversimplified analyses, and understated impacts.

Like all engineers, traffic engineers must understand and comply with professional ethics codes. Primary codes of ethics for traffic engineers are those of the National Society of Professional Engineers and the American Society of Civil Engineers. The most up-to-date versions of each are available on-line. In general, good professional ethics requires that traffic engineers work only in their areas of expertise; do all work completely and thoroughly; be completely honest with the general public, employers, and clients; comply with all applicable codes and standards; and work to the best of their ability. In traffic engineering, the pressure to understate negative impacts of projects, sometimes brought to bear by clients who wish a project to proceed and employers who wish to keep clients happy, is a particular concern. As in all engineering professions, the pressure to minimize costs must give way to basic needs for safety and reliability.

Experience has shown that the greatest risk to a project is an incomplete analysis. Major projects have been upset because an impact was overlooked or analysis oversimplified. Sophisticated developers and experienced professionals know that the environmental impact process calls for a fair and complete statement of impacts and a *policy decision by the reviewers* on accepting the impacts, given an overall good analysis report. The process does not require zero impacts; it does, however, call for clear and complete disclosure of impacts so that policy makers can make informed decisions. Successful challenges to major projects are almost always based on flawed analysis, not on disagreements with policy makers. Indeed, such disagreements are not a valid basis for a legal challenge to a

project. In the case of the Westway Project proposed in the 1970s for the west side of Manhattan, one of the bases for legal challenge was that the impact of project construction on striped bass in the Hudson River had not been properly identified or disclosed.

The traffic engineer also has a responsibility to protect the community from liability by good practice. There are many areas in which agencies charged with traffic and transportation responsibilities can be held liable. These include (but are not limited to):

- Placing control devices that do not conform to applicable standards for their physical design and placement.

- Failure to maintain devices in a manner that ensures their effectiveness; the worst case of this is a "dark" traffic signal in which no indication is given due to bulb or other device failure.

- Failure to apply the most current standards and guidelines in making decisions on traffic control, developing a facility plan or design, or conducting an investigation.

- Implementing traffic regulations (and placing appropriate devices) without the proper legal authority to do so.

A historic standard has been that "due care" be exercised in the preparation of plans, and that determinations made in the process be reasonable and "not arbitrary." It is generally recognized that professionals must make value judgments, and the terms "due care" and "not arbitrary" are continually under legal test.

The fundamental ethical issue for traffic engineers is to provide for the public safety through positive programs, good practice, knowledge, and proper procedure. The negative (albeit important) side of this is the avoidance of liability problems.

1.2 Transportation Systems and their Function

Transportation systems are a major component of the U.S. economy and have an enormous impact on the shape of the society and the efficiency of the economy

- 221.5 Million Registered Vehicles
- 191.0 Million Licensed Drivers
- 2.46 Trillion Vehicle-Miles Traveled
- 3.92 Million Miles of Paved Highway
- $61.6 Billion in State Road User Taxes Collected
- $30.3 Billion in Federal User Taxes in the Highway Trust Fund
- 41,471 Fatalities in 6.3 Million Police-Reported Accidents
- 98% of all Person-Trips Made by Highway

Figure 1.1: Fundamental Highway Traffic Statistics (2000)

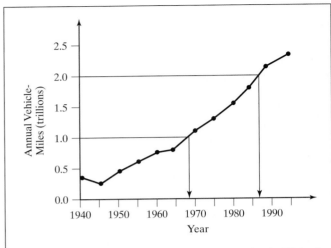

Figure 1.2: Annual Vehicle-Miles Traveled in the United States (1940–2000)

in general. Figure 1.1 illustrates some key statistics for the U.S. highway system for the base year 2000.

America moves on its highways. While public transportation systems are of major importance in large urban areas such as New York, Boston, Chicago, and San Francisco, it is clear that the vast majority of person-travel as well as a large proportion of freight traffic is entirely dependent on the highway system. The system is a major economic force in its own right: Over $90 billion per year is collected by state and federal governments directly from road users in the form of focused user taxes and fees. Such taxes and fees include excise taxes on gasoline and other fuels, registration fees, commercial vehicles fees, and others. The total, however, does not include state, local, and federal general levies that also affect road users. The general state and local sales taxes on vehicle purchases, fuels, parts and labor, etc. are *not* included in this total. Further, well over $100 billion per year is expended by all units of government to plan, build, maintain, and operate highways.

Moreover, the American love affair with the automobile has grown consistently since the 1920s, when Henry Ford's Model T made the car accessible to the average wage earner. This growth has survived wars, gasoline embargoes, depressions, recessions, and almost everything else that has happened in society. As seen in Figure 1.2, annual vehicle-miles traveled reached the 1 trillion mark in 1968 and the 2 trillion mark in 1987. If the trend continues, the 3 trillion mark is not too far in our future.

This growth pattern is one of the fundamental problems to be faced by traffic engineers. Given the relative maturity of our highway systems and the difficulty

faced in trying to add system capacity, particularly in urban areas, the continued growth in vehicle-miles traveled leads directly to increased congestion on our highways. The inability to simply build additional capacity to meet the growing demand creates the need to address alternative modes, fundamental alterations in demand patterns, and management of the system to produce optimal results.

1.2.1 The Nature of Transportation Demand

Transportation demand is directly related to land-use patterns and to available transportation systems and facilities. Figure 1.3 illustrates the fundamental relationship, which is circular and ongoing. Transportation demand is generated by the types, amounts, and intensity of land use, as well as its location. The daily journey to work, for example, is dictated by the locations of the worker's residence and employer and the times that the worker is on duty.

Transportation planners and traffic engineers attempt to provide capacity for observed or predicted travel demand by building transportation systems. The improvement of transportation systems, however, makes the adjacent and nearby lands more accessible and, therefore, more attractive for development. Thus, building new

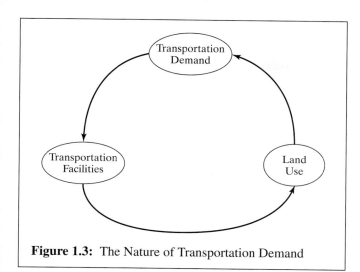

Figure 1.3: The Nature of Transportation Demand

transportation facilities leads to further increases in land-use development, which (in turn) results in even higher transportation demands. This circular, self-reinforcing characteristic of traffic demand creates a central dilemma: building additional transportation capacity invariably leads to incrementally increased travel demands.

In many major cities, this has led to the search for more efficient transportation systems, such as public transit and car-pooling programs. In some of the largest cities, providing additional system capacity on highways is no longer an objective, as such systems are already substantially choking in congestion. In these places, the emphasis shifts to improvements within existing highway rights-of-way and to the elimination of bottleneck locations (without adding to overall capacity). Other approaches include staggered work hours and work days to reduce peak hour demands, and even more radical approaches involve development of satellite centers outside of the central business district (CBD) to spatially disperse highly directional demands into and out of city centers.

On the other hand, demand is not constrained by capacity in all cities, and the normal process of attempting to accommodate demand as it increases is feasible in these areas. At the same time, the circular nature of the travel/demand relationship will lead to congestion if care is not taken to manage both capacity and demand to keep them within tolerable limits.

It is important that the traffic engineer understand this process. It is complex and cannot be stopped at any moment in time. Demand-prediction techniques (not covered in this text) must start and stop at arbitrary points in time. The real process is on-going, and as new or improved facilities are provided, travel demand is constantly changing. Plans and proposals must recognize both this reality and the professional's inability to precisely predict its impacts. *A 10-year traffic demand forecast that comes within approximately ±20% of the actual value is considered a significant success.* The essential truth, however, is that traffic engineers cannot simply build their way out of congestion.

If anything, we still tend to underestimate the impact of transportation facilities on land-use development. Often, the increase in demand is hastened by development occurring simply as a result of the planning of a new facility.

One of the classic cases occurred on Long Island, New York. As the Long Island Expressway was built, the development of suburban residential communities lurched forward in anticipation. While the expressway's link to Exit 7 was being constructed, new homes were being built at the anticipated Exit 10, even though the facility would not be open to that point for several years. The result was that as the expressway was completed section by section, the 20-year anticipated demand was being achieved within a few years, or even months. This process has been repeated in many cases throughout the nation.

1.2.2 Concepts of Mobility and Accessibility

Transportation systems provide the nation's population with both mobility and accessibility. The two concepts are strongly interrelated but have distinctly different elements. *Mobility* refers to the ability to travel to many different destinations, while *accessibility* refers to the ability to gain entry to a particular site or area.

Mobility gives travelers a wide range of choices as to where to go to satisfy particular needs. Mobility allows shoppers to choose from among many competing shopping centers and stores. Similarly, mobility provides the traveler with many choices for all kinds of trip purposes, including recreational trips, medical trips, educational trips, and even the commute to work. The range of available choices is enabled by having an effective transportation network that connects to many alternative trip destinations within a reasonable time, with relative ease, and at reasonable cost.

Accessibility is a major factor in the value of land. When land can be accessed by many travelers from many potential origins, it is more desirable for development and, therefore, more valuable. Thus, proximity of land to major highways and public transportation facilities is a major factor determining its value.

Mobility and accessibility may also refer to different portions of a typical trip. Mobility focuses on the through portion of trips and is most affected by the effectiveness of through facilities that take a traveler from one general area to another. Accessibility requires the ability to make a transfer from the transportation system to the particular land parcel on which the desired activity is taking place. Accessibility, therefore, relies heavily on transfer facilities, which include parking for vehicles, public transit stops, and loading zones.

As is discussed in Chapter 3, most transportation systems are structured to separate mobility and access functions, as the two functions often compete and are not necessarily compatible. In highway systems, mobility is provided by high-type facilities, such as freeways, expressways, and primary and secondary arterials. Accessibility is generally provided by local street networks. Except for limited-access facilities, which serve only through vehicles (mobility), most other classes of highway serve both functions to some degree. Access maneuvers, however (e.g., parking and unparking a vehicle, vehicles entering and leaving off-street parking via driveways, buses stopping to pick up or discharge passengers, trucks stopped to load and/or unload goods), retard the progress of through traffic. High-speed through traffic, on the other hand, tends to make such access functions more dangerous.

A good transportation system must provide for both mobility and accessibility and should be designed to separate the functions to the extent possible to ensure both safety and efficiency.

1.2.3　People, Goods, and Vehicles

The most common unit used by the traffic engineer is "vehicles." Highway systems are planned, designed, and operated to move vehicles safely and efficiently from place to place. Yet the movement of vehicles is not the objective; the goal is the movement of the people and goods that occupy vehicles.

Modern traffic engineering now focuses more on people and goods. While lanes must be added to a freeway to increase its capacity to carry vehicles, its person-capacity can be increased by increasing the average vehicle occupancy. Consider a freeway lane with a capacity of 2,000 vehicles per hour (veh/h). If each vehicle carries one person, the lane has a capacity of 2,000 persons/hour as well. If the average car occupancy is increased to 2.0 persons/vehicle, the capacity in terms of people is doubled to 4,000 persons/hour. If the lane were established as an exclusive bus lane, the vehicle-capacity might be reduced to 1,000 veh/h due to the larger size and poorer operating characteristics of buses as compared with automobiles. However, if each bus carries 50 passengers, the people-capacity of the lane is increased to 50,000 persons per hour.

The efficient movement of goods is also vital to the general economy of the nation. The benefits of centralized and specialized production of various products are possible only if raw materials can be efficiently shipped to manufacturing sites and finished products can be efficiently distributed throughout the nation and the world for consumption. While long-distance shipment of goods and raw materials is often accomplished by water, rail, or air transportation, the final leg of the trip to deliver a good to the local store or the home of an individual consumer generally takes place on a truck using the highway system. Part of the accessibility function is the provision of facilities that allow trucks to be loaded and unloaded with minimal disruption to through traffic and the accessibility of people to a given site.

The medium of all highway transportation is the vehicle. The design, operation, and control of highway systems relies heavily on the characteristics of the vehicle and of the driver. In the final analysis, however, the objective is to move people and goods, not vehicles.

1.2.4 Transportation Modes

While the traffic engineer deals primarily with highways and highway vehicles, there are other important transportation systems that must be integrated into a cohesive national, regional, and local transportation network. Table 1.1 provides a comprehensive listing of various transportation modes and their principal uses.

The traffic engineer deals with all of these modes in a number of ways. All over-the-road modes—automobile, bus transit, trucking—are principal users of highway systems. Highway access to rail and air terminals is critical to their effectiveness, as is the design of specific transfer facilities for both people and freight. General access, internal circulation, parking, pedestrian areas, and terminals for both people and freight are all projects requiring the expertise of the traffic engineer.

Moreover, the effective integration of multimodal transportation systems is a major goal in maximizing efficiency and minimizing costs associated with all forms of travel.

1.3 Highway Legislation and History in the United States

The development of highway systems in the United States is strongly tied to federal legislation that supports and regulates much of this activity. Key historical and legislative actions are discussed in the sections that follow.

1.3.1 The National Pike and the States' Rights Issue

Before the 1800s, roads were little more than trails cleared through the wilderness by adventurous travelers and explorers. Private roadways began to appear in the latter part of the 1700s. These roadways ranged in quality and length from cleared trails to plank roadways. They were built by private owners, and fees were charged for their use. At points where fees were to be collected, a barrier usually consisting of a single crossbar was mounted on a swiveling stake, referred to as a "pike." When the fee was collected, the pike would be swiveled or turned, allowing the traveler to proceed. This early process gave birth to the

term "turnpike," often used to describe toll roadways in modern times.

The National Pike

In 1811, the construction of the first national roadway was begun under the direct supervision of the federal government. Known as the "national pike" or the "Cumberland Road," this facility stretched for 800 miles from Cumberland MD in the east, to Vandalia IL in the west. A combination of unpaved and plank sections, it was finally completed in 1852 at a total cost of $6.8 million. A good deal of the original route is now a portion of U.S. Route 40.

Highways as a States' Right

The course of highway development in the United States, however, was forever changed as a result of an 1832 Supreme Court case brought by the administration of President Andrew Jackson. A major proponent of states' rights, the Jackson Administration petitioned the court claiming that the U.S. constitution did not specifically define transportation and roadways as federal functions; they were, therefore, the responsibility of the individual states. The Supreme Court upheld this position, and the principal administrative responsibility for transportation and highways was forevermore assigned to state governments.

The Governmental Context

If the planning, design, construction, maintenance, and operation of highway systems is a state responsibility, what is the role of federal agencies—for example, the U.S. Department of Transportation and its components, such as the Federal Highway Administration, the National Highway Safety Administration, and others in these processes?

The federal government asserts its overall control of highway systems through the power of the purse strings. The federal government provides massive funding for the construction, maintenance, and operation of highway and other transportation systems. States are not *required* to follow federal mandates and standards but must do so to qualify for federal funding of projects.

Table 1.1: Transportation Modes

Mode	Typical Function	Approximate Range of Capacities*
Urban People-Transportation Systems		
Automobile	Private personal transportation; available on demand for all trips.	1–6 persons/vehicle; approx. 2,000 veh/h per freeway lane; 400–700 veh/h per arterial lane.
Taxi/For-Hire Vehicles	Private or shared personal transportation; available by prearrangement or on call.	1–6 persons/vehicle; total capacity limited by availability.
Local Bus Transit	Public transportation along fixed routes on a fixed schedule; low speed with many stops.	40–70 persons/bus; capacity limited by schedule; usually 100–5,000 persons/h/route.
Express Bus Transit	Public transportation along fixed routes on a fixed schedule; higher speed with few intermediate stops.	40–50 persons/bus (no standees); capacity limited by schedule.
Para-transit	Public transportation with flexible routing and schedules, usually available on call.	Variable seating capacity depends upon vehicle design; total capacity dependent on number of available vehicles.
Light Rail	Rail service using 1–2 car units along fixed routes with fixed schedules.	80–120 persons/car; up to 15,000 persons/h/route.
Heavy Rail	Heavy rail vehicles in multi-car trains along fixed routes with fixed schedules on fully separated rights-of-way in tunnels, on elevated structures, or on the surface.	150–300 persons/car depending on seating configuration and standees; up to 60,000 persons per track.
Ferry	Waterborne public transportation for people and vehicles along fixed routes on fixed schedules.	Highly variable with ferry design and schedule.
Intercity People-Transportation Systems		
Automobile	Private transportation available on demand for all trip purposes.	Same as urban automobile.
Intercity Bus	Public transportation along a fixed intercity route on a fixed (and usually limited) schedule. Provides service to a central terminal location in each city.	40–50 passengers per bus; schedules highly variable.

*Ranges cited represent typical values, not the full range of possibilities.

(Continued)

Table 1.1: Transportation Modes (*Continued*)

Mode	Typical Function	Approximate Range of Capacities*
Intercity People-Transportation Systems (Cont.)		
Railroad	Passenger intercity-rail service on fixed routes on a fixed (and usually limited) schedule. Provides service to a central terminal location or locations within each city.	500–1,000 passengers per train, depending upon configuration; schedules highly variable.
Air	A variety of air-passenger services from small commuter planes to jumbo jets on fixed routes and fixed schedules.	From 3–4 passengers to 500 passengers per aircraft, depending upon size and configuration. Schedules depend upon destination and are highly variable.
Water	Passenger ship service often associated with on-board vacation packages on fixed routes and schedules.	Ship capacity highly variable from several hundred to 3,500 passengers; schedules often extremely limited.
Urban and Intercity Freight Transportation		
Long-Haul Trucks	Single-, double-, and triple tractor-trailer combinations and large single-unit trucks provide over-the-road intercity service, by arrangement.	
Local Trucks	Smaller trucks provide distribution of goods and services throughout urban areas.	Hauling capacity of all freight modes varies widely with the design of the vehicle (or pipeline) and limitations on fleet size and schedule availability.
Railroad	Intercity haulage of bulk commodities with some local distribution to locations with rail sidings.	
Water	International and intercity haulage of bulk commodities on a variety of container ships and barges.	
Air Freight	International and intercity haulage of small and moderately sized parcels and/or time-sensitive and/or high-value commodities where high cost is not a disincentive.	
Pipelines	Continuous flow of fluid or gaseous commodities; intercity and local distribution networks possible.	

*Ranges cited represent typical values, not the full range of possibilities.

Thus, the federal government does not force a state to participate in federal-aid transportation programs. If it chooses to participate, however, it must follow federal guidelines and standards. As no state can afford to give up this massive funding source, the federal government imposes strong control of policy issues and standards.

The federal role in highway systems has four major components:

1. Direct responsibility for highway systems on federally owned lands, such as national parks and Native American reservations.

2. Provision of funding assistance in accord with current federal-aid transportation legislation.

3. Development of planning, design, and other relevant standards and guidelines that must be followed to qualify for receipt of federal-aid transportation funds.

4. Monitoring and enforcing compliance with federal standards and criteria, and the use of federal-aid funds.

State governments have the primary responsibility for the planning, design, construction, maintenance, and operation of highway systems. These functions are generally carried out through a state department of transportation or similar agency. States have:

1. Full responsibility for administration of highway systems.

2. Full responsibility for the planning, design, construction, maintenance, and operation of highway systems in conformance with applicable federal standards and guidelines.

3. The right to delegate responsibilities for local roadway systems to local jurisdictions or agencies.

Local governments have general responsibility for local roadway systems as delegated in state law. In general, local governments are responsible for the planning, design, construction, maintenance, and control of local roadway systems. Often, assistance from state programs and agencies is available to local governments in fulfilling these functions. At intersections of state highways with local roadways, it is generally the state that has the responsibility to control the intersection.

Local organizations for highway functions range from a full highway or transportation department to local police to a single professional traffic or city engineer.

There are also a number of special situations across the United States. In New York State, for example, the state constitution grants "home rule" powers to any municipality with a population in excess of 1,000,000 people. Under this provision, New York City has full jurisdiction over all highways within its borders, including those on the state highway system.

1.3.2 Key Legislative Milestones

Federal-Aid Highway Act of 1916

The Federal-Aid Highway Act of 1916 was the first allocation of federal-aid highway funds for highway construction by the states. It established the "A-B-C System" of primary, secondary, and tertiary federal-aid highways, and provided 50% of the funding for construction of highways in this system. Revenues for federal aid were taken from the federal general fund, and the act was renewed every two to five years (with increasing amounts dedicated). No major changes in funding formulas were forthcoming for a period of 40 years.

Federal-Aid Highway Act of 1934

In addition to renewing funding for the A-B-C System, this act authorized states to use up to 1.5% of federal-aid funds for planning studies and other investigations. It represented the entry of the federal government into highway planning.

Federal-Aid Highway Act of 1944

This act contained the initial authorization of what became the National System of Interstate and Defense Highways. No appropriation of funds occurred, however, and the system was not initiated for another 12 years.

Federal-Aid Highway Act of 1956

The authorization and appropriation of funds for the implementation of the National System of Interstate and Defense Highways occurred in 1956. The act also set the federal share of the cost of the Interstate System at 90%, the first major change in funding formulas since 1916. Because of the major impact on the amounts of federal funds to be spent, the act also created the *Highway Trust Fund* and enacted a series of road-user taxes to provide it with revenues. These taxes included excise taxes on motor fuels, vehicle purchases, motor oil, and replacement parts. Most of these taxes, except for the federal fuel tax, were dropped during the Nixon Administration. The monies housed in the Highway Trust Fund may be disbursed only for purposes authorized by the current federal-aid highway act.

Federal-Aid Highway Act of 1970

Also known as the Highway Safety Act of 1970, this legislation increased the federal subsidy of non-Interstate highway projects to 70% and required all states to implement highway safety agencies and programs.

Federal-Aid Highway Act of 1983

This act contained the "Interstate trade-in" provision that allows states to "trade in" federal-aid funds designated for urban Interstate projects for alternative transit systems. This historic provision was the first to allow road-user taxes to be used to pay for public transit improvements.

ISTEA and TEA-21

The single largest overhaul of federal-aid highway programs occurred with the passage of the Intermodal Surface Transportation Efficiency Act (ISTEA) in 1991 and its successor, the Transportation Equity Act for the 21st Century (TEA-21) in 1998.

Most importantly, these acts combined federal-aid programs for all modes of transportation and greatly liberalized the ability of state and local governments to make decisions on modal allocations. Key provisions of ISTEA included:

1. Greatly increased local options in the use of federal-aid transportation funds.

2. Increased the importance and funding to Metropolitan Planning Organizations (MPOs) and required that each state maintain a state transportation improvement plan (STIP).

3. Tied federal-aid transportation funding to compliance with the Clean Air Act and its amendments.

4. Authorized $38 billion for a 155,000-mile National Highway System.

5. Authorized an additional $7.2 million to complete the Interstate System and $17 billion to maintain it as part of the National Highway System.

6. Extended 90% federal funding of Interstate-eligible projects.

7. Combined all other federal-aid systems into a single surface transportation system with 80% federal funding.

8. Allowed (for the first time) the use of federal-aid funds in the construction of toll roads.

TEA-21 followed in kind, increasing funding levels, further liberalizing local options for allocation of funds, further encouraging intermodality and integration of transportation systems, and continuing the link between compliance with clean-air standards and federal transportation funding.

The creation of the National Highway System answered a key question that had been debated for years: what comes after the Interstate System? The new, expanded NHS is not limited to freeway facilities and is over three times the size of the Interstate System, which becomes part of the NHS.

1.3.3 The National System of Interstate and Defense Highways

The "Interstate System" has been described as the largest public works project in the history of mankind. In 1919, a young army officer, Dwight Eisenhower, was tasked with moving a complete battalion of troops and military equipment from coast to coast on the nation's highways to determine their utility for such movements in a time of potential war. The trip took months and left the young officer with a keen appreciation for the need to develop a national roadway system. It was no accident that the Interstate System was initiated in the administration of President Dwight Eisenhower, nor that the system now bears his name.

After the end of World War II, the nation entered a period of sustained prosperity. One of the principal signs of that prosperity was the great increase in auto ownership along with the expanding desire of owners to use their cars for daily commuting and for recreational travel. Motorists groups, such as the American Automobile Association (AAA), were formed and began substantial lobbying efforts to expand the nation's highway systems. At the same time, the over-the-road trucking industry was making major inroads against the previous rail monopoly on intercity freight haulage. Truckers also lobbied strongly for improved highway systems. These substantial pressures led to the inauguration of the Interstate System in 1956.

The System Concept

Authorized in 1944 and implemented in 1956, the National System of Interstate and Defense Highways is a 42,500-mile national system of multilane, limited-access facilities. The system was designed to connect all standard metropolitan statistical areas (SMSAs) with 50,000 or greater population with a continuous system of limited-access facilities. The allocation of 90% of the cost of the system to the federal government was justified on the basis of the potential military use of the system in wartime.

System Characteristics

Key characteristics of the Interstate System include the following:

1. All highways have at least two lanes for the exclusive use of traffic in each direction.
2. All highways have full control of access.
3. The system must form a closed loop: all Interstate highways must begin and end at a junction with another Interstate highway.
4. North-south routes have odd two-digit numbers (e.g., I-95).
5. East-west routes have even two-digit numbers (e.g., I-80).

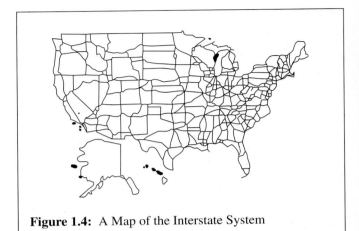

Figure 1.4: A Map of the Interstate System

6. Interstate routes serving as bypass loops or acting as a connector to a primary Interstate facility have three-digit route numbers, with the last two digits indicating the primary route.

A map of the Interstate System is shown in Figure 1.4.

Status and Costs

By 1994, the system was 99.4% complete. Most of the unfinished sections were not expected to ever be completed for a variety of reasons. The total cost of the system was approximately $125 billion.

The impact of the Interstate System on the nation cannot be understated. The system facilitated and enabled the rapid suburbanization of the United States by providing a means for workers to commute from suburban homes to urban jobs. The economy of urban centers suffered as shoppers moved in droves from traditional central business districts (CBDs) to suburban malls.

The system also had serious negative impacts on some of the environs through which it was built. Following the traditional theory of benefit-cost, urban sections were often built through the low-income parts of communities where land was the cheapest. The massive Interstate highway facilities created physical barriers, partitioning many communities, displacing residents, and separating others from their schools, churches, and local shops. Social unrest resulted in several parts of the country, which eventually resulted in important modifications to the public hearing process and in the ability of local opponents to legally stop many urban highway projects.

Between 1944 and 1956, a national debate was waged over whether the Interstate System should be built into and out of urban areas, or whether all Interstate facilities should terminate in ring roads built around urban areas. Proponents of the ring-road option (including Robert Moses) argued that building these roadways into and out of cities would lead to massive urban congestion. On the other side, the argument was that most of the road users who were paying for the system through their road user taxes lived in urban areas and should be served. The latter view prevailed, but the predicted rapid growth of urban congestion also became a reality.

1.4 Elements of Traffic Engineering

There are a number of key elements of traffic engineering:

1. Traffic studies and characteristics
2. Performance evaluation
3. Facility design
4. Traffic control
5. Traffic operations
6. Transportation systems management
7. Integration of intelligent transportation system technologies

Traffic studies and characteristics involve measuring and quantifying various aspect of highway traffic. Studies focus on data collection and analysis that is used to characterize traffic, including (but not limited to) traffic volumes and demands, speed and travel time, delay, accidents, origins and destinations, modal use, and other variables.

Performance evaluation is a means by which traffic engineers can rate the operating characteristics of individual sections of facilities and facilities as a whole in relative terms. Such evaluation relies on measures of performance quality and is often stated in terms of "levels of service." Levels of service are letter grades, from A to F, describing how well a facility is operating using specified performance criteria. Like grades in a course, A is very good, while F connotes failure (on some level). As part of performance evaluation, the *capacity* of highway facilities must be determined.

Facility design involves traffic engineers in the functional and geometric design of highways and other traffic facilities. Traffic engineers, per se, are not involved in the structural design of highway facilities but should have some appreciation for structural characteristics of their facilities.

Traffic control is a central function of traffic engineers and involves the establishment of traffic regulations and their communication to the driver through the use of traffic control devices, such as signs, markings, and signals.

Traffic operations involves measures that influence overall operation of traffic facilities, such as one-way street systems, transit operations, curb management, and surveillance and network control systems.

Transportation systems management (TSM) involves virtually all aspects of traffic engineering in a focus on optimizing system capacity and operations. Specific aspects of TSM include high-occupancy vehicle priority systems, car-pooling programs, pricing strategies to manage demand, and similar functions.

Intelligent transportation systems (ITS) refers to the application of modern telecommunications technology to the operation and control of transportation systems. Such systems include automated highways, automated toll-collection systems, vehicle-tracking systems, in-vehicle GPS and mapping systems, automated enforcement of traffic lights and speed laws, smart control devices, and others. This is a rapidly emerging family of technologies with the potential to radically alter the way we travel as well as the way in which transportation professionals gather information and control facilities. While the technology continues to expand, society will grapple with the substantial "big brother" issues that such systems invariably create.

This text contains material related to all of these components of the broad and complex profession of traffic engineering.

1.5 Modern Problems for the Traffic Engineer

We live in a complex and rapidly developing world. Consequently, the problems that traffic engineers are involved in evolve rapidly.

Urban congestion has been a major issue for many years. Given the transportation demand cycle, it is not always possible to solve congestion problems through expansion of capacity. Traffic engineers therefore are involved in the development of programs and strategies to manage demand in both time and space and to discourage growth where necessary. A real question is not "how much capacity is needed to handle demand?" but rather, "how many vehicles and/or people can be allowed to enter congested areas within designated time periods?"

Growth management is a major current issue. A number of states have legislation that ties development permits to level-of-service impacts on the highway and transportation system. Where development will cause substantial deterioration in the quality of traffic service, either such development will be disallowed or the developer will be responsible for general highway and traffic improvements that mitigate these negative impacts. Such policies are more easily dealt with in good economic times. When the economy is sluggish, the issue will often be a clash between the desire to reduce congestion and the desire to encourage development as a means of increasing the tax base.

Reconstruction of existing highway facilities also causes unique problems. The entire Interstate System has been aging, and many of its facilities have required major reconstruction efforts. Part of the problem is that reconstruction of Interstate facilities receives the 90% federal subsidy, while routine maintenance on the same facility is primarily the responsibility of state and local governments. Deferring routine maintenance on these facilities in favor of major reconstruction efforts has resulted from federal funding policies over the years. Major reconstruction efforts have a substantial major burden not involved in the initial construction of these facilities: maintaining traffic. It is easier to build a new facility in a dedicated right-of-way than to rebuild it while continuing to serve 100,000 or more vehicles per day. Thus, issues of long-term and short-term construction detours as well as the diversion of traffic to alternate routes require major planning by traffic engineers.

Recently, the issue of security of transportation facilities has come to the fore. The creation of facilities and processes for random and systematic inspection of trucks and other vehicles at critical locations is a major challenge, as is securing major public transportation systems such as railroads, airports, and rapid transit systems.

The list goes on and on. The point is that traffic engineers cannot expect to practice their profession only in traditional ways on traditional projects. Like any professional, the traffic engineer must be ready to face current problems and to play an important role in any situation that involves transportation and/or traffic systems.

1.6 Standard References for the Traffic Engineer

In order to remain up to date and aware, the traffic engineer must keep up with modern developments through membership and participation in professional organizations, regular review of key periodicals, and an awareness of the latest standards and criteria for professional practice.

Key professional organizations for the traffic engineer include the Institute of Transportation Engineers (ITE), the Transportation Research Board (TRB), the Transportation Group of the American Society of Civil Engineers (ASCE), ITS America, and others. All of these provide literature and maintain journals, and have local, regional, and national meetings. TRB is a branch of the National Academy of Engineering and is a major source of research papers and reports.

Like many engineering fields, the traffic engineering profession has many manuals and standard references, most of which will be referred to in the chapters of this text. Major references include

- *Traffic Engineering Handbook* [1]
- *Uniform Vehicle Code and Model Traffic Ordinance* [2]
- *Manual on Uniform Traffic Control Devices* [3]
- *Highway Capacity Manual* [4]
- *A Policy on Geometric Design of Highways and Streets* (The AASHTO Green Book) [5]

A few of these have had major updates and revisions since 2000, including References 1, 3, 4, and 5. Most standards such as these are updated frequently, usually on a 5- or 10-year cycle, and the traffic engineer must be aware of how changes in standards, criteria, methodology, and other aspects will affect the practice of the profession.

Other manuals abound and often relate to specific aspects of traffic engineering. These references document the current state of the art in traffic engineering, and those most frequently used should be part of the professional's personal library.

There are also a wide variety of internet sites that are of great value to the traffic engineer. Specific sites are not listed here, as they change rapidly. All of the professional organizations, as well as equipment manufacturers, maintain Web sites. The federal DOT, FHWA, NHTSA, and private highway-related organizations maintain Web sites. The entire *Manual on Uniform Traffic Control Devices* is available on-line through the FHWA Web site.

Because traffic engineering is a rapidly changing field, the reader cannot assume that every standard and analysis process included in this text is current, particularly as the time since publication increases. While the authors will continue to produce periodic updates, the traffic engineer must keep abreast of latest developments as a professional responsibility.

1.7 Metric versus U.S. Units

In the preface to the second edition of this text, it was indicated that the third edition would be in metric units. At the time, legislation was in place to require the conversion of all highway agencies to metric units over a short time period. Since then, the government has once again backed off this stance. Thus, at the current time, there are states continuing to use U.S. units, states continuing to use metric units (they had already converted), and an increasing number of states moving back to U.S. units after conversion to the metric system. Some of the key references, such as the Highway Capacity Manual, have been produced in both metric and U.S. unit versions. Others, like the AASHTO Green Book, contain both metric and U.S. standards.

Metric and U.S. standards are not the same. A standard 12-ft lane converts to a standard 3.6-m lane, which is narrower than 12 feet. Standards for a 70-mi/h

design speed convert to standards for a 120-km/h design speed, which are not numerically equivalent. This is because even units are used in both systems rather than the awkward fractional values that result from numerically equivalent conversions. That is why a metric set of wrenches for use on a foreign car is different from a standard U.S. wrench set.

Because more states are on the U.S. system than on the metric system (with more moving back to U.S. units) and because the size of the text would be unwieldy if dual units were included, this text continues to be written using standard U.S. units.

1.8 Closing Comments

The profession of traffic engineering is a broad and complex one. Nevertheless, it relies on key concepts and analyses and basic principles that do not change greatly over time. This text emphasizes both the basic principles and current (in 2003) standards and practices. The reader must keep abreast of changes that influence the latter.

References

1. Pline, J., Editor, *Traffic Engineering Handbook*, 5th Edition, Institute of Transportation Engineers, Washington DC, 1999.

2. *Uniform Vehicle Code and Model Traffic Ordinance*, National Committee on Uniform Traffic Laws and Ordinance, Washington DC, 1992.

3. *Manual on Uniform Traffic Control Devices*, Millennium Edition, Federal Highway Administration, Washington DC, 2000. (Available on the FHWA Web site—www.fhwa.gov.)

4. *Highway Capacity Manual*, 4th Edition, Transportation Research Board, Washington DC, 2000.

5. *A Policy on Geometric Design of Highways and Streets*, 4th Edition, American Association of State Highway and Traffic Officials, Washington DC, 2001.

PART 1

Components of the Traffic System and their Characteristics

CHAPTER
2
Road User and Vehicle Characteristics

2.1 Overview of Traffic Stream Components

To begin to understand the functional and operational aspects of traffic on streets and highways it is important to understand how the various elements of a traffic system interact. Further, the characteristics of traffic streams are heavily influenced by the characteristics and limitations of each of these elements. There are five critical components that interact in a traffic system:

- Road users—drivers, pedestrians, bicyclists, and passengers
- Vehicles—private and commercial
- Streets and highways
- Traffic control devices
- The general environment

This chapter provides an overview of critical road user and vehicle characteristics. Chapter 3 focuses on the characteristics of streets and highways, while

Chapter 4 provides an overview of traffic control devices and their use.

The general environment also has an impact on traffic operations, but this is difficult to assess in any given situation. Such things as weather, lighting, density of development, and local enforcement policies all play a role in affecting traffic operations. These factors are most often considered qualitatively, with occasional supplemental quantitative information available to assist in making judgments.

2.1.1 Dealing with Diversity

Traffic engineering would be a great deal simpler if the various components of the traffic system had uniform characteristics. Traffic controls could be easily designed if all drivers reacted to them in exactly the same way. Safety could be more easily achieved if all vehicles had uniform dimensions, weights, and operating characteristics.

Drivers and other road users, however, have widely varying characteristics. The traffic engineer must deal with elderly drivers as well as 18-year-olds, aggressive drivers and timid drivers, and drivers subject to myriad distractions both inside and outside their vehicles. Simple subjects like reaction time, vision characteristics, and walking speed become complex because no two road users are the same.

Most human characteristics follow the normal distribution (see Chapter 8). The normal distribution is characterized by a strong central tendency (i.e., most people have characteristics falling into a definable range). For example, most pedestrians crossing a street walk at speeds between 3.0 and 5.0 ft/s. However, there are a few pedestrians that walk either much slower or much faster. A normal distribution defines the proportions of the population expected to fall into these ranges. Because of variation, it is not practical to design a system for "average" characteristics. If a signal is timed, for example, to accommodate the average speed of crossing pedestrians, about half of all pedestrians would walk at a slower rate and be exposed to unacceptable risks.

Thus, most standards are geared to the "85th percentile" (or "15th percentile") characteristic. In general terms, a percentile is a value in a distribution for which the stated percentage of the population has a characteristic that is less than or equal to the specified value. In terms of walking speed, for example, safety demands that we accommodate slower walkers. The 15th percentile walking speed is used, as only 15% of the population walks slower than this. Where driver reaction time is concerned, the 85th percentile value is used, as 85% of the population has a reaction time that is numerically equal to or less than this value. This approach leads to design practices and procedures that safely accommodate 85% of the population. What about the remaining 15%? One of the characteristics of normal distributions is that the extreme ends of the distribution (the highest and lowest 15%) extend to plus or minus infinity. In practical terms, the highest and lowest 15% of the distribution represent very extreme values that could not be effectively accommodated into design practices. Qualitatively, the existence of road users who may possess characteristics not within the 85th (or 15th) percentile is considered, but most standard practices and criteria do not directly accommodate them. Where feasible, higher percentile characteristics can be employed.

Just as road-user characteristics vary, the characteristics of vehicles vary widely as well. Highways must be designed to accommodate motorcycles, the full range of automobiles, and a wide range of commercial vehicles, including double- and triple-back tractor-trailer combinations. Thus, lane widths, for example, must accommodate the largest vehicles expected to use the facility.

Some control over the range of road-user and vehicle characteristics is maintained through licensing criteria and federal and state standards on vehicle design and operating characteristics. While these are important measures, the traffic engineer must still deal with a wide range of road-user and vehicle characteristics.

2.1.2 Addressing Diversity through Uniformity

While traffic engineers have little control over driver and vehicle characteristics, design of roadway systems and traffic controls is in the core of their professional practice. In both cases, a strong degree of uniformity of approach is desirable. Roadways of a similar type and function should have a familiar "look" to drivers; traffic control devices should be as uniform as possible. Traffic engineers strive to provide information to drivers in uniform ways. While this does not assure uniform reactions from drivers, it at least narrows the range of behavior, as drivers become accustomed to and familiar with the cues traffic engineers design into the system.

Chapters 3 and 4 will deal with roadways and controls, respectively, and will treat the issue of uniformity in greater detail.

2.2 Road Users

Human beings are complex and have a wide range of characteristics that can and do influence the driving task. In a system where the driver is in complete control of vehicle operations, good traffic engineering requires a keen understanding of driver characteristics. Much of the task of traffic engineers is to find ways to provide drivers with information in a clear, effective manner that induces safe and proper responses.

The two driver characteristics of utmost importance are visual acuity factors and the reaction process. The two overlap, in that reaction requires the use of vision for most driving cues. Understanding how information is received and processed is a key element in the design of roadways and controls.

There are other important characteristics as well. Hearing is an important element in the driving task (i.e., horns, emergency vehicle sirens, brakes squealing, etc.). While noting this is important, however, no traffic element can be designed around audio cues, as hearing-impaired and even deaf drivers are licensed. Physical strength may have been important in the past, but the evolution of power-steering and power-braking systems has eliminated this as a major issue, with the possible exception of professional drivers of trucks, buses, and other heavy vehicles.

Of course, one of the most important human factors that influences driving is the personality and psychology of the driver. This, however, is not easily quantified and is difficult to consider in design. It is dealt with primarily through enforcement and licensing procedures that attempt to remove or restrict drivers who periodically display inappropriate tendencies, as indicated by accident and violation experience.

2.2.1 Visual Characteristics of Drivers

When drivers initially apply for, or renew, their licenses, they are asked to take an eye test, administered either by the state motor vehicle agency or by an optometrist or ophthalmologist who fills out an appropriate form for the motor vehicle agency. The test administered is a standard chart-reading exercise that measures *static visual acuity*—that is, the ability to see small stationary details clearly.

Visual Factors in Driving

While certainly an important characteristic, static visual acuity is hardly the only visual factor involved in the driving task. The *Traffic Engineering Handbook* [1] provides an excellent summary of visual factors involved in driving, as shown in Table 2.1.

Many of the other factors listed in Table 2.1 reflect the dynamic nature of the driving task and the fact that most objects to be viewed by drivers are in relative motion with respect to the driver's eyes.

As static visual acuity is the only one of these many visual factors that is examined as a prerequisite to issuing a driver's license, traffic engineers must expect and deal with significant variation in many of the other visual characteristics of drivers.

Fields of Vision

Figure 2.1 illustrates three distinct fields of vision, each of which is important to the driving task [2]:

- *Acute or clear vision cone*—3° to 10° around the line of sight; legend can be read only within this narrow field of vision.

- *Fairly clear vision cone*—10° to 12° around the line of sight; color and shape can be identified in this field.

- *Peripheral vision*—This field may extend up to 90° to the right and left of the centerline of the pupil, and up to 60° above and 70° below the line of sight. Stationary objects are generally not seen in the peripheral vision field, but the movement of objects through this field is detected.

These fields of vision, however, are defined for a stationary person. In particular, the peripheral vision field narrows, as speed increases, to as little as 100° at 20 mi/h and to 40° at 60 mi/h.

The driver's visual landscape is both complex and rapidly changing. Approaching objects appear to expand in size, while other vehicles and stationary objects are in relative motion both to the driver and to each other. The typical driver essentially samples the available visual information and selects appropriate cues to make driving decisions.

The fields of vision affect a number of traffic engineering practices and functions. Traffic signs, for example, are placed so that they can be read within the acute vision field without requiring drivers to change their line of sight. Thus, they are generally placed within a 10° range of the driver's expected line of sight, which is assumed to be in line with the highway alignment. This leads to signs that are intended to be read when they are

Table 2.1: Visual Factors in the Driving Task

Visual Factor	Definition	Sample Related Driving Task(s)
Accommodation	Change in the shape of the lens to bring images into focus.	Changing focus from dashboard displays to roadway.
Static Visual Acuity	Ability to see small details clearly.	Reading distant traffic signs.
Adaptation	Change in sensitivity to different levels of light.	Adjust to changes in light upon entering a tunnel.
Angular Movement	Seeing objects moving across the field of view.	Judging the speed of cars crossing drivers' paths.
Movement in Depth	Detecting changes in visual image size.	Judging speed of an approaching vehicle.
Color	Discrimination between different colors.	Identifying the color of signals.
Contrast Sensitivity	Seeing objects that are similar in brightness to their background.	Detecting dark-clothed pedestrians at night.
Depth Perception	Judgment of the distance of objects.	Passing on two-lane roads with oncoming traffic.
Dynamic Visual Acuity	Ability to see objects that are in motion relative to the eye.	Reading traffic signs while moving.
Eye Movement	Changing the direction of gaze.	Scanning the road environment for hazards.
Glare Sensitivity	Ability to resist and recover from the effects of glare.	Reduction in visual performance due to headlight glare.
Peripheral Vision	Detection of objects at the side of the visual field.	Seeing a bicycle approaching from the left.
Vergence	Angle between the eyes' line of sight.	Change from looking at the dashboard to the road.

(Used with permission of Institute of Transportation Engineers, Dewar, R., "Road Users," *Traffic Engineering Handbook*, 5th Edition, Chapter 2, Table 2.2, pg. 8, 1999.)

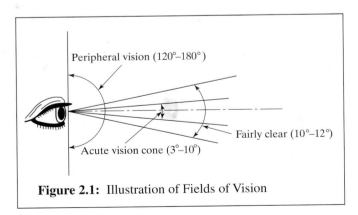

Figure 2.1: Illustration of Fields of Vision

a significant distance from the driver; in turn, this implies how large the sign and its lettering must be in order to be comprehended at that distance. Objects or other vehicles located in the fairly clear and peripheral vision fields may draw the driver's attention to an important event occurring in that field, such as the approach of a vehicle on an intersection street or driveway or a child running into the street after a ball. Once noticed, the driver may turn his/her head to examine the details of the situation.

Peripheral vision is the single most important factor when drivers estimate their speed. The movement of objects through the peripheral vision field is the driver's single most important indicator of speed. Old studies have demonstrated time and again that drivers deprived of peripheral vision (using blinders in experimental cases) and deprived of a working speedometer have little idea of how fast they are traveling.

2.2.2 Important Visual Deficits

There are a number of visual problems that can affect driver performance and behavior. Unless the condition causes a severe visual disability, drivers affected by various visual deficits often continue to drive. Reference 3 contains an excellent overview and discussion of these.

Some of the more common problems involve cataracts, glaucoma, peripheral vision deficits, ocular muscle imbalance, depth perception deficits, and color blindness. Drivers who have eye surgery to correct a problem may experience temporary or permanent impairments. Other diseases, such as diabetes, can have a significant negative impact on vision if not controlled. Some conditions, like cataracts and glaucoma, if untreated, can lead to blindness.

While color blindness is not the worst of these conditions, it generally causes some difficulties for the affected driver, since color is one of the principal means to impart information. Unfortunately, one of the most common forms of color blindness involves the inability to discern the difference between red and green. In the case of traffic signals, this could have a devastating impact on the safety of such drivers. To ameliorate this difficulty to some degree, some blue pigment has been added to green lights and some yellow pigment has been added to red lights, making them easier to discern by color blind drivers. Also, the location of colors on signal heads has long been standardized, with red on the top and green on the bottom of vertical signal heads. On horizontal heads, red is on the left and green on the right. Arrow indications are either located on a separate signal head or placed below or to the right of ball indications on a mixed signal head.

2.2.3 Perception–Reaction Time

The second critical driver characteristic is perception-reaction time (PRT). During perception and reaction, there are four distinct processes that the driver must perform [4]:

- *Detection*. In this phase, an object or condition of concern enters the driver's field of vision, and

the driver becomes consciously aware that something requiring a response is present.

- *Identification*. In this phase, the driver acquires sufficient information concerning the object or condition to allow the consideration of an appropriate response.

- *Decision*. Once identification of the object or condition is sufficiently completed, the driver must analyze the information and make a decision about how to respond.

- *Response*. After a decision has been reached, the response is now physically implemented by the driver.

The total amount of time that this process takes is called the perception–reaction time (PRT). In some of the literature, the four phases are referred to as perception, identification, emotion, and volition, leading to the term "PIEV time." This text will use PRT, but the reader should understand that this is equivalent to PIEV time.

Design Values

Like all human characteristics, perception–reaction times vary widely amongst drivers, as do a variety of other factors, including the type and complexity of the event perceived and the environmental conditions at the time of the response.

Nevertheless, design values for various applications must be selected. The American Association of State Highway and Transportation Officials (AASHTO) mandates the use of 2.5 seconds for most computations involving braking reactions [5], based upon a number of research studies [6–9]. This value is believed to be approximately a 90th percentile criterion (i.e., 90% of all drivers will have a PRT as fast or faster than 2.5 s).

For signal timing purposes, the Institute of Transportation Engineers [10] recommends a PRT time of 1.0 second. Because of the simplicity of the response and the preconditioning of drivers to respond to signals, the PRT time is significantly less than that for a braking response on an open highway. While this is a lower value, it still represents an approximately 85th percentile for the particular situation of responding to a traffic signal.

AASHTO criteria, however, recognize that in certain more complex situations, drivers may need considerably more time to react than 1.0 or 2.5 seconds. Situations where drivers must detect and react to unexpected events, or a difficult-to-perceive information source in a cluttered highway environment, or a situation in which there is a likelihood of error involving either information reception, decisions, or actions, all would result in increased PRT times. Some of the examples cited by AASHTO of locations where such situations might exist include complex interchanges and intersections where unusual movements are encountered and changes in highway cross-sections such as toll plazas, lane drops, and areas where the roadway environment is cluttered with visual distractions. Where a collision avoidance maneuver is required, AASHTO criteria call for a PRT of 3.0 seconds for stops on rural roads and 9.1 seconds for stops on urban roads. Where collision avoidance requires speed, path, and/or direction changes, AASHTO recommends a PRT of between 10.2–11.2 seconds on rural roads, 12.1–12.9 seconds on suburban roads, and 14.0–14.5 on urban roads. Complete AASHTO criteria are given in Exhibit 3-3, page 116 of Reference 5.

Expectancy

The concept of expectancy is important to the driving task and has a significant impact on the perception–reaction process and PRT. Simply put, drivers will react more quickly to situations they *expect* to encounter as opposed to those that they *do not expect* to encounter. There are three different types of expectancies:

- *Continuity.* Experiences of the immediate past are generally expected to continue. Drivers do not, for example, expect the vehicle they are following to suddenly slow down.
- *Event.* Things that have not happened previously will not happen. If no vehicles have been observed entering the roadway from a small driveway over a reasonable period of time, then the driver will assume that none will enter now.
- *Temporal.* When events are cyclic, such as a traffic signal, the longer a given state is observed, drivers will assume that it is more likely that a change will occur.

The impact of expectancy on PRT is illustrated in Figure 2.2. This study by Olsen, et al. [*11*] in 1984 was a controlled observation of student drivers reacting to a similar hazard when they were unaware that it would appear, and again where they were told to look for it. In a third experiment, a red light was added to the dash to initiate the braking reaction. The PRT under the "expected" situation was consistently about 0.5 seconds faster than under the "unexpected" situation.

Given the obvious importance of expectancy on PRT, traffic engineers must strive to avoid designing "unexpected" events into roadway systems and traffic controls. If there are all right-hand ramps on a given freeway, for example, left-hand ramps should be avoided if at all possible. If absolutely required, guide signs must be very carefully designed to alert drivers to the existence and location of the left-hand ramp, so that when they reach it, it is no longer "unexpected."

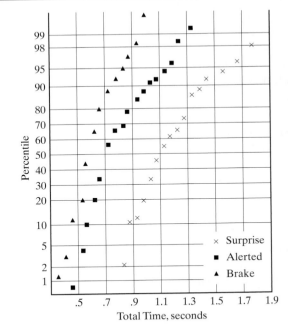

Figure 2.2: Comparison of Perception–Reaction Times Between Expected and Unexpected Events (Used with permission of Transportation Research Board, National Research Council, Olson, P., et al., "Parameters Affecting Stopping Sight Distance," *NCHRP Report 270*, Washington DC, 1984.)

Other Factors Affecting PRT

In general, PRTs increase with a number of factors, including (1) age, (2) fatigue, (3) complexity of reaction, and (4) presence of alcohol and/or drugs in the driver's system. While these trends are well documented, they are generally accounted for in recommended design values, with the exception of the impact of alcohol and drugs. The latter are addressed primarily through enforcement of ever-stricter DWI/DUI laws in the various states, with the intent of removing such drivers from the system, especially where repeated violations make them a significant safety risk. Some of the more general affects of alcohol and drugs, as well as aging, on driver characteristics are discussed in a later section.

Reaction Distance

The most critical impact of perception–reaction time is the distance the vehicle travels while the driver goes through the process. In the example of a simple braking reaction, the PRT begins when the driver first becomes aware of an event or object in his or her field of vision and ends when his or her foot is applied to the brake. During this time, the vehicle continues along its original course at its initial speed. Only after the foot is applied to the brake pedal does the vehicle begin to slow down in response to the stimulus.

The reaction distance is simply the PRT multiplied by the initial speed of the vehicle. As speed is generally in units of mi/h and PRT is in units of seconds, it is convenient to convert speeds to ft/s for use:

$$\frac{1\text{ mi} * \left(\dfrac{5,280\text{ ft}}{\text{mi}}\right)}{1\text{ h} * \left(\dfrac{3,600\text{ s}}{\text{h}}\right)} = 1.46666\ldots\frac{\text{ft}}{\text{s}} = 1.47\frac{\text{ft}}{\text{s}}$$

Thus, the reaction distance may be computed as:

$$d_r = 1.47\, S\, t \qquad (2\text{-}1)$$

where: d_r = reaction distance, ft

S = initial speed of vehicle, mi/h

t = reaction time, s

The importance of this factor is illustrated in the following sample problem: A driver rounds a curve at a speed of 60 mi/h and sees a truck overturned on the roadway ahead. How far will the driver's vehicle travel before the driver's foot reaches the brake? Applying the AASHTO standard of 2.5 s for braking reactions:

$$d_r = 1.47 * 60 * 2.5 = 220.5\text{ ft}$$

The vehicle will travel 220.5 ft (approximately 11–12 car lengths) before the driver even engages the brake. The implication of this is frightening. If the overturned truck is closer to the vehicle than 220.5 ft when noticed by the driver, not only will the driver hit the truck, he or she will do so at full speed—60 mi/h. Deceleration begins only when the brake is engaged—*after* the perception–reaction process has been completed.

2.2.4 Pedestrian Characteristics

One of the most critical safety problems in any highway and street system involves the interactions of vehicles and pedestrians. A substantial number of traffic accidents and fatalities involve pedestrians. This is not surprising, as in any contact between a pedestrian and a vehicle, the pedestrian is at a significant disadvantage.

Virtually all of the interactions between pedestrians and vehicles occur as pedestrians cross the street at intersections and at mid-block locations. At signalized intersections, safe accommodation of pedestrian crossings is as critical as vehicle requirements in establishing an appropriate timing pattern. Pedestrian walking speed in crosswalks is the most important factor in the consideration of pedestrians in signal timing.

At unsignalized crossing locations, gap-acceptance behavior of pedestrians is another important consideration. "Gap acceptance" refers to the clear time intervals between vehicles encroaching on the crossing path and the behavior of pedestrians in "accepting" them to cross through.

Walking Speeds

Table 2.2 shows 50th percentile walking speeds for pedestrians of various ages. It should be noted that these speeds were measured as part of a controlled experiment [12] and not specifically at intersection or mid-block

Table 2.2: 50th Percentile Walking Speeds for Pedestrians of Various Ages

Age (years)	50th Percentile Walking Speed (ft/s)	
	Males	**Females**
2	2.8	3.4
3	3.5	3.4
4	4.1	4.1
5	4.6	4.5
6	4.8	5.0
7	5.0	5.0
8	5.0	5.3
9	5.1	5.4
10	5.5	5.4
11	5.2	5.2
12	5.8	5.7
13	5.3	5.6
14	5.1	5.3
15	5.6	5.3
16	5.2	5.4
17	5.2	5.4
18	4.9	N/A
20–29	5.7	5.4
30–39	5.4	5.4
40–49	5.1	5.3
50–59	4.9	5.0
60+	4.1	4.1

(Compiled from Eubanks, J. and Hill, P., *Pedestrian Accident Reconstruction and Litigation*, 2nd Edition, Lawyers & Judges Publishing Co., Tucson, AZ, 1999.)

crosswalks. Nevertheless, the results are interesting. The standard walking speed used in timing signals is 4.0 ft/s, with 3.5 ft/s recommended where older pedestrians are predominant. Most studies indicate that these standards are reasonable and will accommodate 85% of the pedestrian population.

One problem with standard walking speeds involves physically impaired pedestrians. A study of pedestrians with various impairments and assistive devices concluded that average walking speeds for virtually all categories were lower than the standard 4.0 ft/s used in signal timing [13]. Table 2.3 includes some of the results of this study. These and similar results of other studies suggest that more consideration needs to be given to the needs of handicapped pedestrians.

Gap Acceptance

When a pedestrian crosses at an uncontrolled (either by signals, STOP, or YIELD signs) location, either at an intersection or at a mid-block location, the pedestrian must select an appropriate "gap" in the traffic stream through which to cross. The "gap" in traffic is measured as the time lag between two vehicles in any lane encroaching on the pedestrian's crossing path. As the pedestrian waits to cross, he or she views gaps and decides whether to "accept" or "reject" the gap for a safe crossing. Some studies have used a gap defined as the distance between the pedestrian and the approaching vehicle at the time the pedestrian begins his or her crossing. An early study [14] using the latter approach resulted in an 85th percentile gap of approximately 125 ft.

Table 2.3: Walking Speeds for Physically Impaired Pedestrians

Impairment/Assistive Device	Average Walking Speed (ft/s)
Cane/Crutch	2.62
Walker	2.07
Wheelchair	3.55
Immobilized Knee	3.50
Below-Knee Amputee	2.46
Above-Knee Amputee	1.97
Hip Arthritis	2.44–3.66
Rheumatoid Arthritis (Knee)	2.46

(Compiled from Perry, J., *Gait Analysis*, McGraw-Hill, New York, NY, 1992.)

Gap acceptance behavior, however, is quite complex and varies with a number of other factors, including the speed of approaching vehicles, the width of the street, the frequency distribution of gaps in the traffic stream, waiting time, and others. Nevertheless, this is an important characteristic that must be considered due to its obvious safety implications. Chapter 18, for example, presents warrants for (conditions justifying) the imposition of traffic signals. One of these is devoted entirely to the safety of pedestrian crossings and to the frequency of adequate gaps in the traffic stream to permit safe crossings.

Pedestrian Comprehension of Controls

One of the problems in designing controls for pedestrians is generally poor understanding of and poor adherence to such devices. One questionnaire survey of 4,700 pedestrians [15] detailed many problems of misunderstanding. The proper response to a flashing "DON'T WALK" signal, for example, was not understood by 50% of road users, who thought it meant they should return to the curb from which they started. The meaning of this signal is to not start crossing while it is flashing; it is safe to complete a crossing if the pedestrian has already started to do so. Another study [16] found that violation rates for the solid "DON'T WALK" signal were higher than 50% in most cities, that the use of the flashing "DON'T WALK" for pedestrian clearance was not well understood, and that pedestrians tend not to use pedestrian-actuated signals. Chapter 20 (on signal timing) discusses some of the problems associated with pedestrian-actuation buttons and their use that compromise both pedestrian comprehension and the efficiency of the signalization. Since this study was completed, the flashing and solid "DON'T WALK" signals have been replaced by the Portland orange "raised hand" symbol.

Thus, the task of providing for a safe environment for pedestrians is not an easy one. The management and control of conflicts between vehicles and pedestrians remains a difficult one.

2.2.5 Impacts of Drugs and Alcohol on Road Users

The effect of drugs and alcohol on drivers has received well-deserved national attention for many years, leading to substantial strengthening of DWI/DUI laws and enforcement. These factors remain, however, a significant contributor to traffic fatalities and accidents. Drivers, however, are not the only road users who contribute to the nation's accident and fatality statistics. Consider that in 1996, 47.3% of fatal pedestrian accidents involved either a driver or a pedestrian with detectable levels of alcohol in their systems. For this group, 12.0% of the drivers and 32.3% of the pedestrians had blood-alcohol levels above 0.10%, the legal definition of "drunk" in many states. More telling is that 7% of the drivers and 6% of the pedestrians had detectable alcohol levels below this limit.

The importance of these isolated statistics is to make the following point: legal limits for DWI/DUI do not define the point at which alcohol and/or drugs influence the road user. Recognizing this is important for individuals to ensure safe driving, and is now causing many states to reduce their legal limits on alcohol to 0.08%, and for some to consider "zero tolerance" criteria (0.01%) for new drivers for the first year or two they are licensed.

Figure 2.3 is a summary of various studies on the effects of drugs and alcohol on various driving factors.

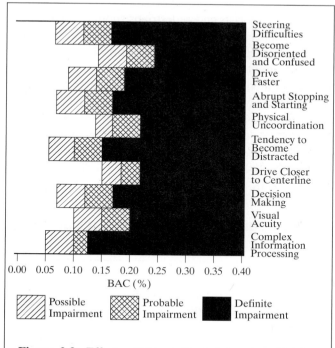

Figure 2.3: Effects of Blood-Alcohol Level on Driving Tasks (Used with permission of Institute of Transportation Engineers, Blaschke, J.; Dennis, M.; and Creasy, F., "Physical and Psychological Effects of Alcohol and Other Drugs on Drivers," *ITE Journal*, 59, Washington DC, 1987.)

Note that for many factors, impairment of driver function begins at levels well below the legal limits—for some factors at blood-alcohol levels as low as 0.05%.

What all of these factors add up to is an impaired driver. This combination of impairments leads to longer PRT times, poor judgments, and actions that can and do cause accidents. Since few of these factors can be ameliorated by design or control (although good designs and well-designed controls help both impaired and unimpaired drivers), enforcement and education are critical elements in reducing the incidence of DWI/DUI and the accidents and deaths that result.

The statistics cited in the opening paragraph of this section also highlight the danger caused by pedestrians who are impaired by drugs or alcohol. In the case of impaired pedestrians, the danger is primarily to themselves. Nevertheless, if crossing a street or highway is required, "walking while impaired" is also quite dangerous. Again, enforcement and education are the major weapons in combating the problem, as there is not a great deal that can be done through design or control to address the issue.

Both motorists and pedestrians should also be aware of the impact of common prescription and over-the-counter medications on their performance capabilities. Many legitimate medications have effects that are similar to those of alcohol and/or illegal drugs. Users of medications should always be aware of the side effects of what they use (a most frequent effect of many drugs is drowsiness), and to exercise care and good judgment when considering whether or not to drive. Some legitimate drugs can have a direct impact on blood-alcohol levels and can render a motorist legally intoxicated without "drinking."

2.2.6 Impacts of Aging on Road Users

As life expectancy continues to rise, the number of older drivers has risen dramatically over the past several decades. Thus, it becomes increasingly important to understand how aging affects driver needs and limitations and how these should impact design and control decisions. Reference 17 is an excellent compilation sponsored by the National Academy of Sciences on a wide range of topics involving aging drivers.

Many visual acuity factors deteriorate with age, including both static and dynamic visual acuity, glare

sensitivity and recovery, night vision, and speed of eye movements. Such ailments as cataracts, glaucoma, macular degeneration, and diabetes are also more common as people age, and these conditions have negative impacts on vision.

The increasing prevalence of older drivers presents a number of problems for both traffic engineers and public officials. At some point, deterioration of various capabilities must lead to revocation of the right to drive. On the other hand, driving is the principal means of mobility and accessibility in most parts of the nation, and the alternatives for those who can no longer drive are either limited or expensive. The response to the issue of an aging driver population must have many components, including appropriate licensing standards, consideration of some license restrictions on older drivers (for example, a daytime only license), provision of efficient and affordable transportation alternatives, and increased consideration of their needs, particularly in the design and implementation of control devices and traffic regulations. Older drivers may be helped, for example, by such measures as larger lettering on signs, better highway lighting, larger and brighter signals, and other measures. Better education can serve to make older drivers more aware of the types of deficits they face and how to best deal with them. More frequent testing of key characteristics such as eyesight may help ensure that prescriptions for glasses and/or contact lenses are frequently updated.

2.2.7 Psychological, Personality, and Related Factors

In the past few years, traffic engineers and the public in general have become acquainted with the term "road rage." Commonly applied to drivers who lose control of themselves and react to a wide variety of situations violently, improperly, and almost always dangerously, the problem (which has always existed) is now getting well-deserved attention. "Road rage," however, is a colloquial term, and is applied to everything from a direct physical assault by one road user on another to a variety of aggressive driving behaviors.

According to the testimony of Dr. John Larsen to the House Surface Transportation Subcommittee on July 17,

1997 (as summarized in Chapter 2 of Reference 1), the following attitudes characterize aggressive drivers:

- The desire to get to one's destination as quickly as possible, leading to the expression of anger at other drivers/pedestrians who impede this desire.
- The need to compete with other fast cars.
- The need to respond competitively to other aggressive drivers.
- Contempt for other drivers who do not drive, look, and act as they do on the road.
- The belief that it is their right to "hit back" at other drivers whose driving behavior threatens them.

"Road rage" is the extreme expression of a driver's psychological and personal displeasure over the traffic situation he or she has encountered. It does, however, remind traffic engineers that drivers display a wide range of behaviors in accordance with their own personalities and psychological characteristics.

Once again, most of these factors cannot be addressed directly through design or control decisions and are best treated through vigorous enforcement and educational programs.

2.3 Vehicles

In the year 2000, there were 217,293,000 registered vehicles in the United States, a number that represents more than one vehicle per licensed driver. The characteristics of these vehicles vary as widely as those of the motorists who drive them.

In general, motor vehicles are classified by AASHTO [5] into four main categories:

- *Passenger cars*—all passenger cars, SUVs, minivans, vans, and pickup trucks.
- *Buses*—intercity motor coaches, transit buses, school buses, and articulated buses
- *Trucks*—single-unit trucks, tractor-trailer, and tractor-semi-trailer combination vehicles
- *Recreational vehicles*—motor homes, cars with various types of trailers (boat, campers, motorcycles, etc.)

Motorcycles and bicycles also use highway and street facilities but are not isolated as a separate category, as their characteristics do not usually limit or define design or control needs.

There are a number of critical vehicle properties that must be accounted for in the design of roadways and traffic controls. These include:

- Braking and deceleration
- Acceleration
- Low-speed turning characteristics
- High-speed turning characteristics

In more general terms, the issues associated with vehicles of vastly differing size, weight, and operating characteristics sharing roadways must also be addressed by traffic engineers.

2.3.1 Concept of the Design Vehicle

Given the immense range of vehicle types using street and highway facilities, it is necessary to adopt standard vehicle characteristics for design and control purposes. For geometric design, AASHTO has defined 20 "design vehicles," each with specified characteristics. The 20 design vehicles are defined as follows:

P	=	passenger car
SU	=	single-unit truck
BUS-40	=	intercity bus with a 40-ft wheelbase
BUS-45	=	intercity bus with a 45-ft wheelbase
CITY-BUS	=	transit bus
S-BUS36	=	conventional school bus for 65 passengers
S-BUS40	=	large school bus for 84 passengers
A-BUS	=	articulated bus
WB-40	=	intermediate semi-trailer (wheelbase = 40 ft)
WB-50	=	intermediate semi-trailer (wheelbase = 50 ft)
WB-62	=	interstate semi-trailer (wheelbase = 62 ft)

WB-65   =   interstate semi-trailer
              (wheelbase = 65 ft)

WB-67D   =   double trailer combination
              (wheelbase = 67 ft)

WB-100T   =   triple semi-trailer/trailers
              (wheelbase = 100 ft)

WB-109D   =   turnpike double semi-trailer/trailer
              (wheelbase = 109 ft)

MH   =   motor home

P/T   =   passenger car and camper

P/B   =   passenger car and boat trailer

MH/B =   motor home and boat trailer

TR/W =   farm tractor with one wagon

Wheelbase dimensions are measured from the frontmost axle to the rearmost axle, including both the tractor and trailer in a combination vehicle.

Design vehicles are primarily employed in the design of turning roadways and intersection curbs, and are used to help determine appropriate lane widths, and such specific design features as lane-widening on curves. Key to such usage, however, is the selection of an appropriate design vehicle for various types of facilities and situations. In general, the design should consider the largest vehicle likely to use the facility with reasonable frequency.

In considering the selection of a design vehicle, it must be remembered that all parts of the street and highway network must be accessible to emergency vehicles, including fire engines, ambulances, emergency evacuation vehicles, and emergency repair vehicles, among others. Therefore the single-unit truck is usually the minimum design vehicle selected for most local street applications. The mobility of hook-and-ladder fire vehicles is enhanced by having rear-axle steering that allows these vehicles to negotiate sharper turns than would normally be possible for combination vehicles, so the use of a single-unit truck as a design vehicle for local streets is not considered to hinder emergency vehicles.

The passenger car is used as a design vehicle only in parking lots, and even there, access to emergency vehicles must be considered. For most other classes or types of highways and intersections, the selection of a design vehicle must consider the expected vehicle mix. In general, the design vehicle selected should easily accommodate 95% or more of the expected vehicle mix.

The physical dimensions of design vehicles are also important considerations. Design vehicle heights range from 4.25 ft for a passenger car to 13.5 ft for the largest trucks. Overhead clearances of overpass and sign structures, electrical wires, and other overhead appurtenances should be sufficient to allow the largest anticipated vehicles to proceed. As all facilities must accommodate a wide variety of potential emergency vehicles, use of 14.0 ft for minimum clearances is advisable for most facilities.

The width of design vehicles ranges from 7.0 ft for passenger cars to 8.5 ft for the largest trucks (excluding special "wide load" vehicles such as a tractor pulling a prefabricated or motor home.) This should influence the design of such features as lane width and shoulders. For most facilities, it is desirable to use the standard 12-ft lane width. Narrower lanes may be considered for some types of facilities when necessary, but given the width of modern vehicles, 10 ft is a reasonable minimum for virtually all applications.

2.3.2 Turning Characteristics of Vehicles

There are two conditions under which vehicles must make turns:

- Low-speed turns (≤ 10 mi/h)
- High-speed turns (> 10 mi/h)

Low-speed turns are limited by the characteristics of the vehicle, as the minimum radius allowed by the vehicle's steering mechanism can be supported at such speeds. High-speed turns are limited by the dynamics of side friction between the roadway and the tires, and by the superelevation (cross-slope) of the roadway.

Low-Speed Turns

AASHTO specifies minimum design radii for each of the design vehicles, based on the centerline turning radius and minimum inside turning radius of each vehicle. While the actual turning radius of a vehicle is controlled by the front wheels, rear wheels do not follow the same path. They "off-track" as they are dragged through the turning movement.

Reference 5 contains detailed low-speed turning templates for all AASHTO design vehicles. An example (for a WB-40 combination vehicle) is shown in Figure 2.4.

of the "lane" occupied by the vehicle as it turns. The path of the inside rear wheel is not circular, and has a variable radius.

Turning templates provide illustrations of the many different dimensions involved in a low-speed turn. In designing for low-speed turns, the minimum design turning radius is the minimum centerline radius plus one-half of the width of the front of the vehicle.

Minimum design turning radii range from 24.0 ft for a passenger car to a high of 60.0 ft for the WB-109D double tractor-trailer combination vehicle. Depending upon the specific design vehicle, the minimum inside curb radius is generally considerably smaller than the minimum design turning radius, reflecting the variable radius of the rear-inside wheel's track. In designing intersections, off-tracking characteristics of the design vehicle should be considered when determining how far from travel lanes to locate (or cut back) the curb. In a good design, the outside wheel of the turning design vehicle should be able to negotiate its path without "spilling over" into adjacent lanes as the turn is negotiated. This requires that the curb setback must accommodate the maximum off-tracking of the design vehicle.

High-Speed Turns

When involved in a high-speed turn on a highway curve, centripetal forces of momentum are exerted on the vehicle to continue in a straight path. To hold the curve, these forces are opposed by side friction and superelevation.

Superelevation is the cross-slope of the roadway, always with the lower edge in the direction of the curve. The sloped roadway provides an element of horizontal support for the vehicle. Side-friction forces represent the resistance to sliding provided across the plane of the surface between the vehicle's tires and the roadway. From the basic laws of physics, the relationship governing vehicle operation on a curved roadway is:

$$\frac{0.01e + f}{1 - 0.01ef} = \frac{S^2}{gR} \qquad (2\text{-}2)$$

where: e = superelevation rate, %
f = coefficient of side friction

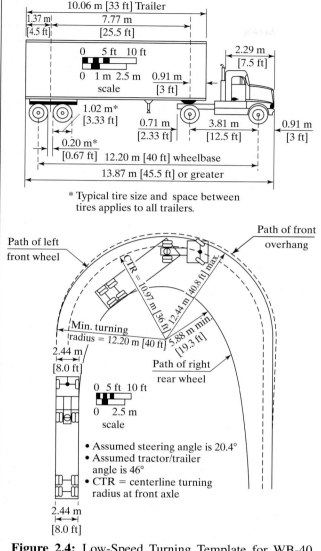

Figure 2.4: Low-Speed Turning Template for WB-40 Combination Vehicles (Used with permission of American Association of State Highway and Transportation Officials, *A Policy on Geometric Design of Highways and Streets,* 4th Edition, 2001, Washington DC, Exhibit 2–13, pg. 31.)

Note that the minimum turning radius is defined by the track of the front outside wheel. The combination vehicle, however, demonstrates considerable "off-tracking" of the rear inside wheel, effectively widening the width

S = speed of the vehicle, ft/s

R = radius of curvature, ft

g = acceleration rate due to gravity, 32.2 ft/s^2

The superelevation rate is the total rise in elevation across the travel lanes of the cross-section (ft) divided by the width of the travel lanes (ft), expressed as a percentage (i.e., multiplied by 100). AASHTO [5] expresses superelevation as a percentage in its 2001 criteria, but many other publications still express the superelevation rate as a decimal proportion.

Equation 2-2 is simplified by noting that the term "0.01ef" is extremely small, and may be ignored for the normal range of superelevation rates and *side-friction factors*. It is also convenient to express vehicle speed in mi/h. Thus:

$$\frac{0.01e + f}{1} = \frac{(1.47S)^2}{32.2R}$$

$$0.01e + f = \frac{0.067S^2}{R} = \frac{S^2}{15R}$$

This yields the more traditional relationship used to depict vehicle operation on a curve:

$$R = \frac{S^2}{15(0.01e + f)} \qquad (2\text{-}3)$$

where all terms are as previously defined, except that "S" is the speed in mi/h rather than ft/s as in Equation 2-2.

The normal range of superelevation rates is from a minimum of approximately 0.5% to support side drainage to a maximum of 12%. As speed increases, higher superelevation rates are used. Where icing conditions are expected, the maximum superelevation rate is generally limited to 8% to prevent a stalled vehicle from sliding towards the inside of the curve.

Coefficients of side friction for design are based upon wet roadway conditions. They vary with speed and are shown in Table 2.4.

Theoretically, a road can be banked to fully oppose centripetal force without using side friction at all. This is, of course, generally not done, as vehicles travel at a range of speeds and the superelevation rate required in many cases would be excessive. High-speed turns on a flat pavement may be fully supported by side friction as

Table 2.4: Coefficient of Side Friction (f) for Wet Pavements at Various Speeds

Speed (mi/h)	30	40	50	60	70
f	0.16	0.15	0.14	0.12	0.10

well, but this generally limits the radius of curvature or speed at which the curve may be safely traversed.

Chapter 3 treats the design of horizontal curves and the relationships among superelevation, side friction, curve radii, and design speed in greater detail.

Equation 2-3 can be used in a number of ways. In design, a minimum radius of curvature is computed based on maximum values of e and f. For example, if a roadway has a design speed of 65 mi/h, and the maximum values are e = 8% and f = 0.11, the minimum radius is computed as:

$$R = \frac{65^2}{15(0.01*8 + 0.11)} = 1{,}482.5 \text{ ft}$$

It can also be used to solve for a maximum safe speed, given a radius of curvature and maximum values for e and f. If a highway curve with radius of 800 ft has a superelevation rate of 6%, the maximum safe speed can be estimated. However, doing so requires that the relationship between the *coefficient of side friction, f,* and speed, as indicated in Table 2.4, be taken into account. Solving Equation 2-3 for S yields:

$$S = \sqrt{15R(0.01e + f)} \qquad (2\text{-}4)$$

For the example given, the equation is solved for the given values of e (6%) and R (800 ft) using various values of f from Table 2.4. Computations continue until there is closure between the computed speed and the speed associated with the coefficient of side friction selected. Thus:

$$S = \sqrt{15*800*(0.06 + f)}$$

$$S = \sqrt{15*800*(0.06 + 0.10)}$$

$$= 43.8 \text{ mi/h} \ (70 \text{ mi/h assumed})$$

$$S = \sqrt{15*800*(0.06 + 0.12)}$$

$$= 46.5 \text{ mi/h (60 mi/h assumed)}$$
$$S = \sqrt{15*800*(0.06 + 0.14)}$$
$$= 49.0 \text{ mi/h (50 mi/h assumed)}$$
$$S = \sqrt{15*800*(0.06 + 0.15)}$$
$$= 50.2 \text{ mi/h (40 mi/h assumed)}$$

The correct result is obviously between 49.0 and 50.2 mi/h. If straight-line interpolation is used:

$$S = 49.0 + (50.2 - 49.0)*\left[\frac{(50.0 - 49.0)}{(50.2 - 49.0) + (50.2 - 40)}\right]$$
$$= 49.1 \text{ mi/h}$$

Thus, for the curve as described, 49.1 mi/h is the maximum safe speed at which it should be negotiated.

It must be noted that this is based on the design condition of a wet pavement and that higher speeds would be possible under dry conditions.

2.3.3 Braking Characteristics

Another critical characteristic of vehicles is their ability to stop (or decelerate) once the brakes have been engaged. Again, basic physics relationships are used. The distance traveled during a stop is the average speed during the stop multiplied by the time taken to stop, or:

$$d_b = \left(\frac{S}{2}\right)*\left(\frac{S}{a}\right) = \frac{S^2}{2a} \qquad (2\text{-}5)$$

where: d_b = braking distance, ft
S = initial speed, ft/s
a = deceleration rate, ft/s^2

It is convenient, however, to express speed in mi/h, yielding:

$$d_b = \frac{(1.47 S)^2}{2a} = \frac{1.075 S^2}{a}$$

where S is the speed in mi/h. Note that the 1.075 factor is derived from the more exact conversion factor between mi/h and ft/s (1.4666. . . .). It is often also useful

to express this equation in terms of the coefficient of forward rolling or skidding friction, F, where $F = a/g$, and g is the acceleration due to gravity, 32.2 ft/s^2. Then:

$$d_b = \frac{\left(\dfrac{1.075 S^2}{32.2}\right)}{\left(\dfrac{a}{32.2}\right)} = \frac{S^2}{30F}$$

where F = coefficient of forward rolling or skidding friction. When the effects of grade are considered, and where a braking cycle leading to a reduced speed other than "0" are considered, the equation becomes:

$$d_b = \frac{S_i^2 - S_f^2}{30(F \pm 0.01G)} \qquad (2\text{-}6)$$

where: G = grade, %
S_i = initial speed, mi/h
S_f = final speed, mi/h

When there is an upgrade, a "+" is used; a "−" is used for downgrades. This results in shorter braking distances on upgrades, where gravity helps deceleration, and longer braking distances on downgrades, where gravity is causing acceleration.

In previous editions of Reference 5, braking distances were based on coefficients of forward skidding friction on wet pavements. In the latest standards, however, a standard deceleration rate of 11.2 ft/s^2 is adopted as a design rate. This is viewed as a rate that can be developed on wet pavements by most vehicles. It is also expected that 90% of drivers will decelerate at higher rates. This, then, suggests a standard friction factor for braking distance computations of $F = 11.2/32.2 = 0.348$, and Equation 2-6 becomes:

$$d_b = \frac{S_i^2 - S_f^2}{30(0.348 \pm 0.01G)} \qquad (2\text{-}7)$$

Consider the following case: Once the brakes are engaged, what distance is covered bringing a vehicle traveling at 60 mi/h on a 3% downgrade to a complete stop ($S_f = 0$ mi/h). Applying Equation 2-7:

$$d_b = \frac{60^2 - 0^2}{30(0.348 - 0.01*3)} = 377.4 \text{ ft}$$

The braking distance formula is also a favorite tool of accident investigators. It can be used to estimate the initial speed of a vehicle using measured skid marks and an estimated final speed based on damage assessments. In such cases, actual estimated values of F are used, rather than the standard design value recommended by AASHTO. Thus, Equation 2-6 is used.

Consider the following case: An accident investigator estimates that a vehicle hit a bridge abutment at a speed of 20 mi/h, based on his or her assessment of damage. Leading up to the accident location, he or she observes skid marks of 100 ft on the pavement ($F = 0.35$) and 75 ft on the grass shoulder ($F = 0.25$). There is no grade. An estimation of the speed of the vehicle at the beginning of the skid marks is desired.

In this case, Equation 2-6 is used to find the initial speed of the vehicle (S_i) based upon a known (or estimated) final speed (S_f). Each skid must be analyzed separately, starting with the grass skid (for which a final speed has been estimated). Then:

$$d_b = 75 = \frac{S_i^2 - 20^2}{30(0.25)}$$

$$S_i = \sqrt{(75*30*0.25) + 20^2} = \sqrt{962.5}$$

$$= 31.0 \text{ mi/h}$$

This is the estimated speed of the vehicle at the *start* of the grass skid; it is also the speed of the vehicle at the *end* of the pavement skid. Then:

$$d_b = 100 = \frac{S_i^2 - 962.5}{30*0.35}$$

$$S_i = \sqrt{(100*30*0.35) + 962.5} = \sqrt{2012.5}$$

$$= 44.9 \text{ mi/h}$$

It is, therefore, estimated that the speed of the vehicle immediately before the pavement skid was 44.9 mi/h. This, of course, can be compared with the speed limit to determine whether excessive speed was a factor in the accident.

2.3.4 Acceleration Characteristics

The flip side of deceleration is acceleration. Passenger cars are able to accelerate at significantly higher rates than commercial vehicles. Table 2.5 shows typical maximum acceleration rates for a passenger car with a weight-to-horsepower ratio of 30 lbs/hp and a tractor-trailer with a ratio of 200 lbs/hp.

Acceleration is highest at low speeds and decreases with increasing speed. The disparity between passenger cars and trucks is significant. Consider the distance required for a car and a truck to accelerate to 20 mi/h. Converting speed from mi/h to ft/s:

$$d_a = \left(\frac{1.47\,S}{a}\right) * \left(\frac{1.47\,S}{2}\right) = 1.075\left(\frac{S^2}{a}\right) \quad (2\text{-}8)$$

where: d_a = acceleration distance, ft
   S = speed at the end of acceleration (from a stop), mi/h
   a = acceleration rate, ft/s^2

Once again, note that the 1.075 factor is derived using the more precise factor for converting mi/h to ft/s (1.46666 ......). Then:

For a passenger car to accelerate to 20 mi/h at a rate of 7.5 ft/s^2:

$$d_a = 1.075\left(\frac{20^2}{7.5}\right) = 57.3 \text{ ft}$$

Table 2.5: Acceleration Characteristics of a Typical Car versus a Typical Truck on Level Terrain

Speed Range (mi/h)	Acceleration Rate (ft/s²) for:	
	Typical Car (30 lbs/hp)	**Typical Truck (200 lbs/hp)**
0–20	7.5	1.6
20–30	6.5	1.3
30–40	5.9	0.7
40–50	5.2	0.7
50–60	4.6	0.3

(Compiled from *Traffic Engineering Handbook*, 5th Edition, Institute of Transportation Engineers, Washington DC, 2000, Chapter 3, Tables 3.9 and 3.10.)

For a truck to accelerate to 20 mi/h at a rate of 1.6 ft/s^2:

$$d_a = 1.075\left(\frac{20^2}{1.6}\right) = 268.8 \text{ ft}$$

The disparity is striking. If a car is at a "red" signal behind a truck, the truck will significantly delay the car. If a truck is following a car in a standing queue, a large gap between the two will occur as they accelerate.

Unfortunately, there is not much that can be done about this disparity in terms of design and control. In the analysis of highway capacity, however, the disparity between trucks and cars in terms of acceleration and in terms of their ability to sustain speeds on upgrades leads to the concept of "passenger car equivalency." Depending on the type of facility, severity and length of grade, and other factors, one truck may consume as much roadway capacity as six to seven or more passenger cars. Thus, the disparity in key operating characteristics of trucks and passenger cars is taken into account in design by providing additional capacity as needed.

2.4 Total Stopping Distance and Applications

The total distance to bring a vehicle to a full stop, from the time the need to do so is first noted, is the sum of the reaction distance, d_r, and the braking distance, d_b. If Equation 2-1 (for d_r) and Equation 2-7 (for d_b) are combined, the total stopping distance becomes:

$$d = 1.47S_i\, t + \frac{S_i^2 - S_f^2}{30(0.348 \pm 0.01G)} \qquad (2\text{-}9)$$

where: d = total stopping distance, ft
  S_i = initial speed, mi/h
  S_f = final speed, mi/h
  t = reaction time, s
  G = grade, %

The concept of total stopping distance is critical to many applications in traffic engineering. Three of the more important applications are discussed in the sections that follow.

2.4.1 Safe Stopping Sight Distance

One of the most fundamental principles of highway design is that the driver must be able to see far enough to avoid a potential hazard or collision. Thus, on all roadway sections, the driver must have a sight distance that is at least equivalent to the total stopping distance required at the design speed.

Essentially, this requirement addresses this critical concern: A driver rounding a horizontal curve and/or negotiating a vertical curve is confronted with a downed tree, an overturned truck, or some other situation that completely blocks the roadway. The only alternative for avoiding a collision is to stop. The design must be such that every point along its length, the driver has a clear line of vision for at least one full stopping distance. By ensuring this, the driver can never be confronted with the need to stop without having sufficient distance to do so.

Consider a section of rural freeway with a design speed of 70 mi/h. On a section of level terrain, what safe stopping distance must be provided? Equation 2-9 is used with a final speed (S_f) of "0" and the AASHTO standard reaction time of 2.5 s. Then:

$$d = 1.47 * 70 * 2.5 + \frac{70^2 - 0^2}{30(0.348)}$$
$$= 257.3 + 469.3 = 726.6 \text{ ft}$$

This means that for the entire length of this roadway section drivers must be able to see at least 726.6 ft ahead. Providing this safe stopping sight distance will limit various elements of horizontal and vertical alignment, as discussed in Chapter 3.

What could happen, for example, if a section of this roadway provided a sight distance of only 500 ft? It would now be possible that a driver would initially notice an obstruction when it is only 500 ft away. If the driver were approaching at the design speed of 70 mi/h, a collision would occur. Again, assuming design values of reaction time and forward skidding friction, Equation 2-9 could be solved for the collision speed (i.e., the final speed of the deceleration cycle), using a known deceleration distance of 500 ft:

$$500 = 1.47 * 70 * 2.5 + \frac{70^2 - S_f^2}{30(0.348)}$$

$$500 - 257.3 = 242.7 = \frac{70^2 - S_f^2}{10.44}$$

$$2{,}533.8 = 4{,}900 - S_f^2$$

$$S_f = \sqrt{4{,}900 - 2{,}533.8} = 48.6 \text{ mi/h}$$

If the assumed conditions hold, a collision at 48.6 mi/h would occur. Of course, if the weather were dry and the driver had faster reactions than the design value (remember, 90% of drivers do), the collision might occur at a lower speed, and might be avoided altogether. The point is that such a collision *could* occur if the sight distance were restricted to 500 ft.

2.4.2 Decision Sight Distance

While every point and section of a highway must be designed to provide at least safe stopping sight distance, there are some sections that should provide greater sight distance to allow drivers to react to potentially more complex situations than a simple stop. Previously, reaction times for collision avoidance situations were cited [5].

Sight distances based upon these collision-avoidance decision reaction times are referred to as "decision sight distances." AASHTO recommends that decision sight distance be provided at interchanges or intersection locations where unusual or unexpected maneuvers are required; changes in cross-section such as lane drops and additions, toll plazas, and intense-demand areas where there is substantial "visual noise" from competing information (e.g., control devices, advertising, roadway elements).

The decision sight distance is found by using Equation 2-9, replacing the standard 2.5 s reaction time for stopping maneuvers with the appropriate collision avoidance reaction time for the situation.

Consider the decision sight distance required for a freeway section with a 60 mi/h design speed approaching a busy urban interchange with many competing information sources. The approach is on a 3% downgrade. For this case, AASHTO suggests a reaction time up to 14.5 s to allow for complex path and speed changes in response to conditions. The decision sight distance is still based on the assumption that a worst case would require a complete stop. Thus, the decision sight distance would be:

$$d = 1.47 * 60 * 14.5 + \frac{60^2 - 0^2}{30(0.348 - 0.01 * 3)}$$

$$= 1{,}278.9 + 377.4 = 1{,}656.3 \text{ ft}$$

AASHTO criteria for decision sight distances do not assume a stop maneuver for the speed/path/direction changes required in the most complex situations. The criteria, which are shown in Table 2.6, replace the braking distance in these cases with maneuver distances consistent with maneuver times between 3.5 and 4.5 s. During the maneuver time, the initial speed is assumed to be in effect. Thus, for maneuvers involving speed, path, or direction change on rural, suburban, or urban roads, Equation 2-10 is used to find the decision sight distance.

$$d = 1.47(t_r + t_m)S_i \qquad (2\text{-}10)$$

where: t_r = reaction time for appropriate avoidance maneuver, s

t_m = maneuver time, s

Thus, in the sample problem posed previously, AASHTO would not assume that a stop is required. At 60 mi/h, a maneuver time of 4.0 s is used with the 14.5 s reaction time, and:

$$d = 1.47 * (14.5 + 4.0) * 60 = 1{,}631.7 \text{ ft}$$

The criteria for decision sight distance shown in Table 2.6 are developed from Equations 2-9 and 2-10 for the decision reaction times indicated for the five defined avoidance maneuvers.

2.4.3 Other Sight Distance Applications

In addition to safe stopping sight distance and decision sight distance, AASHTO also sets criteria for (1) passing

Table 2.6: Decision Sight Distances Resulting From Equations 2-9 and 2-10

Design Speed (mi/h)	Assumed Maneuver Time (s)	Decision Sight Distance for Avoidance Maneuver (ft)				
		A (Equation 2-9)	B (Equation 2-9)	C (Equation 2-10)	D (Equation 2-10)	E (Equation 2-10)
Reaction Time (s)		3	9.1	11.2	12.9	14.5
30	4.5	219	488	692	767	838
40	4.5	330	688	923	1023	1117
50	4.0	460	908	1117	1242	1360
60	4.0	609	1147	1341	1491	1632
70	3.5	778	1406	1513	1688	1852
80	3.5	966	1683	1729	1929	2117

A: Stop on a rural road
B: Stop on an urban road
C: Speed/path/direction change on a rural road
D: Speed/path/direction change on a suburban road
E: Speed/path/direction change on an urban road

sight distance on two-lane rural highways and (2) intersection sight distances for various control options. These are covered in other chapters of this text. See Chapter 16 for a discussion of passing sight distance on two-lane highways and Chapter 18 for intersection sight distances.

2.4.4 Change (Yellow) and Clearance (All Red) Intervals for a Traffic Signal

The yellow interval for a traffic signal is designed to allow a vehicle that cannot comfortably stop when the green is withdrawn to enter the intersection legally. Consider the situation shown in Figure 2.5.

In Figure 2.5, d is the safe stopping distance. At the time the green is withdrawn, a vehicle at d or less feet from the intersection line will not be able to stop, assuming normal design values hold. A vehicle further away than d would be able to stop without encroaching into the intersection area. The yellow signal is timed to allow a vehicle that cannot stop to traverse distance d at the approach speed (S). A vehicle may legally enter the intersection on yellow.

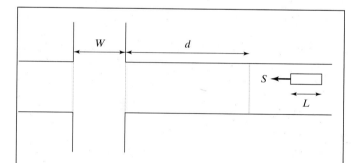

Figure 2.5: Timing Yellow and All-Red Intervals at a Signal

Having entered the intersection legally, the all-red period must allow the vehicle to cross the intersection width (W) and clear the back end of the vehicle (L) past the far intersection line.

Thus, the yellow interval must be timed to allow a vehicle to traverse the safe stopping distance. Consider a case in which the approach speed to a signalized intersection is 40 mi/h. How long should the yellow interval be?

The safe stopping distance is computed using a standard reaction time of 1.0 s for signal timing and level grade:

$$d = 1.47 * 40 * 1.0 + \frac{40^2 - 0^2}{30(0.348)}$$

$$= 58.8 + 153.3 = 212.1 \text{ ft}$$

The length of the yellow signal is the time it takes an approaching vehicle to traverse 212.1 ft at 40 mi/h, or:

$$y = \frac{212.1}{1.47 * 40} = 3.6 \text{ s}$$

In actual practice, the yellow interval is computed using a direct time-based algorithm and a standard deceleration rate. The principle, however, is the same. This example shows how the concept of safe stopping distance is incorporated into signal timing methodologies, which are discussed in detail in Chapter 20.

2.5 Closing Comments

This chapter has summarized some of the key elements of driver, pedestrian, and vehicle characteristics that influence highway design and traffic control. Together with the characteristics of the roadway itself, these elements combine to create traffic streams. As will be seen, the characteristics of traffic streams are the result of interactions among and between these elements. The characteristics of human road users and their vehicles have a fundamental impact on traffic streams.

References

1. Dewar, Robert, "Road Users," *Traffic Engineering Handbook*, 5th Edition, Institute of Transportation Engineers, Washington DC, 1999.

2. Ogden, K.W., *Safer Roads: A Guide to Road Safety Engineering*, University Press, Cambridge, England, 1996.

3. Allen, Merrill, *et al.*, *Forensic Aspects of Vision and Highway Safety*, Lawyers and Judges Publishing Co., Inc.,Tucson, AZ, 1996.

4. Olson, Paul, *Forensic Aspects of Driver Perception and Response*, Lawyers and Judges Publishing Co., Inc., Tucson, AZ, 1996.

5. *A Policy on Geometric Design of Highways and Streets*, 4th Edition, American Association of State Highway and Transportation Officials, Washington DC, 2001.

6. Johansson, G. and Rumar, K., "Driver's Brake Reaction Times," *Human Factors*, Vol. 13, No. 1, Human Factors and Ergonomics Society, February 1971.

7. *Report of the Massachusetts Highway Accident Survey*, Massachusetts Institute of Technology, Cambridge, MA, 1935.

8. Normann, O.K., "Braking Distances of Vehicles from High Speeds," *Proceedings of the Highway Research Board*, Vol. 22, Highway Research Board, Washington DC, 1953.

9. Fambro, D.B., *et al.*, "Determination of Safe Stopping Distances," *NCHRP Report 400*, Transportation Research Board, Washington DC, 1997.

10. *Determination of Vehicle Signal Change and Clearance Intervals*, Publication IR-073, Institute of Transportation Engineers, Washington DC, 1994.

11. *Human Factors*, Vol. 28, No. 1, Human Factors and Ergonomics Society, 1986.

12. Eubanks, J.J. and Hill, P.L., *Pedestrian Accident Reconstruction and Litigation*, 2nd Edition, Lawyers and Judges Publishing Co, Inc., Tucson, AZ, 1998.

13. Perry, J., *Gait Analysis*, McGraw-Hill, New York, NY, 1992.

14. Sleight, R.B., "The Pedestrian," *Human Factors in Traffic Safety Research*, John Wiley and Sons, Inc., New York, NY, 1972.

15. Tidwell, J.E. and Doyle, D., *Driver and Pedestrian Comprehension of Pedestrian Laws and Traffic Control Devices*, AAA Foundation for Traffic Safety, Washington DC, 1993.

16. Herms, B.F., "Pedestrian Crosswalk Study: Accidents in Painted and Unpainted Crosswalks," *Pedestrian Protection*, Highway Research Record 406, Transportation Research Board, Washington DC, 1972.

17. "Transportation in an Aging Society," *Special Report 218*, Transportation Research Board, Washington DC, 1988.

Problems

2-1. A driver takes 3.2 s to react to a complex situation while traveling at a speed of 55 mi/h. How far does the vehicle travel before the driver initiates a physical response to the situation (i.e., putting his or her foot on the brake)?

2-2. A driver traveling at 60 mi/h rounds a curve on a level grade to see a truck overturned across the roadway at a distance of 400 ft. If the driver is able to decelerate at a rate of 10 ft/s^2, at what speed will the vehicle hit the truck? Plot the result for reaction times ranging from 0.50 to 5.00 s in increments of 0.5 s. Comment on the results.

2-3. A car hits a tree at an estimated speed of 35 mi/h on a 3% downgrade. If skid marks of 100 ft are observed on dry pavement $(F = 0.45)$, followed by 250 ft $(F = 0.20)$ on a grass-stabilized shoulder, estimate the initial speed of the vehicle just before the pavement skid was begun.

2-4. Drivers must slow down from 70 mi/h to 60 mi/h to negotiate a severe curve on a rural highway. A warning sign for the curve is clearly visible for a distance of 100 ft. How far in advance of the curve must the sign be located in order to ensure that vehicles have sufficient distance to safely decelerate? Use the standard reaction time and deceleration rate recommended by AASHTO for basic braking maneuvers.

2-5. How long should the "yellow" signal be for vehicles approaching a traffic signal on a 2% downgrade at a speed of 35 mi/h? Use a standard reaction time of 1.0 s and the standard AASHTO deceleration rate.

2-6. What is the safe stopping distance for a section of rural freeway with a design speed of 80 mi/h on a 4% upgrade?

2-7. What minimum radius of curvature may be designed for safe operation of vehicles at 70 mi/h if the maximum rate of superelevation (e) is 6% and the maximum coefficient of side friction (f) is 0.10?

Roadways and their Geometric Characteristics

3.1 Highway Functions and Classification

Roadways are a major component of the traffic system, and the specifics of their design have a significant impact on traffic operations. There are two primary categories of service provided by roadways and roadway systems:

- Accessibility
- Mobility

"Accessibility" refers to the direct connection to abutting lands and land uses provided by roadways. This accessibility comes in the form of curb parking, driveway access to off-street parking, bus stops, taxi stands, loading zones, driveway access to loading areas, and similar features. The access function allows a driver or passenger to depart the transport vehicle to enter the particular land use in question. "Mobility" refers to the through movement of people, goods, and vehicles from Point A to Point B in the system.

The essential problem for traffic engineers is that the specific design aspects that provide for good access—parking, driveways, loading zones, etc.—tend to retard through movement, or mobility. Thus, the two major

services provided by a roadway system are often in conflict. This leads to the need to develop roadway systems in a hierarchal manner, with various classes of roadways specifically designed to perform specific functions.

3.1.1 Trip Functions

The American Association of State Highway and Transportation Officials (AASHTO) defines up to six distinct travel movements that may be present in a typical trip:

- Main movement
- Transition
- Distribution
- Collection
- Access
- Termination

The *main movement* is the through portion of trip, making the primary connection between the area of origin and the area of destination. *Transition* occurs when a vehicle transfers from the through portion of the trip to the remaining functions that lead to access and termination.

A vehicle might, for example, use a ramp to transition from a freeway to a surface arterial. The *distribution* function involves providing drivers and vehicles with the ability to leave a major through facility and get to the general area of their destinations. *Collection* brings the driver and vehicle closer to the final destination, while *access* and *termination* result in providing the driver with a place to leave his or her vehicle and enter the land use sought. Not all trips will involve all of these components.

The hierarchy of trip functions should be matched by the design of the roadways provided to accomplish them. A typical trip has two terminals, one at the origin, and one at the destination. At the origin end, the access function provides an opportunity for a trip-maker to enter a vehicle and for the vehicle to enter the roadway system. The driver may go through a series of facilities, usually progressively favoring higher speeds and through movements, until a facility—or set of facilities—is found that will provide the primary through connection. At the destination end of the trip, the reverse occurs, with the driver progressively moving toward facilities favoring access until the specific land parcel desired is reached.

3.1.2 Highway Classification

All highway systems involve a hierarchal classification by the mix of access and mobility functions provided. There are four major classes of highways that may be identified:

- Limited-access facilities
- Arterials
- Collectors
- Local streets

The *limited-access facility* provides for 100% through movement, or mobility. No direct access to abutting land uses is permitted. *Arterials* are surface facilities that are designed primarily for through movement but permit some access to abutting lands. *Local streets* are designed to provide access to abutting land uses with through movement only a minor function, if provided at all. The *collector* is an intermediate category between arterials and local streets. Some measure of both mobility and access is provided. The term "collector" comes from a common use of such facilities to collect vehicles from a number of local streets and deliver them to the nearest arterial or limited access facility.

Figure 3.1 illustrates the traditional hierarchy of these categories.

The typical trip starts on a local street. The driver seeks the closest collector available, using it to access the nearest arterial. If the trip is long enough, a freeway or limited-access facility is sought. At the destination end of the trip, the process is repeated in reverse order. Depending upon the length of the trip and specific characteristics of the area, not all component types of facilities need be included in every trip.

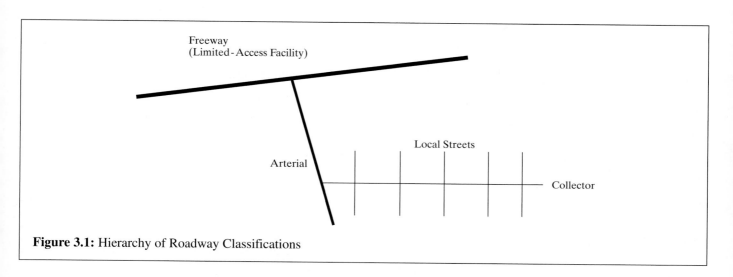

Figure 3.1: Hierarchy of Roadway Classifications

Table 3.1: Through Service Provided by Various Roadway Categories

Roadway Class	Percent Through Service
Freeways (Limited-Access Facilities)	100
Arterials	60–80
Collectors	40–60
Local Streets	0–40

Table 3.1 shows the range of through (or mobility) service provided by the major categories of roadway facility.

Many states have their own classification systems that often involve subcategories. Table 3.2 provides a general description of frequently-used subcategories in highway classification.

It is emphasized that the descriptions in Table 3.2 are presented as typical. Each highway agency will have its own highway classification system, and many have features that are unique to the agency. The traffic engineer should be familiar with highway classification systems, and be able to properly interpret any well-designed system.

3.1.3 Preserving the Function of a Facility

Highway classification systems enable traffic engineers to stratify the highway system by functional purpose. It is important that the intended function of a facility be reinforced through design and traffic controls.

Figure 3.2, for example, illustrates how the design and layout of streets within a suburban residential subdivision can reinforce the intended purpose of each facility.

The character of local streets is assured by incorporating sharp curvature into their design, and through the use of cul-de-sacs. No local street has direct access to an arterial; collectors within the subdivision provide the only access to arterials. The nature of collectors can be strengthened by not having any residence front on the collector.

The arterials have their function strengthened by limiting the number of points at which vehicles can enter or leave the arterial. Other aspects of an arterial, not obvious here, that could also help reinforce their function include:

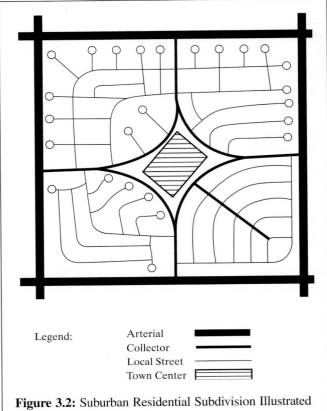

Legend:
Arterial
Collector
Local Street
Town Center

Figure 3.2: Suburban Residential Subdivision Illustrated

- Parking prohibitions
- Coordinated signals providing for continuous progressive movement at appropriate speeds
- Median dividers to limit midblock left turns, and
- Speed limits appropriate to the facility and its environment.

In many older cities, it is difficult to separate the functions served by various facilities due to basic design and control problems. The historic development of many older urban areas has led to open-grid street systems. In such systems, local streets, collectors, and surface arterials all form part of the grid. Every street is permitted to intersect every other street, and all facilities provide some land access. Figure 3.3 illustrates this case. The only thing that distinguishes an arterial in such a system is its width and provision of progressive signal timing to encourage through movement.

Table 3.2: Typical Rural and Urban Roadway Classification Systems

Subcategory	Rural	Urban
Freeways		
Interstate Freeways	All freeways bearing interstate designation.	All freeways bearing interstate designation.
Other Freeways	All other facilities with full control of access.	All other facilities with full control of access.
Expressways	Facilities with substantial control of access, but having some at-grade crossings or entrances.	Facilities with substantial control of access, but having some at-grade crossings or entrances.
Arterials		
Major or Principal Arterials	Serving significant corridor movements, often between areas with populations over 25,000 to 50,000. High-type design and alignment prevail.	Principal service for through movements, with very limited land-access functions that are incidental to the mobility function. High-type design prevails.
Minor Arterials	Provide linkage to significant traffic generators, including towns and cities with populations below the range for principal arterials; serve shorter trip lengths than principal arterials.	Principal service for through movements, with moderate levels of access service also present.
Collectors		
Major Collectors	Serve generators of intra-county importance not served by arterials; provide connections to arterials and/or freeways.	No subcategories usually used for urban collectors.
Minor Collectors	Link locally important generators with their rural hinterlands; provide connections to major collectors or arterials.	Provide land access and circulation service within residential neighborhoods and/or commercial/industrial areas; collect trips from local generators and channel them to nearby arterials; distribute trips from arterials to their ultimate destination.
Local Roads		
Subcategory	Rural	Urban
Residential	No subcategories generally used in rural classification schemes.	Provide land access and circulation within residential neighborhoods.
Commercial	Provide access to adjacent lands of all types; serve travel over relatively short distances.	Provide land access and circulation in areas of commercial development.
Industrial		Provide land access and circulation in areas of industrial development.

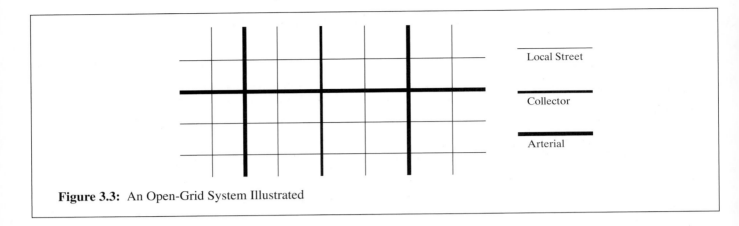

Figure 3.3: An Open-Grid System Illustrated

Such systems often experience difficulties when development intensifies, and all classes of facility, including arterials, are subjected to heavy pedestrian movements, loading and unloading of commercial vehicles, parking, and similar functions. Because local streets run parallel to collectors and arterials, drivers experiencing congestion on arterials often reroute themselves to nearby local streets, subjecting them to unwanted and often dangerous heavy through flows.

The importance of providing designs and controls that are appropriate to the intended function of a facility cannot be understated. Chapters 28 and 29 provide a more detailed discussion of techniques for doing so.

3.2   Highway Design Elements

Highways are complex physical structures involving compacted soil, sub-base layers of aggregate, pavements, drainage structures, bridge structures, and other physical elements.

From an operational viewpoint, it is the geometric characteristics of the roadway that primarily influence traffic flow and operations. Three main elements define the geometry of a highway section:

- Horizontal alignment
- Vertical alignment
- Cross-sectional elements

Virtually all standard practices in geometric highway design are specified by the American Association of State Highway and Transportation Officials in the current version of *Policy on Geometric Design of Highways and Streets* [*1*]. The latest edition of this key reference (at this writing) was published in 2001. Because of severe restrictions on using material directly from Reference 1, this text presents general design practices that are most frequently based upon AASHTO standards.

3.2.1   Introduction to Horizontal Alignment

The horizontal alignment refers to a plan view of the highway. The horizontal alignment includes tangent sections and the horizontal curves and other transition elements that join them.

Highway design is generally initiated by laying out a set of tangents on topographical and development maps of the service area. Selection of an appropriate route, and specific location of these tangent lines involves many considerations and is a complex task. Some of the more important considerations include:

- Forecast demand volumes, with known or projected origin-destination patterns
- Patterns of development
- Topography
- Natural barriers
- Subsurface conditions

- Drainage patterns
- Economic considerations
- Environmental considerations
- Social considerations

The first two items deal with anticipated demand on the facility and the specific origins and destinations that are to be served. The next four are important engineering factors that must be considered. The last three are critically important. Cost is always an important factor, but it must be compared with quantifiable benefits.

Environmental impact statements are required of virtually all highway projects, and much effort is put into providing remedies for unavoidable negative impacts on the environment. Social considerations are also important and cover a wide range of issues. It is particularly important that highways be built in ways that do not disrupt local communities, either by dividing them, enticing unwanted development, or causing particularly damaging environmental impacts. While this text does not deal in detail with this complex process of decision making, the reader should be aware of its existence and of the influence it has on highway programs in the United States.

3.2.2 Introduction to Vertical Alignment

Vertical alignment refers to the design of the facility in the profile view. Straight grades are connected by vertical curves, which provide for transition between adjacent grades. The "grade" refers to the longitudinal slope of the facility, expressed as "feet of rise or fall" per "longitudinal foot" of roadway length. As a dimensionless value, the grade may be expressed either as a decimal or as a percentage (by multiplying the decimal by 100).

In vertical design, attempts are made to conform to the topography, wherever possible, to reduce the need for costly excavations and landfills as well as to maintain aesthetics. Primary design criteria for vertical curves include:

- Provision of adequate sight distance at all points along the profile
- Provision of adequate drainage
- Maintenance of comfortable operations
- Maintenance of reasonable aesthetics

The specifics of vertical design usually follow from the horizontal route layout and specific horizontal design. The horizontal layout, however, is often modified or established in part to minimize problems in the vertical design.

3.2.3 Introduction to Cross-Sectional Elements

The third physical dimension, or view, of a highway that must be designed is the cross-section. The cross-section is a cut across the plane of the highway. Within the cross-section, such elements as lane widths, superelevation (cross-slope), medians, shoulders, drainage, embankments (or cut sections), and similar features are established. As the cross-section may vary along the length of a given facility, cross-sections are generally designed every 100 ft along the facility length and at any other locations that form a transition or change in the cross-sectional characteristics of the facility.

3.2.4 Surveying and Stationing

In the field, route surveyors define the geometry of a highway by "staking" out the horizontal and vertical position of the route and by similarly marking of the cross-section at intervals of 100 ft.

While this text does not deal with the details of route surveying, it is useful to understand the conventions of "stationing" which are used in the process. Stationing of a new or reconstructed route is generally initiated at the western or northern end of the project. "Stations" are established every 100 ft, and are given the notation $xxx + yy$. Values of "xxx" indicate the number of hundreds of feet of the location from the origin point. The "yy" values indicate intermediate distances of less than 100 ft.

Regular stations are established every 100 ft, and are numbered $0 + 00$, $100 + 00$, $200 + 00$, etc. Various elements of the highway are "staked" by surveyors at these stations. If key points of transition occur between full stations, they are also staked and would be given a notation such as $1200 + 52$, which signifies a location 1,252 ft. from the origin. This notation is used to describe points along a horizontal or vertical alignment in subsequent sections of this chapter.

Reference 2 is a text in route surveying, which can be consulted for more detailed information on the subject.

3.3 Horizontal Alignment of Highways

3.3.1 Geometric Characteristics of Horizontal Curves

Radius and Degree of Curvature

All highway horizontal curves are circular (i.e., they have a constant radius). The severity of a circular horizontal curve is measured by the *radius* or by the *degree of curvature*, which is a related measure. Degree of curvature is most often used, as higher values depict sharper, or more severe, curves. Conversely, larger radii depict less severe curves.

Figure 3.4 illustrates two ways of defining degree of curvature. The *chord definition* is illustrated in Figure 3.4 (a). The degree of curvature is defined as the central angle subtending a 100-ft chord on the circular curve. The *arc definition* is illustrated in Figure 3.4 (b), and is the most frequently used. In this definition, the degree of curvature is defined as the central angle subtending a 100-ft arc.

Using the arc definition, it is possible to derive the relationship between the radius (R) and the degree of curvature (D). The ratio of the circumference of the circle to 360° is set equal to the ratio of 100 ft to $D°$. Then:

$$\frac{2\pi R}{360} = \frac{100}{D}$$

$$D = \frac{100(360)}{2\pi R}$$

Noting that $\pi = 3.141592654\ldots$, then:

$$D = \frac{36,000}{2(3.1415915)R} = \frac{5,729.58}{R} \qquad (3\text{-}1)$$

where: D = degree of curvature, degrees
         R = radius of curvature, ft

Thus, for example, a circular curve with a radius of 2,000 ft has a degree of curvature of:

$$D = \frac{5,729.58}{2,000} = 2.865°$$

It should be noted that for up to 4° curves, there is little difference between the arc and the chord definition of degree of curvature. This text, however, will use only the arc definition illustrated in Figure 3.4 (b).

Review of Trigonometric Functions

The geometry of horizontal curves is described mathematically using trigonometric functions. A brief review of these functions is included as a refresher for those who may not have used trigonometry for some time. Figure 3.5 illustrates a right triangle, from which the definitions of trigonometric functions are drawn.

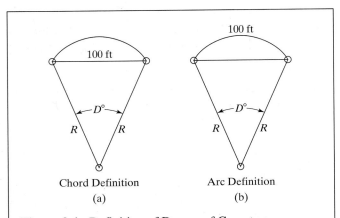

Figure 3.4: Definition of Degree of Curvature

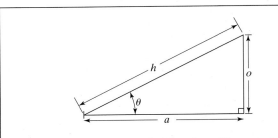

Figure 3.5: Trigonometric Functions Illustrated

In Figure 3.5:

- o = length of the opposite leg of the right triangle
- a = length of the adjacent leg of the right triangle
- h = hypotenuse of the right triangle

Using the legs of the right triangle, the following trigonometric functions are defined:

- Sine θ = o/h
- Cosine θ = a/h
- Tangent θ = o/a

From these primary functions, several derivative functions are also defined:

- Cosecant θ = 1/Sine θ = h/o
- Secant θ = 1/Cosine θ = h/a
- Cotangent θ = 1/Tangent θ = a/o

and:

- Exsecant θ = Secant θ − 1
- Versine θ = 1 − Cosine θ

Trigonometric functions are tabulated in many mathematics texts and are generally included on most calculators and in virtually all spreadsheet software. When using spreadsheet software or calculators, the user must determine whether angles are entered in *degrees* or *radians*. In a full circle, there are 2π radians and 360°. Thus, one radian is equal to $360/2(3.141592654) = 57.3°$.

Critical Characteristics of Circular Horizontal Curves

Figure 3.6 depicts a circular horizontal curve connecting two tangent lines. The following points are defined:

- *P.I.* = point of intersection; point at which the two tangent lines meet
- *P.C.* = point of curvature; point at which the circular horizontal curve begins
- *P.T.* = point of tangency; point at which the circular horizontal curve ends
- T = length of tangent, from the *P.C.* to the *P.I.* and from the *P.I.* to the *P.T.*, in feet

- E = external distance, from point 5 to the *P.I.* in Figure 3.6, in feet
- M = middle ordinate distance, from point 5 to point 6 in Figure 3.6, in feet
- *L.C.* = long chord, from the *P.C.* to the *P.T.*, in feet
- Δ = external angle of the curve, sometimes referred to as the angle of deflection, in degrees
- R = the radius of the circular curve, in feet

A number of geometric characteristics of the circular curve are of interest in deriving important relationships:

- Radii join tangent lines at right (90°) angles at the *P.C.* and *P. T.*
- A line drawn from the *P.I.* to the center of the circular curve bisects $\angle 412$ and $\angle 432$ (numbers refer to Figure 3.6)
- $\angle 412$ equals $180 − \Delta$; thus, $\angle 413$ and $\angle 312$ must be half this, or $90 − \Delta/2$ as shown in Figure 3.6
- Triangle 412 is an isosceles triangle. Thus, $\angle 142 = \angle 124$, and the sum of these, plus $\angle 412 \ (180 − \Delta)$, must be 180°. Therefore, $\angle 142 = \angle 124 = \Delta/2$
- $\angle 346$ and $\angle 326$ must be $90 − \Delta/2$, and the central angle, $\angle 432$, is equal to Δ
- The long chord (*L.C.*) and the line from point 1 to point 3 meet at a right (90°) angle

Given the characteristics shown in Figure 3.6, some of the key relationships for horizontal curves may be derived.

Length of the Tangent (T) Consider the tangent of triangle 143:

$$\mathrm{Tan}(\Delta/2) = \frac{T}{R}$$

Then:

$$T = R\,\mathrm{Tan}(\Delta/2) \tag{3-2}$$

Length of the Middle Ordinate (M) The length of the middle ordinate is found by subtracting line segment 3-6

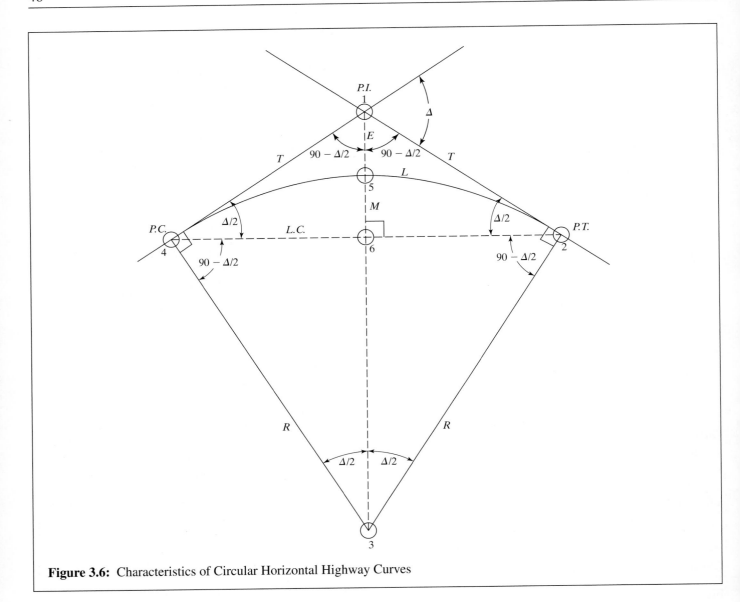

Figure 3.6: Characteristics of Circular Horizontal Highway Curves

from the radius, which is line segment 3-6-5. Then, considering triangle 362:

$$Cos(^\Delta/_2) = \frac{\text{seg } 36}{R}$$

$$\text{seg } 36 = R\, Cos(^\Delta/_2)$$

Then:

$$M = R - R\, Cos(^\Delta/_2) = R[1 - Cos(^\Delta/_2)] \quad (3\text{-}3)$$

Length of the External Distance (E)   Consider triangle 162. Then:

$$Sin(^\Delta/_2) = \frac{E + M}{T}$$

$$E = T\, Sin(^\Delta/_2) - M$$

Substituting the appropriate equations for T and M:

$$E = R\, Tan(^\Delta/_2)\, Sin(^\Delta/_2) - R[1 - Cos(^\Delta/_2)]$$

By manipulating the trigonometric functions, this may be rewritten as:

$$E = R\frac{Sin(^\Delta/_2)}{Cos(^\Delta/_2)}Sin(^\Delta/_2) - R + R\,Cos(^\Delta/_2)$$

$$E = R\frac{Sin^2(^\Delta/_2)}{Cos(^\Delta/_2)} - R - R\,Cos(^\Delta/_2)$$

$$E = \frac{R\,Sin^2(^\Delta/_2) - R\,Cos(^\Delta/_2) + R\,Cos^2(^\Delta/_2)}{Cos(^\Delta/_2)}$$

$$E = \frac{R[1 - Cos^2(^\Delta/_2)] - R\,Cos(^\Delta/_2) + R\,Cos(^\Delta/_2)}{Cos(^\Delta/_2)}$$

$$E = \frac{R[1 - Cos(^\Delta/_2)]}{Cos(^\Delta/_2)}$$

and:

$$E = R\left[\left(\frac{1}{Cos(^\Delta/_2)}\right) - 1\right] \qquad (3\text{-}4)$$

Length of the Curve (L)   The length of the curve derives directly from the arc definition of degree of curvature. A central angle equal to the degree of curvature subtends an arc of 100 ft, while the actual central angle (Δ) subtends the length of the curve (L). Thus:

$$\frac{L}{100} = \frac{\Delta}{D}$$

$$L = 100\left(\frac{\Delta}{D}\right) \qquad (3\text{-}5)$$

Length of the Long Chord (L.C.)   Note that the long chord is bisected by line segment 3-6-5. Then, considering triangle 364:

$$Sin(^\Delta/_2) = \frac{L.C./2}{R}$$

$$L.C. = 2\,R\,Sin(^\Delta/_2) \qquad (3\text{-}6)$$

An Example

Two tangent lines meet at Station 3,200 + 15. The radius of curvature is 1,200 ft, and the angle of deflection is 14°. Find the length of the curve, the stations for the P.C. and P.T., and all other relevant characteristics of the curve (L.C., M, E). Figure 3.7 illustrates the case.

Using the relationships discussed previously, all of the key measures for the curve of Figure 3.7 may be found:

$$D = \frac{5{,}729.58}{1200} = 4.77°$$

$$L = 100\left(\frac{14}{4.77}\right) = 293.5 \text{ ft}$$

$$L.C. = 2(1{,}200)\,Sin(^{14}/_2) = 292.5 \text{ ft}$$

$$T = 1{,}200\,Tan(^{14}/_2) = 147.3 \text{ ft}$$

$$M = 1{,}200[1 - Cos(^{14}/_2)] = 8.9 \text{ ft}$$

$$E = 1{,}200\left[\frac{1}{Cos(^{14}/_2)} - 1\right] = 9.0 \text{ ft}$$

Obviously, this is a fairly short curve, due to a small angle of deflection (14°). The question also asked for station designations for the P.C. and P.T. The computed characteristics are used to find these values. The

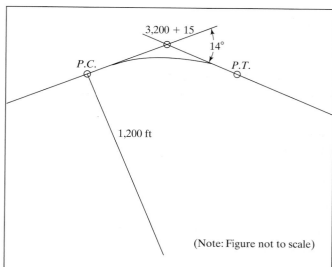

Figure 3.7: Horizontal Curve for Example

station of the *P.I.* is given as 3,200 + 15, which indicates that it is 3,215 ft from the beginning of the project.

The *P.C.* is found as the *P.I.* − *T*. This is 3,215 − 147.3 = 3,067.7, which is station 3,000 + 67.7. The station of the *P.T.* is found as the *P.C.* + *L*. This is 3,067.7 + 293.5 = 3361.2, which is station 3,300.0 + 61.2. All station units are in feet.

Superelevation of Horizontal Curves

Most highway curves are "superelevated", or banked, to assist drivers in resisting the effects of centripetal force. Superelevation is quantified as a percentage, computed as follows:

$$e = \left(\frac{\text{total rise in pavement from edge to edge}}{\text{width of pavement}} \right) \times 100 \quad (3\text{-}7)$$

As noted in Chapter 2, the two factors that keep a vehicle on a highway curve are side friction between the tires and the pavement, and the horizontal element of support provided by a banked or "superelevated" pavement. The speed of a vehicle and the radius of curvature are related to the superelevation rate (*e*) and the coefficient of side friction (*f*), by the equation:

$$R = \frac{S^2}{15(0.01e + f)} \quad (3\text{-}8)$$

where: R = radius of curvature, ft
S = speed of vehicle, mi/h
e = rate of superelevation, %
f = coefficient of side friction

In design, these values become limits: *S* is the design speed for the facility; *e* is the maximum rate of superelevation permitted; and *f* is a design value of the coefficient of side friction representing tires in reasonable condition on a wet pavement. The resulting value of *R* is the minimum radius of curvature permitted for these conditions.

Maximum Superelevation Rates

AASHTO [1] recommends the use of maximum superelevation rates between 4% and 12%. For design purposes, only increments of 2% are used. Maximum rates adopted vary from region to region based upon factors such as climate, terrain, development density, and frequency of slow-moving vehicles. Some of the practical considerations involved in setting this range, and for selection of an appropriate rate include:

1. Twelve percent (12%) is the maximum superelevation rate in use. Drivers feel uncomfortable on sections with higher rates, and driver effort to maintain lateral position is high when speeds are reduced on such curves.

2. Where snow and ice are prevalent, a maximum value of 8% is generally used. Many agencies use this as an upper limit regardless, due to the effect of rain or mud on highways.

3. In urban areas, where speeds may be reduced frequently due to congestion, maximum rates of 4%–6% are often used.

4. On low-speed urban streets or at intersections, superelevation may be eliminated.

It should be noted that on open highway sections, there is generally a minimum superelevation maintained, even on straight sections. This is to provide for cross-drainage of water to the appropriate roadside(s) where sewers or drainage ditches are present for longitudinal drainage. This minimum rate is usually in the range of 1.5% for high-type surfaces and 2.0% for low-type surfaces.

Side-Friction Factors (Coefficient of Side Friction)

Design values of the side-friction factor vary with design speed. Design values represent wet pavements and tires in reasonable but not top condition. Values also represent frictional forces that can be comfortably achieved; they do not represent, for example, the maximum side friction that is achieved the instant before skidding.

Design values for the coefficient of side friction (*f*) vary more-or-less linearly with speed from 0.080 at 80 mi/hr to 0.175 at 15 mi/h, as shown in Figure 3.8.

Determining Design Values of Superelevation

Once a maximum superelevation rate and a design speed are set, the minimum radius of curvature can be found using Equation 3-8. This can be expressed as a maximum degree of curvature using Equation 3-1.

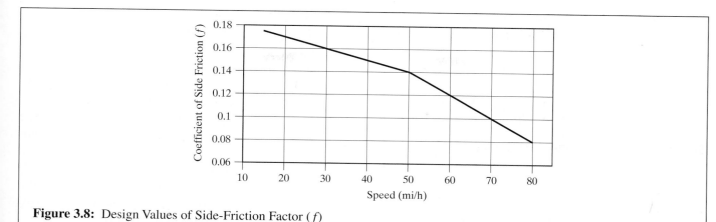

Figure 3.8: Design Values of Side-Friction Factor (f)

Consider a roadway with a design speed of 60 mi/h, for which a maximum superelevation rate of 6% has been selected. What are the minimum radius of curvature and/or maximum degree of curve that can be included on this facility?

For a design speed of 60 mi/h, Figure 3.8 indicates a design value for the coefficient of side friction (f) of 0.120. Then:

$$R_{min} = \frac{S^2}{15(0.01e_{max} + f_{des})} = \frac{60^2}{15(0.01 * 6 + 0.120)}$$

$$= 1{,}333.33 \text{ ft}$$

$$D_{max} = \frac{5{,}729.58}{R_{min}} = \frac{5{,}729.58}{1{,}333.33} = 4.3°$$

While this limits the degree of curvature to a maximum of 4.3° for the facility, it does not determine the appropriate rate of superelevation for degrees of curvature less than 4.3° (or a radius greater than 1,333.33 ft). The actual rate of superelevation for any curve with less than the maximum degree of curvature (or more than the minimum radius) is found by solving Equation 3-8 for e using the design speed for S and the appropriate design value of f. Then:

$$e = 100\left[\left(\frac{S^2_{des}}{15R}\right) - f_{des}\right] \qquad (3\text{-}9)$$

For the highway described above, what superelevation rate would be used for a curve with a radius of 1,500 ft? Using Equation 3-9:

$$e = 100\left[\left(\frac{60^2}{15 * 1{,}500}\right) - 0.120\right] = 4.0\%$$

Thus, while the maximum superelevation rate for this facility was set at 6%, a superelevation rate of 4.0% would be used for a curve with a radius of 1,500 ft, which is *larger* than the minimum radius for the design constraints specified for the facility. AASHTO standards [1] contain many curves and tables yielding results of such analyses for various specified constraints, for ease of use in design.

Achieving Superelevation

The transition from a tangent section with a normal superelevation for drainage to a superelevated horizontal curve occurs in two stages:

- *Tangent Runoff*: The outside lane of the curve must have a transition from the normal drainage superelevation to a level or flat condition prior to being rotated to the full superelevation for the horizontal curve. The length of this transition is called the tangent runoff and is noted as L_t.

- *Superelevation Runoff*: Once a flat cross-section is achieved for the outside lane of the curve, it must be rotated (with other lanes) to the full superelevation rate of the horizontal curve. The length of this transition is called the superelevation runoff and is noted as L_s.

For most undivided highways, rotation is around the centerline of the roadway, although rotation can also be accomplished around the inside or outside edge of the roadway as well. For divided highways, each directional roadway is separately rotated, usually around the inside or outside edge of the roadway.

Figure 3.9 illustrates the rotation of undivided two-lane, four-lane, and six-lane highways around the centerline, although the slopes shown are exaggerated for clarity. The rotation is accomplished in three steps:

1. The outside lane(s) are rotated from their normal cross-slope to a flat condition.

2. The outside lane(s) are rotated from the flat position until they equal the normal cross-slope of the inside lanes.

3. All lanes are rotated from the condition of step 2 to the full superelevation of the horizontal curve.

The tangent runoff is the distance taken to accomplish step 1, while the superelevation runoff is the distance taken to accomplish steps 2 and 3. The tangent and superelevation runoffs are, of course, implemented for the transition from tangent to horizontal curve and for the reverse transition from horizontal curve back to tangent.

In effect, the transition from a normal cross-slope to a fully superelevated section is accomplished by creating a grade differential between the rotation axis and the pavement edge lines. To achieve safe and comfortable operations, there are limitations on how much of a differential may be accommodated. The recommended minimum length of superelevation runoff is given as:

$$L_r = \frac{w * n * e_d * b_w}{\Delta} \qquad (3\text{-}10)$$

where: L_r = minimum length of superelevation runoff, ft
  w = width of a lane, ft
  n = number of lanes being rotated
  e_d = design superelevation rate, %
  b_w = adjustment factor for number of lanes rotated
  Δ = maximum relative gradient, %

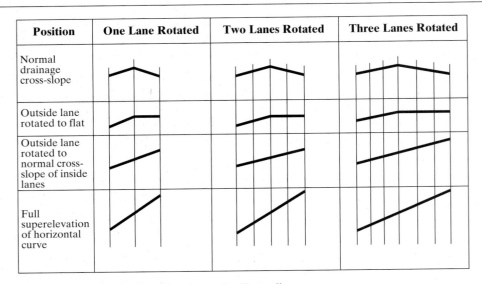

Position	One Lane Rotated	Two Lanes Rotated	Three Lanes Rotated
Normal drainage cross-slope			
Outside lane rotated to flat			
Outside lane rotated to normal cross-slope of inside lanes			
Full superelevation of horizontal curve			

Figure 3.9: Achieving Superelevation by Rotation Around a Centerline

AASHTO-recommended values for the maximum relative gradient, Δ, are shown in Table 3.3. The adjustment factor, b_w, depends upon the number of lanes being rotated. A value of 1.00 is used when one lane is being rotated, 0.75 when two lanes are being rotated, and 0.67 when three lanes are being rotated.

Consider the example of a four-lane highway, with a superelevation rate of 4% achieved by rotating two 12-ft lanes around the centerline. The design speed of the highway is 60 mi/h. What is the appropriate minimum length of superelevation runoff? From Table 3.3, the maximum relative gradient for 60 mi/h is 0.45%; the adjustment factor for rotating two lanes is 0.75. Thus:

$$L_r = \frac{w * n * e_d * b_w}{\Delta}$$

$$L_r = \frac{12 * 2 * 4 * 0.75}{0.45} = 160 \text{ ft}$$

Note that while it is a four-lane cross-section being rotated, $n = 2$, as rotation is around the centerline. Where separate pavements on a divided highway are rotated around an edge, the full number of lanes on the pavement would be used.

The length of the tangent runoff is related to the length of the superelevation runoff, as follows:

$$L_t = \frac{e_{NC}}{e_d} L_r \qquad (3\text{-}11)$$

where: L_t = length of tangent runoff, ft
L_r = length of superelevation runoff, ft
e_{NC} = normal cross-slope, %
e_d = design superelevation rate, %

If, in the previous example, the normal drainage cross-slope was 1%, then the length of the tangent runoff would be:

$$L_t = \left(\frac{1}{4}\right)160 = 40 \text{ ft}$$

The total transition length between the normal cross-section to the fully superelevated cross-section is the sum of the superelevation and tangent runoffs, or (in this example) 160 + 40 = 200 ft.

To provide drivers with the most comfortable operation, from 60% to 90% of the total runoff is achieved on the tangent section, with the remaining runoff achieved on the horizontal curve. AASHTO allows states or other jurisdictions to set a constant percentage split anywhere within this range as a matter of policy, but also gives criteria for optimal splits based on design speed and the number of lanes rotated. For design speeds between 15 and 45 mi/h, 80% (one lane rotated) or 90% (two or more lanes rotated) of the runoff is on the tangent section. For higher design speeds, 70% (one lane rotated), 80% (two lanes rotated) or 85% (three or more lanes rotated) of the runoff is on the tangent section.

Table 3.3: Maximum Relative Gradients (Δ) for Superelevation Runoff

Design Speed (mi/h)	Maximum Relative Gradient (%)	Design Speed (mi/h)	Maximum Relative Gradient (%)
15	0.78	50	0.50
20	0.74	55	0.47
25	0.70	60	0.45
30	0.66	65	0.43
35	0.62	70	0.40
40	0.58	75	0.38
45	0.54	80	0.35

(Used with permission of the American Association of State Highway and Transportation Officials, *A Policy on Geometric Design of Highways and Streets*, 4th Edition, condensed from Table 3-17, Pg. 170, Washington DC, 2001.)

Where a spiral transition curve (see next section) is used between the tangent and horizontal curves, the superelevation is achieved entirely on the spiral. If possible, the tangent and superelevation runoff may be accomplished on the spiral.

3.3.2 Spiral Transition Curves

While not impossible, it is difficult for drivers to travel immediately from a tangent section to a circular curve with a constant radius. A spiral transition curve begins with a tangent (degree of curve, $D = 0$) and gradually and uniformly increases the degree of curvature (decreases the radius) until the intended circular degree of curve is reached.

Use of a spiral transition provides for a number of benefits:

- Provides an easy path for drivers to follow: centrifugal and centripetal forces are increased gradually
- Provides a desirable arrangement for superelevation runoff
- Provides a desirable arrangement for pavement widening on curves (often done to accommodate off-tracking of commercial vehicles)
- Enhances highway appearance

The latter is illustrated in Figure 3.10, where the visual impact of a spiral transition curve is obvious. Spiral transition curves are not always used, as construction is difficult and construction cost is generally higher than for a simple circular curve. They are recommended for high-volume situations where degree of curvature exceeds 3°. The geometric characteristics of spiral transition curves are complex; they are illustrated in Figure 3.11.

The key variables in Figure 3.11 are defined as:

$T.S.$ = transition station from tangent to spiral
$S.C.$ = transition station from spiral to circular curve
$C.S.$ = transition station from circular curve to spiral
$S.T.$ = transition station from spiral to tangent
Δ = angle of deflection (central angle) of original circular curve without spiral
Δ_s = angle of deflection (central angle) of circular portion of curve with spiral

δ = angle of deflection for spiral portion of curve
L_s = length of the spiral, ft

Without going through the very detailed derivations for many of the terms included in Figure 3.11, some of the key relationships are described below.

Length of Spiral, L_s

The length of the spiral can be set in one of two ways: (a) L_s is set equal to the length of the superelevation runoff, as described in the previous section; (b) the length of the spiral can be determined as [6]:

$$L_s = \left(\frac{3.15 S^3}{RC} \right) \qquad (3\text{-}12)$$

where: L_s = length of the spiral, ft
S = design speed of the curve, mi/h
R = radius of the circular curve, ft
C = rate of increase of lateral acceleration, ft/s^3

The values of C commonly used in highway design range between 1 and 3 ft/s^3. When a value of 1.97 is used (a common standard value adopted by highway agencies), the equation becomes:

$$L_s = 1.6 \frac{S^3}{R} \qquad (3\text{-}13)$$

Angle of Deflection (Central Angle) for the Spiral, δ

The angle of deflection for the spiral reflects the average degree of curvature along the spiral. As the degree of curvature is uniformly increased from 0 to D, the average degree of curvature for the spiral is $D/2$. Thus, the angle of deflection for the spiral is:

$$\delta = \frac{L_s D}{200} \qquad (3\text{-}14)$$

where: δ = spiral angle of deflection, degrees
L_s = length of the spiral, ft
D = degree of curve for the circular curve, degrees

Figure 3.10: The Visual Impact of a Spiral Transition Curve (Used with permission of Yale University Press, C. Tunnard and B. Pushkarev, *Manmade America*, New Haven, CT, 1963.)

Angle of Deflection (Central Angle) for Circular Portion of Curve with Spiral Easement, Δ_s

By definition (see Figure 3.11):

$$\Delta_s = \Delta - 2\delta \qquad (3\text{-}15)$$

where: Δ_s = angle of deflection for circular curve with spiral, degrees or radians

Δ = angle of deflection for circular curve without spiral, degrees or radians

δ = angle of deflection for the spiral, degrees or radians

Length of Tangent Distance, T_s, between *P.I.* and *T.S.* (and *P.I.* and *S.T.*)

As shown on Figure 3.11, this is the distance between the point of intersection (*P.I.*) and the points at which the spiral curve transitions to or from the tangent. This distance is needed to appropriately station the curves. A very complicated derivation, the resulting equation is:

$$T_s = R \operatorname{Tan}\left(\frac{\Delta}{2}\right) + \left[R \operatorname{Cos}(\delta) - R + \frac{L_s^2}{6R} \right] \times$$

$$\operatorname{Tan}\left(\frac{\Delta}{2}\right) + [L_s - R \operatorname{Sin}(\delta)] \qquad (3\text{-}16)$$

where: T_s = distance between *P.I.* and *T.S.* (also *P.I.* and *S.T.*), ft

R = radius of circular curve, ft

Δ = angle of deflection for circular curve without spiral, degrees or radians

δ = angle of deflection for spiral, degrees or radians

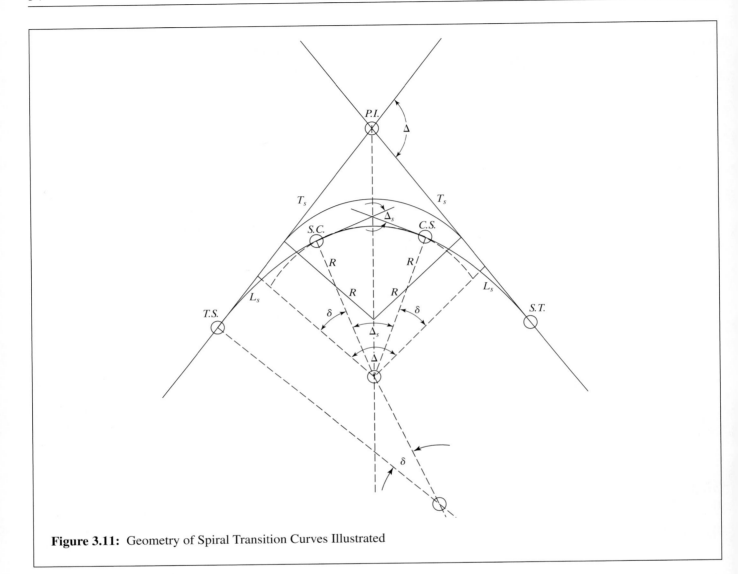

Figure 3.11: Geometry of Spiral Transition Curves Illustrated

A Sample Problem

A 4° curve is to be designed on a highway with two 12-ft lanes and a design speed of 60 mi/h. A maximum super-elevation rate of 6% has been established, and the appropriate side-friction factor for 60 mi/h is found from Figure 3.8 as 0.120. The normal drainage cross-slope on the tangent is 1%. Spiral transition curves are to be used. Determine the length of the spiral and the appropriate stations for the *T.S., S.C., C.S.,* and *S.T.* The angle of deflection for the original tangents is 38°, and the *P.I.* is at station 1,100 + 62. The segment has a two-lane cross section.

Solution: The radius of curvature for the circular portion of the curve is found from the degree of curvature as (Equation 3-1):

$$R = \frac{5,729.58}{D} = \frac{5,729.58}{4} = 1,432.4 \text{ ft}$$

The length of the spiral may now be computed using Equation 3-13:

$$L_s = 1.6\left(\frac{S^3}{R}\right) = 1.6\left(\frac{60^3}{1,432.4}\right) = 241.3 \text{ ft}$$

The minimum length of the spiral can also be determined as the length of the superelevation runoff. For a 60-mi/h design speed and a radius of 1,432.4 ft, the superelevation rate is found using Equation 3-9:

$$e = 100\left[\left(\frac{60^2}{15*1,432.4}\right) - 0.12\right] = 4.8\%$$

The length of the superelevation and tangent runoffs are computed from Equations 3-10 and 3-11 respectively. For 60 mi/h, the design value of Δ is 0.45 (Table 3-3). The adjustment factor for two lanes being rotated is 0.75. Then:

$$L_e = \frac{w*n*e_d*b_w}{\Delta} = \frac{12*2*4.8*0.75}{0.45} = 192 \text{ ft}$$

$$L_t = \frac{e_{NC}}{e_d}L_r = \left(\frac{1}{4.8}\right)192 = 40 \text{ ft}$$

The spiral must be at least as long as the superelevation runoff, or 192 ft. The result from Equation 3-13 is 241.3 ft, so this value controls. In fact, at 241.3 ft the minimum length of the spiral is sufficient to encompass both the superelevation runoff of 192 ft *and* the tangent runoff of 40 ft. Normally, the length of the spiral would be rounded, perhaps to 250 ft, which will be assumed for this problem.

The angle of deflection for the spiral is computed from Equation 3-14:

$$\delta = \frac{L_s D}{200} = \frac{250*4}{200} = 5°$$

The angle of deflection for the circular portion of the curve is (Equation 3-15):

$$\Delta_s = \Delta - 2\delta = 38 - 2(5) = 28°$$

The length of the circular portion of the curve, L_c, is found from Equation 3-6:

$$L_c = 100\left(\frac{\Delta_s}{D}\right) = 100\left(\frac{28}{4}\right) = 700 \text{ ft}$$

The distance between the *P.I.* and the *T.S.* is:

$$T_s = 1,432.4 \text{ Tan}\left(\frac{38}{2}\right) + \left[1,432.4 \text{ Cos}(5) - 1,432.4\right.$$

$$\left. + \frac{250^2}{6(1,432.4)}\right]\text{Tan}\left(\frac{38}{2}\right)$$

$$+ \left[250 - 1,432.4 \text{ Sin}(5)\right] = 619.0 \text{ ft}$$

From these results, the curve may now be stationed:

$$T.S. = P.I. - T_s = 1,162 - 619.0$$
$$= 543.0 = 500 + 43.0$$

$$S.C. = T.S. + L_s = 543.1 + 250$$
$$= 793.0 = 700 + 93.0$$

$$C.S. = S.C. + L_c = 793.0 + 700 = 1,493.0$$
$$= 1,400 + 93.0$$

$$S.T. = C.S. + L_s = 1,493.0 + 250$$
$$= 1,743.0 = 1,700 + 43.0$$

3.3.3 Sight Distance on Horizontal Curves

One of the most fundamental design criteria for all highway facilities is that a minimum sight distance equal to the safe stopping distance must be provided at every point along the roadway.

On horizontal curves, sight distance is limited by roadside objects (on the inside of the curve) that block drivers' line of sight. Roadside objects such as buildings, trees, and natural barriers disrupt motorists' sight lines. Figure 3.12 illustrates a sight restriction on a horizontal curve.

Figure 3.13 illustrates the effect of horizontal curves on sight distance. Sight distance is measured along the arc of the roadway, using the centerline of the inside travel lane. The middle ordinate, M, is taken as the distance from the centerline of the inside lane to the nearest roadside sight blockage.

The formula for the middle ordinate was given previously as:

$$M = R\left[1 - \text{Cos}\left(\frac{\Delta}{2}\right)\right]$$

The length of the circular curve has also been defined previously. In this case, however, the length of the

Figure 3.12: A Sight Restriction on a Horizontal Curve

curve is set equal to the required stopping sight distance. Then:

$$L = d_s = 100\left(\frac{\Delta}{D}\right)$$

$$\Delta = \frac{d_s D}{100}$$

Substituting in the equation for M:

$$M = R\left[1 - \text{Cos}\left(\frac{d_s D}{200}\right)\right]$$

The equation can be expressed uniformly using either the degree of curvature, D, or the radius of curvature, R:

$$M = \frac{5,729.58}{D}\left[1 - \text{Cos}\left(\frac{d_s D}{200}\right)\right]$$

$$M = R\left[1 - \text{Cos}\left(\frac{28.65\, d_s}{R}\right)\right] \qquad (3\text{-}17)$$

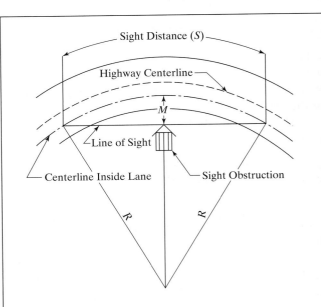

Figure 3.13: Sight Restrictions on Horizontal Curves (Used with permission of American Association of State Highway and Transportation Officials, *A Policy on Geometric Design of Highways and Streets*, 4th Edition, Exhibit 3-58, pg. 231, Washington DC, 2001.)

Remember (see Chapter 2) that the safe stopping distance used in each of these equations may be computed as:

$$d_s = 1.47\,S\,t + \frac{S^2}{30(0.348 \pm 0.01G)}$$

where: d_s = safe stopping distance, ft

$\quad S$ = design speed, mi/h

$\quad t$ = reaction time, secs

$\quad G$ = grade, %

A Sample Problem:

A 6° curve (measured at the centerline of the inside lane) is being designed for a highway with a design speed of 70 mi/h. The grade is level, and driver reaction time will be taken as 2.5 seconds, the AASHTO standard for highway braking reaction. What is the closest any roadside object may be placed to the centerline of the inside lane of the roadway?

Solution: The safe stopping distance, d_s, is computed as:

$$d_s = 1.47(70)(2.5) + \frac{70^2}{30(0.348 + 0.01*0)}$$

$$= 257.3 + 469.3 = 726.6 \text{ ft}$$

The minimum clearance at the roadside is given by the middle ordinate for a sight distance of 726.6 ft:

$$M = \frac{5{,}729.58}{6}\left[1 - \text{Cos}\left(\frac{726.6 * 6}{200}\right)\right] = 68.3 \text{ ft}$$

Thus, for this curve, no objects or other sight blockages on the inside roadside may be closer than 68.3 ft to the centerline of the inside lane.

3.3.4   Compound Horizontal Curves

A compound horizontal curve consists of two or more consecutive horizontal curves in a single direction with different radii. Figure 3.14 illustrates such a curve.

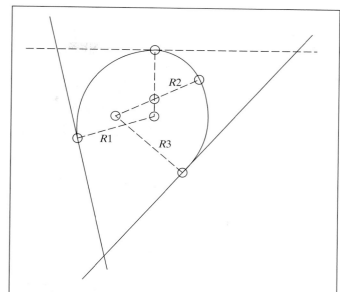

Figure 3.14: Compound Horizontal Curve Illustrated

Some general criteria for such curves include:

- Use of compound curves should be limited to cases in which physical conditions require it.
- Whenever two consecutive curves are connected on a highway segment, the larger radii should not be more than 1.5 times the smaller. A similar criteria is that the degrees of curvature should not differ by more than 5°.
- Whenever two consecutive curves in the same direction are separated by a short tangent (<200 ft), they should be combined in a compound curve.
- A compound curve is merely a series of simple horizontal curves subject to the same criteria as isolated horizontal curves.
- AASHTO relaxes some of these criteria for compound curves for ramp design.

3.3.5   Reverse Horizontal Curves

A reverse curve consists of two consecutive horizontal curves in opposite directions. Such a curve is illustrated in Figure 3.15. Two horizontal curves in opposite directions should always be separated by a tangent of at least 200 ft. Use of spiral transition curves is a significant assist to drivers negotiating reverse curves.

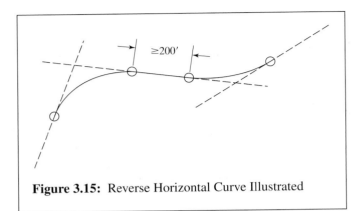

Figure 3.15: Reverse Horizontal Curve Illustrated

3.4 Vertical Alignment of Highways

The vertical alignment of a highway is the *profile* design of the facility in the vertical plane. The vertical alignment is composed of a series of vertical tangents connected by vertical curves. Vertical curves are in the shape of a *parabola*. This provides for a natural transition from a tangent to a curved section as part of the curve characteristics. Therefore, there is no need to investigate or provide transition curves, such as the spiral for horizontal curves.

The longitudinal slope of a highway is called the *grade*. It is generally stated as a percentage.

In vertical design, attempts are made to conform to the topography wherever possible to reduce the need for costly excavations and landfills as well as to maintain aesthetics. Primary design criteria for vertical curves include:

- Provision of adequate sight distance at all points along the profile
- Provision of adequate drainage
- Maintenance of comfortable operations
- Maintenance of reasonable aesthetics

3.4.1 Grades

Vertical tangents are characterized by their longitudinal slope, or grade. When expressed as a percent, the grade indicates the relative rise (or fall) of the facility in the longitudinal direction as a percentage of the length of the section under study. Thus, a 4% grade of 2,000 ft involves

a vertical rise of $2,000 * (4/100) = 80$ ft. Upgrades have positive slopes and percent grades, while downgrades have negative slopes and percent grades.

Maximum recommended grades for use in design depend upon the type of facility, the terrain in which it is built, and the design speed. Figure 3.16 presents a general overview of usual practice. These criteria represent a balance between the operating comfort of motorists and passengers and the practical constraints of design and construction in more severe terrains.

The principal operational impact of a grade is that trucks will be forced to slow down as they progress up the grade. This creates gaps in the traffic stream that cannot be effectively filled by simple passing maneuvers. Figure 3.17 illustrates the effect of upgrades on the operation of trucks with a weight-to-horsepower ratio of 200 lbs/hp, which is considered to be operationally typical of the range of commercial vehicles on most highways. It depicts deceleration behavior with an assumed entry speed of 70 mi/h.

Because of the operation of trucks on grades, simple maximum grade criteria are not sufficient for design. An example is shown in Figure 3.18. Trucks entering an upgrade with an assumed speed of 70 mi/h begin to slow. The length of the upgrade determines the extent of deceleration. For example, a truck entering a 5% upgrade at 70 mi/h slows to 50 mi/h after 2,000 ft and 32 mi/h after 4,000 ft. Eventually, the truck reaches its "crawl speed." The crawl speed is that constant speed that the truck can maintain for any length of grade (of the given steepness). Using the same example, a truck on a 5% upgrade has a crawl speed of 26 mi/h that is reached after approximately 7,400 ft.

Thus, the interference of trucks with general highway operations is related not only to the steepness of the grade but to its length as well. For most design purposes, grades should not be longer than the "critical length." For grades entered at 70 mi/h, the critical length is generally defined as the length at which the speed of trucks is 15 mi/h less than their speed upon entering the grade. When trucks enter an upgrade from slower speed, a speed reduction of 10 mi/h may be used to define the critical length of grade.

Figure 3.19 shows the relationship between length of grade, percent grade, and speed reduction for 200 lb/hp

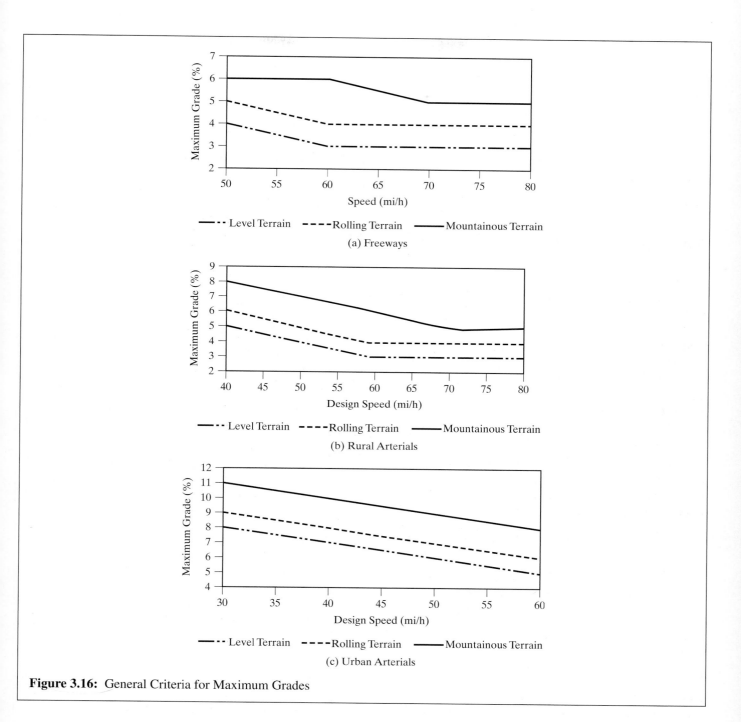

Figure 3.16: General Criteria for Maximum Grades

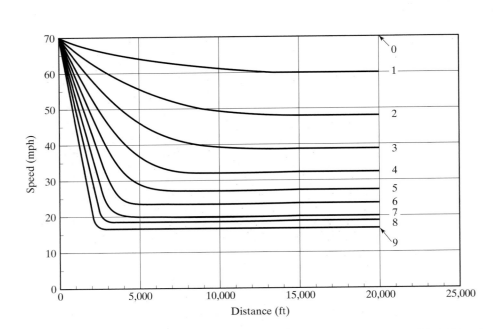

Figure 3.17: Deceleration of Typical Trucks (200 lbs/hp) on Upgrades (Used with permission of American Association of State Highway and Transportation Officials, *A Policy on Geometric Design of Highways and Streets*, 4th Edition, Exhibit 3-59, pg. 237, Washington DC, 2001.)

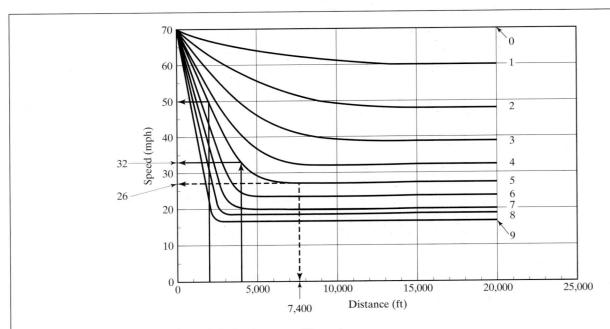

Figure 3.18: An Example of Truck Behavior on an Upgrade

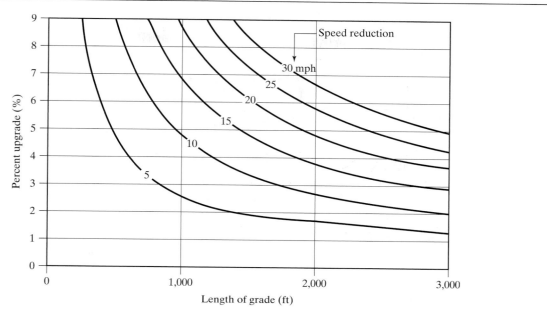

Figure 3.19: Critical Lengths of Grade for a Typical Truck (200 lbs/hp) (Used with permission of American Association of State Highway and Transportation Officials, *A Policy on Geometric Design of Highways and Streets*, 4th Edition, Exhibit 3-63, pg. 245, Washington DC, 2001.)

trucks entering a grade at 70 mi/h. These curves can be used to determine critical length of grade. It should be noted that terrain may make it impossible to limit grades to the critical length or shorter.

A Sample Problem:

A rural freeway in rolling terrain has a design speed of 60 mi/h. What is the longest and steepest grade that should be included on the facility?

Solution: From Figure 3.16 (a), for a freeway facility with a design speed of 60 mi/h in rolling terrain, the maximum allowable grade is 4%. Entering Figure 3.19 with 4% on the vertical axis, moving to the "15 mi/h" curve, the critical length of grade is seen to be approximately 1,900 ft.

   Again, it must be emphasized that terrain sometimes makes it impossible to consistently follow maximum grade design criteria. This is particularly true for desirable maximum grade lengths. Where the terrain is rising for significant distances, the profile of the roadway must do so as well. It is, however, true that grades longer than the critical length will generally operate poorly, and the addition of a climbing lane may be warranted in such situations.

3.4.2 Geometric Characteristics of Vertical Curves

As noted previously, vertical curves are in the shape of a parabola. In general, there are two types of vertical curves:

- Crest vertical curves
- Sag vertical curves

For crest vertical curves, the entry tangent grade is greater than the exit tangent grade. While traveling along a crest vertical curve, the grade is constantly declining. For sag vertical curves, the opposite is true: the entry tangent grade is lower than the exit tangent grade, and while traveling along the curve, the grade is constantly increasing. Figure 3.20 illustrates the various types of vertical curves.

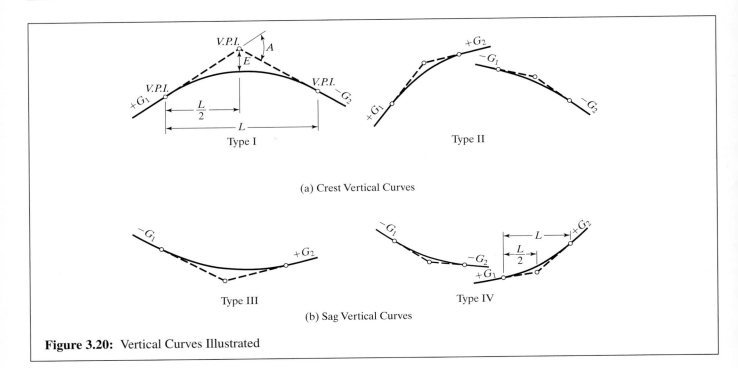

Figure 3.20: Vertical Curves Illustrated

The terms used in Figure 3.20 are defined as:

> $V.P.I.$ = vertical point of intersection
>
> $V.P.C.$ = vertical point of curvature
>
> $V.P.T.$ = vertical point of tangency
>
> G_1 = approach grade, %
>
> G_2 = departure grade, %
>
> L = length of vertical curve, in *hundreds* of ft

Length, and all stationing, on a vertical curve is measured in the plan view, (i.e., along a level axis). Two other useful variables are defined as follows:

$$A = G_2 - G_1$$

$$r = \frac{G_2 - G_1}{L} \tag{3-18}$$

where: A = algebraic change in grade, percent
r = rate of change in grade per 100 ft

The general form of a parabola is:

$$y = ax^2 + bx + c$$

For the purposes of describing a vertical curve, let:

> $y = Y_x$ = elevation of the vertical curve at a point
>
> x = distance from $V.P.C.$ in hundreds of ft
>
> $c = Y_0$ = elevation of the $V.P.C.$,
>   which occurs where $x = 0$ hundreds of ft

Then:

$$Y_x = ax^2 + bx + Y_0$$

Also, consider that the slope of the curve at any point x is the first derivative of this equation, or:

$$\frac{dY}{dx} = 2ax + b$$

When $x = 0$, the slope is equal to the entry grade, G_1. Thus:

$$\frac{dY}{dx} = G_1 = 2a(0) + b$$

$$b = G_1$$

The second derivative of the equation is equal to the rate of change in slope along the grade, or:

$$\frac{d^2Y}{dx^2} = 2a = r = \frac{G_2 - G_1}{L}$$

$$a = \frac{G_2 - G_1}{2L}$$

Thus, the final form of the equation for a vertical curve is given as:

$$Y_x = \left(\frac{G_2 - G_1}{2L}\right)x^2 + G_1 x + Y_0 \qquad (3\text{-}19)$$

The location of the high point (on a crest vertical curve) or the low point (on a sag vertical curve) is at a point where the slope (or first derivative) is equal to "zero." Note that for curves in which both grades are either up or down, no such point will exist on the curve. Taking the first derivative of the final curve:

$$\frac{dY_x}{dx} = 0 = \left(\frac{G_2 - G_1}{L}\right)x + G_1$$

$$x = \frac{-G_1 L}{G_2 - G_1} \qquad (3\text{-}20)$$

In all of these equations, care must be taken to address the sign of the grade. A negative grade has a minus $(-)$ sign that must be accounted for in the equation. Double negatives become positives in the equation.

A Sample Problem:

A vertical curve of 600 ft (L = 6) connects a +4% grade to a −2% grade. The elevation of the V.P.C. is 1,250 ft. Find the elevation of the P.V.I., the high point on the curve, and the V.P.T.

Solution: The elevation of the V.P.I. is found from the elevation of the V.P.C., the approach grade, and the length of the vertical curve. The V.P.I. is located on the extension of the approach grade at a point $^1/_2$ L into the curve, or:

$$Y_{V.P.I.} = Y_{V.P.C.} + G_1\left(\frac{L}{2}\right) = 1{,}250 + 4\left(\frac{6}{2}\right)$$

$$= 1{,}250 + 12 = 1{,}262 \text{ ft}$$

Following the format of Equation 3-19, the equation for this particular vertical curve is:

$$Y_x = \left(\frac{-2 - 4}{2*6}\right)x^2 + 4x + 1{,}250$$

$$= -\left(\frac{1}{2}\right)x^2 + 4x + 1{,}250$$

The elevation of the V.P.T. is the elevation of the curve at the end of its length of 600 ft, or where $x = 6$:

$$Y_{V.P.T.} = -\left(\frac{1}{2}\right)6^2 + 4(6) + 1{,}250$$

$$= -18 + 24 + 1{,}250 = 1{,}256 \text{ ft}$$

The high point of the curve occurs at a point where:

$$x = \frac{-G_1 L}{G_2 - G_1} = \frac{-4(6)}{-2 - 4} = \frac{-24}{-6} = 4(100 \text{ ft})$$

Then:

$$Y_{high} = -\left(\frac{1}{2}\right)4^2 + 4(4) + 1{,}250 = 1{,}258 \text{ ft}$$

3.4.3   Sight Distance on Vertical Curves

The minimum length of vertical curve is governed by sight-distance considerations. On vertical curves, sight distance is measured from an assumed eye height of 3.5 ft and an object height of 2.0 ft. Figure 3.21 shows a situation in which sight distance is limited by vertical curvature.

For crest vertical curves, the daylight sight line controls minimum length of vertical curves. The minimum length of a crest vertical curve is given by Equation 3-21 for cases in which the stopping sight distance is less than the length of the curve ($d_s < L$) or Equation 3-22 for cases in which the stopping sight distance is greater than the length of the vertical curve ($d_s > L$). The two equations yield equal results when $d_s = L$.

$$L = \frac{|G_2 - G_1|*d_s^2}{2{,}158} \qquad \text{for } d_s < L \qquad (3\text{-}21)$$

$$L = 2d_s - \left(\frac{2{,}158}{|G_2 - G_1|}\right) \qquad \text{for } d_s > L \qquad (3\text{-}22)$$

Figure 3.21: Sight Distance Limited by a Crest Vertical Curve

For sag vertical curves, the sight distance is limited by the headlamp range during nighttime driving conditions. Again, two equations result. Equation 3-23 is used when $d_s < L$, and Equation 3-24 is used when $d_s > L$. Once again, both equations yield the same results when $d_s = L$.

$$L = \frac{|G_2 - G_1| * d_s^2}{400 + 3.5\,d_s} \quad \text{for } d_s < L \qquad (3\text{-}23)$$

$$L = 2\,d_s - \left(\frac{400 + 3.5\,d_s}{|G_2 - G_1|}\right) \quad \text{for } d_s > L \quad (3\text{-}24)$$

where: L = minimum length of vertical curve, ft
   d_s = required stopping sight distance, ft
   G_2 = departure grade, %
   G_1 = approach grade, %

A Sample Problem:

What is the minimum length of vertical curve that must be provided to connect a 5% grade with a 2% grade on a highway with a design speed of 60 mi/h? Driver reaction time is the AASHTO standard of 2.5 s for simple highway stopping reactions.

Solution: This vertical curve is a *crest* vertical curve, as the departure grade is less than the approach grade.

The safe stopping distance is computed assuming that the vehicle is on a 2% upgrade. This results in a worst-case stopping distance:

$$d_s = 1.47(60)(2.5) + \frac{60^2}{30(0.348 + 0.01*2)}$$

$$= 220.5 + 326.1 = 546.6 \text{ ft}$$

Rounding off this number, a stopping sight distance requirement of 547 ft will be used. The first computation is made assuming that the stopping sight distance is less than the resulting length of curve. Using Equation 3-21 for this case:

$$L = \frac{|2 - 5|\,547^2}{2{,}158} = 416.0 \text{ ft}$$

From this result, it is clear that the initial assumption that $d_s < L$ was not correct. Equation 3-22 is now used:

$$L = 2*547 - \left(\frac{2{,}158}{|2 - 5|}\right) = 1{,}094 - 719.3$$

$$= 374.7 \text{ ft}$$

In this case, $d_s > L$, as assumed, and the 375 ft (rounded) is taken as the result.

3.4.4 Other Minimum Controls on Length of Vertical Curves

There are two other controls on the minimum length of *sag vertical curves only* vertical curves. For driver comfort, the minimum length of vertical curve is given by:

$$L = \frac{|G_2 - G_1|S^2}{46.5} \qquad (3\text{-}25)$$

For general appearance, the minimum length of vertical curve is given by:

$$L = 100|G_2 - G_1| \qquad (3\text{-}26)$$

where: S = design speed, mi/h
   all other variables as previously defined

Neither of these controls would enter into the example done previously, as it involved a crest vertical curve.

Equations 3-21 through 3-24 for minimum length of vertical curve are based upon stopping sight distances only. Consult AASHTO standards [1] directly for similar criteria based upon passing sight distance (for two-lane roadways) and for sag curves interrupted by overpass structures that block headlamp paths for night vision.

3.4.5   Some Design Guidelines for Vertical Curves

AASHTO gives a number of common-sense guidelines for the design of highway profiles, which are summarized below:

1. A smooth grade line with gradual changes is preferred to a line with numerous breaks and short grades.

2. Profiles should avoid the "roller-coaster" appearance, as well as "hidden dips" in the alignment.

3. Undulating grade lines involving substantial lengths of momentum (down) grades should be carefully evaluated with respect to operation of trucks.

4. Broken-back grade lines (two consecutive vertical curves in the same direction separated by a short tangent section) should be avoided wherever possible.

5. On long grades, it may be preferable to place the steepest grades at the bottom, lightening the grade on the ascent. If this is difficult, short sections of lighter grades should be inserted periodically to aid operations.

6. Where at-grade intersections occur on roadway sections with moderate to steep grades, the grade should be reduced or flattened through the intersection area.

7. Sag vertical curves in cuts should be avoided unless adequate drainage is provided.

3.5   Cross-Section Elements of Highways

The cross-section of a highway includes a number of elements critical to the design of the facility. The cross-section view of a highway is a 90° cut across the facility from roadside to roadside. The cross-section includes the following features:

- Travel lanes
- Shoulders
- Side slopes
- Curbs
- Medians and median barriers
- Guardrails
- Drainage channels

General design practice is to specify the cross-section at each station (i.e., at points 100 ft apart and at intermediate points where a change in the cross-sectional design occurs). The important cross-sectional features are briefly discussed in the sections that follow.

3.5.1   Travel Lanes and Pavement

Paved travel lanes provide the space that moving (and sometimes parked) vehicles occupy during normal operations. The standard width of a travel lane is 12 ft (metric standard is 3.6 m), although narrower lanes are permitted when necessary. The minimum recommended lane width is 9 ft. Lanes wider than 12 ft are sometimes provided on curves to account for the off-tracking of the rear wheels of large trucks. Narrow lanes will have a negative impact on the capacity of the roadway and on operations [7]. In general, 9-ft and 10-ft lanes should be avoided wherever possible. Nine-foot (9-ft) lanes are acceptable only on low-volume, low-speed rural or residential roadways, and 10-ft lanes are acceptable only on low-speed facilities.

All pavements have a cross-slope that is provided (1) to provide adequate drainage, and (2) to provide superelevation on curves (see Section 3.3 of this chapter). For high-type pavements (portland cement concrete, asphaltic concrete), normal drainage cross-slopes range from 1.5% to 2.0%. On low-type pavements (penetration surfaces, compacted earth, etc.), the range of drainage cross-slopes is between 2% and 6%.

How the drainage cross-slope is developed depends upon the type of highway and the design of other drainage facilities. A pavement can be drained to *both* sides of the roadway or to one side. Where water is drained to both sides of the pavement, there must be drainage ditches or culverts and pipes on both sides of the pavement. In some cases, water drained to the roadside is simply absorbed into the earth; studies testing whether the soil is adequate to handle maximum expected water loads must be conducted before adopting this approach. Where more than one lane is drained to one side of the roadway, each successive lane should have a cross-slope that is 0.5% steeper than the previous lane. Figure 3.22 illustrates a typical cross-slope for a four-lane pavement.

On superelevated sections, cross-slopes are usually sufficient for drainage purposes, and a slope differential between adjacent lanes is not needed. Superelevated sections, of course, must drain to the inside of the horizontal curve, and the design of drainage facilities must accommodate this.

3.5.2 Shoulders

AASHTO defines shoulders in the following way: "*A shoulder is the portion of the roadway contiguous with the traveled way that accommodates stopped vehicles, emergency use, and lateral support of sub-base, base, and surface courses (of the roadway structure).*" [Ref 1, pg. 316]. Shoulders vary widely in both size and physical appearance. For some low-volume rural roads in difficult terrain, no shoulders are provided. Normally, the shoulder width ranges from 2 ft to 12 ft. Most shoulders

are "stabilized" (i.e. treated with some kind of material that provides a reasonable surface for vehicles). This can range from a fully-paved shoulder to shoulders stabilized with penetration or stone surfaces or simply grass over compacted earth. For safety, it is critical that the joint between the traveled way and the shoulder be well maintained.

Shoulders are generally considered necessary on rural highways serving a significant mobility function, on all freeways, and on some types of urban highways. In these cases, a minimum width of 10 ft is generally used, as this provides for stopped vehicles to be about 2 ft clear of the traveled way. The narrowest 2-ft shoulders should be used only for the lowest classifications of highways. Even in these cases, 6–8 ft is considered desirable.

Shoulders serve a variety of functions, including:

- Providing a refuge for stalled or temporarily stopped vehicles
- Providing a buffer for accident recovery
- Contributing to driving ease and driver confidence
- Increasing sight distance on horizontal curves
- Improving capacity and operations on most highways
- Provision of space for maintenance operations and equipment
- Provision of space for snow removal and storage
- Provision of lateral clearance for signs, guardrails, and other roadside objects
- Improved drainage on a traveled way
- Provision of structural support for the roadbed

Reference 8 provides an excellent study of the use of roadway shoulders.

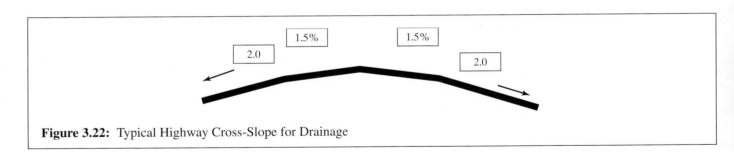

Figure 3.22: Typical Highway Cross-Slope for Drainage

Table 3.4: Recommended Cross-Slopes for Shoulders

Type of Surface	Recommended Cross-Slope (%)
Bituminous	2.0–6.0
Gravel or stone	4.0–6.0
Turf	6.0–8.0

Table 3.4 shows recommended cross-slopes for shoulders, based on the type of surface. No shoulder should have a cross-slope of more than 7:1, as the probability of rollover is greatly increased for vehicles entering a more steeply sloped shoulder.

3.5.3 Side-Slopes for Cuts and Embankments

Where roadways are located in cut sections or on embankments, side-slopes must be carefully designed to provide for safe operation. In urban areas, sufficient right-of-way is generally not available to provide for natural side-slopes, and retaining walls are frequently used.

Where natural side-slopes are provided, the following limitations must be considered:

- A 3:1 side-slope is the maximum for safe operation of maintenance and mowing equipment.

- A 4:1 side-slope is the maximum desirable for accident safety. Barriers should be used to prevent vehicles from entering a side-slope area with a steeper slope.

- A 2:1 side-slope is the maximum on which grass can be grown, and only then in good climates.

- A 6:1 side-slope is the maximum that is structurally stable for where sandy soils are predominate.

Table 3.5 shows recommended side-slopes for various terrains and heights of cut and/or fill.

3.5.4 Guardrail

One of the most important features of any cross-section design is the use and placement of guardrail. "Guardrail" is intended to prevent vehicles from entering a dangerous area of the roadside or median during an accident or intended action.

Roadside guardrail is provided to prevent vehicles from entering a cross-slope steeper than 4:1, or from colliding with roadside objects such as trees, culverts, lighting standards, sign posts, etc. Once a vehicle hits a section of guardrail, the physical design also guides the vehicle into a safer trajectory, usually in direction of traffic flow.

Median guardrail is primarily provided to prevent vehicles from encroaching into the opposing lane(s) of traffic. It also prevents vehicles from colliding with median objects. The need for median guardrail depends upon the design of the median itself. If the median is 20 ft or wider and if there are no dangerous objects in the median, guardrail is usually not provided, and the median is not curbed. Wide medians can effectively serve as accident recovery areas for encroaching drivers.

Table 3.5: Recommended Side-Slopes for Cut and Fill Sections

Height of Cut Or Fill (ft)	Terrain		
	Level or Rolling	Moderately Steep	Steep
0–4	6:1	4:1	4:1
4–10	4:1	3:1	2:1
10–15	3:1	2.5:1	1.75:1[*]
15–20	2:1	2:1	1.5:1[*]
>20	2:1	1.5:1[*]	1.5:1[*]

[*]Avoid where soils are subject to erosion.

Narrower medians generally require some type of barrier, as the potential for encroaching vehicles to cross the entire median and enter the opposing traffic lanes is significant.

Figure 3.23 illustrates common types of guardrail in current use. The barriers shown in Figure 3.23 are *all configured for median use* (i.e., they are designed to protect encroachment from either side of the barrier).

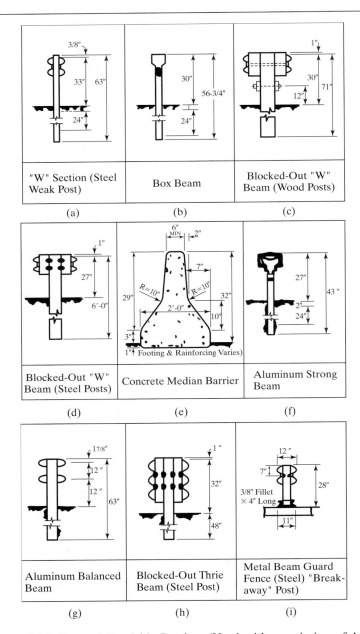

Figure 3.23: Common Types of Median and Roadside Barriers (Used with permission of American Association of State Highway and Transportation Officials, *A Policy on Geometric Design of Highways and Streets*, 2nd Edition, Figure IV-8, pg. 398, Washington DC, 1984.)

For roadside use, the same designs are used with only one collision surface.

The major differences in the various designs are the flexibility of guardrail upon impact and the strength of the barrier in preventing a vehicle from crossing through the barrier.

The box-beam design, for example, is quite flexible. Upon collision, several posts of the box-beam will give way, allowing the beam to flex as much as 10 to 15 ft. The colliding vehicle is gently straightened and guided back towards the travel lane over a length of the guardrail. Obviously, this type of guardrail is not useful in narrow medians, as it could well deflect into the opposing traffic lanes.

The most inflexible design is the concrete median or roadside barrier. These blocks are almost immovable, and it is virtually impossible to crash through them. Thus, they are used in narrow roadway medians (particularly on urban freeways), and on roadsides where virtually no deflection would be safe. On collision with such a barrier, the vehicle is straightened out almost immediately, and the friction of the vehicle against the barrier brings it to a stop.

The details of guardrail design are critical. End treatments must be carefully done. A vehicle colliding with a blunt end of a guardrail section is in extreme danger. Thus, most W-beam and box-beam guardrails are bent away from the traveled way, with their ends buried in the roadside. Even with this done, vehicles can (with some difficulty) hit the buried end and "ramp up" the guardrail with one or more wheels. Concrete barriers have sloped ends, but are usually protected by impact-attenuating devices, such as sand or water barrels or mechanical attenuators.

Connection of guardrail to bridge railings and abutments is also important. As most guardrails deflect, they cannot be isolated from fixed objects, as they could conceivably "guide" a vehicle into a dangerous collision with such an object. Thus, where guardrails meet bridge railings or abutments, they are anchored onto the railing or abutment itself to ensure that encroaching vehicles are guided away from the object.

3.6 Closing Comments

This chapter has provided a brief overview of the critical functional and geometric characteristics of highways. There are many more details involved in highway

geometry than those illustrated herein. The current AASHTO standard—*A Policy on Geometric Design of Highways and Streets*—should be consulted directly for a more detailed presentation of specific design practices and policies.

References

1. *A Policy on Geometric Design of Highways and Streets*, 4th Edition, American Association of State Highway and Transportation Officials, Washington DC, 2001.

2. Kavanagh, B.F., *Surveying With Construction Applications*, 4th Edition, Prentice-Hall, Upper Saddle River, NJ, 2001.

3. MacAdam, C.C., Fancher, P.S., and Segal, L., *Side Friction for Superelevation on Horizontal Curves*, Report No. FHWA-RD-86-024, U.S. Department of Transportation, Federal Highway Administration, McLean, VA, August 1985.

4. Moyer, R.A., "Skidding Characteristics of Automobile Tires on Roadway Surfaces and Their Relation to Highway Safety," *Bulletin No. 120*, Iowa Engineering Experiment Station, Ames, Iowa, 1934.

5. Hajela, G.P., *Compiler, Résumé of Tests on Passenger Cars on Winter Driving Surfaces*, National Safety Council, Chicago, IL, 1968.

6. Shortt, W.H., "A Practical Method for Improvement of Existing Railroad Curves," *Proceedings: Institution of Civil Engineering*, Vol 76, Institution of Civil Engineering, London ENG, 1909.

7. *Highway Capacity Manual*, 4th Edition, National Research Council, Transportation Research Board, Washington DC, 2000.

8. Zegeer, C.V., Stewart, R., Council, F.M., and Neuman, T.R., "Roadway Widths for Low-Traffic Volume Roads," *National Cooperative Highway*

Research Report 362, National Research Council, Transportation Research Board, Washington DC, 1994.

Problems

3-1. The point of intersection (*P.I.*) of two tangent lines is Station 1500 + 20. The radius of curvature is 900 feet, and the angle of deflection is 52°. Find the length of the curve, the stations for the *P.C.* and *P.T.*, and all other relevant characteristics of the curve (*LC, M, E*).

3-2. A 7° curve is to be designed on a highway with a design speed of 50 mi/h. Spiral transition curves are to be used. Determine the length of the spiral and the appropriate stations for the *T.S.*, *S.C.*, *C.S.*, and *S.T.* The angle of deflection for the original tangents is 45°, and the *P.I.* is at station 1200 + 80. The segment consists of two 12-ft lanes.

3-3. A 5° curve (measured at the centerline of the inside lane) is being designed for a highway with a design speed of 60 mi/h. The curve is on a 2% upgrade, and driver reaction time may be taken as 2.5 seconds. What is the closest any roadside object may be placed to the centerline of the inside lane of the roadway while maintaining adequate stopping sight distance?

3-4. What is the appropriate superelevation rate for a curve with a 1,000-ft radius on highway with a design speed of 55 mi/h? The maximum design superelevation is 6% for this highway.

3-5. What length of superelevation runoff should be used to achieve a superelevation rate of 10%? The design speed is 70 mi/h, and a three-lane cross section (12-ft lanes) is under consideration. Superelevation will be achieved by rotating all three lanes around the inside edge of the pavement.

3-6. Find the maximum allowable grade and critical length of grade for each of the following facilities:

(a) A rural freeway in mountainous terrain with a design speed of 60 mi/h

(b) A rural arterial in rolling terrain with a design speed of 45 mi/h

(c) An urban arterial in level terrain with a design speed of 40 mi/h

3-7. A vertical curve of 1,000 ft is designed to connect a grade of +4% to a grade of −5%. The *V.P.I.* is located at station 1,500 + 55 and has a known elevation of 500 ft. Find the following:

(a) The station of the *V.P.C.* and the *V.P.T.*

(b) The elevation of the *V.P.C.* and the *V.P.T.*

(c) The elevation of points along the vertical curve at 100-ft intervals

(d) The location and elevation of the high point on the curve

3-8. Find the minimum length of curve for the following scenarios:

Entry Grade	Exit Grade	Design Speed	Reaction Time
3%	7%	55 mi/h	2.5 s
−5%	2%	60 mi/h	2.5 s
2%	−3%	70 mi/h	2.5 s

3-9. A vertical curve is to be designed to connect a −4% grade to a +1% grade on a facility with a design speed of 70 mi/h. For economic reasons, a minimum-length curve will be provided. A driver-reaction time of 2.5 seconds may be used in sight distance determinations. The *V.P.I.* of the curve is at station 5,100 + 22 and has an elevation of 210 ft. Find the station and elevation of the *V.P.C.* and *V.P.T.*, the high point of the curve, and at 100-ft intervals along the curve.

4

Introduction to Traffic Control Devices

Traffic control devices are the media by which traffic engineers communicate with drivers. Virtually every traffic law, regulation, or operating instruction must be communicated through the use of devices that fall into three broad categories:

- Traffic markings
- Traffic signs
- Traffic signals

The effective communication between traffic engineer and driver is a critical link if safe and efficient traffic operations are to prevail. Traffic engineers have no direct control over any individual driver or group of drivers. If a motorman violated a RED signal while conducting a subway train, an automated braking system would force the train to stop anyway. If a driver violates a RED signal, only the hazards of conflicting vehicular and/or pedestrian flows would impede the maneuver. Thus, it is imperative that traffic engineers design traffic control devices that communicate uncomplicated messages clearly, in a way that encourages proper observance.

This chapter introduces some of the basic principles involved in the design and placement of traffic control devices. Subsequent chapters cover the details of specific applications to freeways, multilane and two-lane highways, intersections, and arterials and streets.

4.1 The Manual on Uniform Traffic Control Devices

The principal standard governing the application, design, and placement of traffic control devices is the current edition of the *Manual on Uniform Traffic Control Devices* (MUTCD) [*1*]. The Federal Highway Administration publishes a national MUTCD which serves as a minimum standard and a model for individual state MUTCDs. Many states simply adopt the federal manual by statute. Others develop their own manuals. In the latter case, the state MUTCD must meet all of the minimum standards of the federal manual, but it may impose additional or more stringent standards. As is the case with most federal mandates in transportation, compliance is enforced through partial withholding of federal-aid highway funds from states deemed in violation of federal MUTCD standards.

4.1.1   History and Background

One of the principal objectives of the MUTCD is to establish *uniformity* in the use, placement, and design of traffic control devices. Communication is greatly enhanced when the same messages are delivered in the same way and in similar circumstances at all times. Consider the potential confusion if each state designed its own STOP sign, with different shapes, colors, and legends.

Varying device design is not a purely theoretical issue. As late as the early 1950s, two-color (red, green) traffic signals had the indications in different positions in different states. Some placed the "red" ball on top; others placed the "green" ball on top. This is a particular problem for drivers with color blindness, the most common form of which is the inability to distinguish "red" from "green." Standardizing the order of signal lenses was a critical safety measure, guaranteeing that even colorblind drivers could interpret the signal by position of the light in the display. More recently, small amounts of blue and yellow pigment have been added to "green" and "red" lenses to enhance their visibility to color-blind drivers.

Early traffic control devices were developed in various locales with little or no coordination on their design, much less their use. The first centerline appeared on a Michigan roadway in 1911. The first electric signal installation is thought to have occurred in Cleveland, Ohio in 1914. The first STOP sign was installed in Detroit in 1915, where the first three-color traffic signal was installed in 1920.

The first attempts to create national standards for traffic control devices occurred during the 1920s. Two separate organizations developed two manuals in this period. In 1927, the American Association of State Highway Officials (AASHO, the forerunner of AASHTO), published the *Manual and Specification for the Manufacture, Display, and Erection of U.S. Standard Road Markings and Signs.* It was revised in 1929 and 1931. This manual addressed only rural signing and marking applications. In 1930, the National Conference on Street and Highway Safety (NCSHS) published the *Manual on Street Traffic Signs, Signals, and Markings,* which addressed urban applications.

In 1932, the two groups formed a merged Joint Committee on Uniform Traffic Control Devices and

published the first complete MUTCD in 1935, revising it in 1939. This group continued to have responsibility for subsequent editions until 1972, when the Federal Highway Administration formally assumed responsibility for the manual.

The latest edition of the MUTCD (at this writing) is the *Millennium Edition*, published, as the name suggests, in 2000. Work on this edition had started in the late 1980s, and its completion was later than originally expected (by two or three years). This chapter is based on the millennium edition.

Table 4.1 summarizes the editions of the MUTCD, and the formal revisions that have been issued between the release of fully updated manuals.

For an excellent history of the MUTCD and its development, consult a series of articles by Hawkins. [2–5].

4.1.2   General Principles of the MUTCD

The MUTCD states that the purpose of traffic control devices is "to promote highway safety and efficiency by providing for orderly movement of all road users on streets and highways, throughout the Nation." [Ref. 1, pg. 1A-1]. It also defines five requirements for a traffic control device to be effective in fulfilling that mission. A traffic control device must:

1. Fulfill a need
2. Command attention
3. Convey a clear, simple message
4. Command respect of road users
5. Give adequate time for a proper response

In addition to the obvious meanings of these requirements, some subtleties should be carefully noted. The first strongly implies that superfluous devices *should not* be used. Each device must have a specific purpose and must be needed for the safe and efficient flow or traffic. The fourth requirement reinforces this. Respect of drivers is commanded only when drivers are conditioned to expect that all devices carry meaningful and important messages. Overuse or misuse of devices encourages drivers to ignore them—it is like "crying

Table 4.1: Evolution of the MUTCD

Year	Title	Revisions Issued
1927	Manual and Specification for the Manufacture, Display, and Erection of U.S. Standard Road Markers and Signs	4/29, 12/31
1930	Manual on Street Traffic Signs, Signals, and Markings	None
1935	Manual on Uniform Traffic Control Devices for Streets and Highways	2/39
1943	Manual on Uniform Traffic Control Devices for Streets and Highways—War Emergency Edition	None
1948	Manual on Uniform Traffic Control Devices for Streets and Highways	9/54
1961	Manual on Uniform Traffic Control Devices for Streets and Highways	None
1971	Manual on Uniform Traffic Control Devices for Streets and Highways	11/71, 4/72, 3/73, 10/73, 6/74, 6/75, 9/76, 12/77
1978	Manual on Uniform Traffic Control Devices for Streets and Highways	12/79, 12/83, 9/84, 3/86
1988	Manual on Uniform Traffic Control Devices for Streets and Highways	1/90, 3/92, 9/93, 11/94, 12/96, 6/96, 1/00
2000	Manual on Uniform Traffic Control Devices for Streets and Highways—Millennium Edition	6/01

(Used with permission of the Federal Highway Administration, U.S. Department of Transportation, *Manual on Uniform Traffic Control Devices*, Millennium Edition, 2000, Table I-1.)

wolf" too often. In such an atmosphere, drivers may not pay attention to those devices that are really needed.

Items 2 and 3 affect the design of a device. Commanding attention requires proper visibility and a distinctive design that attracts the driver's attention in what is often an environment filled with visual distractions. Standard use of color and shape coding plays a major role in attracting this attention. Clarity and simplicity of message is critical; the driver is viewing the device for only a few short seconds while traveling at what may be a high speed. Again, color and shape coding is used to deliver as much information as possible. Legend, the hardest element of a device to understand, must be kept short and as simple as possible.

Item 5 affects the placement of devices. A STOP sign, for example, is always placed at the stop line, but must be visible for at least one safe stopping distance.

Guide signs requiring drivers to make lane changes must be placed well in advance of the diverge area to give drivers sufficient distance to execute the required maneuvers.

4.1.3 Contents of the MUTCD

The MUTCD addresses three critical aspects of traffic control devices. It contains:

1. Detailed standards for the physical design of the device, specifying shape, size, colors, legend types and sizes, and specific legend.

2. Detailed standards and guidelines on where devices should be located with respect to the traveled way.

3. Warrants, or conditions, that justify the use of a particular device.

The most detailed and definitive standards are for the physical design of the device. Little is left to judgment, and virtually every detail of the design is fully specified. Colors are specified by specific pigments and legend by specific fonts. Some variance is permitted with respect to size, with minimum sizes specified and optional larger sizes for use when needed for additional visibility.

Placement guidelines are also relatively definitive but often allow for some variation within prescribed limits. Placement guidelines sometimes lead to obvious problems. One frequent problem involves STOP signs. When placed in the prescribed position, they may wind up behind trees or other obstructions where their effectiveness is severely compromised. Figure 4.1 shows such a case, in which a STOP sign placed at the prescribed height and lateral offset at the stop line winds up virtually hidden by a tree. Common sense must be exercised in such cases if the device is to be effective.

Warrants are given with various levels of specificity and clarity. Signal warrants, for example, are detailed and relatively precise. This is necessary, as signal installations represent a significant investment, both in initial investment, and in continuing operating and maintenance costs. The warrants for STOP and YIELD signs,

however, are far more general and leave substantial latitude for the exercise of professional judgment.

Chapter 16 deals with the selection of an appropriate form of intersection control and covers the warrants for signalization, two-way and multiway STOP signs, and YIELD signs in some detail. Because of the cost of signals, much study has been devoted to the defining of conditions warranting their use. Proper implementation of signal and other warrants in the MUTCD requires appropriate engineering studies to be made to determine the need for a particular device or devices.

4.1.4 Legal Aspects of the MUTCD

The Millennium Edition of the MUTCD provides guidance and information in four different categories:

1. *Standard.* A standard is a statement of a required, mandatory, or specifically prohibitive practice regarding a traffic control device. Typically, standards are indicated by the use of the term "shall" or "shall not" in the statement.

2. *Guidance.* Guidance is a statement of recommended, but not mandatory, practice in typical situations. Deviations are allowed if engineering judgment or a study indicates that a deviation is appropriate. Guidance is generally indicated by use of the word "should" or "should not."

3. *Option.* An option is a statement of practice that is a permissive condition. It carries no implication of requirement or recommendation. Options often contain allowable modifications to a Standard or Guidance. An option is usually stated using the word "may" or "may not."

4. *Support.* This is a purely information statement provided to supply additional information to the traffic engineer. The words "shall," "should," or "may" do not appear in these statements (nor do their negative counterparts).

The four types of statements given in the MUTCD have legal implications for traffic agencies. Violating a standard leaves the jurisdictional agency exposed to liability for any accident that occurs because of the violation.

Figure 4.1: Placement of STOP Sign with Visibility Impaired by Tree

Thus, placing a nonstandard STOP sign would leave the jurisdictional agency exposed to liability for any accident occurring at the location. Guidelines, when violated, also leave some exposure to liability. Guidelines should be modified only after an engineering study has been conducted and documented, justifying the modification(s). Without such documentation, liability for accidents may also exist. Options and Support carry no implications with respect to liability.

It should also be understood that jurisdiction over traffic facilities is established as part of each state's vehicle and traffic law. That law generally indicates what facilities fall under the direct jurisdiction of the state (usually designated state highways and all intersections involving such highways) and specifies the state agency exercising that jurisdiction. It also defines what roadways would fall under control of county, town, and other local governments. Each of those political entities, in turn, would appoint or otherwise specify the local agency exercising jurisdiction.

Many traffic control devices must be supported by a specific law or ordinance enacted by the appropriate level of government. Procedures for implementing such laws and ordinances must also be specified. Many times (such as in the case of speed limits and parking regulations), public hearings and/or public notice must be given before imposition. For example, it would not be legal for an agency to post parking prohibitions during the night and then ticket or tow all parked vehicles without having provided adequate advance public notice, which is most often accomplished using local or regional newspapers.

This chapter presents some of the principles of the MUTCD, and generally describes the types of devices and their typical applications. Chapter 15 goes into greater detail concerning the use of traffic control devices on freeways, multilane, and two-lane highways. Chapter 19 contains additional detail concerning use of traffic control devices at intersections.

4.1.5 Communicating with the Driver

The driver is accustomed to receiving a certain message in a clear and standard fashion, often with redundancy. A number of mechanisms are used to convey messages. These mechanisms make use of recognized human limitations, particularly with respect to eyesight. Messages are conveyed through the use of:

- *Color*. Color is the most easily visible characteristic of a device. Color is recognizable long before a general shape may be perceived and considerably before a specific legend can be read and understood. The principal colors used in traffic control devices are red, yellow, green, orange, black, blue, and brown. These are used to code certain types of devices and to reinforce specific messages whenever possible.

- *Shape*. After color, the shape of the device is the next element to be discerned by the driver. Particularly in signing, shape is an important element of the message, either identifying a particular type of information that the sign is conveying or conveying a unique message of its own.

- *Pattern*. Pattern is used in the application of traffic markings. In general, double solid, solid, dashed, and broken lines are used. Each conveys a type of meaning with which drivers become familiar. The frequent and consistent use of similar patterns in similar applications contributes greatly to their effectiveness and to the instant recognition of their meaning.

- *Legend*. The last element of a device that the driver comprehends is its specific legend. Signals and markings, for example, convey their entire message through use of color, shape, and pattern. Signs, however, often use specific legend to transmit the details of the message being transmitted. Legend must be kept simple and short, so that drivers do not divert their attention from the driving task, yet are able to see and understand the specific message being given.

Redundancy of message can be achieved in a number of ways. The STOP sign, for example, has a unique shape (octagon), a unique color (red), and a unique one-word legend (STOP). Any of the three elements alone is sufficient to convey the message. Each provides redundancy for the others.

Redundancy can also be provided through use of different devices, each reinforcing the same message. A left-turn lane may be identified by arrow markings on the pavement, a "This Lane Must Turn Left" sign, and a protected left-turn signal phase indicated by a green arrow. Used together, the message is unmistakable.

The MUTCD provides a set of standards, guidelines, and general advice on how to best communicate various traffic rules and regulations to drivers. The MUTCD, however, is a document that is always developing. The traffic engineer must always consult the latest version of the manual (with all applicable revisions) when considering traffic control options.

4.2 Traffic Markings

Traffic markings are the most plentiful traffic devices in use. They serve a variety of purposes and functions and fall into three broad categories:

- Longitudinal markings
- Transverse markings
- Object markers and delineators

Longitudinal and transverse markings are applied to the roadway surface using a variety of materials, the most common of which are paint and thermoplastic. Reflectorization for better night vision is achieved by mixing tiny glass beads in the paint or by applying a thin layer of glass beads over the wet pavement marking as it is placed. The latter provides high initial reflectorization, but the top layer of glass beads is more quickly worn. When glass beads are mixed into the paint before application, some level of reflectorization is preserved as the marking wears. Thermoplastic is a naturally reflective material, and nothing need be added to enhance drivers' ability to see them at night.

In areas where snow and snow-plowing is not a problem, paint or thermoplastic markings can be augmented by pavement inserts with reflectors. Such inserts greatly improve the visibility of the markings at night. They are visible in wet weather (often a problem with markings) and resistant to wear. They are generally not used where plowing is common, as they can be dislodged or damaged during the process.

Object markers and delineators are small object-mounted reflectors. Delineators are small reflectors mounted on lightweight posts and are used as roadside markers to help drivers in proper positioning during inclement weather, when standard markings are not visible.

4.2.1 Colors and Patterns

Five marking colors are in current use: yellow, white, red, blue, and black. In general, they are used as follows:

- *Yellow* markings separate traffic traveling in opposite directions.
- *White* markings separate traffic traveling in the same direction, and are used for all transverse markings.
- *Red* markings delineate roadways that shall not be entered or used by the viewer of the marking.
- *Blue* markings are used to delineate parking spaces reserved for persons with disabilities.
- *Black* markings are used in conjunction with other markings on light pavements. To emphasize the pattern of the line, gaps between yellow or white markings are filled in with black to provide contrast and easier visibility.

A solid line prohibits or discourages crossing. A double solid line indicates maximum or special restrictions. A broken line indicates that crossing is permissible. A dotted line uses shorter line segments than a broken line. It provides trajectory guidance and often is used as a continuation of another type of line in a conflict area.

Normally, line markings are 4 to 6 inches wide. Wide lines, which provide greater emphasis, should be at least twice the width of a normal line. Broken lines normally consist of 10-ft line segments and 30-ft gaps. Similar dimensions with a similar ratio of line segments to gaps may be used as appropriate for prevailing traffic speeds and the need for delineation. Dotted lines usually consist of 2-ft line segments and 4-ft (or longer) gaps. MUTCD suggests a maximum segment-to-gap ratio of 1:3 for dotted lines.

4.2.2 Longitudinal Markings

Longitudinal markings are those markings placed parallel to the direction of travel. The vast majority of longitudinal markings involve centerlines, lane lines, and pavement edge lines.

Longitudinal markings provide guidance for the placement of vehicles on the traveled way cross-section and basic trajectory guidance for vehicles traveling along the facility. The best example of the importance of longitudinal markings is the difficulty in traversing a newly paved highway segment on which lane markings have not yet been repainted. Drivers do not automatically form neat lanes without the guidance of longitudinal markings; rather, they tend to place themselves somewhat randomly on the cross-section, encountering many difficulties. Longitudinal markings provide for organized flow and optimal use of the pavement width.

Centerlines

The yellow centerline marking is critically important and is used to separate traffic traveling in opposite directions. Use of centerlines on all types of facilities is not mandated by the MUTCD. The applicable standard is:

> Centerline markings shall be placed on all paved urban arterials and collectors that have a traveled way of 20 ft or more and an ADT of 6,000 veh/day or greater. Centerline markings shall also be placed on all paved, two-way streets or highways that have 3 or more traffic lanes. [MUTCD. pg. 3B–01].

Further guidance indicates that placing centerlines is recommended for urban arterials and streets with ADTs of 4,000 or more and on rural highways with a width in excess of 18 ft and ADT $>3,000$ vehicles/day. Caution should be used in placing centerlines on pavements of 16 ft or less, which may increase the incidence of traffic encroaching beyond the traveled way.

On two-lane, two-way rural highways, centerline markings supplemented by signs are used to regulate passing maneuvers. A double-solid yellow center marking indicates that passing is not permitted in either direction. A solid yellow line with a dashed yellow line indicates that passing is permitted from the dashed side only. Where passing is permissible in both directions, a single dashed yellow centerline is used. Chapter 14 contains additional detail on the use and application of centerlines on two-lane, two-way, rural highways.

There are other specialized uses of yellow markings. Figure 4.2(a) illustrates the use of double dashed-yellow markings to delineate a reversible lane on an arterial. Signing and/or lane-use signals would have to supplement these to denote the directional use of the lane. Figure 4.2(b) shows the markings used for two-way left-turn lanes on an arterial.

Lane Markings

The typical lane marking is a single white dashed line separating lanes of traffic in the same direction. MUTCD standards require the use of lane markings on all freeways and Interstate highways and recommend their use on all highways with two or more adjacent traffic lanes in a single direction. The dashed lane line indicates that lane changing is permitted. A single solid white lane line is used to indicate that lane-changing is discouraged but not illegal. Where lane-changing is to be prohibited, a double-white solid lane line is used.

Edge Markings

Edge markings are a required standard on freeways, expressways, and rural highways with a traveled way of 20 ft or more in width and an ADT of 6,000 veh/day or greater. They are recommended for rural highways with ADTs over 3,000 veh/day and a 20-ft or wider traveled way.

When used, right-edge markings are a single normal solid white line; left-edge markings are a single normal solid yellow line.

Other Longitudinal Markings

The MUTCD provides for many options in the use of longitudinal markings. Consult the manual directly for further detail. The manual also provides standards and guidance for other types of applications, including freeway and non-freeway merge and diverge areas, lane drops, extended markings through intersections, and other situations.

Chapter 15 contains additional detail on the application of longitudinal markings on freeways, expressways,

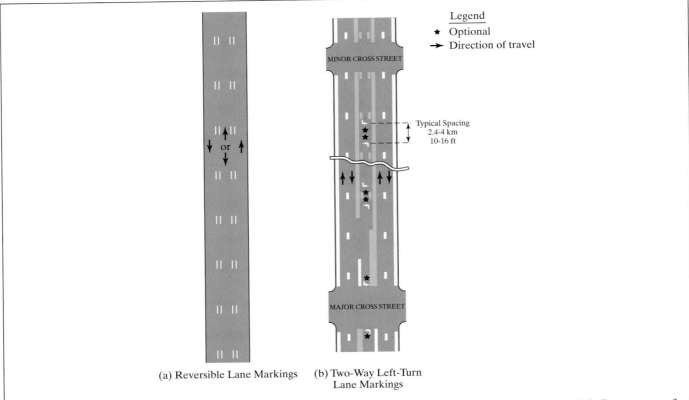

(a) Reversible Lane Markings   (b) Two-Way Left-Turn Lane Markings

Figure 4.2: Special Purpose Center Markings (Used with permission of Federal Highway Administration, U.S. Department of Transportation, *Manual of Uniform Traffic Control Devices*, Millennium Edition, Figures 3B-6 and 3B-7, Washington DC, 2000.)

and rural highways. Chapter 19 includes additional discussion of intersection markings.

4.2.3 Transverse Markings

Transverse markings, as their name implies, include any and all markings with a component that cuts across a portion or all of the traveled way. When used, all transverse markings are white.

STOP Lines

STOP lines are not mandated by the MUTCD. In practice, STOP lines are almost always used where marked crosswalks exist, and in situations where the appropriate location to stop for a STOP sign or traffic signal is not clear. When used, it is recommended that the width of the line be 12 to 24 inches. When used, STOP lines must extend across all approach lanes.

Crosswalk Markings

While not mandated by the MUTCD, it is recommended that crosswalks be marked at all intersections at which "substantial" conflict between vehicles and pedestrians exists. They should also be used at points of pedestrian concentration and at locations where pedestrians might not otherwise recognize the proper place and/or path to cross. A marked crosswalk should be 6 ft or more in width.

Figure 4.3 shows the three types of crosswalk markings in general use. The most frequently used is composed of two parallel white lines. Cross-hatching

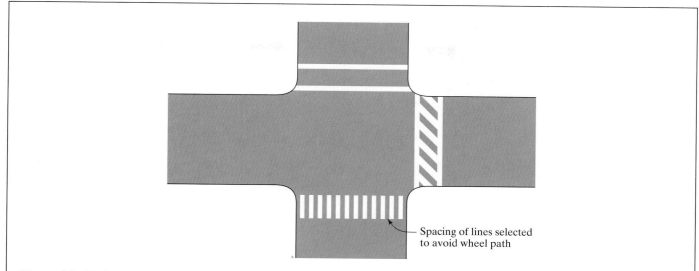

— Spacing of lines selected
to avoid wheel path

Figure 4.3: Typical Crosswalk Markings (Used with permission of Federal Highway Administration, U.S. Department of Transportation, *Manual of Uniform Traffic Control Devices*, Millennium Edition, Figure 3B-16, pg. 3B–17, Washington DC, 2000.)

may be added to provide greater focus in areas with heavy pedestrian flows. The use of parallel transverse markings to identify the crosswalk is another option used at locations with heavy pedestrian flows.

The manual also contains a special pedestrian crosswalk marking for signalized intersections where a full pedestrian phase is included. Consult the manual directly for details of this particular marking.

Parking Space Markings

Parking space markings are not purely transverse, as they contain both longitudinal and transverse elements. They are officially categorized as transverse markings, however, in the MUTCD. They are always optional and are used to encourage efficient use of parking spaces. Such markings can also help prevent encroachment of parked vehicles into fire-hydrant zones, loading zones, taxi stands and bus stops, and other specific locations at which parking is prohibited. They are also useful on arterials with curb parking, as they also clearly demark the parking lane, separating it from

travel lanes. Figure 4.4 illustrates typical parking lane markings.

Note that the far end of the last marked parking space should be at least 20 ft away from the nearest crosswalk marking (30 ft on a signalized intersection approach.)

Word and Symbol Markings

The MUTCD prescribes a number of word and symbol markings that may be used, often in conjunction with signs and/or signals. These include arrow markings indicating lane-use restrictions. Such arrows (with accompanying signs) are mandatory where a through lane becomes a left- or right-turn-only lane approaching an intersection.

Word markings include "ONLY," used in conjunction with lane use arrows, and "STOP," which can be used only in conjunction with a STOP line and a STOP sign. "SCHOOL" markings are often used in conjunction with signs to demark school and school-crossing zones. The MUTCD contains a listing of all authorized

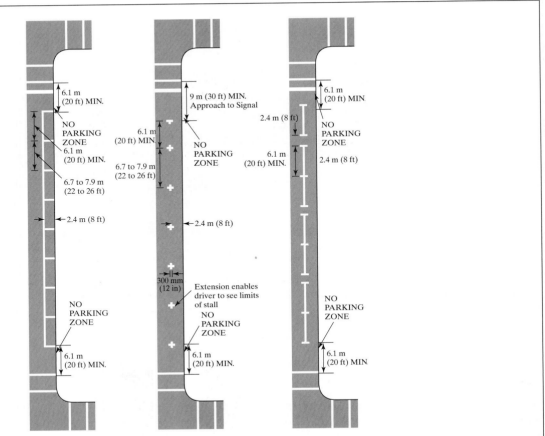

Figure 4.4: Typical Parking Space Markings (Used with permission of Federal Highway Administration, U.S. Department of Transportation, *Manual on Uniform Traffic Control Devices,* Millennium Edition, Figure 3B-17, pg. 3B–19, Washington DC, 2000.)

word markings and allows for discretionary use of unique messages where needed.

Other Transverse Markings

Consult the MUTCD directly for examples of other types of markings, including preferential lane markings, curb markings, roundabout and traffic circle markings, and speed-hump markings. Chapter 19 contains a detailed discussion of the use of transverse and other markings at intersections.

4.2.4 Object Markers

Object markers are used to denote obstructions either in or adjacent to the traveled way. Object markers are mounted on the obstruction in accordance with MUTCD standards and guidelines. In general, the lower edge of the marker is mounted a minimum of 4 ft above the surface of the nearest traffic lane (for obstructions 8 ft or less from the pavement edge) or 4 ft above the ground (for obstructions located further away from the pavement edge).

Typical Type 1 Object Markers

OM1-1   OM1-2   OM1-3

Nine yellow retroreflectors with 3-in minimum diameter on a yellow or black diamond panel of 18 in or more on a side; or an all-yellow retroreflective diamond panel of the same size.

Typical Type 2 Object Markers

OM2-1V   OM2-2V   OM2-1H   OM2-2H

Three yellow retroreflectors with 3-in minimum diameter arranged horizontally or vertically on white panel of at least 6 × 12 in; or an all-yellow retroreflective panel of the same size.

Typical Type 3 Object Markers

A striped marker measuring 12 × 36 in with alternating black and yellow stripes sloping downward at an angle of 45° toward the side of the obstruction on which traffic is to pass.

Figure 4.5: Object Markers Illustrated (Used with permission of Federal Highway Administration, U.S. Department of Transportation, *Manual on Uniform Traffic Control Devices,* Millennium Edition, pg. 3C-2, Washington DC, 2000.)

There are three types of object markers used, as illustrated in Figure 4.5. Obstructions within the roadway *must* be marked using a Type 1 or Type 3 marker. The Type 3 marker, when used, must have the alternating yellow and black stripes sloped downward at a 45° angle towards the side on which traffic is to pass the obstruction. When used to mark a roadside obstruction, the in-side edge of the marker must be in line with the inner edge of the obstruction.

4.2.5 Delineators

Delineators are reflective devices mounted at a 4-ft height on the side(s) of a roadway to help denote its alignment. They are particularly useful during in-clement weather, where pavement edge markings may not be visible. When used on the right side of the road-way, delineators are white; when used on the left side of the roadway, delineators are yellow. The back of delin-eators may have red reflectors to indicate wrong-way travel on a one-direction roadway.

Delineators are mandated on the right side of freeways and expressways and on at least one side of interchange ramps, with the exception of tangent sections where raised pavement markers are used continuously on all lane lines, where whole routes (or substantial portions thereof) have large tangent sections, or where delineators are used to lead into all curves. They may also be omitted where there is continuous roadway lighting between in-terchanges. Delineators may be used on an optional basis on other classes of roads. Figure 4.6 illustrates a typical installation of delineators.

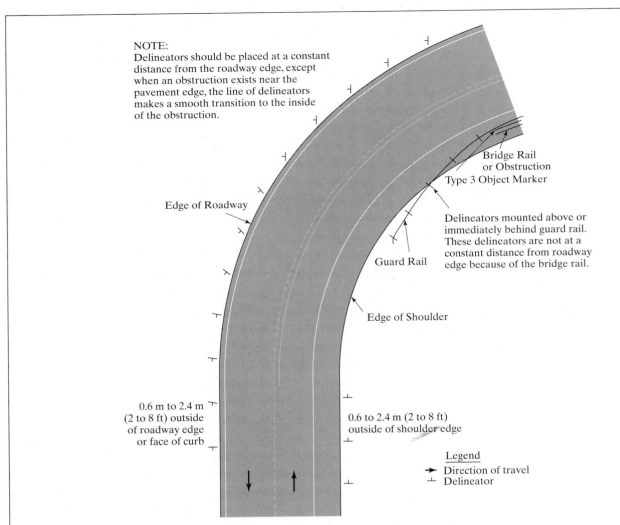

NOTE:
Delineators should be placed at a constant distance from the roadway edge, except when an obstruction exists near the pavement edge, the line of delineators makes a smooth transition to the inside of the obstruction.

Bridge Rail
or Obstruction
Type 3 Object Marker

Delineators mounted above or immediately behind guard rail. These delineators are not at a constant distance from roadway edge because of the bridge rail.

Edge of Roadway

Guard Rail

Edge of Shoulder

0.6 m to 2.4 m (2 to 8 ft) outside of roadway edge or face of curb

0.6 to 2.4 m (2 to 8 ft) outside of shoulder edge

Legend
→ Direction of travel
⊥ Delineator

Figure 4.6: Typical Use of Delineators on a Highway Curve (Used with permission of Federal Highway Administration, U.S. Department of Transportation, *Manual on Uniform Traffic Control Devices*, Millennium Edition, Figure 3D-1, pg. 3D-5, Washington DC, 2000.)

4.3 Traffic Signs

The MUTCD provides specifications and guidelines for the use of literally hundreds of different signs for myriad purposes. In general, traffic signs fall into one of three major categories:

- *Regulatory signs.* Regulatory signs convey information concerning specific traffic regulations. Regulations may relate to right-of-way, speed limits, lane usage, parking, or a variety of other functions.

- *Warning signs.* Warning signs are used to inform drivers about upcoming hazards that they might not see or otherwise discern in time to safely react.

- *Guide signs.* Guide signs provide information on routes, destinations, and services that drivers may be seeking.

It would be impossible to cover the full range of traffic signs and applications in a single chapter. The sections that follow provide a general overview of the various types of traffic signs and their use.

4.3.1 Regulatory Signs

Regulatory signs shall be used to inform road users of selected traffic laws or regulations and indicate the applicability of the legal requirements. Regulatory signs shall be installed at or near where the regulations apply. The signs shall clearly indicate the requirements imposed by the regulations and shall be designed and installed to provide adequate visibility and legibility in order to obtain compliance. [MUTCD, Millennium Edition, Pg 2B–1].

Drivers are expected to be aware of many general traffic regulations, such as the basic right-of-way rule at intersections and the state speed limit. Signs, however, should be used in all cases where the driver cannot be expected to know the applicable regulation.

Except for some special signs, such as the STOP and YIELD sign, most regulatory signs are rectangular, with the long dimension vertical. Some regulatory signs are square. These are primarily signs using symbols instead of legend to impart information. The use of symbol signs generally conforms to international practices established at a 1971 United Nations conference on traffic safety. The background color of regulatory signs, with a few exceptions, is white, while legend or symbols are black. In symbol signs, a red circle with a bar through it signifies a prohibition of the movement indicated by the symbol.

The MUTCD contains many pages of standards for the appropriate size of regulatory signs and should be consulted directly on this issue.

Regulatory Signs Affecting Right-of-Way

The regulatory signs in this category have special designs reflecting the extreme danger that exists when one is ignored. These signs include the STOP and YIELD signs, which assign right-of-way at intersections, and WRONG WAY and ONE WAY signs, indicating directional flow. The STOP and YIELD signs have unique shapes, and they use a red background color to denote danger. The WRONG WAY sign also uses a red background for this purpose. Figure 4.7 illustrates these signs.

Figure 4.7: Regulatory Signs Affecting Right-of-Way (Used with permission of Federal Highway Administration, U.S. Department of Transportation, *Manual on Uniform Traffic Control Devices,* Millennium Edition, pgs. 2B–7, 2B–32, and 2B–36, Washington DC, 2000.)

The "4-Way" and "All Way" panels are mounted below a STOP sign where multiway STOP control is in use. Consult Chapter 16 for a detailed presentation and discussion of warrants for use of STOP and YIELD signs at intersections.

Speed Limit Signs

One of the most important issues in providing for safety and efficiency of traffic movement is the setting of appropriate speed limits. To be effective, a speed limit must be communicated to the driver and should be sufficiently enforced to engender general observance.

There are a number of different types of speed limits that may be imposed:

- Linear speed limits
- Areawide (statutory) speed limits
- Night speed limits
- Truck speed limits
- Minimum speed limits

Speed limits may be stated in terms of standard U.S. units (mi/h) or in metric units (km/h). Current law allows each state to determine the system of units to be used. Where the metric system is used, supplementary panels are used to indicate the appropriate units.

Figure 4.8 shows a variety of speed signs in common use. While most signs consist of black lettering on a white background, night speed limits are posted using the reverse of this: white lettering on a black background. The "metric" panel is a warning sign and uses a yellow background.

Linear speed limits apply to a designated section of roadway. Signs should be posted such that no driver can enter the roadway without seeing a speed limit sign within approximately 1,000 ft. This is not an MUTCD standard, but reflects common practice.

Area speed limits apply to all roads within a designated area (unless otherwise posted). A state statutory speed limit is one example of such a regulation. Cities, towns, and other local governments may also enact ordinances establishing a speed limit throughout their jurisdiction. Areawide speed limits should be posted on every facility at the boundary entering the jurisdiction for which the limit is established.

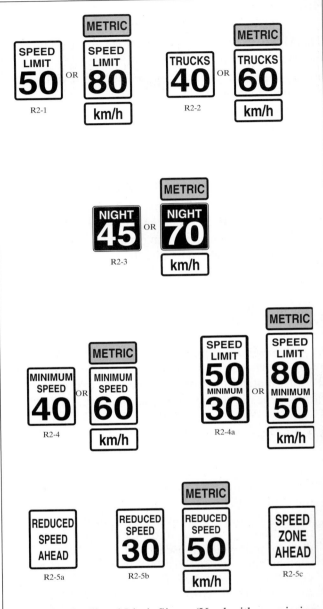

Figure 4.8: Speed Limit Signs (Used with permission of Federal Highway Administration, U.S. Department of Transportation, *Manual on Uniform Traffic Control Devices,* Millennium Edition, pgs. 2B-7 and 2B-15, Washington DC, 2000.)

Figure 4.9: Turn Prohibition Signs (Used with permission of Federal Highway Administration, U.S. Department of Transportation, *Manual on Uniform Traffic Control Devices,* Millennium Edition, pg. 2B-14, Washington DC, 2000.)

The "reduced speed" or "speed zone ahead" signs should be used wherever engineering judgment indicates a need to warn drivers of a reduced speed limit for compliance. When used, however, the sign must be followed by a speed limit sign posted at the beginning of the section in which the reduced speed limit applies.

Consult Chapter 15 for a discussion of criteria for establishing an appropriate speed limit on a highway or roadway section.

Turn Prohibition Signs

Where right and/or left turns are to be prohibited, one or more of the turn prohibition signs shown in Figure 4.9 are used. In this category, international symbol signs are preferred. The traditional red circle with a bar is placed over an arrow indicating the movement to be banned.

Lane-Use Signs

Lane-use control signs are used wherever a given movement or movements are restricted and/or prohibited from designated lanes. Such situations include left-turn- and right-turn-only lanes, two-way left-turn lanes on arterials, and reversible lanes. Lane-use signs, however, may also be used to clarify lane usage even where no regulatory restriction is involved. Where lane usage is complicated, advance lane-use control signs may be used as well. Figure 4.10 illustrates these signs.

Two-way left-turn lane signing must be supplemented by the appropriate markings for such a lane, as illustrated previously. Reversible lane signs must be posted as overhead signs, placed over the lane or lanes that are reversible. Roadside signs may supplement

(a) Sample Lane-Use Signs

(b) Sample Advanced Lane-Use Signs

(c) Two-Way Left-Turn Lane Signs

(d) Sample Reversible Lane Signs

Figure 4.10: Lane-Use Control Signs (Used with permission of Federal Highway Administration, U.S. Department of Transportation, *Manual on Uniform Traffic Control Devices*, Millennium Edition, pgs. 2B-19 and 2B-22, Washington DC, 2000.)

overhead signs. In situations where signing may not be sufficient to ensure safe operation of reversible lanes, overhead signals should be used.

Parking Control Signs

Curb parking control is one of the more critical aspects of urban network management. The economic viability of business areas often depends upon an adequate and convenient supply of on-street and off-street parking. At the same time, curb parking often interferes with through traffic and occupies space on the traveled way that might otherwise be used to service moving traffic. Chapter 11 provides a detailed coverage of parking issues and programs. It is imperative that curb parking regulations be clearly signed, and strict enforcement is often necessary to achieve high levels of compliance.

When dealing with parking regulations and their appropriate signing, three terms must be understood:

- *Parking*. A "parked" vehicle is a stationary vehicle located at the curb with the engine not running; whether or not the driver is in the vehicle is not relevant to this definition.
- *Standing*. A "standing" vehicle is a stationary vehicle located at the curb with the engine running and the driver in the car.
- *Stopping*. A "stopping" vehicle is one that makes a momentary stop at the curb to pick up or discharge a passenger; the vehicle moves on immediately upon completion of the pick-up or discharge, and the driver does not leave the vehicle.

In legal terms, most jurisdictions maintain a common hierarchal structure of prohibitions. "No Stopping" prohibits stopping, standing, and parking. "No Standing" prohibits standing and parking, but permits stopping. "No Parking" prohibits parking, but permits standing and stopping.

Parking regulations may also be stated in terms of a prohibition or in terms of what is permitted. Where a sign is indicating a prohibition, red legend on a white background is used. Where a sign is indicating a permissive situation, green legend on a white background is used. Figure 4.11 illustrates a variety of parking-control signs in common use.

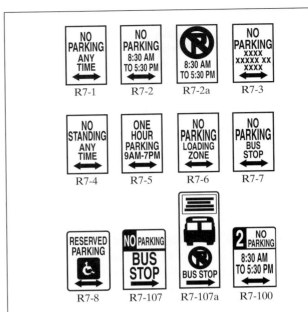

Figure 4.11: Sample Parking-Control Signs (Used with permission of Federal Highway Administration, U.S. Department of Transportation, *Manual on Uniform Traffic Control Devices*, Millennium Edition, pg. 2B–38, Washington DC, 2000.)

Parking signs must be carefully designed and placed to ensure that the often complex regulations are understood by the majority of drivers. The MUTCD recommends that the following information be provided on parking-control signs, in order from top to bottom of the sign:

- The restriction or prohibition (or condition permitted in the case of a permissive sign)
- The times of the day that it is applicable (if not every day)
- The days of the week that it is applicable (if not every day)

Parking-control signs should always be placed at the boundaries of the restricted area and at intermediate locations as needed. At locations where the parking restriction changes, two signs should be placed on a single support, each with an arrow pointing in the direction of application. Where areawide restrictions are in effect, the restriction should be signed at all street locations crossing into the restricted area.

In most local jurisdictions, changes in parking regulations must be disclosed in advance using local newspapers and/or other media and/or by placing posters throughout the affected area warning of the change. It is not appropriate, for example, to place new parking restrictions overnight and then ticket or remove vehicles now illegally parked without adequate advance warning.

Other Regulatory Signs

The millennium edition of the MUTCD provides standards and guidelines for over 100 different regulatory signs. Some of the most frequently used signs have been discussed in this section, but they are merely a sample of the many such signs that exist. New signs are constantly under development as new types of regulations are introduced. Consult the MUTCD directly for additional regulatory signs and their applications.

4.3.2   Warning Signs

Warning signs call attention to unexpected conditions on or adjacent to a highway or street and to situations that might not be readily apparent to road users. Warning signs alert road users to conditions that might call for a reduction of speed or an action in the interest of safety and efficient traffic operations. [MUTCD, Millennium Edition, Pg 2C-1]

Most warning signs are diamond-shaped, with black lettering or symbols on a yellow background. A pennant shape is used for the "No Passing Zone" sign, used in conjunction with passing restrictions on two-lane, two-way rural highways. A rectangular shape is used for some arrow indications. A circular shape is used for railroad crossing warnings.

The MUTCD specifies minimum sizes for various warning signs on different types of facilities. For the standard diamond-shaped sign, minimum sizes range from 30 inches by 30 in to 36 in by 36 in. Larger signs are generally permitted.

The MUTCD indicates that warning signs shall be used only in conjunction with an engineering study or based on engineering judgment. While this is a fairly loose requirement, it emphasizes the need to avoid overuse of such signs. A warning sign should be used only to alert drivers of conditions that they could not normally expected to discern on their own. Overuse of warning signs encourages drivers to ignore them, which could lead to dangerous situations.

When used, warning signs must be placed far enough in advance of the hazard to allow drivers adequate time to perform the required adjustments. Table 4.2 gives the recommended advance placement distances for three conditions, defined as follows:

- *Condition A: High judgment required.* Applies where the road user must use extra time to adjust speed and change lanes in heavy traffic due to a complex driving situation. Typical applications are warning signs for merging, lane drop, and similar situations. A PIEV time of 6.7 to 10.0 s is assumed plus 4.5 s for each required maneuver.

- *Condition B: Stop condition.* Applies in cases where the driver may be required to come to a stop before the hazard location. Typical applications are stop ahead, yield ahead, and signal ahead warnings. The AASHTO standard PIEV time of 2.5 s is applied.

- *Condition C: Deceleration to the listed advisory speed for the condition.* Applies in cases where the road user must decelerate to a posted advisory speed to safely maneuver through the hazard. A 1.6 s PIEV time is assumed with a deceleration rate of 10 ft/s^2.

In all cases, sign visibility of 175 ft is assumed, based on sign-design standards.

Warning signs may be used with supplementary panels indicating either the distance to the hazard or an advisory speed. The advisory speed is the recommended safe speed through the hazardous area and is determined by an engineering study of the location. While no specific guideline is given, common practice is to use an advisory speed panel wherever the safe speed through the hazard is 10 mi/h or more less than the posted or statutory speed limit.

Table 4.2: Guidelines for Advance Placement of Warning Signs

Posted or 85% Speed (mi/h)	Condition A "High Judgment"	Condition B "Stop Condition"	Advanced Placement Distance (ft) Condition C: Decelerate to:				
			10 mi/h	20 mi/h	30 mi/h	40 mi/h	50 mi/h
20	175	NA[1]	NA[1]	—	—	—	—
25	250	NA[1]	100	NA[1]	—	—	—
30	325	100	150	100	—	—	—
35	400	150	200	175	NA[1]	—	—
40	475	225	275	250	175	—	—
45	550	300	350	300	250	NA[1]	—
50	625	375	425	400	325	225	—
55	700	450	500	475	400	300	NA[1]
60	775	550	575	550	500	400	300
65	850	650	650	625	575	500	375

[1]No suggested minimum distance; placement location dependent upon site conditions and other signing to provide adequate advance warning for the driver.
(Used with permission of Federal Highway Administration, U.S. Department of Transportation, *Manual on Uniform Traffic Control Devices,* Millennium Edition, Table 2C-4, pg. 2C-7, Washington DC, 2000.)

Warning signs are used to inform drivers of a variety of potentially hazardous circumstances, including:

- Changes in horizontal alignment
- Intersections
- Advance warning of control devices
- Converging traffic lanes
- Narrow roadways
- Changes in highway design
- Grades
- Roadway surface conditions
- Railroad crossings
- Entrances and crossings
- Miscellaneous

Figure 4.12 shows some sample warning signs from these categories.

While not shown here, the MUTCD contains other warning signs in special sections of the manual related to work zones, school zones, and railroad crossings. The practitioner should consult these sections of the MUTCD directly for more specific information concerning these special situations.

4.3.3 Guide Signs

Guide signs provide information to road users concerning destinations, available services, and historical/recreational facilities. They serve a unique purpose in that drivers who are familiar or regular users of a route will generally not need to use them; they provide critical information, however, to unfamiliar road users. They serve a vital safety function: a confused driver approaching a junction or other decision point is a distinct hazard.

Guide signs are rectangular, with the long dimension horizontal, and have white lettering and borders.

Figure 4.12: Sample Warning Signs (Used with permission of Federal Highway Administration, U.S. Department of Transportation, *Manual on Uniform Traffic Control Devices,* Millennium Edition, pgs. 2C-9, 2C-15, 2C-19, 2C-31, Washington DC, 2000.)

The background varies by the type of information contained on the sign. Directional or destination information is provided by signs with a green background; information on services is provided by signs with a blue background; cultural, historical, and/or recreational information is provided by signs with a brown background. Route markers, included in this category, have varying shapes and colors depending on the type and jurisdiction of the route.

The MUTCD provides guide-signing information for three types of facilities: conventional roads, freeways, and expressways. Guide signing is somewhat different from other types in that overuse is generally not a serious issue, unless it leads to confusion. Clarity and consistency of message is the most important aspect of guide signing. Several general principles may be applied:

1. If a route services a number of destinations, the most important of these should be listed. Thus, a highway serving Philadelphia as well as several lesser suburbs would consistently list Philadelphia as the primary destination.

2. No guide sign should list more than three (four may be acceptable in some circumstances) destinations on a single sign. This, in conjunction with the first principle, makes the selection of priority destinations a critical part of effective guide signing.

3. Where roadways have both a name and a route number, both should be indicated on the sign if space permits. In cases where only one may be listed, the route number takes precedence. Road maps show route numbers prominently,

while not all facility names are included. Unfamiliar drivers are, therefore, more likely to know the route number than the facility name.

4. Wherever possible, advance signing of important junctions should be given. This is more difficult on conventional highways, where junctions may be frequent and closely spaced. On freeways and expressways, this is critical, as high approach speeds make advance knowledge of upcoming junctions a significant safety issue.

5. Confusion on the part of the driver must be avoided at all cost. Sign sequencing should be logical and should naturally lead the driver to the desired route selections. Overlapping sequences should be avoided wherever possible. Left-hand exits and other unusual junction features should be signed extremely carefully.

The size, placement, and lettering of guide signs vary considerably, and the manual gives information on numerous options. A number of site-specific conditions affect these design features, and there is more latitude and choice involved than for other types of highway signs. The MUTCD should be consulted directly for this information.

Figure 4.13: Route Markers Illustrated (Used with permission of Federal Highway Administration, U.S. Department of Transportation, *Manual on Uniform Traffic Control Devices*, Millennium Edition, pg. 2D-7, Washington DC, 2000.)

Route Markers

Figure 4.13 illustrates route markers that are used on all numbered routes. The signs have unique designs that signify the type of route involved. Interstate highways have a unique shield shape, with red and blue background and white lettering. The same design is used for designated "business loops." Such loops are generally a major highway that is not part of the interstate system but one that serves the business area of a city from an interchange on the interstate system. U.S. route markers consist of black numerals on a white shield that is placed on a square sign with a black background. State route markers are designed by the individual states and, therefore, vary from state to state. All county route markers, however, follow a standard design, with yellow lettering on a blue background and a unique shape. The name of the county is placed on the route marker. Routes in national parks and/or national forests also have a unique shape and have white lettering on a brown background.

Route markers may be supplemented by a variety of panels indicating cardinal directions or other special purposes. Special purpose panels include JCT, ALT or ALTERNATE, BY-PASS, BUSINESS, TRUCK, TO, END, and TEMPORARY. Auxiliary panels match the colors of the marker they are supplementing.

Chapter 15 includes a detailed discussion of the use of route markers and various route marker assemblies for freeways, expressways, and conventional roads.

Destination Signs—Conventional Roads

Destination signs are used on conventional roadways to indicate the distance to critical destinations along the route and to mark key intersections or interchanges. On conventional roads, destination signs use an all-capital white legend on a green background. The distance in miles to the indicated destination may be indicated to the right of the destination.

Destination signs are generally used at intersections of U.S. or state numbered routes with Interstate, U.S., state numbered routes, or junctions forming part of a route to such a numbered route. Distance signs are usually placed on important routes leaving a municipality of a major junction with a numbered route.

Local street name signs are recommended for all suburban and urban junctions as well as for major intersections at rural locations. Local street name signs are categorized as conventional roadway destination signs. Figure 4.14 illustrates a selection of these signs.

Destination Signs—Freeways and Expressways

Destination signs for freeways and expressways are similar, although there are different requirements for size and placement specified in the MUTCD. They differ from conventional road guide signs in a number of ways:

- Destinations are indicated in initial capitals and small letters.

- Numbered routes are indicated by inclusion of the appropriate marker type on the guide sign.

- Exit numbers are included as auxiliary panels located at the upper right or left corner of the guide sign.

- At major junctions, diagrammatic elements may be used on guide signs.

As for conventional roadways, distance signs are frequently used to indicate the mileage to critical destinations along the route. Every interchange and every significant at-grade intersection on an expressway is extensively signed with advance signing as well as with signing at the junction itself.

The distance between interchanges has a major impact on guide signing. Where interchanges are widely separated, advance guide signs can be placed as much as five or more miles from the interchange and may be repeated several times as the interchange is approached.

In urban and suburban situations, where interchanges are closely spaced, advanced signing is more difficult to achieve. Advance signing usually gives information

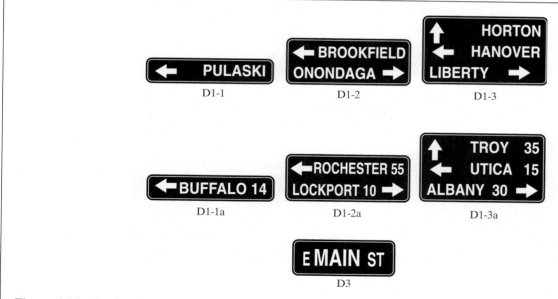

Figure 4.14: Destination Signs for Conventional Roads (Used with permission of Federal Highway Administration, U.S. Department of Transportation, *Manual on Uniform Traffic Control Devices*, Millennium Edition, pgs. 2D–26 and 2D–30, Washington DC, 2000.)

only concerning the *next* interchange, to avoid confusion caused by overlapping signing sequences. The only exception to this is a distance sign indicating the distance to the next several interchanges. Thus, in urban and suburban areas with closely spaced interchanges, the advance

sign for the next interchange is placed at the last off-ramp of the previous interchange.

A wide variety of sign types are used in freeway and expressway destination signing. A few of these are illustrated in Figure 4.15.

(a) Advance Interchange Sign

(b) Next Exit Panel

(c) Supplemental Multiple Exit Sign

(d) Gore Area Exit Sign

(e) Exit Direction Sign

(f) Pull-Through Sign

(g) Diagrammatic Advance Sign for a Left-Hand Exit

Figure 4.15: Sample Freeway and Expressway Destination Signs (Used with permission of Federal Highway Administration, U.S. Department of Transportation, *Manual on Uniform Traffic Control Devices,* Millennium Edition, pgs. 2E–7, 2E–21, 2E–42, 2E–44, 2E–47, 2E–50, Washington DC, 2000.)

Figure 4.15(a) shows a typical advance exit sign. These are placed at various distances from the interchange in accordance with the overall signing plan. The number and placement of advance exit sign is primarily dependent on interchange spacing. Figure 4.15(b) depicts a "next exit" panel. These may be placed below an exit or advance exit sign. The supplemental multiple exit sign of Figure 4.15(c) is used where separate ramps exist for the two directions of the connecting roadway. It is generally placed after the last advance exit sign and before the interchange itself. The gore area exit sign of Figure 4.15(d) is placed in the gore area, and is the last sign associated with a given ramp connection. Such signs are usually mounted on breakaway sign posts to avoid serious damage to vehicles straying into the gore area. Figure 4.15(e) is an exit direction sign, which is posted at the location of the diverge and includes an exit number panel. The "pull-through" sign of Figure 4.15(f) is used primarily in urban or other areas with closely-spaced interchanges. It is generally mounted on overhead supports next to the exit direction sign. It reinforces the direction for drivers intending to continue on the freeway. The final illustration, Figure 4.15(g), is a diagrammatic advance exit sign. These are very useful where the diverge junction is other than a simple one-lane, right-side off-ramp. It informs drivers, in a relatively straightforward way, of what lanes should be used for different destinations.

Chapter 15 contains a detailed discussion of guide signing for freeways, expressways, and conventional roadways.

Service Guide Signs

Another important type of information drivers require is directions to a variety of motorists' services. Drivers, particularly those who are unfamiliar with the area, need to be able to easily locate such services as fuel, food, lodging, medical assistance, and similar services. The MUTCD provides for a variety of signs, all using white legend and symbols on a blue background, to convey such information. In many cases, symbols are used to indicate the type of service available. On freeways, large signs using text messages may be used with exit number auxiliary panels. The maximum information is provided by freeway signs that indicate the actual brand names of available services (gas companies, restaurant names,

etc.). Figure 4.16 illustrates some of the signs used to provide motorist service information.

There are a number of guidelines for specific service signing. No service should be included that is more than three miles from the freeway interchange. No specific services should be indicated where drivers cannot easily reenter the freeway at the interchange.

Specific services listed must also conform to a number of criteria regarding hours of operations and specific functions provided. All listed services must also be in compliance with all federal, state, and local laws and regulations concerning their operation. Consult the MUTCD directly for the details of these requirements.

Service guide signs on conventional highways are similar to those of Figure 4.16 but do not use exit numbers or auxiliary exit number panels.

Recreational and Cultural-Interest Guide Signs

Information on historic, recreational, and/or cultural-interest areas or destinations is given on signs with white legend and/or symbols on a brown background. Symbols are used to depict the type of activity, but larger signs with word messages may be used as well. Figure 4.17 shows some examples of these signs. The millennium edition of the MUTCD has introduced many more acceptable symbols and should be consulted directly for illustrations of these.

Mileposts

Mileposts are small 6 × 9-in vertical white-on-green panels indicating the mileage along the designated route. These are provided to allow the driver to estimate his/her progress along a route, and provide a location system for accident reporting and other emergencies that may occur along the route. Distance numbering is continuous within a state, with "zero" beginning at the south or west state lines or at the southern-most or western-most interchange at which the route begins. Where routes overlap, mileposts are continuous only for *one* of the routes. In such cases, the first milepost beyond the overlap should indicate the total mileage traveled along the route that is *not* continuously numbered and posted.

On some freeways, markers are placed every tenth of a mile for a more precise location system.

(a) General Service Information Signs

(b) Specific Service Information Signs

Figure 4.16: Service Information Signs (Used with permission of Federal Highway Administration, U.S. Department of Transportation, *Manual on Uniform Traffic Control Devices*, Millennium Edition, Figure 2E–34, pg. 2E–72, pg. 2F–4, Washington DC, 2000.) * Exit number panel is optional.

4.4   Traffic Signals

The MUTCD defines nine types of traffic signals:

- Traffic control signals
- Pedestrian signals
- Emergency vehicle traffic control signals
- Traffic control signals for one-lane, two-way facilities
- Traffic control signals for freeway entrance ramps
- Traffic control signals for moveable bridges
- Lane-use control signals
- Flashing beacons
- In-roadway lights

The most common of these is the traffic control signal, used at busy intersections to direct traffic to alternately stop and move.

4.4.1   Traffic Control Signals

The MUTCD specifies two critical standards with respect to traffic control signals:

- A traffic control signal shall be operated in either a steady-state (stop and go) mode or a flashing mode at all times. [MUTCD, Millennium Edition, pg. 4D-1].
- STOP signs shall not be used in conjunction with any traffic control signal operation except in either of the two following cases: (1) If the signal indication for an approach is a flashing red at all times; and (2) If a minor street or driveway is located within or adjacent to the area controlled by the traffic control signal, but does not require separate traffic signal control because an extremely low potential for conflict exists. [MUTCD, Millennium Edition, pg. 4D–1].

The first mandate requires that signals be operated *at all times*. No traffic signal should ever be "dark," that

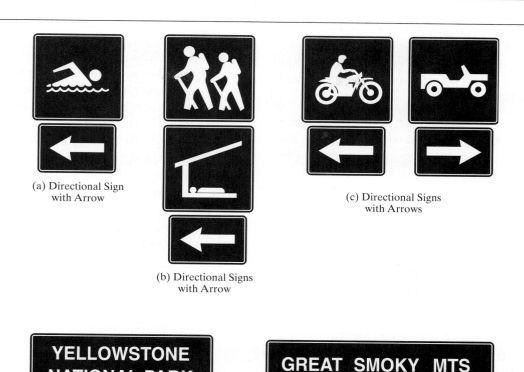

(a) Directional Sign
with Arrow

(b) Directional Signs
with Arrow

(c) Directional Signs
with Arrows

(d) Text Legend Cultural-Interest Signs

Figure 4.17: Recreational and Cultural Interest Signs (Used with permission of Federal Highway Administration, U.S. Department of Transportation, *Manual on Uniform Traffic Control Devices*, Millennium Edition, Fig 2H–1, pg. 2H–5, Fig 2H–2, pg. 2H–7, Washington DC, 2000.)

is, showing no indications. This is particularly confusing to drivers and can result in accidents. Any accidents occurring while a signal is in the dark mode are the legal responsibility of the agency operating the signal in most states. When signals are inoperable, signal heads should be bagged or taken down to avoid such confusion. In power outages, police or other authorized agents should be used to direct traffic at all signalized locations.

The second mandate is actually related to the first and addresses a common past practice—turning off signals at night and using STOP control during these hours. The problem is that during daytime hours, the driver may be confronted with a green signal *and* a STOP sign. This is extremely confusing and is now prohibited.

The use of STOP signs in conjunction with permanent operation of a red flashing light is permissible, as the legal interpretation of a flashing red signal is the same as

that of a STOP sign. The second exception addresses cases where a minor driveway might be near a signalized intersection but would not require separate signal control.

Signal Warrants

Traffic signals, when properly installed and operated at appropriate locations, provide a number of significant benefits:

- With appropriate physical designs, control measures, and signal timing, the capacity of critical intersection movements is increased.

- The frequency and severity of accidents is reduced for certain types of crashes, including right-angle, turn, and pedestrian accidents.

- When properly coordinated, signals can provide for nearly continuous movement of through traffic along an arterial at a designated speed under favorable traffic conditions.

- They provide for interruptions in heavy traffic streams to permit crossing vehicular and pedestrian traffic to safely cross.

At the same time, misapplied or poorly designed signals can cause excessive delay, signal violations, increased accidents (particularly rear-end accidents), and drivers rerouting their trips to less appropriate routes.

The MUTCD provides very specific warrants for the use of traffic control signals. These warrants are far more detailed than those for other devices, due to their very high cost (relative to other control devices) and the negative impacts of their misapplication. Thus, the manual is clear that traffic control signals shall be installed only at locations where an engineering study has indicated that one or more of the specified warrants has been met, and that application of signals will improve safety and/or capacity of the intersection. The manual goes further; if a study indicates that an existing signal is in place at a location that does not meet any of the warrants, it should be removed and replaced with a less severe form of control.

The Millennium Edition of the MUTCD details eight different warrants, any one of which may indicate that installation of a traffic control signal is appropriate. Chapter 16 contains a detailed treatment of these warrants and their application as part of an overall process for determining the appropriate form of intersection control for any given situation.

Signal Indications

The MUTCD defines the meaning of each traffic control signal indication as follows:

- *Green ball.* A steady green circular indication allows vehicular traffic facing the ball to enter the intersection to travel straight through the intersection or to turn right or left, except when prohibited by lane-use controls or physical design. Turning vehicles must yield the right-of-way to opposing through vehicles and to pedestrians legally in a conflicting crosswalk. In the absence of pedestrian signals, pedestrians may proceed to cross the roadway within any legally marked or unmarked crosswalk.

- *Yellow ball.* The steady yellow circular indication is a transition between the Green Ball and the Red Ball indication. It warns drivers that the related green movement is being terminated or that a red indication will immediately follow. In general, drivers are permitted to enter the intersection on "yellow," but are prohibited from doing so on the "red" that follows it. In the absence of pedestrian signals, pedestrians may not begin to cross the street on a "yellow" indication.

- *Red ball.* The steady red circular indication requires all traffic (vehicular and pedestrian) facing it to stop at the STOP line, crosswalk line (if no STOP line exists), or at the conflicting pedestrian path (if no crosswalk or STOP line exists). All states allow right-turning traffic to proceed with caution after stopping, unless specifically prohibited by signing or statute. Some states allow left-turners from one one-way street turning into another to proceed with caution after stopping, but this is far from a universal statute.

- *Flashing ball.* A flashing "yellow" allows traffic to proceed with caution through the intersection. A flashing "red" has the same meaning as a STOP sign—the driver may proceed with caution after coming to a complete stop.

- *Arrow indications.* Green, yellow, and red arrow indications have the same meanings as ball indications, except that they apply only to the movement designated by the arrow. A green left-turn arrow is only used to indicate a protected left turn (i.e., a left turn made on a green arrow will not encounter an opposing vehicular through movement). Such vehicles, however, may encounter pedestrians legally in the conflicting crosswalk and must yield to them. A green right-turn arrow is shown only when there are no pedestrians legally in the conflicting crosswalk. Yellow arrows warn drivers that the green arrow is about to terminate. The yellow arrow may be followed by a green ball indication where the protected left- and/or right-turning movement is followed by a permitted movement. A "permitted" left turn is made against an opposing vehicular flow. A "permitted" right turn is made against a conflicting pedestrian flow. It is followed by a red arrow where the movement must stop.

The MUTCD provides additional detailed discussion on how and when to apply various sequences and combinations of indications.

Signal Faces and Visibility Requirements

In general, a signal face should have three to five signal lenses (see Figure 4.18), with some exceptions allowing for a sixth to be shown. Two lens sizes are provided for: 8-in. diameter and 12-in. lenses. The manual requires that 12-in. lenses be used:

- Where road users view both traffic control and lane-use control signal heads simultaneously.

- Where the nearest signal face is between 120 ft and 150 ft beyond the STOP line, unless a near-side supplemental signal face is provided.

- Where signal faces are located more than 150 ft from the STOP line.

- Where minimum sight distances (see Table 4.3) cannot be met.

- For all arrow signal indications.

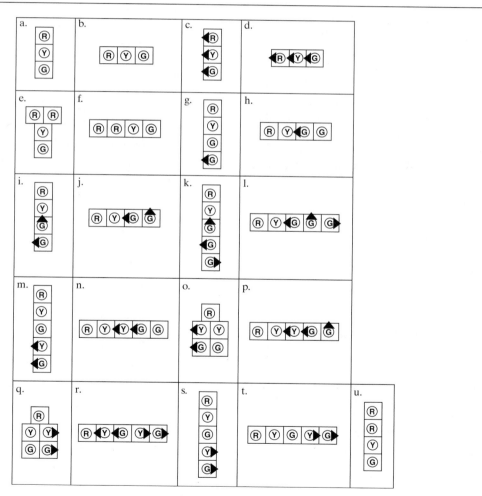

Figure 4.18: Typical Signal Face Arrangements (Used with permission of Federal Highway Administration, U.S. Department of Transportation, *Manual of Uniform Traffic Control Devices*, Millennium Edition, Figure 4D–3, pg. 4D–31, Washington DC, 2000.)

It further recommends that 12-in lenses *should* be used:

- Where 85th percentile approach speeds exceed 40 mi/h
- Where the traffic control signal might be unexpected
- On all approaches without curbs and gutters where only post-mounted signals are used.
- Where there is a significant percentage of elderly drivers

The red signal lens must be the same size or larger than other lenses. Thus, a 12-in red lens can be used in conjunction with 8-in green and yellow lenses. However, where green and yellow lenses are 12 inches, the red lens must also be 12 inches.

Table 4.3: Minimum Sight Distances for Signal Faces

85th Percentile Speed (mi/h)	Minimum Sight Distance (ft)
20	175
25	215
30	270
35	325
40	390
45	460
50	540
55	625
60	715

(Used with permission of Federal Highway Administration, U.S. Department of Transportation, *Manual on Uniform Traffic Control Devices,* Millennium Edition, Table 4D-1, pg. 4D-23, Washington DC, 2000.)

Table 4.3 shows the minimum visibility distances required for signal faces. A minimum of two signal faces must be provided for the major movement on each approach, even if the major movement is a turning movement. This requirement provides some measure of redundancy in case of an unexpected bulb failure.

Where the minimum visibility distances of Table 4.3 cannot be provided, 12-in. lenses must be used, and placement of appropriate "Signal Ahead" warning signs is required. The warning signs may be supplemented by a "hazard identification beacon."

The arrangement of lenses on a signal face is also limited to approved sequences. In general, the red ball must be at the top of a vertical signal face or at the left of a horizontal signal face, followed by the yellow and green. Where arrow indications are on the same signal face as ball indications, they are located on the bottom of a vertical display or right of a horizontal display. Figure 4.18 shows the most commonly used lens arrangements. The MUTCD contains detailed discussion of the applicability of various signal face designs.

Figure 4.19 illustrates the preferred placement of signal faces. At least one of the two required signal faces for the major movement must be located between 40 and 150 ft of the STOP line, unless the physical design of the intersection prevents it. Horizontal placement should be within 20° of the centerline of the approach, facing straight ahead.

Figure 4.20 illustrates the standard for vertical placement of signal faces that are between 40 and 53 ft from the STOP line. The standard prescribes the maximum height of the top of the signal housing above the pavement.

Operational Restrictions

Continuous operation of traffic control signals is critical for safety. No signal face should ever be "dark" (i.e., with no lens illuminated). In cases where signalization is not deemed necessary at night, signals must be operated in the flashing mode ("yellow" for one street and "red" for the other). Signal operations must also be designed to allow flashing operation to be maintained even when the signal controller is undergoing maintenance or replacement.

When being installed, signal faces should be bagged and turned to make it obvious to drivers that they are not in operation. Signals should be made operational as soon as possible after installation—again, to minimize possible confusion to drivers.

Bulb maintenance is a critical part of safe signal operation, as a burned-out bulb can make a signal face

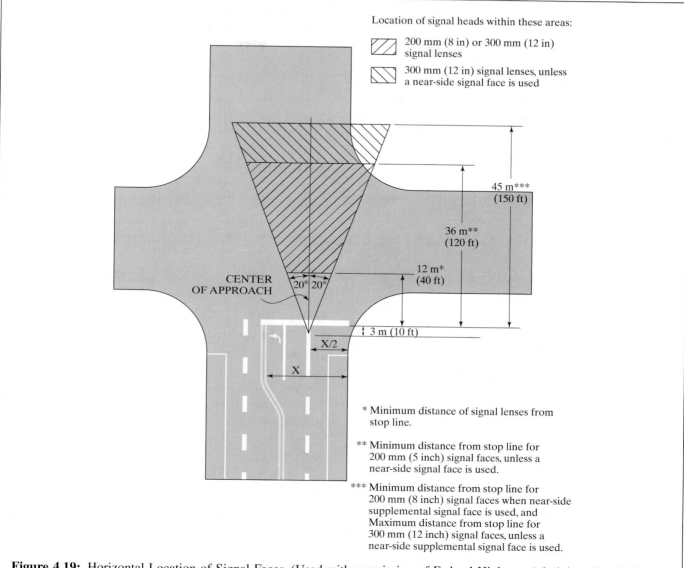

Location of signal heads within these areas:

⬛ 200 mm (8 in) or 300 mm (12 in) signal lenses

⬛ 300 mm (12 in) signal lenses, unless a near-side signal face is used

45 m*** (150 ft)

36 m** (120 ft)

12 m* (40 ft)

CENTER OF APPROACH

20° 20°

3 m (10 ft)

X/2

X

* Minimum distance of signal lenses from stop line.

** Minimum distance from stop line for 200 mm (5 inch) signal faces, unless a near-side signal face is used.

*** Minimum distance from stop line for 200 mm (8 inch) signal faces when near-side supplemental signal face is used, and Maximum distance from stop line for 300 mm (12 inch) signal faces, unless a near-side supplemental signal face is used.

Figure 4.19: Horizontal Location of Signal Faces (Used with permission of Federal Highway Administration, U.S. Department of Transportation, *Manual of Uniform Traffic Control Devices,* Millennium Edition, Figure 4D-2, pg. 4D-26, Washington DC, 2000.)

appear to be "dark" during certain intervals. A regular bulb-replacement schedule must be maintained. It is common to replace signal bulbs regularly at about 75%–80% of their expected service life to avoid burn-out problems. Other malfunctions can lead to other nonstandard indications appearing, although most controllers are programmed to fall back to the flashing mode in the event of most malfunctions. Most signal agencies maintain a contract with a private maintenance organization that requires rapid response (in the order of 15–30 minutes) to any reported malfunction. The agency can also operate its own maintenance group under similar rules. Any accident occurring during a signal malfunction can lead to legal liability for the agency with jurisdiction.

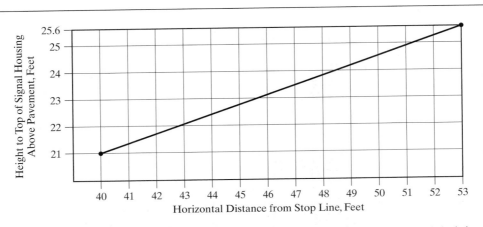

Figure 4.20: Vertical Placement of Signal Faces (Used with permission of Federal Highway Administration, U.S. Department of Transportation, *Manual of Uniform Traffic Control Devices,* Millennium Edition, Figure 4D-1, pg. 4D-25, Washington DC, 2000.)

4.4.2 Pedestrian Signals

The millennium edition of the MUTCD has mandated the use of new pedestrian signals that had been introduced as options over the past several years. The use of the older "WALK" and "DON'T WALK" designs has been discontinued in favor of the following indications:

- *Walking man (steady).* The new "WALK" indication is the image of a walking person in the color white. This indicates that it is permissible for a pedestrian to enter the crosswalk to begin crossing the street.

- *Upraised hand (flashing).* The new "DON'T WALK" indication is an upraised hand in the color Portland orange. In the flashing mode, it indicates that no pedestrian may enter the crosswalk to begin crossing the street but that those already crossing may continue safely.

- *Upraised hand (steady).* In the steady mode, the upraised hand indicates that no pedestrian should begin crossing and that no pedestrian should still be in the crosswalk.

In previous manuals, a flashing "WALK" indication was an option that could be used to indicate that right-turning vehicles may be conflicting with pedestrians legally in the crosswalk. The new manual does not permit a flashing WALKING MAN, effectively discontinuing this practice.

Figure 4.21 shows the new pedestrian signals. Note that both the UPRAISED HAND and WALKING MAN symbols can be shown in outline form or as a solid image. They may be located side-by-side on a single-section signal or arranged vertically on two-section signal. The UPRAISED HAND is on the left, or on top in these displays. When not illuminated, neither symbol should be readily visible to pedestrians at the far end of the crosswalk.

Chapters 18 and 19 discuss the use and application of pedestrian signals in the context of overall intersection control and operation. They include a discussion of when and where pedestrian signals are mandated as part of a signalization design.

4.4.3 Other Traffic Signals

The MUTCD provides specific criteria for the design, placement, and use of a number of other types of signals, including:

- Beacons
- In-roadway lights
- Lane-use control signals
- Ramp control signals (or ramp meters)

Beacons are generally used to identify a hazard or call attention to a critical control device, such as a speed

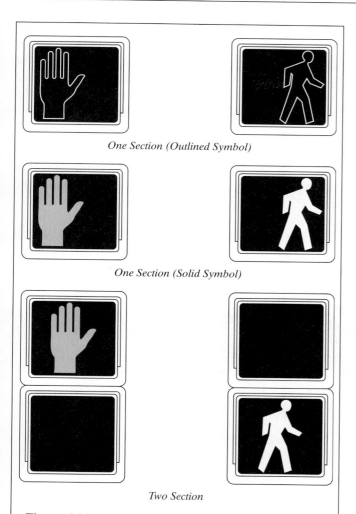

One Section (Outlined Symbol)

One Section (Solid Symbol)

Two Section

Figure 4.21: Pedestrian Signal Indications (Used with permission of Federal Highway Administration, U.S. Department of Transportation, *Manual of Uniform Traffic Control Devices*, Millennium Edition, Figure 4E-1, pg. 4E-4, Washington DC, 2000.)

limit sign, a STOP or YIELD sign, or a DO NOT ENTER sign. Lane-use control signals are used to control reversible lanes on bridges, in tunnels, and on streets and highways.

4.4.4 Traffic Signal Controllers

Modern traffic signal controllers are a complex combination of hardware and software that implements signal timing and ensures that signal indications operate consistently

and continuously in accordance with the programmed signal-timing. Each signalized intersection has a controller dedicated to implementing the signal-timing plan at that intersection. In addition, master controllers coordinate the operation of many signals, allowing signals along an arterial or in a network to be coordinated to provide progressive movement and/or other arterial or network control policies.

Individual traffic controllers may operate in the *pretimed* or *actuated* mode. In pretimed operation, the sequence and timing of every signal indication is preset and is repeated in each signal cycle. In actuated operation, the sequence and timing of some or all of the green indications may change on a cycle-by-cycle basis in response to detected vehicular and pedestrian demand. Chapter 18 discusses the timing and design of pre-timed signals. Chapter 20 discusses the timing and design of semi-actuated and fully actuated signals.

Traffic controllers implement signal timing designs. They are connected to *display hardware*, which consists of various traffic control and pedestrian signal faces that inform drivers and pedestrians of when they may legally proceed. Such display hardware also includes various types of supporting structures. Where individual intersection control and/or the signal system has demand-responsive elements, controllers must be connected with properly placed *detectors* that provide information on vehicle and/or pedestrian presence that interacts with controllers to determine and implement a demand-responsive signal timing pattern. Chapter 19 contains a more detailed discussion of street hardware, detectors, and their placement as part of an intersection design.

The *Manual of Traffic Control Devices* [6] and the *Traffic Detector Handbook* [7] are standard traffic engineering references that provide significant detail on all elements of traffic signal hardware.

Standards for Traffic Signal Controllers

The National Electrical Manufacturers Association (NEMA) is the principal trade group for the electronics industry in the United States. Its Traffic Control Systems Section sets manufacturing guidelines and standards for traffic control hardware [8]. The group's philosophy is to encourage industry standards that:

- Are based upon proven designs
- Are downward compatible with existing equipment

- Reflect state-of-the-art reliability and performance

- Minimize the potential for malfunctions

NEMA standards are not product designs but rather descriptions and performance criteria for various product categories. The standards cover such products as solid-state controllers, load switches, conflict monitors, loop detectors, flashers, and terminals and facilities. Ref. [9] provides a good overview of what the NEMA standards mean and imply.

NEMA does not preclude any manufacturer from making and selling a nonconforming product. Many funding agencies, however, require the use of hardware conforming to the latest NEMA standards.

Two other standards also exist. In New York State, standards for the Type 170 controller have been developed and extensively applied. In California, standards for the Type 2070 controller have been similarly developed and implemented. Both have similar features, but are implemented using different hardware and software architecture.

Electro-Mechanical Controllers

Historically, pretimed controllers were electro-mechanical devices marked by simplicity, reliability, and reasonable cost. The primary functioning elements of the electro-mechanical controller are illustrated in Figure 4.22.

The controller shown is a "three-dial" controller. It permits the implementation of three different pretimed

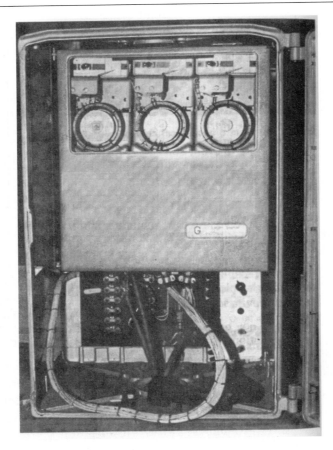

Figure 4.22: Electro-Mechanical Controller Inside a Standard Cabinet (Courtesy of General Traffic Equipment.)

signal timings for different times of the day. A clock setting determines when each of the timings is in effect. Typically, a signal timing would be developed and implemented for an AM peak period, a PM peak period, and an off-peak period.

Each timing dial is a cylinder that contains 100 "slots" into which tabbed keys may be inserted. The cycle length is determined by the speed at which the cylinder rotates. One cycle is one complete rotation of the appropriate timing dial. Keys, inserted into the slots, trip cam switches as they rotate, causing signal indications to change. In effect, these controllers allowed a pretimed setting of each indication to the nearest 100th of the cycle length.

The actual signal indication occurs because the teeth on a given cam are broken out, allowing a contact to close a circuit so current can flow to designated bulbs in the signal heads. The cams are broken out in a pattern that allows the logical sequence of green-yellow-red first on one phase and then another. Since each cam generally controls one set of signal faces, a given signal phase often involves two cams (often referred to as rings), one controlling direction 1 and the other controlling direction 2. Most complete signal timings can be worked out using two rings. While this definition of "rings" is no longer necessary, the concept of control rings remains critical to the understanding of signal timing design and operation. Chapter 18 presents extensive discussion of the ring concept.

While the technology is old, the reliability and cost of electro-mechanical controllers were desirable features. Many such controllers remain in operation in traffic control systems.

Modern Controllers

As indicated previously, modern controllers are a complex combination of hardware and software. Many are modular, so that individual modules can be easily replaced in the field and taken back to the "shop" for repairs or modification.

Because signal technology advances significantly and rapidly, the details of any specific illustrations will become rapidly dated. Students should consult manufacturers' Web sites for the most up-to-date information on signal controllers.

4.5 Special Types of Control

While not covered in this chapter, the MUTCD contains significant material covering special control situations, including:

- School zones
- Railroad crossings
- Construction and maintenance zones
- Pedestrian and bicycle controls

These situations invariably involve a combination of signing, markings, and/or signals for fully effective control. Consult the MUTCD directly for details on these and other applications not covered herein.

4.6 Summary and Conclusion

This chapter has provided an introduction and overview to the design, placement, and use of traffic control devices. The MUTCD is not a stagnant document, and updates and revisions are constantly being issued. Thus, it is imperative that users consult the latest version of the manual, and all of its formal revisions. For convenience, the MUTCD can be accessed on-line at mutcd.fhwa.gov/kno-millennium.htm or through the federal highway administration home Web site—www.fhwa.dot.com. This is a convenient way of using the manual, as all updates and revisions are always included. Similarly, virtually every signal manufacturer has a Web site that can be accessed to review detailed specifications and characteristics of controllers and other signal hardware and software. A directory of web sites related to traffic control devices may be found at www.traffic-signals.com.

The use of the manual is discussed and illustrated in a number of other chapters of this text: Chapter 16 for signal warrants and warrants for STOP and YIELD control; Chapter 15 for application of traffic control devices on freeways and rural highways; and Chapter 19 for application of control devices as part of intersection design.

References

1. *Manual of Uniform Traffic Control Devices*, Millennium Edition, Federal Highway Administration, U.S. Department of Transportation, Washington DC, 2000.

2. Hawkins, H.G., "Evolution of the MUTCD: Early Standards for Traffic Control Devices," *ITE Journal*, Institute of Transportation Engineers, Washington DC, July 1992.

3. Hawkins, H.G., "Evolution of the MUTCD: Early Editions of the MUTCD," *ITE Journal*, Institute of Transportation Engineers, Washington DC, August 1992.

4. Hawkins, H.C., "Evolution of the MUTCD: The MUTCD Since WWII," *ITE Journal*, Institute of Transportation Engineers, Washington DC, November 1992.

5. Hawkins, H.C., "Evolution of the MUTCD Mirrors American Progress Since the 1920's," *Roads and Bridges*, Scranton Gillette, Communications Inc., Des Plaines, IL, July 1995.

6. Kell, J. and Fullerton, I., *Manual of Traffic Signal Design*, 2nd Edition, Institute of Transportation Engineers, Prentice Hall Inc., Englewood Cliffs, NJ, 1991.

7. *Traffic Detector Handbook*, 2nd Edition, JHK & Associates, Institute of Transportation Engineers, Washington DC, nd.

8. Parris, C., "NEMA and Traffic Control," *ITE Journal*, Institute of Transportation Engineers, Washington DC, August 1986.

9. Parris, C., "Just What Does a NEMA Standard Mean?" *ITE Journal*, Institute of Transportation Engineers, Washington DC, July 1987.

Problems

4-1. Define the following terms with respect to their meaning in the millennium edition of the MUTCD: standard, guideline, option, and support.

4-2. Describe how color, shape, and legend are used to convey and reinforce messages given by traffic control devices.

4-3. Why should overuse of regulatory and warning signs be avoided? Why is this not a problem with guide signs?

4-4. How far from the point of a hazard should the following warning signs be placed?

    (a) A "STOP ahead" warning sign on a road with a posted speed limit of 50 mi/h.

    (b) A "curve ahead" warning sign with an advisory speed of 30 mi/h on a road with a posted speed limit of 45 mi/h.

    (c) A "merge ahead" warning sign on a ramp with an 85th percentile speed of 35 mi/h.

4-5. Select a one-mile stretch of freeway in your vicinity. Drive one direction of this facility with a friend or colleague. The passenger should count and note the number and type of traffic signs encountered. Are any of them confusing? Suggest improvements as appropriate. Comment on the overall quality of the signing in the test section.

4-6. Select one signalized and one STOP or YIELD controlled intersection in your neighborhood. Note the placement of all devices at each intersection. Do they appear to meet MUTCD standards? Is visibility of all devices adequate? Comment on the effectiveness of traffic controls at each intersection.

CHAPTER
5
Traffic Stream Characteristics

Traffic streams are made up of individual drivers and vehicles interacting with each other and with the physical elements of the roadway and its general environment. Because both driver behavior and vehicle characteristics vary, individual vehicles within the traffic stream do not behave in exactly the same manner. Further, no two traffic streams will behave in exactly the same way, even in similar circumstances, because driver behavior varies with local characteristics and driving habits.

Dealing with traffic, therefore, involves an element of variability. A flow of water through channels and pipes of defined characteristics will behave in an entirely predictable fashion, in accord with the laws of hydraulics and fluid flow. A given flow of traffic through streets and highways of defined characteristics will vary with both time and location. Thus, the critical challenge of traffic engineering is to plan and design for a medium that is not predictable in exact terms—one that involves both physical constraints and the complex behavioral characteristics of human beings.

Fortunately, while exact characteristics vary, there is a reasonably consistent range of driver and, therefore, traffic stream behavior. Drivers on a highway designed for a safe speed of 60 mi/h may select speeds in a broad range (perhaps 45–65 mi/h); few, however, will travel at 80 mi/h or at 20 mi/h.

In describing traffic streams in quantitative terms, the purpose is to both understand the inherent variability in their characteristics and to define normal ranges of behavior. To do so, key parameters must be defined and measured. Traffic engineers will analyze, evaluate, and ultimately plan improvements in traffic facilities based on such parameters and their knowledge of normal ranges of behavior.

This chapter focuses on the definition and description of the parameters most often used for this purpose and on the characteristics normally observed in traffic streams. These parameters are, in effect, the traffic engineer's measure of reality, and they constitute a language with which traffic streams are described and understood.

5.1 Types of Facilities

Traffic facilities are broadly separated into two principal categories:

- Uninterrupted flow
- Interrupted flow

Uninterrupted flow facilities have no external interruptions to the traffic stream. Pure uninterrupted flow exists primarily on freeways, where there are no intersections at grade, traffic signals, STOP or YIELD signs, or other interruptions external to the traffic stream itself. Because such facilities have full control of access, there are no intersections at grade, driveways, or any forms of direct access to abutting lands. Thus, the characteristics of the traffic stream are based solely on the interactions among vehicles and with the roadway and the general environment.

While pure uninterrupted flow exists only on freeways, it can also exist on sections of surface highway, most often in rural areas, where there are long distances between fixed interruptions. Thus, uninterrupted flow may exist on some sections of rural two-lane highways and rural and suburban multilane highways. As a very general guideline, it is believed that uninterrupted flow can exist in situations where the distance between traffic signals and/or other significant fixed interruptions is more than two miles.

It should be remembered that the term "uninterrupted flow" refers to a type of facility, not the quality of operations on that facility. Thus, a freeway that experiences breakdowns and long delays during peak hours is still operating under uninterrupted flow. The causes for the breakdowns and delay are not external to the traffic stream but are caused entirely by the internal interactions within the traffic stream.

Interrupted flow facilities are those that incorporate fixed external interruptions into their design and operation. The most frequent and operationally significant external interruption is the traffic signal. The traffic signal alternatively starts and stops a given traffic stream, creating a platoons of vehicles progressing down the facility. Other fixed interruptions include STOP and YIELD signs, unsignalized at-grade intersections, driveways, curb parking maneuvers, and other land-access operations. Virtually all urban surface streets and highways are interrupted flow facilities.

The major difference between uninterrupted and interrupted flow facilities is the impact of time. On uninterrupted facilities, the physical facility is available to drivers and vehicles at all times. On a given interrupted flow facility, movement is periodically barred by "red" signals. The signal timing, therefore, limits access to particular segments of the facility in time. Further, rather than a continuously moving traffic stream, at traffic signals, the traffic stream is periodically stopping and starting again.

Interrupted flow is, therefore, more complex than uninterrupted flow. While many of the traffic flow parameters described in this chapter apply to both types of facilities, this chapter focuses primarily on the characteristics of uninterrupted flow. Many of these characteristics may also apply within a moving platoon of vehicles on an interrupted flow facility. Specific characteristics of traffic interruptions and their impact on flow are discussed in detail in Chapter 17.

5.2 Traffic Stream Parameters

Traffic stream parameters fall into two broad categories. *Macroscopic parameters* describe the traffic stream as a whole; *microscopic parameters* describe the behavior of individual vehicles or pairs of vehicles within the traffic stream.

The three principal macroscopic parameters that describe a traffic stream are (1) volume or rate of flow, (2) speed, and (3) density. Microscopic parameters include (1) the speed of individual vehicles, (2) headway, and (3) spacing.

5.2.1 Volume and Rate of Flow

Traffic volume is defined as the number of vehicles passing a point on a highway, or a given lane or direction of a highway, during a specified time interval. The unit of measurement for volume is simply "vehicles," although it is often expressed as "vehicles per unit time." Units of time used most often are "per day" or "per hour."

Daily volumes are used to establish trends over time, and for general planning purposes. Detailed design or control decisions require knowledge of hourly volumes for the peak hour(s) of the day.

Rates of flow are generally stated in units of "vehicles per hour," but represent flows that exist for periods of time less than one hour. A volume of 200 vehicles observed over a 15-minute period may be expressed as a rate of $200 \times 4 = 800$ vehicles/hour, even though 800 vehicles would not be observed if the full hour were counted. The 800 vehicles/hour becomes a rate of flow that exists for a 15-minute interval.

Daily Volumes ✳

As noted, daily volumes are used to document annual trends in highway usage. Forecasts based upon observed trends can be used to help plan improved or new facilities to accommodate increasing demand.

There are four daily volume parameters that are widely used in traffic engineering:

- *Average annual daily traffic (AADT)*. The average 24-hour volume at a given location over a full 365-day year; the number of vehicles passing a site in a year divided by 365 days (366 days in a leap year).

- *Average annual weekday traffic (AAWT)*. The average 24-hour volume occurring on weekdays over a full 365-day year; the number of vehicles passing a site on weekdays in a year divided by the number of weekdays (usually 260).

- *Average daily traffic (ADT)*. The average 24-hour volume at a given location over a defined time period less than one year; a common application is to measure an ADT for each month of the year.

- *Average weekday traffic (AWT)*. The average 24-hour weekday volume at a given location over a defined time period less than one year; a common application is to measure an AWT for each month of the year.

All of these volumes are stated in terms of vehicles per day (veh/day). Daily volumes are generally not differentiated by direction or lane but are totals for an entire facility at the designated location.

Table 5.1 illustrates the compilation of these daily volumes based upon one year of count data at a sample location.

The data in Table 5.1 generally comes from a permanent count location (i.e., a location where automated detection of volume and transmittal of counts electronically to a central computer is in place). Average weekday traffic (AWT) for each month is found by dividing the total monthly weekday volume by the number of weekdays in the month (Column 5 ÷ Column 2). The average daily traffic is the total monthly volume divided by the number of days in the month (Column 4 ÷ Column 3). Average annual daily traffic is the total observed volume for the year divided by 365 days/year. Average annual weekday traffic is the total observed volume on weekdays divided by 260 weekdays/year.

The sample data of Table 5.1 gives a capsule description of the character of the facility on which it was measured. Note that ADTs are significantly higher than AWTs in each month. This suggests that the facility is serving a recreational or vacation area, with traffic strongly peaking on weekends. Also, both AWTs and ADTs are highest during the summer months, suggesting that the facility serves a warm-weather recreational/vacation area. Thus, if a detailed study were needed to provide data for an upgrading of this facility, the period to focus on would be weekends during the summer.

Hourly Volumes

Daily volumes, while useful for planning purposes, cannot be used alone for design or operational analysis purposes. Volume varies considerably over the 24 hours of the day, with periods of maximum flow occurring during the morning and evening commuter "rush hours." The single hour of the day that has the highest hourly volume is referred to as the *peak hour*. The traffic volume within this hour is of greatest interest to traffic engineers for design and operational analysis usage. The peak-hour volume is generally stated as a *directional* volume (i.e., each direction of flow is counted separately).

Highways and controls must be designed to adequately serve the peak-hour traffic volume in the peak direction of flow. Since traffic going one way during the

Table 5.1: Illustration of Daily Volume Parameters

1. Month	2. No. of Weekdays In Month (days)	3. Total Days in Month (days)	4. Total Monthly Volume (vehs)	5. Total Weekday Volume (vehs)	6. AWT 5/2 (veh/day)	7. ADT 4/3 (veh/day)
Jan	22	31	425,000	208,000	9,455	13,710
Feb	20	28	410,000	220,000	11,000	14,643
Mar	22	31	385,000	185,000	8,409	12,419
Apr	22	30	400,000	200,000	9,091	13,333
May	21	31	450,000	215,000	10,238	14,516
Jun	22	30	500,000	230,000	10,455	16,667
Jul	23	31	580,000	260,000	11,304	18,710
Aug	21	31	570,000	260,000	12,381	18,387
Sep	22	30	490,000	205,000	9,318	16,333
Oct	22	31	420,000	190,000	8,636	13,548
Nov	21	30	415,000	200,000	9,524	13,833
Dec	22	31	400,000	210,000	9,545	12,903
Total	**260**	**365**	**5,445,000**	**2,583,000**	—	—

$$AADT = 5,445,000/365 = 14,918 \text{ veh/day}$$

$$AAWT = 2,583,000/260 = 9,935 \text{ veh/day}$$

morning peak is going the opposite way during the evening peak, *both* sides of a facility must generally be designed to accommodate the peak directional flow during the peak hour. Where the directional disparity is significant, the concept of reversible lanes is sometimes useful. Washington DC, for example, makes extensive use of reversible lanes (direction changes by time of day) on its many wide boulevards and some of its freeways.

In design, peak-hour volumes are sometimes estimated from projections of the AADT. Traffic forecasts are most often cast in terms of AADTs based on documented trends and/or forecasting models. Because daily volumes, such as the AADT, are more stable than hourly volumes, projections can be more confidently made using them. AADTs are converted to a peak-hour volume in the peak direction of flow. This is referred to as the "directional design hour volume" (DDHV), and is found using the following relationship:

$$DDHV = AADT * K * D \qquad (5-1)$$

where: K = proportion of daily traffic occurring during the peak hour

D = proportion of peak hour traffic traveling in the peak direction of flow.

For design, the K factor often represents the proportion of AADT occurring during the *30th peak hour* of the year. If the 365 peak hour volumes of the year at a given location are listed in descending order, the 30th peak hour is 30th on the list and represents a volume that is exceeded in only 29 hours of the year. For rural facilities, the 30th peak hour may have a significantly lower volume than the worst hour of the year, as critical peaks may occur only infrequently. In such cases, it is not considered economically feasible to invest large amounts of capital in providing additional capacity that will be used in only 29 hours of the year. In urban cases, where traffic is frequently at capacity levels during the daily commuter peaks, the 30th peak hour is often not substantially different from the highest peak hour of the year.

Factors K and D are based upon local or regional characteristics at existing locations. Most state highway departments, for example, continually monitor these proportions, and publish appropriate values for use in various areas of the state. The K factor decreases with increasing development density in the areas served by the facility. In high-density areas, substantial demand during off-peak periods exists. This effectively lowers the proportion of traffic occurring during the peak hour of the day. The volume generated by high-density development is generally larger than that generated by lower-density areas. Thus, it is important to remember that a high proportion of traffic occurring in the peak hour does not suggest that the peak-hour volume is large.

The D factor tends to be more variable and is influenced by a number of factors. Again, as development density increases, the D factor tends to decrease. As density increases, it is more likely to have substantial bi-directional demands. Radial routes (i.e, those serving movements into and out of central cities or other areas of activity), will have stronger directional distributions (higher D values) than those that are circumferential, (i.e., going around areas of central activity). Table 5.2 indicates general ranges for K and D factors. These are purely illustrative; specific data on these characteristics should be available from state or local highway agencies, or should be locally calibrated before application.

Consider the case of a rural highway that has a 20-year forecast of AADT of 30,000 veh/day. Based upon the data of Table 5.2, what range of directional design hour volumes might be expected for this situation? Using the values of Table 5.2 for a rural highway, the K factor ranges from 0.15 to 0.25, and the D factor ranges from 0.65 to 0.80. The range of directional design hour volumes is, therefore:

$$DDHV_{LOW} = 30,000*0.15*0.65 = 2,925 \text{ veh/h}$$

$$DDHV_{HIGH} = 30,000*0.25*0.80 = 6,000 \text{ veh/h}$$

The expected range in DDHV is quite large under these criteria. Thus, determining appropriate values of K and D for the facility in question is critical in making such a forecast.

This simple illustration points out the difficulty in projecting future traffic demands accurately. Not only does volume change over time, but the basic characteristics of volume variation may change as well. Accurate projections require the identification of causative relationships that remain stable over time. Such relationships are difficult to discern in the complexity of observed travel behavior. Stability of these relationships over time cannot be guaranteed in any event, making volume forecasting an approximate process at best.

Subhourly Volumes and Rates of Flow

While hourly traffic volumes form the basis for many forms of traffic design and analysis, the variation of traffic within a given hour is also of considerable interest. The quality of traffic flow is often related to short-term fluctuations in traffic demand. A facility may have sufficient capacity to serve the peak-hour demand, but short-term peaks of flow within the hour may exceed capacity and create a breakdown.

Volumes observed for periods of less than one hour are generally expressed as equivalent hourly rates of flow. For example, 1,000 vehicles counted over a 15-minute

Table 5.2: General Ranges for K and D Factors

Facility Type	Normal Range of Values	
	K-Factor	**D-Factor**
Rural	0.15–0.25	0.65–0.80
Suburban	0.12–0.15	0.55–0.65
Urban:		
Radial Route	0.07–0.12	0.55–0.60
Circumferential Route	0.07–0.12	0.50–0.55

Table 5.3: Illustration of Volumes and Rates of Flow

Time Interval	Volume for Time Interval (vehs)	Rate of Flow for Time Interval (vehs/h)
5:00–5:15 PM	1,000	1,000/0.25 = 4,000
5:15–5:30 PM	1,100	1,100/0.25 = 4,400
5:30–5:45 PM	1,200	1,200/0.25 = 4,800
5:45–6:00 PM	900	900/0.25 = 3,600
5:00–6:00 PM	Σ = 4,200	

interval could be expressed as 1,000 vehs/0.25 h = 4,000 veh/h. The rate of flow of 4,000 veh/h is valid for the 15-minute period in which the volume of 1,000 vehs was observed. Table 5.3 illustrates the difference between volumes and rates of flow.

The full hourly volume is the sum of the four 15-minute volume observations, or 4,200 veh/h. The rate of flow for each 15-minute interval is the volume observed for that interval divided by the 0.25 hours over which it was observed. In the worst period of time, 5:30–5:45 PM, the rate of flow is 4,800 veh/h. This is a *flow rate*, not a volume. The actual volume for the hour is only 4,200 veh/h.

Consider the situation that would exist if the capacity of the location in question were exactly 4,200 vehs/h. While this is sufficient to handle the full-hour demand indicated in Table 5.3, the demand *rate of flow* during two of the 15-minute periods noted (5:15–5:30 PM and 5:30–5:45 PM) exceeds the capacity. The problem is that while demand may vary within a given hour, capacity is constant. In each 15-minute period, the capacity is 4,200/4 or 1,050 vehs. Thus, within the peak hour shown, queues will

develop in the half-hour period between 5:15 and 5:45 PM, during which the demand exceeds the capacity. Further, while demand is less than capacity in the first 15-minute period (5:00–5:15 PM), the unused capacity cannot be used in a later period. Table 5.4 compares the demand and capacity for each of the 15-minute intervals. The queue at the end of each period can be computed as the queue at the beginning of the period plus the arriving vehicles minus the departing vehicles.

Even though the capacity of this segment over the full hour is equal to the peak-hour demand volume (4,200 veh/h), at the end of the hour, there remains a queue of 50 vehicles that has not been served. While this illustration shows that a queue exists for three out of four 15-minute periods within the peak hour, the dynamics of queue clearance may continue to negatively affect traffic for far longer.

Because of these types of impacts, it is often necessary to design facilities and analyze traffic conditions for a period of maximum rate of flow within the peak hour. For most practical purposes, 15 minutes is considered to be the minimum period of time over

Table 5.4: Queuing Analysis for the Data of Table 5.3

Time Interval	Arriving Vehicles (vehs)	Departing Vehicles (vehs)	Queue Size at End of Period (vehs)
5:00–5:15 PM	1,000	1,050	0
5:15–5:30 PM	1,100	1,050	0 + 1,100 − 1,050 = 50
5:30–5:45 PM	1,200	1,050	50 + 1,200 − 1,050 = 200
5:45–6:00 PM	900	1,050	200 + 900 − 1,050 = 50

which traffic conditions are statistically stable. While rates of flow can be computed for any period of time and researchers often use rates for periods of one to five minutes, rates of flow for shorter periods often represent transient conditions that defy consistent mathematical representations. In recent years, however, use of five-minute rates of flow has increased, and there is some thought that these might be sufficiently stable for use in design and analysis. Despite this, most standard design and analysis practices continue to use the 15-minute interval as a base period.

The relationship between the hourly volume and the maximum rate of flow within the hour is defined by the *peak hour factor*, as follows:

$$PHF = \frac{\text{hourly volume}}{\text{max. rate of flow}}$$

For standard 15-minute analysis period, this becomes:

$$PHF = \frac{V}{4 * V_{m15}} \qquad (5\text{-}2)$$

where: V = hourly volume, vehs
  V_{m15} = maximum 15-minute volume within the hour, vehs
  PHF = peak-hour factor

For the illustrative data in Tables 5-3 and 5-4:

$$PHF = \frac{4{,}200}{4 * 1{,}200} = 0.875$$

The maximum possible value for the *PHF* is 1.00, which occurs when the volume in each interval is constant. For 15-minute periods, each would have a volume of exactly one quarter of the full hour volume. This indicates a condition in which there is virtually no variation of flow within the hour. The minimum value occurs when the entire hourly volume occurs in a single 15-minute interval. In this case, the *PHF* becomes 0.25, and represents the most extreme case of volume variation within the hour. In practical terms, the *PHF* generally varies between a low of 0.70 for rural and sparsely developed areas to 0.98 in dense urban areas.

The peak-hour factor is descriptive of trip generation patterns and may apply to an area or portion of a street and highway system. When the value is known, it can be used to estimate a maximum flow rate within an hour based on the full-hour volume:

$$v = \frac{V}{PHF} \qquad (5\text{-}3)$$

where: v = maximum rate of flow within the hour, veh/h
  V = hourly volume, veh/h
  PHF = peak-hour factor.

This conversion is frequently used in the techniques and methodologies covered throughout this text.

5.2.2 Speed and Travel Time

Speed is the second macroscopic parameter describing the state of a traffic stream. Speed is defined as a rate of motion in distance per unit time. Travel time is the time taken to traverse a defined section of roadway. Speed and travel time are inversely related:

$$S = \frac{d}{t} \qquad (5\text{-}4)$$

where: S = speed, mi/h or ft/s
  d = distance traversed, mi or ft
  t = time to traverse distance d, h or s

In a moving traffic stream, each vehicle travels at a different speed. Thus, the traffic stream does not have a single characteristic value, but rather a distribution of individual speeds. The traffic stream, taken as a whole, can be characterized using an average or typical speed.

There are two ways in which an average speed for a traffic stream can be computed:

- *Time mean speed* (TMS). The average speed of all vehicles passing a point on a highway or lane over some specified time period.
- *Space mean speed* (SMS). The average speed of all vehicles occupying a given section of highway or lane over some specified time period.

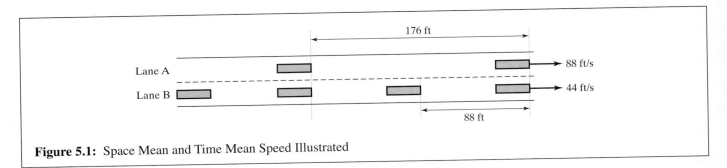

Figure 5.1: Space Mean and Time Mean Speed Illustrated

In essence, time mean speed is a point measure, while space mean speed describes a length of highway or lane. Figure 5.1 shows an example illustrating the differences between the two average speed measures.

To measure time mean speed (TMS), an observer would stand by the side of the road and record the speed of each vehicle as it passes. Given the speeds and the spacing shown in Figure 5.1, a vehicle will pass the observer in lane A every 176/88 = 2.0 s. Similarly, a vehicle will pass the observer in lane B every 88/44 = 2.0 s. Thus, as long as the traffic stream maintains the conditions shown, for every n vehicles traveling at 88 ft/s, the observer will also observe n vehicles traveling at 44 ft/s. The TMS may then be computed as:

$$TMS = \frac{88.0n + 44.0n}{2n} = 66.0 \text{ ft/s}$$

To measure space mean speed (SMS), an observer would need an elevated location from which the full extent of the section may be viewed. Again, however, as long as the traffic stream remains stable and uniform, as shown, there will be twice as many vehicles in lane B as there are in lane A. Therefore, the SMS is computed as:

$$SMS = \frac{(88.0n) + (44*2n)}{3n} = 58.7 \text{ mi/h}$$

In effect, space mean speed accounts for the fact that it takes a vehicle traveling at 44.0 ft/s twice as long to traverse the defined section as it does a vehicle traveling at 88.0 ft/s. The space mean speed weights slower vehicles more heavily, based on the amount of time they occupy a highway section. Thus, the space mean speed is usually lower than the corresponding time mean speed,

in which each vehicle is weighted equally. The two speed measures may conceivably be equal if all vehicles in the section are traveling at exactly the same speed.

Both the time mean speed and space mean speed may be computed from a series of measured travel times over a specified distance using the following relationships:

$$TMS = \frac{\sum_i \left(d / t_i \right)}{n} \tag{5-5}$$

$$SMS = \frac{d}{\left(\sum_i t_i / n \right)} = \frac{nd}{\sum_i t_i} \tag{5-6}$$

where: TMS = time mean speed, ft/s
SMS = space mean speed, ft/s
d = distance traversed, ft
n = number of observed vehicles
t_i = time for vehicle "i" to traverse the section, s

TMS is computed by finding each individual vehicle speed and taking a simple average of the results. SMS is computed by finding the average travel time for a vehicle to traverse the section and using the average travel time to compute a speed. Table 5.5 shows a sample problem in the computation of time mean and space mean speeds.

5.2.3 Density and Occupancy

Density

Density, the third primary measure of traffic stream characteristics, is defined as the number of vehicles occupying

Table 5.5: Illustrative Computation of TMS and SMS

Vehicle No.	Distance d (ft)	Travel Time t (s)	Speed (ft/s)
1	1,000	18.0	1,000/18 = 55.6
2	1,000	20.0	1,000/20 = 50.0
3	1,000	22.0	1,000/22 = 45.5
4	1,000	19.0	1,000/19 = 52.6
5	1,000	20.0	1,000/20 = 50.0
6	1,000	20.0	1,000/20 = 50.0
Total	**6,000**	**119**	**303.7**
Average	**6,000/6 = 1,000**	**119/6 = 19.8**	**303.7/6 = 50.6**

$$\text{TMS} = 50.6 \text{ ft/s}$$
$$\text{SMS} = 1,000/19.8 = 50.4 \text{ ft/s}$$

a given length of highway or lane, generally expressed as vehicles per mile or vehicles per mile per lane.

Density is difficult to measure directly, as an elevated vantage point from which the highway section under study may be observed is required. It is often computed from speed and flow rate measurements. (See Section 5.3 of this chapter).

Density, however, is perhaps the most important of the three primary traffic stream parameters, because it is the measure most directly related to traffic demand. Demand does not occur as a rate of flow, even though traffic engineers use this parameter as the principal measure of demand. Traffic is generated from various land uses, injecting a number of vehicles into a confined roadway space. This process creates a density of vehicles. Drivers select speeds that are consistent with how close they are to other vehicles. The speed and density combine to give the observed rate of flow.

Density is also an important measure of the quality of traffic flow, as it is a measure of the proximity of other vehicles, a factor which influences freedom to maneuver and the psychological comfort of drivers.

Occupancy

While density is difficult to measure directly, modern detectors can measure *occupancy*, which is a related parameter. Occupancy is defined as the proportion of time that a detector is "occupied," or covered, by a vehicle in a defined time period. Figure 5.2 illustrates.

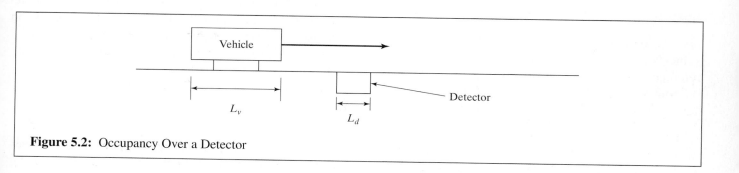

Figure 5.2: Occupancy Over a Detector

In Figure 5.2, L_v is the average length of a vehicle (ft), while L_d is the length of the detector (which is normally a magnetic loop detector). If "occupancy" over a given detector is "O," then density may be computed as:

$$D = \frac{5{,}280 * O}{L_v + L_d} \qquad (5\text{-}7)$$

The lengths of the average vehicle and the detector are added, as the detector is generally activated as the front bumper engages the front boundary of the detector and is deactivated when the rear bumper clears the back boundary of the detector.

Consider a case in which a detector records an occupancy of 0.200 for a 15-minute analysis period. If the average length of a vehicle is 28 ft, and the detector is 3 ft long, what is the density?

$$D = \frac{5{,}280 * 0.200}{28 + 3} = 34.1 \text{ veh/mi/ln}$$

The occupancy is measured for a specific detector in a specific lane. Thus, the density estimated from occupancy is in units of vehicles per mile per lane. If there are adjacent detectors in additional lanes, the density in each lane may be summed to provide a density in veh/mi for a given direction of flow over several lanes.

5.2.4   Spacing and Headway: Microscopic Parameters

While flow, speed, and density represent macroscopic descriptors for the entire traffic stream, they can be related to microscopic parameters that describe individual vehicles within the traffic stream, or specific pairs of vehicles within the traffic stream.

Spacing

Spacing is defined as the distance between successive vehicles in a traffic lane, measured from some common reference point on the vehicles, such as the front bumper or front wheels. The *average* spacing in a traffic lane can be directly related to the density of the lane:

$$D = \frac{5{,}280}{d_a} \qquad (5\text{-}8)$$

where: D = density, veh/mi/ln

d_d = average spacing between vehicles in the lane, ft

Headway

Headway is defined as the time interval between successive vehicles as they pass a point along the lane, also measured between common reference points on the vehicles. The *average* headway in a lane is directly related to the rate of flow:

$$v = \frac{3{,}600}{h_a} \qquad (5\text{-}9)$$

where: v = rate of flow, veh/h/ln

h_a = average headway in the lane, s

Use of Microscopic Measures

Microscopic measures are useful for many traffic analysis-purposes. Because a spacing and/or a headway may be obtained for every pair of vehicles, the amount of data that can be collected in a short period of time is relatively large. A traffic stream with a volume of 1,000 vehs over a 15-minute time period results in a *single* value of rate of flow, space mean speed, and density when observed. There would be, however, 1,000 headway and spacing measurements, assuming that all vehicle pairs were observed.

Use of microscopic measures also allows various vehicle types to be isolated in the traffic stream. Passenger car flows and densities, for example, could be derived from isolating spacing and headway for pairs of passenger cars following each other. Heavy vehicles could be similarly isolated and studied for their specific characteristics. Chapter 12 illustrates such a process for calibrating basic capacity analysis variables.

Average speed can also be computed from headway and spacing measurements as:

$$S = \frac{\left(d_a \big/ h_a\right)}{1.47} = 0.68\left(d_a \big/ h_a\right) \qquad (5\text{-}10)$$

where: S = average speed, mi/h

$\quad\quad d_a$ = average spacing, ft

$\quad\quad h_a$ = average headway, s

A Sample Problem:

Traffic in a congested multilane highway lane is observed to have an average spacing of 200 ft, and an average headway of 3.8 s. Estimate the rate of flow, density and speed of traffic in this lane.

Solution:

$$v = \frac{3,600}{3.8} = 947 \text{ veh/h/ln}$$

$$D = \frac{5,280}{200} = 26.4 \text{ veh/mi/ln}$$

$$S = 0.68 \left(200 \Big/ 3.8 \right) = 35.8 \text{ mi/h}$$

5.3 Relationships among Flow Rate, Speed, and Density

The three macroscopic measures of the state of a given traffic stream—flow, speed, and density—are related as follows:

$$v = S * D \quad\quad\quad (5\text{-}11)$$

where: v = rate of flow, veh/h or veh/h/ln

$\quad\quad S$ = space mean speed, mi/h

$\quad\quad D$ = density, veh/mi or veh/mi/ln

Space mean speed and density are measures that refer to a specific *section* of a lane or highway, while flow rate is a point measure. Figure 5.3 illustrates the relationship. The space mean speed and density measures must apply to the same defined section of roadway. Under stable flow conditions (i.e., the flow entering and leaving the section are the same; no queues are forming within the section), the rate of flow computed by Equation 5-11 applies to *any* point within the section. Where unstable operations exist (a queue is forming within the section), the computed flow rate represents an average for all points within the section.

If a freeway lane were observed to have a space mean speed of 55 mi/h and a density of 25 veh/mi/ln, the flow rate in the lane could be estimated as:

$$v = 55 * 25 = 1,375 \text{ veh/h/ln}$$

As noted previously, this relationship is most often used to estimate density, which is difficult to measure directly, from measured values of flow rate and space mean speed. Consider a freeway lane with a measured space mean speed of 60 mi/h and a flow rate of 1,000 veh/h/ln. The density could be estimated from Equation 5-11 as:

$$D = \frac{v}{S} = \frac{1,000}{60} = 16.7 \text{ veh/mi/ln}$$

Equation 5-11 suggests that a given rate of flow (v) could be achieved by an infinite number of speed (S) and density (D) pairs having the same product. Thankfully, this is not what happens, as it would make the mathematical interpretation of traffic flow unintelligible. There are additional relationships between pairs of these variables that restrict the number of combinations

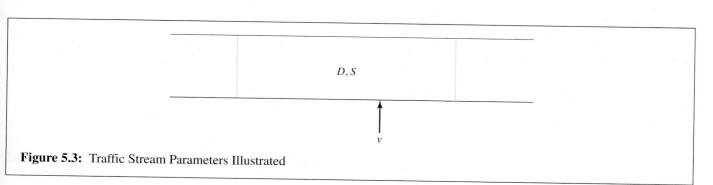

Figure 5.3: Traffic Stream Parameters Illustrated

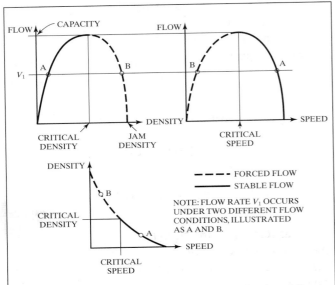

Figure 5.4: Relationships Among Flow, Speed, and Density (Used with permission of Transportation Research Board, National Research Council, from *Highway Capacity Manual*, 3rd Edition, *Special Report 209*, pgs. 1–7, Washington DC, 1994.)

that can and do occur in the field. Figure 5.4 illustrates the general form of these relationships.

The exact shape and calibration of these relationships depends upon prevailing conditions, which vary from location to location and even over time at the same location.

Note that a flow rate of "0 veh/h" occurs under two very different conditions. When there are no vehicles on the highway, density is "0 veh/mi" and no vehicles can be observed passing a point. Under this condition, speed is unmeasurable and is referred to as "free-flow speed," a theoretical value that exists as a mathematical extension of the relationship between speed and flow (or speed and density). In practical terms, free-flow speed can be thought of as the speed a single vehicle could achieve when there are no other vehicles on the road and the motorist is driving as fast as is practicable given the geometry of the highway and its environmental surroundings.

A flow of "0 veh/h" also occurs when there are so many vehicles on the road that all motion stops. This occurs at a very high density, called the "jam density," and

no flow is observed, as no vehicle can pass a point to be counted when all vehicles are stopped.

Between these two extreme points on the relationships, there is a peaking characteristic. The peak of the flow-speed and flow-density curves is the maximum rate of flow, or the *capacity* of the roadway. Its value, like everything else about these relationships, depends upon the specific prevailing conditions at the time and location of the calibration measurements. Operation at capacity, however, is very unstable. At capacity, with no usable gaps in the traffic stream, the slightest perturbation caused by an entering or lane-changing vehicle, or simply a driver hitting the brakes, causes a chain reaction that cannot be damped. The perturbation propagates upstream and continues until sufficient gaps in the traffic stream allow the event to be effectively dissipated.

The dashed portion of the curves represents *unstable* or *forced* flow. This effectively represents flow within a queue that has formed behind a breakdown location. A breakdown will occur at any point where the arriving flow rate exceeds the downstream capacity of the facility. Common points for such breakdowns include on-ramps on freeways, but accidents and incidents are also common, less predictable causes for the formation of queues. The solid line portion of the curves represents *stable* flow (i.e. moving traffic streams that can be maintained over a period of time).

Except for capacity flow, any flow rate may exist under two conditions:

1. A condition of relatively high speed and low density (on the stable portion of flow relationships)

2. A condition of relatively low speed and high density (on the unstable portion of flow relationships)

Obviously, traffic engineers would prefer to keep all facilities operating on the stable side of the curves.

Because a given volume or flow rate may occur under two very different sets of operating conditions, these variables cannot completely describe flow conditions, nor can they be used as measures of the quality of traffic flow. Values of speed and/or density, however, would define unique points on any of the relationships of Figure 5.4, and both describe aspects of quality that can be perceived by drivers and passengers.

Over the years, various researchers have studied speed-flow-density relationships and have attempted to develop many mathematical descriptions for these curves. In the 1930s, Bruce Greenshields conducted the first formal studies of traffic flow. He hypothesized that the speed-density relationship was linear [1]. Later, Ellis [2] suggested piecewise linear relationships. Using two or three linear relationships for various portions of the speed-density curve, this approach created discontinuous curves, with the critical discontinuity in the vicinity of capacity. Greenberg [3] hypothesized a logarithmic curve for speed-density, while Underwood [4] used an exponential model for this relationship. Edie [5] suggested a discontinuous relationship for speed-density using both exponential and logarithmic relationships, and May [6] suggested a bell-shaped curve. Reference 6 reports on a classic study in which all of these mathematical descriptions were compared using a single set of data from the Merritt Parkway in Connecticut, circa 1958. For the study data, the Edie hypothesis produced the best "fit" to field observations.

All of these historic studies focused on calibration of the speed-density relationship. This is considered to be the basic behavioral relationship—drivers selecting speeds based on their proximity to other vehicles (and the geometric and general environment of the roadway). Flow rate results from this relationship. Mathematically, once the speed-density relationship has been established, the speed-flow and flow-density relationships may be derived. Consider Greenshield's linear speed-density model, selected for its simplicity. Assume that a speed-density study has resulted in the following calibrated relationship:

$$S = 55.0 - 0.45D$$

Knowing the general relationship, $v = S*D$, speed-flow and flow-density relationships can be derived by substitution:

$$S = 55.0 - 0.45\left(\frac{v}{S}\right)$$

$$v = 122.2\,S - 2.22\,S^2$$

$$\left(\frac{v}{D}\right) = 55.0 - 0.45\,D$$

$$v = 55.0\,D - 0.45\,D^2$$

As indicated, a linear speed-density hypothesis leads to parabolic speed-flow and flow-density relationships.

Mathematical models for speed-density may also be manipulated to determine (1) free-flow speed, (2) jam density, and (3) capacity. Free-flow speed occurs when the density is "0 veh/h." Thus:

$$S = 55.0 - 0.45*0 = 55.0 \text{ mi/h}$$

Similarly, jam density occurs when speed is "0 mi/h," or:

$$S = 0 = 55.0 - 0.45\,D$$

$$D = \frac{55.0}{0.45} = 122.2 \text{ veh/mi/ln}$$

Capacity is found by determining the peak of the speed-flow or flow-density curves. The peak occurs when the first derivative of the relationship is 0. Using the flow-density curve:

$$v = 55.0\,D - 0.45\,D^2$$

$$\frac{dv}{dD} = 0 = 55.0 - 0.90\,D$$

$$D = \frac{55.0}{0.90} = 61.1 \text{ veh/mi/ln}$$

Capacity, therefore, occurs when the density is 61.1 veh/mi/ln (which is exactly one-half the jam density for a linear curve). Using the speed-density curve, the speed at which a density of 61.1 veh/mi/ln is achieved may be determined:

$$S = 55.0 - 0.45*61.1 = 27.5 \text{ mi/h}$$

which is exactly one-half the free-flow speed for a linear relationship. Now, the basic flow equation is used to find the flow rate that results from this combination of speed and density:

$$v = S*D = 27.5*61.1 = 1{,}680 \text{ veh/h/ln}$$

This is the capacity of the section, based on the calibrated linear speed-density relationship for the section.

It should be noted that there is no consensus as to which mathematical description best describes an uninterrupted-flow traffic stream. Indeed, studies suggest that the best form may vary by location and even over time at a given location. The linear model of Bruce Greenshields, originally calibrated in the 1930s, does not represent modern traffic behavior particularly well. The Fourth Edition of the *Highway Capacity Manual* [7] uses a very different model for speed-flow as the basis for its analysis methodologies for uninterrupted flow facilities. These are based upon a more recent and comprehensive study of freeway flow characteristics [8]. Chapter 12 contains a more detailed discussion of the results of this study and their application to capacity and level of service analysis.

References

1. Greenshields, B., "A Study of Highway Capacity," *Proceedings of the Highway Research Board*, Vol. 14, Transportation Research Board, National Research Council, Washington DC, 1934.

2. Ellis, R., "Analysis of Linear Relationships in Speed-Density and Speed-Occupancy Curves," *Final Report*, Northwestern University, Evanston, IL, December 1964.

3. Greenberg, H., "An Analysis of Traffic Flows," *Operations Research*, Vol. 7, ORSA, Washington DC, 1959.

4. Underwood, R., "Speed, Volume, and Density Relationships," *Quality and Theory of Traffic Flow*, Yale Bureau of Highway Traffic, Yale University, New Haven, CT, 1961.

5. Edie, L., "Car-Following and Steady-State Theory for Non-Congested Traffic," *Operations Research*, Vol. 9, ORSA, Washington DC, 1961.

6. Duke, J., Schofer, J., and May Jr., A., "A Statistical Analysis of Speed-Density Hypotheses," *Highway Research Record 154*, Transportation Research Board, National Research Council, Washington DC, 1967.

7. *Highway Capacity Manual*, 4th Edition, Transportation Research Board, National Research Council, Washington DC, 2000.

8. Scheon, J., *et al.*, "Speed-Flow Relationships for Basic Freeway Sections," *Final Report*, NCHRP Project 3-45, JHK & Associates, Tucson, AZ, May 1995.

Problems

5-1. A volume of 900 veh/h is observed at an intersection approach. Find the peak rate of flow within the hour for the following peak-hour factors: 1.00, 0.90, 0.80, 0.70. Plot and comment on the results.

5-2. A traffic stream displays average vehicle headways of 2.2 s at 50 mi/h. Compute the density and rate of flow for this traffic stream.

5-3. A freeway detector records an occupancy of 0.255 for a 15-minute period. If the detector is 3.5 ft long, and the average vehicle has a length of 20 ft, what is the density implied by this measurement?

5-4. At a given location, the space mean speed is measured as 40 mi/h and the rate of flow as 1,600 pc/h/ln. What is the density at this location for the analysis period?

5-5. The AADT for a given highway section is 25,000 veh/day. If the route is classified as an urban radial roadway, what range of directional design hour volumes would be expected?

5-6. The following travel times were measured for vehicles as they traversed a 2.0-mile segment of highway. Compute the time mean speed (TMS) and space mean speed (SMS) for this data. Why is SMS always lower than TMS?

Vehicle	Travel Time(s)
1	156
2	144
3	144
4	168
5	126
6	132

5-7. The following counts were taken on an intersection approach during the morning peak hour. Determine (a) the hourly volume, (b) the peak rate of flow within the hour, and (c) the peak hour factor.

Time Period	Volume
8:00–8:15 AM	150
8:15–8:30 AM	155
8:30–8:45 AM	165
8:45–9:00 AM	160

5-8. The following traffic count data were taken from a permanent detector location on a major state highway.

1. Month	2. No. of Weekdays In Month (days)	3. Total Days in Month (days)	4. Total Monthly Volume (vehs)	5. Total Weekday Volume (vehs)
Jan	22	31	200,000	170,000
Feb	20	28	210,000	171,000
Mar	22	31	215,000	185,000
Apr	22	30	205,000	180,000
May	21	31	195,000	172,000
Jun	22	30	193,000	168,000
Jul	23	31	180,000	160,000
Aug	21	31	175,000	150,000
Sep	22	30	189,000	175,000
Oct	22	31	198,000	178,000
Nov	21	30	205,000	182,000
Dec	22	31	200,000	176,000

From this data, determine (a) the AADT, (b) the ADT for each month, (c) the AAWT, and (d) the AWT for each month. From this information, what can be discerned about the character of the facility and the demand it serves?

5-9. A study of freeway flow at a particular site has resulted in a calibrated speed-density relationship as follows:

$$S = 57.5(1 - 0.008 \, D)$$

For this relationship, determine: (a) the free-flow speed, (b) jam density, (c) the speed-flow relationship, (d) the flow-density relationship, and (e) capacity.

5-10. Answer all of the questions in Problem 5-9 for the following calibrated speed-density relationship:

$$S = 61.2e^{-0.015D}$$

CHAPTER
6

Intelligent Transportation Systems

In the work leading to the first Interstate Surface Transportation Efficiency Act (ISTEA) legislation in 1991, considerable attention was paid to the advanced technologies that collectively define the elements of "intelligent transportation systems" (ITS). The legislation itself contained directives on ITS, up to and including automated highway demonstrations.

Concurrently, work has been progressing on such systems in many countries, and an international dialog was vital for the efficient incorporation of new technologies and systems. *ITS America* became a focal point of this, consistent with the intent of the ISTEA legislation.

The ITS emphasis was based on advances over the past few decades in traffic system surveillance and control projects, variable message signing, signal optimization, and simulation.

Perhaps even more important, the emphasis was put in place at a time of a true revolution in computing and communications. Computational cost was decreasing, computer memory expanding beyond prior concepts, and microprocessors—the heart of the computer—were being integrated into systems ranging from automobiles to dishwashers. At the same time, the availability

of a global positioning system (GPS) enabled the use of geographic information systems (GIS) so that individuals could locate themselves rather precisely and do it in terms of defined networks, on a global scale. Wireless communication was also becoming widespread, so that the cell phones have become commonplace. These devices allow travelers to call for assistance, get advisories, report incidents, and be located (within the bounds within which the data base can be used, under existing law).

This explosion in enabling technologies then focused attention on how *transportation systems* could be enhanced, building on past work but taking advantage of new, relatively inexpensive and pervasive technological tools.

In the early years of this new focus, there was much attention to standards-setting by government and *ITS America*: what are the common protocols, interchange formats, and such that should be adopted and used to assure seamless systems?

At the same time, there were major advances in the private sector that acted to preempt the orderly introduction of such standards: cell phone features and technology are driven by an extremely competitive market; commercial vehicle routing systems became a sales feature in a competitive market and are tailored to customer needs by the private sector; locator and emergency assistance became selling points, as did routing systems; traffic advisories based on Internet-based camera networks became features of regional radio, another competitive private sector market. Cell phone networks expanded explosively.

This duality continues to exist. The authors see in it an emerging new paradigm in which (a) the individual has access to information delivered by the private sector and can make more informed decisions that serve personal goals, while (b) the agencies and other units in the public sector that have historically "controlled" the networks become providers of raw information and have less direct control over the network because of the level of information available to the individual travelers, who can continually adapt based on personal goals. At the same time, the public sector can—and has—introduce(d) efficiencies by use of automated toll collection, variable information advisories, enforcement of HOV rules and traffic rules generally (e.g., red light violations are now routinely handled with automated monitoring systems, with tickets sent to the vehicle owner along with the photographic evidence of the violation).

The authors also see as inevitable the pricing of transportation being used to influence the temporal demand pattern to make best use of available capacity, with automated billing and collection an essential part of "congestion pricing." This influence can in turn affect mode choice, because the system users have to make tradeoffs involving convenience, need to travel at given times, and cost.

With the growing concern over security in the face of terrorism, ITS technology will be extremely relevant, and the range of applications—including cargo and passenger inspection—will surely expand.

Following this overview, this chapter is devoted primarily to the range of ITS applications in common use today and their potential.

6.1 The Range of ITS Applications

A fundamental question is: What constitutes an intelligent transportation system application and what does not?

To some, much of what has been done over the past few decades fits the definition and simply has a new name and expanded emphasis—freeway surveillance and control, route advisories, traffic control centers, and so forth. An important distinction, however, is that the longer-term view is that of ITS providing the user with information for decisions, in an environment in which enabling technologies are reaching unprecedented sophistication and availability and in which the public/private sector roles are shifting rather dynamically.

Still, the individual projects and research do seem to sound familiar, although there are new, timely issues such as dynamic routing assignment models and algorithms and a new emphasis on applications in several modes—auto, transit, and freight.

If one were to go over the program of the Annual Meeting of the Transportation Research Board (TRB) over the past few years, one would see papers[1]on:

- Traffic signal optimization [1], data needs and quality [2,3], and fuzzy control algorithms [4]
- Bus transit [5,6]
- Dynamic assignment and dynamic routing [7]
- Commercial fleet management [8]
- Trucking [9]
- Specific technical problems [10]

Further, one would note a variety of other ITS themes, including Advanced Traveler Information Systems (ATIS) and Advanced Traffic Management Systems (ATMS), among others.

The *ITS America Annual Meeting and Exposition* is another important forum for ITS applications, as are meetings of the Institute of Transportation Engineers (ITE). Relevant publications for survey articles and technology advertisements include *ITS International* (www.itsinternational.com) and *Traffic Technology International*.

---

[1]Most of these papers appear in the *Transportation Research Record*, based on the peer reviews conducted by the various TRB committees.

6.2   Network Optimization

Chapter 24 of this text addresses the subject of traffic signal optimization on surface streets; freeway surveillance and control is briefly discussed in Chapter 12. These subjects are central themes in ITS applications, perhaps with greater emphasis placed on communication with the drivers and on the autonomous decisions then made by those drivers.

There has also been a growing emphasis on color-coded maps of the freeway system, available from traffic control centers and used by local TV stations. These graphically depict the trouble spots in the network at any given time.

Some TV and radio stations also depend heavily on a network of "Web-cams" to observe conditions in the network and inform the public. Once this was the domain of the agencies operating the facilities and employing video cameras; it is now relatively common for TV and radio to depend both on these and on privately sponsored "Web-cams." Some agencies feature such displays on their Web sites or are used by TV and radio stations on their Web sites.

Given the information flowing to drivers from such systems, by in-vehicle routing systems, and by traffic advisory services (often based on cell phones), an interesting challenge emerges for the transportation professional: how are the drivers going to reroute themselves and how does this affect the network loading and hence travel times?

This challenge can be stated as trying to estimate the actions of drivers based on their current location, the origin-destination that underlies their initial route selection, and their (generally unknown) personal decision rules. Therefore, the technical problems of (a) dynamic traffic reassignment and (b) estimation of O/D patterns from observed data take on greater importance and, in fact, blend into the traffic optimization problem.

A related issue is the type of data available from which to estimate the underlying O/D pattern. Some work has assumed that only traffic count observations are available, with turn information. Other work has focused on "probe vehicles" that transmit information from which their exact routing is known. And, of course, with the rather pervasive use of cell phones—and subject to privacy issues related to accessing the data—the route information on many vehicles could be obtained from cell phone locations over time.

This, in turn, raises another issue: who collects and disseminates such information? If it is a private firm, such as the cell phone provider, it is a nontrivial computing task for which compensation would logically be expected. If it is an agency, then the question of access to the private sector's data must be addressed.

There are an emerging set of models and model platforms in which the assignment/routing tools are combined with (or more precisely, generally interface with) the signal optimization tools.

6.3   Sensing Traffic Using Virtual Detectors

The mainstay of traffic detection for decades has been the magnetic sensing of vehicles by a loop installed into the road surface. While this system is relatively reliable, complete intersection coverage is expensive and the desired detail (vehicle classification, extent of queue, speed estimation) is lacking.

To a large extent, the newest tool is a standard or infrared video camera covering an area such as an intersection approach, coupled with software that creates "virtual detectors" that are defined and located by the user, to observe counts at points, occupancy at points or areas (including queue extent), and speeds. Some capability to classify vehicles may also be included.

Using such a tool, the transportation professional can "locate" numerous "detectors" essentially by drawing them on top of the intersection image and depending upon the software to process the data. Refer to Figure 6.1 for an illustration of virtual detector placement.

AutoScope® is one of the pioneering video imaging systems; there are now a number of such products in the international ITS community.

The use of infrared imaging allows vehicles to be detected in a variety of weather conditions. The use of sophisticated algorithms based on coverage of the underlying

Figure 6.1: Software-Based Virtual Detector (Note: Six user-defined detectors are shown.)

pavement image allows data to be collected from stationary traffic as well as moving traffic.

The literature now shows variants of these concepts, including laser systems for traffic detection as well as a new generation of radar detectors.

6.4 In-Vehicle Routing, and Personal Route Information

Decades ago, a concept for roadside computers that would communicate with drivers and provide routing information was sketched out by the Federal Highway Administration (FHWA). But the relatively high cost and relatively slow speeds of the technology then available made this concept unrealizable.

In recent years, the phenomenal growth in computing speeds and memory, coupled with radical decreases in the associated costs, has made such systems feasible. The implementation is not by roadside computers but rather by systems that use GPS to establish the vehicle's position,

access a data base of network travel times (static or dynamic), and compute best routes. Drivers can tailor the computation to their perception of "best"—travel time, distance, etc. In-vehicle displays, accompanied by audio instructions, provide the driver with the detailed routing.

It is now commonplace for new vehicles (particularly high-end) to have such systems as standard or optional equipment. Packages for adding the feature to existing vehicles are also on the market.

Again, the underlying theme is the individual's autonomous decision making, in an information-rich environment. It is the private sector competitive market forces that are driving the evolution of these systems—simply put, they are a selling point.

There are complications, in that there is some literature that suggests such devices may attract the driver's attention for up to 20% of the task time, and thus lead to greater accident potential.

But routing and mapping software is available in a number of formats, both in-vehicle and otherwise. Figure 6.2 illustrates map and route displays on a notebook computer and on a Palm™ personal device.

Figure 6.2: Map and Route Information on Two Personal Devices (Source: *ITS International,* Sept–Oct 2001, reprinted with permission (pg. 6 of ITS Profiles insert).)

And most readers are familiar with the map-routing features at Web portals such as Yahoo!®.

6.5   The Smart Car

The authors believe that the commercial market will be a major driver of ITS innovations, simply because market pressures require products that are differentiated, if only in the short term. Anti-lock braking (ALB) was one such feature. In-vehicle route guidance, based on a static or active data base, is another, more recent one. Although not yet integrally a part of the vehicle, cell phones were initially extensively marketed for the personal safety and security they provide to the stranded or lost driver.

Conceptually, the modern car is computer based for such essential features as fuel regulation and efficiency. Sensors add valued features to the car, and some active control mechanisms address specific safety needs. It is easy to envision a car with:

- GPS and on-board communications, including route guidance

- Airbags that inflate at different rates, depending upon the severity of the incident

- Anti-collision sensors that alert drivers to surrounding traffic risks and obstacles while in motion, and to obstacles while backing up

- More efficient transmissions

- Active anti-roll control, particularly for vehicles with a high center of gravity

- Active systems to stabilize vehicles in turns

- Sensors to detect under-inflated tires, and tires that maintain their shape in case of a blowout

- Greater fuel efficiency, with a goal of zero emissions through active control and different fuels

While it might be extreme to state that the car *is* the driver's environment, generations of vehicles have been marketed on that premise.

6.6   Commercial Routing and Delivery

Perhaps routing systems are of even greater importance to trucking and service vehicles. This is a specialty market, with software available that can compute long-haul routing and urban routing and can provide answers that take into account the set of scheduled pickup and delivery points. Dynamic rerouting, as new pickups are added en route, is feasible.

Today, it is commonplace for package delivery services (FedEx, UPS, and others) to offer real-time package tracking to customers on their Web sites. Using bar-code scanning technology and wireless communication, packages are tracked in detail from origin to destination. At the delivery point, the driver uses a computer-based pad to record delivery time and often the receiver's signature. This information is available in virtually real time to the sender.

Clearly, the package delivery services found a differentiating service feature that has rapidly been adopted in a highly competitive industry. What was special a few years ago has now become the expected standard of service.

The same data allows the service providers to obtain a wealth of data on the productivity (and down time) of their vehicles and drivers as well as on the cost of delivery in various areas.

6.7   Electronic Toll Collection

At the time of this writing, it is still appropriate to treat electronic toll collection (ETC) and the "smart card" as distinct topics (next section), but the distinction is blurring as the field advances. The concept of one debit card to be used seamlessly on different transit services and also on toll roads is exceptionally attractive to the public and is evolving rapidly in the United States and elsewhere.

One of the highly visible systems in the United States is the "E-Z Pass" system used in a number of jurisdictions in New York, New Jersey, and, more recently, in Delaware. The great appeal is that one in-vehicle device can be used on many different facilities, with the driver not concerned about which agency operates which facility.

Underlying the tag's popularity is of course the design of special toll lanes that accept only the E-Z Pass and have much smaller queues (if any). Another major feature is discounts for tag users, compared with tolls paid at the standard lanes.

At the time E-Z Pass was introduced, a great concern was whether the public would accept and use the system. In 2001, there were some 2 million E-Z Pass tags in use, and—in order to encourage reasonable speeds through the toll booths—the penalty that was presented for violations was suspension of the right to use E-Z Pass.

When the E-Z Pass system was introduced on the New Jersey Turnpike in 2000, the immediate effect was a 40% usage of the special lanes, thanks, in good part, to the extensive number of users in the region.

In Florida, a tag system known as E-Pass is in use. As of this writing, an extensive operational test of a public/private effort named the "Orlando Regional Alliance for Next Generation Electronic payment Systems" (ORANGES). This system is intended to create a seamless, multimodal electronic payment system that includes toll roads, bus transit riders, and parking services.

There are, in fact, a variety of ETC systems in use or selling planned throughout the world, including projects in Europe, Brazil, Canada, Australia, China, Malaysia, and Thailand.

Three issues and trends have been emerging since the initiation of the ETC systems:

1. Data can be collected from the tags to estimate travel times on sections of road and between toll stations. The same data can be used for noting routing patterns. At the present time, one of the uses of travel-time data is providing the public with estimates of the condition of the system, a benign application. But the potential for using the same tags for speed enforcement exists, although it currently would be viewed as counterproductive in encouraging card use. The ability to establish routings and to know starting times, down to a personal level, also raises questions of privacy, an issue that is a companion to such an information-rich environment.

2. A new generation of toll "stations" is emerging that do not have (or need) the physical infrastructure of existing toll plazas and can sense vehicles at prevailing speeds.

3. The pervasiveness of the tags enables congestion pricing to be considered as technically feasible. Indeed, various discounting options already in effect can lay the groundwork for public acceptance and perhaps enthusiasm for such systems.

6.8   The Smart Card

Systems using "smart cards" have generally implemented seamless travel using a common card on public transportation, thus making the trips easier. In New York City, the MetroCard (perhaps not fully enabled yet as a smart card) allows users free transfers between bus and subway if the transfer is made within a certain time. Other systems use variable pricing with distance and time of day.

As already cited, the ORANGES project is extending the use of smart cards from public transportation to parking and to toll roads.

One should look forward and consider at least three interesting applications of future generations of smart cards:

1. The opportunity to encourage seamless and efficient use of all transportation modes.

2. The ability to introduce variable cost with time of day and with distance, thus moving to a wider use of the "congestion pricing" or "road user cost" concepts.

3. The extensive data base that is being created, and the potential to use that data to obtain both

traffic statistics (volume, speeds, travel times) and O/D estimates for use in:

- Building historical patterns, by season, weather, and other factors

- Observing trends and changes in the historical patterns

- Use in planning, in scheduling, and even in revising transit routes

- Real-time use of the data in traffic advisories and control, as well as in response to incidents

6.9 Congestion Pricing

The general topic of congestion pricing has already been introduced. Indeed, it is a reality in some modest ways (Washington DC metro area) and in major plans for trucking and other road user costing in Europe and elsewhere. Figure 6.3 shows a congestion pricing zone planned in the center of London for implementation in 2003.

The congestion pricing issue in the United States and elsewhere is one charged with political impact, concerns over equity among user groups, and equal or historical access to facilities or routes.

While a good case can be made for the logic and effectiveness of such demand-based pricing, the public policy issues are nontrivial. Indeed, the move to congestion pricing that seems inevitable to the authors will surely focus the public's attention on the underlying role of transportation in their economic and personal lives.

Figure 6.3: Proposed Congestion Pricing Zone in London, England (Source: *Tolltrans, Traffic Technology International Supplement 2001*, pg. 26. Used with permission.)

Questions raised over the years on the importance of freight movement to the economy of a region, on the true costs of various modes, and about embedded subsidies will once again draw attention, in a context of (relatively) rapid implementation.

The present text is not the place to define the "right" approach to these issues; that lies in the future. But it is important that transportation professionals recognize the issues that will become increasingly a part of their lives.

6.10 Dynamic Assignment

The traditional transportation planning models focus on assigning traffic to networks in specified time periods, based on historical or forecast traffic loads. They generally assume that the system is fully operational (no incidents), while allowing that some segments may hit their capacity limits. These models also allow for some of the traffic to be assigned outside the design period, if it cannot enter the network.

A far more challenging problem is that of dynamic assignment: traffic exists on the network, and an incident or major event disrupts the capacity of one or more links. How shall travelers reroute themselves, given that they have launched themselves on trips and routings based upon their historical knowledge? Travelers deal with this problem every day and use radio, cell phones, information from variable message signs, and other inputs to make their decisions. The issue with dynamic assignment is: How does one *model* and *anticipate* what people will do and/or advise them on what they should do? To address this requires a new generation of comprehensive models that are not only computationally fast (so as to be of value in real time) but also take into account (1) the planned routings (which in turn are based on the underlying O/D pattern) and (2) the decision rules employed by a range of travelers.

There is extensive and evolving literature on the development and use of such models. References [11–14] are only a sampling. For present purposes, suffice it to say that such models are central to providing informed advice and to anticipating the effects of disruptions.

6.11 Traffic Enforcement

ITS technology provides considerable opportunities to enforce traffic rules and regulations and to thereby enhance safety. Reference was already made to the pros and cons of using travel time data from ETC tags for enforcement.

One area in which technology has been used is the detection of vehicles that pass red lights. This is a clear safety issue and one that has gained public acceptance and support. The essence of such systems, already in use in many locations, is that a video camera observes both the signal indication and the moving vehicle with sufficient detail to see and recover the license plate. A ticket is then sent to the registered owner for what was done with his or her vehicle, without citing the person for the infraction.

The monitoring of vehicle license plates for the purposes of identifying stolen vehicles, vehicles with outstanding tickets, and those sought for security or police matters is another application that is generating interest in the international community. Of particular interest is use of such systems at border crossings. Another clear application is assigning road-user costs. Other candidate enforcement activities include (1) emissions monitoring and (2) weight in motion, to ascertain axle loads.

6.12 Bus Transit and Paratransit

Bus transit ITS applications include vehicle location systems, signal priority at intersections and on arterials, and electronic toll collection. The value of smart cards has already been noted as an effective tool in enabling use of public transportation. With a growing emphasis on multimodal transportation, the matching of schedules for bus to rail and for bus to bus is also an important item.

Perhaps most fundamental in the emphasis on multimodal transportation is the concept of a *minimally acceptable transit net* to service a corridor or area. A single bus route is likely to lack ridership, for the simple reason that the number of origin-destination pairs that can be served is limited and service frequency is often unattractive. Solutions that depend upon using auto to get to bus transit seem self-defeating, in that people are already in their personal vehicles (although we recognize that some park-n-ride scenarios are valid). But single routes—or even uncoordinated routes—tend to lead to a self-fulfilling prophecy: ridership is low because of limited appeal and value, and low ridership justifies the lack of need.

Therefore, one approach is to use the O/D patterns of the area to define a minimally acceptable transit net, one that can have enough interlocking routes so that the overall service is viable. The transportation adequacy can be first measured against whether such a net exists, and later against whether it attracts riders. The role of ITS in this approach is manifold, ranging from the data base needed to coordinated schedules to seamless travel using smart cards that allow discounted transfers for multibus or multimode trips.

Paratransit is often an alternative or supplement to conventional bus networks, for trip purposes or O/D pairs that cannot be met efficiently with fixed bus routes, even when they are interlocking. ITS technology can enable trips to be combined efficiently, provide variable fares based upon traveler flexibility and trip departure times, and track vans and taxis for availability.

6.13 Emerging Issues

At the risk of being dated by events, the authors suggest that the following issues should be considered in anticipating the future:

1. The emergence of congestion pricing as a priority, given that ITS technologies are creating the enabling infrastructure. Beyond the technological issues, the market, economic, and elasticity issues need in-depth research.

2. The need for high-quality O/D data, and the ability of the cell phone network to provide detailed trip information, will cause major attention to focus on this issue and the related personal privacy issues.

3. An emphasis on models that specifically address the departure-time decisions and mode decisions of the travelers, so that the "demand profile" is not fixed but is subject to traveler knowledge of network loading and travel cost.

Complicating this is that the demand profile changes interactively with the decisions made.

4. The national priority in the United States on antiterrorism and security, and the same priority in the international community, will set priorities for certain inspection and detection technologies and will influence the design of ITS systems.

6.14 Summary

This chapter has in some ways raised more issues than it answers, and it skips some details of specific ITS systems. This is done intentionally, because (1) the field is moving rapidly and any "snapshot" its present state is sure to be dated rapidly, perhaps even by the publication date of the text, and (2) the real issue is for the reader to be prepared to expand his or her view of providing transportation service in a highly competitive market in which computing, communications, and Web services are being used in novel ways.

Furthermore, the evolving roles of private and public sectors—in some ways, structure vis-à-vis market responsiveness—should draw the reader's attention. Today's "right answer" can be swept away by what the enabling technologies make available.

And there is another fundamental issue for the reader to consider: providing transportation information may simply be a selling point for manufacturers of automobiles, other vehicles, computers, personal data assistants tailored to various users (e.g., package delivery services), cell phones, wireless communication devices, and Web services. That is, while valued by the user, the underlying objective is making products that are both more attractive and differentiated (at least in the short term, until the competition copies success). In this view, transportation data and information is not an end in its own right—the traditional view in our profession—but is rather a product enhancement. In addition, competitive private sector forces may provide a data-rich environment for transportation professionals as a by-product of their work, *and at an innovative pace driven by that work and its market.* And this pace far exceeds the traditional pace of public-sector planning and innovation, and the competing value of standardization.

References

1. Stamatiadis, C. and Gartner, R., "Accelerated Progression Optimzation in Large Urban Grid Networks," TRB 79th Annual Meeting, January 9–13, 2000, Washington DC.

2. Chen, M., "Determining the Number of Probe Vehicles for Freeway Travel Time Estimation Using Microscopic Simulation," TRB 79th Annual Meeting, January 9–13, 2000, Washington DC.

3. Sunkari, S.R., Charara, H., and Urbanik, T., "Automated Turning Movement Counts for Shared Lane Configurations at Signalized Diamond Interchanges," TRB 79th Annual Meeting, January 9–13, 2000, Washington DC.

4. Nittymaki, J. and Granberg, M., "The Development of Fuzzy Control Algorithms for a Signalized Pedestrian Crossing—Fuzzy Similarity and Calibration of Membership Functions," TRB 79th Annual Meeting, January 9–13, 2000, Washington DC.

5. Ding, Y., Chien, S., and Zayas, N.A., "Analysis of Bus Transit Operations with Enhanced CORSIM Case Study: Bus Route #39 of New Jersey Transit," TRB 79th Annual Meeting, January 9–13, 2000, Washington DC.

6. Cevallos, F. and Willis, A., "Essential Transit Software Applications," TRB 79th Annual Meeting, January 9–13, 2000, Washington DC.

7. Chabini, I. and Canugapati, S., "Design and Implementation of Parallel Dynamic Shortest Path Algorithms for Intelligent Transportation Systems Applications," TRB 79th Annual Meeting, January 9–13, 2000, Washington DC.

8. Mahmassani, H.S., Kim, Y., and Jaillet, P., "Local Optimization Approaches to Solve Dynamic Commercial Fleet Management Problems," TRB 79th Annual Meeting, January 9–13, 2000, Washington DC.

9. Golob, T.F., "Trucking Industry Demand for Information Technology: A Multivariate Discrete Choice Model," TRB 79th Annual Meeting, January 9–13, 2000, Washington DC.

10. Zabilansky, I.J. and Yankielun, N.F., "Measuring Scour Under Ice in Real Time," TRB 79th Annual Meeting, January 9–13, 2000, Washington DC.

11. Ran, B., Rouphail, N.M., and Tarko, A., "Toward a Class of Link Travel Time Functions for Dynamic Assignment Models on Signalized Networks," *Transportation Research, Part B: Methodological* 31:4, 1997.

12. Van der Zijpp, N.J. and Lindveld, C.D.R., "Estimation of Origin-Destination Demand for Dynamic Assignment with Simultaneous Route and Departure Time Choice," *Transportation Research Record 1771*.

13. Astarita, V., Er-Rafia, K., Florian, M., Mahut, M., and Velan, S., "Comparison of Three Methods for Dynamic Network Loading," *Transportation Research Record 1771*.

14. Abdelghany, A.F., Mahmassani, H.S., and Chiu, Y-C., "Spatial Microassignment of Travel Demand with Activity Trip Chains," *Transportation Research Record 1777*.

Problems

6-1. Consider the feasibility of designing a Web site that offers the public the opportunity to have several service providers bid on paratransit trips. Users might indicate that they need to get to the airport by a certain time and have flexibility in travel time duration (or not) and in multiple pickups (or not). The user might also give advance notice of several days or only hours. Service providers would then bid for the particular trip or for a set of trips specified by the user. Consider how to market the service, what options it should provide the user, and why it might (or might not) be attractive to service providers. Write a paper not to exceed 10 pages on such a system.

6-2. Refer to the second-from-last paragraph in this chapter and prepare for (1) a class discussion of this issue and/or (2) a 5- to 10-page paper on this issue, as specified by the course instructor.

6-3. Refer to the last paragraph in this chapter and prepare for (1) a class discussion of this issue and/or (2) a 5- to 10-page paper on this issue, as specified by the course instructor.

6-4. Address the first "emerging issue" listed in Section 6.13, with particular attention to the research needed to advance the state of the art to where it must be to make informed decisions. This may require a literature search and thus be a major course project. But two timelines—the desire for informed congestion pricing implementations and the advance planning/research needed—may be inconsistent.

6-5. Address the second "emerging issue" listed in Section 6.13 of that name, with emphasis on how much data can be obtained and how it can be used in routing and assignment algorithms. At the same time, address the privacy issues and how personal privacy can realistically be assured. Prepare for (1) a class discussion of this issue and/or (2) a 5- to 10-page paper on this issue, as specified by the course instructor.

6-6. Address the third "emerging issue" in Section 6.13, with emphasis on how closely existing advanced models for dynamic assignment truly address the issue. Prepare for (1) a class discussion of this issue and/or (2) a 5- to 10-page paper on this issue, as specified by the course instructor.

6-7. Address the fourth "emerging issue" listed in Section 6.13, with emphasis on the most acute needs in face of the range of threats that must be considered. Take reasonable account of the reality that passive action—detection and then remedy—may be neither wise or cost-effective.

PART 2
Traffic Studies and Programs

CHAPTER

7

Statistical Applications in Traffic Engineering

Because traffic engineering involves the collection and analysis of large amounts of data for performing all types of traffic studies, it follows that statistics is also an important element in traffic engineering. Statistics helps us determine how much data will be required, as well as what meaningful inferences can confidently be made based on that data.

Statistics is required whenever it is not possible to directly observe or measure all of the values needed. If a room contained 100 people, the average weight of these people could be measured with 100% certainty by weighing each one and computing the average. In traffic, this is often not possible. If the traffic engineer needs to know the average speed of all vehicles on a particular section of roadway, not all vehicles could be observed. Even if all speeds could be measured over a specified time period (a difficult accomplishment in most cases), speeds of vehicles arriving before or after the study period, or on a different day than the sample day, would be unknown. In effect, no matter how many speeds are measured, there are always more that are not known. For all practical and statistical purposes, the number of vehicles using a particular section of roadway over time is infinite.

Because of this, traffic engineers often observe and measure the characteristics of a finite *sample* of vehicles in a *population* that is effectively infinite. The mathematics of statistics is used to estimate characteristics that cannot be established with absolute certainty, and to assess the degree of certainty that does exist. When this is done, statistical analysis is used to address the following questions:

- How many samples are required (i.e., how many individual measurements must be made)?
- What confidence should I have in this estimate (i.e., how sure can I be that this *sample* measurement has the same characteristics as the *population*)?
- What statistical distribution best describes the observed data mathematically?
- Has a traffic engineering design resulted in a change in characteristics of the population? (For example, has a new speed limit resulted in reduced speeds?)

This chapter explores the statistical techniques used in answering these critical questions and provides

131

some common examples of their use in traffic engineering. This chapter is not, however, intended as a substitute for a course in statistics. For some basic references, see [1–4]. For an additional traffic reference addressing statistical tests, see [5]. The review that follows assumes that either (1) the students have had a previous course in statistics or (2) the lecturer will supplement or expand the materials as needed by an individual class.

7.1   An Overview of Probability Functions and Statistics

Before exploring some of the more complex statistical applications in traffic engineering, some basic principles of probability and statistics that are relevant to these analyses are reviewed.

7.1.1   Discrete versus Continuous Functions

Discrete functions are made up of discrete variables— that is, they can assume only specific whole values and not any value in between. Continuous functions, made up of continuous variables, on the other hand, can assume any value between two given values. For example, Let N = the number of children in a family. N can equal 1, 2, 3, etc, but not 1.5, 1.6, 2.3. Therefore it is a discrete variable. Let H = the height of an individual. H can equal 5 ft, 5.5 ft, 5.6 ft, etc., and, therefore, is a continuous variable.

Examples of discrete probability functions are the Bernoulli, binomial, and Poisson distributions, which will be discussed in the following sections. Some examples of continuous distributions are the normal, exponential, and chi-square distributions.

7.1.2   Randomness and Distributions Describing Randomness

Some events are very predictable, or should be predictable. If you add mass to a spring or a force to a beam, you can expect it to deflect a predictable amount. If you depress the gas pedal a certain amount and you are on level terrain, you expect to be able to predict the speed of the vehicle. On the other hand, some events may

be totally random. The emission of the next particle from a radioactive sample is said to be completely random.

Some events may have very complex mechanisms and *appear* to be random for all practical purposes. In some cases, the underlying mechanism cannot be perceived, while in other cases we cannot afford the time or money necessary for the investigation.

Consider the question of who turns north and who turns south after crossing a bridge. Most of the time, we simply say there is a probability p that a vehicle will turn north, and we treat the outcome as a random event. However, if we studied who was driving each car and where each driver worked, we might expect to make the estimate a very predictable event, for each and every car. In fact, if we kept a record of their license plates and their past decisions, we could make very predictable estimates. The events—to a large extent—are not random. Obviously, it is not worth that trouble because the random assumption serves us well enough. That, of course, is the crux of engineering: model the system as simply (or as precisely) as possible (or necessary) *for all practical purposes*. Albert Einstein was once quoted as saying "Make things as simple as possible, but no simpler."

In fact, a number of things are modeled as random *for all practical purposes*, given the investment we can afford. Most of the time, these judgments are just fine and are very reasonable but, as with every engineering judgment, they can sometimes cause errors.

7.1.3   Organizing Data

When data is collected for use in traffic studies, the raw data can be looked at as individual pieces of data or grouped into classes of data for easier comprehension. Most data will fit into a common distribution. Some of the common distributions found in traffic engineering are the normal distribution, the exponential distribution, the chi-square distribution, the Bernoulli distribution, the binomial distribution, and the Poisson distribution.

As part of the process for determining which distribution fits the data, one often summarizes the raw data into classes and creates a frequency distribution table (often the data is collected without even recording the individual data points). This makes the data more easily readable and understood. Consider Table 7.1, which lists the unorganized data of the heights in inches of 100 students in

Table 7.1: Heights of Students in Engineering 101 (Inches)

62.3	67.5	73	73	63	69.5	70	63.2	74	70.2
72	67	64	67.5	66.1	64.6	67.5	74	66	68
66	61	67	70.5	67	70.1	60.5	67.9	70	61.4
67.5	68	67.5	64.6	69.8	63	66	64.9	68	67.5
73.9	66	66.2	69.2	66.5	67.8	71	69.1	69.4	70.5
64.5	67.4	72.5	61.8	63.7	69	67	68	67.9	64.5
67	73.5	67.4	67.4	69.3	70	66.8	62	64	68
66	64	63	71	66	63.9	70	69	68	67
69.3	69.4	68	64.5	69	66.5	67.2	66.5	63.9	69
71	67	70	67	70.5	71	64.2	70.5	67	67.5

the Engineering 101 class. You could put the data into an array format, which means listing the data points in order from either lowest to highest or highest to lowest. This will give you some feeling for the character of the data, but it is still not very helpful, particularly when you have a large number of data points.

It is more helpful to summarize the data into defined categories, and display it as a frequency distribution table, as in Table 7.2.

Table 7.2: Frequency Distribution Table: Heights of Students in Engineering 101

Height Group (inches)	Number of Observations
60–62	5
63–65	18
66–68	42
69–71	27
72–74	8

Even though the details of the individual data points are lost, the grouping of the data adds much clarity to the character of the data. From this frequency distribution table, plots can be made of the frequency histogram, the frequency distribution, the relative frequency distribution, and the cumulative frequency distributions.

7.1.4 Common Statistical Estimators

In dealing with a distribution, there are two key characteristics that are of interest. These are discussed in the following subsections.

Measures of Central Tendency

Measures of central tendency are measures that describe the center of data in one of several different ways. The *arithmetic mean* is the average of all observed data. The true underlying mean of the population, μ, is an exact number that we do not know, but can estimate as:

$$\bar{x} = \frac{1}{N} \sum_{i=1}^{N} x_i \tag{7-1}$$

where: $\bar{x}$ = arithmetic average or mean of observed values

x_i = ith individual value of statistic

N = sample size, number of values x_i

Consider the following example: Estimate the mean from the following sample speeds in mi/h: (53, 41, 63, 52, 41, 39, 55, 34). Using Equation 7-1:

$$\bar{x} = \frac{1}{8}(53 + 41 + 63 + 52 + 41 + 39 + 55 + 34)$$

$$= 47.25$$

Since the original data had only two significant digits, the more correct answer is 47 mi/h.

For grouped data, the average value of all observations in a given group is considered to be the midpoint value of the group. The overall average of the entire sample may then be found as:

$$\bar{x} = \frac{\sum_{j} f_j m_j}{N} \tag{7-2}$$

where: f_j = number of observations in group j

$\quad\quad\quad m_j$ = middle value of variable in group j

$\quad\quad\quad N$ = total sample size or number of observations

For the height data above of Table 7.2:

$$\bar{x} = \frac{61(5) + 64(18) + 67(42) + 70(27) + 73(8)}{100}$$

$$= 67.45 = 67$$

The *median* is the middle value of all data when arranged in an array (ascending or descending order). The median divides a distribution in half: half of all observed values are higher than the median, half are lower. For nongrouped data, it is the middle value; for example, for the set of numbers (3, 4, 5, 5, 6, 7, 7, 7, 8), the median is 6. It is the fifth value (in ascending or descending order) in an array of 9 numbers. For grouped data, the easiest way to get the median is to read the 50% percentile point off a cumulative frequency distribution curve (see Chapter 9).

The *mode* is the value that occurs most frequently—that is, the most common single value. For example, in nongrouped data, for the set of numbers (3, 4, 5, 5, 6, 7, 7, 7, 8), the mode is 7. For the set of numbers (3, 3, 4, 5, 5, 5, 6, 7, 8, 8, 8, 9), both 5 and 8 are modes, and the data is said to be bimodal. For grouped data, the mode is estimated as the peak of the frequency distribution curve (see Chapter 9).

For a perfectly symmetrical distribution, the mean, median, and mode will be the same.

Measures of Dispersion

Measures of dispersion are measures that describe how far the data spread from the center.

The *variance and standard deviation* are statistical values that describe the magnitude of variation around the mean, with the variance defined as:

$$s^2 = \frac{\sum (x_i - \bar{x})^2}{N - 1} \tag{7-3}$$

where: s^2 = variance of the data

$\quad\quad\quad N$ = sample size, number of observations

All other variables as previously defined.

The standard deviation is the square root of the variance. It can be seen from the equation that what you are measuring is the distance of each data point from the mean. This equation can also be rewritten (for ease of use) as:

$$s^2 = \frac{1}{N} \sum_{i=1}^{N} x_i^2 - \left(\frac{N}{N-1} \right) \bar{x}^2 \tag{7-4}$$

For grouped data, the standard deviation is found from:

$$s = \sqrt{\frac{\sum fm^2 - N(\bar{x})^2}{N - 1}} \tag{7-5}$$

where all variables are as previously defined. The standard deviation (STD) may also be estimated as:

$$s_{est} = \frac{P_{85} - P_{15}}{2} \tag{7-6}$$

where: P_{85} = 85th percentile value of the distribution (i.e., 85% of all data is at this value or less).

$\quad\quad\quad P_{15}$ = 15th percentile value of the distribution (i.e., 15% of all data is at this value or less).

The xth *percentile* is defined as that value below which x% of the outcomes fall. P_{85} is the 85th percentile, often used in traffic speed studies; it is the speed that encompasses 85% of vehicles. P_{50} is the 50th percentile speed or the median.

The *coefficient of variation* is the ratio of the standard deviation to the mean and is an indicator of the spread of outcomes relative to the mean.

The distribution or the underlying shape of the data is of great interest. Is it normal? Exponential? But the engineer is also interested in anomalies in the shape of the distribution (e.g., skewness or bimodality). Skewness is defined as the (mean − mode)/std. If a distribution is negatively skewed, it means that the data is concentrated to the left of the most frequent value (i.e., the mode). When a distribution is positively skewed, the data is concentrated to the right of the mode. The engineer should look for the underlying reasons for skewness in a distribution. For instance, a negatively skewed speed distribution may indicate a problem such as sight distance or pavement condition that is inhibiting drivers from selecting higher travel speeds.

7.2   The Normal Distribution and Its Applications

One of the most common statistical distributions is the *normal distribution*, known by its characteristic bell-shaped curve (illustrated in Figure 7.1). The normal distribution is a continuous distribution. Probability is indicated by the area under the probability density function $f(x)$ between specified values, such as $P(40 < x < 50)$.

The equation for the normal distribution function is:

$$f(x) = \frac{1}{\sigma\sqrt{2\pi}}e^{-\left[\frac{(x-\mu)^2}{2\sigma^2}\right]} \qquad (7\text{-}7)$$

where: x = normally distributed statistic

μ = true mean of the distribution

σ = true standard deviation of the distribution

π = 3.14

The probability of any occurrence between values x_1 and x_2 is given by the area under the distribution function between the two values. The area may be found by integration between the two limits. Likewise, the mean, μ, and the variance, σ^2, can be found through integration. The normal distribution is the most common distribution, because any process that is the sum of many parts tends to be normally distributed. Speed, travel time, and delay are all commonly described using the normal distribution. The function is completely defined by two parameters: the mean and the variance. All other values in Equation 7-6, including π, are constants. The notation for a normal distribution is $x: N[\mu, \sigma^2]$, which means that the variable x is normally distributed with a mean of μ and a variance of σ^2.

7.2.1   The Standard Normal Distribution

For the normal distribution, the integration cannot be done in closed form due to the complexity of the equation for $f(x)$; thus, tables for a "standard normal" distribution, with zero mean ($\mu = 0$) and unit variance ($\sigma^2 = 1$), are constructed. Table 7.3 presents tabulated values of the standard normal distribution. The standard normal is denoted $z: N[0,1]$. Any value of x on any normal distribution, denoted $x: N[\mu, \sigma^2]$, can be converted to an equivalent value of z on the standard normal distribution. This can also be done in reverse when needed. The translation of an arbitrary normal distribution of values of x to equivalent values of z on the standard normal distribution is accomplished as:

$$z = \frac{x - \mu}{\sigma} \qquad (7\text{-}8)$$

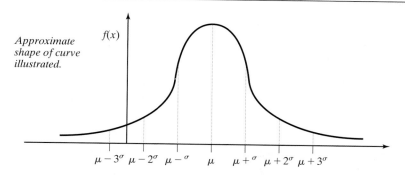

$f(x)$

Approximate shape of curve illustrated.

$\mu - 3\sigma \quad \mu - 2\sigma \quad \mu - \sigma \qquad \mu \qquad \mu + \sigma \quad \mu + 2\sigma \quad \mu + 3\sigma$

Figure 7.1: Shape of the Normal Distribution Function

Table 7.3: Tabulated Values of the Standard Normal Distribution

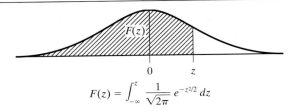

$$F(z) = \int_{-\infty}^{z} \frac{1}{\sqrt{2\pi}} \, e^{-z^2/2} \, dz$$

z	.00	.01	.02	.03	.04	.05	.06	.07	.08	.09
.0	.5000	.5040	.5080	.5120	.5160	.5199	.5239	.5279	.5319	.5359
.1	.5398	.5438	.5478	.5517	.5557	.5596	.5636	.5675	.5714	.5753
.2	.5793	.5832	.5871	.5910	.5948	.5987	.6026	.6064	.6103	.6141
.3	.6179	.6217	.6255	.6293	.6331	.6368	.6406	.6443	.6480	.6517
.4	.6554	.6591	.6628	.6661	.6700	.6736	.6772	.6808	.6844	.6879
.5	.6913	.6950	.6985	.7019	.7054	.7083	.7123	.7157	.7190	.7224
.6	.7257	.7291	.7324	.7357	.7389	.7422	.7454	.7486	.7517	.7549
.7	.7580	.7611	.7642	.7673	.7704	.7734	.7764	.7794	.7823	.7852
.8	.7881	.7910	.7939	.7967	.7995	.8023	.8051	.8078	.8106	.8133
.9	.8159	.8186	.8212	.8238	.8264	.8289	.8315	.8340	.8365	.8389
1.0	.8413	.8438	.8461	.8485	.8508	.8531	.8554	.8577	.8599	.8621
1.1	.8643	.8665	.8686	.8708	.8729	.8749	.8770	.8790	.8810	.8830
1.2	.8849	.8869	.8888	.8907	.8925	.8944	.8962	.8980	.8997	.9015
1.3	.9032	.9049	.9066	.9082	.9099	.9115	.9131	.9147	.9162	.9177
1.4	.9192	.9207	.9222	.9236	.9251	.9265	.9279	.9292	.9306	.9319
1.5	.9332	.9345	.9357	.9370	.9382	.9394	.9406	.9418	.9429	.9441
1.6	.9432	.9463	.9474	.9484	.9495	.9505	.9515	.9525	.9535	.9545
1.7	.9554	.9564	.9573	.9582	.9591	.9599	.9608	.9616	.9625	.9633
1.8	.9641	.9649	.9658	.9664	.9671	.9678	.9686	.9693	.9699	.9706
1.9	.9713	.9719	.9726	.9732	.9738	.9744	.9750	.9756	.9716	.9767
2.0	.9772	.9778	.9783	.9788	.9793	.9798	.9803	.9808	.9812	.9817
2.1	.9812	.9826	.9830	.9834	.9838	.9842	.9846	.9854	.9854	.9857
2.2	.9861	.9864	.9868	.9871	.9875	.9878	.9881	.9884	.9887	.9890
2.3	.9893	.9896	.9898	.9901	.9904	.9906	.9909	.9911	.9913	.9916
2.4	.9918	.9920	.9922	.9925	.9927	.9929	.9931	.9932	.9934	.9936
2.5	.9938	.9940	.9941	.9943	.9945	.9946	.9948	.9949	.9951	.9952
2.6	.9953	.9955	.9956	.9937	.9959	.9960	.9961	.9962	.9963	.9964
2.7	.9965	.9966	.9967	.9968	.9969	.9970	.9971	.9972	.9973	.9974
2.8	.9974	.9975	.9976	.9977	.9977	.9978	.9979	.9979	.9980	.9981
2.9	.9981	.9982	.9982	.9983	.9984	.9984	.9985	.9985	.9986	.9986
3.0	.9987	.9987	.9987	.9988	.9988	.9989	.9989	.9989	.9990	.9990
3.1	.9990	.9991	.9991	.9991	.9992	.9992	.9992	.9992	.9993	.9993
3.2	.9993	.9993	.9994	.9994	.9994	.9994	.9994	.9995	.9995	.9995
3.3	.9995	.9995	.9995	.9996	.9996	.9996	.9996	.9996	.9996	.9997
3.4	.9997	.9997	.9997	.9997	.9997	.9997	.9997	.9997	.9997	.9998

where: z = equivalent statistic on the standard normal distribution, z: $N[0,1]$

x = statistic on any arbitrary normal distribution, x: $N[\mu, \sigma^2]$ other variables as previously defined

Figure 7.2 shows the translation for a distribution of spot speeds that has a mean of 55 mi/h and standard deviation of 7 mi/h to equivalent values of z.

Consider the following example: For the spot speed distribution of Figure 7.2, x: $N[55,49]$, what is the probability that the next observed speed will be 65 mph or less? Translate and scale the x-axis as shown in Figure 7.2. The equivalent question for the standard normal distribution, z: $N[0,1]$, is found using Equation 7-8: Determine the probability that the next value of z will be less than:

$$z = \frac{65 - 55}{7} = 1.43$$

Entering Table 7.3 on the vertical scale at 1.4 and on the horizontal scale at 0.03, the probability of having a value of z less than 1.43 is 0.9236 or 92.36%.[1]

Another type of application frequently occurs: For the case just stated, what is the probability that the speed of the next vehicle is between 55 and 65 mph?

The probability that the speed is less than 65 mph has already been computed. We can now find the probability that the speed is less than 55 mph, which is equivalent to $z = (55 - 55)/7 = 0.00$, so that the probability is 0.50 or 50% exactly.[2] The probability of being between 55 and 65 mph is just the difference of the two probabilities: $(0.9236 - 0.5000) = 0.4236$ or 42.36%.

In similar fashion, using common sense and the fact of symmetry, one can find probabilities less than 0.5000, even though they are not tabulated directly in Figure 7.2.

For the case stated above, find the probability that the next vehicle's speed is less than 50 mph. Translating to the z-axis, we wish to find the probability of a value being less than $z = (50 - 55)/7 = -0.71$. Negative values of z are not given in Table 7.3, but by symmetry it can be seen that the desired shaded area is the same size as the area *greater* than $+0.71$. Still, we can only find the shaded area less than $+0.71$ (it is 0.7611). However, knowing that the area (or total probability) under the curve is 1.00, the remaining area (i.e., the desired quantity) is therefore $(1.0000 - 0.7611) = 0.2389$ or 23.89%.

From these illustrations, three important procedures have been presented: (1) the conversion of values

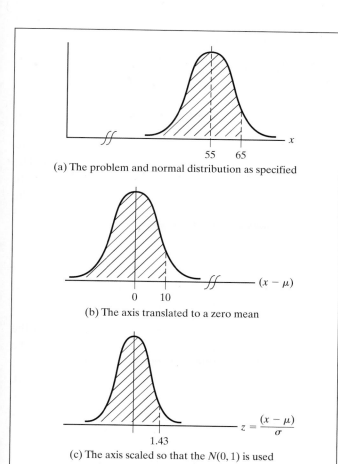

(a) The problem and normal distribution as specified

(b) The axis translated to a zero mean

$$z = \frac{(x - \mu)}{\sigma}$$

(c) The axis scaled so that the $N(0, 1)$ is used

Figure 7.2: Translating a Normal Distribution to the Standard Normal

---

[1] Probabilities are numbers between zero and one, inclusive, and not percentages. However, many people talk about them as percentages.

[2] Of course, given that the normal distribution is symmetric and the mean in this case is 55 mph, the probability of 0.50 could also have been found by inspection.

from any arbitrary normal distribution to the standard normal distribution, (2) the use of the standard normal distribution to determine the probability of occurrences, and (3) the use of Figure 7.2 to find probabilities less than both positive and negative values of z, and between specified values of z.

7.2.2　Important Characteristics of the Normal Distribution Function

The foregoing exercises allow one to compute relevant areas under the normal curve. Some numbers occur frequently in practice, and it is useful to have those in mind. For instance, what is the probability that the next observation will be within one standard deviation of the mean, given that the distribution is normal? That is, what is the probability that x is in the range $(\mu \pm 1.00\sigma)$? By a similar process to those illustrated above, we can find that this probability is 68.3%.

The following ranges have frequent use in statistical analysis involving normal distributions:

- 68.3% of the observations are within $\mu \pm 1.00\sigma$
- 95.0% of the observations are within $\mu \pm 1.96\sigma$
- 95.5% of the observations are within $\mu \pm 2.00\sigma$
- 99.7% of the observations are within $\mu \pm 3.00\sigma$

The total probability under the normal curve is 1.00, and the normal curve is symmetric around the mean. It is also useful to note that the normal distribution is asymptotic to the x-axis and extends to values of $\pm\infty$. These critical characteristics will prove to be useful throughout the text.

7.3　Confidence Bounds

What would happen if we asked everyone in class (70 people) to collect 50 samples of speed data and to compute their own estimate of the mean. How many estimates would there be? What distribution would they have? There would be 70 estimates and the histogram of these 70 means would look normally distributed. Thus the "estimate of the mean" is itself a random variable that is normally distributed.

Usually we compute only one estimate of the mean (or any other quantity), but in this class exercise we are confronted with the reality that there is a range of outcomes. We may, therefore, ask how good is our estimate of the mean. How confident are we that our estimate is correct? Consider that:

1. The estimate of the mean quickly tends to be normally distributed.

2. The expected value (the true mean) of this distribution is the unknown fixed mean of the original distribution.

3. The standard deviation of this new distribution of means is the standard deviation of the original distribution divided by the square root of the number of samples, N. (This assumes independent samples and infinite population.)

The standard deviation of this distribution of the means is called the standard error of the mean (E), where:

$$E = \sigma/\sqrt{N} \qquad (7-9)$$

where the sample standard deviation, s, is used to estimate σ, and all variables are as previously defined. The same characteristics of any normal distribution apply to this distribution of means as well. In other words, the single value of the estimate of the mean, $\bar{x}_n$, approximates the true mean population, μ, as follows:

$\mu = \bar{x} \pm E$, with 68.3% confidence

$\mu = \bar{x} \pm 1.96\,E$, with 95% confidence

$\mu = \bar{x} \pm 3.00\,E$, with 99.7% confidence

The $\pm$ term (E, 1.96E, or 3.00E, depending upon the confidence level) in the above equation is also called the *tolerance* and is given the symbol e.

Consider the following: 54 speeds are observed, and the mean is computed as 47.8 mph, with a standard deviation of 7.80 mph. What are the 95% confidence bounds?

$$P\left[47.8 - 1.96*(7.80/\sqrt{54})\right] \leq \mu$$
$$\leq \left[47.8 + 1.96*(7.80/\sqrt{54})\right] = 0.95 \text{ or}$$
$$P(45.7 \leq \mu \leq 49.9) = 0.95$$

Thus, it is said that there is a 95% chance that the true mean lies between 45.7 and 49.9 mph. Further, while not proven here, any random variable consisting of sample means tends to be normally distributed for reasonably large n, regardless of the original distribution of individual values.

7.4   Sample Size Computations

We can rewrite the equation for confidence bounds to solve for N, given that we want to achieve a specified tolerance and confidence. Resolving the 95% confidence bound equation for N gives:

$$N \geq \frac{1.96^2\, s^2}{e^2} \qquad (7\text{-}10)$$

where 1.96^2 is used only for 95% confidence. If 99.7% confidence is desired, then the 1.96^2 would be replaced by 3^2.

Consider another example: With 99.7% and 95% confidence, estimate the true mean of the speed on a highway, plus or minus 1 mph. We know from previous work that the standard deviation is 7.2 mph. How many samples do we need to collect?

$$N = \frac{3^2 * 7.2^2}{1^2} \approx 467 \text{ samples for 99.7\% confidence,}$$

and

$$N = \frac{1.96^2 * 7.2^2}{1^2} \approx 200 \text{ samples for 95\% confidence}$$

Consider further that a spot speed study is needed at a location with unknown speed characteristics. A tolerance of ±0.2 mph and a confidence of 95% is desired. What sample size is required? Since the speed characteristics are unknown, a standard deviation of 5 mi/h (a most common result in speed studies) is assumed. Then for 95% confidence, $N = (1.96^2 * 5^2)/0.2^2 = 2,401$ samples. This number is unreasonably high. It would be too expensive to collect such a large amount of data. Thus the choices are to either reduce the confidence or increase the tolerance. A 95% confidence level is considered the minimum that is acceptable; thus, in this case, the tolerance would be increased. With a tolerance of 0.5 mi/h:

$$N = \frac{1.96^2 * 5^2}{0.5^2} = 384 \text{ vehicles}$$

Thus the increase of just 0.3 mi/h in tolerance resulted in a decrease of 2,017 samples required. Note that the sample size required is dependent on s, which was assumed at the beginning. After the study is completed and the mean and standard deviation are computed, N should be rechecked. If N is greater (i.e., the actual s is greater than the assumed s) then more samples may need to be taken.

Another example: An arterial is to be studied, and it is desired to estimate the mean travel time to a tolerance of ±5 seconds with 95% confidence. Based on prior knowledge and experience, it is estimated that the standard deviation of the travel times is about 15 seconds. How many samples are required?

Based on an application of Equation 7-10, $N = 1.96^2(15^2)/(5^2) = 34.6$, which is rounded to 35 samples.

As the data is collected, the s computed is 22 seconds, not 15 seconds. If the sample size is kept at $n = 35$, the confidence bounds will be $\pm 1.96(22)/\sqrt{35}$ or about ±7.3 seconds. If the confidence bounds must be kept at ±5 seconds, then the sample size must be increased so that $N \geq 1.96^2(22^2)/(5^2) = 74.4$ or 75 samples. Additional data will have to be collected to meet the desired tolerance and confidence level.

7.5   Addition of Random Variables

One of the most common occurrences in probability and statistics is the summation of random variables, often in the form $Y = a_1 X_1 + a_2 X_2$ or in the more general form:

$$Y = \sum a_i X_i \qquad (7\text{-}11)$$

where the summation is over i, usually from 1 to n.

It is relatively straightforward to prove that the expected value (or mean) μ_Y of the random variable Y is given by:

$$\mu_Y = \sum a_i \mu_{xi} \qquad (7\text{-}12)$$

and that if the random variables x_i are independent of each other, the variance $s^2{}_Y$ of the random variable Y is given by:

$$\sigma_Y^2 = \sum a_i^2 \sigma_{xi}^2 \qquad (7\text{-}13)$$

The fact that the coefficients, a_i, are multiplied has great practical significance for us in all our statistical work.

A Sample Problem: Adding Travel Times

A trip is composed of three parts, each with its own mean and standard deviation as shown below. What is the mean, variance, and standard deviation of the total trip time?

Trip Components	Mean	Standard Deviation
1. Auto	7 min	2 min
2. Commuter Rail	45 min	6 min
3. Bus	15 min	3 min

Solution: The variance may be computed by squaring the standard deviation. The total trip time is the sum of the three components. Equations 7-12 and 7-13 may be applied to yield a mean of 67 minutes, a variance of $(1^2 \times 2^2) + (1^2 \times 6^2) + (1^2 \times 3^2) = 49$ min^2, and, therefore, a standard deviation of 7 minutes.

A Sample Problem: Parking Spaces

Based on observations of condos with 100 units, it is believed that the mean number of parking spaces desired is 70, with a standard deviation of 4.6. If we are to build a larger complex with 1,000 units, what can be said about the parking?

Solution: If X = the number of parking spaces for a condo with 100 units, then the mean (μ_x) of this distribution of values has been estimated to be 70 spaces, and the standard deviation (σ_x) has been estimated to be 4.6 spaces. The assumption must be made that the number of parking spaces needed is proportional to the size of the condo development (i.e., that a condo with 1,000 units will need 10 times as many parking spaces as one with 100 units). In Equations 7-12 and 7-13, 10 becomes the value of a. Then, if Y = the number of parking spaces needed for a condo with 1,000 units:

$$\mu_y = 10(70) = 700 \text{ spaces}$$

and:

$$\sigma_y^2 = (10^2)(4.6^2) = 2116$$

$$\sigma_y = 46 \text{ spaces}$$

Note that this estimates the *average* number of spaces needed by a condo of size 1,000 units, but it does not address the specific need for parking at any specific development of this size. The estimate is also based on the assumption that parking needs will be proportional to the size of the development, which may not be true. If available, a better approach would have been to collect information on parking at condo developments of varying size to develop a relationship between parking and size of development. The latter would involve regression analysis, a more complex type of statistical analysis.

7.5.1 The Central Limit Theorem

One of the most impressive and useful theorems in probability is that the sum of n similarly distributed random variables tends to the normal distribution, no matter what the initial, underlying distribution is. That is, the random variable $Y = \Sigma X_i$, where the X_i have the same distribution, tends to the normal distribution.

The words "tends to" can be read as "tends to look like" the normal distribution. In mathematical terms, the actual distribution of the random variable Y approaches the normal distribution asymptotically.

Sum of Travel Times

Consider a trip made up of 15 components, all with the same underlying distribution, each with a mean of 10 minutes and standard deviation of 3.5 minutes. The underlying distribution is unknown. What can you say about the total travel time?

While there might be an odd situation to contradict this, $n = 15$ should be quite sufficient to say that the distribution of total travel times tends to look normal. From Equation 7-12, the mean of the distribution of total travel times is found by adding 15 terms $(a_i \mu_i)$ where $a_i = 1$ and $\mu_i = 10$ minutes, or

$$\mu_y = 15 * (1 * 10) = 150 \text{ minutes}$$

The variance of the distribution of total travel times is found from Equation 7-13 by adding 15 terms $(a_i^2 \sigma_i^2)$ where a_i is again 1, and σ_i is 3.5 minutes. Then:

$$\sigma_y^2 = 15 * (1 * 3.5^2) = 183.75 \text{ minutes}^2$$

The standard deviation, σ_y is, therefore, 13.6 minutes.

If the total travel times are taken to be normally distributed, 95% of all observations of total travel time will lie between the mean (150 minutes) $\pm$ 1.96 standard deviations (13.6 minutes), or:

$$X_y = 150 \pm 1.96\,(13.6)$$

Thus, 95% of all total travel times would be expected to fall within the range of 123 to 177 minutes (values rounded to the nearest minute).

Hourly Volumes

Five-minute counts are taken, and they tend to look rather smoothly distributed but with some skewness (asymmetry). Based on many observations, the mean tends to be 45 vehicles in the five-minute count, with a standard deviation of seven vehicles. What can be said of the hourly volume?

The hourly volume is the sum of 12 five-minute distributions, which should logically be basically the same if traffic levels are stable. Thus, the hourly volume will tend to look normal, and will have a mean computed using Equation 7-12, with $a_i = 1$, $\mu_i = 45$ vehicles, and $n = 12$, or $12*(1*45) = 540$ veh/h. The variance is computed using Equation 7-13, with $a_i = 1$, $\sigma_i = 7$, and $n = 12$, or $12*(1^2*7^2) = 588$ (veh/h)^2. The standard deviation is 24.2 veh/h. Based on the assumption of normality, 95% of hourly volumes would be between $540 \pm 1.96\,(24.2) = 540 \pm 47$ veh/h (rounded to the nearest whole vehicle).

Note that the summation has had an interesting effect. The σ/μ ratio for the five-minute count distribution was $7/45 = 0.156$, but for the hourly volumes it was $47/540 = 0.087$. This is due to the summation, which tends to remove extremes by canceling "highs" with "lows" and thereby introduces stability. The mean of the sum grows in proportion to n, but the standard deviation grows in proportion to the square root of n.

Sum of Normal Distributions

Although not proven here, it is true that the sum of any two normal distributions is itself normally distributed. By extension, if one normal is formed by n_1 summations of one underlying distribution and another normal is formed by n_2 summations of another underlying distribution, the sum of the total also tends to the normal.

Thus, in the foregoing travel-time example, not all of the elements had to have exactly the same distribution as long as subgroupings each tended to the normal.

7.6 The Binomial Distribution Related to the Bernoulli and Normal Distributions

7.6.1 Bernoulli and the Binomial Distribution

The *Bernoulli distribution* is the simplest discrete distribution, consisting of only two possible outcomes: yes or no, heads or tails, one or zero, etc. The first occurs with probability p, and therefore the second occurs with probability $(1 - p = q)$. This is modeled as:

$$P(X = 1) = p$$
$$P(X = 0) = 1 - p = q$$

In traffic, it represents any basic choice—to park or not to park; to take this route or that; to take auto or transit (for one individual). It is obviously more useful to look at more than one individual, however, which leads us to the binomial distribution. The *binomial distribution* can be thought of in two common ways:

1. Observe N outcomes of the Bernoulli distribution, make a record of the number of events that have the outcome "1," and report that number as the outcome X.
2. There binomial distribution is characterized by the following properties:

   - There are N events, each with the same probability p of a positive outcome and $(1-p)$ of a negative outcome.

   - The outcomes are independent of each other.

   - The quantity of interest is the total number X of positive outcomes, which may logically vary between 0 and N.

   - N is a finite number.

The two ways are equivalent, for most purposes.

Consider a situation in which people may choose "transit" or "auto" where each person has the same probability $p = 0.25$ of choosing transit, and each person's decision is independent of that of all other persons. Defining "transit" as the positive choice for the purpose of this example and choosing $N = 8$, note that:

1. Each person is characterized by the Bernoulli distribution, with $p = 0.25$.

2. There are $2^8 = 256$ possible combinations of choices, and some of the combinations not only yield the same value of X but also have the same probability of occurring. For instance, the value of $X = 2$ occurs for both

TTAAAAAA

and

TATAAAAA

and several other combinations, each with probability of $p^2(1 - p)^6$, for a total of 28 such combinations.

Stated without proof is the result that the probability $P(X = x)$ is given by:

$$P(X = x) = \frac{N!}{(N - x)!x!}p^x(1 - p)^{N-x} \quad (7\text{-}14)$$

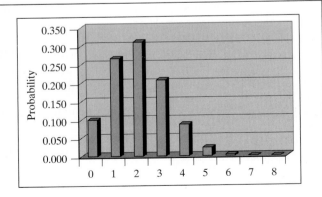

Figure 7.3: Plot of Binomial Results for $N = 8$ and $p = 0.25$

Table 7.4: Tabulated Probabilities for a Binomial Distribution with $N = 8$ and $p = 0.25$

Outcome X People Using Transit	Probability of X
0	0.100
1	0.267
2	0.311
3	0.208
4	0.087
5	0.023
6	0.004
7	0.000
8	0.000

with a mean of Np and a variance of Npq where $q = 1 - p$. The derivation may be found in any standard probability text.

Figure 7.3 shows the plot of the binomial distribution for $p = 0.25$ and $N = 8$. Table 7.4 tabulates the probabilities of each outcome. The mean may be computed as $\mu = Np = 8 \times 0.25 = 2.00$ and the standard deviation as $\sigma = Npq = 8 \times 0.25 \times 0.75 = 1.5$.

There is an important concept that the reader should master, in order to use statistics effectively throughout this text. Even though on average two out of eight people will choose transit, there is absolutely no guarantee what the next eight randomly selected people will choose, even if they follow the rules (same p, independent decisions, etc). In fact, the number could range anywhere from $X = 0$ to $X = 8$. And we can expect that the result $X = 1$ will occur 10.0% of the time, $X = 4$ will occur 8.7% of the time, and $X = 2$ will occur only 31.1% of the time.

This is the crux of the variability in survey results. If there were 200 people in the senior class and each student surveyed eight people from the subject population, we would get different results. In general, our results, if plotted and tabulated, would conform to Figure 7.3 and Table 7.4 but would not mimic them perfectly. Likewise, if we average our results, the result would probably be close to 2.00, but would almost surely not be identical to it.

7.6.2 Asking People Questions: Survey Results

Consider that the commuting population has two choices—$X = 0$ for auto and $X = 1$ for public transit. The probability p is generally unknown, and it is usually of great interest. Assuming the probability is the same for all people (to our ability to discern, at least), then each person is characterized by the Bernoulli distribution.

If we ask $n = 50$ people for their value of X, the resulting distribution of the random variable Y is binomial and may tend to look like the normal. Figure 7.3 shows this exact distribution for $p = 0.25$. Without question, this distribution looks normal. Applying some "quick facts" and noting that the expected value (that is, the mean) is 12.5 (50×0.25), the variance is 9.375 ($50 \times 0.25 \times 0.75$), and the standard deviation is 3.06, one can expect 95% of the results to fall in the range 12.5 ± 6.0 or between 6.5 and 18.5.

If $n = 200$ had been selected, then the mean of Y would have been 50 when $p = 0.25$ and the standard deviation would have been 6.1, so that 95% of the results would have fallen in the range of 38 to 62. Figure 7.4 illustrates the resulting distribution.

7.6.3 The Binomial and the Normal Distributions

The central limit theorem informs us that the sum of Bernoulli distributions (i.e., the binomial distribution) tends to the normal distribution. The only question is: How fast? A number of practitioners in different fields use a rule of thumb that says "for large n and small p" the normal approximation can be used without restriction. This is incorrect and can lead to serious errors.

The most notable case in which the error occurs is when rare events are being described, such as auto accidents per million miles traveled or aircraft accidents. Consider Figure 7.5 on the next page, which is an *exact* rendering of the actual binomial distribution for $p = 0.7(10)^{-6}$ and two values of n—namely $n = 10^6$ and $n = 2(10)^6$, respectively. Certainly p is small and n is large in these cases, and, just as clearly, they do *not* have the characteristic symmetric shape of the normal distribution.

It can be shown that in order for there to be some chance of symmetry—that is, in order for the normal distribution to approximate the binomial distribution—the condition that $np/(1 - p) \geq 9$ is necessary. Clearly, neither of the cases in Figure 7.5 satisfies such a condition.

7.7 The Poisson Distribution

The Poisson distribution is known in traffic engineering as the "counting" distribution. It has the clear physical meaning of a number of events X occurring in a specified counting interval of duration T and is a one-parameter distribution with:

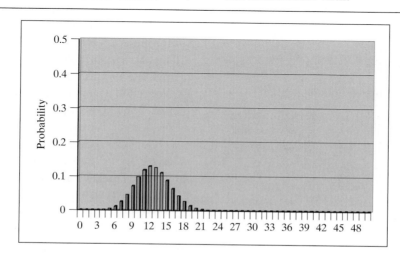

Figure 7.4: Binomial Distribution for $N = 200$ and $p = 0.25$

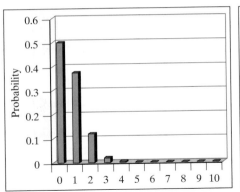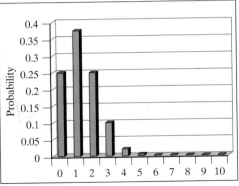

Figure 7.5: Binomial Does Not Always Look Normal for Low p and High N

$$P(X = x) = e^{-m}\frac{m^x}{x!} \qquad (7\text{-}15)$$

with mean $\mu = m$ and variance $\sigma^2 = m$.

The fact that one parameter m specifies both the mean and the variance is a limitation, in that if we encounter field data where the variance and mean are clearly different, the Poisson does not apply.

The Poisson distribution often applies to observations per unit of time, such as the arrivals per five-minute period at a toll booth. When headway times are exponentially distributed with mean $\mu = 1/\lambda$, the number of arrivals in an interval of duration T is Poisson distributed with mean $\mu = m = \lambda T$.

Applying the Poisson distribution is done the same way we applied the Binomial distribution earlier. For example, say there is an average of five accidents per day on the Florida freeways. Table 7.5 shows the proba-

Table 7.5: A Poisson Distribution with $m = 5$

Number of Accidents (x)	Probability $P(x)$	Cumulative Probability $P(X \le x)$
0	.007	.007
1	.034	.041
2	.084	.125
3	.140	.265
4	.175	0.440
5	.175	0.615
6	.146	0.761

bility of there being 0, 1, 2, etc. accidents on a given day on the freeways in Florida. The last column shows the cumulative probabilities or the probability $P(X \le x)$.

7.8 Hypothesis Testing

Very often traffic engineers must make a decision based on sample information. For example, is a traffic control effective or not? To test this, we formulate a hypothesis, H_0, called the null hypothesis and then try to disprove it. The null hypothesis is formulated so that there is no difference or no change, and then the opposite hypothesis is called the alternative hypothesis, H_1.

When testing a hypothesis, it is possible to make two types of errors: (1) We could reject a hypothesis that should be accepted (e.g., say an effective control is not effective). This is called a *Type I error*. The probability of making a Type I error is given the variable name, α. (2) We could accept a false hypothesis (e.g., say an ineffective control is effective). This is called a *Type II error*. A Type II error is given the variable name β.

Consider this example: An auto inspection program is going to be applied to 100,000 vehicles, of which 10,000 are "unsafe" and the rest are "safe." Of course, we do not know which cars are safe and which are unsafe.

We have a test procedure, but it is not perfect, due to the mechanic and test equipment used. We know that 15% of the unsafe vehicles are determined to be safe, and 5% of the safe vehicles are determined to be unsafe, as seen in Figure 7.6.

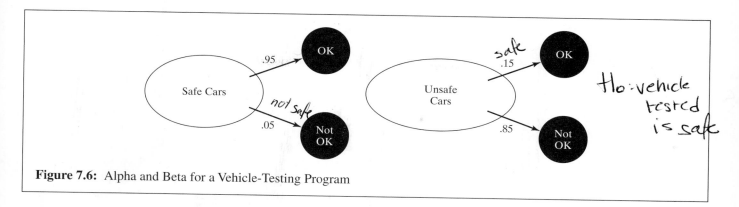

Figure 7.6: Alpha and Beta for a Vehicle-Testing Program

We would define: H_0: The vehicle being tested is "safe," and H_1: the vehicle being tested is "unsafe." The Type I error, rejecting a true null hypothesis (false negative), is labeling a safe vehicle as "unsafe." The probability of this is called the level of significance, α, and in this case $\alpha = 0.05$. The Type II error, failing to reject a false null hypothesis (false positive), is labeling an unsafe vehicle as "safe." The probability of this, β, is 0.15. In general, for a given test procedure, one can reduce Type I error only by living with a higher Type II error, or vice versa.

7.8.1 Before-and-After Tests with Two Distinct Choices

In a number of situations, there are two clear and distinct choices, and the hypotheses seem almost self-defining:

- Auto inspection (acceptable, not acceptable)
- Disease (have the disease, don't)
- Speed reduction of 5 mph (it happened, it didn't)
- Accident reduction of 10% (it happened, it didn't)
- Mode shift by five percentage points (it happened, it didn't)

Of course, there is the distinction between *the real truth* (reality, unknown to us) and *the decision* we make, as already discussed and related to Type I and Type II errors. That is, we can decide that some cars in good working order need repairing and we can decide that some unsafe cars do not need repairing.

There is also the distinction that people may not want to reduce the issue to a binary choice or might not be able to do so. For instance, if an engineer expects a 10% decrease in the accident rate, should we test "H_0: no change" against "H_1: 10% decrease" and not allow the possibility of a 5% change? Such cases are addressed in the next section. For the present section, we will concentrate on binary choices.

Application: Travel Time Decrease

Consider a situation in which the existing travel time on a given route is known to average 60 minutes, and experience has shown the standard deviation to be about 8 minutes. An "improvement" is recommended that is expected to reduce the true mean travel time to 55 minutes.

This is a rather standard problem, with what is now a fairly standard solution. The logical development of the solution follows.

The first question we might ask ourselves is whether we can consider the mean and standard deviation of the initial distribution to be truly known or whether they must be estimated. Actually, we will avoid this question simply by focusing on whether the after situation has a true mean of 60 minutes or 55 minutes. Note that we do not know the shape of the travel time distribution, but the central limit theorem tells us: a new random variable Y, formed by averaging several travel time observations, will tend to the normal distribution if enough observations are taken. Figure 7.7 shows the shape of Y for two different hypotheses, which we now form:

H_0: The true mean of Y is 60 minutes

H_1: The true mean of Y is 55 minutes

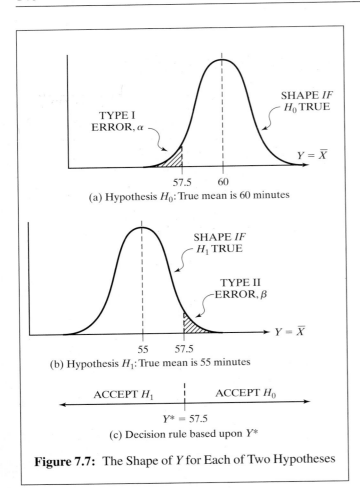

(a) Hypothesis H_0: True mean is 60 minutes

(b) Hypothesis H_1: True mean is 55 minutes

(c) Decision rule based upon Y^*

Figure 7.7: The Shape of Y for Each of Two Hypotheses

Figure 7.7 also shows a logical decision rule: if the actual observation Y falls to the right of a certain point, Y^*, then accept H_0; if the observation falls to the left of that point, then accept H_1. Finally, Figure 7.7 shows shaded areas that are the probabilities of Type I and Type II errors. Note that:

1. The n travel time observations are all used to produce the one estimate of Y.

2. If the point Y^* is fixed, then the only way the Type I and Type II errors can be changed is to increase n, so that the shapes of the two distributions become narrower because the standard deviation of Y involves the square root of n in its denominator.

3. If the point Y^* is moved, the probabilities of Type I and Type II errors vary, with one increasing while the other decreases.

To complete the definition of the test procedure, the point Y^* must be selected and the Type I and Type II errors determined. It is common to require that the Type I error (also known as the level of significance, α) be set at 0.05, so that there is only a 5% chance of rejecting a true null hypothesis. In the case of two alternative hypotheses, it is common to set both the Type I and Type II errors to 0.05, unless there is very good reason to imbalance them (both represent risks, and the two risks—repairing some cars needlessly versus having unsafe cars on the road, for instance—may not be equal).

Inspecting Figure 7.7, Y^* will be set at 57.5 min in order to equalize the two probabilities. The only way these errors can be equal is if the value of Y^* is set at exactly half the distance between 55 and 60 min. The symmetry of the assumed normal distribution requires that the decision point be equally distant from both 55 and 60 min, assuming that the standard deviation of both distributions (before and after) remains 8 min.

To ensure that both errors are not only equal but have an equal value of 0.05, Y^* must be 1.645 standard deviations away from 60 minutes, based on the standard normal table. Therefore, $n \geq (1.645^2)(8^2)/2.5^2$ or 28 observations, where 8 = the standard deviation, 2.5 = the tolerance (57.5 mph is 2.5 mph away from both 55 and 60 mph), and 1.645 corresponds to the z statistic on the standard normal distribution for a beta value of 0.05 (which corresponds to a probability of $z \leq 95\%$).

The test has now been established with a decision point of 57.5 min. If the "after" study results in an average travel time of under 57.5 min, we will accept the hypothesis that the true average travel time has been reduced to 55 min. If the result of the "after" study is an average travel time of more than 57.5 min, the null hypothesis—that the true average travel time has stayed at 60 min—is accepted.

Was all this analysis necessary to make the commonsense judgment to set the decision time at 57.5 min—halfway between the existing average travel time of 60 min and the desired average travel time of 55 min? The answer is in two forms: the analysis provides the logical basis for making such a decision. This is useful. The analysis also provided the minimum sample size required for the "after" study to restrict both alpha and beta errors to 0.05. This is the most critical result of the analysis.

Application: Focus on the Travel Time Difference

The preceding illustration assumed that we would focus on whether the underlying true mean of the "after" situation was either 60 minutes or 55 minutes. What are some of the practical objections that people could raise?

Certainly one objection is that we implicitly accepted at face value that the "before" condition truly had an underlying true mean of 60 minutes. Suppose, to overcome that, we focus on the difference between before and after observations.

The n_1 "before" observations can be averaged to yield a random variable Y_1 with a certain mean μ_1 and a variance of σ_1^2/n_1. Likewise, the n_2 "after" observations can be averaged to yield a random variable Y_2 with a (different?) certain mean μ_2 and a variance of σ_2^2/n_2. Another random variable can be formed as $Y = (Y_2 - Y_1)$, which is itself normally distributed and that has an underlying mean of $(\mu_2 - \mu_1)$ and variance $\sigma^2 = \sigma_2^2/n_2 + \sigma_1^2/n_1$. This is often referred to as the normal approximation. Figure 7.8 shows the distribution of Y, assuming two hypotheses.

What is the difference between this example and the previous illustration? The focus is directly on the difference and does not implicitly assume that we know the initial mean. As a result, "before" samples are required. Also, there is more uncertainty, as reflected in the larger variance. There are a number of practical observations stemming from using this result: it is common that the "before" and "after" variances are equal, in which case the total number of observations can be minimized if $n_1 = n_2$. If the variances are not known, the estimators s_i^2 are used in their place.

If the "before" data was already taken in the past and n_1 is therefore fixed, it may not be possible to reduce the total variance enough (by just using n_2) to achieve a desired level of significance, such as $\alpha = 0.05$. Comparing with the previous problem, note that if both variances are 8^2 and $n_1 = n_2$ is specified, then $n_1 \geq 2*(1.645^2)(8^2)/(2.5^2)$ or 55 and $n_2 \geq 55$. The total required is 110 observations. The fourfold increase is a direct result of

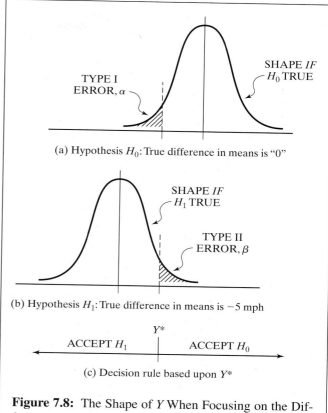

(a) Hypothesis H_0: True difference in means is "0"

(b) Hypothesis H_1: True difference in means is -5 mph

(c) Decision rule based upon Y^*

Figure 7.8: The Shape of Y When Focusing on the Differences

focusing on the difference of -5 mph rather than the two binary choices (60 or 55 minutes).

7.8.2 Before-and-After Tests with Generalized Alternative Hypothesis

It is also common to encounter situations in which the engineer states the situation as "there was a decrease" or "there was a change" versus "there was not," but does not state or claim the magnitude of the change. In these cases, it is standard practice to set up a null hypothesis of "there was no change" ($\mu_1 = \mu_2$) and an alternative hypothesis of "there was a change" ($\mu_1 \neq \mu_2$). In such cases, a level of significance of 0.05 is generally used.

Figure 7.9 shows the null hypotheses for two possible cases, both having a null hypothesis of "no change." The first case implicitly considers that if there were a change, it would be negative—that is, either there was no change or there was a decrease in the mean. The second

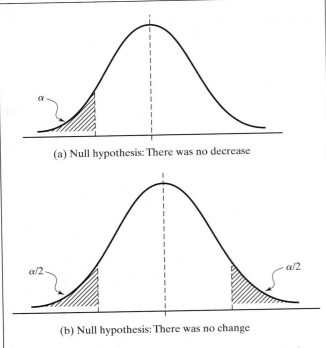

(a) Null hypothesis: There was no decrease

(b) Null hypothesis: There was no change

Figure 7.9: Hypothesis Testing for a Generalized Alternative Hypothesis

case does not have any sense (or suspicion) about the direction of the change, if it exists. Note that:

1. The first is used when physical reasoning leads one to suspect that if there were a change, it would be a decrease.[3] In such cases, the Type I error probability is concentrated where the error is most likely to occur, in one tail.

2. The second is used when physical reasoning leads one to simply assert "there was a change" without any sense of its direction. In such cases, the Type I error probability is spread equally in the two tails.

In using the second case often we might hope that there was no change, and really not want to reject the

null hypothesis. That is, not rejecting the null hypothesis in this case is a measure of success. There are, however, other cases in which we wish to prove that there is a difference. The same logic can be used, but in such cases, rejecting the null hypothesis is "success."

An Application: Travel Time Differences

Let's assume, we have made some improvements and suspect that there is a decrease in the true underlying mean travel time. Figure 7.9(a) applies. Using information from the previous illustration, let us specify that we wish a level of significance $a = 0.05$. The decision point depends upon the variances and the n_i. If the variances are as stated in the prior illustration and $n_1 = n_2 = 55$, then the decision point $Y^* = -2.5$ mph, as before.

Let us now go one step further. The data is collected, and $Y = -3.11$ results. The decision is clear: reject the null hypothesis of "there is no decrease." But what risk did we take?

Consider the following:

- Under the stated terms, had the null hypothesis been valid, we were taking a 5% chance of rejecting the truth. The odds favor (by 19 to 1, in case you are inclined to wager with us) not rejecting the truth in this case.
- At the same time, there is no stated risk of accepting a false hypothesis H_1, for the simple reason that no such hypothesis was stated.
- The null hypothesis was rejected because the value of Y was higher than the decision value of 2.5 mph. Since the actual value of -3.11 is considerably higher than the decision value, one could ask about the confidence level associated with the rejection. The point $Y = -3.11$ is 2.033 standard deviations away from the zero point, as can be seen from:

$$\sigma_Y = \sqrt{\frac{\sigma_1^2}{n_1} + \frac{\sigma_2^2}{n_2}}$$

$$= \sqrt{\frac{8^2}{55} + \frac{8^2}{55}} = 1.53$$

---

[3]The same logic—and illustration—can be used for cases of suspected *increases*, just by a sign change.

and $z = 3.11/1.53 = 2.033$ standard deviations. Entering the standard normal distribution table with $z = 2.033$ yields a probability of 0.9790. This means that if we had been willing to take only a 2% chance of rejecting a valid H_0, we still would have rejected the null hypothesis; we are 98% confident that our rejection of the null hypothesis is correct. This test is called the Normal Approximation and is only valid when $n_1 \geq 30$ and $n_2 \geq 30$.

Since this reasoning is a little tricky, let us state it again: If the null hypothesis had been valid, you were initially willing to take a 5% chance of rejecting it. The data indicated a rejection. Had you been willing to take only a 2% chance, you still would have rejected it.

One-Sided Versus Two-Sided Tests

The material just discussed appears in the statistics literature as "one-sided" tests, for the obvious reason that the probability is concentrated in one tail (we were considering only a decrease in the mean). If there is no rationale for this, a "two-sided" test should be executed, with the probability split between the tails. As a practical matter, this means that one does not use the probability tables with a significance level of 0.05, but rather with $0.05/2 = 0.025$.

7.8.3 Other Useful Statistical Tests

The *t*-Test

For small sample sizes ($N < 30$), the normal approximation is no longer valid. It can be shown that if x_1 and x_2 are from the same population, the statistic t is distributed according to the tabulated t distribution, where:

$$t = \frac{x_1 - x_2}{s_p \sqrt{1/n_1 + 1/n_2}} \qquad (7\text{-}16)$$

and s_p is a pooled standard deviation, which equals:

$$s_p = \sqrt{\frac{(n_1 - 1)s_1^2 + (n_2 - 1)s_2^2}{n_1 + n_2 - 2}} \qquad (7\text{-}17)$$

The t distribution depends upon the degrees of freedom, f, which refers to the number of independent pieces of data that form the distribution. For the t distribution, the non-independent pieces of data are the two means, x_1 and x_2. Thus: *dependent*

$$f = N_1 + N_2 - 2 \qquad (7\text{-}18)$$

Once the t statistic is determined, the tabulated values of Table 7.6 yield the probability of a t value being greater than the computed value. In order to limit the probability of a Type I error to 0.05, the difference in the means will be considered significant only if the probability is less than or equal to 0.05—that is, if the calculated t value falls in the 5% area of the tail, or in other words, if there is less than a five percent chance that such a difference could be found in the same population. If the probability is greater than 5% that such a difference in means could be found in the same population, then the difference would be considered not significant.

Consider the following example: Ten samples of speed data are taken both before and after a change in the speed limit is implemented. The mean and standard deviations found were:

Before		After
35	$\overline{x}$	32
4.0	s	5.0
10	N	10

Then:

$$s_p = \sqrt{\frac{(10 - 1)4^2 + (10 - 1)5^2}{10 + 10 - 2}} = 4.53,$$

$$t = \frac{35 - 32}{4.53\sqrt{1/10 + 1/10}} = 1.48, \text{ and}$$

$$f = 10 + 10 - 2 = 18$$

The probability that $t \geq 1.48$, with 18 degrees of freedom, falls between 0.05 and 0.10. Thus, it is greater than 0.05, and we conclude that there is not a significant difference in the means.

Table 7.6: Upper Percentage Points of the t-Distribution*

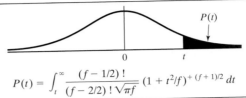

$$P(t) = \int_t^\infty \frac{(f-1/2)!}{(f-2/2)!\sqrt{\pi f}}(1+t^2/f)^{+(f+1)/2}\,dt$$

Deg. of Freedom f	Probability of a Value Equal to or Greater Than t										
	0.40	0.30	0.25	0.20	0.15	0.10	0.05	0.025	0.01	0.005	0.0003
1	0.325	0.727	1.000	1.376	1.936	3.078	6.314	12.706	31.821	63.657	636.619
2	0.289	0.617	0.816	1.061	1.386	1.886	2.290	4.303	6.965	9.925	31.598
3	0.277	0.584	0.765	0.978	1.250	1.638	2.353	3.182	4.541	5.841	12.924
4	0.271	0.569	0.741	0.941	1.190	1.533	2.132	2.776	3.747	4.604	8.610
5	0.267	0.559	0.727	0.920	1.156	1.476	2.015	2.571	3.365	4.032	6.869
6	0.265	0.553	0.718	0.906	1.134	1.440	1.943	2.447	3.143	3.707	5.959
7	0.263	0.549	0.711	0.896	1.119	1.415	1.895	2.365	2.998	3.499	5.608
8	0.262	0.546	0.706	0.889	1.108	1.397	1.860	2.306	2.896	3.355	5.401
9	0.261	0.544	0.703	0.883	1.100	1.383	1.833	2.262	2.821	3.250	4.781
10	0.260	0.542	0.700	0.879	1.093	1.372	1.812	2.228	2.764	3.169	4.587
11	0.260	0.540	0.697	0.876	1.088	1.363	1.796	2.201	2.718	3.106	4.437
12	0.259	0.539	0.695	0.873	1.083	1.356	1.782	2.179	2.681	3.055	4.318
13	0.259	0.538	0.694	0.870	1.079	1.350	1.771	2.160	2.650	3.012	4.221
14	0.258	0.537	0.692	0.866	1.076	1.345	1.761	2.143	2.624	2.977	4.140
15	0.258	0.536	0.691	0.866	1.074	1.341	1.753	2.131	2.602	2.947	4.073
16	0.258	0.535	0.690	0.865	1.071	1.337	1.746	2.120	2.583	2.921	4.015
17	0.257	0.534	0.689	0.865	1.069	1.333	1.740	2.110	2.567	2.898	3.965
18	0.257	0.534	0.688	0.862	1.067	1.330	1.734	2.101	2.552	2.878	3.922
19	0.257	0.533	0.688	0.861	1.066	1.328	1.729	2.093	2.539	2.861	3.883
20	0.257	0.533	0.687	0.860	1.064	1.325	1.725	2.086	2.528	2.845	3.850
21	0.257	0.532	0.686	0.859	1.063	1.323	1.721	2.080	2.518	2.831	3.819
22	0.256	0.532	0.686	0.858	1.061	1.321	1.717	2.074	2.508	2.819	3.792
23	0.256	0.532	0.685	0.858	1.060	1.319	1.714	2.069	2.500	2.807	3.767
24	0.256	0.531	0.685	0.857	1.059	1.318	1.711	2.064	2.492	2.797	3.745
25	0.256	0.531	0.684	0.856	1.058	1.316	1.708	2.060	2.485	2.787	3.735
26	0.256	0.531	0.684	0.856	1.058	1.315	1.706	2.056	2.479	2.779	3.707
27	0.256	0.531	0.684	0.855	1.057	1.314	1.703	2.052	2.473	2.771	3.690
28	0.256	0.530	0.683	0.855	1.056	1.313	1.701	2.048	2.467	2.763	3.674
29	0.256	0.530	0.683	0.854	1.055	1.311	1.699	2.045	2.462	2.756	3.659
30	0.256	0.530	0.683	0.854	1.055	1.310	1.697	2.042	2.457	2.750	3.646
40	0.255	0.529	0.681	0.851	1.050	1.303	1.684	2.021	2.423	2.704	3.551
60	0.254	0.527	0.679	0.848	1.046	1.296	1.671	2.000	2.390	2.660	3.460
120	0.254	0.526	0.677	0.845	1.041	1.289	1.658	1.980	2.358	2.617	3.373
Infinity	0.253	0.524	0.674	0.842	1.036	1.282	1.645	1.960	2.326	2.576	3.291

*Values of "t" are in the body of the table.
(Used with permission of U.S. Naval Ordinance Test Station, Crow, Davis, and Maxfield, *Statistics Manual*, Dover, NJ, 1960, Table 3.)

The F-Test

In using the t-test, and in other areas as well, there is an implicit assumption made that the $\sigma_1 = \sigma_2$. This may be tested with the F distribution, where:

$$F = \frac{s_1^2}{s_2^2}$$

(by definition the larger s is always on top) (7-19)

It can be proven that this F value is distributed according to the F distribution, which is tabulated in Table 7.7. The F distribution is tabulated according to the degrees of freedom in each sample, thus $f_1 = n_1 - 1$ and $f_2 = n_2 - 1$. Since the f distribution in Table 7.7 gives the shaded area in the tail, like the t distribution, the decision rules are as follows:

- If [Prob $F \geq$ F] ≤ 0.05, then the difference is significant.

- If [Prob $F \geq$ F] > 0.05, then the difference is not significant.

Consider the following problem: Based on the following data, can we say that the standard deviations come from the same population?

Before		After
30	$\bar{x}$	35
5.0	s	4.0
11	N	21

Thus:

$$F = \frac{5^2}{4^2} = 1.56; \quad f_1 = 11 - 1 = 10;$$

$$f_2 = 21 - 1 = 20$$

The f distribution is tabulated for various probabilities, as follows, based on the given degrees of freedom; thus:

when
$$\begin{aligned} p &= 0.10, & F &= 1.94 \\ p &= 0.05, & F &= 2.35 \\ p &= 0.025, & F &= 2.77 \end{aligned}$$

The F values are increasing, and the probability $[F \geq 1.56]$ must be greater than 0.10 given this trend; thus, the difference in the standard deviations is not significant. The assumption that the standard deviations are equal, therefore, is valid.

Paired Differences

In some applications, notably simulation, where the environment is controlled, data from the "before" and "after" situations can be paired and only the differences are important. In this way, the entire statistical analysis is done directly on the differences, and the overall variation can be much lower, because of the identity tags.

An example application of paired differences is presented in Table 7.8, which shows the results of two methods for measuring speed. Both methods were applied to the same vehicles, and Table 7.8, considers the data first with no attempt at pairing and then with the data paired.

It is relatively easy to see what appears to be a significant difference in the data. The question is, could it have been statistically detected without the pairing? Using the approach of a one-sided test on the null hypothesis of "no increase," the following computations result:

$$s_1 = 7.74$$
$$s_2 = 7.26$$
$$N_1 = N_2 = 15$$

Then:

$$s_Y = \sqrt{\frac{7.74^2}{15} + \frac{7.26^2}{15}} = 2.74$$

For a significance level of 0.05 (or 95% confidence), in a one-tailed test, the decision point is 1.65 standard deviations, or $1.65 * 2.74 = 4.52$ mi/h. Given that the observed difference in average speeds is only 4.30 mi/h, the null hypothesis of "no increase" *cannot* be rejected, and the test indicates that the two measurement techniques yield *statistically equal* speeds.

Table 7.7: Upper Percentage Points of the f Distribution[*]

$$P(F) = \int_{F}^{\infty} \frac{(f_1 + f_2 - 2)/2\,!}{(f_1 - 2)/2\,!\,(f_2 - 2)/2\,!}\, f^{f_1/2} f^{f_2/2} F^{(f_1 - 2)/2} (f_2 + f_1 F)^{-(f_1 + f_2)/2}\, dF$$

(a) All $P(F) = 0.10$

Deg. of Freedom f_2 Denom.	Degrees of Freedom f_1 (Numerator)								
	1	5	10	20	30	40	60	120	Infinity
1	39.86	57.24	60.20	61.24	62.26	62.53	62.79	63.06	63.33
5	4.06	3.45	3.30	3.21	3.17	3.16	3.14	3.12	3.10
10	3.28	2.52	2.32	2.20	2.16	2.13	2.11	2.08	2.06
20	2.97	2.16	1.94	1.79	1.74	1.71	1.68	1.64	1.61
30	2.89	2.05	1.82	1.64	1.61	1.57	1.54	1.50	1.46
40	2.84	2.00	1.76	1.61	1.57	1.51	1.47	1.42	1.38
60	2.79	1.95	1.71	1.54	1.51	1.44	1.40	1.35	1.29
120	2.75	1.90	1.65	1.48	1.45	1.37	1.32	1.26	1.19
Infinity	2.71	1.85	1.00	1.42	1.39	1.30	1.24	1.17	1.00

(b) All $P(F) = 0.05$

Deg of Freedom f_2 Denom.	Degrees of Freedom f_1 (Numerator)								
	1	5	10	20	30	40	60	120	Infinity
1	161.45	230.16	241.88	248.01	250.09	251.14	252.20	253.25	254.32
5	6.61	5.05	4.74	4.56	4.50	4.46	4.43	4.40	4.36
10	4.96	3.33	2.98	2.77	2.70	2.66	2.62	2.58	2.54
20	4.35	2.71	2.35	2.12	2.04	1.99	1.95	1.90	1.84
30	4.17	2.53	2.16	1.93	1.84	1.79	1.74	1.68	1.62
40	4.08	2.45	2.08	1.84	1.79	1.69	1.64	1.58	1.51
60	4.00	2.37	1.99	1.75	1.65	1.59	1.53	1.47	1.39
120	3.92	2.29	1.91	1.66	1.55	1.50	1.43	1.35	1.23
Infinity	3.84	2.21	1.83	1.57	1.46	1.39	1.32	1.22	1.00

(c) All $P(F) = 0.025$

Deg of Freedom f_2 Denom.	Degrees of Freedom f_1 (Numerator)								
	1	5	10	20	30	40	60	120	Infinity
1	647.79	921.85	968.63	993.10	1,001.40	1,005.60	1,009.80	1,014.00	1,018.30
5	10.01	7.15	6.62	6.33	6.23	6.18	6.12	6.07	6.02
10	6.94	4.24	3.72	3.42	3.31	3.26	3.20	3.14	3.08
20	5.87	3.29	2.77	2.46	2.35	2.29	2.22	2.16	2.09
30	5.57	3.03	2.51	2.20	2.07	2.01	1.94	1.87	1.79
40	5.42	2.90	2.39	2.07	1.94	1.88	1.80	1.72	1.64
60	5.29	2.79	2.27	1.94	1.82	1.74	1.67	1.58	1.48
120	5.15	2.67	2.16	1.82	1.69	1.61	1.53	1.43	1.31
Infinity	5.02	2.57	2.05	1.71	1.57	1.48	1.39	1.27	1.00

[*]Values of F in the body of table. (Condensed from Crow, Davis, and Maxfield, *Statistics Manual*, 1960.)

Table 7.8: Example Showing the Benefits of Paired Testing

Vehicle	Method 1 Speed (mi/h)	Method 2 Speed (mi/h)	Difference by Vehicle (D) (mi/h)	D^2
1	55	60	5	25
2	47	55	8	64
3	70	74	4	16
4	62	67	5	25
5	49	53	4	16
6	67	71	4	16
7	52	57	5	25
8	57	60	3	9
9	58	61	3	9
10	45	48	3	9
11	68	70	2	4
12	52	57	5	25
13	51	56	5	25
14	58	64	6	36
15	62	65	3	9
MEAN =	56.9	61.2	4.3	
STD =	7.74	7.26	1.5	

However, with the inspection of paired differences, it is seen that *every* measurement resulted in an increase when Method 2 is applied. Application of the same statistical approach, but using the paired differences yields:

$$s = 1.50$$
$$N = 15$$

The standard error of the mean for this case is:

$$E = \frac{1.50}{\sqrt{15}} = 0.387$$

For a 0.05 level of significance, the decision point is $Y^* = 1.65 * 0.378 = 0.64$ mi/h. With an observed difference of 4.3 mi/h, the null hypothesis of "no increase" is clearly rejected. The two measurement methodologies do yield different results, with Method 2 resulting in higher speeds than Method 1.

Chi-Square Test: Hypotheses on an Underlying Distribution $f(x)$

One of the early problems stated was a desire to "determine" the underlying distribution, such as in a speed study. The most common test to accomplish this is the Chi-square (χ^2) goodness-of-fit test.

In actual fact, the underlying distribution will not be determined. Rather, a hypothesis such as "H_0: The underlying distribution is normal" will be made, and we test that it is not rejected, so we may then act as if the distribution were in fact normal.

The procedure is best illustrated by an example. Consider data on the height of 100 people showed in Table 7.9. To simplify the example, we will test the hypothesis that this data is *uniformly* distributed (i.e., there are equal numbers of observations in each group).

In order to test this hypothesis, the goodness-of-fit test is done by following these steps:

1. Compute the theoretical frequencies, f_i, for each group. Since a uniform distribution is assumed and there are 10 categories with 100 total observations, $f_i = 100/10 = 10$ for all groups.

2. Compute the quantity:

$$\chi^2 = \sum_{i=1}^{N} \frac{(n_i - f_i)^2}{f_i} \qquad (7\text{-}20)$$

where the summation is done over N data categories or groups. These computations are shown in Table 7.10.

Table 7.9: Sample Height Data for Chi-Square Test

Height Category	Number of People
5.0 ft–5.2 ft	4
5.2 ft–5.4 ft	6
5.4 ft–5.6 ft	11
5.6 ft–5.8 ft	14
5.8 ft–6.0 ft	21
6.0 ft–6.2 ft	18
6.2 ft–6.4 ft	16
6.4 ft–6.6 ft	5
6.6 ft–6.8 ft	4
6.8 ft–7.0 ft	1
Total	**100**

3. As shown in any standard statistical text, the quantity χ^2 is chi-squared distributed, and we expect low values if our hypothesis is correct. (If the observed samples exactly equal the expected, then the quantity is zero.) Therefore, refer to a table of the chi-square distribution (Table 7.11) and look up the number that we would not exceed more than 5% of the time (i.e., $\alpha = 0.05$). To do this, we must also have the number of degrees of freedom, designated df, and defined as $df = N - 1 - g$, where N is the number of categories and g is the number of things we estimated from the data in defining the hypothesized distribution. For the uniform distribution, only the sample size was needed to estimate the theoretical frequencies, thus "0" parameters were used in computing χ^2 ($g = 0$). Therefore, for this case, $df = 10 - 1 - 0 = 9$.

4. Table 7.11 is now entered with $\alpha = 0.05$ and $df = 9$. A decision value of $\chi^2 = 16.92$ is found. As the value obtained is $43.20 > 16.92$, the hypothesis that the underlying distribution is uniform must be rejected.

In this case, a rough perusal of the data would have led one to believe that the data were certainly *not* uniform, and the analysis has confirmed this. The distribution

Table 7.10: Computation of χ^2 for the Sample Problem

Height Category (ft)	Observed Frequency n	Theoretical Frequency F	χ^2
5.0–5.2	4	10	$(4-10)^2/10 = 3.60$
5.2–5.4	6	10	$(6-10)^2/10 = 1.60$
5.4–5.6	11	10	$(11-10)^2/10 = 0.10$
5.6–5.8	14	10	$(14-10)^2/10 = 1.60$
5.8–6.0	21	10	$(21-10)^2/10 = 12.10$
6.0–6.2	18	10	$(18-10)^2/10 = 6.40$
6.2–6.4	16	10	$(16-10)^2/10 = 3.60$
6.4–6.6	5	10	$(5-10)^2/10 = 2.50$
6.6–6.8	4	10	$(4-10)^2/10 = 3.60$
6.8–7.0	1	10	$(1-10)^2/10 = 8.10$
Total	**100**	**100**	**43.20**

Table 7.11: Upper Percentage Points on the Chi-Square Distribution

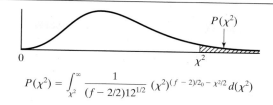

$$P(\chi^2) = \int_{\chi^2}^{\infty} \frac{1}{(f-2/2)12^{1/2}} \, (\chi^2)^{(f-2)/2_0 - \chi^{2/2}} \, d(\chi^2)$$

df	.995	.990	.975	.950	.900	.750	.500	.250	.100	.050	.025	.010	.005
1	3927×10^{-2}	1571×10^{-7}	9821×10^{-7}	3932×10^{-8}	0.01579	0.1015	0.4549	1.323	2.706	3.841	5.024	6.635	7.879
2	0.01003	0.02010	0.05064	0.1026	.2107	.5754	1.386	2.773	4.605	5.991	7.378	9.210	10.60
3	.07172	.1148	.2158	.3578	.5844	1.213	2.366	4.108	6.251	7.815	9.348	11.34	12.34
4	.2070	.2971	.4844	.7107	1.064	1.923	3.357	5.585	7.779	9.488	11.14	13.28	14.86
5	.4117	.5543	.8312	1.145	1.610	2.675	4.351	6.626	9.236	11.07	12.83	15.09	16.75
6	.6757	.8721	1.237	1.635	2.204	3.455	5.348	7.841	10.64	12.59	14.45	16.81	18.55
7	.9893	1.259	1.690	2.167	2.833	4.255	6.346	9.037	12.02	14.07	16.01	18.48	20.28
8	1.344	1.646	2.180	2.733	3.199	5.071	7.344	10.22	13.36	15.51	17.63	20.09	21.98
9	1.735	2.088	2.700	3.325	4.168	5.899	8.343	11.39	14.68	16.92	19.02	21.67	23.59
10	2.150	2.558	3.247	3.940	4.865	6.737	9.342	12.55	15.99	18.31	20.48	23.21	25.19
11	2.603	3.053	3.816	4.575	5.578	7.584	10.34	13.70	17.28	19.68	21.92	24.72	26.76
12	3.074	3.571	4.404	5.226	6.304	8.458	11.34	14.85	18.55	21.03	23.34	26.22	28.30
13	3.565	4.107	5.009	5.892	7.042	9.299	12.34	15.98	19.81	22.36	24.74	27.69	29.82
14	4.075	4.660	5.629	6.571	7.790	10.17	13.34	17.12	21.06	23.68	26.12	29.14	31.32
15	4.601	5.229	6.262	7.261	8.547	11.04	14.34	18.25	22.31	25.00	27.49	30.58	32.80
16	5.142	5.812	6.908	7.962	9.312	11.91	15.34	19.37	23.54	26.30	28.85	32.00	34.27
17	5.697	5.408	7.564	8.672	10.09	12.79	16.34	20.49	24.77	27.59	30.19	33.41	35.72
18	6.265	7.015	8.231	9.390	10.86	13.68	17.34	21.60	25.99	28.87	31.53	34.81	37.16
19	6.844	7.644	8.907	10.12	11.65	14.56	18.34	22.72	27.20	30.14	32.85	36.19	38.58
20	7.434	8.260	9.591	10.85	12.44	15.45	19.34	23.83	28.41	31.41	34.17	37.57	40.00
21	8.034	8.897	10.28	11.59	13.24	16.34	20.34	24.93	29.62	32.67	35.48	38.93	41.40
22	8.643	9.542	10.98	12.34	14.04	17.24	21.34	26.04	30.81	33.92	36.78	40.29	42.60
23	9.260	10.20	11.69	13.09	-14.85	18.14	22.34	27.14	32.01	35.17	38.08	41.64	44.18
24	9.886	10.86	12.40	13.85	15.66	19.04	23.24	28.24	33.20	36.42	39.36	42.98	45.58
25	10.52	11.52	13.12	14.61	16.47	19.94	24.34	29.34	34.38	37.65	40.65	44.31	46.93
26	11.16	12.20	13.84	15.38	17.29	20.84	25.34	30.43	35.56	38.89	41.92	45.64	48.29
27	11.81	12.88	14.57	16.15	18.11	21.75	26.34	31.53	36.74	40.11	43.19	46.96	49.64
28	12.46	13.56	15.31	16.93	18.94	22.66	27.34	32.62	37.92	41.34	44.46	48.28	50.99
29	13.12	14.26	16.05	17.71	19.77	23.57	28.34	33.71	39.09	42.58	45.72	49.59	52.34
30	13.79	14.95	16.79	18.49	20.60	24.48	29.34	34.80	43.26	43.77	46.98	50.89	53.67
40	20.71	22.16	24.43	26.51	29.05	33.66	39.34	45.62	51.80	55.76	59.34	63.69	66.77
50	27.99	29.71	32.36	34.76	37.69	42.94	49.33	56.33	63.17	67.50	71.42	76.15	79.49
60	35.53	37.48	40.48	43.19	46.46	52.29	59.33	66.98	79.08	79.08	83.30	88.38	91.95
70	43.28	45.44	48.76	51.74	55.33	61.70	69.33	77.58	85.53	90.53	95.02	100.42	104.22
80	51.17	53.54	57.15	60.39	64.28	71.14	79.33	88.13	96.58	101.88	106.63	112.33	116.32
90	59.20	61.75	65.65	69.13	73.29	80.62	89.33	98.65	107.56	113.14	118.14	124.12	128.30
100	67.33	70.00	74.22	77.93	82.36	90.13	99.33	109.14	118.50	124.34	129.56	135.81	140.17
	-2.576	-2.326	1.960	-1.645	-1.28	-0.6745	0.0000	+0.6745	+1.282	+1.645	+1.960	+2.326	576

(Source: E. L. Crow, F. A. Davis, and M. W. Maxwell, *Statistics Manual*, Dover Publications, Mineola, NY, 1960.)

actually appears to be closer to normal, and this hypothesis can be tested. Determining the theoretical frequencies for a normal distribution is much more complicated, however. An example applying the χ^2 test to a normal hypothesis is discussed in detail in Chapter 9, using spot speed data.

Note that in conducting goodness-of-fit tests, it is possible to show that several different distributions could represent the data. How could this happen? Remember that the test does not prove what the underlying distribution is in fact. At best, it does not oppose the desire to assume a certain distribution. Thus, it is possible that different hypothesized distributions for the same set of data may be found to be statistically acceptable.

A final point on this hypothesis testing: the test is not directly on the hypothesized distribution, but rather on the expected versus the observed number of samples. Table 7.10 shows this in very plain fashion: the computations involve the expected and observed number of samples; the actual distribution is important only to the extent that it influences the probability of category. That is, the actual test is between two histograms. It is therefore important that the categories be defined in such a way that the "theoretic histogram" truly reflects the essential features and detail of the hypothesized distribution. That is one reason why categories of different size are sometimes used: the categories should each match the fast-changing details of the underlying distribution.

7.9   Summary and Closing Comments

Traffic studies cannot be done without using some basic statistics. In using statistics, the engineer is often faced with the need to act despite the lack of certainty, which is why statistical analysis is employed. This chapter is meant to review basic statistics for the reader to be able to understand and perform everyday traffic studies.

There are a number of very important statistical techniques not presented in this chapter that are useful for traffic engineers. They include contingency tables, correlation analysis, analysis of variance (ANOVA), regression analysis, and cluster analysis. The reader is referred to reference [4] or other basic statistical texts [1–3, 5].

References

1. Vardeman, Stephen B. and Jobe, J. Marcus, *Data Collection and Analysis*, Duxbury Press, Boston, 2001.

2. Chao, Lincoln L., *Statistics Methods and Analyses*, McGraw-Hill, New York, 1974.

3. Mendenhall, *et al.*, *Mathematical Statistics with Applications*, PWS-KENT, Boston, 1990.

4. Crow, *et al.*, *Statistics Manual*, Dover Publications, New York, 1960.

5. *Statistics with Applications to Highway Traffic Analyses*, 2nd Ed., ENO Foundation for Transportation, Connecticut, 1978.

Problems

7-1. Experience suggests that spot speed data at a given location is normally distributed with a mean of 57 mph and a standard deviation of 7.6 mph. What is the speed below which 85% of the vehicles are traveling?

7-2 Travel time data is collected on an arterial, and with 30 runs, an average travel time of 152 seconds is computed over the 2.00-mile length, with a computed standard deviation of 17.3 seconds. Compute the 95% confidence bounds on your estimate of the mean. Was it necessary to make any assumption about the shape of the travel time distribution?

7-3. Vehicle occupancy data is taken in a high occupancy vehicle (HOV) lane on a freeway, with the following results:

Vehicle Occupancy	Number of Vehicles Observed
2	120
3	40
4	30
5	10

(a) Compute the estimated mean and standard deviation of the vehicle occupancy in the HOV Lane. Compute the 95% confidence bounds on the estimate of the mean.

(b) When the HOV lane in question has an hourly volume of 900 vph, what is your estimate of how many people on average are being carried in the lane? Give a 95% confidence range.

(c) If we observe the lane tomorrow and observe a volume of 900 vph, what is the range of persons moved we can expect in that hour? Use a range encompassing 95% of the likely outcomes. List any assumptions or justifications not made in part (b) but necessary in part (c) if any.

7-4. Run a compliance study at a STOP sign in a residential area, dividing the work amongst a group of three to four students, so that each group has about $N = 200$ observations. Estimate the fraction who are fully compliant, and compute the estimated 95% confidence bounds on this fraction, where $s^2 = p(1 - p)/N$.

7-5. Consider a survey of n people asking whether they plan to use Route A or Route B, with an anticipated fraction p choosing Route A.

If $p = 0.30$, plot the 95% confidence bounds and the expected value of X as a function of n, for $n = 50$ to $n = 800$.

7-6. Two different procedures are used to measure "delay" on the same intersection approach at the same time. Both are expressed in terms of delay per vehicle. The objective is to see which procedure can be best used in the field. Assume that the following data are available from the field.

(a) A statistician, asked whether the data "are the same," conducts tests on the mean and the variance to determine whether they are the same at

a level of significance of 0.05. Use the appropriate tests on the hypothesis that the difference in the means is zero and the variances are equal. Do the computations. Present the conclusions.

Delay (sec/veh)	
Procedure 1	Procedure 2
8.4	7.2
9.2	8.1
10.9	10.3
13.2	10.3
12.7	11.2
10.8	7.5
15.3	10.7
12.3	10.5
19.7	11.9
8.0	8.7
7.4	5.9
26.7	18.6
12.1	8.2
10.7	8.5
10.1	7.5
12.0	9.5
11.9	8.1
10.0	8.8
22.0	19.8
41.3	36.4

(b) Would it have been more appropriate to use the "paired-t" test in part (a) rather than a simple t test? Why?

7-7. Based on long standing observation and consensus, the speeds on a curve are observed to average 57 mph with a 6 mph standard deviation. This is taken as "given." Some active controls are put in place (signing, flashing lights, etc.), and the engineer is sure that the average will fall to at least 50 mph, with the standard deviation probably about the same.

(a) Formulate a null and alternative hypothesis based on taking "after" data only, and determine the required sample size n so that the Type I and Type II errors are each 0.05.

(b) Assume that the data has been taken for the required N observations and that the average is 52.2 mph and the standard deviation is 6.0 mph.

What is your decision? What error may have occurred, and what is its probability?

(c) Resolve part (b) with a mean of 52.2 mph and a standard deviation of 5.4 mph.

CHAPTER

8

Volume Studies and Characteristics

8.1 Introduction to Traffic Studies

The starting point for most traffic engineering is the current state of facilities and traffic along with a prediction or anticipation of future demand. The former requires that a wide variety of data and information be assembled that adequately describe the current status of systems, facilities, and traffic.

Traffic engineers collect data for many reasons and applications:

- *Managing the physical system.* Physical inventories of physical system elements are always needed. These include traffic control devices, lighting fixtures, and roadways. These inventories help assess which items need to be replaced or repaired, and on what anticipated schedule.

- *Investigating trends over time.* Traffic engineers need trend data to help forecast future transportation needs. Parametric studies on traffic volumes, speeds, and densities help quantify current demand and assess operating quality of traffic facilities. Accident studies reveal locations with

problems that must be addressed and mitigated. Establishing trends allows the traffic engineer to make needed improvements *before* an obvious deficiency in the system manifests itself.

- *Understanding the needs and choices of the public and industry.* The traffic engineer must have a good measure of how and why people travel for planning and development purposes. Studies of how travelers make mode choices, time of trip decisions, and other judgments is critical to understanding the nature of travel demand. Studies of parking and goods delivery characteristics help plan facilities to effectively handle these demands.

- *Calibrating basic relationships or parameters.* Fundamental measures, such as perception-reaction time, discharge headways at a signalized intersection, headway and spacing relationships on freeways and other uninterrupted flow facilities, and other key parameters and relationships must be properly quantified and calibrated to existing conditions. Such measures are incorporated into a variety of predictive and assessment models on which much of traffic engineering is based.

- *Assessing the effectiveness of improvements.* When improvements of any kind are implemented, follow-up studies are needed to confirm their effectiveness, and to allow for adjustments if all objectives are not fully met.

- *Assessing potential impacts.* An essential part of traffic engineering is the ability to predict and analyze projected traffic impacts of new developments and to provide traffic input to air pollution models.

- *Evaluating facility or system performance.* All traffic facilities and systems must be periodically studied to determine whether they are delivering the intended quantity and quality of access and/or mobility service to the public.

Traffic studies provide the underpinning of all traffic planning, design, and analysis. If the data is not correct, then the traffic engineering that is based on it must be flawed. Some of the tasks involved in data collection, reduction, analysis, and presentation are somewhat mundane. Focused attention must be paid to these tasks, however, if the final outcome of traffic engineering is to be effective.

8.1.1 Modern Technology

The rapid advancement of the information technologies in recent years has greatly enhanced the traffic engineer's ability to collect, summarize, and analyze information. Technologies such as wide-area sensing, image identification processes, wireless communications, satellite-based location systems, mapping technologies, and the like continue to revolutionize both the quantity and quality of data that can be collected, stored in recoverable formats, and analyzed.

Such technology is expensive, however, and there are still myriad small and focused studies that make use of traditional technologies and even manual implementation.

The potential of new technologies is, however, truly amazing. Tag systems for automated toll collection—such as E-Z Pass, now used on the east coast from New York to Washington DC—could also be used to track vehicle movements throughout the system by placing high-speed sensors at periodic intervals along major highways. The potential to gain tremendous insight into travel characteristics and origin-destination patterns is immense. So too, however, are the legal and moral questions raised in the use of such technologies. The issue of "big brother" watching the detailed daily movements of private individuals is one of great social significance.

Global positioning systems can be used to track vehicle fleets, such as trucks and rail cars. They can also be used as individual vehicle guidance systems, now available on most cars and most likely to become standard equipment within a decade. The next level of technology will be to overlay these positioning and guidance systems with current information on traffic conditions, including accidents and incidents, to provide real-time assessments of the best routes available between points A and B.

The rate of progress in information and data technologies far outstrips the pace at which textbooks and other literature can be updated. Traffic engineers, therefore, have a responsibility to keep up with the state of the art in traffic data collection, reduction, and analysis developments so that they may apply the best possible tools to obtain needed information concerning traffic systems.

8.1.2 Types of Studies

It would be literally impossible to list all of the studies in which traffic engineers get involved. Some of the most common include:

1. *Volume studies.* Traffic counts are the most basic of traffic studies and are the primary measure of demand; virtually all aspects of traffic engineering require volume as an input, including highway planning and design, decisions on traffic control and operations, detailed signal timing, and others.

2. *Speed studies.* Speed characteristics are strongly related to safety concerns and are needed to assess the viability of existing speed regulations and/or to set new ones on a rational basis.

3. *Travel-time studies.* Travel times along significant sections of roadway or throughout entire

systems constitute a major measure of quality of service to motorists and passengers; many demand-forecasting models also require good and accurate travel-time measures.

4. *Delay studies.* Delay is a term that has many meanings, as will be discussed in later chapters; in essence, it is the part or parts of travel time that users find particularly annoying, such as stopping at a traffic signal or because of a mid-block obstruction.

5. *Density studies.* Density is rarely directly measured but is computed from measurements of flow rate and speed at important locations; modern sensor technology can measure "occupancy" over a detector, which can be converted to a density.

6. *Accident studies.* As traffic safety is the primary responsibility of the traffic engineer, the focused study of accident characteristics, in terms of systemwide rates, relationships to causal factors and at specific locations, is a critically important function.

7. *Parking studies.* These involve inventories of parking supply and a variety of counting methodologies to determine accumulations within defined areas and duration distributions; interview techniques are also important in determining attitudinal factors in the use of parking facilities and the selection among alternative destinations.

8. *Goods movement and transit studies.* Inventories of existing truck-loading facilities and transit systems are important descriptors of the transportation system. As these elements can be significant causes of congestion, proper planning and operational policies are a significant need.

9. *Pedestrian studies.* Pedestrians are a part of the traffic system traffic engineers must manage; their characteristics in using crosswalks at signalized and unsignalized locations constitute a required input to many analyses; interview approaches can be used to assess behavioral patterns and to obtain more detailed information.

10. *Calibration studies.* Traffic engineering uses a variety of basic and not-so-basic models and relationships to describe and analyze traffic; studies are needed to provide for proper calibration of these to ensure that they are reasonably representative of the conditions they claim to describe.

11. *Observance studies.* Studies on the effectiveness of various traffic controls are needed to assess how well controls have been designed and implemented; rates of observance and violation are critical inputs in the evaluation of control measures.

This text has major chapters on volume studies; speed, travel time, and delay studies; accident studies; and parking studies. Other types of studies are noted in passing, but the engineer should consult other sources for detailed descriptions of procedures and methodologies in these areas.

8.2 Volume Characteristics

The most fundamental measurement in traffic engineering is counting—counting vehicles, passengers, and/or people. Various automated and manual counting techniques are used to produce estimates of:

- Volume
- Rate of flow
- Demand
- Capacity

Sometimes these are used in conjunction with other measures or conditions. The four parameters listed are closely related, and all are expressed in terms of the same or similar units. They are *not*, however, the same.

1. *Volume* is the number of vehicles (or persons) passing a point during a specified time period, which is usually one hour, but need not be.

2. *Rate of flow* is the rate at which vehicles (or persons) pass a point during a specified time period less than one hour, expressed as an equivalent hourly rate.

3. *Demand* is the number of vehicles (or persons) that desire to travel past a point during a specified period (also usually one hour). Demand is frequently higher than actual volumes where congestion exists. Some trips divert to alternative routes, while other trips are simply not made.

4. *Capacity* is the maximum rate at which vehicles can traverse a point or short segment during a specified time period. It is a characteristic of the roadway. Actual volume can never be observed at levels higher than the true capacity of the section. However, such results may appear, because capacity is most often estimated using standard analysis procedures of the *Highway Capacity Manual* [1]. These estimates may indeed be too low for some locations.

The field measurement of each of these, however, relies upon counting vehicles at appropriate locations. This chapter presents basic methodologies for counting studies ranging from isolated single-location studies to ongoing statewide counting programs, as well as information on common characteristics of traffic volumes.

8.2.1 Volume, Demand, and Capacity

It has been noted that volume, demand, and capacity are three different measures, even though all are expressed in the same units and may relate to the same location. In practical terms, volume is what *is*, demand is what motorists would like *to be*, and capacity is the physical limit of what *is possible*. In very simple terms, if vehicles were counted at any defined location for one hour:

- Volume would be the number of vehicles counted passing the study location in the hour.

- Demand would be the volume plus the vehicles of motorists wishing to pass the site during the study hour who were prevented from doing so by congestion. The latter would include motorists in queue waiting to reach the study location, motorists using alternative routes to avoid the congestion around the study location, and motorists deciding not to travel at all due to the existing congestion.

- Capacity would be the maximum volume that could be accommodated by the highway at the study location.

Consider the illustration of Figure 8.1. It shows a classic "bottleneck" location on a freeway, in this case consisting of a major merge area. For each approaching leg, and for the downstream freeway section, the actual volume (v), the demand (d), and the capacity (c) of the segment are given. Capacity is the primary constraint on the facility. As shown in Figure 8.1, the capacity is

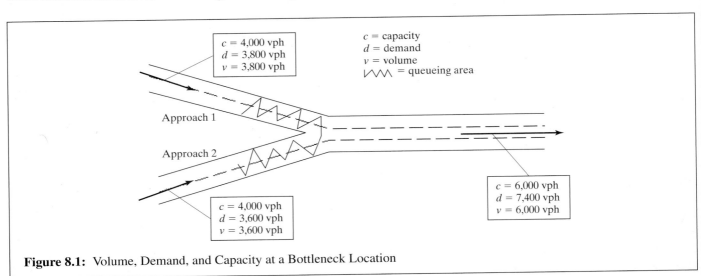

Figure 8.1: Volume, Demand, and Capacity at a Bottleneck Location

2,000 veh/h/ln, so that the capacity of the two-lane approach legs are 4,000 veh/h each, while the capacity of the downstream freeway, which has three lanes, is 6,000 veh/h.

Assuming that the stated capacities are correct, no volume in excess of these capacities can ever be counted. Simply put, you can't carry six gallons of water in a five-gallon bucket. Therefore, it is informative to consider what would be observed for the situation as described. On Approach 1, the true demand is 3,800 veh/h and the capacity is 4,000 veh/h. On Approach 2, the true demand is 3,600 veh/h and the capacity is also 4,000 veh/h. There is no capacity deficiency on either approach. Downstream of the merge, however, the capacity is 6,000 veh/h, but the sum of the approaching demands is 3,800 + 3,600 = 7,400 veh/h. This exceeds the capacity of the segment.

Given this scenario, what can we expect to observe?

- Any volume count downstream of the merge cannot exceed 6,000 veh/h for as long as the illustrated conditions exist. A count of 6,000 veh/h is expected.

- Because of the capacity deficiency downstream of the merge, a queue of vehicles will begin to form and propagate upstream on both approaches.

- If a count of entering vehicles on both approaches is taken upstream of the forming queues, the true demand would be counted on each approach, assuming that there has been no diversion of vehicles to alternative routes.

- If a count of approaching vehicles is taken within the forming queues, they would be unstable in both time and space, but their total should not exceed 6,000 veh/h, the capacity of the downstream freeway section.

A final question is also interesting: Given that queues are observed on both approaches, is it reasonable to assume that the downstream count of 6,000 veh/h is a *direct measurement* of the capacity of the section? This issue involves a number of subtleties.

The existence of queues on both approaches certainly suggests that the downstream section has experienced

capacity flow. Capacity, however, is defined as the maximum flow rate that can be achieved under stable operating conditions (i.e., without breakdown). Thus, capacity would most precisely be the flow rate for the period immediately preceding the formation of queues. After the queues have formed, flow is in the "queue discharge" mode, and the flow rates and volumes measured may be equal to, less than, or even more than capacity. In practical terms, however, the queue discharge capacity may be more important than the stable-flow value, which is, in many cases, a transient that cannot be maintained for long periods of time.

As can be seen from this illustration, volume (or rate of flow) can be counted anywhere and a result achieved. In a situation where queuing exists, it is reasonable to assume that downstream flows represent either capacity or queue discharge conditions. Demand, however, is much more difficult to address. While queued vehicles can be added to counts, this is not necessarily a measure of true demand. True demand contains elements that go well beyond queued vehicles at an isolated location. Determining true demand requires an estimation of how many motorists changed their routes to avoid the subject location. It also requires knowledge of motorists who either traveled to alternative destinations or who simply decided to stay home (and not travel) as a result of congestion.

Figure 8.2 illustrates the impact of a capacity constraint on traffic counts. Part (a) shows a plot of demand and capacity. The demand shown would be observable if it were not clear that a capacity constraint is present. Part (b) shows what will actually occur. Volume can never rise to a level higher than capacity. Thus, actual counts peak at capacity. Since not all vehicles arriving can be accommodated, the peak period of flow is essentially lengthened until all vehicles can be served. The result is that observed counts will indicate that the peak flow rate is approximately the same as capacity and that it occurs over an extended period of time. The volume distribution looks as if someone took the demand distribution and flattened it out with their hand.

The difference between observed volume counts and true demand can have some interesting consequences when the difference is not recognized and

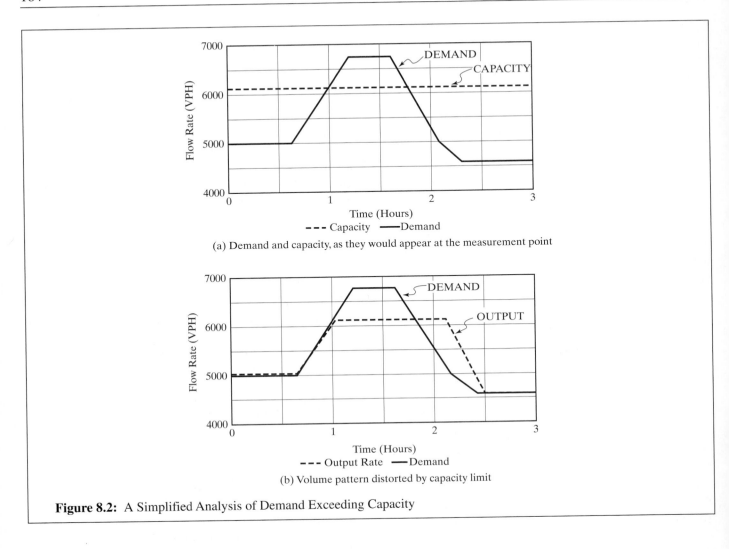

Figure 8.2: A Simplified Analysis of Demand Exceeding Capacity

included in planning and design. Figure 8.3 illustrates an interesting case of a freeway section consisting of four ramps.

From Figure 8.3, the arriving demand volume on Segment 3 of 3,700 veh/h, exceeds its capacity of 3,400 veh/h. From the point of view of counts taken within each segment, 3,400 veh/h will be observed in Segment 3. Because only 3,400 veh/h are output from Segment 3, the downstream counts in Segments 4 and 5 will be lower than their true demand. In this case, the counts shown reflect a proportional distribution of volume to the various ramps, using the same distribution as reflected in the demand values. The capacities of Segments 4 and 5 are not exceeded by these counts. Upstream counts in Segments 1 and 2 will

be unstable and will reflect the transient state of the queue during the count period.

Assume that as a result of this study, a decision is made to add a lane to Segment 3, essentially increasing its capacity to a value larger than the demand of 3,700 veh/h. Once this is done, the volume now discharged into Segment 4 is 3,200 veh/h—more than the capacity of 3,000 veh/h. This secondary bottleneck, often referred to as a "hidden bottleneck," was not apparent in the volume data originally obtained. It was not obvious, because the existing demand was constrained from reaching the segment due to an upstream bottleneck. Such a constraint is often referred to as "demand starvation."

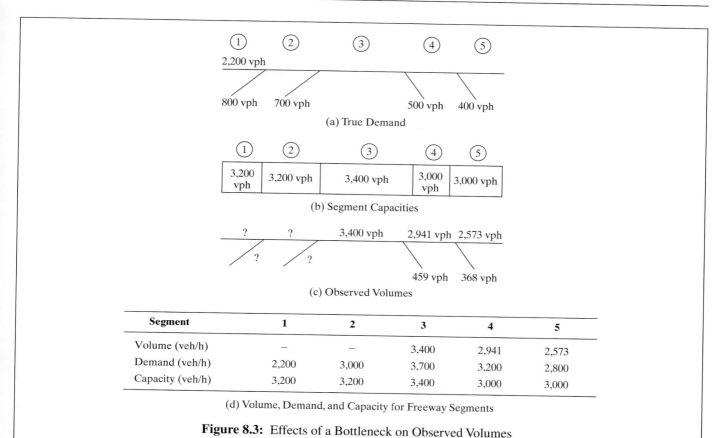

(a) True Demand

(b) Segment Capacities

(c) Observed Volumes

Segment	1	2	3	4	5
Volume (veh/h)	–	–	3,400	2,941	2,573
Demand (veh/h)	2,200	3,000	3,700	3,200	2,800
Capacity (veh/h)	3,200	3,200	3,400	3,000	3,000

(d) Volume, Demand, and Capacity for Freeway Segments

Figure 8.3: Effects of a Bottleneck on Observed Volumes

In designing corrective highway improvements, it is critical that all downstream points be properly evaluated to identify such hidden bottlenecks. In the case illustrated, the improvement project would have to address *both* the existing and hidden bottlenecks to achieve a successful result.

The case of a freeway bottleneck is relatively simple to analyze, because the number of entry and exit points are limited and the number of available alternative routes is generally small. On arterials, however, the situation is far more complex, as every intersection represents a diversion opportunity and the number of alternative routes is generally quite large. Thus, the demand response to an arterial bottleneck is much harder to discern. An arterial may also have a number of overlapping bottlenecks, further complicating the analysis. If several consecutive signalized intersections, for example, are failing, it is difficult to trace the impacts. Is an upstream signal apparently failing because it is inadequate, or because a queue from the downstream signal has blocked the intersection?

On arterial systems, it is often impossible to measure existing demand, except in cases where there are no capacity constraints. Later in this chapter, a method for discerning intersection approach demand from observed volumes in a capacity-constrained case is discussed. It applies, however, only to an isolated breakdown and does not account for the effects of diversion. Congestion in a surface street network severely distorts demand patterns, and observed volumes are more a reflection of capacity constraints than true demand.

The student is cautioned that the terms "volume" and "demand" are often used imprecisely, even in the technical literature. It is often necessary to discern the true meaning of these terms from the context in which they are used.

In the final analysis, volume counts always result in an observation of "volume." Depending upon the circumstances, observed volumes may be equivalent to demand, to capacity, or to neither. The traffic engineer must, however, gain sufficient insight through counting studies and programs to recognize which situation exists and to properly incorporate this into the interpretation of the study data and the development of improvement plans.

8.2.2 Volume Patterns and Characteristics

If traffic distributed itself uniformly amongst the $365 \times 24 = 8{,}760$ hours of the year, there is not a location in the nation that would experience congestion or significant delay. The problem for traffic engineers, of course, is that there are strong peaks during a typical day, fueled primarily by commuters going to and from work. Depending upon the specific region and location, the peak hour of the day typically contains from 10 to 15% of the 24-hour volume. In remote or rural areas, the percentage can go much higher, but the volumes are much lower in these surroundings.

The traffic engineer, therefore, must deal with the travel preferences of our society in planning, designing, and operating highway systems. In some dense urban areas, policies to induce spreading of the peak have been attempted, including the institution of flex-hours or days and/or variable pricing policies for toll and parking facilities. Nevertheless, the traffic engineer must still face the fundamental problem: traffic demand varies in time in ways that are quite inefficient. Demand varies by time of day, by day of the week, by month or season of the year, and in response to singular events (both planned and unplanned) such as construction detours, accidents or other incidents, and even severe weather. Modern intelligent transportation system (ITS) technologies will increasingly try to manage demand on a real-time basis by providing information on routes, current travel times, and related conditions directly to drivers. This is a rapidly growing technology sector, but its impacts have not yet been well documented.

One of the many reasons for doing volume studies is to document these complex variation patterns and to evaluate the impact of ITS technologies and other measures on traffic demand.

Hourly Traffic Variation Patterns: The Phenomenon of the Peak Hour

When hourly traffic patterns are contemplated, we have been conditioned to think in terms of two "peak hours" of the day: morning and evening. Dominated by commuters going to work in the morning (usually between 7 AM and 10 AM) and returning in the evening (usually between 4 PM and 7 PM), these patterns tend to be repetitive and more predictable than other facets of traffic demand. This so-called "typical" pattern holds only for weekday travel, and modern evidence may suggest that this pattern is not as typical as we have been inclined to accept.

Figure 8.4 shows a number of hourly variation patterns documented in the *Highway Capacity Manual* [1], compiled from References [2] and [3]. In part (a) of Figure 8.4, hourly distributions for rural highways are depicted. Only the weekday pattern on a local rural route displays the expected AM and PM peak patterns. Intercity, recreational, and local weekend traffic distributions have only a single, more dispersed peak occurring across the mid- to late afternoon. In part (b), weekday data from four urban sites is shown in a single direction. Sites 1 and 3 are in the opposite direction from Sites 2 and 4, which are only two blocks apart on the same facility. While Sites 2 and 4 show clear AM peaks, traffic after the peak stays relatively high and surprisingly uniform for most of the day. Sites 1 and 3, in the opposite direction, show evening peaks, with Site 3 also displaying considerable off-peak hour traffic volume. Only Site 1 shows a strong PM peak with significantly less traffic during other portions of the day.

The absence of clear AM and PM peaks in many major urban areas is a spreading phenomenon. On one major facility, the Long Island Expressway (I-495) in New York, a recent study showed that on a typical weekday, only one peak was discernible in traffic volume data—and it lasted for 10 to 12 hours per day. This characteristic is a direct result of system capacity constraints. Everyone who would like to drive during the normal peak hours cannot be accommodated. Because of this, individuals begin to make travel choices that allow them to increasingly travel during the "off-peak" hours. This process continues until off-peak periods are virtually impossible to separate from peak periods.

Figure 8.4(b) displays another interesting characteristic of note. The outer lines of each plot show the 95% confidence intervals for hourly volumes over the course of one year. Traffic engineers depend on the basic repeatability of peak hour traffic demands. The variation

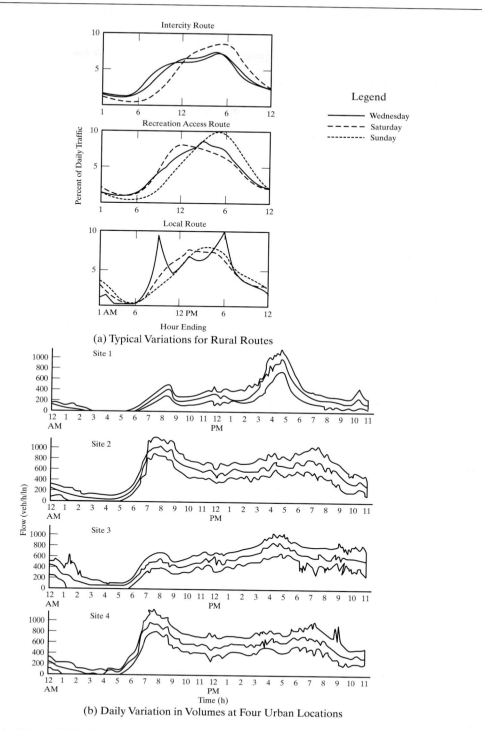

(a) Typical Variations for Rural Routes

(b) Daily Variation in Volumes at Four Urban Locations

Figure 8.4: Examples of Hourly Volume Variation Patterns (Used with permission of Transportation Research Board, *Highway Capacity Manual*, 4th Edition, Washington DC, 2000, Exhibits 8-6 and 8-7, pgs. 8-6 and 8-7, repeated from Refs 2 and 3.)

in these volumes in Figure 8.4(b), however, is not insignificant. During the course of any given year, there are 365 peak hours at any location, one for each day of the year. The question for the traffic engineer is: Which one should be used for planning, design, and operations?

Figure 8.5 shows plots of peak hour volumes (as a percentage of AADT) in decreasing order for a variety of facilities in Minnesota. In all cases, there is clearly a "highest" peak hour of the year. The difference between this highest peak and the bulk of the year's peak hours, however, depends upon the type of facility. The recreational route has greatest disparity. This is not unexpected, as traffic on such a route will tend to have enormous peaks during the appropriate season on weekends, with far less traffic on a "normal" day. The main rural route has less of a disparity, as at least some component of traffic consists of regular commuters. Urban roadways show far less of a gap between the highest hour and the bulk of peak hours.

It is interesting to examine the various peak hours for the types of facilities illustrated in Figure 8.5, which represents data from various facilities in Minnesota. Table 8.1 tabulates the percentage of AADT occurring within designated peak hours for the facility types represented.

The choice of which peak hour to use as a basis for planning, design, and operations is most critical for the recreational access route. In this case, the highest hour of the year carries twice the traffic as the 200th peak hour of the year, and 1.36 times that of the 30th hour of the year. In the two urban cases, the highest hour of the year is only 1.2 times the 200th highest hour.

Historically, the 30th highest hour has been used in rural planning, design, and operations. There are two primary arguments for such a policy: (1) the target demand would be exceeded only 29 times per year and (2) the 30th peak hour generally marks a point where subsequent peak hours have similar volumes. The latter defines a point on many relationships where the curve begins to "flatten out," a range of demands where it is deemed economic to invest in additional roadway capacity.

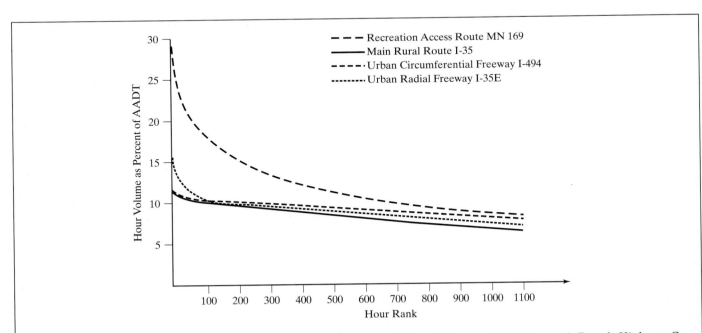

Figure 8.5: Peak Hours as a Percentage of AADT (Used with permission of Transportation Research Board, *Highway Capacity Manual*, 4th Edition, Washington DC, 2000. Exhibit 8-8, pg. 8-8.)

Table 8.1: Key Values from Figure 8.5

Type of Facility	Percent of AADT Occurring in the ___ Peak Hour			
	1st	30th	100th	200th
Recreational Access	30.0%	22.0%	18.0%	15.0%
Main Rural	15.0%	13.0%	10.0%	9.0%
Urban Circumferential Freeway	11.5%	10.5%	10.0%	9.5%
Urban Radial Freeway	11.5%	10.5%	10.0%	9.5%

In urban settings, the choice of a design hour is far less clear and has far less impact. Typical design hours selected range from the 30th highest hour to the 100th highest hour. For the facilities of Figure 8.5, this choice represents a range from 10.5% to 10.0% of AADT. With an AADT of 80,000 veh/day, for example, this range is a difference of only 400 veh/h in demand.

Subhourly Variation Patterns: Flow Rates versus Volumes

In Chapter 5, it was noted that peaking of traffic flows within the peak hour often needed to be considered in design and operations. The peak hour factor (PHF) was defined as a means of quantifying the difference between a maximum flow rate and the hourly volume within the peak hour. Figure 8.6 shows the difference among 5-minute, 15-minute, and peak hourly flow rates from a freeway location in Minnesota.

Flow rates can be measured for almost any period of time. For research purposes, periods from one to five minutes have frequently been used. Very small increments of time, however, become impractical at some point. In a two-second interval, the range of volumes in a given lane would be limited to "0" or "1," and flow rates would be statistically meaningless.

For most traffic engineering applications, 15 minutes is the standard time period used, primarily based on the belief that this is the shortest period of time over which flow rates are "statistically stable." Statistically stable implies that reasonable relationships can be calibrated among flow parameters, such as flow rate, speed, and density. In recent years, there is some thought that 5-minute flow rates might qualify as statistically stable, particularly on freeway facilities. Practice, however, continues to use 15 minutes as the standard period for flow rates.

The choice, however, has major implications. In Figure 8.6, the highest 5-minute rate of flow is 2,200 veh/h/ln; the highest 15-minute rate of flow is 2,050 veh/h/ln; the peak hour volume is 1,630 veh/h/ln. Selecting a 15-minute base period for design and analysis means that, in this case, the demand flow rate (assuming no capacity constraints) would be 2,050 veh/h/ln. This value is 7% lower than the peak 5-minute flow rate and 20% higher than the peak hour volume. In real design terms, these differences could translate into a design with one more or fewer lanes or differences in other geometric and control features. The use of 15-minute flow periods also implies that breakdowns of a shorter duration do not cause the kinds of instabilities that accompany breakdowns extending for 15 minutes or more.

Daily Variation Patterns

Traffic volumes also conform to daily variation patterns that are caused by the type of land uses and trip purposes served by the facility. Figure 8.7 illustrates some typical relationships.

The recreational access route displays strong peaks on Fridays and Sundays. This is a typical pattern for such routes, as motorists leave the city for recreational areas on Fridays, returning on Sundays. Mondays through Thursdays have far less traffic demand, although Monday is somewhat higher than other weekdays due to some vacationers returning after the weekend rather than on Sunday. The suburban freeway obviously caters to commuters. Commuter trips are virtually a mirror image of recreational trips, with peaks occurring on weekdays and lower demand on weekends. The main rural route in this exhibit has a pattern similar to the recreational route, but with less variation between the weekdays and weekends. The route serves

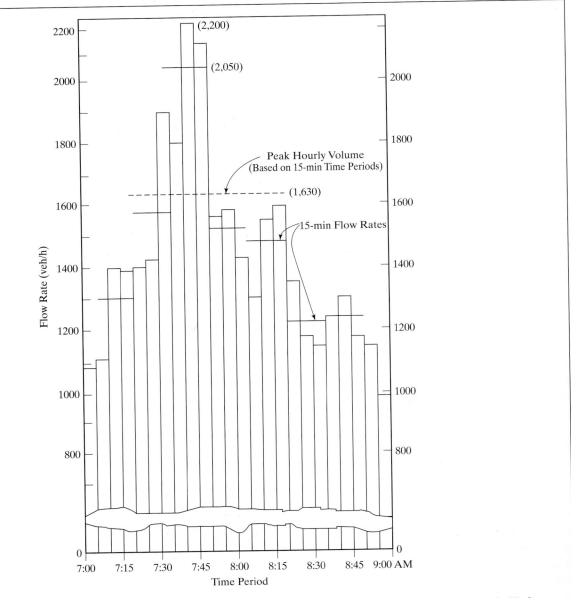

Figure 8.6: Variations of Flow Within the Peak Hour (Used with permission of Transportation Research Board, *Highway Capacity Manual*, 4th Edition, Washington DC, 2000, Figure 8.10, pg. 8.10.)

both recreational and commuter trips, and the mix tends to dampen the amount of variation observed.

Monthly or Seasonal Variation Patterns

Figure 8.8 illustrates typical monthly volume variation patterns. Recreational routes will have strong peaks occurring during the appropriate seasons (i.e., summer for beaches, winter for skiing). Commuter routes often show similar patterns with less variability. In Figure 8.8, recreational routes display monthly ADT's that range from 77% to 158% of the AADT. Commuter routes, while showing similar peaking

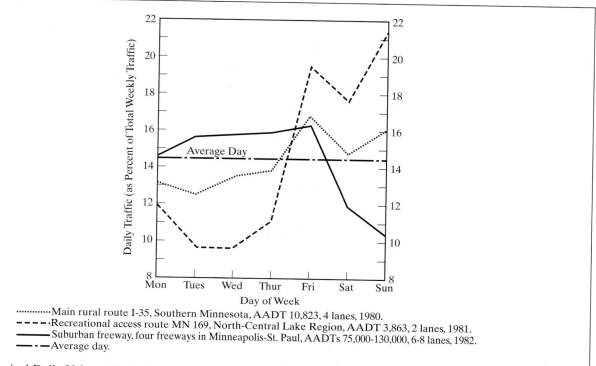

Figure 8.7: Typical Daily Volume Variation Patterns (Used with permission of Transportation Research Board, *Highway Capacity Manual*, 4th Edition, Washington DC, 2000, Exhibit 8-4, pg. 8-5.)

periods, have monthly ADT's ranging from 82% to 119% of AADT.

It might be expected that commuter routes would show a trend opposite to recreational routes (i.e., if recreational routes are peaking in the summer, then commuter routes should have less traffic during those periods). The problem is that few facilities are purely recreational or commuter; there is always some mix present. Further, much recreational travel is done by inhabitants of the region in question; the same motorists may be part of both the recreational and commuter demand during the same months. There are, however, some areas in which commuter traffic does clearly decline during summer recreational months. The distributions shown here are illustrative; different distributions are possible, and they do occur in other regions.

Some Final Thoughts on Volume Variation Patterns

One of the most difficult problems in traffic engineering is that we are continually planning and designing for a demand that represents a peak flow rate within a peak hour on a peak day during a peak season. When we are successful, the resulting facilities are underutilized most of the time.

It is only through the careful documentation of these variation patterns, however, that the traffic engineer can know the impact of this underutilization. Knowing the volume variation patterns governing a particular area or location is critical to finding appropriate design and control measures to optimize operations. It is also important to document these patterns so that estimates of an AADT can be discerned from data taken for much shorter time periods. It is simply impractical to count every location for a full year to determine AADT and related demand factors. Counts taken over a shorter period of time can, however, be adjusted to reflect a yearly average or a peak occurring during another part of the year, if the variation patterns are known and well documented. These concepts will be illustrated and applied in the sections that follow.

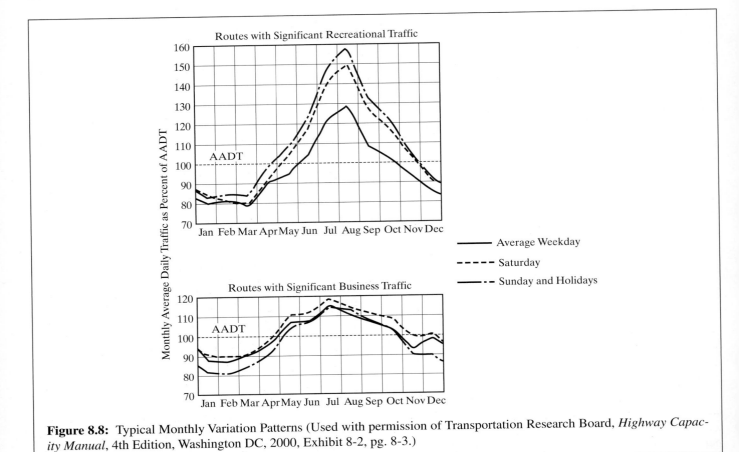

Figure 8.8: Typical Monthly Variation Patterns (Used with permission of Transportation Research Board, *Highway Capacity Manual*, 4th Edition, Washington DC, 2000, Exhibit 8-2, pg. 8-3.)

8.3   Field Techniques for Volume Studies

There are a wide variety of techniques and equipment that can be used to count vehicles as they pass a point. Despite this, most studies are conducted using relatively simple techniques and basic technology. Volume can literally be counted using a satellite. While this is technically feasible, the expense involved generally does not warrant such sophistication.

Many counts are still conducted by hand using field observers with mechanical hand-counters. There are two primary reasons for using manual count techniques for volume studies. Often, a quick count is needed on a deadline. If the need is to obtain PM peak hour vehicle counts at an intersection today or tomorrow, the expense and time spent in gathering and placing automated equipment at the site is often not justified. The second reason for manual counts is that there are a number of parameters that cannot be easily discerned using automated procedures. These include:

- Vehicle occupancy
- Turning movements at intersections
- Vehicle classification

There are multivariate detectors that can classify vehicles by length. If data is desired that separates taxis from passenger cars, however, manual observation is needed.

Road tubes with mechanical or electronic counters are still frequently used. The last option is the use of permanent detectors with remote monitoring of the site.

This option is rarely economic unless a multiparametric study is desired, or if permanent detectors are already in place for other reasons, such as the operation of actuated signals.

8.3.1 Manual Count Techniques

Manual counts are relatively inexpensive and can generally be set up in a short period of time. They also involve some unique challenges, including the coordination of personnel, accurate observation and recording of data, and organization of data in a manner that can be easily understood and interpreted.

Equipment for Manual Counts

There is a variety of equipment available to assist manual counts. The simplest of these involve mechanical hand-counters that can be used individually, or as part of a set. Figure 8.9 shows a single hand counter and four board-mounted counters for use in intersection counts.

Hand-counters usually have four individual registers, as shown in Figure 8.9, but can be acquired with up to six. On a single counter, an observer is able to keep track of four different items. This allows an observer to classify volumes by vehicle type or record lane or movement volumes separately. When placed on a board, four counters can be operated to observe the four approaches to a typical intersection. Simultaneous observation of four approaches is difficult, unless volumes are relatively light. With experience, however, a single observer can record a number of different items at the same time. The details of

intersection counts are discussed in greater detail in a later section.

The major disadvantage of hand counters is that the data must be manually recorded in the field at periodic intervals. This disrupts the count periodically and calls for close coordination between observers to ensure that the data collected is consistent. For the most part, therefore, it is advantageous to use equipment capable of recording and storing the data electronically in a machine-readable form that can be directly downloaded to a computer using a data interface. Figure 8.10 illustrates an electronic counter for intersection studies.

Recording Volume Data During Manual Counts

When manual counts are conducted using hand-counters, observers must periodically enter the data on field sheets. This, of course, requires that they stop counting for some period of time while they manually transfer the data from the counter(s) to paper. The count is therefore not continuous when such methodologies are used.

In order to obtain continuous count information on a common basis, short breaks are introduced into the counting procedure. Such breaks must be systematic and uniform for all observers. The system revolves

Figure 8.10: Illustrative Intersection Counter with Data Interface (Used with permission of JAMAR Technologies, Inc.)

Figure 8.9: Hand-Counters Illustrated (Used with permission of the Denominator Company.)

around the *count period* of the study (i.e., the unit of time for which volumes are to be observed and recorded).

Common count periods are 5 minutes, 15 minutes, and 60 minutes, although (as will be seen) other times can be—and occasionally must be—used.

Short breaks are generally arranged in one of two ways: (1) a portion of each counting period is set aside for a short break or (2) every other count period is used as a short break. In the first case, counts may be taken for 4 minutes out of every 5-minute counting period (or 13 out of 15, or any other convenient combination). Assuming that the rate of flow during the short break is the same as for the observed portion of the counting period, the total count during each counting period can be estimated as:

$$V_{cp} = V_i * CF$$

$$CF = \frac{\text{Counting Period (min)}}{\text{Actual Count Time (min)}} \quad (8\text{-}1)$$

where:

V_{cp} = estimated count for the full counting period, vehs

V_i = actual count for period i, a portion of the counting period, vehs

CF = count adjustment factor

For example, if a survey involved counting 12 minutes out of every 15-minute counting period, the count adjustment factor (CF) would be 15/12 = 1.25. An observed count (for 12 minutes) of 150 vehicles would be adjusted to reflect an estimated count for 15 minutes: 150 × 1.25 = 188 vehs. Note that such estimates are always rounded to the nearest whole vehicle.

When alternating periods are used as breaks, direct observations of the full counting period volume are obtained for every other period. Straight-line interpolation is then used to estimate the count during the break periods.

In practice, it is often necessary to combine the two procedures. Consider the example illustrated in Table 8.2. A single observer is used to count two lanes of traffic on an urban arterial. The observer, who can only count and

Table 8.2: Data from an Illustrative Volume Study

Period	Time (PM)	Actual Counts (vehs) Lane 1	Lane 2	Expanded Counts (× 5/4 = 1.25) Lane 1	Lane 2	Estimated Counts (vehs) Lane 1	Lane 2	Estimated Flow Rates (vehs) Lane 1	Lane 2
1	5:00	24		30.0		30	43	360	516
2	5:05		36		45.0	33	45	396	540
3	5:10	28		35.0		35	47	420	564
4	5:15		39		48.8	36	49	432	588
5	5:20	30		37.5		38	54	456	648
6	5:25		47		58.8	41	59	492	708
7	5:30	36		45.0		45	61	540	732
8	5:35		50		62.5	44	63	528	756
9	5:40	34		42.5		43	61	516	732
10	5:45		48		60.0	46	60	552	720
11	5:50	40		50.0		50	59	600	708
12	5:55		46		57.5	55	58	660	696
Total		**192**	**266**	**240.0**	**332.6**	**496**	**659**		
% in Lane		41.9%	58.1%	41.9%	58.1%	42.9%	57.1%		

classify one lane at a time, observes each lane in alternating periods. The counts must also be made on the basis of 4 out of 5 minutes, so that the observer can manually record the data for each counting period. The result is that an alternating period count is obtained for each lane, but each period was observed for only 4 out of 5 minutes.

Note that these computations were conducted using a spreadsheet. Because the estimation of counting period volumes is a two-step process (expand from 4 to 5 minutes; interpolate for alternating periods), then rounding off to the nearest integer. Rounding is done only at the second step to avoid compounding round-off errors. Also, the count in Period 1 for Lane 2 and Period 12 for Lane 1 could not be interpolated. Because these periods were the first and last respectively, the counts must be *extrapolated* using the trends of the adjacent count periods. This is clearly a somewhat inexact process involving some judgment. For the first period in Lane 2, the trend is relatively stable—a difference of two vehicles in each period exists for Periods 2, 3, and 4.

For the last period in Lane 1, the trend is less clear. From Period 8 to 9, a decrease of 1 vehicle is noted; from Period 9 to 10, an increase of 3 vehicles is noted; from Period 10 to 11, the increase is 4 vehicles. Extrapolating the most immediate trend seems to suggest that the count in Period 12 should be 5 more than the count in Period 11. These extrapolations, however, distort the balance of flows in the two lanes. Actual counts indicate that the balance is 41.9% in Lane 1 and 58.1% in Lane 2. The extrapolations shift the balance to 42.9% and 57.1% respectively. One might argue that the extrapolations should have been based upon keeping the lane distribution constant. This, however, ignores that the lane distribution changes from period to period in any event, and was never directly observed in any one period, given the alternating period system utilized.

It should also be noted that the count procedure employed in this example yields an actual count of 4 minutes out of every 10 minutes for each lane of the facility. Only 40% of the time (and presumably, the counts) has been observed; the remainder has been estimated through expansion and interpolation or extrapolation. It is desirable to keep the percentage of actual observations as high as possible, but practical considerations of personnel and equipment impact this as well.

Personnel Considerations

When conducting manual counts, practical human constraints must be taken into account. Some necessary considerations include:

- All personnel must be properly trained concerning the details of the study and the expected results. This includes making them familiar with field sheets and equipment to be used as well as their specific counting assignments.
- No study should place so many observers at a location that they distract drivers and cause traffic disruptions.
- Unless professional counters are used, a typical observer cannot be expected to count and classify more than one heavy movement or two light movements simultaneously.
- As times must be carefully coordinated at all locations and among all counting personnel, some on-line communication system must be maintained. The time of counts and short breaks must be the same; they are generally centrally timed, with a coordinator telling all personnel when to start and stop counts.
- Some relief personnel must be maintained so that others may take longer breaks when needed.

Field sheets are a major component in controlling the quality of data. Field sheets should be designed to meet the specific needs of the study and are intended to foster easy transfer of data in a consistent fashion. However, all field sheets should have some common elements:

- Location of count must be clearly identified.
- Specific movements, lanes, and/or classification included on the field sheet must be noted.
- Weather, roadway conditions, and traffic incidents or accidents occurring during the count should be noted.

- Observer's name must be noted.
- Date and time of study and of specific counts must be noted.
- Counting periods must be identified and linked to clock time.
- Each sheet should have page x of y noted.

These items identify what is on the sheet. The observer's identity is critical if questions arise concerning the data. Remember, a study may involve many observers, possibly at widely spaced locations. The initial results will be a large pile of field sheets. These must eventually be compiled and presented in a clear and concise format that a traffic engineer can use to help interpret existing conditions and propose improvements where needed.

8.3.2 Portable Count Techniques

All count equipment using pneumatic road tubes or other temporary detection equipment is referred to as "portable." This is because the equipment used can be transported from location to location to conduct various counts as needed.

The most frequently used portable equipment involves the use of pneumatic road tubes. Tubes are fastened across the pavement. As vehicles pass over the tube, an air (or pneumatic pulse) is created in the tube. The pulse can be sensed by a variety of counters connected to the tube. Historically, counters were electromechanical and were capable of accumulating total counts and/or of summarizing them on a punch tape at prescribed intervals. Modern equipment is smaller and lighter and fully electronic, having data interchange and even microwave communications capability. Counters are available which:

- Accumulate a total count for the period of the study.
- Accumulate a total count, recording the total on a daily basis.
- Accumulate a total count, recording the accumulated total at prescribed intervals.

The latter are the most useful, as summaries can be obtained for every hour or even every 15 minutes. Figure 8.11 illustrates counters in use with road tubes. Most modern counters can also be used in conjunction with magnetic loop or other types of detectors.

When road tubes are used, counters are not observing the number of vehicles directly. Rather, they record the number of road-tube actuations. For the most part, this is the number of *axles* that cross the tube. It is not sufficient, however, to merely divide the number of axles by two to estimate vehicle counts. Commercial vehicles may have three, four, five, or more axles. The conversion of axle counts provided by road tubes to vehicle counts requires that the average number of axles per vehicle be measured, at least for some sample period within the study.

Assume that a 10-hour road-tube count resulted in 8,000 actuations. An estimate of the number of vehicles is desired. Table 8.3 shows the results of a sample vehicle classification count of 500 vehicles within the 10-hour study period.

Assuming that this sample is representative, the average number of axles per vehicle at this location is 1,163/500 = 2.33 axles per vehicle. Thus, 8,000 actuations are normally caused by 8,000/2.33 = 3,433 vehicles.

Roadside counters can be hooked up to other types of detectors. Loop detectors, for example, count vehicles directly, while tape switches count axles. The traffic engineer must be aware of what is being counted and must be prepared to institute a logical procedure for adjusting results as needed.

Road tubes can be applied in a variety of ways. If a single road tube is stretched across both directions of travel, a total count of axles in both directions is obtained. The most prevalent use of road tubes is to gather directional counts. This is accomplished by using separate road tubes for each direction of travel. A sequence of road tubes can be used to obtain lane counts, but this is not a common application.

Table 8.3: Sample Classification Count for Volume Study

Number of Axles	Number of Vehicles Observed	Total Number of Axles in Group
2	390	780
3	70	210
4	30	120
5	7	35
6	3	18
Total	**500**	**1,163**

Figure 8.11: Portable Electronic Counter in Use with Road Tubes

Figure 8.12 illustrates the alternatives for setting up road tubes for traffic counts. When used to obtain lane counts, the tubes must be close enough to avoid problems with vehicles changing lanes. Lane counts are obtained by successive subtraction of the results from each counter.

In addition to counting axles, road tubes have some other characteristics that should be noted: (1) if not fastened tautly to the pavement, road tubes will "whip" when crossed and eventually break; all data collection ceases with breakage; (2) when used across multiple lanes, some simultaneous actuations will occur, resulting in slight undercounts. Road tube installations must be periodically checked to ensure that they are still operable. Sample manual counts can be compared to machine counts to estimate the undercount due to simultaneous actuations; however, this is rarely done, as the undercount is related to traffic volumes and is not a significant problem for most applications.

8.3.3 Permanent Counts

As part of every statewide counting program, there are locations at which counting is carried out for 24 hours a day, 365 days a year. These locations use a variety of permanent detectors and data communications interfaces to deliver the results to a remote computer system for storage, manipulation, and analysis.

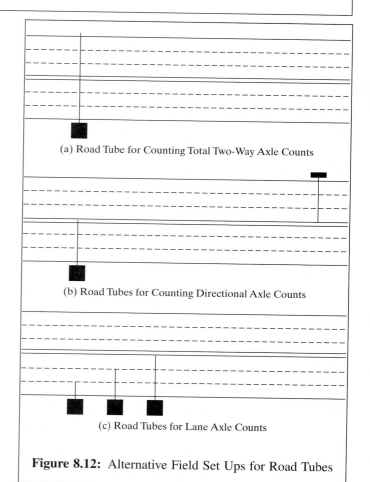

(a) Road Tube for Counting Total Two-Way Axle Counts

(b) Road Tubes for Counting Directional Axle Counts

(c) Road Tubes for Lane Axle Counts

Figure 8.12: Alternative Field Set Ups for Road Tubes

Few permanent count locations are installed specifically for the purpose of conducting traffic counts. Modern actuated signals, toll collection systems, and traffic management systems all contain detectors that can also be employed as permanent counters. While most of these detectors use the data for on-line, real-time monitoring, some can be designated as permanent count sites and the data is then transmitted to and stored at central computer facilities. The vast majority of detectors used are magnetic-loop detectors, although other technologies are available and used in some locations.

The technology of traffic data collection is advancing rapidly. As has been noted, there are portable multiparametric detectors that can be used to count vehicles, observe speeds, obtain occupancy/density data, and classify vehicles by length. The use of video-imaging and virtual detection is also rapidly advancing. The primary impediment to widespread usage of such sophisticated equipment in normal traffic engineering applications is its cost. The use of such systems is, however, already relatively widespread for research purposes.

8.4   Intersection Volume Studies

There is no single location more complex in a traffic system than an at-grade intersection. At a typical four-leg intersection, there are 12 separate movements—left, through, and right from each leg. If a count of intersection volumes is desired, with each movement classified by cars, taxis, trucks, and buses, each count period requires the observation of $12 \times 4 = 48$ separate pieces of data.

When intersections are counted manually (and they often are), observers must be positioned to properly see the movements they are counting. It is doubtful that an inexperienced counter could observe and classify more than one major or two minor movements simultaneously. For heavily used multilane approaches, it may be necessary to use separate observers for different lanes. In manual intersection studies, short-break and alternating-period approaches are almost always combined to reduce the number of observers needed. Rarely, however, can an intersection be counted with fewer than four observers, plus one crew chief to time count periods and breaks.

8.4.1 Arrival versus Departure Volumes: A Key Issue for Intersection Studies

At most intersections, volumes are counted as they depart the intersection. This is done both for convenience and because turning movements cannot be fully resolved until vehicles exit the intersection. While this approach is fine where there is no capacity constraint (i.e. an unstable build-up of queues on the approach), it is not acceptable where demand exceeds the capacity of the approach. In such cases, it is necessary to observe *arrival* volumes, as these are a more accurate reflection of demand.

At signalized intersections, "unstable queue build-up" is detected when vehicles queued during a red interval are not fully cleared during the next green interval. At unsignalized intersections, "unstable queue build-up" can be identified by queues that become larger during each successive counting period.

Direct observation of arrival volumes at an intersection is difficult, as the queue is dynamic. As the queue grows and declines, the point of "arrival" changes. Therefore, the technique used to count arrival volumes is to count departure volumes and the number of queued vehicles at periodic intervals. For signalized approaches, the size of the queue would be recorded *at the beginning of each red phase*. This identifies the "residual queue" of vehicles that arrived during the previous signal cycle but were not serviced. For unsignalized approaches, the queue is counted at the end of each count period. When such an approach is followed, the arrival volume is estimated as:

$$V_{ai} = V_{di} + N_{qi} - N_{q(i-1)} \qquad (8\text{-}2)$$

where: V_{ai} = arrival volume during period i, vehs

V_{di} = departure volume during period i, vehs

N_{qi} = number of queued vehicles at the end of period i, vehs.

$N_{q(i-1)}$ = number of queued vehicles at the end of period $i - 1$, vehs

Estimates of arrival volume using this procedure identify only the localized arrival volume. This procedure *does not* identify diverted vehicles or the number of trips that were not made due to general congestion levels.

Table 8.4: Estimating Arrival Volumes from Departure Counts: An Example

Time Period (PM)	Departure Count (vehs)	Queue Length (vehs)	Arrival Volume (vehs)
4:00–4:15	50	0	50
4:15–4:30	55	0	55
4:30–4:45	62	5	62 + 5 = 67
4:45–5:00	65	10	65 + 10 − 5 = 70
5:00–5:15	60	12	60 + 12 − 10 = 62
5:15–5:30	60	5	60 + 5 − 12 = 53
5:30–5:45	62	0	62 − 5 = 57
5:45–6:00	55	0	55
Total	**469**		**469**

Thus, while arrival volumes do represent localized demand, they do not measure diverted or repressed demand. Table 8.4 shows sample study data using this procedure to estimate arrival volumes.

Note that the study is set up so that the first and last count periods do not have residual queues. Also, the total departure and arrival count are the same, but the conversion from departures to arrivals causes a shift in the distribution of volumes by time period. Based on departure counts, the maximum 15-minute volume is 65 vehs, or a flow rate of 65/0.25 = 260 veh/h. Using arrival counts, the maximum 15-minute volume is 70, or a flow rate of 70/0.25 = 280 veh/h. The difference is important, as the higher arrival flow rate (assuming that the study encompasses the peak period) represents a value that would be valid for use in planning, design, or operations.

8.4.2 Special Considerations for Signalized Intersections

At signalized intersections, count procedures are both simplified and more complicated at the same time. For manual observers, the signalized intersection simplifies counting, as not all movements are flowing at the same time. An observer who can normally count only one through movement at a time could actually count two such movements in the same count period by selecting, for example, the EB and NB through movements. These two operate during different cycles of the signal.

Count periods at signalized intersections, however, must be equal multiples of the cycle length. Further, actual counting times (exclusive of breaks) must also be equal multiples of the cycle length. This is to guarantee that all movements get the same number of green phases within a count period. Thus, for a 60-second signal cycle, a 4-out-of-5-minute counting procedure may be employed. For a 90-second cycle, however, neither 4 nor 5 minutes are equal multiples of 90 seconds (1.5 minutes). For a 90-second cycle, a counting process of 12 out of 15 minutes would be appropriate, as would 4.5 out of 6 minutes.

Actuated signals present special problems, as both cycle lengths and green splits vary from cycle to cycle. Count periods are generally set to encompass a minimum of five signal cycles, using the maximum cycle length as a guide. The actual counting sequence is arbitrarily chosen to reflect this principle, but it is not possible to assure equal numbers of phases for each movement in each count period. This is not viewed as a major difficulty, as the premise of actuated signalization is that green times should be allocated proportionally to vehicle demands present during each cycle.

8.4.3 Presentation of Intersection Volume Data

Intersection volume data may be summarized and presented in a variety of ways. Simple tabular arrays can summarize counts for each count period by movement.

Breakdowns by vehicle type are also most easily depicted in tables. More elaborate graphic presentations are most often prepared to depict peak-hour and/or full-day volumes. Figures 8.13 and 8.14 illustrate common forms for display of peak-hour or daily data. The first is a graphic intersection summary diagram that allows simple entry of data on a predesigned graphic form. The second is an intersection flow diagram in which the thickness of flow lines is based on relative volumes.

8.5   Limited Network Volume Studies

Consider the following proposition: A volume study is to be made covering the period from 6 AM to 12 Midnight on the street network comprising midtown Manhattan (i.e., from 14th Street to 59th Street, 1st Avenue to 12th Avenue). While this is a very big network, including over 500 street links and 500 intersections, it is not the entire city of New York, nor is it a statewide network.

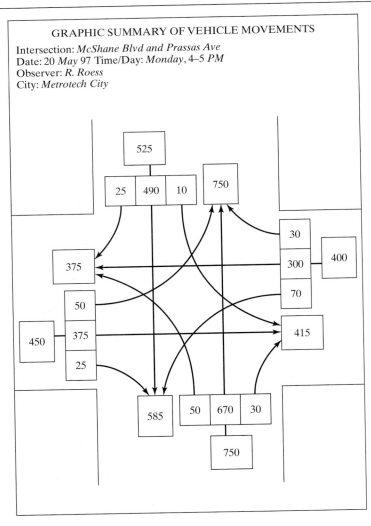

Figure 8.13: Graphic Intersection Summary Diagram

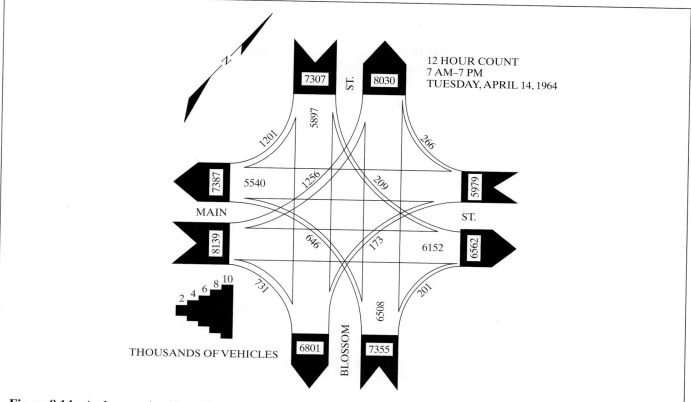

Figure 8.14: An Intersection Flow Diagram (Used with permission of Institute for Transportation Engineers, *Transportation and Traffic Engineering Handbook*, 1st Edition, Washington DC, 1976, pg. 410.)

Nevertheless, the size of the network is daunting for a simple reason: it is virtually impossible to acquire and train sufficient personnel to count all of these locations at the same time. Further, it would be impractically expensive to try and acquire sufficient portable counting equipment to do so. To conduct this study, it will be necessary to employ *sampling* techniques (i.e., not all locations within the study area will be counted at the same time, nor even on the same day). Statistical manipulation based on these samples will be required to produce an hourly volume map of the network for each hour of the intended survey period, or for an average peak period.

Such "limited" networks exist in both small towns and large cities and around other major trip generators, such as airports, sports facilities, shopping malls, and other activity centers. Volume studies on such networks involve individual planning and some knowledge of basic characteristics, such as location of major generators and the nature of traffic on various facilities (local versus through users, for example). The establishment of a reasonable sampling methodology will require judgment based on such local familiarity.

Sampling procedures rely on the assumption that entire networks, or identifiable subportions of networks, have similar demand patterns in time. If these patterns can be measured at a few locations, the pattern can be superimposed on sample measurements from other locations in the network. To implement such a procedure, two types of counts are conducted:

- *Control counts*. Control counts are taken at selected representative locations to measure and quantify demand variation patterns in time. In general, control counts must be maintained continuously throughout the study period.

- *Coverage counts*. Coverage counts are taken at all locations for which data is needed. They are conducted as samples, with each location being counted for only a portion of the study period, in accordance with a preestablished sampling plan.

These types of counts and their use in volume analysis are discussed in the sections that follow.

8.5.1 Control Counts

Because control counts will be used to expand and adjust the results of coverage counts throughout the network under study, it is critical that representative control-count locations be properly selected. The hourly and daily variation patterns observed at a control count must be representative of a larger portion of the network if the sampling procedure is to be accurate and meaningful. It should be remembered that volume variation patterns are generated by land-use characteristics and by the type of traffic, particularly the percentages of through vs. locally generated traffic in the traffic stream. With these principles in mind, there are some general guidelines that can be used in the selection of appropriate control-count locations:

1. There should be one control-count location for every 10 to 20 coverage-count locations to be sampled.

2. Different control-count locations should be established for each class of facility in the network—local streets, collectors, arterials, etc., as different classes of facilities serve different mixes of through and local traffic.

3. Different control-count locations should be established for portions of the network with markedly different land-use characteristics.

These are only general guidelines. The engineer must exercise judgment and use his or her knowledge of the area under study to identify appropriate control-count locations.

8.5.2 Coverage Counts

All locations at which sample counts will be taken are called *coverage counts*. All coverage counts (and control counts as well) in a network study are taken at midblock locations to avoid the difficulty of separately recording turning movements. Each link of the network is counted at least once during the study period. Intersection turning movements may be approximately inferred from successive link volumes, and, when necessary, supplementary intersection counts can be taken. Counts at midblock locations allow for the use of portable automated counters, although the duration of some coverage counts may be too short to justify their use.

8.5.3 An Illustrative Study

The types of computations involved in expanding and adjusting sample network counts is best described by a simple example. Figure 8.15 shows one segment of a larger

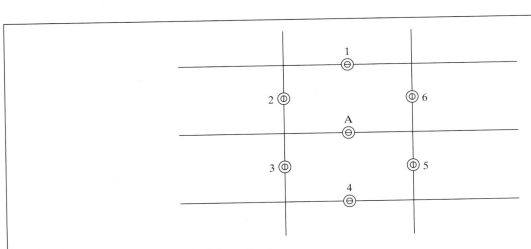

Figure 8.15: A Sample Network Volume Study

network that has been identified as having reasonably uniform traffic patterns in time. The network segment has seven links, one of which has been established as a control-count location. Each of the other six links are coverage-count locations at which sample counts will be conducted. The various proposed study procedures all assume that there are only two field crews or automated counters that can be employed simultaneously in this segment of the network. A study procedure is needed to find the volume on each link of the network between 12:00 Noon and 8:00 PM on a typical weekday. Three different approaches will be discussed. They are typical and not the only approaches that could be used. However, they illustrate all of the expansion and adjustment computations involved in such studies.

A One-Day Study Plan

It is possible to complete the study in a single day. One of the two available crews or set-ups would be used to count Control Location A for the entire 8-hour period of the study. The second crew or set-up would be used to count each of Coverage Locations 1–6 for one hour. The sample data and analysis for this approach is shown in Table 8.5.

Note that full-hour data is shown. This data reflects expansion of actual counts for break periods. If machine-counts were conducted, they would also reflect the conversion of axle-counts to vehicle-counts.

In Table 8.5(b), the control-count data is used to quantify the hourly variation pattern observed. It is now assumed that this pattern applies to all of coverage locations within the network. Thus, a count of 840 vehicles at location 1 would represent 0.117 (or 11.7%) of the 8-hour total at this location. The 8-hour total can then be estimated as 840/0.117 = 7,179 vehicles. Moreover, the peak-hour volume can be estimated as 0.163 × 7,179 = 1,170 vehicles, as the hourly distribution shows that the highest volume hour contains 0.163 (or 16.3%) of the 8-hour volume. Note that this expansion of data results in estimates of 8-hour and peak-hour volumes at each of the seven count locations that represent *the day on which the counts were taken*. Daily and seasonal variations have not been eliminated by this study technique. Volumes for the entire network, however, have been estimated for common time periods.

A Multi-Day Study

In the one-day study approach, each coverage location was counted for one hour. Based on hourly variation patterns documented at the control location, these counts were expanded into 8-hour volume estimates. Hourly variation patterns, however, are not as stable as variations over larger periods of time. For this reason, it could be argued that a better approach would be to count each coverage location for a full eight hours.

Given the limitation to two simultaneous counts due to personnel and/or equipment, such a study would take place over six days. One crew would monitor the control location for the entire period of the study, while the second would count at one coverage location for eight hours on each of six days.

The data and computations associated with a 6-day study are illustrated in Table 8.6. In this case, hourly patterns do not have to be modeled, because each coverage location is counted for every hour of the study period. Unfortunately, the counts are spread over six days, over which volume may vary considerably at any given location. In this case, the control data is used to quantify the underlying *daily* variation pattern. This data is used to *adjust* the coverage data.

Daily volume variations are quantified in terms of adjustment factors defined as follows: the volume for a given day multiplied by the factor yields a volume for the average day of the study period. Stated mathematically:

$$V_a = V_i F_{vi} \qquad (8\text{-}3)$$

where V_a = volume for the average day of the study period, vehs

V_i = volume for day i

F_{vi} = adjustment factor for day i

Using data from the control location, at which the average volume will be known, adjustment factors for each day of the study may be computed as:

$$F_{vi} = {V_a}/{V_i} \qquad (8\text{-}4)$$

Table 8.5: Data and Computations for a One-Day Network Volume Study

Control-Count Data Location A		Coverage-Count Data		
Time (PM)	Count (vehs)	Location	Time (PM)	Count (vehs)
12–1	825	1	12–1	840
1–2	811	2	1–2	625
2–3	912	3	2–3	600
3–4	975	4	4–5	390
4–5	1,056	5	5–6	1,215
5–6	1,153	6	6–7	1,440
6–7	938			
7–8	397			

(a) Data From a One-Day Study

Time (PM)	Count (vehs)	Proportion of 8-Hour Total
12–1	825	825/7,067 = 0.117
1–2	811	811/7,067 = 0.115
2–3	912	912/7,067 = 0.129
3–4	975	975/7,067 = 0.138
4–5	1,056	1,056/7,067 = 0.149
5–6	1,153	1,153/7,067 = 0.163
6–7	938	938/7,067 = 0.133
7–8	397	397/7,067 = 0.056
Total	**7,067**	**1.000**

(b) Computation of Hourly Volume Proportions From Control-Count Data

Location	Time (PM)	Count (Vehs)	Estimated 8-Hr Volume (Vehs)	Estimated Peak Hour Volume (Vehs)
1	12–1	840	840/0.117 = 7,179	× 0.163 = 1,170
2	1–2	625	625/0.115 = 5,435	× 0.163 = 886
3	2–3	600	600/0.129 = 4,651	× 0.163 = 758
4	4–5	390	390/0.149 = 2,617	× 0.163 = 427
5	5–6	1,215	1,215/0.163 = 7,454	× 0.163 = 1,215
6	6–7	1,440	1,440/0.133 = 10,827	× 0.163 = 1,765

(c) Expansion of Hourly Counts

Table 8.6: Data and Computations for a Six-Day Study Option

Control-Count Data Location A		Coverage-Count Data		
Day	**8-Hour Count (vehs)**	**Coverage Location**	**Day**	**8-Hour Count (Vehs)**
Monday 1	7,000	1	Monday 1	6,500
Tuesday	7,700	2	Tuesday	6,200
Wednesday	7,700	3	Wednesday	6,000
Thursday	8,400	4	Thursday	7,100
Friday	7,000	5	Friday	7,800
Monday 2	6,300	6	Monday 2	5,400

(a) Data for a Six-Day Study

Day	8-Hour Count (Vehs)	Adjustment Factor
Monday 1	7,000	7,350/7,000 = 1.05
Tuesday	7,700	7,350/7,700 = 0.95
Wednesday	7,700	7,350/7,700 = 0.95
Thursday	8,400	7,350/8,400 = 0.88
Friday	7,000	7,350/7,000 = 1.05
Monday 2	6,300	7,350/6,300 = 1.17
Total	44,100	
Average	44,100/6 = 7,350	

(b) Computation of Daily Adjustment Factors

Station	Day	8-Hour Count (Vehs)	Adjusted 8-Hour Count (Vehs)
1	Monday 1	6,500	× 1.05 = 6,825
2	Tuesday	6,200	× 0.95 = 5,890
3	Wednesday	6,000	× 0.95 = 5,700
4	Thursday	7,100	× 0.88 = 6,248
5	Friday	7,800	× 1.05 = 8,190
6	Monday 2	5,400	× 1.17 = 6,318

(c) Adjustment of Coverage Counts

where all terms are as previously defined. Factors for the sample study are calibrated in Table 8.6(b). Coverage counts are adjusted using Equation 8-3 in Table 8.6 (c).

The results represent the average 8-hour volumes for all locations for the 6-day period of the study. Seasonal variations are not accounted for, nor are weekend days, which were excluded from the study.

A Mixed Approach: A Three-Day Study

The first two approaches can be combined. If a one-day study is not deemed appropriate due to the estimation of 8-hour volumes based on one-hour observations, and the 6-day study is too expensive, a 3-day study program can be devised in which each coverage location is counted for four hours on one of three days. The control location would have to be counted for the entire 3-day study period; results would be used to calibrate the distribution of volume by 4-hour period and by day.

In this approach, 4-hour coverage counts must be (1) expanded to reflect the full 8-hour study period, and (2) adjusted to reflect the average day of the 3-day study period. Table 8.7 illustrates the data and computations for the 3-day study approach.

Note that in expanding the 4-hour coverage counts to eight hours, the proportional split of volume varied from day to day. The expansions used the proportion appropriate to the day of the count. Since the variation was not great, however, it would have been equally justifiable to use the average hourly split for all three days.

Again, the results obtained represent the particular 3-day period over which the counts were conducted. Volume variations involving other days of the week or seasonal factors are not considered.

The three approaches detailed in this section are illustrative. Expansion and adjustment of coverage counts based upon control observations can be organized in many different ways, covering any network size and study period. The selection of control locations involves much judgment, and the success of any particular study depends upon the quality of the judgment exercised in designing the study. The traffic engineer must design each study to achieve the particular information goals at hand.

8.5.4 Estimating Vehicle Miles Traveled (VMT) on a Network

One output of most limited-network volume studies is an estimate of the total vehicle-miles traveled (VMT) on the network during the period of interest. The estimate is done roughly by assuming that a vehicle counted on a link travels the entire length of the link. This is a reasonable assumption, as some vehicles traveling only a portion of a link will be counted while others will not, depending upon whether they cross the count location. Using the sample network of the previous section, the 8-hour volume results of Table 8.7 and assuming all links are 0.25 miles long, Table 8.8 illustrates the estimation of VMT. In this case, the estimate is the average 8-hour VMT for the three days of the study. It cannot be expanded into an estimate of *annual* VMT without knowing more about daily and seasonal variation patterns throughout the year.

8.5.5 Display of Network Volume Results

As was the case with intersection volume studies, most detailed results of a limited network study are presented in tabular form, some of which have been illustrated herein. For peak hours or for daily total volumes, it is often convenient to provide a network flow map. This is similar to an intersection flow diagram in that the thickness of flow lines is proportional to the volume. An example of such a map is shown in Figure 8.16.

8.6   Statewide Counting Programs

States generally have a special interest in observing trends in AADT, shifts within the ADT pattern, and vehicle-miles traveled. These trends are used in statewide planning and for the programming of specific highway improvement projects. In recent years, there has been growing interest in person-miles traveled (PMT) and in statistics for other modes of transportation. Similar programs at the local and/or regional level are desirable for non-state highway systems, although the cost is often prohibitive.

Following some general guidelines, as in Reference [4] for example, the state road system is divided

Table 8.7: Data and Computations for a Three-Day Study Option

Time (PM)	Monday Count (Vehs)	Monday % of 8 Hours	Tuesday Count (Vehs)	Tuesday % of 8 Hours	Wednesday Count (Vehs)	Wednesday % of 8 Hours	Avg % of 8 Hours
12–4	3,000	42.9%	3,200	42.7%	2,800	43.8%	43.1%
4–8	4,000	57.1%	4,300	57.3%	3,600	56.2%	56.9%
Total	**7,000**	**100.0%**	**7,500**	**100.0%**	**6,400**	**100.0%**	**100.0%**

(a) Control Data and Calibration of Hourly Variation Pattern

Day	8-Hour Control-Count Location A (Vehs)	Adjustment Factor
Monday	7,000	6,967/7,000 = 1.00
Tuesday	7,500	6,967/7,500 = 0.93
Wednesday	6,400	6,967/6,400 = 1.09
Total	**20,900**	
Average	**20,900/3 = 6,967**	

(b) Calibration of Daily Variation Factors

Station	Day	Time (PM)	Count (Vehs)	8-Hour Expanded Count (Vehs)	8-Hour Adjusted Counts (Vehs)
1	Monday	12–4	2,213	2,213/0.429 = 5,159	× 1.00 = 5,159
2	Monday	4–8	3,000	3,000/0.571 = 5,254	× 1.00 = 5,254
3	Tuesday	12–4	2,672	2,672/0.427 = 6,258	× 0.93 = 5,820
4	Tuesday	4–8	2,500	2,500/0.573 = 4,363	× 0.93 = 4,058
5	Wednesday	12–4	3,500	3,500/0.438 = 7,991	× 1.09 = 8,710
6	Wednesday	4–8	3,750	3,750/0.562 = 6,673	× 1.09 = 7,274

(c) Expansion and Adjustment of Coverage Counts

into functional classifications. Within each classification, a pattern of control count locations and coverage count locations is established so that trends can be observed. Statewide programs are similar to limited network studies, except that the network involved is the entire state highway system and the time frame of the study is continuous (i.e., 365 days a year, every year).

Some general principles for statewide programs are:

1. The objective of most statewide programs is to conduct a coverage count every year on every 2-mile segment of the state highway system, with the exception of low-volume roadways (AADT < 100 veh/day). Low-volume roadways usually comprise about 50% of state system mileage and are classified as tertiary local roads.

Table 8.8: Estimation of Vehicle-Miles Traveled on a Limited Network: An Example

Station	8-Hour Count (vehs)	Link Length (mi)	Link VMT (veh-miles)
A	6,967	0.25	1,741.75
1	5,159	0.25	1,289.75
2	5,254	0.25	1,313.50
3	5,820	0.25	1,455.00
4	4,058	0.25	1,014.50
5	8,710	0.25	2,177.50
6	7,274	0.25	1,818.50
Network Total			**10,810.50**

2. The objective of coverage counts is to produce an annual estimate of AADT for each coverage location.

3. One control-count location is generally established for every 20 to 50 coverage-count locations, depending upon the characteristics of the region served. Criteria for establishing control locations are similar to those used for limited networks.

4. Control-count locations can be either *permanent counts* or *major or minor control counts,* which use representative samples. In both cases, control-count locations must monitor and calibrate daily variation patterns and monthly or seasonal variation patterns for the full 365-day year.

5. All coverage counts are for a minimum period of 24 to 48 hours, eliminating the need to calibrate hourly variation patterns.

At permanent count locations, fixed detection equipment with data communications technology is used to provide a continuous flow of volume information.

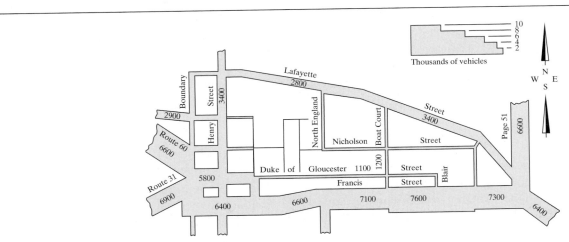

Figure 8.16: A Typical Network Flow Map (Used with permission of Wilbur Smith and Associates, *Traffic, Parking, and Transit—Colonial Williamsburg,* Columbia, South Carolina, 1963.)

Major and minor control counts are generally made using portable counters and road tubes. Major control counts are generally made for one week during each month of the year. Minor control counts are generally made for one 5-day (weekdays only) period in each season.

8.6.1 Calibrating Daily Variation Factors

The illustrative data in Table 8.9 is obtained from a permanent count location. At a permanent count location, data exists for all 52 weeks of the year (i.e., for 52 Sundays, 52 Mondays, 52 Tuesdays, etc.). (Note that in a 365-day year, one day will occur 53 times).

Daily variation factors are calibrated based upon the average volumes observed during each day of the week. The base value for factor calibration is the average of the seven daily averages, which is a rough estimate of the AADT (but not exact, due to the 53rd piece of data for one day of the week). The factors can be plotted, as illustrated in Figure 8.17, and display a clear variation pattern that can be applied to coverage count results.

Note that the sum of the seven daily adjustment factors *does not* add up to 7.00 (the actual total is 7.11). This is because of the way in which the factors are defined and computed. The daily averages are in the denominator of the calibration factors. In effect, the average factor is inverse to the average daily volume, so that the totals would not be expected to add to 7.00.

Daily adjustment factors can also be computed from the results of major and/or minor control counts. In a major control count, there would be 12 weeks of data, one week from each month of the year. The daily averages, rather than representing 52 weeks of data, reflect 12 representative weeks of data. The calibration computations, however, are exactly the same.

8.6.2 Calibrating Monthly Variation Factors

Table 8.10 illustrates the calibration of monthly variation factors (MF) from permanent count data. The monthly factors are based upon monthly ADT's that have been observed at the permanent count location.

Note that the sum of the 12 monthly variation patterns is not 12.00 (the actual sum is 12.29), as the monthly ADTs are in the denominator of the calibration.

Table 8.9 is based on permanent count data, such that the monthly ADTs are directly measured. One 7-day count in each month of the year would produce similar values, except that the ADT for each month would be estimated based on a single week of data, not the entire month. This type of procedure can yield a bias when the week in which the data were collected varies from month to month. In effect, an ADT for a given month is most likely to be observed in the middle of the month (i.e., the 14th to the 16th of any month). This statement is based upon the assumption that the volume trend within each month is unidirectional (i.e., volume grows

Table 8.9: Calibration of Daily Variation Factors

Day	Yearly Average Volume for Day (vehs/day)	Daily Adjustment Factor DF
Monday	1,820	1,430/1,820 = 0.79
Tuesday	1,588	1,430/1,588 = 0.90
Wednesday	1,406	1,430/1,406 = 1.02
Thursday	1,300	1,430/1,300 = 1.10
Friday	1,289	1,430/1,289 = 1.11
Saturday	1,275	1,430/1,275 = 1.12
Sunday	1,332	1,430/1,332 = 1.07
Total	**10,010**	
Estimated AADT	**1,430**	

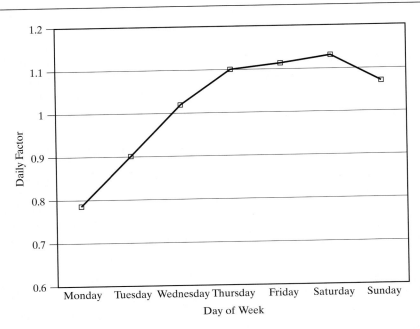

Figure 8.17: Calibration and Plot of Daily Variation Factors

throughout the month or declines throughout the month). Where a peak or low point exists within the month, this statement is not true.

Figure 8.18 illustrates a plot of 12 calibrated monthly variation factors, but one week of data is taken from each month. The daily variation factors are plotted against the midpoint of the week in which the data for the month were taken.

This graph may now be entered at the middle of each month (the 15th), and adjusted factors read from

Table 8.10: Calibration of Monthly Variation Factors

Month	Total Traffic (vehs)	ADT for Month (veh/day)	Monthly Factor (AADT/ADT)
January	19,840	/31 = 640	797/640 = 1.25
February	16,660	/28 = 595	797/595 = 1.34
March	21,235	/31 = 685	797/685 = 1.16
April	24,300	/30 = 810	797/810 = 0.98
May	25,885	/31 = 835	797/835 = 0.95
June	26,280	/30 = 876	797/876 = 0.91
July	27,652	/31 = 892	797/892 = 0.89
August	30,008	/31 = 968	797/968 = 0.82
September	28,620	/30 = 954	797/954 = 0.84
October	26,350	/31 = 850	797/850 = 0.94
November	22,290	/30 = 743	797/743 = 1.07
December	21,731	/31 = 701	797/701 = 1.14
Total	**290,851**	**AADT = 290,851/365 = 797 veh/day**	

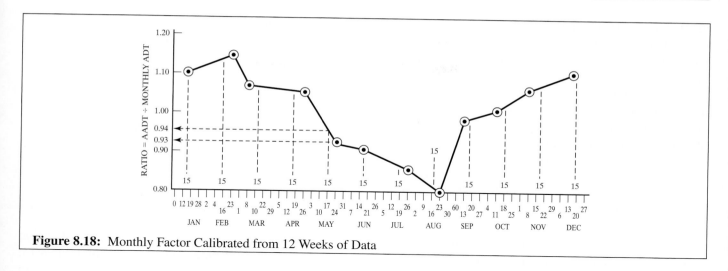

Figure 8.18: Monthly Factor Calibrated from 12 Weeks of Data

the vertical axis. For example, in May the computed factor was 0.93, while the plot indicates that a factor computed for the middle of that month would have resulted in a factor of 0.96. Adjusting the factors in this manner results in a more representative computation based on monthly midpoints.

8.6.3 Grouping Data from Control Count Locations

On state highway networks and systems, particularly in rural areas, it is possible for a broad region to have similar, if not the same, daily and/or monthly adjustment factors. In such regions, spatially contiguous control stations on the same classification of highway may be combined to form a single control group. The average

factors for the group may then be applied over a wide area with similar variation patterns. In general, a statistical standard is applied to such groupings: Contiguous control counts on similar highway types may be grouped if the factors at the individual locations do not differ by more than ± 0.10 from the average for the group.

Consider the example shown in Table 8.11. The daily variation factors for four consecutive control counts on a state highway have been calibrated as shown. It has been hypothesized that the four represent regions with similar daily variation patterns. Average factors have, therefore, been computed for the four grouped stations.

The bold-faced factors indicate cases that violate the statistical rule for grouping (i.e., differences between these factors and the average for the group are

Table 8.11: A Trial Grouping of Four Contiguous Control Stations: Daily Variation Factors

Day	Daily Factor (DF) for Station Number:				Average Daily Factor (DF)
	1	2	3	4	
Monday	1.05	1.00	1.06	0.92	1.01
Tuesday	1.10	1.02	1.06	**0.89**	1.02
Wednesday	1.10	1.05	1.11	0.97	1.06
Thursday	1.06	1.06	1.03	1.00	1.04
Friday	1.01	1.03	1.00	0.91	0.99
Saturday	**0.85**	0.94	0.90	**1.21**	0.98
Sunday	0.83	0.90	0.84	**1.10**	0.92

more than ± 0.10). This suggests that the proposed grouping is not appropriate. One might be tempted to remove Stations 1 and 4 from the group and combine only Stations 2 and 3. The proper technique, however, is to remove one station from the group at a time, as the resulting average factors will change. In this case, a cursory observation indicates that Station 4 does not really display a daily variation pattern similar to the others. This station has its peak traffic (DF < 1.00) occurring during the week, while the other stations have their peak traffic on weekends. Thus, Station 4 is deleted from the proposed grouping and new averages are computed, as illustrated in Table 8.12.

Now, all factors at individual stations are within ± 0.10 of the average for the group. This would be an appropriate grouping of control stations.

8.6.4 Using the Results

It should be noted that groups for daily factors and groups for monthly factors do not have to be the same. It is convenient if they are, however, and it is not at all unlikely that a set of stations grouped for one type of factor would also be appropriate for the other.

The state highway agency will use its counting program to generate basic trend data throughout the state. It will also generate, for contiguous portions of each state highway classification, a set of daily and monthly variation factors that can be applied to any coverage count within the influence area of the subject control grouping. An example of the type of data that would be made available is shown in Table 8.13.

Using these tables, any coverage count for a period of 24 hours or more can be converted to an estimate of the AADT using the following relationship:

$$AADT = V_{24ij} * DF_i * MF_j \qquad (8\text{-}5)$$

where $AADT$ = average annual daily traffic, vehs/day

V_{24ij} = 24-hour volume for day i in month j, vehs

DF_i = daily adjustment factor for day i

MF_j = monthly adjustment factor for month j

Consider a coverage count taken at a location within the area represented by the factors of Table 8.13. A count of 1,000 vehicles was observed on a Tuesday in July. From Table 8.13, the daily factor for Tuesdays (DF) is 1.121, and the factor for July (MF) is 0.913. Then:

$$AADT = 1,000 * 1.121 * 0.913 = 1,023 \text{ vehs/day}$$

8.6.5 Estimating Annual Vehicle-Miles Traveled

Given estimates of AADT for every two-mile segment of each category of roadway in the state system (excluding low-volume roads), estimates of annual vehicle-miles traveled can be assembled. For each segment, the annual vehicle-miles traveled is estimated as:

$$VMT_{365} = AADT * L * 365 \qquad (8\text{-}6)$$

Table 8.12: A Second Trial Grouping of Control Stations Daily Variation Factors

Day	Daily Factor (DF) for Station:			Average Daily Factor (DF)
	1	2	3	
Monday	1.05	1.00	1.06	1.04
Tuesday	1.10	1.02	1.06	1.06
Wednesday	1.10	1.05	1.11	1.09
Thursday	1.06	1.06	1.03	1.05
Friday	1.01	1.03	1.00	1.01
Saturday	0.85	0.94	0.90	0.90
Sunday	0.83	0.90	0.84	0.86

Table 8.13: Typical Daily and Monthly Variation Factors for a Contiguous Area on a State Highway System

Daily Factors (DF)		Monthly Factors (MF)			
Day	Factor	Month	Factor	Month	Factor
Monday	1.072	January	1.215	July	0.913
Tuesday	1.121	February	1.191	August	0.882
Wednesday	1.108	March	1.100	September	0.884
Thursday	1.098	April	0.992	October	0.931
Friday	1.015	May	0.949	November	1.026
Saturday	0.899	June	0.918	December	1.114
Sunday	0.789				

where VMT_{365} = annual vehicle-miles traveled over the segment,

$AADT$ = AADT for the segment, vehs/day, and

L = length of the segment, mi

For any given roadway classification or system, the segment VMTs can be summed to give a regional or statewide total. The question of the precision or accuracy of such estimates is interesting, given that none of the low-volume roads are included and that a real statewide total would need to include inputs for all non-state systems in the state. Regular counting programs at the local level are, in general, far less rigorous than state programs.

There are two other ways that are commonly used to estimate VMT:

- Use the number of registered vehicles with reported annual mileages, adjusting for out-of-state travel.
- Use fuel tax receipts by category of fuel (which relates to categories of vehicles), and estimate VMT using average fuel consumption ratings for different types of vehicles.

There is interest in improving statewide VMT estimating procedures, and a number of significant research efforts have been sponsored on this topic in recent years. There is also growing interest in nationwide person-miles traveled (PMT) estimates, with appropriate modal categories.

8.7 Specialized Counting Studies

There are a number of instances in which simple counting of vehicles at a point, or at a series of points, is not sufficient to provide the information needed. Three examples of specialized counting techniques are (1) origin and destination counts, (2) cordon counts, and (3) screen-line counts.

8.7.1 Origin and Destination Counts

There are many instances in which normal point counts of vehicles must be supplemented with knowledge of the origins and destinations of the vehicles counted. In major regional planning applications, origin and destination studies involve massive home-interview efforts to establish regional travel patterns. In traffic applications, the scope of origin and destination counts are more limited. Common applications include:

- Weaving-area studies
- Freeway studies
- Major activity center studies

Proper analysis of weaving-area operations requires that volume be broken down into two weaving and two non-weaving flows that are present. A total count is insufficient to evaluate performance. In freeway corridors, it is often important to know where vehicles enter and exit the freeway. Alternative routes, for example, cannot be accurately assessed without knowing the underlying pattern of origins and destinations. At major activity centers (sports facilities, airports, regional shopping centers, etc.), traffic planning of access and egress also requires knowledge of where vehicles are coming from when entering the center or going to when leaving the center.

Many ITS technologies hold great promise for providing detailed information on origins and destinations. Automated toll-collection systems can provide data on where vehicles enter and leave toll facilities. Automated license-plate reading technology is used in traffic enforcement and could be used to track vehicle paths through a traffic system. While these technologies continue to advance rapidly, their use in traditional traffic data collection has been much slower due to the privacy issues that such use raises.

Historically, one of the first origin-destination count techniques was called a *lights-on study*. This method was often applied in weaving areas, where vehicles arriving on one leg could be asked to turn on their lights. With the advent of daytime running lights, this methodology is no longer viable.

Conventional traffic origin and destination counts rely primarily on one of three approaches:

- License-plate studies
- Postcard studies
- Interview studies

In a license-plate study, observers (or automated equipment) record the license plate numbers as they pass designated locations. This is a common method used to track freeway entries and exits at ramps. Postcard studies involve handing out color- or otherwise coded cards as vehicles enter the system under study and collecting them as vehicles leave. In both license-plate and postcard studies, the objective is to match up vehicles at their origin and at their destination. Interview studies involve stopping vehicles (with the approval and assistance of police), and asking a short series of questions concerning their trip, where it began, where it is going, and what route will be followed.

Major activity centers are more easily approached, as one end of the trip is known (everyone is at the activity center). Here, interviews are easier to conduct, and license-plate numbers of parked vehicles can be matched to home locations using data from the state Department of Motor Vehicles.

When attempting to match license-plate observations or postcards, sampling becomes a significant issue. If a sample of drivers is recorded at each entry and exit location, then the probability of finding matches is diminished considerably. If 50% of the entering vehicles at Exit 2 are observed, and 40% of the exiting vehicles at Exit 5 are observed, then the number of matches of vehicles traveling from Exit 2 to Exit 5 would be $0.50*0.40 = 0.20$ or 20%. When such sampling techniques are used, separate counts of vehicles at all entry and exit points must be maintained to provide a means of expanding the sample data.

Consider the situation illustrated in Figure 8.19. It shows a small local downtown street network with four

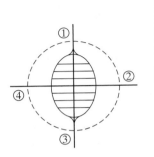

Destination Station	Origin Station				Row Sum T_j	Vol V_j
	1	**2**	**3**	**4**		
1	50	8	20	17	**95**	250
2	10	65	21	10	**106**	310
3	15	12	38	15	**80**	200
4	13	14	18	42	**87**	375
Col Sum T_i	**88**	**99**	**97**	**84**		
Volume V_i	210	200	325	400		**1,135**

Figure 8.19: Data from an Origin-Destination Count Using License-Plate Matching

entry roadways and four exit roadways. Thus, there are $4*4 = 16$ possible origin-destination pairs for vehicles accessing or traveling through the area. The data shown reflect both the observed origins and destinations (using license-plate samples) and the full-volume counts observed on each entry and exit leg.

If the columns and rows are totaled, the sums should be equal to the observed total volumes, assuming that a 100% sample of license plates was obtained at each location. This is obviously not the case. Thus, the origin-destination volumes must be expanded to reflect the total number of vehicles counted. This can be done in two ways: (1) origin-destination cells can be expanded so that the row totals are correct (i.e., match the measured volume), or (2) origin-destination cells can be expanded so that the column totals are correct. Unfortunately, these two approaches will lead to two different sets of origin-destination volumes.

In practice, the average of the two approaches is adopted. This creates an iterative process, as the initial adjustment will still result in column and row totals that are not the same as the measured volumes. Iteration is continued until all row and column totals are within $\pm 10\%$ of the measured volumes.

The cell volumes, representing matched trips from Station i to Station j, are adjusted using factors based upon column closure and row closure:

$$T_{ijN} = T_{ij(N-1)} * \left(\frac{F_i + F_j}{2} \right)$$

$$F_i = \frac{V_i}{T_i}$$

$$F_j = \frac{V_j}{T_j} \qquad (8\text{-}7)$$

where
T_{ijN} = number of trips from Station i to Station j after the Nth iteration of the data (trips)

$T_{ij(N-1)}$ = number of trips from Station i to Station j after the $(N - 1)$th iteration of the data (trips)

T_i = sum of matched trips from Station i (trips)

T_j = sum of matched trips to Station j (trips)

V_i = observed total volume at Station i (vehs)

V_j = observed total volume at Station j (vehs)

F_i = adjustment factor for Origin i

F_j = adjustment factor for Destination j

The actual data of Figure 8.19 serves as the 0th iteration. Each adjustment cycle results in new values of T_{ij}, T_i, T_j, F_i, and F_j. The observed total volumes, of course, remain constant.

Table 8.14 shows the results of several iterations, with the final O-D counts accepted when all adjustment

Table 8.14: Sample O-D Data Expansion

Destination Station	Origin Station				T_j	V_j	F_j
	1	2	3	4			
1	50	8	20	17	**95**	250	2.63
2	10	65	21	10	**106**	310	2.92
3	15	12	38	15	**80**	200	2.50
4	13	14	18	42	**87**	375	4.31
T_i	**88**	**99**	**97**	**84**	**368**		
V_i	210	200	325	400		**1135**	
F_i	2.39	2.02	3.35	4.76			

(a) Field Data and Factors for Iteration 0

(Continued)

Table 8.14: Sample O-D Data Expansion (*Continued*)

Destination Station	Origin Station				T_j	V_j	F_j
	1	**2**	**3**	**4**			
1	125	19	60	63	**267**	250	0.93
2	27	161	66	38	**292**	310	1.06
3	37	27	111	54	**229**	200	0.87
4	44	44	69	190	**347**	375	1.08
T_i	**234**	**251**	**306**	**345**	**1135**		
V_i	210	200	325	400		**1135**	
F_i	0.90	0.80	1.06	1.16			

(b) Initial Expansion of O-D Matrix (Iteration 0)

Destination Station	Origin Station				T_j	V_j	F_j
	1	**2**	**3**	**4**			
1	115	16	60	66	**257**	250	0.97
2	26	150	70	42	**288**	310	1.08
3	33	23	107	55	**218**	200	0.92
4	44	41	74	213	**372**	375	1.01
T_i	**218**	**230**	**311**	**376**	**1135**		
V_i	210	200	325	400		**1135**	
F_i	0.96	0.87	1.05	1.06			

(c) First Iteration of O-D Matrix

Destination Station	Origin Station				T_j	V_j	F_j
	1	**2**	**3**	**4**			
1	111	15	61	67	**254**	250	0.98
2	27	146	75	45	**293**	310	1.06
3	31	21	105	54	**211**	200	0.95
4	43	39	76	220	**378**	375	0.99
T_i	**212**	**221**	**317**	**386**	**1136**		
V_i	210	200	325	400		**1135**	
F_i	0.99	0.90	1.03	1.04			

(d) Second Iteration of O-D Matrix

factors are greater than or equal to 0.90 or less than or equal to 1.10. In this case, the initial expansion of O-D counts was iterated twice to obtain the desired accuracy.

8.7.2 Cordon Counts

A cordon is an imaginary boundary around a study area of interest. It is generally established to define a CBD or other major activity center where the accumulation of vehicles within the area is of great importance in traffic planning. Cordon volume studies require counting all streets and highways that cross the cordon, classifying the counts by direction and by 15- to 60-minute intervals. In establishing the cordon, several principles should be followed:

- The cordoned area must be large enough to define the full area of interest, yet small enough so that accumulation estimates will be useful for parking and other traffic planning purposes.

- The cordon is established to cross all streets and highways at *midblock* locations, to avoid the complexity of establishing whether turning vehicles are entering or leaving the cordoned area.

- The cordon should be established to minimize the number of crossing points wherever possible. Natural or man-made barriers (e.g., rivers, railroads, limited access highways, and similar features) can be used as part of the cordon.

- Cordoned areas should have relatively uniform land use. Accumulation estimates are used to estimate street capacity and parking needs. Large cordons encompassing different land-use activities will not be focused enough for these purposes.

The accumulation of vehicles within a cordoned area is found by summarizing the total of all counts entering and leaving the area by time period. The cordon counts should begin at a time when the streets are virtually empty. As this condition is difficult to achieve, the study should start with an estimate of vehicles already within the cordon. This can be done by circulating through the area and counting parked and circulating vehicles encountered. Off-street parking facilities can be surveyed to estimate their overnight population.

Note that an estimate of parking and standing vehicles may *not* reflect true parking demand if supply is inadequate and many circulating vehicles are merely looking for a place to park. Also, demand discouraged from entering the cordoned area due to congestion is not evaluated by this study technique.

When all entry and exit counts are summed, the accumulation of vehicles within the cordoned area during any given period may be estimated as:

$$A_i = A_{i-1} + V_{Ei} - V_{Li} \qquad (8\text{-}8)$$

where: A_i = accumulation for time period i, vehs

A_{i-1} = accumulation for time period $i-1$, vehs

V_{Ei} = total volume entering the cordoned area during time period i, vehs

V_{Li} = total volume leaving the cordoned area during time period i, vehs

An example of a cordon volume study and the estimation of accumulation within the cordoned area is shown in Table 8.15. Figure 8.20 illustrates a typical presentation of accumulation data, while Figure 8.21 illustrates an interesting presentation of cordon crossing information.

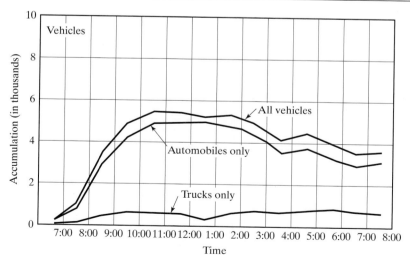

Figure 8.20: Typical Presentation of Accumulation Data (Used with permission of San Diego Area Transportation Study, San Diego CA, 1958.)

Table 8.15: Accumulation Computations for an Illustrative Cordon Study

Time	Vehicles Entering (vehs)	Vehicles Leaving (vehs)	Accumulation (vehs)
4:00–5:00 AM	—	—	250[*]
5:00–6:00 AM	100	20	250 + 100 − 20 = 330
6:00–7:00 AM	150	40	330 + 150 − 40 = 440
7:00–8:00 AM	200	40	440 + 200 − 40 = 600
8:00–9:00 AM	290	80	600 + 290 − 80 = 810
9:00–10:00 AM	350	120	810 + 350 − 120 = 1,040
10:00–11:00 AM	340	200	1,040 + 340 − 200 = 1,180
11:00–12:00 N	350	350	1,180 + 350 − 350 = 1,180
12:00–1:00 PM	260	300	1,180 + 260 − 300 = 1,140
1:00–2:00 PM	200	380	1,140 + 200 − 380 = 960
2:00–3:00 PM	180	420	960 + 180 − 420 = 720
3:00–4:00 PM	100	350	720 + 100 − 350 = 470
4:00–5:00 PM	120	320	470 + 120 − 320 = 270

[*]Estimated beginning accumulation.

8.7.3 Screen-Line Counts

Screen-line counts and volume studies are generally conducted as part of a larger regional origin-destination study involving home interviews as the principal methodology. In such regional planning studies, home interview responses constitute a small but detailed sample that is used to estimate the number of trips per day

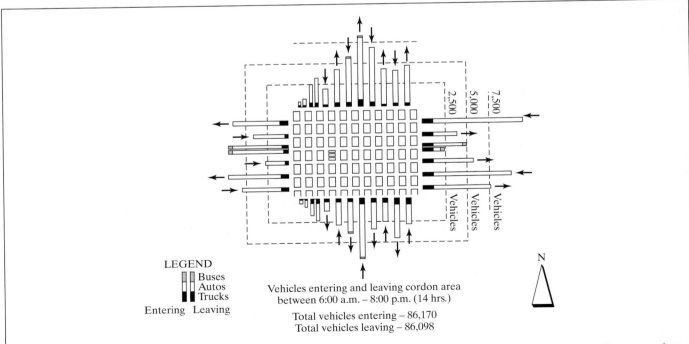

LEGEND

Buses
Autos
Trucks

Entering Leaving

Vehicles entering and leaving cordon area
between 6:00 a.m. – 8:00 p.m. (14 hrs.)

Total vehicles entering – 86,170
Total vehicles leaving – 86,098

N

Figure 8.21: Typical Presentation of Daily Cordon Crossings (Used with permission of San Diego Area Transportation Study, San Diego CA, 1958.)

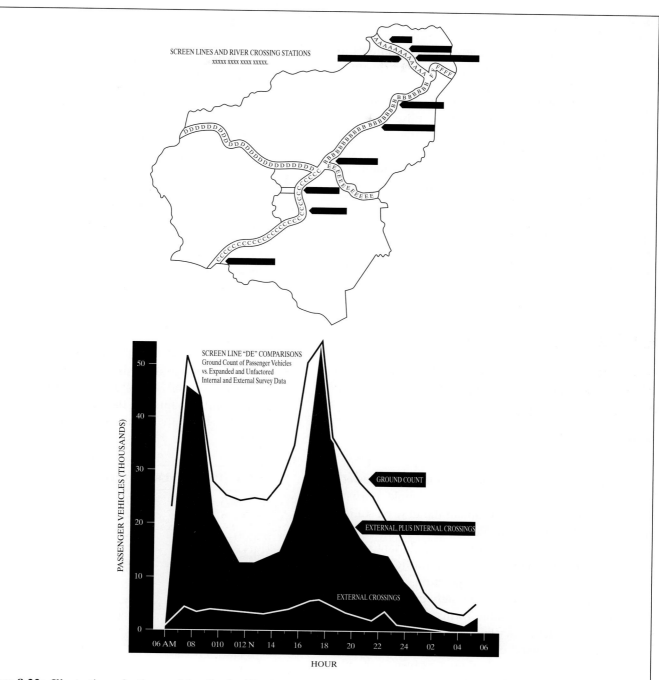

Figure 8.22: Illustration of a Screen-Line Study (Used with permission of Institute of Transportation Engineers, Box, P.C. and Oppenlander, J.C., *Manual of Traffic Engineering Studies*, Washington DC, 1975, Figure 3-35, pg. 43.)

(or some other specified time interval) between defined transportation zones that have been established within the study region. Because home interview samples are small and because additional data is used to estimate trip patterns for those passing through the study area or having only a single trip-end within the study area, it is necessary to use some form of field observations to check on the accuracy of predicted movements.

Screen lines are convenient barriers cutting through the study area with only a limited number of crossing points. Rivers, railroads, limited-access highways, and other features make good screen lines. The zone-to-zone trip estimates of a regional study can be summed in a way that yields the predicted number of trips across the screen line in a defined time period. A screen-line count can then be made to observe the actual number of crossings. The comparison of predicted versus observed crossings provides a means by which predicted zone-to-zone trips can be adjusted.

Figure 8.22 illustrates a study area for which two screen lines have been established. Predicted versus observed crossings are presented in graphic form. The ratio of observed to predicted crossings provides an adjustment factor that can be applied to all zonal trip combinations.

8.8   Closing Comments

The concept is simple: counting vehicles. As reviewed in this chapter, the process is not always simple, nor is the proper use of field results to obtain the desired statistics always straightforward. The field work of volume studies is relatively pedestrian but crucially important. Volume data is one of the primary bases for all traffic engineering analysis, planning, design, and operation.

Volume data must be accurately collected. It must be reduced to understandable forms, and properly analyzed to obtain the prescribed objective of the study. It must then be presented clearly and unambiguously for use by traffic engineers and others involved in the planning and engineering process. No geometric or traffic control design can be effective if it is based on incorrect data related to traffic volumes and true demand. The importance, therefore, of performing volume studies properly cannot be understated.

References

1. *Highway Capacity Manual*, 4th Edition, Transportation Research Board, National Research Council, Washington DC, 2000.

2. *Transportation and Traffic Engineering Handbook*, 2nd Edition, Prentice-Hall, Englewood Cliffs, NJ, 1982.

3. McShane, W. and Crowley, K., "Regularity of Some Detector-Observed Arterial Traffic Volume Characteristics," *Transportation Research Record 596*, Transportation Research Board, National Research Council, Washington DC, 1976.

4. *Traffic Monitoring Guide*, Federal Highway Administration, U.S. Department of Transportation, Washington DC, 1985.

Problems

8-1. The following data was collected during a study of two arterial lanes. Estimate the continuous 15-minute counts for the two-lane roadway as a whole. Find the peak hour, and compute the PHF.

Time (PM)	Lane 1 (Veh)	Lane 2 (Veh)
3:30–3:40	100	–
3:45–3:55	–	120
4:00–4:10	106	–
4:15–4:25	–	124
4:30–4:40	115	–
4:45–4:55	–	130
5:00–5:10	120	–
5:15–5:25	–	146
5:30–5:40	142	–
5:54–5:55	–	140
6:00–6:10	135	–
6:15–6:25	–	130
6:30–6:40	120	–
6:45–6:55	–	110
7:00–7:10	105	–

8-2. What count period would you select for a volume study at an intersection with a signal cycle length of (a) 60 s, (b) 90 s, and (c) 120 s?

8-3. A small-network count was conducted for the network illustrated below using portable machine counts. Because only two field setups were available, the count program was conducted over a period of several days, using Station A as a control location. Using the data presented below, estimate the 12-hour volume (6 AM to 6 PM) at each station for the average day of the study.

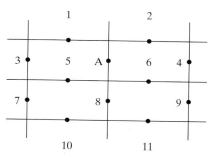

Axle-Counts for Control Station A

Day	Time Period		
	6:00 AM–10:00 AM	10:00 AM–2:00 PM	2:00 PM–6:00 PM
Monday	4,000	2,800	4,400
Tuesday	3,700	2,600	4,000
Wednesday	3,950	2,680	4,200
Thursday	4,200	2,950	4,550

Axle-Counts for Coverage Stations

Station	Day	Time	Count (vehs)
1	Monday	6 AM–9:30 AM	2,100
2	Monday	10 AM–1:30 PM	1,200
3	Monday	2 PM–5:30 PM	930
4	Tuesday	6 AM–9:30 AM	872
5	Tuesday	10 AM–1:30 PM	1,100
6	Tuesday	2 PM–5:30 PM	1,000
7	Wednesday	6 AM–9:30 AM	680
8	Wednesday	10 AM–1:30 PM	1,920
9	Wednesday	2 PM–5:30 PM	1,230
10	Thursday	6 AM–9:30 AM	2,900
11	Thursday	10 AM–1:30 PM	2,000

Sample Vehicle Classification Count

Vehicle Class	Count
2-axle	850
3-axle	75
4-axle	50
5-axle	25

8-4. The following control counts were made at an urban count station to develop daily and monthly variation factors. Calibrate these factors given the data shown below.

24-Hour Daily Volumes

First Week in Month of:	Day of Week			
	Mon	Tue	Wed	Thu
January	2,000	2,200	2,250	2,000
April	1,900	2,080	2,110	1,890
July	1,700	1,850	1,900	1,710
October	2,100	2,270	2,300	2,050

First Week in Month of:	Day of Week		
	Fri	Sat	Sun
January	1,800	1,500	950
April	1,750	1,400	890
July	1,580	1,150	800
October	1,800	1,550	1,010

Standard Monthly Volumes

Third Week in Month of:	Average 24-Hour Count (Vehs)
January	2,250
February	2,200
March	2,000
April	2,100
May	1,950
June	1,850
July	1,800
August	1,700
September	2,000
October	2,100
November	2,150
December	2,300

8-5. The four control stations shown on the next page have been regrouped for the purposes of calibrating daily variation factors. Is the grouping appropriate?

If not, what would an appropriate grouping be? What are the combined daily variation factors for the appropriate group(s)? The stations are located sequentially along a state route.

Daily Variation Factors for Individual Stations

Station	Mon	Tue	Wed	Thu
1	1.05	1.01	0.99	1.10
2	1.10	1.05	0.97	1.10
3	0.99	1.00	0.89	1.01
4	1.04	1.02	1.01	1.09

Daily Variation Factors for Individual Stations

Station	Fri	Sat	Sun
1	1.12	0.91	0.81
2	1.08	0.89	0.88
3	0.89	1.05	1.10
4	1.12	0.89	0.83

8-6. Estimate the annual VMT for a section of the state highway system represented by the variation factors of Table 8.11. The following coverage counts are available for the locations within the section.

Coverage Count Data

Station	Segment Length (Mi)	Coverage Count Date	24-Hour Count (Vehs)
1	2.0	Tue in March	8,765
2	3.0	Mon in September	11,432
3	2.5	Fri in August	15,376
4	4.0	Sat in May	20,010
5	2.0	Wed in December	8,111
6	1.6	Wed in January	10,520

8-7. The following origin and destination results were obtained from sample license plate observations at five locations. Expand and adjust the initial trip-table results to reflect the full population of vehicles during the study period.

Initial Origin and Destination Matches from Sample License-Plate Observations

Destination Station	Origin Station					Total Destination Count (Vehs)
	1	2	3	4	5	
1	50	120	125	210	75	1,200
2	105	80	143	305	100	2,040
3	125	100	128	328	98	1,500
4	82	70	100	125	101	985
5	201	215	180	208	210	2,690
Total Origin Count (Vehs)	**1,820**	**1,225**	**1,750**	**2,510**	**1,110**	**8,415**

CHAPTER
9

Speed, Travel Time, and Delay Studies

9.1 Introduction

Speed, travel time, and delay are all related measures that are commonly used as indicators of performance for traffic facilities. All relate to a factor that is most directly experienced by motorists: how long does it take to get from A to B? Motorists have the obvious desire to complete their trip in the minimum time consistent with safety. The performance of a traffic facility is often described in terms of how well that objective is achieved.

In the *Highway Capacity Manual* [1], for example, average travel speed is used as a measure of effectiveness for arterials and for two-lane rural highways. Control delay is the measure of effectiveness for signalized and STOP-controlled intersections. While freeways use density as a primary measure of effectiveness, speed is an important component of the evaluation of freeway system operation.

Thus, it is important that traffic engineers understand how to measure and interpret data on speed, travel time, and delay in ways that yield a basic understanding

of the quality of operations on a facility, and in ways that directly relate to defined performance criteria. Speed is also an important factor in evaluating high accident locations as well as in other safety-related investigations.

Speed is inversely related to travel time. The reasons and locations at which speeds or travel times would be measured are quite different. Speed measurements are most often taken at a point (or a short section) of roadway under conditions of free flow. The intent is to determine the speeds that drivers select, unaffected by the existence of congestion. This information is used to determine general speed trends, to help determine reasonable speed limits, and to assess safety. Such studies are referred to as "spot speed studies," because the focus is on a designated "spot" on a facility.

Travel time must be measured over a distance. While spot speeds can indeed be measured in terms of travel times over a short measured distance (generally < 1,000 ft), most travel time measurements are made over a significant length of a facility. Such studies are generally done during times of congestion specifically to measure or quantify the extent and causes of congestion.

In general terms, delay is a portion of total travel time. It is a portion of travel time that is particularly identifiable and unusually annoying to the motorist. Delay along an arterial, for example, might include stopped time due to signals, mid-block obstructions, or other causes of congestion.

At signalized and STOP-controlled intersections, delay takes on more importance, as travel time is difficult to define for a point location. Unfortunately, delay at intersections, specifically signalized intersections, has many different definitions, and the traffic engineer must be careful to use measurements and criteria that relate to the same delay definition. Some of the most frequently used forms of intersection delay include:

- *Stopped-time delay*—the time a vehicle spends stopped waiting to proceed through a signalized or STOP-controlled intersection.

- *Approach delay*—adds the delay due to deceleration to and acceleration from a stop to stopped time delay.

- *Time-in-queue delay*—the time between a vehicle joining the end of a queue at a signalized or STOP-controlled intersection and the time it crosses the STOP line to proceed through the intersection.

- *Control delay*—the total delay at an intersection caused by a control device (either a signal or a STOP-sign), including both time-in-queue delay plus delays due to acceleration and deceleration.

Control delay was a term introduced in the 1985 *Highway Capacity Manual*, and it is used as the measure of effectiveness for signalized and STOP-controlled intersections.

Along routes, another definition of delay may be applied: *travel-time delay* is the difference between the actual travel time traversing a section of highway and the driver's expected or desired travel time. It is more of a philosophic approach, as there are no clearly accurate methodologies for determining the expected travel time of a motorist over a given section of highway. For this reason, it is seldom used for assessing congestion along a highway segment.

Because speeds are generally studied at points under conditions of free flow and travel times and delays are generally studied along sections of roadway under congested conditions, the study techniques for each are quite different. The major sections of this chapter detail current methodologies for conducting these studies.

9.2 Spot Speed Studies

Spot speeds have been previously defined as the average speed of vehicles passing a point on a highway. Recalling the basic discussion of speed measures in Chapter 5, this is the *time mean speed*. As the traffic engineer is interested in conducting spot speed studies under conditions of free flow (i.e., observed speeds are not impeded by volume and density conditions), they are generally not conducted when volumes are in excess of 750–1,000 veh/h/ln on freeways or 500 veh/h/ln on other types of uninterrupted flow facilities.

9.2.1 Speed Definitions of Interest

When the speeds of individual vehicles are measured at a given spot or location, the result is a *distribution* of speeds, as no two vehicles will be traveling at exactly the same speed. The results of the study, therefore, must describe the observed distribution of speeds as clearly as possible. There are several key statistics that are used to describe spot speed distributions:

- *Average or time mean speed*:—the average speed of all vehicles passing the study location during the period of the study.

- *Standard deviation*:—in simplistic terms, the standard deviation of speeds is the average difference between observed speeds and the time mean speed during the period of the study.

- *85th percentile speed*:—the speed below which 85% of the vehicles travel.

- *Median*:—the speed that equally divides the distribution of spot speeds; 50% of observed speeds are higher than the median; 50% of observed speeds are lower than the median.

- *Pace*:—a 10-mi/h increment in speeds that encompasses the highest proportion of observed speeds (as compared with any other 10-mi/h increment).

The desired result of a spot speed study is to determine each of these measures and to determine an adequate mathematical description of the entire observed distribution.

9.2.2 Uses of Spot Speed Data

The results of spot speed studies are used for many different purposes by traffic engineers, including:

- Establishing the effectiveness of new or existing speed limits or enforcement practices.

- Determining appropriate speed limits for application.

- Establishing speed trends at the local, state, and national level to assess the effectiveness of national policy on speed limits and enforcement.

- Specific design applications determining appropriate sight distances, relationships between speed and highway alignment, and speed performance with respect to steepness and length of grades.

- Specific control applications for the timing of "yellow" and "all red" intervals for traffic signals, proper placement of signs, and development of appropriate signal progressions.

- Investigation of high-accident locations at which speed is suspected to be a contributing cause to the accident experience.

This list is illustrative. It is not intended to be complete, as there are myriad situations that may require speed data for a complete analysis. Such studies are of significant importance and are among the tasks most commonly conducted by traffic engineers.

9.2.3 Measurement Techniques

Spot speeds are generally measured using one of two techniques:

- Measurement of travel times as vehicles traverse a short measured distance along the highway

- Use of hand-held or fixed-mounted radar meters

For more complex multiparametric studies, there are detectors that can measure speed and other parameters. Video technologies, such as the Autoscope™, are also rapidly advancing and may be applied to traffic studies that extend over a long period of time or that involve permanent installations for remote monitoring of a given location.

Travel Times Traversing a Short Measured Distance

In the early days of speed enforcement, the primary method used by police was to place two road tubes a short distance apart, using a meter to measure the time between successive actuations to determine speed. The short distance between the two road tubes was called a "trap." This is the origin of the colloquial "speed trap," which now is used to describe any police speed monitoring location.

For traffic engineering studies, the simplest, cheapest, and easiest methodology involves manual use of stopwatches to time vehicles as they traverse an easily-recognized trap. While the trap can be marked with wide tape, natural and/or existing boundaries are most often used. Thus, the time for vehicles to traverse the distance between two pavement cracks or two light standards might be used. In such studies, the observer stands at the location of one of the trap boundaries (usually the entry boundary). This allows the observer to view vehicles crossing the boundary without distortion. It guarantees, however, that the observer views the other (usually the exit) boundary at an angle. This creates a systematic measurement error called "parallax," which is illustrated in Figure 9.1. The observer sees the vehicle crossing the exit boundary that is d feet from the entry boundary

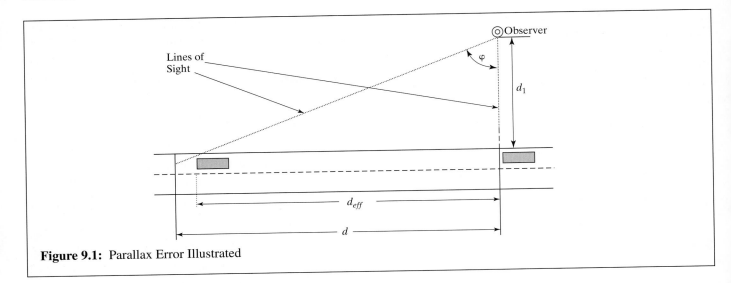

Figure 9.1: Parallax Error Illustrated

when the vehicle is actually only d_{eff} feet from the entry boundary.

While the exhibit shows parallax error occurring from the horizontal angle at which the observer sees the vehicle crossing the exit line, it can also occur in the vertical dimension when measurements are taken from an overpass or other elevated position. The parallax error is easily corrected, as long as the angle of observation, φ, is known. If known, the distance d_{eff} may then be computed as:

$$d_{eff} = d_1 \, \text{Tan}\varphi \qquad (9\text{-}1)$$

The speed of an individual vehicle is then computed as:

$$S_i = \frac{d_{eff}}{t_i} \qquad (9\text{-}2)$$

where: S_i = speed of vehicle i, ft/s

   d_{eff} = actual distance over which travel time of vehicle i is measured, ft

   t_i = times for vehicle i to traverse the distance d_{eff}, s

   d_1 = perpendicular distance of observer from line of travel, ft

   φ = angle between lines of sight, degrees

While parallax is a systematic error and can be corrected as indicated, there is another larger source of error that is incurred when a human observer must depress a manual stopwatch as a vehicle is observed crossing a boundary. A human observer can be a bit early or late in depressing the stopwatch as the vehicle crosses either end of the trap. Fortunately, this error tends to be random, and (with a number of observations) will average out relatively accurately. This error can also be minimized by using traps that are long enough so that such errors will be a small percentage of the actual result.

Assume that a human observer may err by as much as 0.20 s as a vehicle enters or leaves the trap. A total error of 0.20 + 0.20 = 0.40 s could be incurred. If the average time for a vehicle to cross the trap is 1.0 s, the error could be as much as 40% on any given measurement. If a trap three times as long is used, the average travel time becomes 3.0 s. The maximum error in this case would be only 13.3%(0.40/3.0) × 100.

The random error created by human observation of trap travel times can also be eliminated by using a meter to measure the elapsed time between sequential boundaries, as was done in the early days of speed enforcement. Today, however, road tubes are not used. Tapes with wire switches are attached across the pavement at the trap boundaries and connected to a meter.

Radar Meters

The easiest methodology for observing speeds is through the use of either hand-held or vehicle-mounted radar meters. These devices measure speed directly by reflecting an emitted radar wave off an oncoming vehicle and measuring the difference in the frequency between the emitted and reflected radar wave. This difference in frequency is referred to as the "Doppler effect," and is proportional to the speed of the oncoming vehicle.

Radar meters have several practical limitations:

1. As radar waves are in the range of government-regulated frequencies and wave lengths, a Federal Communications Commission (FCC) license must be obtained for each device.

2. While the accuracy of various meters vary, they are generally limited to plus or minus 1–2 mi/h accuracy.

3. They are difficult to conceal, and motorists' general association of radar with a police enforcement technique may cause drivers to slow down, affecting the results of measurements.

4. Accurate measurements are obtained only when the radar wave is reflected directly along the axis of vehicle movement; adjustment requires a good measurement of the angle of wave deflection.

5. Multilane traffic streams are difficult to resolve.

Figure 9.2 shows two examples of modern radar equipment for speed measurements, as well as illustrating the difficulty in resolving multilane traffic streams.

Because the angle of reflection should be kept as shallow as possible, the emitted wave can intercept vehicles in different lanes at widely divergent distances. Where traffic is reasonably busy, this makes it hard to exactly match a speed measurement with a particular vehicle.

Some Practical Measurement Issues

There are a number of practical issues that must be kept in mind when collecting speed data:

- *All field personnel need to make an effort to conceal themselves and their activity.* Once motorists become aware that speeds are being monitored, they assume that it is an enforcement activity and slow down. When this happens, measurements no longer reflect the ambient conditions at the site. Concealment is a particularly difficult issue when radar is being used, as it requires a clear path for the radar waves to travel and be reflected.

- *Under the best of circumstances, it is still virtually impossible to record the speed of every vehicle passing a study site.* Thus, it is critically important that the selected sample be truly representative of the entire traffic stream. Such common failings as trying to measure the fastest and/or slowest speeds, all truck speeds, or other specific element will bias the sample, yielding statistics that do not represent accurately the traffic stream as a whole.

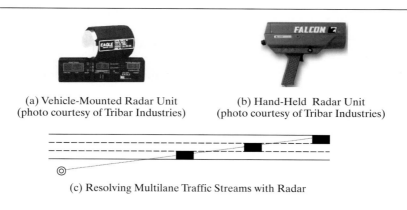

(a) Vehicle-Mounted Radar Unit
(photo courtesy of Tribar Industries)

(b) Hand-Held Radar Unit
(photo courtesy of Tribar Industries)

(c) Resolving Multilane Traffic Streams with Radar

Figure 9.2: Measuring Speeds with Radar Meters

A sample must be randomly selected. Therefore, if it is anticipated that only 1/3 of all vehicle speeds can be observed, then a pattern of observing and recording the speed of every third vehicle in each lane, or a similar procedure, would be adopted.

- *Even in light traffic, vehicle platoons may form.* In platoon flow, only the first vehicle in the platoon is truly selecting his or her speed. Other vehicles are opting to accept the platoon leader's speed. Therefore, in platoon situations, only the first vehicle in the platoon can be observed. As a general guideline, a vehicle is considered to be in platoon if it is spaced less than 200 ft from the lead vehicle at speeds less than 40 mi/h, or 350 ft at speeds of 40 mi/h or higher.

Recording Data in the Field

During field studies, data on individual vehicle speeds is generally not recorded. There are two reasons for this: (1) most measurement techniques (except for the use of two closely-spaced detectors) have limited accuracy ranges for individual measurements (usually limited to ± 1 or ± 2 mi/h) and (2) mathematical descriptions of data (see Chapter 8) cannot describe the occurrence of a single discrete value in a continuous distribution. Rather, continuous mathematical distributions describe the proportion (or probability) of occurrences between "x" and "y" on the distribution. In practical terms, a continuous mathematical distribution cannot describe the probability of a vehicle traveling at a speed of 36.55 mi/h. It can, however, describe the probability of a vehicle traveling at a speed *between* 36 and 37 mi/h.

Because of these characteristics, speed data is generally recorded as the frequency of occurrence within predetermined ranges of speeds. The sample field data sheet of Figure 9.3 illustrates this collection/reduction process. In the left two columns, speed groups of 2 mi/h range are predefined. In practical terms, most studies use speed groups that are between 2 mi/h and 5 mi/h in range. Smaller speed groupings are impractical, and larger speed groupings have a negative effect on the overall accuracy and precision of computations. Individual observations (in this case using a radar meter) are entered as a "tick mark" in the appropriate speed group box, with passenger cars, trucks, and other vehicles considered

separately. After the completion of the field study, the "tick marks" can be tallied and entered into the columns on the right side of the field sheet.

In this case, the sample sizes for trucks and other vehicles are too small for separate analysis, so only the totals for all vehicles are useful. Subsequent sections of this chapter will use the data from Figure 9.3 as a continuing sample problem to illustrate computations on and interpretations of the study results.

The field sheet of Figure 9.3 is also useful if measurements are being made of travel time over a short trap. Each speed boundary can be converted to an equivalent travel time value. For example, if a trap of 1,000 ft were being used, the 40 mi/h speed group boundary may be converted to a travel time value of:

$$t = \frac{1,000}{40 * 1.47} = 17.0 \text{ s}$$

The second set of columns on the left of the field sheet allow for the pre-entry of elapsed times corresponding to each of the speed group limits. In this way, "tick marks" can be entered directly onto the field sheet on the basis of a stopwatch reading or other travel-time device.

9.2.4 Reduction and Analysis of Spot Speed Data

The data summarized in the illustrative field sheet of Figure 9.3 represents, in raw form, the results of the spot speed study. To fully understand and describe the results, a number of techniques are used to both present and analyze the data in a systematic format.

Frequency Distribution Table

The data of Figure 9.3 is reformatted into a frequency distribution table, shown in Table 9.1. This tabular array shows the total number of vehicles observed in each speed group. For the convenience of subsequent use, the table includes one speed group at each extreme for which no vehicles were observed. The "middle speed" (S) of the third column is taken as the midpoint value within the speed group. The use of this value will be discussed in a later section.

LOCATION: _Route 10 @ MP 125.3_
DATE: _July 10, 2003_
TIME: _1:00 - 4:00 PM_

WEATHER CONDITIONS: _Good - Clear, Dry_
ROADWAY SURFACE CONDITIONS: _Asphaltic concrete - good._

SPEED GROUP		TIME GROUP		PASSENGER CARS	TRUCKS	OTHER	TOTALS			
Lower limit (mph)	Upper limit (mph)	Lower limit (secs)	Upper limit (secs)				PC	Trucks	Other	Total
30	32									
32	34									
34	36			II	II	I	2	2	1	5
36	38			III	II		3	2	0	5
38	40			JHT	I	I	5	1	1	7
40	42			JHT JHT	III		10	3	0	13
42	44			JHT JHT JHT III	III		18	3	0	21
44	46			JHT JHT JHT JHT JHT IIII	IIII		29	4	0	33
46	48			JHT JHT JHT JHT JHT JHT JHT II	II	II	42	2	2	46
48	50			JHT JHT JHT JHT JHT JHT JHT JHT JHT JHT	II		60	2	0	62
50	52			JHT JHT JHT JHT JHT JHT JHT II			37	0	0	37
52	54			JHT JHT JHT JHT III			23	1	0	24
54	56			JHT JHT III		I	13	0	1	14
56	58			JHT II	I	I	7	1	1	9
58	60			JHT			5	0	0	5
60	62			II			2	0	0	2
62	64									
64	66									
66	68									
68	70									

METHOD OF MEASUREMENT
x   Radar
_____ Time over measured course length of _____ ft.
   _____ Stop watch/manual
   _____ Road tubes w/timer
   _____ Electronic contact w/timer

Signature _(signed)_   7/10/03

Figure 9.3: Field Data for an Illustrative Spot Speed Study

The fourth column of the table shows the number of vehicles observed in each speed group. This value is known as the *frequency* for the speed group. These values are taken directly from the field sheet of Figure 9.3.

In the fifth column, the percentage of total observations in each speed group is computed as:

$$\% = 100\frac{n_i}{N} \qquad (9\text{-}3)$$

where: n_i = number of observations (frequency) in speed group i

N = total number of observations in the sample

For the 40–42 mi/h speed group, there are 13 observations in a total sample of 283 speeds. Thus, the percent frequency, %, is $100 * (13/283) = 4.6\%$ for this group.

The cumulative percent frequency (cum %) is the percentage of vehicles traveling at or below the highest speed in the speed group:

$$\text{cum}\% = 100\left(\sum_{1-x} n_i \Big/ N\right) \qquad (9\text{-}4)$$

where: x = consecutive number (starting with the lowest speed group) of the speed group for which the cumulative percent frequency is desired

Table 9.1: Frequency Distribution Table for Illustrative Spot Speed Study

Speed Group		Middle Speed S (mi/h)	Observed Freq. in Group n	% Freq. in Group (%)*	Cum. % Freq (%)*	nS**	nS²**
Lower Limit (mi/h)	Upper Limit (mi/h)						
32	34	33	0	0.0%	0.0%	0	0
34	36	35	5	1.8%	1.8%	175	6,125
36	38	37	5	1.8%	3.5%	185	6,845
38	40	39	7	2.5%	6.0%	273	10,647
40	42	41	13	4.6%	10.6%	533	21,853
42	44	43	21	7.4%	18.0%	903	38,829
44	46	45	33	11.7%	29.7%	1,485	66,825
46	48	47	46	16.3%	45.9%	2,162	101,614
48	50	49	62	21.9%	67.8%	3,038	148,862
50	52	51	37	13.1%	80.9%	1,887	96,237
52	54	53	24	8.5%	89.4%	1,272	67,416
54	56	55	14	4.9%	94.3%	770	42,350
56	58	57	9	3.2%	97.5%	513	29,241
58	60	59	5	1.8%	99.3%	295	17,405
60	62	61	2	0.7%	100.0%	122	7,442
62	64	63	0	0.0%	100.0%	0	0
			283	**100.0%**		**13,613**	**661,691**

*All percents computed to two decimal places and rounded to one; this may cause apparent "errors" in cumulative percents due to rounding.
**Computations rounded to the nearest whole number.

For the 40–42 mi/h speed group, the sum of the frequencies for all speed groups having a high-speed boundary of 42 mi/h or less is found as 13 + 7 + 5 + 5 + 0= 30. The cumulative percent frequency is then 100*(30/283) = 10.6%.

The last two columns of the frequency distribution table are simple multiplications that will be used in subsequent computations.

Frequency and Cumulative Frequency Distribution Curves

Figure 9.4 illustrates two standard plots that are generally prepared from the data summarized in the frequency distribution table. These are:

- *Frequency distribution curve.* For each speed group, the % frequency of observations within the group is plotted versus the middle speed of the group (S).

- *Cumulative frequency distribution curve.* For each speed group, the % cumulative frequency of observations is plotted versus the high limit of the speed group.

Note that the two frequencies are plotted versus *different* speeds. The middle speed is used for the frequency distribution curve. The cumulative frequency distribution curve, however, results in a very useful plot of speed versus the percent of vehicles traveling at or below the designated speed. For this reason, the upper limit of the speed group is used as the plotting point.

In both cases, the plots are connected by a *smooth* curve that minimizes the total distance of points falling above the line and those falling below the line (on the vertical axis). A smooth curve is defined as one without

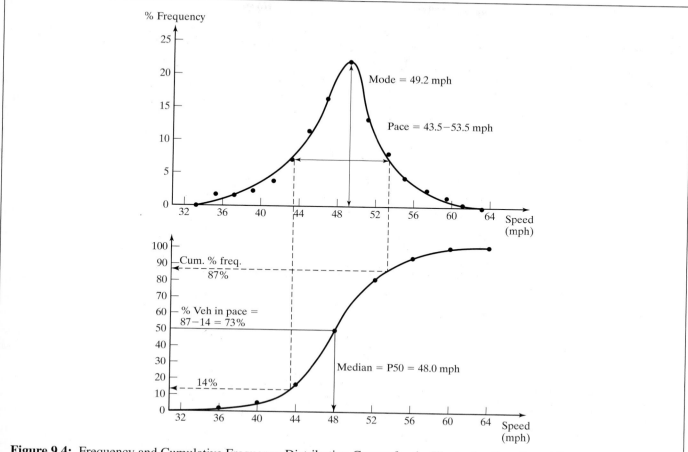

Figure 9.4: Frequency and Cumulative Frequency Distribution Curves for the Illustrative Spot Speed Study

any breaks in the slope of the curve. The "best fit" is done approximately (by eye), generally a lightly sketched curve in freehand. A French curve may then be used to darken the line.

It is also convenient to plot the frequency distribution curve directly above the cumulative frequency distribution curve, using the same horizontal scale. This makes it easier to use the curves to graphically extract critical parameters. The frequency and cumulative frequency distribution curves for the data of Figure 9.3 and Table 9.1 are shown in Figure 9.4.

Figure 9.4 also illustrates the graphic determination of several key variables that help describe the observed distribution. These parameters are defined and their determination explained in the sections that follow.

Common Descriptive Statistics

Common descriptive statistics may be computed from the data in the frequency distribution table or determined graphically from the frequency and cumulative frequency distribution curves. These statistics are used to describe two important characteristics of the distribution:

- *Central tendency*—measures that describe the approximate middle or center of the distribution.

- *Dispersion*—measures that describe the extent to which data spreads around the center of the distribution.

Measures of central tendency include the average or mean speed, the median speed, the modal speed, and the pace. Measures of dispersion include the 85th and 15th percentile speeds and the standard deviation.

The Mean Speed: A Measure of Central Tendency

The average or mean speed of a distribution is usually easily found as the sum of the observed values divided by the number of observations. In a spot speed study, however, individual values of speed are not recorded; rather, the frequency of observations within defined speed groups is known. Computing the mean speed requires the assumption that *the average speed within a given speed group is the middle speed, S, of the group.* This is the reason that speed groups of more than 5 mi/h are never used. This assumption becomes less valid as the size of the speed groups increases. For 2 mi/h speed groups, as in our illustrative study, the assumption is usually quite good. If this assumption is made, the sum of all speeds in a given speed group may be computed as:

$$n_i S_i$$

where: n_i = frequency of observations in speed group i

   S_i = middle speed of speed group i

The sum of all speeds in the distribution may then be found by adding this product for all speed groups:

$$\sum_i n_i S_i$$

The mean or average speed is then computed as the sum divided by the number of observed speeds:

$$\bar{x} = \frac{\sum_i n_i S_i}{N} \qquad (9\text{-}5)$$

where: $\bar{x}$ = average speed for the sample observations, mi/h

   N = total sample size

For the illustrative study data presented in Figure 9.3 and Table 9.1, the average or mean speed is:

$$\bar{x} = \frac{13,613}{283} = 48.1 \text{ mi/h}$$

where $\sum n_i S_i$ is the sum of the next-to-last column of the frequency distribution table of Table 9.1.

The Median Speed: Another Measure of Central Tendency

The median speed is defined as the speed that divides the distribution into equal parts (i.e., there are as many observations of speeds higher than the median as there are lower than the median). It is a positional value and is not affected by the absolute value of extreme observations.

The difference between the median and mean is best illustrated by example. Three speeds are observed: 30 mi/h, 40 mi/h, and 50 mi/h. Their average is $(30 + 40 + 50)/3 = 40$ mi/h. Their median is also 40 mi/h, as it equally divides the distribution, with one speed higher than 40 mi/h and one speed lower than 40 mi/h. Another three speeds are then observed: 30 mi/h, 40 mi/h, and 70 mi/h. Their average is $(30 + 40 + 70)/3 = 46.7$ mi/h. The median, however, is still 40 mi/h, with one speed higher and one speed lower than this observation. The mean is affected by the magnitude of the extreme observations; the median is affected only by the number of such observations.

As individual speeds have not been recorded in the illustrative study, however, the "middle value" is not easily determined from the tabular data of Table 9.1. It is easier to estimate the median graphically using the cumulative frequency distribution curve of Figure 9.4. By definition, the median equally divides the distribution. Therefore, 50% of all observed speeds should be less than the median. This is exactly what the cumulative frequency distribution curve plots. If the curve is entered at 50% on the vertical axis, the median speed is found, as illustrated in Figure 9.4. For the illustrative study:

$$P_{50} = 48.0 \text{ mi/h}$$

where P_{50} is the median or 50th percentile speed.

The Pace: Another Measure of Central Tendency

The pace is a traffic engineering measure not commonly used for other statistical analyses. It is defined as *the 10-mi/h increment in speed in which the highest percentage of drivers is observed.* It is also found graphically using the frequency distribution curve of Figure 9.4. The solution recognizes that the area under the frequency distribution curve between any two speeds approximates the percentage of vehicles traveling between those two speeds, where the total area under the curve is 100%.

The pace is found as follows: A 10-mi/h template is scaled from the horizontal axis. Keeping this template horizontal, place an end on the lower left side of the curve and move slowly along the curve. When the right side of the template intersects the right side of the curve, the pace has been located. This procedure identifies the 10-mi/h increment that intersects the peak of the curve; this contains the most area and, therefore, the highest percentage of vehicles. The pace is shown in Figure 9.4 as:

$$43.5 - 53.5 \text{ mi/h}$$

The Modal Speed: Another Measure of Central Tendency   The mode is defined as the single value of speed that is most likely to occur. As no discrete values were recorded, the modal speed is also determined graphically from the frequency distribution curve. A vertical line is dropped from the peak of the curve, with the result found on the horizontal axis. For the illustrative study, the modal speed is:

$$49.2 \text{ mi/h}$$

The Standard Deviation: A Measure of Dispersion
The most common statistical measure of dispersion in a distribution is the standard deviation. It is a measure of how far data spreads around the mean value. In simple terms, the standard deviation is the average value of the difference between individual observations and the average value of those observations. Where discrete values of a variable are available, the equation for computing the standard deviation is:

$$s = \sqrt{\frac{\sum (x_i - \overline{x})^2}{N - 1}} \qquad (9\text{-}6)$$

where: s = the standard deviation

x_i = observation i

$\overline{x}$ = average of all observations

N = number of observations

The difference between a given data point and the average is a direct measure of the magnitude of dispersion.

These differences are squared to avoid positive and negative differences canceling, and summed for all data points. They are then divided by $N - 1$. One statistical *degree of freedom* is lost because the mean of the distribution is known and used to compute the differences. If there are three numbers and it is known that the differences between the values and the mean for the first two are "3" and "2," then the third or last difference must be "−5," as the sum of all differences must be zero. Only the first "$N - 1$" observations of differences are statistically random. A square root is taken of the results, as the values of the differences were squared to begin the computation.

Because discrete values of speed are not recorded, Equation 9-6 is modified to reflect group frequencies:

$$s = \sqrt{\frac{\sum n_i (S_i - \overline{x})^2}{N - 1}}$$

which may be manipulated into a more convenient form, as follows:

$$s = \sqrt{\frac{\sum n_i S_i^2 - N\overline{x}^2}{N - 1}} \qquad (9\text{-}7)$$

where all terms are as previously defined. This form is most convenient, as the first term is the sum of the last column of the frequency distribution table of Table 9.1. For the illustrative study, the standard deviation is:

$$s = \sqrt{\frac{661,691 - (283)(48.1)^2}{283 - 1}} = 4.96 \text{ mi/h}$$

Most observed speed distributions have standard deviations that are close to 5 mi/h, as this represents most driver behavior patterns reasonably well. Unlike averages and other central speeds, which vary widely from location to location, most speed studies yield similar standard deviations.

The 85th and 15th Percentile Speeds   The 85th and 15th percentile speeds give a general description of the high and low speeds observed by most reasonable drivers. It is generally thought that the upper and lower 15% of the distribution represents speeds that are either too fast or too slow for existing conditions. These values are found graphically from the cumulative frequency distribution curve of Figure 9.4. The curve is entered on the vertical axis at values of 85% and 15%. The respective speeds are found on the horizontal axis, as shown in Figure 9.4. For the illustrative study, these speeds are:

$$P_{85} = 52.7 \text{ mi/h}$$

$$P_{15} = 43.7 \text{ mi/h}$$

The 85th and 15th percentile speeds can be used to roughly estimate the standard deviation of the distribution, although this is not recommended when the data is available for a precise determination:

$$s_{est} = \frac{P_{85} - P_{15}}{2} \qquad (9\text{-}8)$$

where all terms are as previously defined. For the illustrative spot speed study:

$$s_{est} = \frac{52.7 - 43.7}{2} = 4.5 \text{ mi/h}$$

In this case, the estimated value is close to the actual computed value of 4.96 mi/h.

The 85th and 15th percentile speeds give insight to both the central tendency and dispersion of the distribution. As these values get closer to the mean, less dispersion exists and the stronger the central tendency of the distribution becomes.

Percent Vehicles Within the Pace   The pace itself is a measure of the center of the distribution. The percentage of vehicles traveling within the pace speeds is a measure of both central tendency and dispersion. The smaller the percentage of vehicles traveling within the pace, the greater the degree of dispersion in the distribution.

The percent of vehicles within the pace is found graphically using both the frequency distribution and cumulative frequency distribution curves of Figure 9.4. The pace speeds were determined previously from the frequency distribution curves. Lines from these speeds are dropped vertically to the cumulative frequency distribution curve. The percentage of vehicles traveling at or below each of these speeds can then be determined

from the vertical axis of the cumulative frequency distribution curve, as shown. Then:

$$\% \text{ Vehicles under } 53.5 \text{ mi/h} = 87.0\%$$

$$\underline{\% \text{ Vehicles under } 43.5 \text{ mi/h} = 14.0\%}$$

$$\% \text{ Veh between } 43.5 \text{ and } 53.5 \text{ mi/h} = 73.0\%$$

Even though speeds between 34 and 62 mi/h were observed in this study, over 70% of the vehicles traveled at speeds between 43.5 and 53.5 mi/h. This represents normal traffic behavior with a standard deviation of approximately 5 mi/h.

Using the Normal Distribution in the Analysis of Spot Speed Data

Most speed distributions tend to be statistically normal (i.e., they can be reasonably represented by a normal distribution). Chapter 7 contains a detailed description of the normal distribution and its properties that should be reviewed in conjunction with this section.

If observed speeds are assumed to be normally distributed, then several additional analyses of the data may be conducted. Recall that the standard notation $x{:}N[40,25]$ signifies that the variable "x" is normally distributed with a mean of "40" and a variance of "25." The standard deviation is the square root of the variance, or "5" in this case. Recall also that a value of "x" on any normal distribution can be converted to an equivalent value of "z" on the standard normal distribution, where $z{:}N[0,1]$:

$$z_i = \frac{x_i - \mu}{\sigma} \qquad (9\text{-}8)$$

where: $x_i =$ a value on any normal distribution $x{:}N[\mu, \sigma^2]$

$\mu =$ the true mean of the distribution of values x_i

$\sigma =$ the true standard deviation of the distribution of values x_i

$z_i =$ equivalent value on the standard normal distribution $z{:}N[0,1]$

In practical terms, the true values, μ and σ, are unknown. What results from a spot speed study are estimates of the true mean and standard deviation of the distribution based upon a measured sample, $\bar{x}$ and s. A table of values of the standard normal distribution is included in Chapter 7 as Table 7.3, and is used in sample calculations in this section.

Precision and Confidence Intervals When a spot speed study is conducted, a single value of the mean speed is computed. For the illustrative study of this chapter, the mean is 48.1 mi/h, based upon a sample of 283 observations. In effect, this value, based upon a finite number of measured speeds, is being used to estimate the true mean of the underlying distribution of all vehicles traversing the site under uncongested conditions. The number of such vehicles, for all practical and statistical purposes, is infinite. The measured value of $\bar{x}$ is being used as an estimate for μ. The first statistical question that must be answered is: How good is this estimate?

In Chapter 7, the standard error of the mean, E, was introduced and defined. If a variable x is normally distributed:

$$x : N[\mu, \sigma^2]$$

it can be shown that the distribution of sample means (of a set of means with a constant sample size, n) is also normally distributed, as follows:

$$\bar{x}_n : N\left[\mu, \left(\sigma^2/n\right)\right]$$

Assume that 100 speed observations had an average value of 50 mi/h. The speeds are then arranged in 10 groups of 10 speeds, and 10 separate averages are computed (one for each group). The average of the 10 group averages would still be 50 mi/h, as the mean of the distribution of sample means is the same as the mean of the original distribution. The standard deviations, however, would be different, as the grouping and averaging process significantly reduces the occurrence of extreme values. For example, in a distribution with an average speed of 50 mi/h, it is conceivable that some observations of 70 mi/h or more would be obtained. However, at the same site, it is highly unlikely that the average of any 10 observed speeds would be 70 mi/h or higher.

The standard error of the mean, E, is simply the standard deviation of a distribution of sample means with a constant group size of n:

$$E = s/\sqrt{n} \qquad (9\text{-}9)$$

where: E = standard error of the mean

s = standard deviation of the original distribution of individual values

n = number of samples in each group of observations

The characteristics of the normal distribution are also discussed in Chapter 7. These characteristics, together with the standard error of the mean, can be used to quantify the quality of the sample estimate of the true mean of the underlying distribution. In effect, the entire illustrative spot speed study (with its sample size of 283 values) is considered to be a single point on a distribution of sample means, all with a group size of 283. Assuming a normal distribution, it is known that 95% of all values lie between the mean ± 1.96 standard deviations; 99.7% of all values lie between the mean ± 3.00 standard deviations. Thus, it is 95% certain that the sample mean (48.1 mi/h) is within the range of the true mean ± 1.96 standard deviations. The standard deviation is, in this case, the standard error of the mean. Then:

$$\bar{x} = \mu \pm 1.96E \Rightarrow \mu = \bar{x} \pm 1.96E \qquad (9\text{-}10)$$

95% of the time. The percentage is referred to as the confidence interval, while the precision of the measurement is given by the term $1.96\,E$. For the illustrative spot speed study:

$$E = 4.96/\sqrt{283} = 0.295 \text{ mi/h}$$
$$\mu = 48.1 \pm 1.96(0.295) = 48.1 \pm 0.578$$
$$\mu = 47.522 - 48.678 \text{ mi/h}$$

Rounding off the values, it can be stated that we are *95% confident* that the true mean of the underlying speed distribution lies between 47.5 and 48.7 mi/h. For a 99.7% confidence level:

$$\bar{x} = \mu \pm 3.00E \Rightarrow \mu = \bar{x} \pm 3.00E$$
$$\mu = 48.1 \pm 3.00(0.295)$$

$$\mu = 48.1 \pm 0.885$$

$$\mu = 47.215 - 48.985 \text{ mi/h} \qquad (9\text{-}11)$$

Again rounding off these values, it can be stated that we are *99.7%* confident that the true mean of the underlying speed distribution lies between 47.2 and 49.0 mi/h.

These statements provide a quantitative description of the precision of the measurement and the confidence with which the estimate is given. Note that as the confidence level increases, the precision of the estimate decreases (i.e., the range of the estimate increases). Given that speeds are normally distributed, we can be 100% confident that the true mean speed lies between 48.10 mi/h $\pm \infty$.

Such a statement is useless in engineering terms. Because spot speed studies represent a sample of measurements selected from a virtually infinite population, the average can never be measured with complete precision and 100% confidence. The most common approach uses the 95% confidence interval as to compute the precision and confidence of the sample mean as an estimator of the true mean of the underlying distribution.

Estimating the Required Sample Size   While it is useful to know the confidence level and precision of a measured sample mean after the fact, it is more useful to determine what sample size is required to obtain a measurement that satisfies a predetermined precision and confidence level. Given that the *precision* or *tolerance* (*e*) of the estimate is the $\pm$ range around the mean:

$$95\%: \quad e = 1.96E = 1.96\left(\frac{s}{\sqrt{n}}\right)$$

$$99.7\%: \quad e = 3.00E = 3.00\left(\frac{s}{\sqrt{n}}\right)$$

These equations can now be solved for the sample size, *n*. To obtain a desired precision with 95% confidence:

$$n = \frac{3.84s^2}{e^2} \qquad (9\text{-}11)$$

To obtain a desired precision with 99.7% confidence:

$$n = \frac{9.0s^2}{e^2} \qquad (9\text{-}12)$$

where all variables are as previously defined.

Consider the following problem: How many speeds must be collected to determine the true mean speed of the underlying distribution to within ± 1.0 mi/h with 95% confidence? How do the results change if the tolerance is changed to ± 0.5 mi/h and the confidence level to 99.7%?

The first problem is that the standard deviation of the distribution, *s*, is not known, as the study has not yet been conducted. Here, practical use is made of the knowledge that most speed distributions have standard deviations of approximately 5.0 mi/h. This value is assumed, and the results are shown in Table 9.2.

A sample size of 96 speeds is required to achieve a tolerance of ± 1.0 mi/h with 95% confidence. To achieve a tolerance of ± 0.5 mi/h with 99.7% confidence, the required sample size must be almost 10 times greater. For most traffic engineering studies, a tolerance of ± 1.0 mi/h and a confidence level of 95% are quite sufficient.

Before and After Spot Speed Studies

There are many situations in which existing speeds at a given location should be reduced. This occurs in situations where a high accident and/or accident severity rate

Table 9.2: Sample Size Computations Illustrated

Tolerance	Confidence Level	
e (mi/h)	**95%**	**99.7%**
1.0	$n = \dfrac{3.84(5)^2}{(1.0)^2} = 96$	$n = \dfrac{9.0(5)^2}{(1.0)^2} = 225$
0.5	$n = \dfrac{3.84(5)^2}{(0.5)^2} = 384$	$n = \dfrac{9.0(5)^2}{(0.5)^2} = 900$

is found to be related to excessive speed. It also arises where existing speed limits are being exceeded by an inordinate number of drivers.

There are many traffic engineering actions that can help reduce speeds, including lowered speed limits, stricter enforcement measures, warning signs, installation of rumble strips, and others. The major study issue, however, is to demonstrate that speeds have indeed been successfully reduced.

This is not an easy issue. Consider the following scenario: Assume that a new speed limit has been installed at a given location in an attempt to reduce the average speed by 5 mi/h. A speed study is conducted before implementing the reduced speed limit, and another is conducted several months after the new speed limit is in effect. Note that the "after" study is normally conducted after the new traffic engineering measures have been in effect for some time. This is done so that stable driver behavior is observed, rather than a transient response to something new. It is observed that the average speed of the "after" study is 3.5 mi/h less than the average speed of the "before" study. Statistically, there are two questions that must be answered:

- Is the observed reduction in average speeds real?

- Is the observed reduction in average speeds the intended 5 mi/h?

While both questions appear to have obvious answers, they in fact do not. There are two reasons that a reduction in average speeds could have occurred: (1) the observed 3.5-mi/h reduction could occur because the new speed limit caused the true mean speed of the underlying distribution to be reduced. (2) The observed 3.5-mi/h reduction could also occur because two different samples were selected from an underlying distribution that did not change. In statistical terms, the first is referred to as a *significant* reduction in speeds, while the latter is statistically *not significant*.

The second question is equally tricky. Assuming that the observed 3.5-mi/h reduction in speeds is found to be statistically significant, it is necessary to determine whether the true mean speed of the underlying distribution has likely been reduced by 5 mi/h. Statistical testing will be required to answer both questions. Further, it

will not be possible to answer either question with 100% certainty or confidence.

Chapter 7 introduced the concepts and methodologies for before-and-after testing for the significance of observed differences in sample means. The concept of truth tables was also discussed. The statistical tests for the significance of observed differences have four possible results: (1) the actual difference is significant, and the statistical test determines that it is significant; (2) the actual difference is not significant, and the statistical test determines that it is not significant; (3) the actual difference is significant and the statistical test determines that it is not significant; (4) the actual difference is not significant and the statistical test determines that it is significant. The first two outcomes result in an accurate assessment of the situation; the last two represent erroneous results. In statistical terms, outcome (4) is referred to as a Type I or α error, while outcome (3) is referred to as a Type II or β error.

In practical terms, the traffic engineer must avoid making a Type I error. In this case, it will appear that the problem (excessive speed) has been solved, when in fact it has not been solved. This may result in additional accidents, injuries, and/or deaths before the "truth" becomes apparent. If a Type II error is made, additional effort will be expended to entice lower speeds. While this might involve additional expense, it is unlikely to lead to any negative results.

The statistical test applied to assess the significance of an observed reduction in mean speeds is the normal approximation. As discussed in Chapter 7, this test is applicable as long as the "before" and "after" sample sizes are ≥ 30, which will always be the case in properly conducted speed studies. In order to certify that an observed reduction is significant, we wish to be 95% confident that this is so. In other words, we wish to insure that the chance of making a Type I error is less than 5%.

The normal approximation is applied by converting the observed reduction in mean speeds to a value of "z" on the standard normal distribution:

$$z_d = \frac{(\overline{x}_1 - \overline{x}_2) - 0}{s_Y}$$

$$s_Y = \sqrt{\frac{s_1^2}{N_1} + \frac{s_2^2}{N_2}} \qquad (9\text{-}13)$$

where: z_d = standard normal distribution equivalent to the observed difference in sample speeds

$\bar{x}_1$ = mean speed of the "before" sample, mi/h

$\bar{x}_2$ = mean speed of the "after" sample, mi/h

s_Y = pooled standard deviation of the distribution of sample mean differences

s_1 = standard deviation of the "before" sample, mi/h

s_2 = standard deviation of the "after" sample, mi/h

The standard normal distribution table of Table 7.3 (Chapter 7) is used to find the probability that a value equal to or less than z_d occurs when both sample means are from the same underlying distribution. Then:

- If Prob $(z \le z_d) \ge 0.95$, the observed reduction in speeds is *statistically significant*.

- If Prob $(z \le z_d) < 0.95$, the observed reduction in speeds is *not statistically significant*.

In the first case, it means that the observed difference in sample means would be exceeded less than 5% of the time, assuming that the two samples came from the same underlying distribution. Given that such a value was observed, this may be interpreted as being less than 5% probable that the observed difference came from the same underlying distribution and more than 95% probable that it resulted from a change in the underlying distribution.

Note that a *one-sided* test is conducted (i.e., we are testing the significance of an observed *reduction* in sample means, NOT an observed *difference* in sample means). If the observations revealed an increase in sample means, no statistical test is conducted, as it is obvious that the desired result was not achieved.

If the observed reduction is found to be statistically significant, the second question can be entertained (i.e., was the target speed reduction achieved?). This is done using only the results of the "after" distribution. Note that from the normal distribution characteristics, it is 95% probable that the true mean of the distribution is:

$$\mu = \bar{x} \pm 1.96E$$

If the target speed lies within this range, it can be stated that it was successfully achieved.

Consider the following results of a before-and-after spot speed study conducted to evaluate the effectiveness of a new speed limit intended to reduce the average speed at the location to 60 mi/h:

Before Results		After Results
65.3 mi/h	$\bar{x}$	63.0 mi/h
5.0 mi/h	s	6.0 mi/h
50	N	60

A normal approximation test is conducted to determine whether the observed reduction in sample means is statistically significant:

Step 1: Compute the pooled standard deviation

$$s_Y = \sqrt{\frac{5.0^2}{50} + \frac{6.0^2}{60}} = 1.05 \text{ mi/h}$$

Step 2: Compute z_d

$$z_d = \frac{(65.3 - 63.0) - 0}{1.05} = 2.19$$

Step 3: Determine the Prob $(z \le 2.19)$ from Table 7.3

$$\text{Prob } (z \le 2.19) = 0.9857$$

Step 4: Compare Results with the 95% Criteria

As 98.57% > 95%, the results indicate that the observed reduction in sample means was statistically significant.

Given these results, it is now possible to investigate whether or not the target speed of 60 mi/h was successfully achieved in the "after" sample. The 95% confidence interval for the "after" estimate of the true mean of the underlying distribution is:

$$E = 6/\sqrt{60} = 0.7746$$
$$\mu = 63.0 \pm 1.96(0.7746)$$
$$\mu = 63.0 \pm 1.52$$
$$\mu = 61.48 - 64.52 \text{ mi/h}$$

As the target speed of 60 mi/h does not lie in this range, it cannot be stated that it was successfully achieved.

In this case, while a significant reduction of speeds was achieved, it was not sufficient to achieve the target value of 60 mi/h. Additional study of the site would be undertaken and additional measures enacted to achieve additional speed reduction.

The 95% confidence criteria for certifying a significant reduction in observed speeds should be well understood. If a before-and-after study results in a confidence level of 94.5%, it would not be certified as statistically significant. This decision limits the probability of making a Type I error to less than 5%. When we state that the observed difference in mean speeds is not statistically significant in this case, however, it is 94.5% probable that we are making a Type II error. Before expending large amounts of funds on additional speed-reduction measures, a larger "after" speed sample should be taken to see whether or not 95% confidence can be achieved with an expanded data base.

Testing for Normalcy: The Chi-Square Goodness-of-Fit Test

Virtually all of the statistical analyses of this section start with the basic assumption that the speed distribution can be mathematically represented as normal. For completeness, it is therefore necessary to conduct a statistical test to confirm that this assumption is correct. As described in Chapter 7, the chi-square test is used to determine whether the difference between an observed distribution and its assumed mathematical form is significant. For grouped data, the chi-squared statistic is computed as:

$$\chi^2 = \sum_{N_G} \frac{(n_i - f_i)^2}{f_i} \quad (9\text{-}14)$$

where: χ^2 = chi-squared statistic

n_i = frequency of observations in speed group i

f_i = theoretical frequency in speed group i, assuming that the assumed distribution exists

N_G = number of speed groups in the distribution

Table 9.3 shows these computations for the illustrative spot speed study. Speed groups are already specified, and the observed frequencies are taken directly from the field sheet of Table 9.1.

For convenience, the speed groups are listed from highest to lowest. This is to coordinate with the standard normal distribution table (Table 7.3) in this text, which gives probabilities of $z \leq z_d$. The upper limit of the highest group is adjusted to "infinity," as the theoretical normal distribution extends to both positive and negative infinity. The remaining columns of Table 9.3 focus on determining the theoretical frequencies, f_i, and on determining the final value of of χ^2.

The theoretical frequencies are the numbers of observations that would have occurred in the various speed groups *if the distribution were perfectly normal.* To find these values, the probability of an occurrence within each speed group must be determined from the standard normal table (Table 7.3). This is done in columns 4 through 7 of Table 9.3, as follows:

1. The upper limit of each speed group (in mi/h) is converted to an equivalent value of z on the standard normal distribution, using Equation 9-8. This computation is illustrated for the speed group with an upper limit of 60 mi/h:

$$z_{60} = \frac{60.00 - 48.10}{4.96} = 2.40$$

Note that the mean speed and standard deviation of the illustrative spot speed study are used in this computation.

2. Each computed value of z is now looked up in Table 7.3 of Chapter 7. From this, the probability of $z \leq z_d$ is found and entered into column 5 of Table 9.3.

3. Consider the 48–50 mi/h speed group in Table 9.3. From column 5, 0.6480 is the probability of speed $\leq$ 50 mi/h occurring on a normal distribution; 0.4920 is the probability of a speed $\leq$ 48 mi/h occurring. Thus, the probability of an occurrence between 48 and 50 mi/h is 0.6480 − 0.4920 = 0.1560. The probabilities of column 6 are computed via sequential subtractions as shown here. The result is the probability of a speed being in any speed group, assuming a normal distribution.

Table 9.3: Chi-Square Test for Normalcy on Illustrative Spot Speed Data

Speed Group		Observed Frequency n	Upper Limit (Std. Normal) z_d	Prob. $z \leq z_d$ Table 7.3	Prob. of Occurrence in Group	Theoretical Frequency f	Combined Groups n	Combined Groups f	χ^2 Group
Upper Limit (mi/h)	Lower Limit (mi/h)								
∞	60	2	∞	1.0000	0.0082	2.3206			
60	58	5	2.40	0.9918	0.0146	4.1318	7	6.4524	0.0465
58	56	9	2.00	0.9772	0.0331	9.3673	9	9.3673	0.0144
56	54	14	1.59	0.9441	0.0611	17.2913	14	17.2913	0.6265
54	52	24	1.19	0.8830	0.0978	27.6774	24	27.6774	0.4886
52	50	37	0.79	0.7852	0.1372	38.8276	37	38.8276	0.0860
50	48	62	0.38	0.6480	0.1560	44.1480	62	44.1480	7.2188
48	46	46	−0.02	0.4920	0.1548	43.8084	46	43.8084	0.1096
46	44	33	−0.42	0.3372	0.1339	37.8937	33	37.8937	0.6320
44	42	21	−0.83	0.2033	0.0940	26.6020	21	26.6020	1.1797
42	40	13	−1.23	0.1093	0.0577	16.3291	13	16.3291	0.6787
40	38	7	−1.63	0.0516	0.0309	8.7447	7	8.7447	0.3481
38	36	5	−2.04	0.0207	0.0134	3.7922	10	5.8581	2.9285
36	34	5	−2.44	0.0073	0.0073	2.0659			
Total					**1.0000**	**283**	**283**	**283**	**14.3574**

$$\chi^2 = 14.3574$$
$$\text{Degrees of Freedom} = 12 - 3 = 9$$

4. The theoretical frequencies of column 7 are found by multiplying the sample size by the probability of an occurrence in that speed group. Fractional results are permitted for theoretical frequencies.

5. The chi-square test is valid only when all values of the theoretical frequency are 5 or more. To achieve this, the first two and last two speed groups must be combined. The observed frequencies are similarly combined.

6. The value of chi-square for each speed group is computed as shown. The computation for the 40–42 mi/h speed group is illustrated here:

$$\chi_i^2 = \frac{(n_i - f_i)^2}{f_i} = \frac{(13 - 16.3291)^2}{16.3291} = 0.6787$$

These values are summed to yield the final value of x^2 for the distribution, which is 14.3574.

To assess this result, the chi-square table of Table 7.11 (Chapter 7) is used. Probability values are shown on the horizontal axis of the table. The vertical axis shows *degrees of freedom*. For a chi-square distribution, the number of degrees of freedom is the number of data groups (after they are combined to yield theoretical frequencies of 5 or more), minus 3. Three degrees of freedom are lost because the computation of χ^2 requires that three characteristics of the measured distribution be known: the mean, the standard deviation, and the sample size. Thus, for the illustrative spot speed study, the number of degrees of freedom is 12 − 3 = 9.

The values of χ^2 are shown in the body of Table 7.11. For the illustrative data, the value of χ^2 lies between the tabulated values of 11.39 (Prob = 0.25) and 14.68

(Prob = 0.10). Note also that the probabilities shown in Table 7.11 represent the probability of a value being *greater* than or equal to χ^2. Interpolation is used to determine the precise probability level associated with a value of 14.3574 on a chi-square distribution with 9 degrees of freedom:

Value	Probability
11.3900	0.25
14.3574	p
14.6800	0.10

$$p = \text{Prob}(\chi^2 \geq 14.3574)$$
$$= 0.10 + (0.15)\left[\frac{14.6800 - 14.3574}{14.6800 - 11.3900}\right]$$
$$= 0.1147$$

From this determination, it is 11.47% probable that a value of 14.3574 or higher would exist if the distribution were statistically normal. The decision criteria are the same as for other statistical tests (i.e., to say that the data and the assumed mathematical description are *significantly different*, we must be 95% confident that this is true). For tables that yield a probability of a value *less than or equal to* the computed statistic, the probability must be 95% or more to certify a significant difference. This was the case in the normal approximation test. The corresponding decision point using a table with probabilities greater than or equal to the computed statistic is that the probability must be *5% or less* to certify a significant statistical difference. In the case of the illustrative data, the probability of a value of 14.3574 or greater is 11.47%. This is more than 5%. Thus, the data and the assumed mathematical description are *not significantly different*, and its normalcy is successfully demonstrated.

A chi-square test is rarely actually conducted on spot speed results, since they are virtually always normal. If the data is seriously skewed, or takes a shape obviously different from the normal distribution, this will be relatively obvious, and the test can be conducted. It is also possible to compare the data with other types of distributions. There are a number of distributions that

have the same general shape as the normal distribution but have skews to the low or high end of the distribution. It is also possible that a given set of data can be reasonably described using a number of different distributions. This does not negate the validity of a normal description when it occurs. As long as speed data can be described as normal, all of the manipulations described herein are valid.

If a speed distribution is found to be not normal, then other distributions can be used to describe it, and other statistical tests can be performed. These are not covered in this text, and the student is referred to standard statistics textbooks.

9.2.5 Proper Location for Speed Studies

Much of the preceding discussion of spot speed studies has focused on the statistical analysis and interpretation of results. It must be remembered that speed studies are conducted for eminently logical purposes that will influence what traffic engineering measures are implemented in any given case.

The location at which speed measurements are taken must conform to the intended purpose of the study. If approach speeds at a toll plaza are deemed to be too high, measurements should be taken at a point before drivers start to decelerate. Similarly, if excessive speed around a curve is thought to be contributing to off-the-road accidents, speed measurements should be taken in advance of the curve, before deceleration begins. It may also be appropriate, however, to measure speeds at the point where accidents are occurring for comparison with approach speeds. This would allow the traffic engineer to assess whether the problem is excessive approach speed or that drivers are not decelerating sufficiently through the subject geometric element, or a combination of both. A study of intersection approach speeds must also be taken at a point before drivers begin to decelerate. This may be a moving point, given that queues get shorter and longer at different periods of the day.

The guiding philosophy behind spot speed studies is that measurements should include drivers freely selecting their speeds, unaffected by traffic congestion. Thus, spot speed studies are rarely made under conditions of heavy, or even moderate, traffic.

9.3 Travel-Time Studies

Travel-time studies involve significant lengths of a facility or group of facilities forming a route. Information on the travel time between key points within the study area is sought and is used to identify those segments in need of improvements. Travel-time studies are often coordinated with delay observations at points of congestion along the study route.

Travel-time information is used for many purposes, including:

- To identify problem locations on facilities by virtue of high travel times and/or delay.

- To measure arterial level of service, based on average travel speeds and travel times.

- To provide necessary input to traffic assignment models, which focus on link travel time as a key determinant of route selection.

- To provide travel-time data for economic evaluation of transportation improvements.

- To develop time contour maps and other depictions of traffic congestion in an area or region.

9.3.1 Field Study Techniques

Because significant lengths of roadway are involved, it is difficult to remotely observe vehicles as they progress through the study section. The most common techniques for conducting travel time studies involve driving *test cars* through the study section while an observer records elapsed times through the section and at key intermediate points within the section. The observer is equipped with a field sheet predefining the intermediate points for which travel times are desired. The observer uses a stopwatch that is started when the test vehicle enters the study section and records the elapsed time at each intermediate point and when the end of the study section is reached. A second stopwatch is used to measure the length of mid-block and intersection stops. Their location is noted, and if the cause can be identified, it is also noted.

To maintain some consistency of results, test-car drivers are instructed to use one of three driving strategies:

1. *Floating-car technique.* In this technique, the test-car driver is asked to pass as many vehicles as pass the test car. In this way, the vehicle's relative position in the traffic stream remains unchanged, and the test car approximates the behavior of an average vehicle in the traffic stream.

2. *Maximum-car technique.* In this procedure, the driver is asked to drive as fast as is safely practical in the traffic stream without ever exceeding the design speed of the facility.

3. *Average-car technique.* The driver is instructed to drive at the approximate average speed of the traffic stream.

The floating-car and average-car techniques result in estimates of the average travel time through the section. The floating-car technique is generally applied only on two-lane highways, where passing is rare, and the number of passings can be counted and balanced relatively easily. On a multilane freeway, such a driving technique would be difficult at best, and might cause dangerous situations to arise as a test vehicle attempts to "keep up" with the number of vehicles that have passed it. The average-car technique yields similar results with less stress experienced by the driver of the test vehicle.

The maximum-car technique does not result in measurement of average conditions in the traffic stream. Rather, the measured travel times represent the lower range of the distribution of travel times. Travel times are more indicative of a 15th percentile than an average. Speeds computed from these travel times are approximately indicative of the 85th percentile speed.

It is important, therefore, that all test-car runs in a given study follow the same driving strategy. Comparisons of travel times measured using different driving techniques will not yield valid results.

Issues related to sample size are handled similarly to spot speed studies. When specific driving strategies are followed, the standard deviation of the results is somewhat constrained, and fewer samples are needed. This is important. As a practical issue, too many test cars released into the traffic stream over a short period of time will affect its operation, in effect altering the observed results. For most common applications, the number of test-car runs that will yield travel-time measurements with reasonable confidence and precision ranges from a low of 6 to 10 to a high of 50, depending upon the type of facility and the amount of traffic. The latter is difficult to achieve without affecting traffic, and may require that runs be taken over an extended time period, such as during the evening peak hour over several days.

Another technique may be used to collect travel times. Roadside observers can record license plate numbers as vehicles pass designated points along the route. The time of passage is noted along with the license plate number. The detail of delay information at intermediate points is lost with this technique. Sampling is quite difficult, as it is virtually impossible to record every license plate and time. Assume that a sample of 50% of all license plates is recorded at every study location. The probability that a license plate match occurs at two locations is 0.50 × 0.50, or 0.25 (25%). The probability that a license plate match occurs across three locations is 0.50 × 0.50 × 0.50 = 12.5%. Also, as there is no consistent driving strategy among drivers in the traffic stream, many more license plate matches are required than test-car runs to obtain similar precision and confidence in the results.

In some cases, elevated vantage points may be available to allow an entire study section to be viewed. The progress of individual vehicles in the traffic stream can be directly observed. This type of study generally involves videotaping the study section, so that many—or even all—vehicle travel times can be observed and recorded.

An alternative to the use of direct observation is to equip the test vehicle with one of several devices that plots speed against distance as the vehicle travels through the test section. Data can be extracted from the plot to yield checkpoint travel times, and the locations and time of stopped delays can be determined.

Table 9.4 shows a typical field sheet that would be used to collect travel-time data, using a test-car technique.

The sample data sheet of Table 9.4 is for a seven-mile section of Lincoln Highway, which is a major suburban multilane highway of six lanes. Checkpoints are defined in terms of mileposts. As an alternative, intersections or other known geographic markers can be used as identifiers. The elapsed stopwatch time to each checkpoint is noted. Section data refers to the distance

between the previous checkpoint and the checkpoint noted. Thus, for the section labeled MP 16, the section data refers to the section between mileposts 16 and 17. The total stopped delay experienced in each section is noted, along with the number of stops. The "special notes" column contains the observer's determination of the cause(s) of the delays noted. Section travel times are computed as the difference between cumulative times at successive checkpoints.

In this study, the segments ending in mileposts 18 and 19 display the highest delays, and therefore, the highest travel times. If this is consistently shown in *all* or *most* of the test runs, these sections would be subjected to more detailed study. Since the delays are indicated as caused primarily by traffic control signals, their timing and coordination would be examined carefully to see if they can be improved. Double parking is also noted as a cause in one segment. Parking regulations would be reviewed, along with available legal parking supply, as would enforcement practices.

9.3.2 Travel Time Data Along an Arterial: An Example of the Statistics of Travel Times

Given the cost and logistics of travel-time studies (test cars, drivers, multiple runs, multiple days of study, etc), there is a natural tendency to keep the number of observations, N, as small as possible. This case considers a hypothetical arterial on which the true mean running time is 196 seconds over a three-mile section. The standard deviation of the travel time is 15 seconds. The distribution of running times is normal. Note that the discussion is, at this point, limited to *running times*. These do not include stopped delays encountered along the route and are not equivalent to *travel times*, as will be seen.

Given the normal distribution of travel times, the mean travel time for the section is 196 seconds, and 95% of all travel times would fall within 1.96(15) = 29.4 seconds of this value. Thus, the 95% interval for travel times would be between 196 − 29.4 = 166.6 seconds and 196 + 29.4 = 225.4 seconds. The speeds corresponding to these travel times (including the average) are:

$$S_1 = \frac{3 \text{ mi}}{225.4 \text{ s}} * \frac{3600 \text{ s}}{\text{h}} = 47.9 \text{ mi/h}$$

Table 9.4: A Sample Travel Time Field Sheet

Site: <u>Lincoln Highway</u>   *Run No.* <u>3</u>   *Start Location:* <u>Milepost 15.0</u>

Recorder: <u>William McShane</u>   *Date:* <u>Aug 10, 2002</u>   *Start Time:* <u>5:00 PM</u>

| Checkpoint | Cum. Dist. Along Route (mi) | Cum. Trav. Time (min:sec) | Per Section | | | | |
			Stopped Delay (s)	No. of Stops	Section Travel Time (min:sec)	Special Notes	
MP 16	1.0	1:35	0.0	0	1:35		
MP 17	2.0	3:05	0.0	0	1:30		
MP 18	3.0	5:50	42.6	3	2:45	Stops due to signals at: MP17.2 MP17.5 MP18.0	
MP 19	4.0	7:50	46.0	4	2:00	Stops due to signal MP18.5 and double-parked cars.	
MP 20	5.0	9:03	0.0	0	1:13		
MP 21	6.0	10:45	6.0	1	1:42	Stop due to School bus.	
MP 22	7.0	12:00	0.0	0	1:15		
Section Totals	**7.0**		**88.6**	**8**	**12:00**		

$$S_{av} = \frac{3\text{ mi}}{196\text{ s}} * \frac{3600\text{ s}}{\text{h}} = 55.1\text{ mi/h}$$

$$S_2 = \frac{3\text{ mi}}{1666.6\text{ s}} * \frac{3600\text{ s}}{\text{h}} = 64.8\text{ mi/h}$$

Note that the average of the two 95% confidence interval limits is $(47.9 + 64.8)/2 = 56.4$ mi/h, NOT 55.1 mi/h. This discrepancy is due to the fact that the *travel times* are normally distributed and are therefore symmetric. The resulting running speed distribution is skewed. The distribution of speeds, which are inverse to travel times, cannot be normal if the travel times are normal. The 55.1 mi/h value is the appropriate average speed, based on the observed average travel time over the three-mile study section.

So far, this discussion considers only the *running times* of test vehicles through the section. The actual *travel time* results of 20 test-car runs are illustrated in Figure 9.5.

This distribution does not look normal. In fact, it is not normal at all, as the total travel time represents the *sum* of running times (which are normally distributed) and stop time delay that follows another distribution entirely. Specifically, it is postulated that:

No. of Signal Stops	Probability of Occurrence	Duration of Stops
0	0.569	0 s
1	0.300	40 s
2	0.131	80 s

The observations of Table 9.4 result from the combination of random driver selection of running speeds and signal delay effects that follow the relationship specified above.

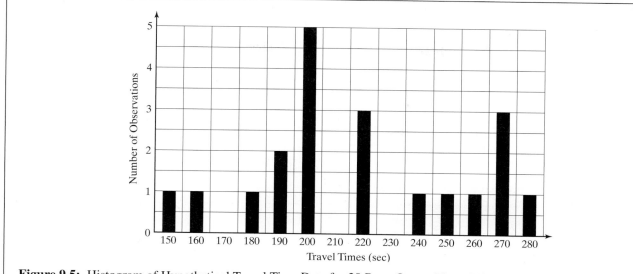

Figure 9.5: Histogram of Hypothetical Travel Time Data for 20 Runs Over a Three-Mile Section

The actual mean travel time of the observations in Figure 9.5 is 218.5 seconds, with a standard deviation of 38.3 seconds. The 95% confidence limits on the average are:

$$218.5 \pm 1.96(38.3/\sqrt{20}) = 218.5 \pm 16.79$$

$$201.71 - 235.29 \text{ s}$$

The speeds associated with these average and limiting travel times are:

$$S_1 = \frac{3\,\text{mi}}{235.29\text{s}} * \frac{3600s}{h} = 45.9 \text{ mi/h}$$

$$S_{av} = \frac{3\,\text{mi}}{218.5\text{s}} * \frac{3600s}{h} = 49.4 \text{ mi/h}$$

$$S_2 = \frac{3\,\text{mi}}{201.71\text{s}} * \frac{3600s}{h} = 53.5 \text{ mi/h}$$

Another way of addressing the average travel time is to add the average running time (196 s) to the average delay time, which is computed from the probabilities noted above as:

$$d_{av} = (0.569 * 0) + (0.300 * 40)$$
$$+ (0.131 * 80) = 22.5\,s$$

The average travel time is then expected to be 196.0 + 22.5 = 218.5 s, which is the same average obtained from the histogram of measurements.

9.3.3 Overriding Default Values: Another Example of Statistical Analysis of Travel-Time Data

Figure 9.6 shows a default curve calibrated by a local highway jurisdiction for average travel speed along four-lane arterials within the jurisdiction. As with all "standard" values, the use of another value is always permissible as long as there are specific field measurements to justify replacing the standard value.

Assume that a case exists in which the default value of travel speed for a given volume, V_1, is 40 mi/h. Based on three travel-time runs over a two-mile section, the measured average travel speed is 43 mi/h. The analysts would like to replace the standard value with the measured value. Is this appropriate?

The statistical issue is whether or not the observed 3 mi/h difference between the standard value and the measured value is *statistically significant*. As a practical matter (in this hypothetical case), practitioners generally believe that the standard values of Figure 9.6 are too low

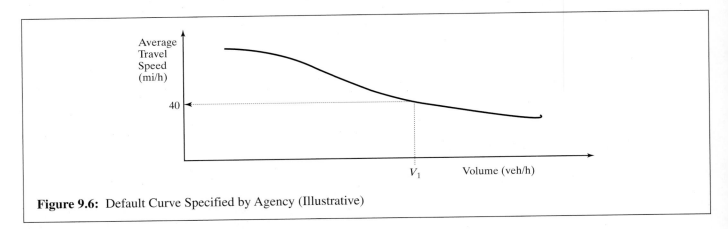

Figure 9.6: Default Curve Specified by Agency (Illustrative)

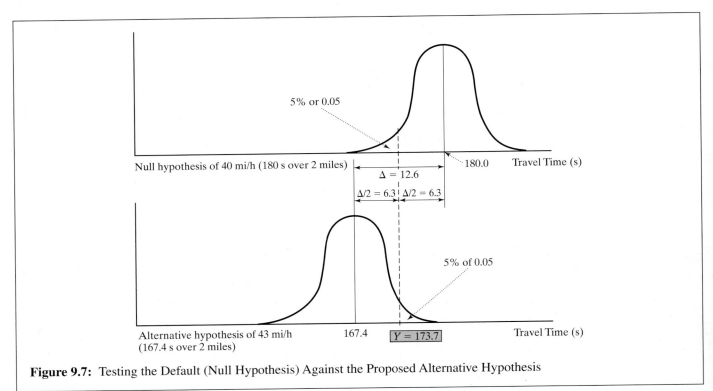

Figure 9.7: Testing the Default (Null Hypothesis) Against the Proposed Alternative Hypothesis

and that higher values are routinely observed. This suggests that a one-sided hypothesis test should be used.

Figure 9.7 shows a probable distribution of the random variable $Y = \Sigma t_i/N$, the estimator of the average travel time through the section. Based on the standard and measured average travel speeds, the corresponding travel times over a two-mile section of the roadway are $(2/40)*3,600 = 180.0$ s and $(2/43)*3,600 = 167.4$ s.

These two values are formulated, respectively, as the null and alternative hypotheses, as illustrated in Figure 9.7. The following points relate to Figure 9.7:

- Type I and Type II errors are equalized and set at 5% (0.05).

- From the standard normal table of Table 7.3 (Chapter 7), the value of z_d corresponding to Prob.

$(z \leq z_d) = 0.95$ (corresponding to a one-sided test with Type I and II errors set at 5%) is 1.645.

- The difference between the null and alternate hypotheses is a travel time of $180.0 - 167.4 = 12.6$, noted as Δ.

- The standard deviation of travel times is known to be 28.0 s.

From Figure 9.7, for the difference between the default and alternative hypotheses to be statistically significant, the value of $\Delta/2$ must be equal to or larger than 1.645 times the standard error for travel times, or:

$$\Delta/2 \geq 1.645\left(s \Big/ \sqrt{N} \right)$$

$$6.3 \geq 1.645\left(28 \Big/ \sqrt{3} \right) = 26.6$$

Obviously, in this case the difference is not significant, and the measured value of 43 mi/h cannot be accepted in place of the default value. This relationship can, of course, be solved for N:

$$N \geq \frac{8,486}{\Delta^2}$$

using the known value of the standard deviation (28). Remember that Δ is stated in terms of the difference in *travel times* over the two-mile test course, not the difference in average travel speeds. Table 9.5 shows the sample size requirements for accepting various alternative average travel speeds in place of the default value. For the alternative hypothesis of 43 mi/h to be accepted, a sample size of $8,486/(12.6)^2 = 54$ would have been required. However, as illustrated in Figure 9.7, had 54 samples

been collected, the alternative hypothesis of 43 mi/h would have been accepted as long as the average travel time was less than 173.7 s (i.e., the average travel speed was greater than $(2/173.7)*3,600 = 41.5$ mi/h. Table 9.5 shows a number of different alternative hypotheses, along with the required sample sizes and decision points for each to be accepted.

While this problem illustrates some of the statistical analyses that can be applied to travel-time data, the student should examine whether the study, as formulated, is appropriate. Should the Type II error be equalized with the Type I error? Does the existence of a default value imply that it should not? Should an alternative value higher than any measured value ever be accepted? (For example, should the alternative hypothesis of 43 mi/h be accepted if the average travel speed from a sample of 54 or more measurements is 41.6 mi/h, which is greater than the decision value of 41.5 mi/h?)

Given the practical range of sample sizes for most travel time studies, it is very difficult to justify overriding default values for individual cases. However, a compendium of such cases—each with individually small sample sizes—can and should motivate an agency to review the default values and curves in use.

9.3.4 Travel-Time Displays

Travel-time data can be displayed in many interesting and informative ways. One method that is used for overall traffic planning in a region is the development of a travel-time contour map, of the type shown in Figure 9.8. Travel times along all major routes entering or leaving a central area are measured. Time contours are then plotted, usually in increments of 15 minutes. The shape of

Table 9.5: Required Sample Sizes and Decision Values for the Acceptance of Various Alternative Hypotheses

Default Value (Average Travel Speed) (mi/h)	Alternative Hypothesis (Average Travel Speed) (mi/h)	Required Sample Size N	Decision Point (Average Travel Speed) Y (mi/h)
40	42	≥ 115	41.0
40	43	≥ 54	41.4
40	44	≥ 32	41.9
40	45	≥ 22	42.4

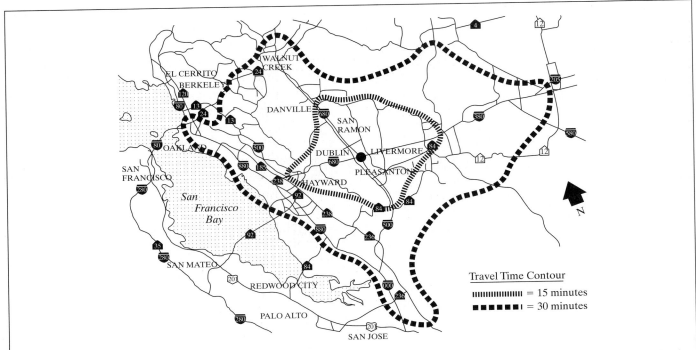

Figure 9.8: A Travel Time Contour Map (Used with permission of Prentice Hall, Inc., from Pline, J., Editor, *Traffic Engineering Handbook*, 4th Edition, Institute of Transportation Engineers, Washington DC, 1992, pg. 69.)

the contours gives an immediate visual assessment of corridor travel times in various directions. The closer together contour lines plot, the longer the travel time to progress any set distance. Such plots can be used for overall planning purposes and for identifying corridors and segments of the system that require improvement.

Travel time along a route can be depicted in different ways as well. Figure 9.9 shows a plot of cumulative time along a route. The slope of the line in any given segment is speed (ft/s), and stopped delays are clearly indicated by vertical lines. Figure 9.10 shows average travel speeds plotted against distance. In both cases, problem areas are clearly indicated, and the traffic engineer can focus on those sections and locations experiencing the most congestion, as indicated by the highest travel times (or lowest average travel speeds).

9.4   Intersection Delay Studies

Some types of delay are measured as part of a travel time study by noting the location and duration of stopped periods during a test run. A complicating feature for all

delay studies lies in the various definitions of delay, as reviewed earlier in the chapter. The measurement technique must conform to the delay definition.

Before 1997, the primary delay measure at intersections was stopped delay. While no form of delay is easy to measure in the field, stopped delay was certainly the easiest. However, the current measure of effectiveness for signalized and STOP-controlled intersections is *total control delay*. Control delay is best defined as time-in-queue delay plus time losses due to deceleration from and acceleration to ambient speed. The 2000 *Highway Capacity Manual* [1] defines a field measurement technique for control delay, using the field sheet shown in Figure 9.11.

The study methodology recommended in the *Highway Capacity Manual* is based on direct observation of vehicles-in-queue at frequent intervals and requires a minimum of two observers. The following should be noted:

1. The method is intended for undersaturated flow conditions, and for cases where the maximum queue is about 20 to 25 vehicles.

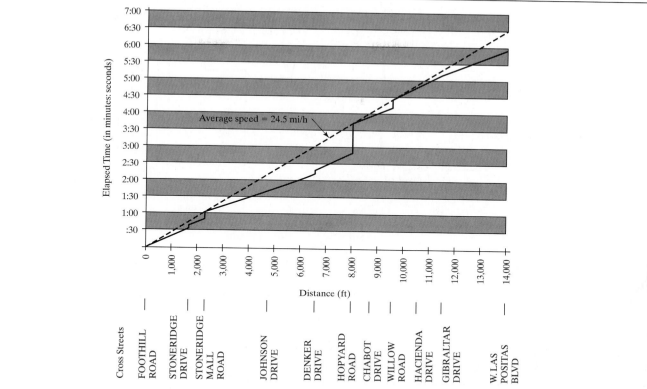

Figure 9.9: A Plot of Elapsed Time versus Distance (Used with permission of Prentice Hall, Inc., from Pline, J., Editor, *Traffic Engineering Handbook*, 4th Edition, Institute of Transportation Engineers, Washington DC, 1992.)

2. The method does not directly measure acceleration-deceleration delay but uses an adjustment factor to estimate this component.

3. The method also uses an adjustment to correct for errors that are likely to occur in the sampling process.

4. Observers must make an estimate of free-flow speed before beginning a detailed survey. This is done by driving a vehicle through the intersection during periods when the light is green and there are no queues and/or by measuring approach speeds at a position where they are unaffected by the signal.

Actual measurements start at the beginning of the red phase of the subject lane group. There should be no overflow queue from the previous green phase when measurements start. The following tasks are performed by the two observers:

Observer 1

- Keeps track of the end of standing queues for each cycle by observing the last vehicle in each lane that stops due to the signal. This count includes vehicles that arrive on green but stop or approach within one car length of queued vehicles that have not yet started to move.

- At intervals between 10 s and 20 s, the number of vehicles in queue are recorded on the field sheet. The regular intervals for these observations should be an integral divisor of the cycle length. Vehicles in queue are those that are included in the queue of stopping vehicles (as defined above) and have not yet exited the intersection. For

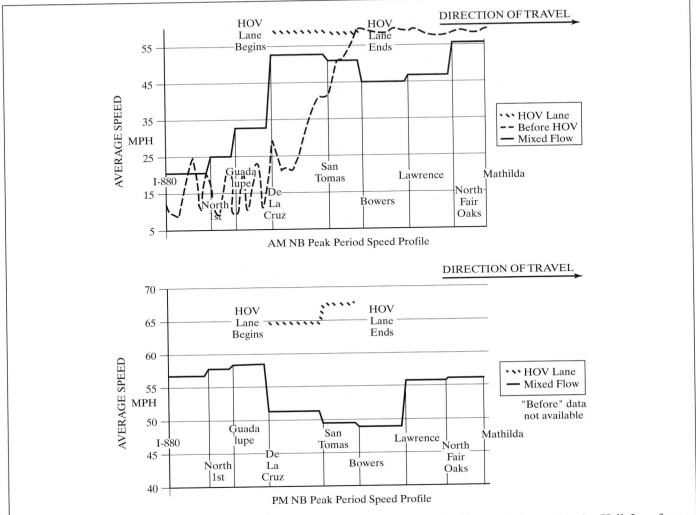

Figure 9.10: Average Travel Speeds Plotted versus Segments of a Route (Used with permission of Prentice Hall, Inc., from Pline, J., Editor, *Traffic Engineering Handbook*, 4th Edition, Institute of Transportation Engineers, Washington DC, 1992.)

through vehicles, "exiting the intersection" occurs when the rear wheels cross the STOP line; for turning vehicles, "exiting" occurs when the vehicle clears the opposing vehicular or pedestrian flow to which it must yield and begins to accelerate.

- At the end of the survey period, vehicle-in-queue counts continue until all vehicles that entered the queue during the survey period have exited the intersection.

Observer 2

- During the entire study period, separate counts are maintained of vehicles arriving during the survey period and of vehicles that stop one or more times during the survey period. Stopping vehicles are counted only once, regardless of how many times they stop.

For convenience, the survey period is defined as an integer number of cycles, although an arbitrary length of

INTERSECTION CONTROL DELAY WORKSHEET

General Information							Information				
Analyst _____							Intersection _____				
Agency or Company _____							Area Type ☐ CBD ☐ Other				
Date Performed _____											
Analysis Time Period _____							Analysis Year _____				

Input Initial Parameters

Number of Lanes, N _____ 　　Total Vehicles Arriving V_T _____

　　　　　　　　　　　　　　　　　　　Stopped Vehicle Count V_{STOP} _____

Survey Count Interval I_s _____ 　Cycle Length D (s) _____

Input Field Data

Clock Time	Cycle Number	Number of Vehicles in Queue Count Interval									
		1	2	3	4	5	6	7	8	9	10
Total											

Figure 9.11: Field Sheet for Signalized Intersection Delay Studies (Used with permission of Transportation Research Board, *Highway Capacity Manual*, 4th Edition, Washington DC, pg. 16-173.)

time (e.g., 15 min) could also be used and would be necessary where an actuated signal is involved.

Each column of the vehicle-in-queue counts is summed; the column sums are then added to yield the total vehicle-in-queue count for the study period. It is than assumed that the average time-in-queue for a counted vehicle is the time interval between counts. Then:

$$T_Q = \left(I_s * \frac{\sum V_{iq}}{V_T} \right) * 0.9 \qquad (9\text{-}15)$$

where: T_Q = average time-in-queue, s/veh

I_s = time interval between time-in-queue counts, s

$\sum V_{iq}$ = sum of all vehicle-in-queue counts, vehs

V_T = total number of vehicles *arriving* during the study period, vehs

0.9 = empirical adjustment factor

The adjustment factor (0.9) adjusts for errors that generally occur when this type of sampling technique is used. Such errors usually result in an overestimate of delay.

A further adjustment for acceleration/deceleration delay requires that two values be computed: (1) the average number of vehicles stopping per lane, per cycle, and (2) the proportion of vehicles arriving that actually stop. These are computed as:

$$V_{SLC} = \frac{V_{STOP}}{N_c * N_L}$$

(9-16)

where: V_{SLC} = number of vehicles stopping per lane, per cycle (veh/ln/cycle)

V_{STOP} = total count of stopping vehicles, vehs

N_c = number of cycles included in the survey

N_L = number of lanes in the survey lane group

$$FVS = \frac{V_{STOP}}{V_T}$$

(9-17)

where FVS = fraction of vehicles stopping
other variables as previously defined

Using the number of stopping vehicles per lane, per cycle, and the measured free-flow speed for the approach in question, an adjustment factor is found in Table 9.6.

The final estimate of control delay is then computed as:

$$d = T_Q + \left(FVS * CF\right)$$

(9-18)

where: d = total control delay, s/veh

CF = correction factor from Table 9.6
other variables as previously defined

Table 9.7 shows a facsimile of a field sheet, summarizing the data for a survey on a signalized intersection approach. The approach has two lanes, and the signal cycle length is 60 seconds. Ten cycles were surveyed, and the vehicle-in-queue count interval is 20 seconds.

The average time-in-queue is computed using Equation 9-15:

$$T_Q = \left(20 * \frac{132}{120}\right) * 0.9 = 19.8 \text{ s/veh}$$

To find the appropriate correction factor from Table 9.6, the number of vehicles stopping per lane per cycle is computed using Equation 9-16:

$$V_{SLC} = \frac{75}{10 * 2} = 3.75 \text{ vehs}$$

Table 9.6: Adjustment Factor for Acceleration/Deceleration Delay

Free-Flow Speed (mi/h)	Vehicles Stopping Per Lane, Per Cycle (V_{SLC})		
	≤7 vehs	8–19 vehs	20–30 vehs
≤37	+5	+2	−1
>37–45	+7	+4	+2
>45	+9	+7	+5

(Used with permission of Transportation Research Board, *Highway Capacity Manual*, 4th Edition, Washington DC, 2000, Exhibit A16-2, pg. 16-91.)

Table 9.7: Sample Data for a Signalized Intersection Delay Study

Clock Time	Cycle Number	Number of Vehicles in Queue		
		+0 secs	+20 secs	+40 secs
5:00 PM	1	4	7	5
5:01 PM	2	6	6	5
5:02 PM	3	3	5	5
5:03 PM	4	2	6	4
5:04 PM	5	5	3	3
5:05 PM	6	5	4	5
5:06 PM	7	6	8	4
5:07 PM	8	3	4	3
5:08 PM	9	2	4	3
5:09 PM	10	4	3	5
	Total	**40**	**50**	**42**
ΣV_{iq} = 132 vehs	V_T = 120 vehs	V_{STOP} = 75		FFS = 35 mi/h

Using this and the measured free-flow speed of 35 mi/h, the correction factor is +5 seconds. The control delay is now estimated using Equations 9-17 and 9-18:

$$FVS = \frac{75}{120} = 0.625$$

$$d = 19.8 + \left(0.625 * 5\right) = 22.9 \text{ s/veh}$$

A similar technique and field sheet can be used to measure stopped time delay as well. In this case, the interval counts include only vehicles stopped within the intersection queue area, not those moving within it. No adjustment for acceleration/deceleration delay would be added.

9.5 Closing Comments

Time is one of the key commodities that motorists and other travelers invest in getting from here to there. Travelers most often wish to minimize this investment by making their trips as short as possible. Travel-time and delay studies provide the traffic engineer with data concerning congestion, section travel times, and point delays. Through careful examination, the causes of congestion, excessive travel times, and delays can be determined and traffic engineering measures developed to ameliorate problems.

Speed is the inverse of travel time. While travelers wish to maximize the speed of their trip, they wish to do so consistent with safety. Speed data provides insight into many factors, including safety, and is used to help time traffic signals, set speed limits, locate signs, and in a variety of other important traffic engineering activities.

References

1. *Highway Capacity Manual*, 4th Edition, Transportation Research Board, National Science Foundation, Washington DC, 2000.

Problems

 9-1. Consider the following spot speed data, collected from a freeway site operating under free-flow conditions:

(a) Plot the frequency and cumulative frequency curves for these data.

Speed Group (mi/h)	Number of Vehicles Observed N
15–20	0
20–25	3
25–30	6
30–35	18
35–40	45
40–45	48
45–50	18
50–55	12
55–60	4
60–65	3
65–70	0

(b) Find and identify on the curves: median speed, modal speed, pace, percent vehicles in pace.

(c) Compute the mean and standard deviation of the speed distribution.

(d) What are the confidence bounds on the estimate of the true mean speed of the underlying distribution with 95% confidence? With 99.7% confidence?

(e) Based on the results of this study, a second is to be conducted to achieve a tolerance of ±1.5 mi/h with 95% confidence. What sample size is needed?

(f) Can this data be appropriately described as normal?

9-2. Consider the following two spot speed samples conducted at a test location to determine the effectiveness of a new speed limit posting at 50 mi/h.

Item	Before	After
Average Speed	55.3 mi/h	52.8 mi/h
STD	5.0 mi/h	5.6 mi/h
Sample Size	100	85

(a) Was the new speed limit effective in reducing average speeds at this location?

(b) Was the new speed limit effective in reducing average speeds to 50 mi/h?

9-3. A series of travel time runs are to be made along an arterial section. Tabulate the number of runs required to estimate the overall average travel time with 95% confidence to within ±2 min, ± 5 min, ± 10 min, for standard deviations of 5, 10, and 15 minutes. Note that a 3 × 3 table of values is desired.

9-4. The following data was collected during a control delay study on a signalized intersection approach. The cycle length of the signal is 90 s.

Clock Time	Cycle Number	Number of Vehicles in Queue					
		+0 s	+15 s	+30 s	+45 s	+60 s	+75 s
8:00 AM	1	2	4	2	1	4	3
	2	3	3	2	1	4	4
8:03 AM	3	3	2	4	2	5	6
	4	1	4	6	3	2	4
8:06 AM	5	5	5	4	5	6	7
	6	5	2	4	4	3	5
8:09 AM	7	4	3	3	2	3	3
	8	2	2	5	3	5	4
8:12 AM	9	1	2	2	4	2	4
	10	3	4	1	1	2	5
8:15 AM	11	3	5	3	4	4	3
	12	2	2	3	4	3	4

$V_T = 200$ vehs $V_{STOP} = 100$ $FFS = 42$ mi/h

(a) Estimate the time spent in queue for the average vehicle.

(b) Estimate the average control delay per vehicle on this approach.

(c) Does the time period of the survey appear to be appropriate? Why or why not?

9-5. The results of a travel time study are summarized in the table that follows. For this data:

(a) Tabulate and graphically present the results of the travel time and delay runs. Show the average travel speed and average running speed for each section.

(b) Note that the number of runs suggested in this Problem (5) is not necessarily consistent with the results of Problem 9-3. Assuming that *each vehicle* makes five runs, how many test vehicles would be needed to achieve a tolerance of ± 3 mi/h with 95% confidence?

Erin Blvd	Recorder: XYZ	Summary of 5 Runs		
Checkpoint Number	Cumulative Section Length (mi)	Cumulative Travel Time (min:sec)	Per Section	
			Delay (s)	No. of Stops
1	–	–	–	–
2	1.00	2:05	10	1
3	2.25	4:50	30	1
4	3.50	7:30	25	1
5	4.00	9:10	42	2
6	4.25	10:27	47	1
7	5.00	11:54	14	1

10

Accidents: Studies, Statistics, and Programs

10.1 Introduction

In the year 2000, 41,821 people were killed in accidents on U.S. highways in a total of 6,394,000 police-reported accidents. As police-reported accidents are generally believed to make up only 50% of all accidents occurring, this implies a staggering total of over 12,000,000 accidents for the year. A more complete set of statistics for the year 2000 is shown in Table 10.1 [1].

To fully appreciate these statistics, some context is needed: More people have been killed in highway accidents in the United States than in all of the wars in which the nation has been involved, from the Revolutionary War through Desert Storm.

A great deal of effort on every level of the profession is focused on the reduction of these numbers. In fact, U.S. highways have become increasingly safe for a variety of reasons. The number of annual fatalities peaked in 1972 with 54,589 deaths in highway-related accidents. While the total number of fatalities has declined by over 23% since 1972, the decline in the fatality rate per 100 million vehicle-miles-traveled (VMT) is even more spectacular. In 1966, the rate

stood at 5.5 fatalities per 100 million VMT; in 2000 the rate was 1.5 fatalities per 100 million VMT, a decline of over 70%.

There are many reasons for this decline. Highway design has incorporated many safety improvements that provide for a more "forgiving" environment for drivers. Better alignments, vast improvements in roadside and guardrail design, use of breakaway supports for signs and lighting, impact-attenuating devices, and other features have made the highway environment safer. Vehicle design has also improved. Federal requirements now dictate such common features as padded dashboards, seat belts and shoulder harnesses, airbags, and energy-absorbing crumple zones. Drivers are more familiar with driving in congested urban environments and on limited-access facilities. Despite vastly improved accident and fatality *rates*, the number of fatalities remains high. This is because Americans continue to drive more vehicle-miles each year. In the year 2000, U.S. motorists drove more than 2.7 *trillion* vehicle-miles. The trend of increasing vehicle usage has continued unchecked for over 35 years, except for a very brief perturbation in the 1970s, during the period of the fuel crisis.

Table 10.1: National Highway Accident Statistics for 2000

Accident Type	Number of Deaths/Injuries	Number of Accidents*	Number of Vehicles Involved
Fatal	41,821	37,409	55,661
Injury	3,189,000	2,070,000	3,759,000
Property Damage Only (PDO)	NA	6,394,000	7,453,000

* Includes only accidents reported by police.
(Compiled from National Traffic Administration, U.S. Department of Transportation, *Traffic Safety Facts 2000.*)

Figure 10.1 shows the trend of total highway fatalities for the years 1966 through 2000. Figure 10.2 shows the trend of fatality rates over the same period.

While the fatality rate trend has been continuously declining, the fatality trend has been less consistent, despite an overall downward trend. Significant decreases in both fatalities and fatality rates in the mid-1970s coincided with the imposition of the national 55-mi/h speed limit. Originally enacted as a measure to reduce fuel consumption, the dramatic decrease in fatalities and fatality rates was a critical, somewhat unanticipated benefit. This is why the 55-mi/h speed limit survived long after the fuel crises of the 1970s had passed.

The statistical connection between the reduction in the national speed limit and improved safety on U.S. highways has been difficult to make. While the 55-mi/h speed limit was a single change that occurred at a definable point in time, other changes in highway design were taking place at the same time. Further, speed limit enforcement practices changed substantially, with more emphasis shifting from lower-speed surface facilities to high-speed facilities such as freeways.

Between 1996 and 1998, the federal government gradually, then completely, eliminated the 55-mi/h speed limit. The national speed limit had been loosely enforced at best. States had to file formal reports on speed behavior each year and were threatened with partial loss of federal-aid highway funds if too high a percentage of drivers were found to be exceeding 55 mi/h. By the mid-1990s, many states were in technical violation of these requirements.

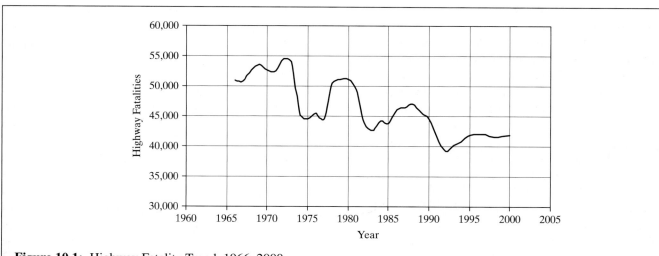

Figure 10.1: Highway Fatality Trend, 1966–2000

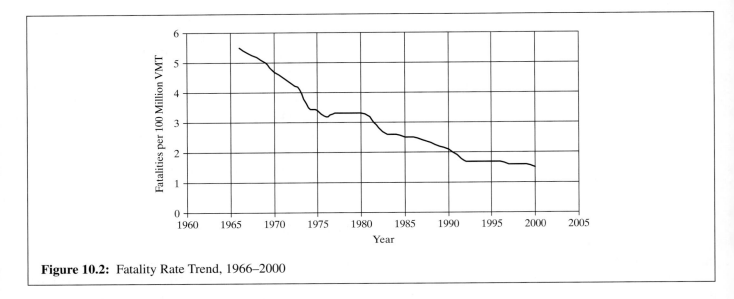

Figure 10.2: Fatality Rate Trend, 1966–2000

Either the restriction had to be changed, or the federal statutes would have to be enforced.

Despite dire predictions, the lifting of the 55-mi/h national speed limit did not lead to dramatic increases in fatalities or fatality rates. The rates, as shown, continued to decline, and the number of fatalities has been relatively stable over the past five years. States, now free to enact their own statutory speed limits, began to systematically increase speed limits on freeways to 65 mi/h or more. Because there has been no clear statistical increase in fatality rates, the clamor against these speed limit increases has declined over the last few years.

While speed limits and their enforcement are clearly aspects that influence highway safety, it is clear that overall safety is provided by a variety of measures working together. Improvements to vehicle and highway design are important components, as are improved traffic control devices and improved techniques for their application.

10.2 Approaches to Highway Safety

Improving highway safety involves consideration of three elements influencing traffic operations: the driver, the vehicle, and the roadway. Unfortunately, the traffic engineer has effective control over only one of these elements—the roadway. Traffic engineers can also play the role of informed advocates for improved driver education and licensing procedures, and for the required incorporation of safety features in vehicle design. The latter factors, however, are subject to the political and legislative process and are not under the direct control of the engineer.

One of the basic references in highway safety is the ITE *Traffic Safety Toolbox* [2]. It lists three basic strategies that may be employed to improve traffic safety:

1. Exposure control
2. Accident risk control
3. Injury control

Another basic reference, *Safer Roads: A Guide to Road Safety Engineering* [3], cites similar strategies, separated into five categories:

1. Exposure control
2. Accident prevention
3. Behavior modification
4. Injury control
5. Post-injury management

10.2.1 Exposure Control

Exposure control is common to both lists and involves strategies that reduce the number of vehicle-miles of travel by motorists. This, of course, has proven to be a very difficult strategy to implement in the United States, given that the automobile is the overwhelming choice of travelers. The march towards ever-increasing national vehicle-miles traveled has continued unabated for over 35 years.

Efforts to reduce auto use and travel cover a wide range of policy, planning, and design issues. Policies and practices that attempt to reduce auto use include:

- Diversion of travel to public transportation modes

- Substitution of telecommunications for travel

- Implementation of policies, taxes, and fees to discourage auto ownership and use

- Reorganization of land uses to minimize travel distances for various trip purposes

- Driver and vehicle restrictions through licensing and registration restrictions

Most of these strategies must take place over long time periods, and many require systemic physical changes to the urban infrastructure and behavioral changes in the traveling public. Some require massive investments (such as providing good public transportation alternatives and changing the urban land use structure), while others have not yet demonstrated the potential to affect large changes in travel behavior.

10.2.2 Accident Risk Control/Accident Prevention

Accident risk control and accident prevention are similar terms with a number of common features. They are not, however, the same. Accident prevention implies actions that reduce the number of accidents that occur for a given demand level. Accident risk control incorporates this, but also includes measures that reduce the *severity* of an accident when it occurs. Reduction of accident severity overlaps accident risk control and injury control strategies.

Accident prevention involves a number of policy measures, including driver and pedestrian training, removal of drivers with "bad" driving records (through the suspension or revocation of licenses), and provision of better highway designs and control devices that encourage good driving practices and minimize the occurrence of driver error.

Risk control, or reduction of severity, often involves the design and protection of roadside and median environments. Proper guardrail and/or impact-attenuating devices will reduce the impact energy transferred to the vehicle in an accident, and can direct the path of a vehicle away from objects or areas that would result in a more serious collision.

10.2.3 Behavior Modification

This category, separately listed in Reference [3], is an important component of strategies for accident prevention and exposure reduction. Affecting mode choice is a major behavior modification action that is hard to successfully achieve. Often, this requires providing very high-class and convenient public transportation alternatives and implementing policies that make public transportation a much more attractive alternative than driving for commuter and other types of trips. This is an expensive process, often involving massive subsidies to keep the cost of public transportation reasonable, coupled with high parking and other fees associated with driving. Use of high-occupancy vehicle lanes and other restricted-use lanes to speed public transportation, providing a visual travel-time differential between public transportation and private automobiles, is another useful strategy.

If drivers and motorists cannot be successfully diverted to alternative modes, driver and pedestrian training programs are a common strategy for behavior modification. Many states offer insurance discounts if a basic driving safety course is completed every three years. There is, however, little statistical evidence that driver training has any measurable effect on accident prevention.

The final strategy in behavior modification is enforcement. This can be very effective, but it is also expensive. Speed limits will be more closely obeyed if enforcement is strict, and the fines for violations are expensive. In recent years, the use of automated systems for ticketing drivers who violate red lights have become quite popular. Automated speed enforcement is also possible with current technologies. The issues involved in automated enforcement are more legal than technical at present. While the license plate of a vehicle running a red light can be automatically recorded, it does not prove who is driving the vehicle. In most states, automated ticketing results in a fine, but does not include "points" on the owner's license, since it cannot be proved that the owner was the driver at the time of the violation.

10.2.4 Injury Control

Injury control focuses on crash survivability of occupants in a vehicular accident. This is primarily affected by better vehicle design that is generally "encouraged by an act of Congress." Vehicle design features that have been implemented with improved crash survivability in mind include:

- Seat belts and shoulder harnesses, and laws requiring their use
- Child-restraint seats and systems, and laws requiring their use
- Anti-burst door locks
- Padded instrument panels
- Energy-absorbing steering posts and crumple zones
- Side door beams
- Air bags
- Head rests and restraints
- Shatterproof glass
- Forgiving interior fittings

10.2.5 Post-Injury Management

While included as part of injury control in Reference [2], this is treated as a separate category in Reference [3]. Traffic fatalities tend to occur during three critical time periods [4]:

- During the accident occurrence, or within minutes of it. Death is usually related to head or heart trauma or extreme loss of blood.

- Within one to two hours of the accident occurrence. In this period, death is usually due to the same causes noted above: head or heart trauma and/or loss of blood.
- Within 30 days of admission to the hospital. Death usually results from cessation of brain activity, organ failure, or infection.

About 50% of traffic fatalities occur in the first category, 35 percent in the second, and 15 percent in the third. There is little that can be done for deaths occurring during the accident or immediately thereafter, and the latter category is difficult to improve in developed countries with high-quality medical care systems. The biggest opportunity for improvement is in the second category.

Deaths within one to two hours of an accident can be reduced by systems that insure speedy emergency medical responses along with high-quality emergency care at the site and during transport to a hospital facility. Such systems involve speedy notification of emergency services, fast dispatch of appropriate equipment to the site, well-trained emergency medical technicians attending to immediate medical needs of victims, and well-staffed and equipped trauma centers at hospitals. Since survival often depends upon quickly stabilizing a victim at the crash site and speedy transport to a trauma center, communications and dispatch systems must be in place to respond to a variety of needs. A simple decision on whether to dispatch an ambulance or a med-evac helicopter is often a life-or-death decision.

10.2.6 Planning Actions to Implement Policy Strategies

Table 10.2 provides a shopping list of planning strategies and actions that can be effective in implementing the various safety approaches discussed.

Note that the list of implementing actions includes a range of measures that traffic engineers normally work within terms of facility design and control. This is not unexpected, as the principal objective of traffic engineering is the provision of safe and efficient traffic operations. The list includes design functions such as interchange design and layout, horizontal and vertical

Table 10.2: Traffic Planning and Operations Measures Related to Highway Safety Strategies

Strategy	Actions to Implement Strategy
Exposure Control	
Reducing transport demand and the amount of road traffic	1. Urban and transport policies, pricing, and regulation 2. Urban renewal (increased density, short distances) 3. Telecommunications (tele-working, tele-shopping) 4. Informatics for pre-trip, on-board information 5. TDM, mobility management (car pools, ride-sharing) 6. Logistics (rail, efficient use of transport fleets)
Promoting safe, comfortable walking and biking	7. Areawide pedestrian and bike networks 8. Land use integrated with public transport
Providing and promoting public transportation	9. Efficient service (bus lanes, fare systems, etc.)
Accident Risk Control	
Through homogenization of the traffic flow	1. Standards for geometric design 2. Classification of links with regard to function 3. Traffic management, pedestrian zones, auto restrictions 4. Traffic calming; speed management
Through separation between traffic streams	5. Grade separation (multilevel interchanges) 6. At-grade separation (traffic signals, roundabouts) 7. Channelization (medians, road markings)
Through traffic control and road management	8. Travel time distribution (staggered hours and holidays) 9. Traffic control (information, warning, flexible signs) 10. Road maintenance and inspection
Injury Control	
Reducing consequences, preventative measures	1. Emergency zones without obstacles; breakaway posts 2. Installation of median and lateral barriers
Reducing consequences, efficient rescue service	3. Establishment of rescue service 4. Emergency operation (traffic regulation, re-routing)
Reinstalling the traffic apparatus	5. Road repair and inspection

(Used with permission of Institute of Transportation Engineers, *Traffic Safety Toolbox: A Primer on Highway Safety*, Washington DC, 1999, Table 2-2, pg. 19.)

alignments, roadside design and protection, and other measures. It includes the full range of control device implementation: markings, signs (warning, regulatory, guide), and traffic signals. It includes making use of modern Intelligent Transportation Systems (ITS), such as motorist information services and rapid dispatch of emergency vehicles.

It also includes broader planning areas that are not the exclusive domain of the traffic engineer. These include public transportation planning and policies to promote its use, as well as major changes in land-use structures and policies. While this text does not deal in detail with these broader issues, the traffic engineer must be aware of them and must actively participate in planning efforts to create appropriate policies and implementation strategies.

The engineering aspects of these implementing actions are treated throughout this text in the appropriate chapters.

10.2.7   National Policy Initiatives

Some aspects of traffic accident and fatality abatement can be addressed through imposition of broad policy initiatives and programs. Such programs generally are initiated in federal or state legislation. At the federal level, compliance is encouraged by tying implementation to receipt of federal-aid-highway funds. Some examples of such policies include:

- State vehicle-inspection programs and requirements
- National speed limit (eliminated in 1996)
- National 21-year-old drinking age
- Reduction in DWI requirements to 0.08 blood content (from 0.10)
- State DUI/DWI programs
- Federal vehicle design standards

Traffic engineers should be involved in creating and implementing these programs and in providing guidance and input to policy-makers through professional and community organizations. As long as traffic accidents and fatalities occur, all levels of government will attempt to deal with the problem programmatically. These programs should be well-founded in research and must concentrate on specific ways to improve safety.

Stricter DUI/DWI policies have been developed in most states over the past decade. Public groups, such as Mothers Against Drunk Driving (MADD) and Students Against Drunk Driving (SADD) have played a major role in forcing federal and state governments to focus on this problem. They provide a compelling case study of the impact of community involvement in a major public policy area.

10.3   Accident Data Collection and Record Systems

There are two national accident data systems that are maintained by the National Highway Traffic Safety Administration (NHTSA). The *Fatality Analysis Reporting System* (FARS) is basically the repository of data on all fatal highway accidents from all 50 states. The primary source of information is police accident reports, which are virtually always filed in the case of a fatal accident. Descriptions of accidents are provided, and as many as 90 coded variables are used in the description. Data can be isolated by geographic location. The *General Estimates System* (GES) is a more general system including information on all types of highway accidents, fatal, injury, and property damage only (PDO) accidents. It uses a sample of police accident reports to estimate national statistics. The sample used is approximately 45,000 police accident reports each year.

The study of traffic accidents is fundamentally different from methods employed to observe other traffic parameters. Because accidents occur infrequently and at unpredictable times and locations, they cannot be directly observed and studied in the field. All accident data comes from secondary sources—primarily police and motorist accident reports. All basic information and data originates in these reports, and a system for collecting, storing, and retrieving this information in a usable and efficient form is an absolute necessity. The information is needed for a wide variety of purposes, including:

1. Identification of locations at which unusually high numbers of accidents occur.

2. Detailed functional evaluations of high-accident locations to determine contributing causes of accidents.

3. Development of general statistical measures of various accident-related factors to give insight into general trends, common causal factors, driver profiles, and other factors.

4. Development of procedures that allow the identification of hazards *before* large numbers of accidents occur.

10.3.1   Accident Reporting

The ultimate basis for all accident information is the individual accident report. These reports come in two types:

- *Motorists' accident reports*—filed by each involved motorist in a traffic accident; required by state law for all accidents with total property damage exceeding a proscribed limit, and for all accidents involving injuries and fatalities.

- *Police accident reports*—filed by an attendant police officer for all accidents at which an officer is present. These would generally include all fatal accidents, most accidents involving a serious injury requiring emergency and/or hospital treatment, and PDO accidents involving major damage. It is estimated that police accident reports are filed for approximately 50% of all traffic accidents that occur.

There are a variety of forms that these reports take, but they have a number of common features that always appear. The centerpiece is a schematic diagram illustrating the accident. While poorly done in many cases, these documents are a principal source of information for traffic engineers. Information on all drivers and vehicles involved is requested, as are notations of probable causal elements. The names of people injured or killed in the accident must be given, along with a general assessment of their condition at the time of the accident. Of course, information on the location, time, and prevailing environmental conditions of the accident are also included.

Police accident reports are the most reliable source of information, as the officer is trained and is a disinterested party. Motorist accident reports reflect the bias of the motorist. Also, there is one police accident report for an accident, but each involved motorist files an accident report, creating several, often conflicting, descriptions of the accident. As noted, the national data systems (FARS, GES) are based entirely on police accident reports.

10.3.2 Manual Filing Systems

While statewide and national computer-based record systems are extremely valuable in developing statistical analyses and a general understanding of the problem, the examination of a particular site requires the details that exist only in the written accident report. Thus, it is still customary to maintain manual files where written police accident reports for a given location can be retrieved and reviewed.

Police accident reports are generally sent and stored in three different locations:

- A copy of each form goes to the state motor vehicle bureau for entry into the state's accident data systems.

- A copy of the form is sent to the central filing location for the municipality or district in which the accident occurred.

- A copy of the form is retained by the officer in his or her precinct as a reference for possible court testimony.

The central file of written accident reports is generally the traffic engineer's most useful source of information for the detailed examination of high-accident locations, while the state's (or municipality's) computer record systems are most useful for the generation and analysis of statistical information.

Location Files

For any individual accident report to be useful, it must be easily retrievable for some time after the occurrence of the accident. Since the traffic engineer must deal with the observed safety deficiencies at a particular location, it is critical that the filing system be organized *by location*. In such a system, the traffic engineer is able to retrieve all of the accident reports for a given location (for some period of time) easily.

There are many ways of coding accident reports for location. For urban systems, a primary file is established for each street or facility. Sub-files are then created for each intersection along the street and for mid-block sections, which are normally identified by a range of street addresses. For example, the following file structure might be set up to store accident reports for all accidents occurring on Main Street:

Main Street
    First Avenue
    Second Avenue
    Third Avenue

Foster Blvd.

Fourth Avenue

Lincoln Road

Fifth Avenue

100–199 Main Street

200–299 Main Street

300–399 Main Street

Etc.

Primary files must select a primary direction. For example, if east-west is the primary direction, all primary files for east-west facilities would include intersection and mid-block sub-files. Primary files for north-south facilities are also needed, but would contain sub-files only for mid-block locations. All intersection accident files would use the east-west artery as the primary locator.

In rural situations, and on freeways and some expressways, the distances between intersections and street addresses are generally too large to provide a useful accident-locating system. Tenth-mile markers and mileposts are used where these exist. Where these are not in place, landmarks and recognizable natural features are used in addition to addresses and crossroads to define accident locations.

Central accident files are generally kept current for one to three years. It is preferable to use a rotating system. At the beginning of each month, all records from the previous (or most distant) year for that month are removed. There is generally a "dead file" maintained for three to five years, after which most records will be discarded or removed to a warehouse location. Use of microfiche can preserve files and allow them to be filed for longer periods of time.

Accident Summary Sheets

A common approach to maintaining records for a longer period of time is to prepare summary sheets of each year's accident records. These may be kept indefinitely, while the individual accident forms are discarded in three to five years.

Figure 10.3 illustrates such a summary sheet, on which all of the year's accidents at one location can be reduced to a single coded sheet. The form retains the basic type of accident, the number and types of vehicles involved, their cardinal direction, weather and roadway conditions, and numbers of injuries and/or fatalities. Of course, such summary information can also be maintained in a computer file.

10.3.3 Computer Record Systems

Two national computer record systems have been noted previously: FARS and GES. Every state—as well as many large municipalities and/or counties—maintains a computerized accident record system. Computer record systems have the advantage of being able to maintain a large number of accident records, keyed to locations. Computer systems are also able to correlate accident records with other data and information. The detail of the individual accident diagram with its accompanying descriptions is often lost, and information is limited to material that can be easily expressed as a series of alphanumeric codes.

Computer record systems serve two vital functions: (1) they produce regular statistical reports at prescribed intervals (usually annually), sorting accident data in ways that provide overall insight into accident trends and problems, and (2) they can produce (on request) large amounts of data on accidents at a specific location or set of locations. Moreover, most computer record systems tie accident files to a number of other statewide information systems, including a highway system network code, traffic volume files from regular counting programs, and project improvement files. These can be correlated to compute statistics and to perform a variety of statistical analyses.

Common types of statistical reports that are available from state computer accident systems include:

- Numbers of accidents by location, type of accident, type of vehicle, driver characteristics, time of day, weather conditions, and other stratifications

- Accident rates by highway location and/or segment, driver characteristics, highway classification, and other variables

- Correlation of types of accidents vs. contributing factors

- Correlation of improvement projects with accident experience

TRAFFIC ACCIDENT RECORD

LOCATION *MARCUS AVE. & NEW HYDE PARK RD.* TSL

VILLAGE *N. NEW HYDE PARK*   PRECINCT

Number	Date	Time	Private Car	Private Car	Private Car	Private Car	Commercial	Omnibus	Bicycle	Pedestrian	Other	Accident Type	Weather	Road Conditions Road Surface	Injured	Fatal
267	1.23.78	1645	N				N					E	C	W		
282	1.24.78	1300	S				S					E	C	slush		
463	2.11.78	1805	S	S	S	S						B	C	W	1	
855	3.11.78	1205	N	N								B	C	D		
1462	5.6.78	1612	W	S								A	C	D	1	
1513	5.11.78	1315	N/W	S								A	C	D		
1528	5.12.78	1645	S	S	S							B	C	D		
1569	5.15.78	1525	S	W								A	C	D	2	
1675	5.24.78	1130	W	S								A	R	W	1	
1801	6.3.78	2039	W	S					Bicycle			A	R	W	1	
1831	6.6.78	1510	N				N/E					A	C	D		
2216	7.8.78	1201	N	W								A	C	D		
2219	7.8.78	0720	S	W								A	C	D	2	
2248	7.10.78	1700	E	E								B	C	D		
2375	7.21.78	1610	S/E								MC/N	A	C	D	1	
2759	8.19.78	2015	N	E								A	C	D		
2787	8.22.78	0850	S	W								A	C	D		
2791	8.22.78	1715	S					School S				B	C	D		
3310	10.1.78	1920	S				W					A	R	W		
3432	10.11.78	1235	S	N/W								A	C	D		
3531	10.19.78	0950	W	W								E	C	D		
4217	12.7.78	0925	S				S					B	C	D		
4234	12.8.78	1030	S	E/S								A	R	W	2	
4281	12.10.78	1900	S	N/W								A	C	D		

Codes: A-right-angle collision; B-rear-end collision; C-head-on collision; D-mv sideswiped (opposite direction); E-mv side-swiped (same direction); F-mv leaving curb; G-mv collided with parked mv; H-mv collided with fixed object; I-mv executing U-turn; K-mv executing improper left turn; L-mv executing improper right-turn; M-left-turn, head-on collision; N-right-turn, head-on collision; O-pedestrian struck by mv; P-unknown; Q-hole in roadway; R-mv backing against traffic; S-operator or occupant fell out of mv; T-person injured while hitchhiking ride on mv; U-mv collided with separated part or object of another mv; V-mv and train collision; W-mv struck by thrown or fallen object; X-parked mv (unattended) rolled into another mv or object; Y-towed mv or trailer broke free of towing vehicle; Z-mv and bicycle in collision.

Figure 10.3: Illustrative Accident Summary Sheet (Courtesy of Nassau County Traffic Safety Board, Mineola, NY.)

Most states also honor requests from traffic professionals for special reports and statistical correlations, though these take somewhat longer to obtain, since such inquiries must be specifically programmed. The regular statistical reports of most states, however, provide a broad range of useful information for the engineer, who should always be aware of the capabilities of the specific state and local systems available for use.

10.4 Accident Statistics

Table 10.1 gave some basic information on traffic accidents in the United States during the year 2000. Accident statistics are measures (or estimates) of the number and severity of accidents. They should be presented in a way that is intended to provide insight into the general state of highway safety and into systematic contributing causes of accidents. These insights can help develop policies, programs, and specific site improvements intended to reduce the number and severity of accidents. Care must be taken, however, in interpreting such statistics, because incomplete or partial statistics can be misleading. Further, it is important to understand what each statistic cited means and (even more important) what it *does not* mean.

Consider a simple statistical statement: In the State of X, male drivers are involved in three times as many accidents as female drivers. This is a simple statement of fact. Its implications, however, are not as obvious as they might appear. Are female drivers safer than male drivers in the State of X? It is really not possible to say, given this fact alone. We would need to know how many licensed male and female drivers exist in the State of X. We would need to know how many miles per year the typical male or female driver in the State of X drives. If male drivers compile three times as many vehicle-miles as do female drivers, then one might expect them to have three times as many accidents. On the other hand, if males and females each account for half the vehicle-miles in the State of X, the higher number of accidents for male drivers would be quite relevant.

It is, therefore, important to understand what various statistics mean and how they are numerically constructed.

10.4.1 Types of Statistics

Accident statistics generally address and describe one of three principal informational elements:

- Accident occurrence
- Accident involvements
- Accident severity

Accident occurrence relates to the numbers and types of accidents that occur, which are often described in terms of rates based on population or vehicle-miles traveled. *Accident involvement* concerns the numbers and types of vehicles and drivers involved in accidents, with population-based rates a very popular method of expression. *Accident severity* is generally dealt with by proxy: the numbers of fatalities and fatality rates are often used as a measure of the seriousness of accidents.

Statistics in these three categories can be stratified and analyzed in an almost infinite number of ways, depending upon the factors of interest to the analyst. Some common types of analyses include:

- Trends over time
- Stratification by highway type or geometric element
- Stratification by driver characteristics (gender, age)
- Stratification by contributing cause
- Stratification by accident type
- Stratification by environmental conditions

Such analyses allow the correlation of accident types with highway types and specific geometric elements, the identification of high-risk driver populations, quantifying the extent of DUI/DWI influence on accidents and fatalities, and other important determinations. Many of these factors can be addressed through policy or programmatic approaches. Changes in the design of guardrails have resulted from the correlation of accident and fatality rates with specific types of installations. Changes in the legal drinking age and in the legal definition of DUI/DWI have resulted partially from statistics showing the very high rate of involvement of this factor in fatal accidents. Improved federal requirements on vehicle safety features (air bags, seat belts and harnesses, energy-absorbing steering columns, padded dashboards) have occurred partially as a result of statistics linking these features to accident severity. All of these changes also involved heavy lobbying of interested groups and special studies demonstrating the impact of specific vehicle and/or highway design changes. These types of statistics, however, direct policy makers to key areas requiring attention and research.

10.4.2 Accident Rates

Simple statistics citing total numbers of accidents, involvements, injuries, and/or deaths can be quite misleading, as they ignore the base from which they arise. An increase in the number of highway fatalities in a specific jurisdiction from one year to the next must be matched against population and vehicle-usage patterns to make any sense. For this reason, many accident statistics are presented in the form of rates.

Population-Based Accident Rates

Accident rates generally fall into one of two broad categories: *population-based* rates, and *exposure-based* rates. Some common bases for population-based rates include:

- Area population
- Number of registered vehicles
- Number of licensed drivers
- Highway mileage

These values are relatively static (they do not change radically over short periods of time) and do not depend upon vehicle usage or the total amount of travel. They are useful in quantifying overall risk to individuals on a comparative basis. Numbers of registered vehicles and licensed drivers may also partially reflect usage.

Exposure-Based Accident Rates

Exposure-based rates attempt to measure the amount of travel as a surrogate for the individual's exposure to potential accident situations. The two most common bases for exposure-based rates are:

- Vehicle-miles traveled
- Vehicle-hours traveled

The two can vary widely depending upon the speed of travel, and comparisons based on mileage can yield different insights from those based on hours of exposure. For point locations, such as intersections, vehicle-miles

or vehicle-hours have very little significance. Exposure rates for such cases are "event-based" using total volume passing through the point to define "events."

True "exposure" to risk involves a great deal more than just time or mileage. Exposure to vehicular or other conflicts that are susceptible to accident occurrence varies with many factors, including volume levels, roadside activity, intersection frequency, degree of access control, alignment, and many others. Data requirements make it difficult to quantify all of these factors in defining exposure. The traffic engineer should be cognizant of these and other factors when interpreting exposure-based accident rates.

Common Bases for Accident and Fatality Rates

In computing accident rates, numbers should be scaled to produce meaningful values. A fatality rate per mile of vehicle-travel would yield numbers with many decimal places before the first significant digit, and would be difficult to conceptualize. The following list indicates commonly used forms for stating accident and fatality rates:

Population-based rates are stated according to:

- Fatalities, accidents, or involvements per 100,000 area population
- Fatalities, accidents, or involvements per 10,000 registered vehicles
- Fatalities, accidents, or involvements per 10,000 licensed drivers
- Fatalities, accidents, or involvements per 1,000 miles of highway

Exposure-based rates are stated according to:

- Fatalities, accidents, or involvements per 100,000,000 vehicle-miles traveled
- Fatalities, accidents, or involvements per 10,000,000 vehicle-hours traveled
- Fatalities, accidents, or involvements per 1,000,000 entering vehicles (for intersections only)

An Example in Computing Accident and Fatality Rates

The following are sample gross accident statistics for a relatively small urban jurisdiction in the Year 2003:

Fatalities:	75
Fatal Accidents:	60
Injury Accidents:	300
PDO Accidents:	2,000
Total Involvements:	4,100
Vehicle-Miles Traveled:	1,500,000,000
Registered Vehicles:	100,000
Licensed Drivers:	150,000
Area Population:	300,000

In general terms, all rates are computed as:

$$Rate = Total * \left({Scale}/{Base} \right) \qquad (10\text{-}1)$$

where: *Total* = total number of accidents, involvements, or fatalities

Scale = scale of the base statistic, as "X" vehicle-miles traveled

Base = total base statistic for the period of the rate

Using this formula, the following fatality rates can be computed using the sample data:

$$Rate\ 1 = 75 * \left({100,000}/{300,000} \right)$$

= 25 deaths per 100,000 population

$$Rate\ 2 = 75 * \left({10,000}/{100,000} \right)$$

= 7.5 deaths per 10,000 registered vehicles

$$Rate\ 3 = 75 * \left({10,000}/{150,000} \right)$$

= 5.0 deaths per 10,000 licensed drivers

$$Rate\ 4 = 75 * \left({100,000,000}/{1,500,000,000} \right)$$

= 5.0 deaths per 100,000,000 veh-mi traveled

Similar rates may also be computed for accidents and involvements but are not shown here.

Accident and fatality rates for a given county, city, or other jurisdiction should be compared against past years, as well as against state and national norms for the analysis year. Such rates may also be subdivided by highway type, driver age and sex groupings, time of day, and other useful breakdowns for analysis.

Severity Index

A widely used statistic for the description of relative accident severity is the severity index (SI), defined as the number of fatalities per accident. For the data of the previous example, there were 75 fatalities in a total of 2,360 accidents. This yields a severity index of:

$$SI = \frac{75}{2360} = 0.0318 \text{ deaths per accident}$$

The severity index is another statistic that should be compared with previous years and state and national norms, so that conclusions may be drawn with respect to the general severity of accidents in the subject jurisdiction.

10.4.3   Statistical Displays and their Use

Graphic and tabular displays of accident statistics can be most useful in transmitting information in a clear and understandable manner. If a picture is worth 1,000 words, then a skillfully prepared graph or table is at least as useful in forcefully depicting facts. Figure 10.4 shows just one example, a graphic depiction of fatality statistics from the present to the early years of motor vehicles and the availability of traffic fatality records.

Death rates have steadily declined since the 1930s, even though number of deaths peaked in the 1960s and 1970s. Declining death rates since the 1970s are due to a number of factors, many of which have already been noted: the imposition of a national 55-mi/h speed limit, improved highway and roadside design (including better guardrail, breakaway posts, and impact-attenuating devices), and improved vehicle design. Increased use of seat belts and harnesses, encouraged by many state laws requiring their use, has also helped cut down on fatalities and fatality rates.

The combined presentation of fatalities, fatality rates, and vehicle-miles traveled on a single graph points out that the most significant underlying problem

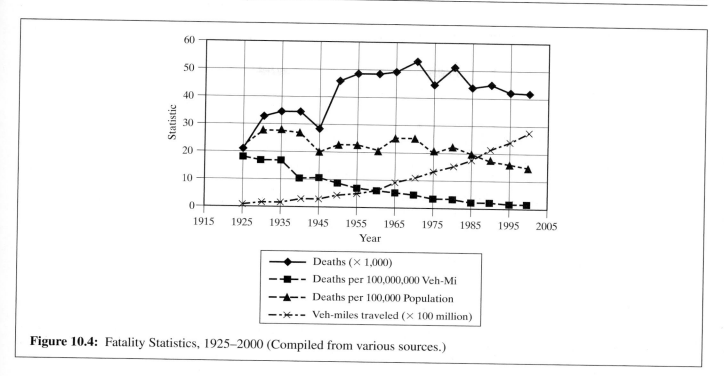

Figure 10.4: Fatality Statistics, 1925–2000 (Compiled from various sources.)

in reducing fatalities is the ever-rising trend in vehicle-miles traveled. While engineers have been able to provide a safer environment for motorists, increased highway travel keeps the total number of fatalities high.

The statistics shown also mirror other factors in U.S. history. During World War II, the increase in vehicle-miles traveled leveled off, and total highway fatalities and population-based fatality rates dipped. Between 1945 and 1950, however, traffic fatalities increased alarmingly as service men and women returned home and onto the nation's highways.

Careful displays of accident statistics can tell a compelling story, identify critical trends, and spotlight specific problem areas. Care should be taken in the preparation of such displays to avoid misleading the reviewer; when the engineer reviews such information, he or she must analyze what the data says, and (more importantly), what it *does not* say.

10.4.4 Identifying High-Accident Locations

A primary function of an accident record system is to regularly identify locations with an unusually high rate of accidents and/or fatalities. Accident spot maps are a tool that can be used to assist in this task.

Figure 10.5 shows a sample accident spot map. Coded pins or markers are placed on a map. Color or shape codes are used to indicate the category and/or severity of the accident. Modern computer technology allows such maps to be electronically generated. To allow this, the system must contain a location code system sufficient to identify specific accident locations.

Computer record systems can also produce lists of accident locations ranked by either total number of accidents occurring or by defined accident or fatality rate. It is useful to examine both types of rankings, as they may yield significantly different results. Some locations with high accident numbers reflect high volumes and have a relatively low accident rate. Conversely, a small number of accidents occurring at a remote location with very little demand can produce a very high accident rate. While statistical rankings give the engineer a starting point, judgment must still be applied in the identification and selection of sites most in need of improvement during any given budget year.

One common approach to determining which locations require immediate attention is to identify those with

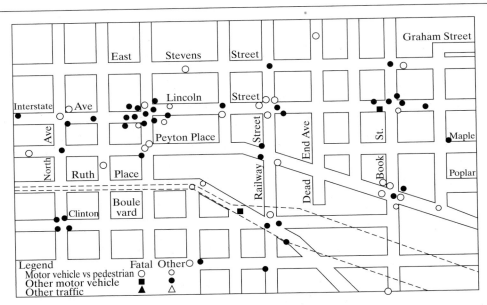

Figure 10.5: Typical Accident Spot Map (Used with permission of Prentice Hall, Inc, *Manual of Traffic Engineering Studies*, Institute of Transportation Engineers, Washington DC, 1994, pg. 400.)

accident rates that are significantly higher than the average for the jurisdiction under study. To say that the accident rate at a specific location is "significantly" higher than the average, only those locations with accident rates in the highest 5% of the (normal) distribution would be selected. In a one-tailed test, the value of z (on the standard normal distribution) for Prob $(z) < 0.95$ is 1.645. The actual value of z for a given accident location is computed as:

$$z = \frac{x_1 - \overline{x}}{s} \qquad (10\text{-}2)$$

where: x_1 = accident rate at the location under consideration

$\overline{x}$ = average accident rate for locations within the jurisdiction under study

s = standard deviation of accident rates for locations within the jurisdiction under study

If the value of z must be at least 1.645 for 95% confidence, the minimum accident rate that would be considered to be significantly higher than the average may be taken to be:

$$x_1 \geq 1.645s + \overline{x} \qquad (10\text{-}3)$$

Locations with a higher accident rate than this value would be selected for specific study and remediation. It should be noted that in comparing average accident rates, similar locations should be grouped (i.e., accident rates for signalized intersections are compared to those for other signalized intersections; mid-block rates are compared to other mid-block rates, etc.).

Consider the following example: A major signalized intersection in a small city has an accident rate of 15.8 per 1,000,000 entering vehicles. The data base for all signalized intersections in the jurisdiction indicates that the average accident rate is 12.1 per 1,000,000 entering vehicles, with a standard deviation of 2.5 per 1,000,000 entering vehicles. Should this intersection be singled out for study and remediation? Using Equation 10-3:

$$15.8 \geq (1.645*2.5) + 12.1 = 16.2$$

For a 95% confidence level, the observed accident rate *does not* meet the criteria for designation as a significantly higher accident rate.

An important factor that tempers statistical identification of high-accident locations is the budget that can be applied to remediation projects in any given year.

Ranking systems are important, as they can help set priorities. Priorities are necessary whenever funding is insufficient to address all locations identified as needing study and remediation. A jurisdiction may have 15 locations that are identified as having significantly higher accident rates than the average. However, if funding is available to address only 8 of them in a given budget year, priorities must be established to select projects for implementation.

10.4.5 Before-and-After Accident Analysis

When an accident problem has been identified, and an improvement implemented, the engineer must evaluate whether or not the remediation has been effective in reducing the number of accidents and/or fatalities. A before-and-after analysis must be conducted. The length of time considered before and after the improvement must be long enough to observe changes in accident occurrence. For most locations, periods ranging from three months to one year are used. The length of the "before" period and the "after" period must be the same.

The normal approximation test is often used to make this determination. This test is more fully discussed in Chapters 7 and 9. The statistic z is computed as:

$$z_1 = \frac{f_B - f_A}{\sqrt{f_A + f_B}} \qquad (10\text{-}4)$$

where: f_A = number of accidents in the "after" period

f_B = number of accidents in the "before" period

z_1 = test statistic representing the reduction in accidents on the standard normal distribution

The standard normal distribution table of Chapter 7 is entered with this value to find the probability of z being equal to or less than z_1. If Prob $[z \leq z_1] \geq 0.95$, the observed reduction in accidents is statistically significant. Note that this is a one-tailed test, and that only an observed *reduction* in accidents would be tested. An increase is a clear sign that the remediation effort failed.

Because the number of accidents may be small, the sample sizes may not be sufficient to justify use of the

normal approximation. It is more accurate to use the Poisson distribution and a modified binomial test. Figure 10.6 shows graphic criteria for rejecting the null hypothesis (i.e., that there has been no change in accident occurrence). The curve is entered with the number of accidents in the "before" period, and the percentage decrease in accidents in the "after" period. If the point plots above the appropriate decision line (one line is for accident reductions, the other for accident increases), the observed change is statistically significant.

Consider the following problem: A signal is installed at a high-accident location to reduce the number of right-angle accidents that are occurring. In the 6-month period prior to installing the signal, 10 such accidents occurred. In the 6-month period following the installation of the signal, 6 such accidents occurred. Was this reduction statistically significant?

Using the normal approximation test, the statistic z is computed as:

$$z = \frac{10 - 6}{\sqrt{10 + 6}} = \frac{4}{4} = 1.00$$

From the standard normal distribution table of Chapter 7, Prob $[z \leq 1.00] = 0.8413 < 0.95$. The reduction in accidents observed is *not* statistically significant.

Given the small sample size involved, it may be better to use the modified binomial test. The percent reduction in accidents for the "after" period is 4/10 or 40%. A point is plotted on Figure 10.6 at (10, 40). This point is clearly below the decision line for accident reductions. Therefore, the observed accident reduction is *not* statistically significant. If the decision curve is entered with 10 "before" accidents, the minimum percentage reduction needed for a "significant" result is 70%.

Technically, there is a serious flaw in the way that most before-and-after accident analyses are conducted. There is generally a base assumption that any observed change in accident occurrence (or severity) is due to the corrective actions implemented. Because the time span involved in most studies is long, however, this may not be correct in any given case.

If possible, a control experiment or experiments should be established. These control experiments involve locations with similar accident experience that have not been treated with corrective measures. The controls

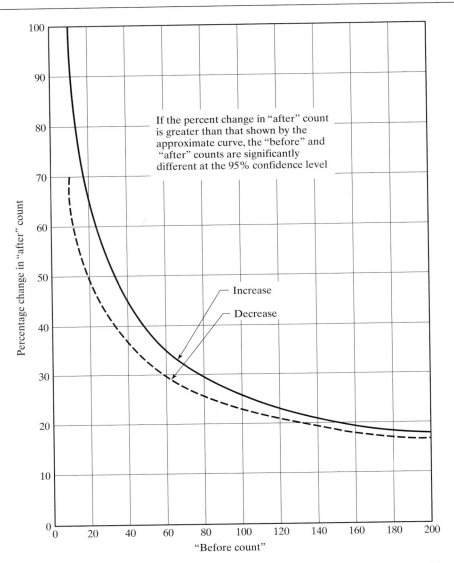

Figure 10.6: Rejection Criteria for Poisson-Distributed Data Based on the Binomial Test (Used with permission of Transportation Research Board, Weed, R., "Research Decision Criteria for Before-and-After Analyses," *Transportation Research Record 1069*, Washington DC, 1986, pg. 11.)

establish the expected change in accident experience due to general environmental causes not influenced by corrective measures. For the subject location, the null hypothesis is that the change in accident experience is not significantly different from the change at observed control locations. Figure 10.7 illustrates this technique.

While desirable from a statistical point of view, the establishment of control conditions is often a practical problem, requiring that some high-accident locations be left untreated during the period of the study. For this reason, many before-after studies of accidents are conducted without such control conditions.

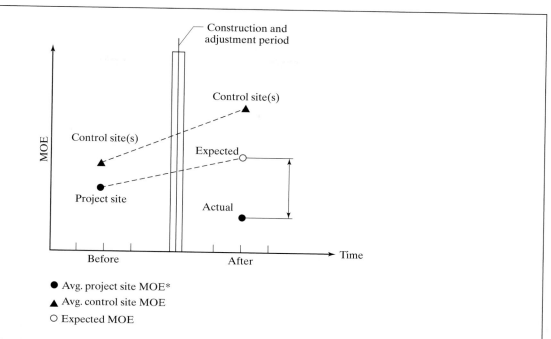

Figure 10.7: Before-and-After Experiment With Control Conditions (Used with permission of Prentice Hall, Inc., *Manual of Traffic Engineering Studies*, Institute of Transportation Engineers, Washington DC, 1994, pg. 368.)

10.5   Site Analysis

One of the most important tasks in traffic safety is the study and analysis of site-specific accident information to identify contributing causes and to develop site remediation measures that will lead to improved safety.

Once a location has been statistically identified as a "high-accident" location, detailed information is required in two principal areas:

1. Occurrence of accidents at the location in question

2. Environmental and physical conditions existing at the location

The analysis of this information must identify the environmental and physical conditions that potentially or actually contribute to the observed occurrence of accidents. Armed with such analyses, engineers may then develop countermeasures to alleviate the problem(s).

The best information on the occurrence of accidents is compiled by reviewing all accident reports for a given location over a specified study period. This can be done using computer accident records, but the most detailed data will be available from the actual police accident reports on file. Environmental and physical conditions are established by a thorough site investigation conducted by appropriate field personnel. Two primary graphical outputs are then prepared:

1. Collision diagram

2. Condition diagram

10.5.1   Collision Diagrams

A collision diagram is a schematic representation of all accidents occurring at a given location over a specified period. Depending upon the accident frequency, the "specified period" usually ranges from one to three years.

Each collision is represented by a set of arrows, one for each vehicle involved, which schematically represents the type of accident and directions of all vehicles. Arrows are generally labeled with codes indicating vehicle types, date and time of accident, and weather conditions.

The arrows are placed on a schematic (not-to-scale) drawing of the intersection with no interior details shown. One set of arrows represents one accident. It should be noted that arrows are not necessarily placed at the exact spot of the accident on the drawing. There could be several accidents that occurred at the same spot, but separate sets of arrows would be needed to depict them. Arrows illustrate the occurrence of the accident, and are placed as close to the actual spot of the accident as possible.

Figure 10.8 shows the standard symbols and codes used in the preparation of a typical collision diagram. Figure 10.9 shows an illustrative collision diagram for an intersection.

The collision diagram provides a powerful visual record of accident occurrence over a significant period of time. In Figure 10.9, it is clear that the intersection has experienced primarily rear-end and right-angle collisions, with several injuries but no fatalities during the study period. Many of the accidents appear to be clustered at night. The diagram clearly points out these patterns, which now must be correlated to the physical and

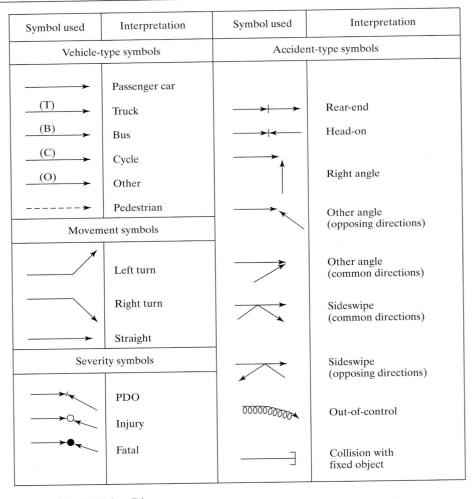

Figure 10.8: Symbols Used in Collision Diagrams

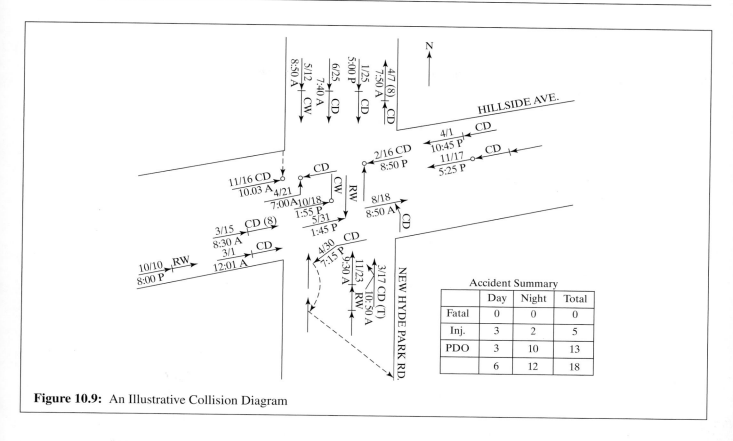

Figure 10.9: An Illustrative Collision Diagram

Accident Summary			
	Day	Night	Total
Fatal	0	0	0
Inj.	3	2	5
PDO	3	10	13
	6	12	18

control characteristics of the site to determine contributing causes and appropriate corrective measures.

10.5.2 Condition Diagrams

A condition diagram describes all physical and environmental conditions at the accident site. The diagram must show all geometric features of the site, the location and description of all control devices (signs, signals, markings, lighting, etc.), and all relevant features of the roadside environment, such as the location of driveways, roadside objects, land uses, etc. The diagram must encompass a large enough area around the location to include all potentially relevant features. This may range from several hundred feet on intersection approaches to .25–.50 mile on rural highway sections.

Figure 10.10 illustrates a condition diagram. It is for the same site and time period as the collision diagram of Figure 10.9. The diagram includes several hundred feet of each approach and shows all driveway

locations and the commercial land uses they serve. Control details include signal locations and timing, location of all stop lines and crosswalks, and even the location of roadside trees, which could conceivably affect visibility of the signals.

10.5.3 Interpretation of Condition and Collision Diagrams

This brief overview chapter cannot fully discuss and present all types of accident site analyses. The objective in analyzing collision and condition diagrams is straightforward: find contributing causes to the observed accidents shown in the collision diagram among the design, control, operational, and environmental features summarized on the condition diagram. Doing so involves virtually all of the traffic engineer's knowledge, experience, and insight, and the application of professional judgment.

Accidents are generally grouped by type. Predominant types of accidents shown in Figure 10.9 are rear-end

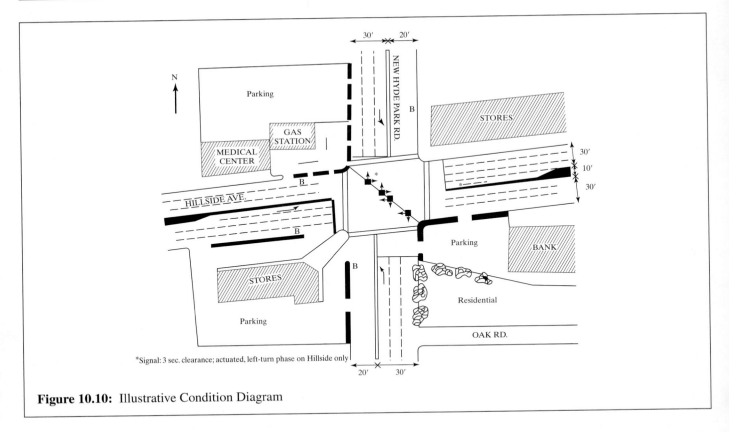

Figure 10.10: Illustrative Condition Diagram

and right-angle collisions. For each type of accident, three questions should be asked:

1. What driver actions lead to the occurrence of such accidents?

2. What existing conditions at the site could contribute to drivers taking such actions?

3. What changes can be made to reduce the chances of such actions taking place?

Rear-end accidents occur when the lead vehicle stops suddenly or unexpectedly and/or when the trailing driver follows too closely for the prevailing speeds and environmental conditions. While "tailgating" by a following driver cannot be easily corrected by design or control measures, there are a number of factors evident in Figures 10.9 and 10.10 that may contribute to vehicles stopping suddenly or unexpectedly.

The condition diagram shows a number of driveways allowing access to and egress from the street at or near the intersection itself. Unexpected movements into or out of these driveways could cause mainline vehicles

to stop suddenly. Because of these driveways, STOP lines are located well back from the sidewalk line, particularly in the northbound direction. Vehicles, therefore, are stopping at positions not normally expected, and following drivers may be surprised and unable to respond in time to avoid a collision. Potential corrective actions include closing some or all of these driveways and moving STOP lines closer to their normal positions.

Other potential causes of rear-end actions include signal timing (insufficient "yellow" and "all red" intervals), signal visibility (do trees block approaching drivers' views), and roadway lighting adequacy (given that most of the accidents occur at night).

Right-angle collisions indicate a breakdown in the right-of-way assignment by the signal. Signal visibility must be checked and the signal timing examined for reasonableness. Again, insufficient "yellow" and "all red" intervals could release vehicles before the competing vehicles have had time to clear the intersection. If the allocation of green is not reasonable, some drivers will "jump" the green or otherwise disregard it.

At this location, some of the causes compound each other. The set-back of STOP lines to accommodate driveways, for example, lengthens the requirements for "all red" clearance intervals and, therefore, amplifies the effect of a shortfall in this factor.

This analysis is illustrative. The number of factors that can affect accident occurrence and/or severity at any given location is large indeed. A systematic approach, however, is needed if all relevant factors are to be identified and dealt with in an effective way. Traffic safety is not an isolated subject for study by traffic engineers. Rather, everything traffic engineers do is linked to a principal objective of safety. The importance of building safety into all traffic designs, control measures, and operational plans is emphasized throughout this text.

10.6 Development of Countermeasures

The ultimate goal of any general or site-specific safety analysis is the development of programmatic or site-specific improvements to mitigate the circumstances leading to those accidents. Each case, however, has its own unique characteristics that must be studied and analyzed in detail. Program development must consider national, regional, and local statistics; site remediation requires detailed collision and condition information.

Programmatic countermeasures are used to attack systemic safety problems that prevail throughout the highway system. These measures generally involve education and/or control of drivers, vehicles, or highway design features. Table 10.3 (on the following page) provides a sampling of systemic problems and the types of programmatic approaches that are used to address them.

Table 10.3 presents only a small sampling of the thousands of legislative and programmatic measures that have been developed to improve highway safety. There are programs that addresses specific user groups, such as pedestrians and child passengers in vehicles. There are programs that address a class of problems, such as impaired driving due to alcohol or drug use. There are programs that address design standards, such as bumper heights and provision of air bags and other safety devices in vehicles and on roadways. The key is that this scale of effort is intended to address systemic and persistent problems that have been studied, and where such study has identified potential ways of improving the situation.

The list of site-specific accident countermeasures is enormous, as the number of accident situations that can arise is, for all intent and purposes, almost infinite. Table 10.4 (on the following pages) summarizes a listing of accident types, probable contributing causes, and potential countermeasures from a 1981 study [5]. Again, the listing is intended to be purely illustrative, as it is certainly not exhaustive. It does provide insight into the linkages between observed accident occurrence and specific measures designed to combat it.

10.7 Closing Comments

This chapter provides a very general overview of the important subject of highway safety and accident studies. The subject is complex and covers a vast range of material. Everything the traffic engineer does, from field studies, to planning and design, to control and operations, is related to the provision of a safe system for vehicular travel. The traffic engineer is not alone in the focus on highway safety, as many other professionals, from urban planners to lawyers to public officials, also have an abiding interest in safe travel.

References

1. *Traffic Safety Facts 2000*, National Highway Traffic Safety Administration, U.S. Department of Transportation, Washington DC, 2001.

2. *The Traffic Safety Toolbox: A Primer on Traffic Safety*, Institute of Transportation Engineers, Washington DC, 1999.

3. Ogden, K.W., *Safer Roads: A Guide to Road Safety Engineering*, Avebury Technical, Brookfield, VT, 1996.

4. Trinca, G., *et al.*, *Reducing Traffic Injury—A Global Challenge*, Royal Australian College of Surgeons, Melbourne, Australia, 1988.

5. Robertson, Hummer, and Nelson (Editors), *Manual of Transportation Engineering Studies*, Institute of Transportation Engineers, Prentice Hall, Inc., Upper Saddle River, NJ, 1994.

Table 10.3: Illustrative Programmatic Safety Approaches

Target Group	Problem Area	Sample Strategies or Programs
Drivers	Ensuring driver competency	1. Implement graduated licensing system. 2. Develop and implement improved competency training and assessment procedures for entry-level drivers. 3. Increase effectiveness of license suspension and revocation procedures. 4. Develop and provide technical aids such as simulators and electronic media for private self-assessment and improvement of driver skills.
Drivers	Reducing impaired driving	1. Implement stronger legislation to reduce drinking and driving. 2. Develop and implement sobriety checkpoints and saturation enforcement blitzes. 3. Develop and implement a comprehensive public awareness campaign.
Drivers	Keeping drivers alert	1. Retrofit rural and other fatigue-prone facilities with shoulder rumble strips. 2. Provide 24-hour "coffee stops" along fatigue-prone facilities.
Vehicles	Increasing safety enhancements in vehicles	1. Implement educational programs on the use of anti-lock brake systems and other vehicle safety features. 2. Strengthen regulations requiring incorporation of safety features on vehicles and their use by motorists and passengers.
Highways	Keeping vehicles on the roadway	1. Improve driver guidance through installation of better pavement markings and delineation. 2. Implement a targeted rumble-strip program. 3. Improve highway maintenance. 4. Develop better guidance to control variance in speed through combinations of geometric, control, and enforcement techniques.
Highways	Reducing the consequences of leaving the roadway	1. Provide improved practices for the selection, installation, and maintenance of upgraded roadside safety hardware (guardrail, impact-attenuating devices, etc.) 2. Implement a program to remove hazardous trees or other natural roadside hazards. 3. Install breakaway sign and lighting posts.
Intersections	Reducing intersection accidents	1. Install and use automated methods to monitor and enforce intersection traffic control. 2. Implement more effective access control strategies with a safety perspective.

(Excerpted from Institute of Transportation Engineers, *The Traffic Safety Toolbox: A Primer on Traffic Safety*, Washington DC, pgs. 8–11.)

Table 10.4: Illustrative Site-Specific Accident Countermeasures

Accident Pattern	Probable Cause[a]	Possible Countermeasures[b]	Accident Pattern	Probable Cause[a]	Possible Countermeasures[b]
Left turn, head-on	A	1–11	Ran off roadway	E	15
	B	3, 6, 12–15		G	15, 19–22
	C	16, 17		H	23
	D	3		K	54
	E	15		U	55–58
Rear-end at unsignalized intersection	A	4, 13, 18		V	14, 53, 59
	E	15		W	60
	F	14		X	6
	G	15, 19–22		Y	61
	H	23	Fixed object	E	15
	I	10, 24		G	20, 22, 55, 62
	J	25		H	23
Rear-end at signalized intersection	A	3, 4, 13, 18		T	53
	G	15, 19–22		U	14, 63
	H	23		Z	58, 64–67
	J	25, 26		AA	68
	K	12, 14, 15, 27–32	Parked or parking vehicle	F	15
	L	16, 17, 33		T	69
	M	34		BB	35
Right angle at signalized intersection	B	6, 12, 14, 15, 35, 36		CC	70
	E	15, 16, 37		DD	45, 50, 71
	H	23		EE	1, 43
	K	14, 27, 32, 38	Sideswipe or head-on	E	15, 72, 73
	L	11, 16, 17, 33, 39, 40		T	53
	N	14		U	1, 55
	O	2, 11		W	60
Right angle at unsignalized intersection	B	6, 10, 12, 14, 15, 24, 35, 36, 41, 42		X	6, 13, 74
	E	15, 16, 37		Y	61
	H	23		FF	38, 75
	N	14	Driveway-related	A	13, 18, 35, 55, 72, 76
	O	10, 43		B	12, 15, 32, 35
	P	44, 45		E	15
Pedestrian-vehicle	B	12, 25, 35, 46		H	23
	E	14, 15, 45, 47		GG	77–81
	H	23		HH	43, 79, 82
	I	10, 25, 26		II	6, 10, 74
	L	11	Train-vehicle	B	12, 14, 24, 83–85
	P	26		E	15
	Q	47, 48		G	62
	R	49		K	23, 54
	S	14, 15, 47, 50		T	36, 42, 53
	T	51, 53		JJ	11
Wet pavement	G	15, 19–22, 62		KK	86
	T	53		LL	87
				MM	88
			Night	K	14, 23, 59
				V	14, 59, 89
				X	14, 53, 59, 89
				FF	44, 90

(Continued)

Table 10.4: Illustrative Site-Specific Accident Countermeasures *(Continued)*

[a]Key to probable causes:

A	Large turn volume	T	Inadequate or improper pavement markings
B	Restricted sight distance	U	Inadequate roadway design for traffic conditions
C	Amber phase too short	V	Inadequate delineation
D	Absence of left-turn phase	W	Inadequate shoulder
E	Excessive speed	X	Inadequate channelization
F	Driver unaware of intersection	Y	Inadequate pavement maintenance
G	Slippery surface	Z	Fixed object in or too close to roadway
H	Inadequate roadway lighting	AA	Inadequate TCDs and guardrail
I	Lack of adequate gaps	BB	Inadequate parking clearance at driveway
J	Crossing pedestrians	CC	Angle parking
K	Poor traffic control device (TCD) visibility	DD	Illegal parking
L	Inadequate signal timing	EE	Large parking turnover
M	Unwarranted signal	FF	Inadequate signing
N	Inadequate advance intersection warning signs	GG	Improperly located driveway
O	Large total intersection volume	HH	Large through traffic volume
P	Inadequate TCDs	II	Large driveway traffic volume
Q	Inadequate pedestrian protection	JJ	Improper traffic signal preemption timing
R	School crossing area	KK	Improper signal or gate warning time
S	Drivers have inadequate warning of frequent midblock crossings	LL	Rough crossing surface
		MM	Sharp crossing angle

[b]Key to possible countermeasures:

1	Create one-way street	30	Install signal back plates
2	Add lane	31	Relocate signal
3	Provide left-turn signal phase	32	Add signal heads
4	Prohibit turn	33	Provide progression through a set of signalized intersections
5	Reroute left-turn traffic	34	Remove signal
6	Provide adequate channelization	35	Restrict parking near corner/crosswalk/driveway
7	Install stop sign	36	Provide markings to supplement signs
8	Revise signal-phase sequence	37	Install rumble strips
9	Provide turning guidelines for multiple left-turn lanes	38	Install illuminated street name sign
10	Provide traffic signal	39	Install multidial signal controller
11	Retime signal	40	Install signal actuation
12	Remove signal	41	Install yield sign
13	Provide lurn lane	42	Install limit lines
14	Install or improve warning sign	43	Reroute through traffic
15	Reduce speed limit	44	Upgrade TCDs
16	Adjust amber phase	45	Increase enforcement
17	Provide all-red phase	46	Reroute pedestrian path
18	Increase curb radii	47	Install pedestrian barrier
19	Overlay pavement	48	Install pedestrian refuge island
20	Provide adequate drainage	49	Use crossing guard at school crossing area
21	Groove pavement	50	Prohibit parking
22	Provide "slippery when wet" sign	51	Install thermoplastic markings
23	Improve roadway lighting	52	Provide signs to supplement markings
24	Provide stop sign	53	Improve or install pavement markings
25	Install or improve pedestrian crosswalk TCDs	54	Increase sign size
26	Provide pedestrian signal	55	Widen lane
27	Install overhead signal	56	Relocate island
28	Install 12-inch signal lenses	57	Close curb lane
29	Install signal visors	58	Install guardrail

(Continued)

Table 10.4: Illustrative Site-Specific Accident Countermeasures *(Continued)*

59	Improve or install delineation	75	Install advance guide sign
60	Upgrade roadway shoulder	76	Increase driveway width
61	Repair road surface	77	Regulate minimum driveway spacing
62	Improve skid resistance	78	Regulate minimum corner clearance
63	Provide proper superelevation	79	Move driveway to side street
64	Remove fixed object	80	Install curb to define driveway location
65	Install barrier curb	81	Consolidate adjacent driveways
66	Install breakaway posts	82	Construct a local service road
67	Install crash cushioning device	83	Reduce grade
68	Paint or install reflectors on obstruction	84	Install train-actuated signal
69	Mark parking stall limits	85	Install automatic flashers or flashers with gates
70	Convert angle to parallel parking	86	Retime automatic flashers or flashers with gates
71	Create off-street parking	87	Improve crossing surface
72	Install median barrier	88	Rebuild crossing with proper angle
73	Remove constriction such as parked vehicle	89	Provide raised markings
74	Install acceleration or deceleration lane	90	Provide illuminated sign

(Reprinted with permission of Prentice Hall, Inc., from Robertson, Hummer, and Nelson (Editors), *Manual of Traffic Engineering Studies*, Institute of Transportation Engineers, Washington DC, 1994, pgs. 214 and 215.)

Problems

10-1. Consider the following data for the year 2001 in a small suburban community:

- Number of accidents   360

    Fatal   10

    Injury   36

    PDO   314

- Number of fatalities   15

- Area population   50,000

- Registered vehicles   35,000

- Annual VMT   12,000,000

- Average speed   30 mi/h

Compute all relevant exposure- and population-based accident and fatality rates for this data. Compare these to national norms for the current year. (Hint: Use the Internet to locate current national norms.)

10-2. A before-after accident study results in 25 accidents during the year before a major improvement to an intersection, and 15 the year after. Is this reduction in accidents statistically significant? What statistical test is appropriate for this comparison? Why?

10-3. Consider the collision and condition diagrams illustrated in Figure 10.11 on the next page. Discuss probable causes of the accidents observed. Recommend improvements, and illustrate them on a revised condition diagram.

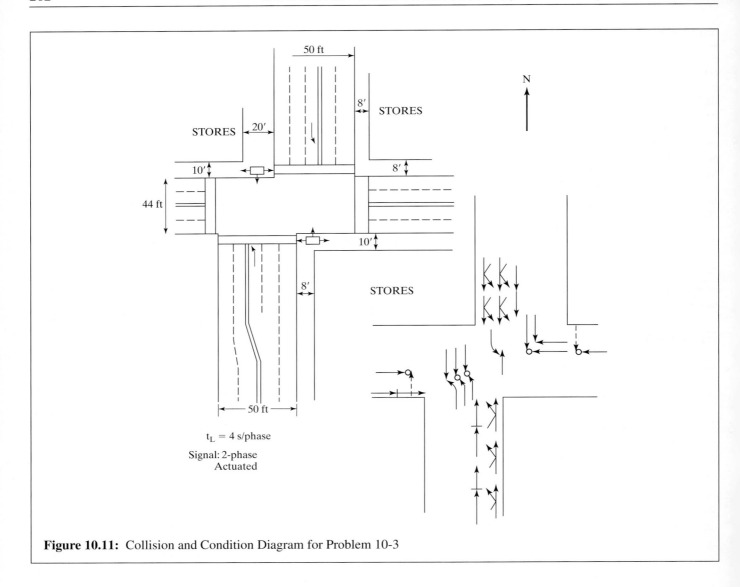

Figure 10.11: Collision and Condition Diagram for Problem 10-3

CHAPTER
11

Parking: Studies, Characteristics, Facilities, and Programs

11.1 Introduction

Every person starts and ends each trip as a pedestrian. With the exception of drive-through facilities now present at such varied destinations as banks and fast-food restaurants, travelers generally leave from their origins as pedestrians and enter their destinations as pedestrians. In terms of trips using private automobiles, the pedestrian portion of the trip starts or ends at a parking space.

At the residential trip end, private vehicles are accessed in private driveways and garages, in on-street parking spaces, or in nearby off-street lots or garages. At the other end of the trip, the location and nature of parking opportunities depends heavily on the land-use function and density as well as on a wide variety of public policy and planning issues.

For land to be productively used, it must be accessible. While public transportation can be a major part of providing accessibility in dense urban areas, for the most part, accessibility is dependent on the supply, convenience, and cost of parking facilities. Major activity centers, from regional shopping malls to sports facilities to airports, rely on significant parking supply to provide site accessibility. Without such supply, these facilities could not operate profitably over a substantial period of time.

The economic survival of most activity centers, therefore, is directly related to parking and other forms of access. Parking supply must be balanced with other forms of access (public transportation), the traffic conditions created by such access, and the general environment of the activity center. While economic viability is most directly related to the availability of parking, the environmental impacts of generated traffic may have negative affects as well.

This chapter attempts to provide an overview of issues related to parking. The coverage is not intended to be exhaustive, and the reader is encouraged to consult the available literature for more complete and detailed treatments of the subject. This chapter will address four key parking issues:

- Parking generation and supply needs
- Parking studies and characteristics
- Parking facility design and location
- Parking programs

Each is covered in the sections that follow.

11.2 Parking Generation and Supply Needs

The key issue in parking is a determination of how many spaces are required for a particular development, and where they should be located. These requirements lead to locally based zoning regulations on minimum numbers of spaces that need to be provided when a development is built.

The need for parking spaces depends upon many factors, some of which are difficult to assess. The type and size of land use(s) in a development is a major factor, but so is the general density of the development environment and the amount and quality of public transportation access available.

11.2.1 Parking Generation

The most comprehensive source of parking generation information is the Institute of Transportation Engineers' *Parking Generation*, 2nd Edition [1]. Unfortunately, this document was published in 1987. At this writing (2002), a third edition is in preparation and is expected within the next year.

Parking generation relates the maximum observed number of occupied parking spaces to *one* underlying variable that is used as a surrogate for the size or activity level of the land use involved. Early studies reported in *Parking Principles* [2], established preferred and alternative variables for establishing parking generation rates. These variables are listed in Table 11.1.

A summary of parking generation rates and relationships, compiled from Reference [1], is shown in Table 11.2.

Table 11.2 shows only a sample of the parking generation data from *Parking Generation*. Data for other uses is included in *Parking Generation*, but is generally backed up with only very small sample sizes. Even for those land uses included, the number of sites used to calibrate the values is

Table 11.1: Typical Parking Generation Specification Units

Type of Land Use	Parking–Related Unit	
	Preferred	**Alternative**
Single-Family Residential	Per Dwelling Unit	Per Dwelling Unit with range by number of bedrooms
Apartment Residential	Per Dwelling Unit with range by number of bedrooms	Per Dwelling Unit
Shopping Center	Per 1,000 sq ft GLA[*]	N/A
Other Retail	Per 1,000 sq ft GFA[**]	N/A
Office	Per Employee	Per 1,000 sq ft GFA[**]
Industrial	Per Employee	Per 1,000 sq ft GFA[**]
Hospital	Per Employee	Per Bed
Medical/Dental	Per Doctor	Per Office
Nursing Home	Per Employee	Per Bed
Hotel/Motel	Per Unit	N/A
Restaurant	Per Seat	Per 1,000 sq ft GFA[**]
Bank	Per 1,000 sq ft GFA[**]	N/A
Public Assembly	Per Seat	N/A
Bowling Alley	Per Lane	Per 1,000 sq ft GFA[**]
Library	Per 1,000 sq ft GFA[**]	N/A

*GLA = gross leaseable area

**GFA = gross floor area

(Used with permission of Transportation Research Board, "Parking Principles," *Special Report 125*, Washington DC, 1971, Table 3.1, pg. 34.)

Table 11.2: Typical Parking Generation Rates

Land Use[a]	Avg Rate	Per	Equation[b]	R^2	No. of Studies
Residential—Low/Mid-Rise Apartment (Sat)	1.21	Dwelling Unit	$P = 1.32\,X - 53.0$	0.962	11
Residential—High-Rise Apartment (wkdy)	0.88	Dwelling Unit	$P = 0.34\,X + 105.0$	0.516	7
Residential—Condominium (wkdy)	1.11	Dwelling Unit	$P = 1.29\,X - 23.0$	0.908	32
Convention Hotel (wkdy)	0.81	Room	$\text{Ln}\,P = 0.71\,\text{Ln}\,X + 1.42$	0.498	22
Motel—w/Restaurant/Lounge (wkdy)	0.89	Room	$P = 0.42\,X + 70.0$	0.496	10
Motel—w/o Restaurant/Lounge (wkdy)	0.51	Room	$P = 0.37\,X + 11.0$	0.854	4
Industrial—Light (wkdy)	1.55	1,000 sq ft GFA	$\text{Ln}\,P = 0.94\,\text{Ln}\,X + 0.54$	0.774	8
Industry—Industrial Park (wkdy)	1.48	1,000 sq ft GFA	$\text{Ln}\,P = 1.38\,\text{Ln}\,X - 2.10$	0.763	5
Industry—Manufacturing (wkdy)	1.59	1,000 sq ft Building area	$P = 1.02\,X + 51.0$	0.961	20
Medical—Hospital (wkdy)	1.79	Bed	$\text{Ln}\,P = 0.95\,\text{Ln}\,X + 0.81$	0.637	20
Medical—Medical/Dental Clinic/Office (wkdy)	4.11	1,000 sq ft Building Area	$\text{Ln}\,P = 0.82\,\text{Ln}\,X + 1.81$	0.958	40
Office—General Office Building (wkdy)	2.79	1,000 sq ft Building Area	$\text{Ln}\,P = 0.93\,\text{Ln}\,X + 1.253$	0.870	207
Office—Office Park (wkdy)	2.52	1,000 sq ft Building Area	$P = 2.58\,X - 14.03$	0.906	24
Shopping—Shopping Center (Sat)	3.97	1,000 sq ft GLA	$\text{Ln}\,P = 1.173\,\text{Ln}\,X + 0.064$	0.939	141
Restaurant—Quality Restaurant (Sat)	15.89	1,000 sq ft GLA	$P = 15.35\,X - 23.0$	0.472	34
Restaurant—Family Restaurant	9.08	1,000 sq ft GLA	$P = 9.54\,X - 2.0$	0.928	11
Recreation—Movie Theater (Sat)	0.26	Seat	$P = 0.50\,X - 322.0$	0.837	11
Recreation—Sports Club (wkdy)	4.37	1,000 sq ft GLA	$\text{Ln}\,P = 0.72\,\text{Ln}\,X + 2.35$	0.574	43
Religion—Church or Synagogue (Sat/Sun)	0.43	Attendee	$P = 0.495\,X - 4.0$	0.849	8

[a]parking generation shown for peak day of the week.

[b]P is the peak number of parking spaces occupied; X is the appropriate underlying variable shown in the "Per" column.

(Compiled from *Parking Generation*, 2nd Edition, Institute of Transportation Engineers, Washington DC, 1987.)

not always impressive, and the R^2 values often connote significant variability in parking characteristics.

For this reason, it is always preferable to base projections of parking needs on locally calibrated values, using similar types of land uses and facilities as a basis.

Consider the case of a small office park, consisting of 25,000 sq ft of office space. What is the peak parking load expected to be at this facility? From Table 11.2 for office parks, the average peak parking occupancy is 2.52 per thousand sq ft of building area, or in this case, $2.52 \times 25 = 63$ parking spaces. A more precise estimate might be obtained using the equation related to facility size:

$$P = 2.58X - 14.03 = 2.58*25 - 14.03$$
$$= 50.5 \text{ spaces}$$

This presents a significant range to the engineer—from 51 to 63 parking spaces needed. Thus, while these general guidelines can provide some insight into parking needs, it is important to do localized studies of parking generation to augment national norms.

More recent data concerning shopping centers is available from a comprehensive 1998 study [3]. Over 400 shopping centers were surveyed, resulting in the establishment of recommended "parking ratios," the number of spaces provided per 1,000 sq ft of GLA. Centers were categorized by total size (in GLA), and by the percentage of total center GLA occupied by movie houses, restaurants, and other entertainment uses. The results are summarized in Table 11.3

The guidelines were established such that the 20th peak parking hour of the year is accommodated (i.e., there are only 19 hours of the year when parking demand would exceed the recommended values). Parking demands accommodate both patrons and employees.

Where movie theaters, restaurants, and other entertainment facilities occupy more than 20% of the GLA, a "shared parking" approach is recommended. Parking requirements would be predicted for shopping facilities, and for movies, restaurants, and entertainment facilities separately. Local studies would be used to establish the amount of overlapping usage that might occur (e.g., spaces used by shoppers in the afternoon would be used by movie patrons in the evening).

Consider the following case: a new regional shopping center with 1,000,000 sq ft of GLA is to be built. It is anticipated that about 15% of the GLA will be occupied by movie theaters, restaurants, or other entertainment facilities. How many parking spaces should be provided? From Table 11.3, the center as described would require a parking ratio of 4.65 spaces per 1,000 sq ft GLA, or $4.65 * 1,000 = 4,650$ parking spaces.

Reference [4] presents a more detailed model for predicting peak parking needs. As the model is more detailed, additional input information is needed in order to apply it. Peak parking demand may be estimated as:

$$D = \frac{NKRP*pr}{O} \qquad (11\text{-}1)$$

where: D = parking demand, spaces

N = size of activity measured in appropriate units (floor area, employment, dwelling units, or other appropriate land-use parameters)

K = portion of destinations that occur at any one time

R = person-destinations per day (or other time period) per unit of activity

P = proportion of people arriving by car

O = average auto occupancy

pr = proportion of persons with primary destination at the designated study location

Consider the case of a 400,000 sq ft retail shopping center in the heart of a CBD. The following estimates have been made:

- Approximately 40% of all shoppers are in the CBD for other reasons ($pr = 0.60$)

- Approximately 70% of shoppers travel to the retail center by automobile ($P = 0.70$)

- Approximate total activity at the center is estimated to be 45 person-destinations per 1,000 sq ft of gross leasable area, of which 20% occur during the peak parking accumulation period ($R = 45$; $K = 0.20$)

- The average auto occupancy of travelers to the shopping center is 1.5 persons per car ($O = 1.5$)

Table 11.3: Recommended Parking Ratios (Parking Spaces per 1,000 sq ft of GLA) from a 1998 Study

Center Size (Total GLA)	Percent Usage by Movie Houses, Restaurants, and Other Entertainment				
	0%	5%	10%	15%	20%
0–399,999	4.00	4.00	4.00	4.15	4.30
400,000–419,999	4.00	4.00	4.00	4.15	4.30
420,000–439,999	4.06	4.06	4.06	4.21	4.36
440,000–459,999	4.11	4.11	4.11	4.26	4.41
460,000–479,999	4.17	4.17	4.17	4.32	4.47
480,000–499,999	4.22	4.22	4.22	4.37	4.52
500,000–519,999	4.28	4.28	4.28	4.43	4.58
520,000–539,999	4.33	4.33	4.33	4.48	4.63
540,000–559,999	4.39	4.39	4.39	4.54	4.69
560,000–579,000	4.44	4.44	4.44	4.59	4.74
580,000–599,999	4.50	4.50	4.50	4.65	4.80
600,000–2,500,000	4.50	4.50	4.50	4.65	4.80

(Used with permission of Urban Land Institute, *Parking Requirements for Shopping Centers*, 2nd Edition, Washington DC, 1999, compiled from Appendix A, Recommended Parking Ratios.)

As the unit of size is 1,000 sq ft of gross leasable area, $N = 400$ for this illustration. The peak parking demand may now be estimated using Equation 11-1 as:

$$D = \frac{400 * 45 * 0.20 * 0.70 * 0.60}{1.5}$$
$$= 1,008 \text{ parking spaces}$$

This is equivalent to $1,008/400 = 2.52$ spaces per 1,000 ft of GLA.

While this technique is analytically interesting, it requires that a number of estimates be made concerning parking activity. For the most part, these would be based on data from similar developments in the localized area or region or on nationwide activity information if no local information is available.

11.2.2   Zoning Regulations

Control of parking supply for significant developments is generally maintained through zoning requirements. Local zoning regulations generally specify the minimum number of parking spaces that must be provided for developments of specified type and size. Zoning regulations also often specify needs for handicapped parking and set minimum standards for loading zones.

Reference [4] contains a substantial list of recommended zoning requirements for various types of development in "suburban" settings. "Suburban" settings have little transit access, no significant ridesharing, and little captive walk-in traffic to reduce parking demands. The recommendations are based upon satisfying the 85th percentile demand (i.e., a level of demand that would be exceeded only 15% of the time), and are summarized in Table 11.4. Zoning requirements are generally set 5% to 10% higher than the 85th percentile demand expectation.

The recommended zoning requirements of Table 11.4 would be significantly reduced in urban areas with good transit access, captive walk-in patrons (people working or living in the immediate vicinity of the development), or organized car-pooling programs. In such areas, the modal split characteristics of users must be determined, and parking spaces may be reduced accordingly. Such a modal split estimate must consider local conditions, as this can vary widely. In a typical small urban community, transit may provide 10% to 15% of total access; in Manhattan (New York City), fewer than 5% of major midtown and downtown access is by private automobile.

Table 11.4: Recommended Parking Space Zoning Requirements in Suburban Settings

Land Use	Unit	Parking Spaces Per Unit	
		Peak Parking Demand[a]	Recommended Zoning Requirement
Residential			
Single Family	Dwelling Unit	2.0	2.0
Multiple Family	Dwelling Unit	2.0	2.0
Efficiency/Studio	Dwelling Unit	1.0	1.0
1 Bedroom Apt.	Dwelling Unit	1.5	1.5
2 or more Bed Apt.	Dwelling Unit	2.0	2.0
Elderly Housing	Dwelling Unit	0.7	0.5 + 1 space/day shift employee
Accessory D.U.	Dwelling Unit	1.0	1.0
Commercial Lodging			
Hotel/Motel	Bedroom	1.2	1.0 + spaces for restaurant, lounge, meeting rooms + 0.25 per day shift employee
Sleeping Room	Bedroom	1.0	1.0 + 2.0 for resident manager
Medical Treatment			
Hospitals	Bed	2.5	**Higher of:** 2.7 *OR* 0.33 + 0.4/employee + 0.2/outpatient + 0.25/staff physician
Medical Center	Bed	5.5	**Higher of:** 6.0 *OR* 0.5 + 0.4/employee + 0.2/outpatient + 0.25/staff physician + 0.33/student
Business Offices			
General Office:	1,000 sq ft GFA	3.0	
≤30,000 sq ft			4.0
>30,000 sq ft			3.3
Banks	1,000 sq ft GFA	3.3	3.6
Branch Drive-In Bank w/walk-In window svcs	1,000 sq ft GFA	3.5	5.6
Retail Services			
General Retail	1,000 sq ft GFA	2.2	2.4
Personal Care	1,000 sq ft GFA	3.5	**Higher of:** 4.0 *OR* 2/treatment station
Coin-Operated Laundries	Wash/Dry Clean Machine	0.5	0.5

(Continued)

Table 11.4: Recommended Parking Space Zoning Requirements in Suburban Settings *(Continued)*

Land Use	Unit	Parking Spaces Per Unit	
		Peak Parking Demand[a]	**Recommended Zoning Requirement**
Retail Goods			
General Retail	1,000 sq ft GFA	3.0	3.3
Convenience Store	1,000 sq ft GFA	4.0	4.4
Hard Goods Store	1,000 sq ft GFA	3.0	2.5 + 1.5/1,000 sq ft interior storage and exterior display/storage
Shopping Centers:	1,000 sq ft GFA		4.7
<400,000 sq ft GLA		4.5	5.2
400,001–600,000		5.0	5.8
>600,000sq ft GLA		5.5	
Food and Beverage			
Quality Restaurant	1,000 sq ft GLA	20.0	22.0 + banquet/meeting room needs
Family Restaurant	1,000 sq ft GLA	11.2	12.3 + banquet/meeting room needs
Fast-Food Restaurant	1,000 sq ft GLA	15.4	16.9 (kitchen, serving counter, waiting areas) + 0.5/seat
Educational			
Elementary/Secondary	Classroom	1.5	1.5 (include classrooms and other rooms used by students and faculty) + 0.25/student of driving age
College/University	Population	n/a	1.0/daytime staff & faculty + 0.5/resident & commuting student
Day-Care Center	Employee	n/a	1.0 + 0.1/licensed enrollment capacity
Cultural, Entertainment and Recreational			
General Public Assembly	Max. Occupancy	n/a	0.25
General Recreation	Max. Occupancy	n/a	0.33
Auditorium, Theater or Stadium	Seat	0.35	0.38
Church	Seat	n/a	0.50
Industrial			
General	Employee	0.60–1.00	1.0 + 1.0/1,000 sq ft GFA
Storage, Wholesale or Utility			
General	1,000 sq ft GFA	n/a	0.50+ required spaces for office or sales areas

[a]Typically, the 85th percentile demand, based on analysis of comparative studies.
(Used with permission of Eno Foundation for Transportation, Weant, R. and Levinson, H., *Parking*, Westport, CT, 1990, reformatted from Table 3.2, pgs. 42 and 43.)

In any parking facility, handicapped spaces must be provided as required by federal and local laws and ordinances. Such standards affect both the number of spaces that must be required and their location. The Institute of Transportation Engineers [4] recommends the following minimum standards for provision of handicapped spaces:

- *Office*—0.02 spaces per 1,000 sq ft GFA
- *Bank*—1 to 2 spaces per bank
- *Restaurant*—0.30 spaces per 1,000 sq ft GFA
- *Retail* (<500,000 sq ft GFA)—0.075 spaces per 1,000 sq ft GFA
- *Retail* (≥ 500,000 sq ft GFA)—0.060 spaces per 1,000 sq ft GFA

In all cases, there is an effective minimum of one handicapped space.

11.3 Parking Studies and Characteristics

There are a number of characteristics of parkers and parking that have a significant influence on planning. Critical to parking supply needs are the duration, accumulation, and proximity requirements of parkers. Duration and accumulation are related characteristics. If parking capacity is thought of in terms of "space-hours," then vehicles parked for a longer duration consume more of that capacity than vehicles parked for only a short period. In any area, or at any specific facility, the goal is to provide enough parking spaces to accommodate the maximum accumulation on a typical day.

11.3.1 Proximity: How Far Will Parkers Walk?

Maximum walking distances that parkers will tolerate vary with trip purpose and urban area size. In general, tolerable walking distances are longer for work trips than for any other type of trip, perhaps because of the relatively long duration involved. Longer walking distances are tolerated for off-street parking spaces as opposed to on-street (or curb) parking spaces. As the urban area population increases, longer walking distances are experienced.

The willingness of parkers to walk certain distances to (or from) their destination to their car must be well understood, as it will have a significant influence over where parking capacity must be provided. Under any conditions, drivers tend to seek parking spaces as close as possible to their destination. Even in cities of large population (1,000,000 to 2,000,000), 75% of drivers park within 0.25 mile of their final destination.

Table 11.5 shows the distribution of walking distances between parking places and final destinations in urban areas. The distribution is based on studies in five different cities (Atlanta, Pittsburgh, Dallas, Denver, and Seattle), as reported in Reference [4].

As indicated in this table, parkers like to be close to their destination. One-half (50%) of all drivers park within 500 ft of their destination. Figure 11.1 shows average walking distances to and from parking spaces vs. the total urban area population.

Again, this data emphasizes the need to place parking capacity in close proximity to the destination(s) served. Even in an urban region of over 10,000,000 population, the average walking distance to a parking place is approximately 900 ft.

Trip purpose and trip duration also affect the walking distances drivers are willing to accommodate. For shopping or other trips where things must be carried, shorter walking distances are sought. For short-term parking, such as to get a newspaper or a take-out order of food, short walking distances are also sought. Drivers will not walk 10 minutes if they are going to be parked for only 5 minutes. In locating parking capacity, general knowledge of parkers' characteristics is important, but local studies would provide a more accurate picture. In many cases, however, application of common sense and professional judgment is also an important component.

11.3.2 Parking Inventories

One of the most important studies to be conducted in any overall assessment of parking needs is an inventory of existing parking supply. Such inventories include observations of the number of parking spaces and their location,

Table 11.5: CBD Walking Distances to Parking Spaced

Distance		% Walking This Distance or Further	
Feet	Miles	Mean	Range
0	0	100	
250	0.05	70	60–80
500	0.10	50	40–60
750	0.14	35	25–45
1,000	0.19	27	17–37
1,500	0.28	16	8–24
2,000	0.38	10	5–15
3,000	0.57	4	0–8
4,000	0.76	3	0–6
5,000+	0.95+	1	0–2

(Used with permission of Eno Foundation for Transportation, Weant, R. and Levinson, H., *Parking*, Westport, CT, 1990, Table 6.3, pg. 98.)

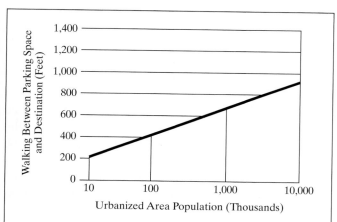

Figure 11.1: Average Walking Distance by Urbanized Area Population (Used with permission of Eno Foundation for Transportation, Weant, R. and Levinson, H., *Parking*, Westport, CT, 1990, Figure 6.5, pg. 98.)

time restrictions on use of parking spaces, and the type of parking facility (e.g., on-street, off-street lot, off-street garage). Most parking inventory data is collected manually, with observers canvassing an area on foot, counting and noting curb spaces and applicable time restrictions, as well as recording the location, type, and capacity of off-street parking facilities. Use of intelligent transportation

system technologies will begin to enhance the quantity of information available and the ease of accessing it. Some parking facilities have begun to use electronic tags (such as EZ Pass) to assess fees. Such a process, however, can also keep track of parking durations and accumulations on a real-time basis. Smart parking meters can provide the same types of information for curb parking spaces.

To facilitate the recording of parking locations, the study area is usually mapped and precoded in a systematic fashion. Figure 11.2 illustrates a simple coding system for blocks and block faces. Figure 11.3 illustrates the field sheets that would be used by observers. Curb parking places are subdivided by parking restrictions and meter duration limits. Where several lines of a field sheet are needed for a given block face, a subtotal is prepared and shown. Where curb spaces are not clearly marked, curb lengths are used to estimate the number of available spaces, using the following guidelines:

- Parallel parking: 23 ft/stall

- Angle parking: 12.0 ft/stall

- 90-degree parking: 9.5 ft/stall

While the parking inventory basically counts the number of spaces available during some period of interest—often the 8- to 11-hour business day—parking supply evaluations must take into account regulatory and time restrictions on those spaces and the average parking duration for the area. Total parking supply can be measured in terms of how many vehicles can be parked during the period of interest within the study area:

$$P = \left(\frac{\sum_n NT}{D} \right) * F \tag{11-2}$$

where: P = parking supply, vehs

N = no. of spaces of a given type and time restriction

T = time that N spaces of a given type and time restriction are available during the study period, hrs

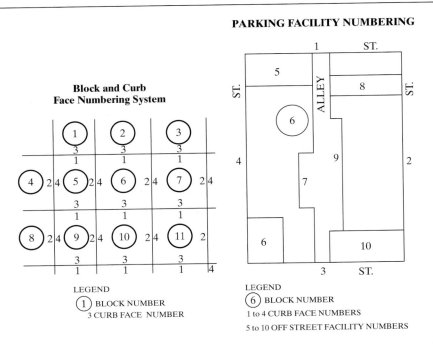

Figure 11.2: Illustrative System for Parking Location Coding (Used with permission of Institute of Transportation Engineers, Box, P. and Oppenlander, J., *Manual of Traffic Engineering Studies*, 4th Edition, Washington DC, 1976, Figs 10.1 and 10.2, pg. 131.)

D = average parking duration during the study period, hrs/veh

F = insufficiency factor to account for turnover—values range from 0.85 to 0.95 and increase as average duration increases

Consider an example in which a 11-hour study of an area revealed that there were 450 spaces available for the full 12 hours, 280 spaces available for 6 hours, 150 spaces available for 7 hours, and 100 spaces available for 5 hours. The average parking duration in the area was 1.4 hours. Parking supply in this study area is computed as:

$$P = \left\{ \frac{[(450*12) + (280*6) + (150*7) + (100*5)]}{1.4} \right\} * 0.90$$

$$= 5,548 \text{ vehs}$$

where an insufficiency factor of 0.90 is used.

This result means that 5,548 vehicles could be parked in the study area over the 11-hour period of the study. It *does not* mean that all 5,548 vehicles could be parked at the same time. This analysis, however, requires that the average parking duration be known. Determining this important factor is discussed in the next section.

Inventory data can be displayed in tabular form, usually similar to that illustrated in Figure 11.3, or can be graphically displayed on coded maps. Maps provide a good overview, but cannot contain the detailed information provided in tabular summaries. Therefore, maps and other graphic displays are virtually always accompanied by tables.

11.3.3 Accumulation and Duration

Parking accumulation is defined as the total number of vehicles parked at any given time. Many parking studies seek to establish the distribution of parking accumulation

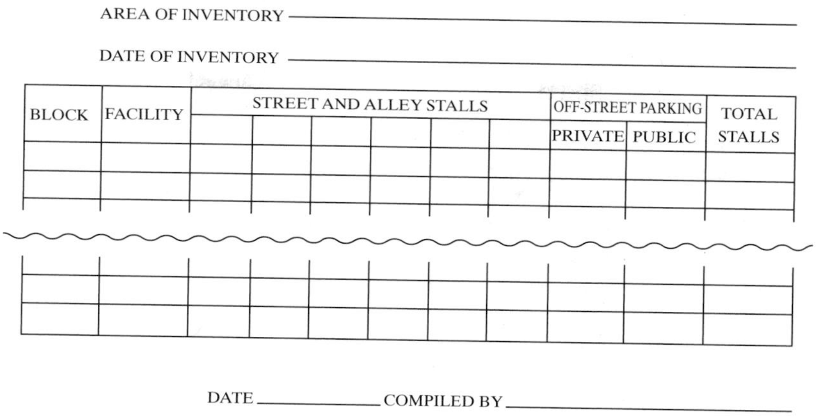

AREA OF INVENTORY _____

DATE OF INVENTORY _____

BLOCK	FACILITY	STREET AND ALLEY STALLS						OFF-STREET PARKING		TOTAL STALLS
								PRIVATE	PUBLIC	

DATE _____ COMPILED BY _____

Figure 11.3: A Parking Inventory Field Sheet (Used with permission of Institute of Transportation Engineers, Box, P. and Oppenlander, J., *Manual of Traffic Engineering Studies*, 4th Edition, Washington DC, 1976, Fig 10.3, pg. 133.)

over time to determine the peak accumulation and when it occurs. Of course, observed parking accumulations are constrained by parking supply; thus, parking demand that is constrained by lack of supply must be estimated using other means.

Nationwide studies have shown that parking accumulation in most cities has increased over time. Total accumulation in an urban area, however, is strongly related to the urbanized area population, as illustrated in Figure 11.4.

Parking duration is the length of time that individual vehicles remain parked. This characteristic is, therefore, a distribution of individual values, and both the distribution and the average value are of great interest.

Like parking accumulation, average parking durations are related to the size of the urban area, with average duration increasing with urban area population, as shown in Figure 11.5. Average duration also varies considerably with trip purpose, as indicated in Table 11.6, which summarizes the results of studies in Boston in 1972 and Charlotte in 1987.

From Table 11.6, it is obvious that durations vary widely from location to location. The Charlotte results are quite different from those obtained in Boston. Thus, local studies of both parking duration and parking accumulation are important elements of an overall approach to the planning and operation of parking facilities.

The most commonly used technique for observing duration and accumulation characteristics of curb parking and surface parking lots is the recording of license plate numbers of parked vehicles. At regular intervals ranging from 10 to 30 minutes, an observer walks a particular route (usually up one block face and down the opposite block face), and records the license plate numbers of vehicles occupying each parking space. A typical field sheet is shown in Figure 11.6.

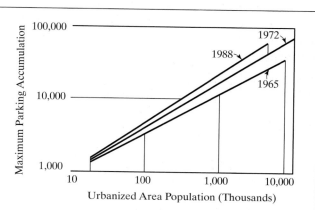

Figure 11.4: Parking Accumulation in Urbanized Areas by Population (Used with permission of Eno Foundation for Transportation, Weant, R., and Levinson, H., *Parking*, Westport, CT, 1990, Figure 6.8, pg. 100.)

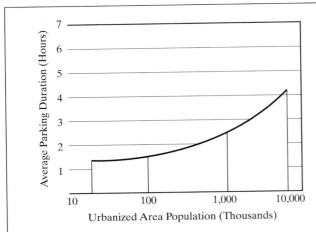

Figure 11.5: Parking Duration versus Urbanized Area Population (Used with permission of Eno Foundation for Transportation, Weant, R. and Levinson, H., *Parking*, Westport, CT, 1990, Figure 6.4, pg. 97.)

Table 11.6: Average Urban Parking Durations by Trip Purpose

Trip Purpose	Average Duration (hours, minutes)	
	Boston (1972)	Charlotte (1987)
Work		
Manager	5h, 30m	
Employee	5h, 59m	
All		8h, 8m
Personal Business	2h, 6m	1h, 5m
Sales/Employment Business	2h, 14m	3h, 32m
Service	2h, 9m	
Recreational	2h, 18m	
Shopping	1h, 57m	1h, 29m
Other	3h, 12m	4h, 17m
All Purposes (Average)	4h, 20m	1h, 41m

(Used with permission of Eno Foundation for Transportation, Weant, R. and Levinson, H., *Parking*, Westport, CT, 1990, Table 6.2, pg. 97.)

Each defined parking space is listed on the field sheet prepared for the specific study, along with any time restrictions associated with it. A variety of special notations can be used to indicate a variety of circumstances, such as "T" for truck, "TK" for illegally parked and ticketed vehicle, and so on. One observer can be expected to observe up to 60 spaces every 15 minutes. Study areas, therefore, must be carefully mapped to allow planning of routes for complete data coverage.

Analysis of the data involves several summaries and computations that can be made using the field sheet information:

- *Accumulation totals.* Each column of each field sheet is summed to provide the total accumulation of parked vehicles within each time period on each observer's route.

- *Duration distribution.* By observing the license plate records of each space, vehicles can be classified as having been parked for one interval, two intervals, three intervals, etc. By examining each line of each field sheet, a duration distribution is created.

- *Violations.* The number of vehicles illegally parked, either because they occupy an illegal

space or have exceeded the legal time restriction of a space, should be noted.

The average parking duration is computed as:

$$D = \frac{\sum\limits_{x}(N_x * X * I)}{N_T} \qquad (11\text{-}3)$$

where: D = average parking duration, h/veh

N_x = number of vehicles parked for x intervals

X = number of intervals parked

I = length of the observation interval, h

N_T = total number of parked vehicles observed

Another useful statistic is the parking turnover rate, *TR*. This rate indicates the number of parkers that, on average, use a parking stall over a period of one hour.

LICENSE PLATE CHECK FIELD DATA SHEET

City ________ Date 10 MAY 1978 Recorded by JONES Side of Street W

Street WRIGHT between 5th and 6th

Codes: 000 last three digits of license number. ✓ for repeat number from prior circuit — for empty space

Space and Regulation	Time circuit begins											
	07	07 30	08	08 30	09	09 30	10	10 30	11	11 30	12 00	
5th												
X-WALK	—	—	—									
NPHC	—	—	—									
I HR M	—	713	✓	✓TK								
" M	631	✓	(✓)	971								
" M	512	34∠	✓	019								
DRIVEWAY	—	—	—	—								
"	—	—	—	613								
I HR M	—	—	418	✓								
" M	117	220	✓	989								
" M	—	148	096	✓								
FIRE HYD	—	—	—	—								
I HR M	042	—	216	✓								
N PHC	—	—	—	774								
X-WALK	—	—	—	—								
6th												

Figure 11.6: A License-Plate Parking Survey Sheet (Used with permission of Institute of Transportation Engineers, Box, P. and Oppenlander, J., *Manual of Traffic Engineering Studies*, 4th Edition, Washington DC, 1976, Fig 10-6, pg. 140.)

It is computed as:

$$TR = \frac{N_T}{P_S * T_S} \tag{11-4}$$

where: TR = parking turnover rate, veh/stall/h

N_T = total number of parked vehicles observed

P_S = total number of legal parking stalls

T_S = duration of the study period, h

The average duration and turnover rate may be computed for each field sheet, for sectors of the study area, and/or for the study area as a whole. Table 11.7 shows a typical field sheet resulting from one observer's route. Table 11.8 shows how data from individual field sheets can be summarized to obtain areawide totals.

Note that the survey includes only the study period. Thus, vehicles parked at 3:00 PM will have a duration that ends at that time, even though they may remain parked for an additional time period outside the study limits. For convenience, only the last three numbers of the license plates are recorded; in most states, the initial two or three letters/numbers represent a code indicating where the plate registration was issued. Thus, these letters/numbers are often repetitive on many plates.

The average duration for the study area, based on the summary of Table 11.8 (b) is:

$$D = \frac{(875*1*0.5) + (490*2*0.5) + (308*3*0.5) + (275*4*0.5) + (143*5*0.5) + (28*6*0.5)}{2119}$$

$$D = 1.12 \text{ h/veh}$$

The turnover rate is:

$$TR = \frac{2119}{1500*7} = 0.20 \text{ veh/stall/h}$$

Table 11.7: Summary and Computations from a Typical Parking Survey Field Sheet

Pkg* Space	8:00	8:30	9:00	9:30	10:00	10:30	11:00	11:30	12:00	12:30	1:00	1:30	2:00	2:30	3:00
1	—	—	861	√	√	—	136	—	140	√	—	-	201	√	070
2	470	√	380	—	—	412	307	—	900	√	√	√	855	999	-
3	-	211	√	√	√	400	√	√	—	—	666	-	√	√	-
4	175	√	√	500	√	222	—	—	616	√	√	√	√	√	-
5	333	—	—	380	√	√	420	√	707	—	—	-	-	-	-
hydrant	—	—	—	—	—	—	—	242TK	—	—	—	-	-	-	-
1-hr	—	—	484	√	909	—	811	√	√	158	√	√	685	√	-
1-hr	301	—	—	525	√	√	696	√	422	—	299	√	√	288	892
1-hr	—	675	895	√	√	703	√	819	—	401	√	√	√	-	412
1-hr	406	—	442	781	882	√	√	√	444	—	903	√	-	893	-
1-hr	—	—	115	√	618	√	818	√	√	906	√	√	-	-	807
2-hr	—	509	√	√	—	705	√	√	√	688	√	696	-	√	√
2-hr	—	—	214	√	√	√	209	—	248	—	797	√	√	-	√
2-hr	101	√	√	√	—	531	√	—	940	√	√	√	628	√	√
2-hr	—	392	√	√	√	251	√	772	—	835	√	√	√	-	-
Accum.	**6**	**7**	**12**	**13**	**11**	**12**	**13**	**10**	**11**	**10**	**12**	**10**	**10**	**7**	**8**

*All data for Block Face 61; timed spaces indicate parking meter limits; √ = same vehicle parked in space.

The maximum observed accumulation occurs at 11:00 AM (from Table 11.8 (a)), and is 1,410 vehicles, which represents use of $(1,410/1,500) * 100 = 94\%$ of available spaces.

For off-street facilities, the study procedure is somewhat altered, with counts of the number of entering and departing vehicles recorded by 15-minute intervals. Accumulation estimates are based on a starting count of occupancy in the facility and the difference between entering and departing vehicles. A duration distribution for off-street facilities can also be obtained if the license-plate numbers of entering and departing vehicles are also recorded.

As noted earlier, accumulation and duration observations cannot reflect repressed demand due to inadequacies in the parking supply. Several findings, however, would serve to indicate that deficiencies exist:

• Large numbers of illegally-parked vehicles

• Large numbers of vehicles parked unusually long distances from primary generators

• Maximum accumulations that occur for long periods of the day and/or where the maximum accumulation is virtually equal to the number of spaces legally available

Even these indications do not reflect trips either not made at all, or those diverted to other locations because of parking constraints. A cordon-count study (see Chapter 8) may be used to estimate the total number of vehicles both parked and circulating within a study area, but trips not made are still not reflected in the results.

11.3.4 Other Types of Parking Studies

A number of other techniques can be used to gain information concerning parked vehicles and parkers. Origins of parked vehicles can be obtained by recording the license plate numbers of parked vehicles and petitioning the state motor vehicle agency for home addresses (which are assumed to be the origins). This technique, which requires special permission from state authorities,

Table 11.8:   Summary Data for an Entire Study Area Parking Survey

Block No.	Accumulation for Interval (1500 Total Stalls)														
	8:00	8:30	9:00	9:30	10:00	10:30	11:00	11:30	12:00	12:30	1:00	1:30	2:00	2:30	3:00
61	6	7	12	13	11	12	13	10	11	10	12	10	10	7	8
62	5	10	15	14	16	18	17	15	15	10	9	9	7	7	8
⋮	⋮	⋮	⋮	⋮	⋮	⋮	⋮	⋮	⋮	⋮	⋮	⋮	⋮	⋮	
180	7	8	13	13	18	14	15	15	11	14	16	10	9	9	6
181	7	5	18	16	12	14	13	11	11	10	10	10	6	6	5
Total	**806**	**900**	**1,106**	**1,285**	**1,311**	**1,300**	**1,410**	**1,309**	**1,183**	**1,002**	**920**	**935**	**970**	**726**	**694**

(a) Summarizing Field Sheets for Accumulation Totals

Block Face No.	Number of Intervals Parked					
	1	2	3	4	5	6
61	28	17	14	8	2	1
62	32	19	20	7	1	3
⋮	⋮	⋮	⋮	⋮	⋮	⋮
180	24	15	12	10	3	0
181	35	17	11	9	4	2
Total	**875**	**490**	**308**	**275**	**143**	**28**

$\Sigma = 2119$ total parkers observed

(b) Summarizing Field Sheets for Duration Distribution

is frequently used at shopping centers, stadiums, and other large trip attractors.

Interviews of parkers are also useful and are most easily conducted at large trip attraction locations. Basic information on trip purpose, duration, distance walked, etc. can be obtained. In addition, however, attitudinal and background parker characteristic information can also be obtained to gain greater insight into how parking conditions affect users.

11.4 Design Aspects of Parking Facilities

Off-street parking facilities are provided as (1) surface lots or (2) parking garages. The latter may be above ground, below ground, or a combination of both. The construction costs of both surface lots and garages vary significantly depending upon location and specific site conditions. In general, surface lots are considerably cheaper than garages. Typically, surface lots cost between $1,500 and $2,500 per space provided. Garages are more complex, and below-ground garages are far more costly than above-ground structures. Typical costs for above-ground garages range from $8,000 to $11,000 per space, while below-ground garages may cost between $16,000 and $20,000 per space. The decision of how to provide off-street parking involves many considerations, including the availability of land, the amount of parking needed, and the cost to provide it.

Reference [4] lists three key objectives in the design of a parking facility:

- A parking facility must be convenient and safe for the intended users.

- A parking facility should be space-efficient and economical to operate.

- A parking facility should be compatible with its environs.

Convenience and safety involve many issues, including proximity to major destinations, adequate access and egress facilities (including reservoir space), a simple and efficient internal circulation system, adequate stall dimensions, and basic security. The latter refers to security against theft of vehicles and security from muggings and other personal crimes. Space efficiency implies that while appropriate circulation, stall, and reservoir space must be provided, parking facilities should be designed to maximize parking capacity and minimize wasted space. The third objective involves issues of architectural beauty and ensuring that the facility and the vehicle-trips it generates do not present a visual or auditory disruption to the environment in its immediate area.

11.4.1   Some Basic Parking Dimensions

Basic parking dimensions are based on one of two "design vehicles." Modern parking facilities often make use of separated parking areas for "small cars" to maximize total parking capacity. Figure 11.7 illustrates the basic criteria for the two design vehicles used in parking facility design:

- Large cars

- Small cars

Parking Stall Width

Parking stalls must be wide enough to encompass the vehicle and allow for door-opening clearance. The minimum door-opening clearance is 22 inches, but this may be increased to 26 inches where turnover rates are high. Only one door-opening clearance is provided per stall, as the parked vehicle and its adjacent neighbor can utilize *the same clearance space.* For large cars, the parking stall width should range between $77 + 22 = 99$ inches (8.25 ft) and $77 + 26 = 103$ inches (8.58 ft).

Reference [5] recommends the use of four parking classes, depending upon turnover rates and typical users. Recommended design guidelines for large-car stall widths are shown in Table 11.9.

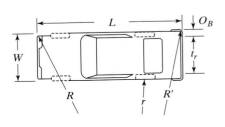

W = overall width, inches
L = overall length, inches
O_R = rear overhang, ft
O_B = body overhang from center of rear tire, ft
t_r = width from center of rear tires, ft

Minimum Turning Radius
  r = inside rear wheel, ft
  R = outside point, front bumper, ft
  R' = outside point, rear bumper, ft

Dimension	Design Vehicle	
	Large Car	Small Car
Width, W (inches)	77	66
Length, L (inches)	215	175
Outside Front Bumper Radius, R (ft)	20.5	18.0
Inside Rear Wheel Radius, r (ft)	12.0	9.6
Rear Width, t_r (ft)	5.1	4.6
Body Overhang, Rear Tire, O_B (ft)	0.63	0.46
Rear Radius, R' (ft)	17.4	15.0

Figure 11.7: Design Vehicles for Parking Design (Used with permission of Eno Foundation for Transportation, Weant, R. and Levinson, H., *Parking*, Westport, CT, 1990, reformatted from Table 8.1, pg. 157.)

For small cars, these guidelines suggest a parking stall width between $66 + 22 = 88$ inches (7.3 ft) and $66 + 26 = 92$ inches (7.7 ft). A 7.6-ft design standard is often applied to small-car stall widths. Reference [5] suggests a design width of 8.0 ft for parking classes A and B (Table 11.9), and 7.5 ft for parking classes C and D.

Parking Stall Length and Depth

Parking stall length is measured parallel to the parking angle. It is generally taken as the length of the design vehicle plus 6 inches for bumper clearance. This implies a length of $215 + 6 = 221$ inches (18.4 ft) for large cars and $175 + 6 = 181$ inches (15.1 ft) for small cars.

The depth of a parking stall is the 90° projection of the design vehicle length and 6-inch bumper clearance. For a 90° parking stall, the length and depth of the stall are equivalent. For other-angle parking, the depth of the stall is smaller than the length.

Aisle Width

Aisles in parking lots must be sufficiently wide to allow drivers to safely and conveniently enter and leave parking stalls in a minimum number of maneuvers, usually one on entry and two on departure. As stalls become narrower, the aisles need to be a bit wider to achieve this. Aisles also carry circulating traffic and accommodate pedestrians walking to or from their vehicles. Aisle width depends upon the angle of parking and upon whether the aisle serves one-way or two-way traffic.

11.4.2 Parking Modules

A "parking module" refers to the basic layout of one aisle with a set of parking stalls on both sides of the aisle. There are many potential ways to lay out a parking module. For 90° stalls, two-way aisles are virtually always used, as vehicles may enter parking stalls conveniently from either approach direction. Where angle parking is used, vehicle may enter a stall in only one direction of travel and must depart in the same direction. In most cases, angle parking is arranged using one-way aisles, and stalls on both sides of the aisle are arranged to permit entries and exits from and to the same direction of travel. Angle stalls can also be arranged such that stalls on one side of the aisle are approached from the opposite direction as those on the other side of the aisle. In such cases, two-way aisles must be provided. Figure 11.8 defines the basic dimensions of a parking module.

Note that Figure 11.8 shows four different ways of laying out a module. One module width applies if both sets of stalls butt up against walls or other horizontal physical barriers. Another applies if both sets of stalls are "interlocked" (i.e., stalls interlock with those of the next adjacent parking module). A third applies if one set of stalls is against a wall, while the other is interlocked. Yet another module reflects only a single set of stalls against a wall. Table 11.10 summarizes critical dimensions for various types of parking modules.

Table 11.9: Stall Width Design Criteria for Various Parking Classifications

Parking Class	Stall Width (ft)	Typical Turnover			Typical Uses
		Low	Medium	High	
A	9.00			X	Retail customers, banks, fast foods, other very high turnover facilities
B	8.75		X	X	Retail customers, visitors
C	8.50	X	X		Visitors, office employees, residential, airport, hospitals
D	8.25	X			Industrial, commuter, universities

(Used with permission of Institute of Transportation Engineers, *Guidelines for Parking Facility Location and Design: A Recommended Practice of the ITE*, Washington DC, 1994, Table 1, pg. 7.)

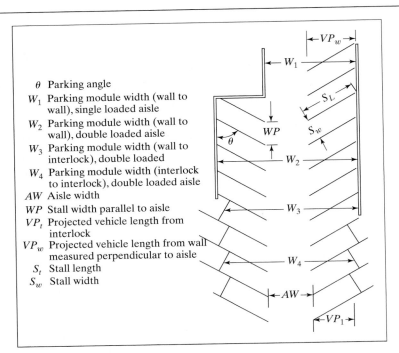

θ Parking angle
W_1 Parking module width (wall to wall), single loaded aisle
W_2 Parking module width (wall to wall), double loaded aisle
W_3 Parking module width (wall to interlock), double loaded
W_4 Parking module width (interlock to interlock), double loaded aisle
AW Aisle width
WP Stall width parallel to aisle
VP_t Projected vehicle length from interlock
VP_w Projected vehicle length from wall measured perpendicular to aisle
S_t Stall length
S_w Stall width

Figure 11.8: Dimensional Elements of Parking Modules (Used with permission of Institute of Transportation Engineers, *Guidelines for Parking Facility Location and Design: A Recommended Practice of the ITE*, Washington DC, 1994, Fig 3, pg. 6.)

Providing separate parking areas for large and small vehicles presents a number of operating problems. Areas must be clearly signed, and circulation must allow drivers to easily get to both types of parking space. As the mix of small and large cars is a variable, there may be times when large cars try to force themselves into small-car stalls and times when small cars occupy large-car stalls. Large cars will have difficulty not only in fitting into small-car stalls but in maneuvering in aisles designed for small cars.

Some designers advocate using a single size for all parking stalls. Ideally, large-car dimensions would be provided for all spaces. A policy colloquially referred to as "one size fits all (OSFA)" is based on uniform stall and modular dimensions that are taken to be the weighted average of small- and large-car criteria. The weighting is on the basis of expected proportions of users in each category. Some traffic engineers have advocated this policy as a means of providing more efficient utilization of scarce off-street parking space. All of the problems, however, associated with large cars attempting to use small-car

spaces and aisles would exist throughout such a facility. Thus, the engineer must carefully weight the negative operating impacts of OSFA against the increased utilization of space OSFA provides.

11.4.3 Separating Small and Large Vehicle Areas

Reference [5] recommends a number of different techniques for separating (or integrating) small and large vehicle stalls where both are provided. Figure 11.9 illustrates various patterns for integrating small and large vehicle stalls.

- *Complete separation of small and large size spaces.* Maximum parking layout efficiency can be obtained by completely separating small and large vehicle parking areas. In this case, all of the reduced dimensions of the small-car requirements can be fully utilized. The downside is that careful

Table 11.10: Parking Module Layout Dimension Guidelines

Basic Layout	Parking Class	S_w Stall Width (ft)	WP Stall Width (ft)	VP_w Stall Depth to Wall (ft)	VP_i Stall Depth to Interlock (ft)	AW Aisle Width (ft)	Modules	
							W_2 Wall to Wall (ft)	W_4 Interlock to Interlock (ft)
Large Cars								
2-Way Aisle–90°	A	9.00	9.00	17.5	17.5	26.0	61.0	61.0
	B	8.75	8.75	17.5	17.5	26.0	61.0	61.0
	C	8.50	8.50	17.5	17.5	26.0	61.0	61.0
	D	8.25	8.25	17.5	17.5	26.0	61.0	61.0
2-Way Aisle–60°	A	9.00	10.4	18.0	16.5	26.0	62.0	59.0
	B	8.75	10.1	18.0	16.5	26.0	62.0	59.0
	C	8.50	9.8	18.0	16.5	26.0	62.0	59.0
	D	8.25	9.5	18.0	16.5	26.0	62.0	59.0
1-Way Aisle–75°	A	9.00	9.3	18.5	17.5	22.0	59.0	57.0
	B	8.75	9.0	18.5	17.5	22.0	59.0	57.0
	C	8.50	8.8	18.5	17.5	22.0	59.0	57.0
	D	8.25	8.5	18.5	17.5	22.0	59.0	57.0
1-Way Aisle–60°	A	9.00	10.4	18.0	16.5	18.0	54.0	51.0
	B	8.75	10.1	18.0	16.5	18.0	54.0	51.0
	C	8.50	9.8	18.0	16.5	18.0	54.0	51.0
	D	8.25	9.5	18.0	16.5	18.0	54.0	51.0
1-Way Aisle–45°	A	9.00	12.7	16.5	14.5	15.0	48.0	44.0
	B	8.75	12.4	16.5	14.5	15.0	48.0	44.0
	C	8.50	12.0	16.5	14.5	15.0	48.0	44.0
	D	8.25	11.7	16.5	14.5	15.0	48.0	44.0
Small Cars*								
2-Way Aisle–90°	A/B	8.0	8.0	15.0	15.0	21.0	51.0	51.0
	C/D	7.5	7.5	15.0	15.0	21.0	51.0	51.0
2-Way Aisle–60°	A/B	8.0	9.3	15.4	14.0	21.0	52.0	50.0
	C/D	7.5	8.7	15.4	14.0	21.0	52.0	50.0
1-Way Aisle–75°	A/B	8.0	8.3	16.0	15.1	17.0	49.0	47.0
	C/D	7.5	7.8	16.0	15.1	17.0	49.0	47.0
1-Way Aisle–60°	A/B	8.0	9.3	15.4	14.0	15.0	46.0	43.0
	C/D	7.5	8.7	15.4	14.0	15.0	46.0	43.0
1-Way Aisle–45°	A/B	8.0	11.3	14.2	12.3	13.0	42.0	38.0
	B/C	7.5	10.6	14.2	12.3	13.0	42.0	38.0

*While various angles are presented, the vast majority of small car layouts are for 90° parking.
(Used with permission of Institute of Transportation Engineers, *Guidelines for Parking Facility Location and Design: A Recommended Practice of the ITE*, Washington DC, 1994, Tables 2 and 3, pgs. 8 and 9.)

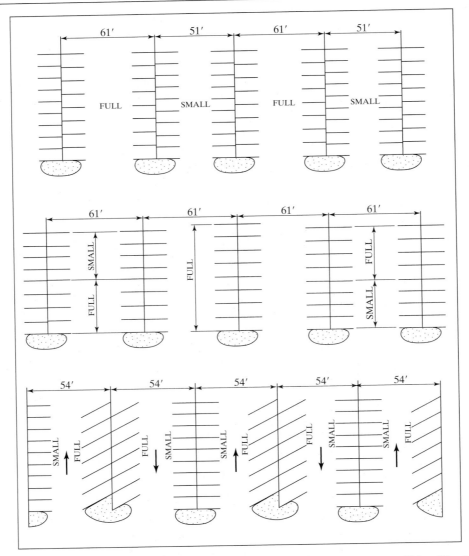

Figure 11.9: Various Layouts for Integrating Small-Car and Large-Car Stalls in Parking Facilities (Used with permission of Institute of Transportation Engineers, *Guidelines for Parking Facility Location and Design: A Recommended Practice of the ITE*, Washington DC, 1994, Fig. 5, pg. 10.)

signing must be used to direct vehicles to the appropriate areas, and one type of vehicle user is virtually guaranteed to be disadvantaged by having spaces further away from primary generators.

- *Mixing small and large size spaces in alternating rows*. In this pattern, large and small car modules

are alternated, as shown at the top of Figure 11.9, producing about a 50–50 split in the number of each type of stall. The advantage is that both types of stall are about equally convenient, and drivers do not have the opportunity to enter a completely "wrong" area for their vehicles. Problems will arise if a 50–50 split is inappropriate for the prevailing mix of vehicles. This pattern can be modified by using a double-alternating layout of two large-car aisles with two small-car aisles.

- *Small and large size spaces in the same row (or module).* This pattern is illustrated in the center of Figure 11.9. A portion of each row is allotted to each type of vehicle. The advantages of this layout are that any mix of small vs. large vehicle spaces can be implemented, and no driver will ever wind up in the "wrong" row. One disadvantage is that all aisles must conform to large-car criteria.

- *Cross-aisle separation.* This layout is illustrated in the bottom of Figure 11.9. Small-car stalls are provided on one side of the aisle, while large-car stalls are provided on the other. In this case, small-car stalls are always placed at a 90° angle, while large-car spaces are at a shallower angle. Use of spaces is somewhat self-enforced, as drivers of large cars find it difficult to maneuver into small-car spaces. If this pattern is used throughout the parking lot, however, the balance of spaces is more in favor of small cars (more than 50%), which may not be appropriate.

While it is preferable to use one layout throughout a parking facility (so as not to confuse drivers), it is always possible to carefully implement more than a single pattern. This may be necessary to obtain maximum utilization of space, but it must be accompanied by careful signing.

11.4.4 Parking Garages

Parking garages are subject to the same stall and module requirements as surface parking lots and have the same requirements for reservoir areas and circulation. The structure of a parking garage, however, presents additional constraints, such as building dimensions and the location of structural columns and other features. Ideal module and stall dimensions must sometimes be compromised to work around these structural features.

Parking garages have the additional burden of providing vertical as well as horizontal circulation for vehicles. This involves a general design and layout that includes a ramp system, at least where self-parking is involved. Some smaller attendant-parking garages use

elevators to move vertically, but this is a slow and often inefficient process.

Ramping systems fall into two general categories:

- *Clearway systems.* Ramps for inter-floor circulation are completely separated from ramps providing entry and exit to and from the parking garage.

- *Adjacent parking systems.* Part or all of the ramp travel is performed on aisles that provide direct access to adjacent parking spaces.

The former provides for easier and safer movement with minimum delays. Such systems, however, preempt a relatively large amount of potential parking space and are therefore usually used only in very large facilities.

Figure 11.10, on the next two pages, illustrates a number of alternative ramp layouts that may be used in parking garages.

In some attendant-park garages and surface lots, mechanical stacking systems are used to increase the parking capacity of the facility. Mechanical systems are generally slow, however, and are most suited to longer-term parking durations, such as the full-day parking needs of working commuters.

There are, of course, many intricate details involved in the design and layout of parking garages and surface parking lots. This text has covered only a few of the major considerations involved. The reader is advised to consult References [4] and [5] directly for additional detail.

11.5 Parking Programs

Every urban governmental unit must have a plan to deal effectively with parking needs and associated problems. Parking is often a controversial issue, as it is of vital concern to the business community in general and to particular businesses that are especially sensitive to parking. Further, parking has enormous financial aspects as well. In addition to the impact of parking on accessibility and the financial health of the community at large, parking facilities are expensive to build and to operate. On the flip side, revenues from parking fees are also enormous.

The public interest in parking falls within the government's general responsibility to protect the health, safety, and welfare of its citizens. Thus, the government has a responsibility to [4]:

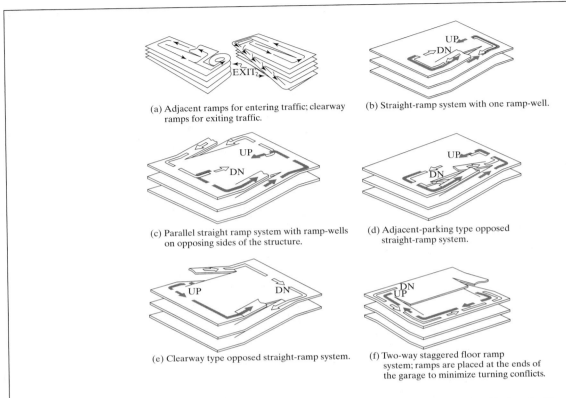

(a) Adjacent ramps for entering traffic; clearway ramps for exiting traffic.

(b) Straight-ramp system with one ramp-well.

(c) Parallel straight ramp system with ramp-wells on opposing sides of the structure.

(d) Adjacent-parking type opposed straight-ramp system.

(e) Clearway type opposed straight-ramp system.

(f) Two-way staggered floor ramp system; ramps are placed at the ends of the garage to minimize turning conflicts.

Figure 11.10: Parking Garage Circulation Systems (Used with permission of Eno Foundation for Transportation, Weant, R. and Levinson, H., *Parking*, Westport, CT, 1990, Figures 9.5-9.16, pgs. 188–192.)

- Establish parking program goals and objectives
- Develop policies and plans
- Establish program standards and performance criteria
- Establish zoning requirements for parking
- Regulate commercial parking
- Provide parking for specific public uses
- Manage and regulate on-street parking and loading
- Enforce laws, regulations, and codes concerning parking, and adjudicate offenses

There are a number of organizational approaches to effectively implementing the public role. Parking can be placed under the authority of an existing department of the government. In small communities, where there is no professional traffic engineer or traffic department, a department of public works might be tasked with parking. In some cases, police departments have been given this responsibility (as an adjunct to their enforcement responsibilities), but this is not considered an optimal solution given that it will be subservient to the primary role of police departments. Where traffic departments exist, responsibility for parking can logically be placed there. In larger municipalities, separate departments can be established for parking. Parking boards may be created with appointed and/or elected members supervising the process. Because of the revenues and costs involved in parking, separate public parking authorities may also be established.

Parking facilities may be operated directly by governmental units or can be franchised to private operators. This is often a critical part of the process and may have

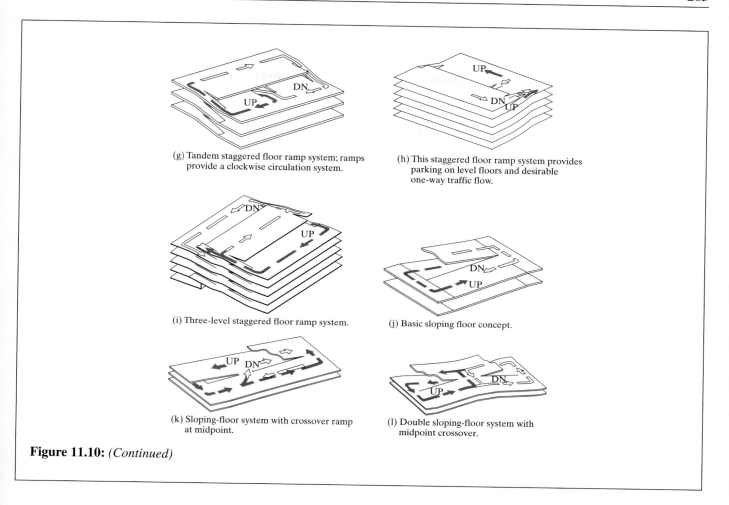

(g) Tandem staggered floor ramp system; ramps provide a clockwise circulation system.

(h) This staggered floor ramp system provides parking on level floors and desirable one-way traffic flow.

(i) Three-level staggered floor ramp system.

(j) Basic sloping floor concept.

(k) Sloping-floor system with crossover ramp at midpoint.

(l) Double sloping-floor system with midpoint crossover.

Figure 11.10: *(Continued)*

a substantial impact on the net revenues from parking that find their way into the public coffers.

Parking policy varies widely depending upon local circumstances. In some major cities, parking supply is deliberately limited, and costs are deliberately kept high as a discouragement to driving. Such a policy works only where there is significant public transportation supply to maintain access to the city's businesses. Where parking is a major part of access, the planning, development, and operation of off-street parking facilities becomes a major issue. Private franchisees are often chosen to build, operate, and manage parking facilities. While this generally provides a measure of expertise and relieves the government of the immediate need to finance and operate such facilities directly, the city must

negotiate and assign a significant portion of parking revenues to the franchisee. Of course, parking lots and garages can be fully private, although such facilities are generally regulated.

Revenues are also earned from parking meter proceeds and from parking violations. Metering programs are implemented for two primary reasons: to regulate turnover rates, and to earn revenue. The former is accomplished through time limits. These limits are established in conjunction with localized needs. Meters at a commuter rail station would, for example, have long-term time limits, as most people would be parking for a full working day. Parking spots near local businesses such as candy stores, barber shops, fast-food restaurants, florists, and similar uses would have relatively short-term time limits to encourage turnover and multiple users. Fees are set based on revenue needs and are influenced by general policy on encouragement or discouragement of parking.

No matter how the effort is organized and managed, parking programs must deal with the following elements:

1. *Planning and policy.* Overall objectives must be established and plans drafted to achieve them; general policy on parking must be set as part of the planning effort.

2. *Curb management.* Curb space must be allocated to curb parking, transit stops, taxi stands, loading areas, and other relevant uses; amounts and locations to be allocated must be set and the appropriate regulations implemented and signed.

3. *Construction, maintenance, and operation of off-street parking facilities.* Whether through private or governmental means, the construction of needed parking facilities must be encouraged and regulated; the financing of such facilities must be carefully planned so as to guarantee feasible operation while providing a revenue stream for the local government.

4. *Enforcement.* Parking and other curb-use regulations must be strictly enforced if they are to be effective; this task may be assigned to local police, or a separate parking violations bureau may be established; adjudication may also be accomplished through a separate traffic court system or through the regular local court system of the community.

To be most effective, parking policies should be integrated into an overall accessibility plan for central areas. Provision and/or improvement of public transportation services may mitigate some portion of parking demands while maintaining the fiscal viability of the city centers.

As was the case with other parking topics, this text can only scratch the surface of the complex issues involved in effective parking programs. The reader is again urged to consult the literature, particularly Reference [4],

for a more complete and detailed coverage of the subject.

11.6 Closing Comments

Without a place to park at both ends of a trip, the automobile would be a very ineffective transportation medium. Because our society relies so heavily on the private automobile for mobility and access, the subject of parking needs and the provision of adequate parking facilities is a critical element of the transportation system.

References

1. *Parking Generation*, 2nd Edition, Institute of Transportation Engineers, Washington DC, 1982.

2. "Parking Principles," *Special Report 125*, Transportation Research Board, Washington DC, 1971.

3. *Parking Requirements for Shopping Centers*, 2nd Edition, Urban Land Institute, Washington DC, 1999.

4. Weant, R. and Levinson, H., *Parking*, Eno Foundation for Transportation, Westport, CT, 1990.

5. *Guidelines for Parking Facility Design and Location: A Recommended Practice*, Institute of Transportation Engineers, Washington DC, April 1994.

Problems

11-1. A high-rise apartment complex with 600 dwelling units is to be built. What is the expected peak parking demand for such a facility, assuming that it is in an area without significant transit access?

11-2. A shopping center with 600,000 sq ft of gross leasable floor area is planned. It is expected that 10% of the floor area will be devoted to movie theaters and restaurants. What peak parking demand would be expected for such a development?

11-3. Based on typical zoning regulations, what number of parking spaces should the developers of Problems 1 and 2 be asked to provide?

11-4. A new office complex will house 2,000 back-office workers for the securities industries. Few external visitors are expected at this site. Each worker will account for 1.0 person-destinations per day. Of these, 35% are expected to occur during the peak hour. Only 7% of the workers will arrive by public transportation. Average car occupancy is 1.3. What peak parking demand can be expected at this facility?

11-5. A parking study has found that the average parking duration in the city center is 35 m, and that the following spaces are available within the 14-hour study period (6 AM–8 PM) with a 90% efficiency factor. How many vehicles may be parked in the study area in one 14-hour day?

Number of Spaces	Time Available
100	6:00 AM–8:00 PM
150	12:00 Nn–8:00 PM
200	6:00 AM–12:00 Nn
300	8:00 AM–6:00 PM

11-6. Consider the license-plate data in Table 11.11 for a study period from 7:00 AM to 2:00 PM. For this data:

Table 11.11: Data for Problem 11-6

Parking Space	7:00	7:30	8:00	8:30	9:00	9:30	10:00	10:30	11:00	11:30	12:00	12:30	1:00	1:30	2:00
1 hr meter	100	√	—	150	√	√	246	385	—	691	√	√	—	810	√
1 hr	—	468	√	630	√	485	—	711	888	927	√	√	108	√	—
1 hr	848	911	√	√	221	747	922	√	—	787	√	452	√	—	289
1 hr	—	—	206	√	242	√	√	—	899	√	205	603	812	√	√
1 hr	—	—	566	665	√	333	848	√	999	—	720	—	802	√	—
1 hr	—	690	—	551	√	√	347	√	265	835	486	√	—	721	855
Hydrant	—	—	—	—	—	—	—	777	—	—	—	—	—	—	—
2 hr meter	—	—	940	√	√	505	608	√	√	√	121	123	√	—	880
2 hr	636	√	√	√	√	—	582	√	√	811	919	√	711	√	√
2 hr	—	399	√	√	401	904	√	√	789	√	556	√	√	√	232
2 hr	—	416	√	√	√	√	√	—	658	√	292	844	493	√	√
2 hr	188	√	√	—	665	558	√	√	√	213	√	—	779	√	√
2 hr	—	—	—	277	√	336	409	√	√	884	√	√	713	895	431
2 hr	—	—	837	√	√	418	575	√	952	√	√	√	√	—	762
2 hr	—	506	√	√	—	786	√	√	√	527	606	√	385	√	√
Hydrant	—	—	—	—	—	518	—	—	—	758	—	—	—	—	—
3 hr	—	079	√	√	√	√	√	√	√	—	441	√	611	√	√
3 hr	256	√	√	√	√	—	295	√	√	338	√	—	499	√	√
3 hr	—	—	848	√	√	√	√	√	—	933	√	√	√	√	√
Bus stop	—	—	—	—	—	740	142	—	—	—	—	—	—	—	—
Bus stop	—	—	—	—	—	915	—	—	—	—	—	—	—	—	—
Bus stop	—	—	—	—	—	—	—	—	—	—	—	—	—	—	—
Bus stop	—	—	—	—	—	—	—	—	—	—	—	—	—	—	818
Bus stop	—	—	—	888	—	175	755	—	—	—	—	—	—	—	397

(a) Find the duration distribution and plot it as a bar chart.

(b) Plot the accumulation pattern.

(c) Compute the average parking duration.

(d) Summarize the overtime and parking violation rates.

(e) Compute the parking turnover rate.

Is there a surplus or deficiency of parking supply on this block? How do you know this?

PART 3
Applications to Freeway and Rural Highway Systems

Capacity and Level-of-Service Analysis for Freeways and Multilane Highways

12.1 Introduction to Capacity and Level-of-Service Concepts

One of the most critical needs in traffic engineering is a clear understanding of *how much* traffic a given facility can accommodate *and under what operating conditions*. These important issues are addressed in highway capacity and level-of-service analysis.

The basis for all capacity and level-of-service analysis is a set of analytic procedures that relate demand or existing flow levels, geometric characteristics, and controls to measures of the resulting quality of operations. These models take many forms. Some are based on regression analysis of significant data bases. Some are based on theoretical algorithms and/or laws of physics. Still others are based on the results of simulation. Their application allows traffic engineers to determine the ultimate traffic-carrying ability of a facility and to estimate operating characteristics at various flow levels.

The U.S. standard for these types of analyses is the *Highway Capacity Manual* (HCM), a publication of the

Transportation Research Board (TRB) of the National Academy of Engineering. Its content is controlled by the Committee on Highway Capacity and Quality of Service (HCQSC) of the TRB. The development of material for the manual is supported by a number of federal agencies through funding for basic and applied research. These agencies include the National Cooperative Highway Research Program (NCHRP) and the Federal Highway Administration (FHWA).

The first edition of the HCM [1] was published by the then Bureau of Public Roads in 1950. Its objective was to provide uniform guidelines for the nation's rapidly-growing highway construction program. The informal group of Bureau personnel that developed the first edition became the founding members of the HCQSC. The second edition [2] was published in 1965. It introduced significant new material on limited access facilities, as well as the level-of-service concept. The third edition [3] was published in 1985, providing refinements to the level-of-service concept and adding material on pedestrian and transit facilities. Introduced for the first time as a loose-leaf document, significant updates to this edition were introduced in 1994 [4] and 1997 [5].

The material in this text is based on the fourth edition of the HCM [6], published in December of 2000. Often referred to as *HCM 2000*, this edition added significant new material on planning applications and corridor and system analysis. It also formally addressed the issue of simulation and its relationship to traditional deterministic highway capacity and level of service models.

The HCM 2000 is also the first manual to be published in two forms: one in standard U.S. units, the other in metric units. The manual was originally developed in metric units in response to legislation requiring all states to convert to a metric system. Late in the development of HCM 2000, the legislation was modified, leaving a split among state highway and transportation departments. As a result, some states now use a metric system, while others have maintained or have returned to standard U.S. units.

This text is presented primarily in standard U.S. units. As the two versions of the HCM 2000 reflect "hard conversion" between the two sets of units, presentation of both forms in this text would have been impractical.

A recent paper by Kittelson [7] provides an excellent history and discussion of the development of the HCM and its key concepts.

12.1.1   The Capacity Concept

The HCM 2000 defines capacity as follows:

> The capacity of a facility is the maximum hourly rate at which persons or vehicles reasonably can be expected to traverse a point or a uniform section of a lane or roadway during a given time period under prevailing roadway, traffic, and control conditions. (HCM 2000, pg. 2-2)

The definition contains a number of significant concepts that must be understood when applying capacity analysis procedures:

- Capacity is defined as a *maximum hourly rate*. For most cases, the rate used is for the peak 15 minutes of the peak hour, although HCM 2000 allows for some discretion in selecting the length of the analysis period. In any analysis, care must be taken to express both the demand and the capacity in terms of the same analysis period.

- Capacity may be expressed in terms of *persons* or *vehicles*. This is critical when transit and pedestrian issues are considered, as well as in the consideration of high-occupancy vehicle lanes and facilities, where the person-capacity is clearly more important than the vehicle-capacity.

- Capacity is defined for *prevailing roadway, traffic, and control* conditions. Roadway conditions refer to the geometric characteristics of the facility, such as the number of lanes, lane widths, shoulder widths, and free-flow speeds. Traffic conditions refer primarily to the composition of the traffic stream, particularly the presence of trucks and other heavy vehicles. Control conditions refer primarily to interrupted flow facilities, where such controls as STOP and YIELD signs and traffic signals have a significant impact on capacity. The important concept is that a change in any of the prevailing conditions causes a change in the capacity of the facility.

- Capacity is defined for a *point* or *uniform section* of a facility. This correlates to the "prevailing conditions" discussed above. A "uniform section" must have consistent prevailing conditions. At any point where these conditions change, the capacity also changes.

- Capacity refers to maximum flows that *can reasonably be expected* to traverse a section. This recognizes that capacity, as are all traffic factors, is subject to variation in both time and space. Thus, capacity is not defined as the single highest flow level ever expected to occur on a facility. Rather it is a value that represents a flow level that can be reasonably achieved repeatedly at a given location and at similar locations throughout the United States. Thus, isolated observations of actual flows in excess of stated capacities is not a contradiction and is, in fact, an expected condition.

It should be noted that, while capacity is an important concept, operating conditions at capacity are generally quite poor, and it is difficult (but not impossible) to maintain capacity operation without breakdowns for long periods of time.

12.1.2   The Level-of-Service Concept

The level-of-service concept was introduced in the 1965 HCM as a convenient way to describe the general quality of operations on a facility with defined traffic, roadway, and control conditions. Using a letter scale from A to F, a terminology for operational quality was created that has become an important tool in communicating complex issues to decision-makers and the general public. The HCM 2000 defines level of service as follows:

> "Level of service (LOS) is a quality measure describing operational conditions within a traffic stream, generally in terms of such service measures as speed and travel time, freedom to maneuver, traffic interruptions, and comfort and convenience." (HCM 2000, pg. 2-2).

The six defined levels of service, A–F, describe operations from best to worst for each type of facility. When originally defined, models did not exist for the prediction of precise quality measures for many types of facilities. This is no longer true. Every facility type now has levels of service defined in terms of a specific *measure of effectiveness*. Table 12.1 shows the measures used to define levels of service in the HCM 2000.

Levels of service are basically step-functions, each representing a range of operating conditions, as illustrated in Figure 12.1. Thus, care must be taken when using level of service as a descriptor. Two similar facilities with the same level of service may differ more than two with different levels of service, depending upon where they are in the defined ranges. Levels of service B and C, for example, may represent very similar conditions if they are both close to the defined boundary between the two levels.

Another critical concept is that level of service is to be defined in terms of parameters that can be perceived by drivers and passengers and that the definitions should reflect that perception.

Thus, volume or flow is never used as a measure of effectiveness, as it is a point measure that cannot be perceived by drivers or passengers from within the traffic stream. On the other hand, it is very difficult to measure driver or passenger perceptions regarding specific threshold values. Thus, most level-of-service criteria are based on the collective judgment of professionals, exercised through the Highway Capacity and Quality of Service Committee and its subcommittees.

Service Flow Rates

Figure 12.1 illustrates the levels of service as ranges or step-functions. The exhibit also illustrates the definition of another related concept—service flow rates. A service flow rate is similar to capacity, except that it represents the maximum flow rate that can be accommodated while maintaining a designated level of service. Service flow rates may be defined for levels of service A–E, but are never defined for level of service F, which represents unstable flow or unacceptably poor service quality. Due to the complexity of some capacity and level-of-service models, service flow rates are difficult to determine for some types of facilities.

Like capacity, service flow rates are defined for prevailing conditions on uniform sections of a facility, and they relate to flow levels that can be reasonably expected to occur at the various levels of service.

Problems With Level of Service

As noted, the level-of-service concept has become a major medium used in describing often complex concepts to decision-makers and the general public. It can, and is, however, often used in ways that were not intended. The Highway Capacity and Quality of Service Committee has been wrestling with many of these issues for a number of years and has even considered abandoning the concept in favor of simply reporting the numerical quality measures on which they are based. One of the most significant problems, the ability to misinterpret changes in level of service due to the step-function nature of the concept, has already been noted. Some of the other key problems are:

- A single set of level-of-service criteria does not recognize differing levels of perception or acceptability in various parts of the United States. What might be acceptable delay at an

Table 12.1: Measures of Effectiveness Defining Levels of Service in HCM 2000

Type of Flow	Type of Facility	Measure of Effectiveness
Uninterrupted Flow	Freeways	
	Basic sections	Density (pc/mi/ln)
	Weaving areas	Density (pc/mi/ln)
	Ramp junctions	Density (pc/mi/ln)
	Multilane Highways	Density (pc/mi/ln)
	Two-Lane Highways	Average Travel Speed (mi/h)
		Percent Time Spent Following (%)
Interrupted Flow	Signalized Intersections	Control Delay (s/veh)
	Unsignalized Intersections	Control Delay (s/veh)
	Urban Streets	Average Travel Speed (mi/h)
	Transit	Service Frequency (veh/day)
		Service Headway (min)
		Passengers/Seat
	Pedestrians	Space (ft²/ped)
	Bicycles	Frequency of (Conflicting) Events (events/h)

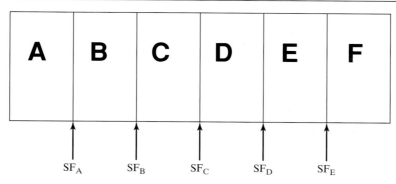

Figure 12.1: Illustration of Levels of Service and Service Flow Rates

intersection in New York City, for example, might not be acceptable at a similar intersection in a small community.

- Some state and local governments have incorporated level-of-service criteria into development legislation. Thus, when revisions to the HCM are made, legal standards are being altered. This is clearly not the intent of the HCQSC, and it is not generally recognized by the legislators enacting these laws.

- In 1985, the adoption of delay as a level-of-service measure for signalized intersections introduced new complexity to the interpretation of level of service F. For some types of facilities, this designation indicates a situation in which demand exceeds capacity. For other types of facilities, it indicates unacceptable service quality.

The availability of models capable of estimating numerical values of measures of effectiveness for all types of facilities in HCM 2000 makes it possible to consider eliminating the level-of-service concept. Measures such as delay, density, speed, and others are continuous measures and better reflect the continuum of operating conditions. Elimination of the concept, however, would place a burden on practicing traffic engineers to explain measures of effectiveness to decision-makers and the public and to explain levels of acceptability in various situations. Because of this, the concept continues to be useful. Professionals, however, should take care to use the concept appropriately and to point out instances in which a level-of-service designation might be misleading.

Additional Performance Measures

The HCM 2000, in addition to the measures of effectiveness that define level-of-service boundaries, also contains methodologies for predicting additional performance measures that might be of interest to traffic engineers. Thus, while uninterrupted flow facilities have levels of service defined by density, predictions of average speed are also provided. For signalized intersections, delay estimates are supplemented by predictions of average queue size.

After considerable discussion and debate, the HCM 2000 does not define levels of service for systems, corridors, or networks. While performance measures are defined, and in some cases estimated, the meaning of a broad-based level of service measure involves a number of new questions that are still under consideration. The most significant involves user perceptions of service quality. A system level-of-service measure, for example, might not relate to any individual's trip experience while using only parts of the system. This discussion, along with others concerning the level-of-service concept, will undoubtedly continue over the next several years, as a fifth edition of the HCM is contemplated and planned.

12.1.3 The *v/c* Ratio and its Use in Capacity Analysis

One of the most important measures resulting from a capacity and/or level-of-service analysis is the *v/c* ratio—the ratio of current or projected demand flow to the capacity of the facility. This ratio is used as a measure of the sufficiency of existing or proposed capacity.

It is, of course, desirable that all facilities be designed to provide sufficient capacity to handle present or projected demands (i.e., that the *v/c* ratio be maintained at a value less than 1.00).

When estimating or considering a *v/c* ratio, care must be taken to understand the origin of the flow rate (*v*) and capacity (*c*) values. The flow rate should represent *demand*. In existing situations, true demand consists of actual *arrival flows* plus traffic that has diverted to alternative routes, other time periods, or alternative destinations due to congestion in the system. It is not possible, for example, for a measured departure flow to exceed the actual capacity of a facility. Yet, it is possible for a measured departure flow to be compared to a capacity estimated using HCM 2000 or other analytic procedures, yielding a *v/c* ratio >1.00. Since actual throughput or departure flow cannot exceed actual capacity, the implication in such cases is that the estimated capacity is too low.

When dealing with future projections, the forecast demand flow is used and compared with the estimated capacity. A *v/c* ratio >1.00 implies that the estimated capacity is not sufficient to handle the forecast demand flows.

When the true ratio of demand flow to capacity (either in the present or the future) is greater than 1.00, the implication is that queuing will occur and propagate upstream of the segment in question. The extent of queues and the time required to clear them depend upon many conditions, including the length of time that the *v/c* ratio exceeds 1.00 and by how much it is exceeded. It also depends upon the demand profile over time, as queues can start to dissipate only when demand flows decrease to levels less than the capacity of the segment. Further, when queuing occurs, drivers tend to seek alternative routes to avoid congestion. Thus, the occurrence of *v/c* > 1.00 often causes a dynamic shift in demand patterns that could significantly impact operations in and around the subject segment.

In any event, the comparison of true demand flows to capacity is a principal objective of capacity and level of service analysis. Thus, in addition to level-of-service criteria, the *v/c* ratio is a major output of such analyses.

12.2 Freeways and Multilane Highways

The procedures described in this chapter cover the analysis of multilane uninterrupted flow. These include basic freeway sections and sections of surface multilane highways with sufficient distances between fixed interruptions (primarily traffic signals) to allow uninterrupted (random, or non-platoon) flow between points of interruption.

12.2.1   Facility Types

Freeways are the only types of facilities providing pure uninterrupted flow. All entries and exits from freeways are made using ramps designed to allow such movements to occur without interruption to the freeway traffic stream. There are no at-grade intersections (either signalized or unsignalized), no driveway access, and no parking permitted within the right-of-way. Full control of access is provided. Freeways are generally classified by the total number of lanes provided in both directions: a six-lane freeway has three lanes in each direction. Common categories are four-, six-, and eight-lane freeways, although some freeway sections in major urban areas may have 10 or more lanes in specific segments.

Multilane surface facilities should be classified and analyzed as urban streets (arterials) if signal spacing is less than one mile. Uninterrupted flow can exist on multilane facilities where the signal spacing is more than two miles. Where signal spacing is between one and two miles, the existence of uninterrupted flow depends on prevailing conditions. There are, unfortunately, no specific criteria to guide traffic engineers in making this determination, which could easily vary over time. In the majority of cases, signal spacings between one and two miles do not result in the complete breakdown of platoon movements unless the signals are not coordinated. Thus, most of these cases are best analyzed as arterials.

Multilane highway segments are classified by the number of lanes and the type of median treatment provided. Surface multilane facilities generally consist of four- or six-lane alignments. They can be *undivided* (i.e., having no median but with a double-solid-yellow marking separating the two directions of flow), or *divided*, with a physical median separating the two directions of flow. In suburban areas, a third median treatment is also used: the two-way left-turn lane. This treatment requires an alignment with an odd number of lanes—most commonly five or seven. The center lane is used as a continuous left-turn lane for both directions of flow.

The median treatment of a surface multilane highway can have a significant impact on operations. A physical median prevents mid-block left turns across the median except at locations where a break in the median barrier is provided. Mid-block left turns can be made at any point on an undivided alignment. Where a two-way left-turn lane is provided, mid-block left turns are permitted without restriction, but vehicles waiting to turn do so in the special lane and do not unduly restrict through vehicles.

In terms of capacity analysis procedures, both basic freeway sections and multilane highways are categorized by the *free-flow speed*. By definition, the free-flow speed is the speed intercept when flow is "zero" on a calibrated speed-flow curve. In practical terms, it is the average speed of the traffic stream when flow rates are less than 1,000 veh/h/ln. Refer to Chapter 5 for a more complete discussion of speed-flow relationships and characteristics.

Figure 12.2 illustrates some common freeway and multilane alignments.

12.2.2   Basic Freeway and Multilane Highway Characteristics

Speed-Flow Characteristics

The basic characteristics of uninterrupted flow were presented in detail in Chapter 5. Capacity analysis procedures for freeways and multilane highways are based on calibrated speed-flow curves for sections with various free-flow speeds operating under *base conditions*. Base conditions for freeways and multilane highways include:

- No heavy vehicles in the traffic stream
- A driver population dominated by regular or familiar users of the facility

(a) A Typical 8-Lane Freeway

(b) A Divided Multilane Rural Highway

(c) A divided Multilane
Suburban Highway

(d) An Undivided Multilane
Suburban Highway

(e) An Undivided Multilane Rural
Highway

Figure 12.2: Typical Freeway and Multilane Highway Alignments (Photo (b) and (e) used with permission of Transportation Research Board, National Research Council, "Highway Capacity Manual," *Special Report 209*, 1994, Illustration 7-2.)

Figures 12.3 and 12.4 show the standard curves calibrated for use in the capacity analysis of basic freeway sections and multilane highways. These exhibits also show the density lines that define levels of service for uninterrupted flow facilities.

Modern drivers maintain high average speeds at relatively high rates of flow on freeways and multilane highways. This is clearly indicated in Figures 12.3 and 12.4. For freeways, the free-flow speed is maintained until flows reach 1,300 to 1,750 pc/h/ln. Multilane highway characteristics are similar. Thus, on most uninterrupted flow facilities, the transition from stable to unstable flow occurs very quickly and with relatively small increments in flow.

Levels of Service

For freeways and multilane highways, the measure of effectiveness used to define levels of service is *density*. The use of density, rather than speed, is based primarily on the shape of the speed-flow relationships depicted in Figures 12.3 and 12.4. Because average speed remains constant through most of the range of flows and because the total difference between free-flow speed and the speed at capacity is relatively small, defining five level-of-service boundaries based on this parameter would be very difficult.

If flow rates vary while speeds remain relatively stable, then density must be varying throughout the range of flows, given the basic relationships that $v = S \times D$. Further, density describes the proximity of vehicles to each other, which is the principal influence on freedom to maneuver. Thus, it is an appropriate descriptor of service quality.

For uninterrupted flow facilities, the density boundary between levels of service E and F is defined as the density at which capacity occurs. The speed-flow curves determine this critical boundary. For freeways, the curves indicate a constant density of 45 pc/mi/ln at capacity for all free-flow speeds. For multilane highways, capacity occurs at densities ranging from 40 to 45 pc/mi/ln, depending on the free-flow speed of the facility.

Other level-of-service boundaries are set judgmentally by the HCQSC to provide reasonable ranges of both density and service flow rates. Table 12.2 shows the defined level-of-service criteria for basic freeway sections and multilane highways.

The general operating conditions for these levels of service can be described as follows:

- *Level of service A* is intended to describe free-flow operations. At these low densities, the operation of

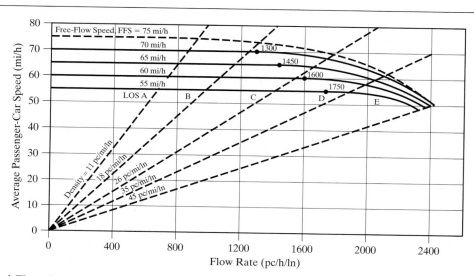

Figure 12.3: Speed-Flow Curves for Basic Freeway Sections (Used with permission of Transportation Research Board, National Research Council, from *Highway Capacity Manual*, Dec 2000, Exhibit 23-3, pg. 23-5.)

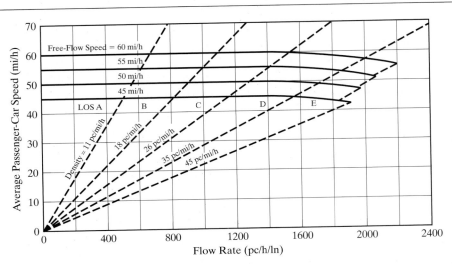

Figure 12.4: Speed-Flow Curves for Multilane Highway Sections (Used with permission of Transportation Research Board, National Research Council, from *Highway Capacity Manual*, Dec 2000, Exhibit 21-3, pg. 21-4.)

Table 12.2: Level-of-Service Criteria for Freeways and Multilane Highways

Level of Service	Density Range for Basic Freeway Sections (pc/mi/ln)	Density Range for Multilane Highways (pc/mi/ln)
A	$\geq 0 \leq 11$	$\geq 0 \leq 11$
B	$>11 \leq 18$	$>11 \leq 18$
C	$>18 \leq 26$	$>18 \leq 26$
D	$>26 \leq 35$	$>26 \leq 35$
E	$>35 \leq 45$	$>35 \leq (40-45)$ depending on FFS
F	>45	$> (40-45)$ depending on FFS

each vehicle is not greatly influenced by the presence of others. Speeds are not affected by flow in this level of service, and operation is at the free-flow speed. Lane changing, merging, and diverging maneuvers are easily accomplished, as many large gaps in lane flow exist. Short-duration lane blockages may cause the level of service to deteriorate somewhat, but do not cause significant disruption to flow. Average spacing between vehicles is a minimum of 480 ft, or approximately 24 car lengths at this level of service.

- At *level of service B*, drivers begin to respond to the existence of other vehicles in the traffic stream, although operation is still at the free-flow speed. Maneuvering within the traffic stream is still relatively easy, but drivers must be more vigilant in searching for gaps in lane flows. The traffic stream still has sufficient gaps to dampen the impact of most minor lane disruptions. Average spacing is a minimum of 293 ft, or approximately 15 car lengths.

- At *level of service C*, the presence of other vehicles begins to restrict maneuverability within the traffic stream. Operations remain at the free-flow speed, but drivers now need to adjust their course to find gaps they can use to pass or merge. A significant increase in driver vigilance is required at this level. While there are still sufficient gaps in the traffic stream to dampen the impact of minor lane blockages, any significant blockage could lead to breakdown and queuing. Average spacing is a minimum of 203 ft, or approximately 10 car lengths.

- *Level of service D* is the range in which average speeds begin to decline with increasing flows. Density deteriorates more quickly with flow in this range. At level of service D, breakdowns can occur quickly in response to small increases in flow. Maneuvering within the traffic stream is now quite difficult, and drivers often have to search for gaps for some time before successfully passing or merging. The ability of the traffic stream to dampen the impact of even minor lane disruptions is severely restricted, and most such blockages result in queue formation unless removed very quickly. Average spacing is a minimum of 151 ft, or approximately seven car lengths.

- *Level of service E* represents operation in the vicinity of capacity. The maximum density limit of level of service E is capacity operation. For such an operation there are few or no usable gaps in the traffic stream, and any perturbation caused by lane-changing or merging maneuvers will create a shock wave in the traffic stream. Even the smallest lane disruptions may cause extensive queuing. Maneuvering within the traffic stream is now very difficult, as other vehicles must give way to accommodate a lane-changing

or merging vehicle. The average spacing is a minimum of 117 ft, or approximately six car lengths.

- *Level of service F* describes operation within the queue that forms upstream of a breakdown point. Such breakdowns may be caused by accidents or incidents, or may occur at locations where arrival demand exceeds the capacity of the section on a regular basis. Actual operating conditions vary widely, and are subject to short-term perturbations. As vehicles "shuffle" through the queue, there are times when they are standing still, and times when they move briskly for short distances. Level of service F is also used to describe the point of the breakdown, where demand flow (v) exceeds capacity (c). In reality, operation at the point of the breakdown is usually good, as vehicles discharge from the queue. Nevertheless, it is insufficient capacity at the point of breakdown that causes the queue, and level of service F provides an appropriate descriptor for this condition.

Service Flow Rates and Capacity

Maximum densities, minimum average speeds, maximum v/c ratios, and maximum service flow rates for the various levels of service for freeways and multilane highways are shown in Tables 12.3 and 12.4.

The values in these tables are taken directly from the curves of Figures 12.3 and 12.4. Service flow rates, however, are rounded to the nearest 10 pc/mi/ln. Maximum service flow rates are stated in terms of pc/mi/ln and reflect the ideal conditions defined previously.

12.3 Analysis Methodologies for Basic Freeway Sections and Multilane Highways

The characteristics and criteria described for freeways and multilane highways in the previous section apply to facilities with base traffic and roadway

Table 12.3: Level-of-Service Criteria for Basic Freeway Sections

Criteria	Level of Service				
	A	**B**	**C**	**D**	**E**
Free-Flow Speed = 75 mi/h					
Maximum density (pc/mi/ln)	11	18	26	35	45
Minimum speed (mi/h)	75.0	74.8	70.6	62.2	53.3
Maximum v/c	0.34	0.56	0.76	0.90	1.00
Maximum service flow rate (pc/h/ln)	820	1,350	1,830	2,170	2,400
Free-Flow Speed = 70 mi/h					
Maximum density (pc/mi/ln)	11	18	26	35	45
Minimum speed (mi/h)	70.0	70.0	68.2	61.5	53.3
Maximum v/c	0.32	0.53	0.74	0.90	1.00
Maximum service flow rate (pc/h/ln)	770	1,260	1,770	2,150	2,400
Free-Flow Speed = 65 mi/h					
Maximum density (pc/mi/ln)	11	18	26	35	45
Minimum speed (mi/h)	65.0	65.0	64.6	59.7	52.2
Maximum v/c	0.30	0.50	0.71	0.89	1.00
Maximum service flow rate (pc/h/ln)	710	1,170	1,680	2,090	2,350
Free-Flow Speed = 60 mi/h					
Maximum density (pc/mi/ln)	11	18	26	35	45
Minimum speed (mi/h)	60.0	60.0	60.0	57.6	51.1
Maximum v/c	0.29	0.47	0.68	0.88	1.00
Maximum service flow rate (pc/h/ln)	660	1,080	1,560	2,020	2,300
Free-Flow Speed = 55 mi/h					
Maximum density (pc/mi/ln)	11	18	26	35	45
Minimum speed (mi/h)	55.0	55.0	55.0	54.7	50.0
Maximum v/c	0.27	0.44	0.64	0.85	1.00
Maximum service flow rate (pc/h/ln)	600	990	1,430	1,910	2,250

Note: The exact mathematical relationship between density and v/c has not always been maintained at LOS boundaries because of the use of rounded values. Density is the primary determinant of LOS. The speed criterion is the speed at maximum density for a given LOS.

(Used with permission of Transportation Research Board, National Research Council, from *Highway Capacity Manual*, Dec 2000, Exhibit 23-2, pg. 23-3.)

conditions. In most cases, base conditions do not exist, and a methodology is required to address the impact of prevailing conditions on these characteristics and criteria.

Analysis methodologies are provided that account for the impact of a variety of prevailing conditions, including:

- Lane widths
- Lateral clearances
- Number of lanes (freeways)
- Type of median (multilane highways)
- Frequency of interchanges (freeways) or access points (multilane highways)

Table 12.4: Level of Service Criteria for Multilane Highways

Criteria	Level of Service				
	A	B	C	D	E
Free-Flow Speed = 60 mi/h					
Maximum density (pc/mi/ln)	11	18	26	35	40
Minimum speed (mi/h)	60.0	60.0	59.4	56.7	55.0
Maximum v/c	0.30	0.49	0.70	0.90	1.00
Maximum service flow rate (pc/h/ln)	660	1,080	1,550	1,980	2,200
Free-Flow Speed = 55 mi/h					
Maximum density (pc/mi/ln)	11	18	26	35	41
Minimum speed (mi/h)	55.0	55.0	54.9	52.9	51.2
Maximum v/c	0.29	0.47	0.68	0.88	1.00
Maximum service flow rate (pc/h/ln)	600	990	1,430	1,850	2,100
Free-Flow Speed = 50 mi/h					
Maximum density (pc/mi/ln)	11	18	26	35	43
Minimum speed (mi/h)	50.0	50.0	50.0	48.9	47.5
Maximum v/c	0.28	0.65	0.65	0.86	1.00
Maximum service flow rate (pc/h/ln)	550	900	1,300	1,710	2,000
Free-Flow Speed = 45 mi/h					
Maximum density (pc/mi/ln)	11	18	26	35	45
Minimum speed (mi/h)	45.0	45.0	45.0	44.4	42.2
Maximum v/c	0.26	0.43	0.62	0.82	1.00
Maximum service flow rate (pc/h/ln)	490	810	1,170	1,550	1,900

Note: The exact mathematical relationship between density and v/c has not always been maintained at LOS boundaries because of the use of rounded values. Density is the primary determinant of LOS. The speed criterion is the speed at maximum density for a given LOS.
(Used with permission of Transportation Research Board, National Research Council, from *Highway Capacity Manual*, Dec 2000, Exhibit 21-2, pg. 21-3.)

- Presence of heavy vehicles in the traffic stream
- Driver populations dominated by occasional or unfamiliar users of a facility

Some of these factors affect the free-flow speed of the facility, while others affect the equivalent demand flow rate on the facility.

12.3.1   Types of Analysis

There are three types of analysis that can be conducted for basic freeway sections and multilane highways:

- Operational analysis
- Service flow rate and service volume analysis
- Design analysis

In addition, the HCM defines "planning analysis." This, however, consists of beginning the analysis with an AADT as a demand input, rather than a peak hour volume. Planning analysis begins with a conversion of an AADT to a directional design hour volume (DDHV) using the traditional procedure as described in Chapter 5.

All forms of analysis require the determination of the free-flow speed of the facility in question. Field measurement and estimation techniques for making this determination are discussed in a later section.

Operational Analysis

The most common form of analysis is *operational analysis*. In this form of analysis, all traffic, roadway, and control conditions are defined for an existing or projected highway section, and the expected level of service and operating parameters are determined.

The basic approach is to convert the existing or forecast demand volumes to an equivalent flow rate under ideal conditions:

$$v_p = \frac{V}{PHF * N * f_{HV} * f_p} \qquad (12\text{-}1)$$

where: v_p = demand flow rate under equivalent ideal conditions, pc/h/ln

PHF = peak-hour factor

N = number of lanes (in one direction) on the facility

f_{HV} = adjustment factor for presence of heavy vehicles

f_p = adjustment factor for presence of occasional or non-familiar users of a facility

This result is used to enter either the standard speed-flow curves of Figure 12.3 (freeways) or 12.4 (multilane highways). Using the appropriate free-flow speed, the curves may be entered on the x-axis with the demand flow rate, v_p, to determine the level of service and the expected average speed. This technique is illustrated in Figure 12.5.

In the example shown, an adjusted demand flow (v_p) of 1,800 pc/h/ln is used for a freeway with a free-flow speed of 65 mi/h. For this condition, the expected speed is determined as 64 mi/h. Density may then be estimated as the flow rate divided by speed. Level of service may be determined on the basis of the computed density, or by examination of the curves: the intersection of 1,800 pc/h/ln with a 65 mi/h free-flow speed obviously falls within the range of level of service D.

Methods for determining the free-flow speed and adjustment factors for heavy vehicles and driver population are presented in later sections.

Service Flow Rate and Service Volume Analysis

It is often useful to determine the service flow rates and service volumes for the various levels of service under prevailing conditions. Various demand levels may then be compared to these estimates for a speedy determination of expected level of service. The service flow rate for a given level of service is computed as:

$$SF_i = MSF_i * N * f_{HV} * f_p \qquad (12\text{-}2)$$

where: SF_i = service flow rate for level of service "i," veh/h

MSF_i = maximum service flow rate for level of service "i," pc/h/ln

N, f_{HV}, f_p as previously defined

The maximum service flow rates for each level of service, MSF_i, are taken from Table 12.3 (for freeways) and Table 12.4 (for multilane highways). The tables are entered with the appropriate free-flow speed. Interpolation may be used to find intermediate values.

Service flow rates are stated in terms of peak flows within the peak hour, usually for a 15-minute analysis period. It is often convenient to convert service flow rates to service volumes over the full peak hour. This is done using the peak-hour factor:

$$SV_i = SF_i * PHF \qquad (12\text{-}3)$$

where: SV_i = service volume over a full peak hour for level of service "i"

SF_i, PHF as previously defined

Design Analysis

In design analysis, an existing or forecast demand volume is used to determine the number of lanes needed to provide for a specified level of service. The number of lanes may be computed as:

$$N_i = \frac{DDHV}{PHF * MSF_i * f_{HV} * f_p} \qquad (12\text{-}4)$$

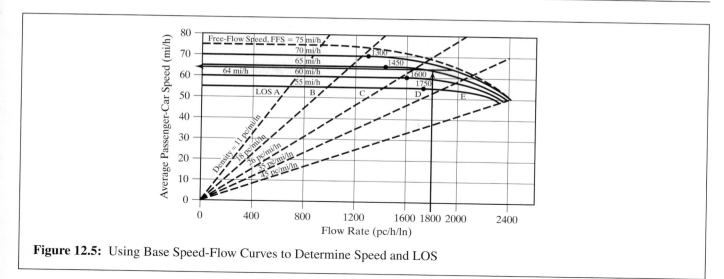

Figure 12.5: Using Base Speed-Flow Curves to Determine Speed and LOS

where: N_i = number of lanes (in one direction) required to provide level of service "i"

$DDHV$ = directional design hour volume, veh/h

MSF_i, f_{HV}, f_p as previously defined

Design analysis for freeways, however, becomes an iterative process. Values of MSF_i depend upon the free-flow speed of the facility. For freeways, as will be seen, the free-flow speed is dependent upon the number of lanes provided. Thus, a number must be assumed, then computed, continuing to iterate until the assumed and computed values agree.

When such iteration is required, it is often more convenient to compute the service flow rate and service volume for the desired level of service for a range of reasonable values of N (usually 2, 3, 4, and possibly 5 lanes). Then the demand volume or flow rate can be compared to the results for a simpler determination of the required number of lanes.

12.3.2   Determining the Free-Flow Speed

The free-flow speed of a facility is best determined by field measurement. Given the shape of speed-flow relationships for freeways and multilane highways, an average speed measured when flow is less than or equal to 1,000 veh/h/ln may be taken to represent the free-flow speed.

It is not always possible, however, to measure the free-flow speed. When new facilities or redesigned facilities are under consideration, it is not possible to measure free-flow speeds. Even for existing facilities, the time and cost of conducting field studies may not be warranted.

Freeways

The free-flow speed of a freeway can be estimated as:

$$FFS = BFFS - f_{LW} - f_{LC} - f_N - f_{ID} \quad (12\text{-}5)$$

where: FFS = free-flow speed of the freeway, mi/h

$BFFS$ = base free-flow speed of the freeway (70 mi/h for urban and suburban freeways, 75 mi/h for rural freeways)

f_{LW} = adjustment for lane width, mi/h

f_{LC} = adjustment for lateral clearance, mi/h

f_N = adjustment for number of lanes, mi/h

f_{ID} = adjustment for interchange density, mi/h

Lane Width Adjustment   The base condition for lane width is an average width of 12 ft or greater. For narrower lanes, the base free-flow speed is reduced by the factors shown in Table 12.5.

Lateral Clearance Adjustment   Base lateral clearance is 6 ft or greater on the right side and 2 ft or greater

Table 12.5: Adjustment to Free-Flow Speed for Lane Width on a Freeway

Lane Width (ft)	Reduction in Free-Flow Speed, f_{LW} (mi/h)
≥12	0.0
11	1.9
10	6.6

(Used with permission of Transportation Research Board, National Research Council, *Highway Capacity Manual*, Dec 2000, Exhibit 23-4, pg. 23-6.)

on the median, or left, side of the basic freeway section. Adjustments for right-side lateral clearances less than 6 ft are given in Table 12.6. There are no adjustments provided for median clearances less than 2 ft, as such conditions are considered rare.

Care should be taken in assessing whether an "obstruction" exists on the right side of the freeway. Obstructions may be continuous, such as a guardrail or retaining wall, or they may be periodic, such as light supports and bridge abutments. In some cases, drivers may become accustomed to some obstructions, and the impact of these on free-flow speeds may be minimal.

Table 12.6: Adjustment to Free-Flow Speed for Lateral Clearance on a Freeway

Right Shoulder Lateral Clearance (ft)	Reduction in Free-Flow Speed, f_{LC} (mi/h)			
	Lanes in One Direction			
	2	3	4	≥5
≥6	0.0	0.0	0.0	0.0
5	0.6	0.4	0.2	0.1
4	1.2	0.8	0.4	0.2
3	1.8	1.2	0.6	0.3
2	2.4	1.6	0.8	0.4
1	2.0	2.0	1.0	0.5
0	3.6	2.4	1.2	0.6

(Used with permission of Transportation Research Board, National Research Council, *Highway Capacity Manual*, Dec 2000, Exhibit 23-5, pg. 23-6.)

Right-side obstructions primarily influence driver behavior in the right lane. Drivers "shy away" from such obstructions, moving further to the left in the lane. Drivers in adjacent lanes may also shift somewhat to the left in response to vehicle placements in the right lane. The overall affect is to cause vehicles to travel closer to each other laterally than would normally be the case, thus making flow less efficient. This is the same affect as for narrow lanes. Since the primary impact is on the right lane, the total impact on free-flow speed declines as the number of lanes increases.

Adjustment for Number of Lanes   This adjustment has been the subject of much discussion. The base condition for number of lanes in one direction on a freeway is five or more lanes. The use of this size freeway as a base has been questioned, as it is a relatively rare occurrence. Eight-lane freeways, providing four lanes in each direction, represent the normal maximum width in most areas of the nation. Further, there is some question as to whether an adjustment to free-flow speed is required for eight-lane and even six-lane freeways. Using such an adjustment implies that no eight-lane, six-lane, or four-lane freeway in an urban or suburban area could ever have a free-flow speed of 70 mi/h, even if all other conditions are ideal.

As a result of the ongoing discussion, the HCQSC modified the application of this adjustment, recommending that no adjustment be applied to rural freeways. The continued application of this adjustment to urban and suburban freeways, however, remains controversial. The adjustment for number of lanes is given in Table 12.7.

Interchange Density Adjustment   Perhaps the most significant impact on freeway free-flow speed is the number and spacing of interchanges. Interchange density is defined as the average number of interchanges per mile over a six-mile section of the facility, taken as three miles upstream and three miles downstream of the point or section under consideration. Note that the interchange density *is not* based on the number of ramps. An interchange may consist of several ramp connections. A typical diamond interchange has four ramps, while a full cloverleaf interchange has eight. To qualify as an interchange,

Table 12.7: Adjustment to Free-Flow Speed for Number of Lanes on a Freeway

Number of Lanes (in one direction)	Reduction in Free-Flow Speed, f_N(mi/h)
≥5	0.0
4	1.5
3	3.0
2	4.5

(*Note:* For all rural highways, $f_N = 0.0$ mi/h.)
(Used with permission of Transportation Research Board, National Research Council, *Highway Capacity Manual*, Dec 2000, Exhibit 23-6, pg. 23-6.)

there must be at least one on-ramp. Thus, a junction with only off-ramps would *not* qualify as an interchange. The base condition for interchange density is 0.50 interchanges/mile, which implies an average interchange spacing of two miles. Adjustments for interchange density are shown in Table 12.8.

Table 12.8: Adjustment to Free-Flow Speed for Interchange Density on a Freeway

Interchanges Per Mi.	Reduction in Free-Flow Speed, f_{ID}(mi/h)
≤0.50	0.0
0.75	1.3
1.00	2.5
1.25	3.7
1.50	5.0
1.75	8.3
2.00	7.5

(Used with permission of Transportation Research Board, National Research Council, *Highway Capacity Manual*, Dec 2000, Exhibit 23-7, pg. 23-7.)

Multilane Highways

The free-flow speed for a multilane highway may be estimated as:

$$FFS = BFFS - f_{LW} - f_{LC} - f_M - f_A \qquad (12-6)$$

where: FFS = free-flow speed of the multilane highway, mi/h
$BFFS$ = base free-flow speed (as discussed below)
f_{LW} = adjustment for lane width, mi/h
f_{LC} = adjustment for lateral clearance, mi/h
f_M = adjustment for type of median, mi/h
f_A = adjustment for access points, mi/h

A base free-flow speed of 60 mi/h may be used for rural and suburban multilane highways, if no field data is available. It may also be estimated using the posted speed limit. The base free-flow speed is approximately 7 mi/h higher than the posted speed limit, for speed limits of 40 and 45 mi/h. For speed limits of 50 and 55 mi/h, the base free-flow speed is approximately 5 mi/h higher than the limit.

Lane Width Adjustment   The base lane width for multilane highways is 12 ft, as was the case for freeways. For narrower lanes, the free-flow speed is reduced by the values shown in Table 12.9.

Table 12.9: Adjustment to Free-Flow Speed for Lane Width on a Multilane Highway

Lane Width (ft)	Reduction in Free-Flow Speed, f_{LW} (mi/h)
≥12	0.0
11	1.9
10	6.6

(Used with permission of Transportation Research Board, National Research Council, *Highway Capacity Manual*, Dec 2000, Exhibit 21-4, pg. 21-5.)

Lateral Clearance Adjustment   For multilane highways, this adjustment is based on the *total lateral clearance*, which is the sum of the lateral clearances on the right side of the roadway and on the left (median) side of the roadway. While this seems like a simple concept, there are some details that must be observed:

- A lateral clearance of 6 ft is the base condition. Thus, no right- or left-side lateral clearance is

Table 12.10: Adjustment to Free-Flow Speed for Total Lateral Clearance on a Multilane Highway

4-Lane Multilane Highways		6-Lane Multilane Highways	
Total Lateral Clearance (ft)	Reduction in Free-Flow Speed, f_{LC} (mi/h)	Total Lateral Clearance (ft)	Reduction in Free-Flow Speed, f_{LC} (mi/h)
≥ 12	0.0	≥ 12	0.0
10	0.4	10	0.4
8	0.9	8	0.9
6	1.3	6	1.3
4	1.8	4	1.7
2	3.6	2	2.8
0	5.4	0	3.9

(Used with permission of Transportation Research Board, National Research Council, *Highway Capacity Manual*, Dec 2000, Exhibit 21-5, pg. 21-6.)

ever taken to be greater than 6 ft, even if greater clearance physically exists. Thus, the base total lateral clearance is 12 ft (6 ft for the right side, 6 ft for the left or median side).

- For an undivided multilane highway, there is no left- or median-side lateral clearance. However, there is a separate adjustment taken for type of median, including the undivided case. To avoid double-counting the impact of an undivided highway, the left or median lateral clearance on an undivided highway is assumed to be 6 ft.

- For multilane highways with two-way left-turn lanes, the left or median lateral clearance is also taken as 6 ft.

- For a divided multilane highway, the left- or median-side lateral clearance may be based on the location of a median barrier, periodic objects (light standards, abutments, etc.) in the median, or the distance to the opposing traffic lane. As noted previously, the maximum value is 6 ft.

The adjustments to free-flow speed for total lateral clearance on a multilane highway are shown in Table 12.10.

Median-Type Adjustment   The median-type adjustment is shown in Table 12.11. A reduction of 1.6 mi/h is

Table 12.11: Adjustment to Free-Flow Speed for Median Type on Multilane Highways

Median Type	Reduction in Free-Flow Speed, f_M (mi/h)
Undivided	1.6
TWLTLs	0.0
Divided	0.0

(Used with permission of Transportation Research Board, National Research Council, *Highway Capacity Manual*, Dec 2000, Exhibit 21-6, pg. 21-6.)

made for undivided configurations, while divided multilane highways, or multilane highways with two-way left-turn lanes, represent base conditions.

Access-Point Density Adjustment   A critical adjustment to base free-flow speed is related to access-point density. Access-point density is the average number of unsignalized driveways or roadways per mile that provide access to the multilane highway on *the right* side of the roadway (for the subject direction of traffic).

Driveways or other entrances with little traffic, or that, for other reasons, do not affect driver behavior, should not be included in the access-point density. Adjustments are shown in Table 12.12.

Table 12.12: Adjustment to Free-Flow Speed for Access-Point Density on a Multilane Highway

Access Density (Access Points/Mi)	Reduction in Free-Flow Speed, f_A (mi/h)
0	0.0
10	2.5
20	5.0
30	7.5
≥ 40	10.0

(Used with permission of Transportation Research Board, National Research Council, *Highway Capacity Manual*, Dec 2000, Exhibit 21-7, pg. 21-7.)

Sample Problems in Free-Flow Speed Estimation

Example 12-1: An Urban Freeway

An old 6-lane urban freeway has the following characteristics: 11-ft lanes; frequent roadside obstructions located 2 ft from the right pavement edge; and an interchange density of 2.00 interchanges/mile (i.e., average interchange spacing of .50 mile). What is the free-flow speed of this freeway?

Solution: The free-flow speed of a freeway may be estimated using Equation 12-5:

$$FFS = BFFS - f_{LW} - f_{LC} - f_N - f_{ID}$$

The following values are used in this computation:

$BFFS = 70$ mi/h (base condition for urban freeways)

$f_{LW} = 1.9$ mi/h (Table 12.5, 11-ft lanes)

$f_{LC} = 1.6$ mi/h (Table 12.6, 2-ft lateral clearance, 3 lanes)

$f_N = 3.0$ mi/h (Table 12.7, 3 lanes in one direction)

$f_{ID} = 7.5$ mi/h (Table 12.8, 2.00 interchanges/mi)

Then:

$$FFS = 70.0 - 1.9 - 1.6 - 3.0 - 7.5 = 56.0 \ mi/h$$

Example 12-2: A Four-Lane Suburban Multilane Highway

A 4-lane undivided multilane highway in a suburban area has the following characteristics: posted speed limit = 50 mi/h; 11-ft lanes; frequent obstructions located 4 ft from the right pavement edge; 30 access points/mi on the right side of the facility. What is the free-flow speed for the direction described?

Solution: The free-flow speed for a multilane highway is computed using Equation 12-6:

$$FFS = BFFS - f_{LW} - f_{LC} - f_M - f_A$$

The base free-flow speed for a multilane highway may be taken as 60 mi/h as a default or may be related to the posted speed limit. In the latter case, for a posted speed limit of 50 mi/h, the base free-flow speed may be taken to be 5 mi/h more than the limit, or $50 + 5 = 55$ mi/h. This is the value that will be used.

Adjustments to the base free-flow speed are as follows:

$f_{LW} = 1.9$ mi/h (Table 12.9, 11-ft lanes)

$f_{LC} = 0.4$ mi/h (Table 12.10, total lateral clearance = 10 ft, 4-lane highway)

$f_M = 1.6$ mi/h (Table 12.11, undivided highway)

$f_A = 7.5$ mi/h (Table 12.12, 30 access points/mi)

Then:

$$FFS = 55.0 - 1.9 - 0.4 - 1.6 - 7.5 = 43.6 \text{ mi/h}$$

Note that in selecting the adjustment for lateral clearance, the total lateral clearance is 4 ft (for the right side) plus an assumed value of 6.0 ft (for the left or median side) of an undivided highway.

12.3.3 Determining the Heavy-Vehicle Factor

The principal adjustment to demand volume is the heavy-vehicle factor, which adjusts for the presence of heavy vehicles in the traffic stream. A *heavy vehicle* is defined as any vehicle with more than four tires touching the pavement during normal operation. Two categories of heavy vehicle are used:

- Trucks and buses
- Recreational vehicles (RVs)

Trucks and buses have similar characteristics and are placed in the same category for capacity analysis purposes. These are primarily commercial vehicles, with the exception of some privately owned small trucks. Trucks vary widely in size and characteristics and range from small panel and single-unit trucks to double-back tractor-trailer combination vehicles. Factors in the HCM 2000 are based upon a typical mix of trucks with an average weight-to-horsepower ratio of approximately 150:1 (150 lbs/hp.)

Recreational vehicles also vary in size and characteristics. Unlike trucks and buses, which are primarily commercial vehicles operated by professional drivers, RVs are mostly privately owned and operated by drivers not specifically trained in their use and who often make only occasional trips in them. RVs include self-contained motor homes and a variety of trailer types hauled by a passenger car, SUV, or small truck. RVs generally have better operating conditions than trucks or buses, and have typical weight-to-horsepower ratios in the range of 75–100 lbs/hp.

The effect of heavy vehicles on uninterrupted multilane flow is the same for both freeways and multilane highways. Thus, the procedures described in this section apply to both types of facility.

The Concept of Passenger Car Equivalents and Their Relationship to the Heavy-Vehicle Adjustment Factor

The heavy-vehicle adjustment factor is based upon the concept of passenger-car equivalents. A *passenger-car equivalent* is the number of passenger cars displaced by one truck, bus, or RV in a given traffic stream under prevailing conditions. Given that two categories of heavy vehicle are used, two passenger car equivalent values are defined:

E_T = passenger car equivalent for trucks and buses in the traffic stream under prevailing conditions

E_R = passenger car equivalent for RV's in the traffic stream under prevailing conditions

The relationship between these equivalents and the heavy-vehicle adjustment factor is best illustrated by example: Consider a traffic stream of 1,000 veh/h, containing 10% trucks and 2% RVs. Field studies indicate that for this particular traffic stream, each truck displaces 2.5 passenger cars (E_T) from the traffic stream, and each RV displaces 2.0 passenger cars (E_R) from the traffic stream. What is the total number of equivalent passenger cars/h in the traffic stream?

Note that from the passenger car equivalent values, it is known that:

1 truck = 2.5 passenger cars

1 RV = 2.0 passenger cars

The number of equivalent passenger cars in the traffic stream is found by multiplying the number of each class of vehicle by its passenger-car equivalent, noting that the passenger-car equivalent of a passenger car is 1.0 by definition. Passenger-car equivalents are computed for each class of vehicle:

Trucks:	$1,000 * 0.10 * 2.5 =$	250 pce/h
RVs:	$1,000 * 0.02 * 2.0 =$	40 pce/h
Cars:	$1,000 * 0.88 * 1.0 =$	880 pce/h
TOTAL:		1,170 pce/h

Thus, the prevailing traffic stream of 1,000 veh/h operates as if it contained 1,170 passenger cars per hour.

By definition, the heavy-vehicle adjustment factor, f_{HV}, converts veh/h to pc/h when divided into the flow rate in veh/h. Thus:

$$V_{pce} = \frac{V_{vph}}{f_{HV}} \qquad (12\text{-}7)$$

where: V_{pce} = flow rate, pce/h
V_{vph} = flow rate, veh/h

In the case of the illustrative computation:

$$1{,}170 = \frac{1{,}000}{f_{HV}}$$

$$f_{HV} = \frac{V_{vph}}{V_{pce}} = \frac{1{,}000}{1{,}170} = 0.8547$$

In the example, the number of equivalent passenger cars per hour for each vehicle type was computed by multiplying the total volume by the proportion of the vehicle type in the traffic stream and by the passenger-car equivalent for the appropriate vehicle type. The number of passenger-car equivalents in the traffic stream may be expressed as:

$$V_{pce} = (V_{vph}*P_T*E_T) + (V_{vph}*P_R*E_R)$$
$$+ (V_{vph}*(1 - P_T - P_R)) \quad (12\text{-}8)$$

where: P_T = proportion of trucks and buses in the traffic stream
P_R = proportion of RVs in the traffic stream
E_T = passenger car equivalent for trucks and buses
E_R = passenger car equivalent for RVs

The heavy-vehicle factor may now be stated as:

$$f_{HV} = \frac{V_{vph}}{V_{pce}}$$

$$= \frac{V_{vph}}{(V_{vph}*P_T*E_T) + (V_{vph}*P_R*E_R) + (V_{vph}*(1 - P_T - P_R))}$$

which may be simplified as:

$$f_{HV} = \frac{1}{1 + P_T(E_T - 1) + P_R(E_R - 1)} \quad (12\text{-}9)$$

For the illustrative computation:

$$f_{HV} = \frac{1}{1 + 0.10(2.5 - 1) + 0.02(2.0 - 1)}$$
$$= \frac{1}{1.170} = 0.8547$$

This, as expected, agrees with the original computation.

Passenger-Car Equivalents for Extended Freeway and Multilane Highway Sections

HCM 2000 specifies passenger-car equivalents for trucks and buses and RVs for extended sections of roadway in general terrain categories, and for specific grade sections of significant impact.

A long section of roadway may be considered as a single extended section if no one grade of 3% or greater is longer than 0.25 miles, and if no grade of less than 3% is longer than 0.5 miles. Such general terrain sections are designated in one of three general terrain categories:

- *Level terrain.* Level terrain consists of short grades, generally less than 2% in severity. The combination of horizontal and vertical alignment permits trucks and other heavy vehicles to maintain the same speed as passenger cars in the traffic stream.

- *Rolling terrain.* Rolling terrain is any combination of horizontal and vertical alignment that causes trucks and other heavy vehicles to reduce their speeds substantially below those of passenger cars, but does not require heavy vehicles to operate at crawl speed for extended distances. Crawl speed is defined as the minimum speed that a heavy vehicle can sustain on a given segment of highway.

- *Mountainous terrain.* Mountainous terrain is severe enough to cause heavy vehicles to operate at crawl speed either frequently or for extended distances.

It should be noted that, in practical terms, mountainous terrain is a rare occurrence. It is difficult to have an extended section of highway that forces heavy vehicles to crawl speed frequently and/or for long distances without violating the limits for extended section analysis. Such situations usually involve longer and steeper grades that would require analysis as a specific grade.

Table 12.13: Passenger-Car Equivalents for Trucks, Buses, and RVs on Extended General Terrain Sections of Freeways or Multilane Highways

Factor	Type of Terrain		
	Level	**Rolling**	**Mountainous**
E_T	1.5	2.5	4.5
E_R	1.2	2.0	4.0

(Used with permission of Transportation Research Board, National Research Council, *Highway Capacity Manual*, Dec 2000, Exhibit 23-8, pg. 23-9.)

Table 12.13 shows passenger-car equivalents for freeways and multilane highways on extended sections of general terrain.

In analyzing extended general sections, it is the alignment of the roadway itself that determines the type of terrain, not the topography of the surrounding landscape. Thus, for example, many urban freeways or multilane highways in a relatively level topography have a rolling terrain based on underpasses and overpasses at major cross streets. Further, since the definitions for each category depend upon the operation of heavy vehicles, the classification may depend somewhat on the mix of heavy vehicles present in any given case.

Passenger-Car Equivalents for Specific Grades on Freeways and Multilane Highways

Any grade of less than 3% that is longer than 0.50 miles and any grade of 3% or steeper that is longer than 0.25 miles must be considered as a specific grade. This is because a long grade may have a significant impact on both heavy-vehicle operation and the characteristics of the entire traffic stream.

HCM 2000 specifies passenger car equivalents for:

- Trucks and buses on specific upgrades (Table 12.14)
- RVs on specific upgrades (Table 12.15)
- Trucks and buses on specific downgrades (Table 12.16)

The passenger car equivalent for RVs on downgrade sections is taken to be the same as that for level terrain sections, or 1.2.

Over time, the operation of heavy vehicles has improved relative to passenger cars. Trucks in particular, now have considerably more power than in the past, primarily due to turbo-charged engines. Thus, the maximum passenger-car equivalent shown in Tables 12.14 and 12.15 is 7.0. In the 1965 HCM, these values were as high as 17.0.

Tables 12.14 through 12.16 indicate the impact of heavy vehicles on the traffic stream. In the worst case, a single truck can displace as many as 7.0–7.5 passenger cars from the traffic stream. This displacement accounts for the both the size of heavy vehicles, and the fact that they cannot maintain the same speed as passenger cars in many situations. The latter is a serious impact that often creates large gaps between heavy vehicles and passenger cars that cannot be continuously filled by passing maneuvers.

Note that in these tables, there are some consistent trends. Obviously, as the grades get steeper and/or longer (either upgrade or downgrade), the passenger car equivalents increase, indicating a harsher impact on the operation of the mixed traffic stream.

A less obvious trend is that the passenger car equivalent in any given situation decreases as the proportion of trucks, buses, and RV's increases. Remember that the values given in Tables 12.14 through 12.16 are passenger-car equivalents (i.e., the number of passenger cars displaced by *one* truck, bus, or RV). The maximum impact of a single heavy vehicle is when it is relatively isolated in the traffic stream; as the flow of heavy vehicles increases, they begin to form their own platoons, within which they can operate more efficiently. The cumulative impact, however, of more heavy vehicles is a reduction in operating quality.

In some cases, the downgrade impact of a heavy vehicle is worse than the same heavy-vehicle situation on a similar upgrade. Downgrade impacts depend on whether or not trucks and other heavy vehicles must shift to low gear to avoid losing control of the vehicle. This is a particular problem for trucks on downgrades steeper than 4%; for lesser downgrades, values of E_T for level terrain are used.

Composite Grades

The passenger-car equivalents given in Tables 12.14 through 12.16 are based on a constant grade of known length. In most situations, however, highway alignment

Table 12.14: Passenger-Car Equivalents for Trucks and Buses on Upgrades

Up Grade (%)	Length (mi)	E_T Percent of Trucks and Buses (%)								
		2	4	5	6	8	10	15	20	25
<2	All	1.5	1.5	1.5	1.5	1.5	1.5	1.5	1.5	1.5
≥2–3	0.00–0.25	1.5	1.5	1.5	1.5	1.5	1.5	1.5	1.5	1.5
	>0.25–0.50	1.5	1.5	1.5	1.5	1.5	1.5	1.5	1.5	1.5
	>0.50–0.75	1.5	1.5	1.5	1.5	1.5	1.5	1.5	1.5	1.5
	>0.75–1.00	2.0	2.0	2.0	2.0	1.5	1.5	1.5	1.5	1.5
	>1.00–1.50	2.5	2.5	2.5	2.5	2.0	2.0	2.0	2.0	2.0
	>1.50	3.0	3.0	2.5	2.5	2.0	2.0	2.0	2.0	2.0
>3–4	0.00–0.25	1.5	1.5	1.5	1.5	1.5	1.5	1.5	1.5	1.5
	>0.25–0.50	2.0	2.0	2.0	2.0	2.0	2.0	1.5	1.5	1.5
	>0.50–0.75	2.5	2.5	2.0	2.0	2.0	2.0	2.0	2.0	2.0
	>0.75–1.00	3.0	3.0	2.5	2.5	2.5	2.5	2.0	2.0	2.0
	>1.00–1.50	3.5	3.5	3.0	3.0	3.0	3.0	2.5	2.5	2.5
	>1.50	4.0	3.5	3.0	3.0	3.0	3.0	2.5	2.5	2.5
>4–5	0.00–0.25	1.5	1.5	1.5	1.5	1.5	1.5	1.5	1.5	1.5
	>0.25–0.50	3.0	2.5	2.5	2.5	2.0	2.0	2.0	2.0	2.0
	>0.50–0.75	3.5	3.0	3.0	3.0	2.5	2.5	2.5	2.5	2.5
	>0.75–1.00	4.0	3.5	3.5	3.5	3.0	3.0	3.0	3.0	3.0
	>1.00	5.0	4.0	4.0	4.0	3.5	2.5	3.0	3.0	3.0
>5–6	0.00–0.25	2.0	2.0	1.5	1.5	1.5	1.5	1.5	1.5	1.5
	>0.25–0.30	4.0	3.0	2.5	2.5	2.0	2.0	2.0	2.0	2.0
	>0.30–0.50	4.5	4.0	3.5	3.0	2.5	2.5	2.5	2.5	2.5
	>0.50–0.75	5.0	4.5	4.0	3.5	3.0	3.0	3.0	3.0	3.0
	>0.75–1.00	5.5	5.0	4.5	4.0	3.0	3.0	3.0	3.0	3.0
	>1.00	6.0	5.0	5.0	4.5	3.5	3.5	3.5	3.5	3.5
>6	0.00–0.25	4.0	3.0	2.5	2.5	2.5	2.5	2.0	2.0	2.0
	>0.25–0.30	4.5	4.0	3.5	3.5	3.5	3.0	2.5	2.5	2.5
	>0.30–0.50	5.0	4.5	4.0	4.0	3.5	3.0	2.5	2.5	2.5
	>0.50–0.75	5.5	5.0	4.5	4.5	4.0	3.5	3.0	3.0	3.0
	>0.75–1.00	6.0	5.5	5.0	5.0	4.5	4.0	3.5	3.5	3.5
	>1.00	7.0	6.0	5.5	5.5	5.0	4.5	4.0	4.0	4.0

(Used with permission of Transportation Research Board, National Research Council, *Highway Capacity Manual*, Dec 2000, Exhibit 29-8, pg. 23-10.)

Table 12.15: Passenger-Car Equivalents for RVs on Upgrades

Grade (%)	Length (mi)	E_R Percentage of RVs (%)								
		2	4	5	6	8	10	15	20	25
≤2	all	1.2	1.2	1.2	1.2	1.2	1.2	1.2	1.2	1.2
>2–3	0.00–0.50	1.2	1.2	1.2	1.2	1.2	1.2	1.2	1.2	1.2
	>0.50	3.0	1.5	1.5	1.5	1.5	1.5	1.2	1.2	1.2
>3–4	0.00–0.25	1.2	1.2	1.2	1.2	1.2	1.2	1.2	1.2	1.2
	>0.25–0.50	2.5	2.5	2.0	2.0	2.0	2.0	1.5	1.5	1.5
	>0.50	3.0	2.5	2.5	2.5	2.0	2.0	2.0	1.5	1.5
>4–5	0.00–0.25	2.5	2.0	2.0	2.0	1.5	1.5	1.5	1.5	1.5
	>0.25–0.50	4.0	3.0	3.0	3.0	2.5	2.5	2.0	2.0	2.0
	>0.50	4.5	3.5	3.0	3.0	3.0	2.5	2.5	2.0	2.0
>5	0.00–0.25	4.0	3.0	2.5	2.5	2.5	2.5	2.0	2.0	1.5
	>0.25–50	6.0	4.0	4.0	4.0	3.5	3.0	2.5	2.5	2.0
	>0.50	6.0	4.5	4.0	4.0	4.0	3.5	3.0	2.5	2.0

(Used with permission of Transportation Research Board, National Research Council, *Highway Capacity Manual*, Dec 2000, Exhibit 23-10, pg. 23-10.)

Table 12.16: Passenger-Car Equivalents for Trucks and Buses on Downgrades

Downgrade (%)	Length (mi)	E_T Percent Trucks and Buses (%)			
		5	10	15	20
<4	all	1.5	1.5	1.5	1.5
≥4–5	≤4	1.5	1.5	1.5	1.5
	>4	2.0	2.0	2.0	1.5
>5–6	≤4	1.5	1.5	1.5	1.5
	>4	5.5	4.0	4.0	3.0
>6	≤4	1.5	1.5	1.5	1.5
	>4	7.5	6.0	5.5	4.5

(Used with permission of Transportation Research Board, National Research Council, *Highway Capacity Manual*, Dec 2000, Exhibit 23-11, pg. 23-11.)

leads to composite grades (i.e., a series of upgrades and/ or downgrades of varying steepness). In such cases, an equivalent uniform grade must be used to determine the appropriate passenger car equivalent values.

Consider the following composite grade profile: 3,000 feet of 3% upgrade followed by 5,000 feet of 5% upgrade. What heavy vehicle equivalents should be used in this case? The general approach is to find a

uniform upgrade, 8,000 ft in length, which has the same impact on the traffic stream as the composite described.

Average Grade Technique One approach to this problem is to find the average grade over the 8,000-ft length of the composite grade. This involves finding the total rise in the composite profile, as follows:

Rise on 3% Grade: 3,000 * 0.03 = 90 ft

Rise on 5% Grade: 5,000 * 0.05 = 250 ft

TOTAL RISE ON COMPOSITE: 340 ft

The average grade is then computed as the total rise divided by the total length of the grade, or:

$$G_{AV} = \frac{340}{8,000} = 0.0425 \; or \; 4.25\%$$

The appropriate values would now be entered to find passenger-car equivalents using a 4.25% grade, 8,000 ft (1.52 mi) in length.

The average grade technique is a good approximation when all subsections of the grade are less than 4%, or when the total length of the composite grade is less than 4,000 ft.

Composite Grade Technique   For more severe grades, a more exact technique is used. In this procedure, a percent grade of 8,000 ft is found that results in the same final operating speed of trucks as the composite described. This is essentially a graphic technique and requires a set of grade performance curves for a "typical" truck, with a weight-to-horsepower ratio of 200. Figure 12.6 shows these performance curves, and Figure 12.7 illustrates their use to find the equivalent grade for the example previously solved using the average-grade technique.

The following steps are followed in Figure 12.7 to find the composite grade equivalent:

- The curves are entered on the x-axis at 3,000 ft, the length of the first portion of the grade. The intersection of a vertical line constructed at this point with the 3% grade curve is found.

- The intersection of 3,000 ft with a 3% grade is projected horizontally back to the y-axis, where a speed of 41 mi/h is determined. This is the speed at which a typical truck is traveling after 3,000 ft on a 3% grade. It is also, however, the speed at which the truck *enters* the 5% segment of the grade.

- The intersection of 41 mi/h with the 5% grade curve is found and projected vertically back to the x-axis. The "length" found is approximately 1,450 ft. Thus, when the truck enters the 5% grade after 3,000 ft of 3% grade, it is as if the truck were on the 5% grade for 1,450 ft.

- The truck will now travel another 5,000 ft on the 5% grade—to an equivalent length of 1,450 + 5,000 = 6,450 ft. A vertical line is constructed at this point, and the intersection with the 5% curve is found. When projected horizontally to the y-axis, it is seen that a typical truck will be traveling at 27 mi/h at the end of the composite grade as described.

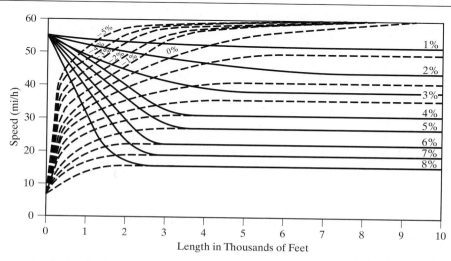

Figure 12.6: Performance of a Typical Truck on Grades (Used with permission of Transportation Research Board, National Research Council, *Highway Capacity Manual*, Dec 2000, Exhibit A23-2, pg. 23–30.)

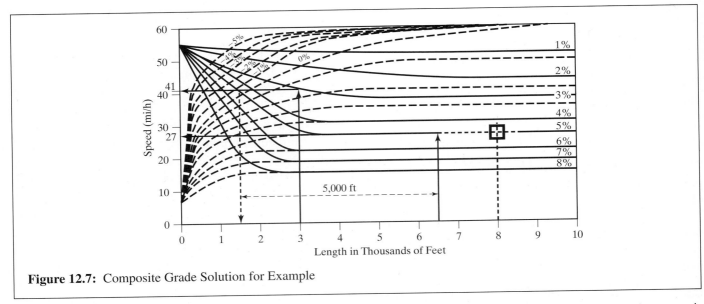

Figure 12.7: Composite Grade Solution for Example

- To find the equivalent grade, the intersection of 27 mi/h and the length of the composite grade (8,000 ft) is found. This is the solution point, indicating that the equivalent grade is a 5% grade of 8,000 ft.

While this methodology is considered "more" exact than the average-grade approach, it embodies a number of simplifications as well. The selection of 200 lbs/hp as the "typical" truck for all cases is certainly one such simplification. Further, the performance curves assume that the truck enters the grade at 55 mi/h and never accelerates to more than 60 mi/h, which are very conservative assumptions. Finally, passenger-car equivalents for all types of heavy vehicles will be selected

based on a composite grade equivalent based on a typical truck.

Despite these simplifications, this is viewed as a more appropriate technique for severe grade profiles than the average-grade technique. A composite grade equivalent can be found using this technique for any number of subsections and may include upgrade and downgrade segments.

It should also be noted that for analysis purposes, the impact of a grade is worst at the end of its steepest (uphill) section. Thus, if 1,000 ft of 4% grade were followed by 1,000 ft of 3% grade, passenger-car equivalents would be found for a 1,000 ft, 4% grade.

Example 12-3: Determination and Use of the Heavy-Vehicle Adjustment Factor

Consider the following situation: A volume of 2,500 veh/h traverses a section of freeway and contains 15% trucks and 5% RVs. The section in question is on a 5% upgrade, 0.75 miles in length. What is the equivalent volume in passenger-car equivalents?

Solution: The solution is started by finding the passenger car equivalent of trucks and RVs on the freeway section described (5% upgrade, 0.75 miles). These are found in Tables 12.14 and 12.15, respectively:

$E_T = 2.5$ (Table 12.14, 15% trucks, >4–5%, >0.50–0.75 mi)
$E_R = 3.0$ (Table 12.15, 5% RV's, >4–5%, >0.50 mi)

In entering values from these tables, care must be taken to observe the boundary conditions.

The heavy-vehicle adjustment factor may now be computed as:

$$f_{HV} = \frac{1}{1 + 0.15(2.5 - 1) + 0.05(3.0 - 1)}$$
$$= \frac{1}{1.325} = 0.7547$$

and the passenger-car equivalent volume may be estimated as:

$$V_{pce} = \frac{V_{vph}}{f_{HV}} = \frac{2,500}{0.7547} = 3,313 \text{ pc/h}$$

The solution can also be found by applying the passenger car equivalents directly:

Truck pces: $2{,}500*0.15*2.5 = 938$

RV pces: $2{,}500*0.05*3.0 = 375$

Pass Cars: $2{,}500*0.80*1.0 = 2{,}000$

TOTAL pces: $3{,}313$

12.3.4   Determining the Driver Population Factor

The base procedures for freeways and multilane highways assume a driver population of commuters or drivers familiar with the roadway and its characteristics. On some recreational routes, the majority of drivers may not be familiar with the route. This can have a significant impact on operations.

This adjustment factor is not well defined and is dependent upon local conditions. In general, the factor ranges between a value of 1.00 (for commuter traffic streams) to 0.85 as a lower limit for other driver populations. Unless specific evidence for a lower value is available, a value of 1.00 is generally used in analysis. Where unfamiliar users dominate a route, field studies comparing their characteristics to those of commuters are suggested to obtain a better estimate of this factor. Where a future situation is being analyzed, and recreational users dominate the driver population, a value of 0.85 is suggested as it represents a "worst-case" scenario.

12.4 Sample Applications

Example 12-4: Analysis of an Older Urban Freeway

Figure 12.8 shows a section of an old freeway in New York City. It is a four-lane freeway (additional service roads are shown in the picture) with the following characteristics:

- Ten-foot travel lanes
- Lateral obstructions at 0 ft at the roadside
- Interchange density = 2.0 interchanges per mile
- Rolling terrain

The roadway has a current peak demand volume of 3,500 veh/h. The peak-hour factor is 0.95, and there are no trucks, buses, or RVs in the traffic stream, as the roadway is classified as a parkway and such vehicles are prohibited.

Figure 12.8: Freeway for Section for Example 12-4 (Used with permission of Transportation Research Board, National Research Council, *Highway Capacity Manual*, Dec 2000, Illustration 12-1, pg. 12-6.)

At what level of service will the freeway operate during its peak period of demand?

Solution:

Step 1:   Determine the Free-Flow Speed of the Freeway

The free-flow speed of the freeway is estimated using Equation 12-5 as:

$$FFS = BFFS - f_{LW} - f_{LC} - f_N - f_{ID}$$

where: $BFFS$ = 70.0 mi/h (urban freeway default)

f_{LW} = 6.6 mi/h (Table 12.5, 10-ft lanes)

f_{LC} = 3.6 mi/h (Table 12.6, 2 lanes, 0-ft obstructions)

f_N = 4.5 mi/h (Table 12.7, 2 lanes)

f_{ID} = 7.5 mi/h (Table 12.8, 2 interchanges/mi)

Thus:

$$FFS = 70.0 - 6.6 - 3.6 - 4.5 - 7.5 = 47.8 \text{ mi/h},$$
say 48 mi/h

Step 2:   Determine the Demand Flow Rate in Equivalent pce Under Base Conditions

The demand volume may be converted to an equivalent flow rate under base conditions using Equation 12-1:

$$v_p = \frac{V}{PHF * N * f_{HV} * f_p}$$

where: V = 3,500 veh/h (given)

PHF = 0.95 (given)

N = 2 lanes (given)

f_{HV} = 1.00 (no trucks, buses, or RVs in the traffic stream)

f_p = 1.00 (assumed commuter driver population)

Then:

$$v_p = \frac{3500}{0.95 * 2 * 1.00 * 1.00} = 1,842 \text{ pc/h/ln}$$

Step 3:   Find the Level of Service and the Speed and Density of the Traffic Stream

Normally, the demand flow of 1,842 veh/h would be used to enter Figure 12.3 to find level of service and speed. However, the figure does not have a curve for a freeway with a free-flow speed of 48 mi/h. Interpolation of a curve between those given is normally acceptable, but in this case an extrapolation will be needed, as the minimum free-flow speed shown is 55 mi/h. This would have to be considered a rough estimate of actual conditions. The estimation is illustrated in Figure 12.9.

From Figure 12.9, a demand flow of 1,842 pc/h/ln on a freeway with a 48 mi/h free-flow speed will result in an expected operating speed of 48.0 mi/h. Even at the relatively high demand flow, the free-flow speed will be maintained. It can also be seen that the level of service for this freeway is E.

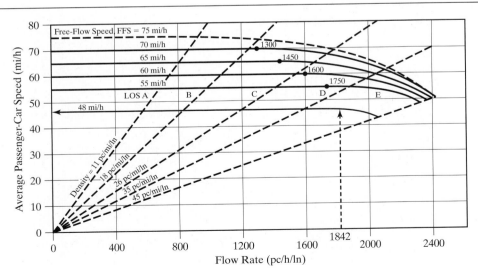

Figure 12.9: Solution to Example 12-4

The density on the freeway may be estimated as the demand flow rate divided by the speed, or:

$$D = \frac{v_p}{S} = \frac{1842}{48} = 38.4 \text{ pc/mi/ln}$$

This value can be used to enter Table 12.2 to confirm the level of service, which is E, falling within the defined boundaries of 35–45 pc/mi/ln for LOS E.

Example 12-5: Analysis of a Multilane Highway Section

A four-lane multilane highway section with a full median carries a peak-hour volume of 2,600 veh/h in the heaviest direction. There are 12% trucks and 2% RVs in the traffic stream. Motorists are primarily regular users of the facility. The section under study is on a 3% sustained grade, 1 mile in length. The *PHF* is 0.88.

Field studies have been conducted to determine that free-flow speed of the facility is 55.0 mi/h.

At what level of service will this facility operate during the peak hour?

Solution: As the free-flow speed has been found from field data, it is not necessary to estimate it using Equation 12-6. The analysis section is a sustained grade. As the peak volume would be expected to travel upgrade during one peak and downgrade during the other, it will be necessary to examine the downgrade as well as the upgrade under peak demand conditions.

Step 1:   Determine the Upgrade Demand Flow Rate in Equivalent pces Under Base Conditions

Equation 12-1 is used to convert the peak hour demand volume to an equivalent flow rate in pces under base conditions:

$$v_p = \frac{V}{PHF * N * f_{HV} * f_p}$$

where: V = 2,600 veh/h (given)

PHF = 0.88 (given)

N = 2 lanes (given)

f_p = 1.00 (regular users)

The heavy-vehicle factor, f_{HV}, is computed using Equation 12-9:

$$f_{HV} = \frac{1}{1 + P_T (E_T - 1) + P_R (E_R - 1)}$$

where: P_T = 0.12 (given)

P_R = 0.02 (given)

E_T = 1.5 (Table 12.14, ≥2–3%, >0.75–1.00 mi, 12% trucks)

E_R = 3.0 (Table 12.15, >2–3%, >0.50 mi, 2% RVs)

Then:

$$f_{HV} = \frac{1}{1 + 0.12(1.5 - 1) + 0.02(3.0 - 1)}$$

$$= \frac{1}{1.10} = 0.909$$

and:

$$v_p = \frac{2,600}{0.88 * 2 * 0.909 * 1.00} = 1,625 \text{ pc/h/ln}$$

Step 2:   Determine the Downgrade Demand Flow Rate in Equivalent pces Under Base Conditions

The downgrade computation follows the same procedure as the upgrade computation, except that the passenger car equivalents for trucks and RVs are selected for the downgrade condition. These are found as follows:

E_T = 1.5 (Table 12.16, <4%, all lengths, 12% trucks)
E_R = 1.2 (Table 12.13, level terrain)

Note that passenger car equivalents for downgrade RVs are found assuming level terrain. Then:

$$f_{HV} = \frac{1}{1 + 0.12(1.5 - 1) + 0.02(1.2 - 1)}$$

$$= \frac{1}{1.064} = 0.940$$

$$v_p = \frac{2,600}{0.88 * 2 * 0.940 * 1.00} = 1,572 \text{ pc/h/ln}$$

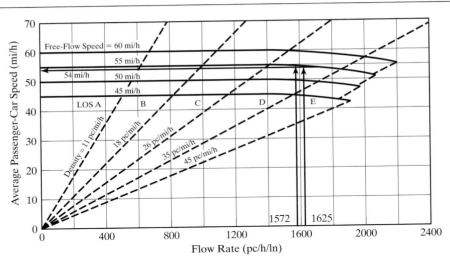

Figure 12.10: Solution to Example 12-5

Step 3: Find the Level of Service and the Speed and Density of the Traffic Stream

Level of service and speed determinations are made using Figure 12.4 for multilane highways, as shown in Figure 12.10. Remember that the free-flow speed was field-measured as 55 mi/h.

From Figure 12.10, it can be seen that the expected speeds for both the upgrade and downgrade sections are approximately 54 mi/h, although it might be argued that the upgrade speed is a fraction less than the downgrade speed. The scale of Figure 12.4 makes it difficult to estimate this small difference. The level of service for both upgrade and downgrade sections is expected to be D during periods of peak demand flow.

The density for upgrade and downgrade sections is estimated as the flow rate divided by the expected speed, or:

$$D_{up} = \frac{1625}{54} = 30.1 \text{ pc/mi/ln}$$

$$D_{dn} = \frac{1572}{54} = 29.1 \text{ pc/mi/ln}$$

both of which are between the limits defined for LOS D—26–35 pc/mi/ln.

Example 12-6: Finding Service Flow Rates and Service Volumes for a Freeway Segment

A six-lane urban freeway has the following characteristics: 12-ft lanes, 6-ft clearances on the right side of the roadway, rolling terrain, an interchange density of 1.0 interchange per mile, and a PHF of 0.92. The traffic consists of 8% trucks and no RVs, and all drivers are regular users of the facility.

The peak hour volume on the facility is currently 3,600 veh/h, which is expected to grow at a rate of 6% a year for the next 20 years.

What is the current level of service on the facility, and what levels of service can be expected in 5 years? In 10 years? In 15 years? In 20 years?

Solution: These questions could be answered by conducting five separate operational analyses, determining the expected LOS for the various demand flows both now and expected in the future target years. It is often easier to solve a problem involving multiple demand levels by simply computing the service flow rates (SF) and service volumes (SV) for the section for each level of service. Then, demand volumes can be easily compared to the results to determine the LOS for each target demand level.

Determine the Free-Flow Speed of the Freeway

Step 1: The free-flow speed of the facility is found using Equation 12-5:

$$FFS = BFFS - f_{LW} - f_{LC} - f_N - f_{ID}$$

where: $BFFS$ = 70 mi/h (urban freeway)

f_{LW} = 0.0 mi/h (Table 12.5, 12-ft lanes)

f_{LC} = 0.0 mi/h (Table 12.6, 6-ft lateral clearance)

f_N = 3.0 mi/h (Table 12.7, 3 lanes)

f_{ID} = 2.5 mi/h (Exh. Table 12.8, 1 int/mi)

$$FFS = 70.0 - 0.0 - 0.0 - 3.0 - 2.5 = 64.5 \text{ mi/h}$$

Step 2: Determine the Maximum Service Flow Rates for Each Level of Service

Maximum service flow (MSF) rates for each level of service are drawn from Table 12.3. Values are obtained for freeways with free-flow speeds of 65 mi/h and 60 mi/h. Straight-line interpolation may be used to obtain maximum service flow rates for the intermediate free-flow speed of 64.5 mi/h. This is illustrated in Table 12.17.

Table 12.17: Values of MSF for Example 12-6

Level of Service	Free-Flow Speed		
	60	64.5	65
A	660	705	710
B	1,080	1,161	1,170
C	1,560	1,668	1,680
D	2,020	2,083	2,090
E	2,300	2,345	2,350

Step 3: Determine the Heavy-Vehicle Factor

The heavy-vehicle factor is computed as:

$$f_{HV} = \frac{1}{1 + P_T(E_T - 1) + P_R(E_R - 1)}$$

where: P_T = 0.08 (given)

P_R = 0.00 (given)

E_T = 2.5 (Exh. 12-18, rolling terrain)

Then:

$$f_{HV} = \frac{1}{1 + 0.08(2.5 - 1)} = \frac{1}{1.12} = 0.893$$

Step 4: Determine the Service Flow Rates and Service Volumes for Each Level of Service

Service flow rates and service volumes are computed using Equations 12-2 and 12-3:

$$SF_i = MSF_i * N * f_{HV} * f_p$$
$$SV_i = SF_i * PHF$$

where: MSF_i = as determined in Step 2

N = 3 (given)

f_{HV} = 0.893 (as computed in Step 3)

f_p = 1.00 (regular users)

PHF = 0.92 (given)

These computations are done in the spreadsheet shown in Table 12.18 on the next page.

The service flow rates (SF) refer to the peak 15-minute interval; service volumes apply to peak-hour volumes.

Step 5: Determine Target-Year Peak-Demand Volumes

The problem statement indicates that present demand is 3,600 veh/h and that this volume will increase by 6% per year for the foreseeable future. Future demand volumes may be computed as:

$$V_j = V_o(1.06^n)$$

where: V_j = peak-hour demand volume in target year j

V_o = peak-hour demand volume in year 0, 3,600 veh/h

N = number of years to target year

Then:

V_o = 3,600 veh/h

V_5 = 3,600(1.06^5) = 4,818 veh/h

V_{10} = 3,600(1.06^{10}) = 6,447 veh/h

V_{15} = 3,600(1.06^{15}) = 8,628 veh/h

V_{20} = 3,600(1.06^{20}) = 11,546 veh/h

Step 6: Determine Target Year Levels of Service

The target year demand volumes are stated as full peak-hour volumes. They are, therefore, compared to the *service volumes* computed in Table 12.18 to determine LOS. The results are shown in Table 12.19 on the next page.

As indicated in Table 12.19, level of service F prevails in target years 10, 15, and 20. In each of these years, demand exceeds capacity. Clearly, the point at which capacity is reached occurs between years 5 and 10.

Table 12.18: Spreadsheet Computation of Service Flow Rates and Service Volumes

Level of Service	MSF (pc/h/ln)	N	f_{HV}	f_p	SF (veh/h)	PHF	SV (veh/h)
A	705	3	0.893	1.000	*1,889*	0.92	*1,738*
B	1,161	3	0.893	1.000	*3,110*	0.92	*2,861*
C	1,668	3	0.893	1.000	*4,469*	0.92	*4,111*
D	2,083	3	0.893	1.000	*5,580*	0.92	*5,134*
E	2,345	3	0.893	1.000	*6,282*	0.92	*5,780*

Table 12.19: Levels of Service for Example 12-6

Target Year	Demand Volume (veh/h)	Level of Service
0	3,600	C
5	4,818	D
10	6,447	F
15	8,628	F
20	11,546	F

Capacity, stated in terms of a full peak hour, is 5,780 veh/h (Table 12.18). The exact year that demand reaches capacity may be found as follows:

$$5,570 = 3,600(1.06^n)$$
$$n = 7.5 \text{ years}$$

Analysis   The results of this analysis indicate that demand will reach the capacity of the freeway in 7.5 years. If no action is taken, users can expect regular breakdowns during the peak hour in this freeway section. To avoid this situation, action must be taken to either reduce demand and/or increase the capacity of the section.

Increasing the capacity of the section suggests adding a lane. Computations would be redone using a four-lane, one-direction cross-section to see whether sufficient capacity was added to handle the 20-year demand forecast. Reduction in demand is more difficult and would involve intensive study of the nature of demand on the freeway section in question. Reduction would require diversion of users to alternative routes or alternative modes, encouraging users to travel at different times or to different destinations, encouraging car-pooling and other actions to increase auto occupancy. Given the constraints of capacity on the current cross-section, it is also unlikely that demand would grow to the levels indicated in later years, as queuing and congestion would reach intolerable levels. In Year 20, the projected demand of 11,546 veh/h is more than twice the capacity of the current cross-section.

As is the case in many uses of the HCM, this analysis identifies and gives insight into a problem. It does not definitively provide a solution—unless engineers are prepared to more than double the current capacity of the facility or modify alternative routes to provide the additional capacity needed. Even these options involve judgments. Capacity and level-of-service analyses of the various alternatives would provide additional information on which to base those judgments, but would not, taken alone, dictate any particular course of action. Economic, social, and environmental issues would obviously also have to be considered as part of the overall process of finding a remedy to the forecasted problem.

Example 12-7: A Design Application

A new freeway is being designed through a rural area. The directional design hour volume (DDHV) has been forecast to be 2,700 veh/h during the peak hour, with a PHF of 0.85 and 15% trucks in the traffic stream. A long section of the facility will have level terrain characteristics, but one 2-mile section involves a sustained grade of 4%. If the objective is to provide level of service C, with a minimum acceptable level of D, how many lanes must be provided?

Solution: The problem calls for the determination of the number of required lanes for three distinct sections of freeway: (1) a level terrain section, (2) a 2-mile, 4% sustained upgrade, and (3) a 2-mile, 4% downgrade.

Step 1: Determine the Free-Flow Speed of the Freeway

This is a design situation. Unless additional information concerning the terrain suggested otherwise, it would be assumed that lane widths (12 ft) and lateral clearances ($\geq$6 ft) conform to modern standards and meet base conditions. No interchange density is noted, but given that this is a rural section of freeway, an interchange density of no more than 0.50 per mile would be assumed. This is also a base condition for freeways.

One adjustment to free-flow speed, however, is the number of lanes in the section, which is what this analysis seeks to determine. From Table 12.7, however, the footnote indicates that for rural freeways, no adjustment for number of lanes is applied.

Thus, there are no adjustments to the base free-flow speed, which, for rural freeways, is generally taken as 75 mi/h. Without adjustments, this is also final free-flow speed.

Step 2: Determine the Maximum Service Flow Rate (MSF) for Levels of Service C and D

As the target level of service is C, with a minimum acceptable level of D, it is necessary to determine the maximum service flow rates that would be permitted if these levels of service are to be maintained. These are found from Table 12.3 for a 75 mi/h free-flow speed:

$$MSF_C = 1{,}830 \text{ pc/h/ln}$$
$$MSF_D = 2{,}170 \text{ pc/h/ln}$$

Step 3: Determine the Number of Lanes Required for the Level, Upgrade, and Downgrade Freeway Sections

The required number of lanes is found using Equation 12-4:

$$N_i = \frac{DDHV}{PHF * MSF_i * f_{HV} * f_p}$$

where: $DDHV$ = 2,700 veh/h (given)

PHF = 0.85 (given)

MSF_C = 1,830 pc/h/ln (determined in Step 2)

f_p = 1.00 (regular users assumed)

Three different heavy-vehicle factors (f_{HV}) must be considered: one for level terrain, one for the upgrade, and one for the downgrade. With no RVs in the traffic stream, the heavy-vehicle factor is computed as:

$$f_{HV} = \frac{1}{1 + P_T (E_T - 1)}$$

where: P_T = 0.15 (given)

E_T (level) = 1.5 (Table 12.13, level terrain)

E_T (up) = 2.5 (Table 12.14, > 3–4%, >1.5 mi)

E_T (down) = 1.5 (Table 12.16, $\geq$ 4–5%, $\leq$4 mi)

Then:

$$f_{HV} \text{ (level, down)} = \frac{1}{1 + 0.15(1.5 - 1)}$$
$$= \frac{1}{1.075} = 0.930$$

$$f_{HV} \text{ (up)} = \frac{1}{1 + 0.15(2.5 - 1)}$$
$$= \frac{1}{1.225} = 0.816$$

and:

$$N \text{(level, down)} = \frac{2{,}700}{0.85 * 1{,}830 * 0.930 * 1.00} = 1.87 \text{ lanes}$$
$$N \text{(up)} = \frac{2{,}700}{0.85 * 1{,}830 * 0.816 * 1.00} = 2.13 \text{ lanes}$$

This suggests that the level and downgrade sections require two lanes in each direction, but that the upgrade requires three lanes. The computed values are minima to provide the target level of service. This suggests that the facility should be constructed as a four-lane freeway with a truck-climbing lane on the sustained upgrade.

Analysis   The results of such an analysis most often result in fractional lanes. An operational analysis could be performed using the DDHV and a four-lane freeway cross-section to determine the resulting LOS. It is at least possible that the level of service will be better than the target.

If this had been an urban or suburban section, the impact of the number of lanes on free-flow speed would have complicated the analysis. A number of lanes would be assumed to determine the FFS and the resulting MSF_C. The number of lanes would then be computed as shown here. When the assumed and computed values agree, the result is accepted. This may require one or two iterations for each such determination.

As level of service of D would have been minimally acceptable, it is useful to examine what LOS would have resulted on the upgrade if only two lanes were provided. Then:

$$v_p = \frac{V}{PHF * N * f_{HV} * f_p} = \frac{2,700}{0.85 * 2 * 0.816 * 1.00}$$
$$= 1,946 \text{ pc/h/ln}$$

As this is less than the MSF_D of 2,170 pc/h/ln determined in Step 2, provision of two lanes on the upgrade would provide LOS D.

Thus, a choice must be made of providing the target LOS C on the upgrade by building a three-lane cross-section with a truck-climbing lane, or accepting the minimal LOS D with a two-lane cross-section. A compromise solution might be to build two lanes, but to acquire sufficient right-of-way

and build all structures so that a climbing lane could be added later.

It should also be noted that the analysis of a three-lane upgrade cross-section is approximate. If one of the lanes is a truck-climbing lane, then there will be substantial segregation of heavy vehicles from passenger cars in the traffic stream. The HCM freeway methodology assumes a mix of vehicles across all lanes.

Again, it must be emphasized that while the results of the analysis provide the engineer with a great deal of information to assist in making a final design decision on the upgrade section, it does not dictate such a decision. Economic, environmental, and social factors would also have to be considered.

12.5 Calibration Issues

The analysis methodologies of the HCM for basic freeway sections and multilane highways rely on defined speed-flow curves for base conditions, and on a variety of adjustment factors applied to determine free-flow speed and the demand flow rate in equivalent pce.

It is important to understand some of the issues involved in calibrating these basic relationships, both as background knowledge and because the HCM allows traffic engineers to substitute locally calibrated relationships and values where they are available.

For interested readers, References 8 and 9 present the research results that were the basis of multilane highway and freeway procedures in the HCM 2000, though both were subject to subsequent modifications. The sections that follow discuss some critical calibration issues involved in these methodologies.

12.5.1   Calibrating Base Speed-Flow Curves

The single most important feature of the freeway and multilane highway methodologies is the set of base speed-flow curves that are used to define and determine levels of service and expected speeds. Speed-flow curves are part of the critical relationships among speed, flow, and density (see Chapter 5). These three critical variables are related:

$$v = S * D \qquad (12\text{-}10)$$

where: v = rate of flow, veh/h or veh/h/ln

   S = space mean speed, mi/h

   D = density, veh/mi or veh/mi/ln

Thus, calibrating a relationship between any two of these variables determines the relationship among all three. As speed and flow are the easier parameters to measure in the field, speed-flow relationships are most often the ones to be calibrated.

Some Historic Studies

The general form of these curves has evolved over the years. One of the earliest studies was conducted by Bruce Greenshields, who hypothesized that the relationship between speed and density was linear [10]. Such a relationship produced continuous parabolic curves for speed-flow and flow-density.

Later, Ellis [11] investigated two- and three-segment linear curves with discontinuities. Greenberg [12] hypothesized a logarithmic curve for speed-density, while Underwood [13] used an exponential form. Edie [14] combined logarithmic and exponential curves for low- and high-density portions of the curve, respectively. As with Ellis, this form produced discontinuities into the relationship. Later, May [15] suggested a bell-shaped curve for speed-density. Reference [15] also provides an interesting study attempting to fit these and other mathematical forms to a single comprehensive set of data from the Eisenhower Expressway in Chicago in the early 1960s.

Historically, these studies are interesting and help focus attention on two important points:

- Early calibration efforts focused on the speed-density relationship
- What is implied by discontinuities in speed-flow-density relationships

The first reflects an understanding that it is the speed-density relationship that directly reflects driver behavior. Traffic demand occurs most immediately as a density, as various land uses generate trips placing numbers of vehicles into a restricted roadway space. Individual drivers select their speed based on many conditions but primarily on their proximity to other vehicles and their perception of safe operation under those conditions. The density, created by trip generation and the driver-selected speed in response primarily to density, results in a rate of flow. Density, however, is rarely measured directly and is most often computed from speed and flow measurements. Thus, most later work has tried to relate the measured values of speed and density directly.

The second introduces a dilemma that is to the present day not completely resolved. Discontinuities in speed-flow-density relationships almost invariably occur in the vicinity of capacity. The result is often a two-segment speed-flow relationship with two values of capacity: one for the stable portion of the curve, and one for the unstable portion of the curve. One such curve, shown in Figure 12.11 from Reference [16], illustrates this characteristic. What it suggests is that capacity, when approached from stable flow, may have a significantly higher value than capacity when approached from unstable flow. In more practical terms, the capacity of a segment after a breakdown may be lower than the capacity when it is operating under stable flow. As will be demonstrated subsequently, this characteristic has a critical impact on how long it takes to recover from a breakdown or queuing situation.

Some of the difficulty in interpreting such discontinuities in older studies results from not having detailed information on how and where the data were collected. This is discussed in the following section with respect to more recent studies.

Modern Freeway and Multilane Highway Characteristics

Figure 12.12 shows the general form of a freeway or multilane highway speed-flow curve that is typical of

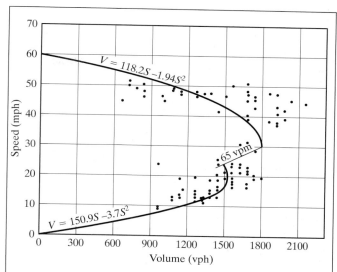

Figure 12.11: Speed-Flow Curve With Discontinuity in the Vicinity of Capacity (Used with permission of Transportation Research Board, National Research Council, Washington DC, J.S. Drake, J.L.Schofer, and A.D. May Jr., "A Statistical Analysis of Speed-Density Hypotheses," *Transportation Research Record 154,* pg. 78, 1967.)

driver behavior on uninterrupted flow facilities. Note that all of the base curves of Figures 12.3 and 12.4 are of this form, except that only parts labeled "Regions 1 and 2" are included. These are the stable flow portions of the curve. The four identified regions of the speed-flow curve in Figure 12.12 may be described as follows:

- *Region 1.* This region shows that drivers maintain a virtually constant speed to relatively high flow rates. This flat portion of the curve defines free-flow speed.
- *Region 2.* In this portion of the curve, still in stable flow, speed begins to decline in response to increasing flow rates. However, the total decline in speed from free-flow speed to the speed at capacity (point 2) is often 5 mi/h or less. Point 1, which indicates the flow rate at which speed begins to decline is often in the range of 1,500–1,700 pc/h/ln. Thus, the path from free-flow speed to capacity (and subsequent breakdown) is often caused by a relatively small increase in flow rate.

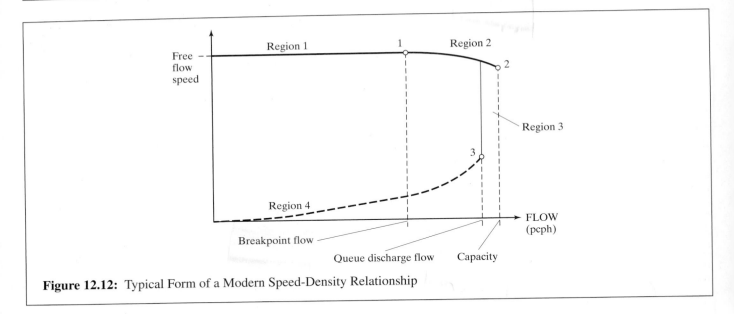

Figure 12.12: Typical Form of a Modern Speed-Density Relationship

- *Region 3*. This portion of the curve is called "queue discharge." Once demand exceeds capacity, a breakdown occurs and a queue propagates upstream of the point of breakdown. Once the queue forms, flow is restricted to what is discharged from the front of the queue. This is a relatively stable value. The variable speed for Region 3 reflects the fact that vehicles discharge from a queue into an uncongested downstream segment. At subsequent downstream points, the flow level will remain constant, but drivers will accelerate after departing the queue.

- *Region 4*. This portion of the curve reflects the unstable operating conditions *within* the queue, upstream of the breakdown.

The shape of this curve represents the characteristics of modern drivers on modern multilane uninterrupted-flow facilities. It reflects the fact that U.S. drivers have had four decades of driving experience on such facilities and that they will aggressively maintain high-speed operation, even at high flow rates.

Issue: Where Should Field Measurements Be Made

In attempting to calibrate a speed-flow curve for a particular facility, it is important to note that the four regions

of the curve *cannot* reflect data collected at a single point. Further, to fully calibrate the curve, capacity must be observed, as must the transition from stable to unstable flow. This can be done only in a situation where sufficient demand is delivered to the section to reach capacity and, indeed, to cause a breakdown and queuing. The downstream section must also remain free of the effect of queuing from other bottlenecks located further downstream, so that queue discharge is not impeded by downstream congestion. The most likely place for these conditions to exist is at a merge area (or on-ramp) where the number of lanes entering the junction exceeds the number of lanes departing it. Figure 12.13 illustrates such a segment.

Because there are four lanes delivering traffic to the merge point and only three lanes departing it, it is at least feasible that more traffic will arrive at the merge than can be discharged from it. General observations should be made to ensure that this is a regular occurrence, so that the study will likely be able to observe the transition from stable flow to breakdown.

Regions 1 and 2 of the speed-flow curve reflect measurements taken under stable flow, before any breakdown occurs. A point downstream of the junction (point 1) is selected such that merging vehicles have fully accelerated to their desired speed. When the breakdown occurs, data collection moves to multiple observation points.

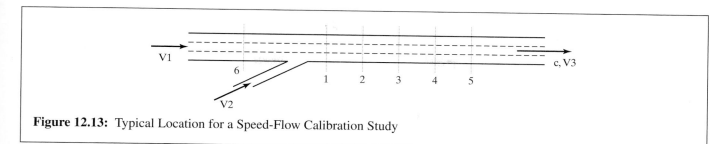

Figure 12.13: Typical Location for a Speed-Flow Calibration Study

Queue discharge flow can be observed at downstream points 1 through 5 (or as many locations as desired). Flow observations should be relatively constant at these points, but speed will increase to desired levels at subsequent downstream points. To observe Region 4, observations must be made within the upstream queue (i.e., at location 6, or perhaps at multiple upstream points within in the range of the queue).

Issue: Ensuring Base Conditions

Base conditions for speed-flow relationships include only two conditions: no heavy vehicles in the traffic stream, and a driver population of regular facility users. While 12-ft lanes and 6-ft minimum lateral clearances are desirable, they are not necessary, because these affect free-flow speed, which is calibrated as part of the study. The number of lanes and interchange density, which also impact on free-flow speed, should also be recorded as part of the data describing the study section.

The most vexing problem in observing base conditions is the presence of heavy vehicles in most traffic streams. While calibration sites should have as little heavy-vehicle presence as possible, it is quite difficult to find sites where none are present.

Thus, the problem is often one of extracting passenger-car–only traffic stream data from a mixed traffic stream. This is done by observing and making use of microscopic traffic stream data, specifically headways. If vehicles are classified only as passenger cars (P) or heavy vehicles (H), then four types of headway may be observed in the traffic stream:

- *P-P:* passenger car following a passenger car

- *P-H:* passenger car following a heavy vehicle

- *H-P:* heavy vehicle following a passenger car

- *H-H:* heavy vehicle following a heavy vehicle

Where the traffic stream contains numerous types of heavy vehicles, headways might be classified only by the type of trailing vehicle, such as:

- *P:* passenger car

- *T:* truck

- *B:* bus

- *R:* recreational vehicle

During each observation period (usually a 15-minute or 5-minute interval), headways would be measured and categorized as noted previously. Average headways for each observation period and each type of headway would be computed. Similarly, average speeds for each type of vehicle in each observation period would be computed.

If it is assumed that the operation of passenger cars is not affected by the presence of heavy vehicles in adjacent lanes, then the base flow rate and speed for each observation period can isolate the behavior of passenger cars by using the relationship between average headways and flow rates:

$$v_{pc} = \frac{3,600}{h_{aPP}} \text{ or } \frac{3,600}{h_{aP}} \qquad (12\text{-}11)$$

where: v_{pc} = rate of flow for passenger cars, pc/h

h_{aPP} = average headway for passenger cars following passenger cars, s

h_{aP} = average of all headways for vehicle pairs in which a passenger car is the following or trailing vehicle, s

The density of the traffic stream, in terms of base passenger cars, can be estimated for each observation period as follows:

$$D_{pc} = \frac{v_{pc}}{S_{aP}} \qquad (12\text{-}12)$$

where: D_{pc} = density, pc/mi

S_{aP} = average (space mean) speed of passenger cars, mi/h

Curve Fitting

Once data has been observed, reduced, and recorded, it is necessary to fit a curve through those points that best represent the trends in the data. Multiple linear and non-linear regression approaches may be used to determine the "best" mathematical descriptions for data in equation form. Most recent studies, however, have observed that the spread of data tends to be large and have pursued simple graphical curve-fitting (i.e., visually estimating the shape and location of the curve) using both the data and the analyst's professional judgment to arrive at a final curve. This is how the base speed-flow curves of Figures 12.3 and 12.4 were calibrated. In developing final text for the HCM 2000, equations describing the graphic curve were added after the fact, so that computer algorithms replicating procedures could be more easily developed.

One Peak or Two?

It was noted that a number of historic speed-flow-density studies derived relationships with sharp discontinuities in the vicinity of capacity. In terms of the typical modern speed-flow relationship, a similar issue arises when the capacity of the stable portion of the curve is compared to queue discharge flows. In both cases, the question is as follows: once a breakdown and queue have been established, is it possible for vehicles to depart the queue at a rate similar to the stable-flow capacity of the facility?

There is, unfortunately, no firm answer to this question. In many cases, the queue discharge flow is observed to be somewhat less than the stable-flow capacity of the

facility section. There have, however, been cases in which the two appear to be equal, and even some cases in which queue discharge flows exceed the stable-flow capacity observed. The importance of the issue is, however, extremely important when analyzing the spatial and time spread of congestion.

Figure 12.14 illustrates a typical situation. A vehicle breakdown occurs on a three-lane section of highway for one-half hour. The demand profile and capacity assumptions for the section are detailed in the exhibit.

Under each scenario, two questions must be answered: (1) how large does the queue become? and (2) how long does it take the queue to dissipate? Deterministic queuing analysis is used to develop these responses. In each time period, the number of vehicles arriving at the section is compared to the number being discharged. The analysis for Scenario 1, which presumes a 200 veh/h/ln difference between capacity and queue discharge flow, is shown in Table 12.20. The analysis for

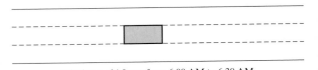

Blockage of 1 Lane from 6:00 AM to 6:30 AM

Blockage of 1 Lane from 6:00 AM to 6:30 AM

Demand Pattern

6–7 AM	6,000 veh/h
7–8 AM	6,000 veh/h
8–9 AM	6,000 veh/h
9–10 AM	5,000 veh/h
>10 AM	4,000 veh/h

Capacity Assumptions

Scenario 1:
Capacity = 2,000 veh/h/ln
Queue Discharge = 1,800 veh/h/ln

Scenario 2:
Capacity = 2,000 veh/h/ln
Queue Discharge = 2,000 veh/h/ln

Figure 12.14: Impact of a Breakdown Illustrated

Table 12.20: Queuing Analysis for Scenario 1

Time	Arrivals (veh)	Capacity (veh)	Queue Size (veh)
6:00–6:30 AM	6,000/2 = 3,000	2 × 1,800/2 = 1,800	3,000 − 1,800 = 1,200
6:30–7:00 AM	6,000/2 = 3,000	3 × 1,800/2 = 2,700	1,200 + 3,000 − 2,700 = 1,500
7:00–8:00 AM	6,000	3 × 1,800 = 5,400	1,500 + 6,000 − 5,400 = 2,100
8:00–9:00 AM	6,000	5,400	2,100 + 6,000 − 5,400 = 2,700
9:00–10:00 AM	5,000	5,400	2,700 + 5,000 − 5,400 = 2,300
>10:00 AM	4,000 veh/h	5,400 veh/h	Queue decreases by 1,400 veh/h

Queue dissipates 2,300/1,400 = 1.64 hrs after 10:00 AM, or 11:38 AM

Scenario 2, which assumes that there is no difference between capacity and queue discharge flow, is shown in Table 12.21.

In Scenario 1, the queue does not dissipate until 11:38 AM. The key point is that the reduced queue discharge flow does not increase to capacity until the queue is gone—which does not occur until 11:38 AM, many hours after the original blockage has been removed. The queue reaches a maximum size of 2,700 vehicles at 9:00 AM.

In Scenario 2, the queue never grows to more than 1,000 vehicles and is fully dissipated by 10:00 AM.

Thus, the issue of whether or not queue discharge flow is less than capacity is more than just a philosophical or theoretical issue. Its impact can be enormous. If the facility is assumed to have three lanes upstream of the breakdown and each vehicle is assumed to occupy 25 ft in queue, compare the maximum lengths of queue developed under each scenario:

$$L_Q \text{ (Scenario 1)} = \left(\frac{2,700}{3}\right) * 25 = 22,500 \text{ ft}$$

$$= \frac{22,500}{5,280} = 4.26 \text{ mi}$$

$$L_Q \text{ (Scenario 2)} = \left(\frac{1000}{3}\right) * 25 = 8,333 \text{ ft}$$

$$= \frac{8,333}{5,280} = 1.58 \text{ mi}$$

Scenario 1 assumes a 10% drop from capacity to queue discharge flow. While this may seem small, it produces a maximum queue over twice as long as a case in which no drop occurs, and it takes almost two hours longer for the queue to clear.

Both cases, however, demonstrate how a phenomenon familiar to many drivers occurs. In the foregoing case, a driver hears a report of the breakdown on a traffic report delivered at 6:00 AM. Another report indicates that the breakdown has been removed at 6:30 AM. The driver arrives at the location at 9:00 AM and is faced with a large queue. On reaching the site of the former breakdown, the driver enters a freely flowing condition downstream. The driver is left wondering why the queue exists, as its cause disappeared hours before. The moral of the story is that breakdowns create a spread of congestion in both time and space, conditions that will exist long after the cause

Table 12.21: Queuing Analysis for Scenario 2

Time	Arrivals (veh)	Capacity (veh)	Queue Size (veh)
6:00–6:30 AM	6,000/2 = 3,000	2 × 2,000/2 = 2,000	3,000 − 2,000 = 1,000
6:30–7:00 AM	6,000/2 = 3,000	3 × 2,000/2 = 3,000	1,000 + 3,000 − 3,000 = 1,000
7:00–8:00 AM	6,000	3 × 2,000 = 6,000	1,000 + 6,000 − 6,000 = 1,000
8:00–9:00 AM	6,000	6,000	1,000 + 6,000 − 6,000 = 1,000
9:00–10:00 AM	5,000	6,000	1,000 + 5,000 − 6,000 = 0

Queue is dissipated at 10:00 AM

of the breakdown itself is removed. The incentive to avoid breakdowns and to develop programs to remove them (when due to accidents or incidents) as quickly as possible is great in terms of traffic quality.

12.5.2 Calibrating Passenger Car Equivalents

The heavy-vehicle adjustment factor is the most significant in converting a demand volume under prevailing conditions to a flow rate in equivalent passenger-car equivalents under base conditions. As noted, the adjustment is based on calibrated passenger-car equivalents for trucks and buses and for RVs under various conditions of terrain. The calibration of these equivalents, is, therefore, of interest. To simplify the presentation, assume that only one category of heavy vehicle and one passenger car equivalent value (E_H) is being used. Recalling several previous relationships:

$$f_{HV} = \frac{v_{vph}}{v_{pce}}$$

$$f_{HV} = \frac{1}{1 + P_H(E_H - 1)}$$

If these equations are manipulated to solve for E_H, the following is obtained:

$$E_H = \left[\frac{\left(\frac{v_{pce}}{v_{vph}}\right) - 1}{P_H} \right] + 1 \qquad (12\text{-}13)$$

The values of v_{pce} and v_{vph} can be related to headway measurements in any traffic stream. If, for example, headways in a traffic stream are categorized by both leading and trailing vehicles in each pair and if only two types of vehicle are used, the following classifications, defined previously, result: P-P, P-H, H-P, and H-H. In this case, from Equation 12-11:

$$v_{pce} = \frac{3,600}{h_{aPP}}$$

The mixed flow rate, v_{vph}, is computed as 3,600 divided by the average of all headways, where the average of all

headways is found from the average headways in each of the above categories:

$$v_{vph} = \frac{3600}{h_a}$$

$$h_a = P_H^2 h_{aHH} + P_H(1 - P_H)h_{aHP}$$
$$+ (1 - P_H)P_H h_{aPH} + (1 - P_H)^2 h_{aPP} \quad (12\text{-}14)$$

where: P_H = proportion of heavy vehicles in the traffic stream

h_{aHH} = average headway for heavy vehicles following heavy vehicles, s

h_{aHP} = average headway for heavy vehicles following passenger cars, s

h_{aPH} = average headway for passenger cars following heavy vehicles, s

h_{aPP} = average headway for passenger cars following passenger cars, s

Inserting all of this into Equation 12-13 results in:

$$E_H = \frac{(1 - P_H)*(h_{aPH} + h_{aHP} - h_{aPP}) + P_H h_{aHH}}{h_{aPP}}$$
$$(12\text{-}15)$$

where all terms are as previously defined.

Where headways are classified only by the type of following or trailing vehicles, the following assumptions are implicit:

$$h_{aPP} = h_{aPH} = h_{aP}$$
$$h_{aHH} = h_{aHP} = h_{aH}$$

When these assumptions are included in Equation 12-15, the relationship simplifies to:

$$E_H = \frac{h_{aH}}{h_{aP}} \qquad (12\text{-}16)$$

This is a convenient form where a number of different categories of heavy vehicles exist.

The entire basis for these computations, however, is the identification of a flow of mixed vehicles that is "equivalent" to a flow of passenger cars only. Once such an equivalence is identified, headways can be used to compute or calibrate values of E_H. The major issue, then, is how to define "equivalent" conditions for use in calibration. The following sections discuss several different approaches.

Driver-Determined Equivalence

Krammas and Crowley [16] used a very straightforward means to establish equivalence. A given traffic stream represents an equilibrium condition in which individual drivers have adjusted their vehicles' operation consistent with their subjective perceptions of optimality. Krammas and Crowley suggest that individual headways within a given traffic stream represent the drivers' view of "equivalent" operational quality or level of service. This is an important concept. As the service flow rate is defined for a given level of service, it is rational to use the concept of level of service as a basis for equivalence.

Using this approach, during each 15-minute period of observations, headways would be classified by type, and equivalents would be computed using either Equation 12-15 or Equation 12-16.

Consider the following example: The data shown in Table 12.22 was obtained during a single 15-minute interval on a freeway. The traffic stream during this interval consisted of 10% trucks and 90% passenger cars.

Table 12.22: Data for Sample Problem in Driver-Determined Equivalence

Type of Headway	Number Observed	Average Headway (s)
P-P	400	3.0
P-T	40	3.4
T-P	40	4.2
T-T	9	4.6

Note that the distribution of headway types is consistent with the distribution of trucks (10%) and passenger cars (90%) in the traffic stream. Using trailing vehicle types, there are 440 passenger cars and 49 trucks in the traffic stream. If lead vehicles are counted, the same distribution arises. With headways classified by both leading and trailing vehicle types, Equation 12-15 is used to determine the value of E_T:

$$E_T = \frac{(1 - 0.10)*(3.4 + 4.2 - 3.0) + (0.10*4.6)}{3.0}$$

$$= 1.53$$

Based on the Krammas and Crowley theory of equivalence, each truck consumes as much space or capacity as 1.53 passenger cars.

It would also be interesting to see how the results would be affected by using the simpler Equation 12-16, where headways are classified only by the type of trailing vehicle. To do this, weighted average headways for trailing passenger cars and trailing trucks would have to be computed from the data in Table 12.22. The weighted average headways are:

$$h_{aT} = \frac{(40*4.2) + (9*4.6)}{(40 + 9)} = 4.27 \text{ s}$$

$$h_{aP} = \frac{(400*3.0) + (40*3.4)}{(400 + 40)} = 3.04 \text{ s}$$

Then, using Equation 12-16:

$$E_T = \frac{4.27}{3.04} = 1.40$$

This equation, however, is based on the assumption that headways depend only on the type of trailing vehicle. From the data in Table 12.22, this is obviously not true in this case—$h_{aPP} \neq h_{aPT}$ and $h_{aTT} \neq h_{aTP}$. Thus, the second computation is more approximate than the first. The results in the two cases, however, differ by only 0.13, indicating that use of Equation 12-16 with headways classified only by the trailing vehicle may be a reasonable approach, particularly where multiple vehicle types exist.

This example, however, illustrates the calibration of *one* value of E_T. In HCM 2000, E_T is shown as varying with type of terrain, the proportion of trucks in the traffic stream, and the length and severity of a sustained grade. Some studies have suggested that these values vary by flow levels and/or level of service as well.

Calibrating a complete set of E_T (and E_R) values would, therefore, require a large data base covering a wide range of underlying conditions. This is both expensive and time-consuming. For practical reasons, most studies of heavy-vehicle equivalence have relied at least partially on simulations to produce the desired data set.

Field studies are then conducted to validate a selection of cells in the multivariable space. Validation results may then be used to adjust equivalents to reflect discrepancies between field data and simulation values.

Equivalence Based on Constant Spacing

Another approach to equivalence is to select headways of various vehicle types such that their *spacing* is constant. This is done by plotting headways of various types of vehicles at a given location (using only the trailing vehicle to classify) vs. their spacing. This is an interesting approach, illustrated in Figure 12.15. Spacing is related to density ($D = 5,280/S_a$), and density is the measure that defines level of service for freeways and multilane highways. Thus, this process results in passenger-car equivalents that define traffic streams of equal density and, therefore, of equal level of service.

The plot of headways vs. spacing can be made from data covering any time period sufficient to define the relationships. The plots would, however, be valid only for the site in question and would relate to the size of the facility (number of lanes), terrain, grade, and length of grade present. Passenger-car equivalents might be shown to vary with spacing (density) as the result of any given set of plots. This approach, however, would not reveal any variation in E_T based on the proportion of trucks in the traffic stream, P_T. Implicit in this approach is the assumption that there is no such variation.

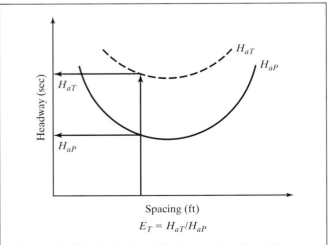

Figure 12.15: Equivalent Headways Based on Constant Spacing

Note also that the equivalent headways selected using this procedure do not have to occur within the same 15-minute period, as the plots are calibrated using a set of points for many 15-minute periods. This is fundamentally different from the Krammas and Crowley approach. It defines "equivalent" in terms of traffic streams having equal average spacings.

Equivalence Based on Constant Speed

It is also possible to prepare a plot of average headways vs. space mean speed for each 15-minute interval. Similar to the approach of Figure 12.15, equivalents may then be based on traffic streams having equal space mean speeds. This is an attractive alternative if speed is the measure defining level of service. Thus, while this method was used to calibrate passenger car equivalents for two-lane rural highways in the 1985 HCM, it is not particularly well suited to such calibrations for freeways or multilane highways. In the 1985 HCM, speed is a secondary service measure for two-lane highways [18].

Macroscopic Calibration of the Heavy-Vehicle Factor

Since f_{HV} is defined as a ratio of equivalent flows in veh/h and pc/h, it seems that a simpler approach would be to measure these values directly and calibrate the factor, rather than using passenger car equivalents.

Figure 12.16 illustrates plots of flow vs. density for similar geometric sites and time periods during which the percentage of trucks varies. Curves are illustrated for 0% trucks (the base condition), and for 10% trucks. Given that level of service for multilane uninterrupted flow facilities is defined by density, equivalent flow levels can be found by holding density constant, as shown in the figure. The ratio of these flows now yields a value of f_{HV} directly.

The data collection for such a calibration is difficult. A family of curves representing different percentages of trucks would be needed as would sites with varying conditions of terrain and/or grade. Finding sites sufficient to produce a curve for truck percentages ranging from 0% to as high as 25%–30% is a significant challenge that makes this approach less practical than others discussed previously.

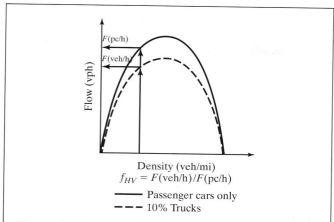

Figure 12.16: Direct Calibration of the Heavy-Vehicle Adjustment Factor Using Flow-Density Relationships

A similar approach can be taken using speed-flow curves, but this would not be as philosophically consistent, as speed is not the measure of service quality use in the HCM.

Additional References

The literature contains numerous theoretical and practical treatments of passenger car equivalents and the heavy-vehicle factor [17–20]. No common concept of equivalence has emerged, nor do all researchers agree on specific calibration approaches. Reference 21 explains the approach to multilane uninterrupted-flow passenger-car equivalents used in HCM 2000.

12.5.3 Calibrating the Driver Population Factor

The adjustment factor for driver population, f_p, virtually always requires local calibration, if a value other than 1.00 (for a commuting traffic stream, or familiar users of a facility) is to be used.

This is because the potential "other" driver populations and their impacts on operations are widely variable, depending on the specific situation in question. Recreational routes frequently involve nonregular users who may not be familiar with local characteristics. However, familiarity, or lack thereof, is not a "yes" or

"no" proposition, but a matter of degrees. Further, recreational and other routes may have a range of driver characteristics, including some familiar or regular users. Geometry also matters. An unfamiliar user of a roadway with severe geometry in a mountainous area will be far more cautious than an unfamiliar user on a roadway that is basically straight and level.

The same facility may have markedly different driver populations on weekdays and weekends if it serves both work-related and recreational destinations.

Calibration is often done by observing capacity operations on a single facility on weekdays and on weekends, or by comparing observed capacities of similar routes with different driver populations. Such calibration is always approximate, as the extension of results from one facility (or a group of facilities) to others cannot ensure accuracy. Nevertheless, this is the best approach available at present and is a reasonable approach to getting a local estimate of a very volatile adjustment factor.

12.5.4 Adjustment Factors to Free-Flow Speed

The HCM 2000 applies some adjustments to the prediction of free-flow speed. These adjustments (for lane width, lateral clearance, number of lanes or median type, and interchange or access-point density) are different from the driver population and heavy-vehicle factors, in that they are subtractive rather than multiplicative. The data on which these factors are based were quite sparse. Lane width and lateral clearance adjustments, for example, were calibrated using some field data and by projecting former adjustments to flow rates for these factors onto the speed axis using standard speed-flow curves.

Free-flow speed for any facility is easily measured as the average speed (of passenger cars uninfluenced by heavy vehicles) when flow rates are low. For freeways and multilane highways, measurements taken when flow is less than 800–1,000 veh/h/ln will provide reasonable estimates. Calibration requires controlled experiments in which one variable (i.e., lane width) is varied while others are held constant. Sites fulfilling this objective will be difficult to find. Another approach is to collect data at a number of sites with varying conditions and use regression analysis to establish a relationship for predicting free-flow speed. The relationship might then be used to examine the impact of individual features on the result.

12.6 Software

There are two major software packages that claim to replicate the methodologies of the HCM 2000. The *Highway Capacity Software* package (HCS) continues to be maintained and available through McTrans Center at the University of Florida, Gainesville. *HiCap 2000* is a new product developed by Catalina Engineering, Inc. of Tucson, Arizona. Catalina Engineering was the primary contractor for NCHRP in developing the final material for HCM 2000.

It should be noted that the Highway Capacity and Quality of Service Committee of the Transportation Research Board *does not* examine, certify, or endorse any software product. The burden of demonstrating that a software package faithfully replicates the current HCM is entirely that of the software producers.

References

1. *Highway Capacity Manual*, Bureau of Public Roads, Washington DC, 1950.

2. *Highway Capacity Manual*, "Special Report 87," Transportation Research Board, Washington DC, 1965.

3. *Highway Capacity Manual*, "Special Report 209," Transportation Research Board, 1985.

4. *Highway Capacity Manual*, "Special Report 209" (as revised in 1994), Transportation Research Board, Washington DC, 1994.

5. *Highway Capacity Manual*, "Special Report 209" (as revised in 1997), Transportation Research Board, Washington DC, 1997.

6. *Highway Capacity Manual*, Transportation Research Board, Washington DC, 2000 (Metric and Standard U.S. versions).

7. Kittelson, W., "Historical Overview of the Committee on Highway Capacity and Quality of Service," *Proceedings of the Fourth International Symposium on Highway Capacity*, Transportation Research Circular E-C018, Maui, Hawaii, June 27–July 1, 2000.

8. Reilly, W., Harwood, D., and Schoen, J., "Capacity and Quality of Flow of Multilane Highways," *Final Report*, JHK & Associates, Tucson AZ, 1988.

9. Schoen, J., May, A. Jr., Reilly, W., and Urbanik, T., "Speed-Flow Relationships for Basic Freeway Sections," *Final Report*, JHK & Associates, Tucson AZ, and Texas Transportation Institute, Texas A&M University, College Station TX, December 1994.

10. Greenshields, B., "A Study of Traffic Capacity," *Proceedings of the Highway Research Board*, Transportation Research Board, Washington DC, 1934.

11. Ellis, R., "Analysis of Linear Relationships in Speed-Density and Speed-Occupancy Curves," *Final Report*, Northwestern University, Evanston IL, December 1964.

12. Greenberg, H., "An Analysis of Traffic Flows," *Operations Research*, Vol. 7, Operations Research Society of America, Washington DC, 1959.

13. Underwood, R., "Speed, Volume, and Density Relationships," *Quality and Theory of Traffic Flow*, Yale Bureau of Highway Traffic, New Haven CT, 1961.

14. Edie, L., "Car-Following and Steady-State Theory for Non-Congested Traffic," *Operations Research*, Vol. 9, Operations Research Society of America, Washington DC, 1961.

15. Duke, J.; Schofer, J.; and May, A. Jr., "A Statistical Analysis of Speed-Density Hypotheses, *Highway Research Record 154*, Transportation Research Board, Washington DC, 1967.

16. Krammas, R. and Crowley, K., "Passenger Car Equivalents for Trucks on Level Freeway Segments," *Transportation Research Record 1194*, Transportation Research Board, Washington DC, 1988.

17. Linzer, E., Roess, R., and McShane, W., "Effect of Trucks, Buses, and Recreational Vehicles on Freeway Capacity and Service Volume," *Transportation Research Record 699*, Transportation Research Board, Washington DC, 1979.

18. Craus, J., Polus, A., and Grinberg, A. "A Revised Method for the Determination of Passenger Car Equivalents," *Transportation Research*, Vol. 14A, No. 4, Pergamon Press, London, England, 1980.

19. Cunagin, W. and Messer, C., "Passenger Car Equivalents for Rural Highways," *Transportation Research Record 905*, Transportation Research Board, Washington DC, 1983.

20. Roess, R. and Messer, C., "Passenger Car Equivalents for Uninterrupted Flow: Revision of the Circular 212 Values," *Transportation Research Record 971*, Transportation Research Board, Washington DC, 1984.

21. Webster, L. and Elefteriadou, A., "A Simulation Study of Truck Passenger Car Equivalents (PCE) on Basic Freeway Sections," *Transportation Research B*, Vol. 33, No. 5, Pergamon Press, London, England, 1999.

Problems

12-1. Estimate the free-flow speed of a four-lane undivided multilane highway having the following characteristics: (a) base free-flow speed = 55 mi/h, (b) average lane width = 11 ft., (c) lateral clearance = 2 ft at both roadsides, (d) access-point density = 20/mi on each side of the roadway.

12-2. Estimate the free-flow speed of a six-lane suburban freeway with 12-ft lanes, lateral clearances of 4 ft, and interchanges spaced at an average of 0.75 miles.

12-3. Find the appropriate composite grade for each of the following grade sequences:

(a) 1,000 ft of 3% grade followed by 1,500 ft of 2% grade, followed by 750 ft of 4% grade.

(b) 2,000 ft of 4% grade, followed by 5,000 ft of 3% grade, followed by 2,000 ft of 5% grade.

(c) 4,000 ft of 5% grade, followed by 3,000 ft of 3% grade.

12-4. A freeway operating in generally rolling terrain has a traffic composition of 10% trucks and 5% RVs. If the observed peak hour volume is 3,500 veh/h, what is the equivalent volume in pc/h?

12-5. Find the upgrade and downgrade service flow rates and service volumes for an eight-lane urban freeway with the following characteristics: (a) 11-ft lanes, (b) 2-ft lateral clearances, (c) interchange density = 2.0/mi, (d) 5% trucks, no recreational vehicles, and (e) driver population consisting of regular facility users. The section in question is on a 4% sustained grade of 1 mile. The PHF is 0.85.

12-6. An existing six-lane divided multilane highway with a field-measured free-flow speed of 45 mi/h serves a peak-hour volume of 4,000 veh/h, with 15% trucks and no RVs. The PHF is 0.90. The highway has rolling terrain. What is the likely level of service for this section?

12-7. A long section of suburban freeway is to be designed on level terrain. A level section of 5 miles is, however, followed by a 5% grade, 1.5 mi in length. If the DDHV is 2,500 veh/h with 10% trucks and 3% RV's, how many lanes will be needed on the (a) upgrade, (b) downgrade, and (c) level terrain section to provide for a minimum of level of service C? Assume that base conditions of lane width and lateral clearance exist and that interchange density is less than 0.50/mi. The PHF = 0.92.

12-8. An old urban four-lane freeway has the following characteristics: (a) 11-ft lanes, (b) no lateral clearances (0 ft), (c) an interchange density of 2.0/mi, (d) 5% trucks, no RVs, (e) PHF = 0.90, and (f) rolling terrain. The present peak-hour demand on the facility is 2,200 veh/h, and anticipated growth is expected to be 3% per year. What is the present level of service expected? What is the expected level of service in 5 years? 10 years? 20 years? To avoid breakdown (LOS F), when will substantial improvements be needed to this facility or alternative routes?

12-9. A freeway with two lanes in one direction has a capacity of 2,000 veh/h/ln under normal stable-flow operation. On a particular morning, one of these lanes is blocked for 15 minutes beginning at 7:00 AM. The arrival pattern of vehicles at this location is as follows: 7–8 AM—4,000 veh/h; 8–9 AM—3,900 veh/h; 9–10 AM—3,500 veh/h; after 10:00 AM—2,800 veh/h.

(a) Assuming that the capacity of this section reduces to 1,800 veh/h/ln under unstable conditions (i.e., immediately upon formation of a queue), how long a queue will be caused by this blockage? When will the maximum queue occur? How long will it take to dissipate the queue from the time of the breakdown?

(b) Reconsider this analysis if the capacity of the section remains 2,000 veh/h/ln after formation of a queue.

12-10. The following headways are observed during a 15-minute period on an urban freeway:

Type of Headway	Number Observed	Average Value (s)
P-P	128	3.1
P-T	32	3.8
T-P	32	4.3
T-T	8	4.9

Compute the effective passenger-car equivalent (pce) for trucks in this case, assuming that all headway types are different. Recompute the pce for trucks, assuming that headway values depend only on the type of following vehicle. Are the results different? Why? Use the "driver selected equivalent" approach in solving this problem.

CHAPTER
13

Weaving, Merging, and Diverging Movements on Freeways and Multilane Highways

13.1 Turbulence Areas on Freeways and Multilane Highways

In Chapter 12, capacity and level-of-service analysis approaches for basic freeway and multilane highway sections were presented and illustrated. Segments of such facilities that accommodate weaving, merging, and/or diverging maneuvers, however, experience additional turbulence as a result of these movements. This additional turbulence in the traffic stream results in operations that cannot be simply analyzed using basic section techniques.

While there are no generally accepted measures of "turbulence" in the traffic stream, the basic distinguishing characteristic of weaving, merging, and diverging segments is the additional lane-changing these maneuvers cause. Other elements of turbulence include the need for greater vigilance on the part of drivers, more

frequent changes in speed, and average speeds that may be somewhat lower than on similar basic sections.

Figure 13.1 illustrates the basic maneuvers involved in weaving, merging, and diverging segments. *Weaving* occurs when one movement must cross the path of another along a length of facility without the aid of signals or other control devices, with the exception of guide and/or warning signs. Such situations are created when a merge area is closely followed by a diverge area. The flow entering on the left leg of the merge and leaving on the right leg of the diverge must cross the path of the flow entering on the right leg of the merge and leaving on the left leg of the diverge. Depending upon the specific geometry of the segment, these maneuvers may require lane changes to be successfully completed. Further, other vehicles in the segment (i.e., those that do not weave from one side of the roadway to the other), may make lane changes to avoid concentrated areas of turbulence within the segment.

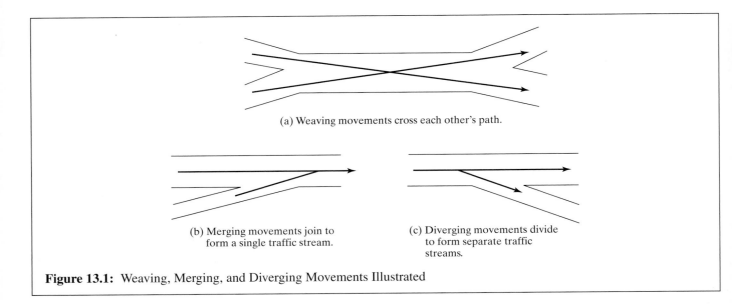

(a) Weaving movements cross each other's path.

(b) Merging movements join to
form a single traffic stream.

(c) Diverging movements divide
to form separate traffic
streams.

Figure 13.1: Weaving, Merging, and Diverging Movements Illustrated

Merging occurs when two separate traffic streams join to form a single stream. Merging can occur at an on-ramp to a freeway or multilane highway, or when two significant facilities join to form one. Merging vehicles often make lane changes to align themselves in lanes appropriate to their desired movement. Non-merging vehicles also make lane changes to avoid the turbulence caused by merging maneuvers in the segment.

Diverging occurs when one traffic stream separates to form two separate traffic streams. This occurs at off-ramps from freeways and multilane highways, but can also occur when a major facility splits to form two separate facilities. Again, diverging vehicles must properly align themselves in appropriate lanes, thus inducing lane-changing; non-diverging vehicles also make lane changes to avoid the turbulence created by diverge maneuvers.

The difference between weaving and separate merging and diverging movements is unclear at best. Weaving occurs when a merge is "closely followed" by a diverge. The exact meaning of "closely followed" is not well defined. The HCM 2000 [1] indicates that the maximum length over which weaving movements are defined is 2,500 ft. Thus, wherever merge and diverge points are separated by more than 2,500 ft, they are treated as isolated merge and diverge movements. Even where the distance between a merge and diverge is less

than 2,500 ft, the classification of the movement depends upon the details of the configuration. For example, a one-lane, right-hand, on-ramp followed by a one-lane, right-hand, off-ramp is considered a weaving section only if the two are connected by a continuous auxiliary lane. If the on-ramp and off-ramp have separate, discontinuous acceleration and deceleration lanes, they are treated as isolated merge and diverge areas, respectively, independent of the distance between them. The 1965 HCM [2] recognized weaving movements over distances up to 8,000 ft, but this was based on a single data point, and lengths greater than 2,500 ft were subsequently removed from consideration as weaving areas.

Even though the nature of lane changing and other turbulence factors is similar in weaving, merging, and diverging segments, the methodologies for analysis treat weaving in a manner distinctly different from merging and diverging. This is primarily an accident of research history, as conceptually, similar procedures would seem to be more appropriate. Research efforts on these subjects have been done at different times using different data bases, as mandated by sponsoring agencies. In writing material for the HCM 2000, however, considerable effort was expended to make the two analytic approaches more consistent, particularly in terms of level-of-service measures and criteria.

HCM methodologies for weaving and for merging/diverging segments are calibrated for freeways. These methods have some limited application to multilane highways with uncontrolled weaving or merging/diverging operations, but must generally be considered more approximate in these cases.

13.2 Level-of-Service Criteria

The measure of effectiveness for weaving, merging, and diverging segments is density. This is consistent with freeway and multilane highway methodologies. Level-of-service criteria are shown in Table 13.1.

For weaving areas, separate criteria are specified for segments on freeways and on multilane highways. Boundary conditions for multilane highways are set at somewhat higher densities than for freeways, reflecting users' lower expectations on multilane highways. This is somewhat inconsistent with the criteria for basic sections, which are the same for freeways and multilane highways, except for the LOS E/F boundary. For weaving areas, the LOS E/F boundary is defined by a maximum density beyond which stable operations are thought to be highly unlikely. While some data does exist to support this boundary, a good

deal of professional judgment (on the part of the Highway Capacity and Quality of Service Committee of the Transportation Research Board) was required to set the values used.

For merge and diverge junctions, only one set of criteria are specified, based on freeways. When applied to multilane highways (as an approximation), the same criteria are applied. Level-of-service boundaries are the same as those used for freeway weaving areas, except for the LOS E/F boundary. For merge and diverge areas, this boundary is crossed when demand exceeds the capacity of the segment.

This reflects an ongoing controversy over how to establish the critical dividing line between levels of service E and F. For uninterrupted flow, there is little difference, except for semantics. The density level for the E/F boundary for weaving areas defines the capacity of the segment (as will be demonstrated) (i.e., a point at which the v/c ratio is 1.00). For merge and diverge sections, the boundary is defined directly in these terms. This issue becomes more important for signalized intersections, where the E/F boundary does not necessarily imply a v/c ratio >1.00, but merely unacceptable operating quality.

Another difference, however, is the spatial relevance of the densities used in determining the level of

Table 13.1: Level-of-Service Criteria for Weaving, Merging, and Diverging Segments

Level of Service	Weaving Areas		Merge or Diverge Areas
	Density Range (pc/mi/ln)		
	On Freeways	On Multilane Highways	On Freeways
A	0–10	0–12	0–10
B	>10–20	>12–24	>10–20
C	>20–28	>24–32	>20–28
D	>28–35	>32–36	>28–35
E	>35–43	>36–40	>35
F	>43	>40	Demand Exceeds Capacity

(Used with permission of Transportation Research Board, National Research Council, *Highway Capacity Manual*, 2000. Compiled from Exhibit 24-2, pg. 24-3 and Exhibit 25-4, pg. 25-5.)

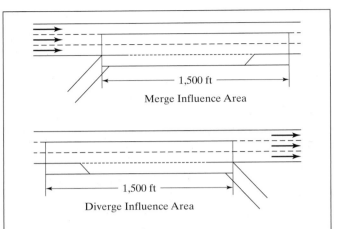

Figure 13.2: Merge and Diverge Influence Areas Illustrated (Used with permission of Transportation Research Board, National Research Council, *Highway Capacity Manual*, 2000, Exhibit 13-14, pg. 13-21.)

service. For weaving areas, the density reflects an average for all vehicles across all lanes of the segment between the entry and exit points of the segment (i.e., the entire weaving area). For merge and diverge areas, densities reflect the "merge/diverge influence area," which consists of lanes 1 and 2 (right and next-to-right lane of the freeway) and the acceleration or deceleration lane 1,500 ft upstream of a diverge or 1,500 ft downstream of a merge. These influence areas are illustrated in Figure 13.2.

In some cases, these definitions cause overlaps. For example if an on-ramp is followed by an off-ramp less than 3,000 ft away, the two 1,500 ft influence areas will at least partially overlap. In such cases, the worst density or level of service is applied. Other overlaps between ramp and weaving segments and/or basic segments are similarly treated.

Despite efforts to make level-of-service definitions consistent among basic segment, weaving, merge, and diverge areas, some discrepancies still remain. Some of these will be highlighted later when capacity is discussed. The principal inconsistency between the weaving and the merge/diverge methodologies arises from the differing definition and view of the level-of-service E/F boundary condition.

13.3 A Common Point: Converting Demand Volumes

Both procedures for weaving areas and for merge/diverge areas rely on algorithms stated in terms of demand flow rates in passenger car units for base conditions. Thus, a point common to both is that all component demand volumes must be converted before proceeding to use either methodology.

$$v_i = \frac{V_i}{PHF * f_{HV} * f_p} \tag{13-1}$$

where: v_i = demand flow rate, pc/h, under equivalent base conditions

V_i = demand volume, veh/h, under prevailing conditions

PHF = peak hour factor

f_{HV} = heavy-vehicle adjustment factor

f_p = driver-population adjustment factor

The heavy-vehicle and driver-population factors are the same ones used for basic freeway and multilane highway segments. They are found using the methods and exhibits discussed in Chapter 12.

13.4 Analysis of Weaving Areas

Weaving areas have been the subject of a great deal of research since the late 1960s, yet many features of current procedures rely heavily on judgment. This is primarily due to the great difficulty in and cost of collecting comprehensive data on weaving operations. Weaving areas cover significant lengths and generally require videotaping from elevated vantage points or time-linked separate observation of entry and exit terminals and visual matching of vehicles. Further, there are a large number of variables affecting weaving operations, and, therefore, a large number of sites reflecting these variables would need to be observed.

The first research study leading up to the third-edition 1985 HCM focused on weaving areas [3]. This was unfortunate, as basic section models would be revised later, causing judgmental modification in weaving models for consistency. It relied on 48 sets of data collected by the then Bureau of Public Roads in the late 1960s and an additional 12 sets collected specifically for the study. The methodology that resulted was complex and iterative. It was later modified as part of a study of all freeway-related methodologies [4] in the late 1970s. In 1980, a set of interim analysis procedures was published by TRB [5], which included the modified weaving analysis procedure. It also contained an independently developed methodology that produced often substantially different results. The latter methodology was documented in a subsequent study [6]. To resolve the differences between these two methodologies, another study was conducted in the early 1980s, using a new data base consisting of 10 sites [7]. This study produced yet a third methodology, substantially different from the first two. As the publication date of the 1985 HCM approached, the three methodologies were judgmentally merged, using the ten 1980s sites for general validation purposes [8]. A number of studies throughout the 1980s and 1990s

continued to examine the various weaving approaches, with no common consensus emerging [9–12].

It was, therefore, no surprise that a new study, relying on some new data but primarily on simulation, was commissioned as part of the research for the HCM 2000 [13]. Unfortunately, the simulation approach was not particularly successful, and it yielded a number of trends that were judged (by the Highway Capacity and Quality of Service Committee of the Transportation Research Board) to be counterintuitive. The method of the HCM 2000, presented here, resulted from a further judgmental modification of earlier procedures [14].

13.4.1 Flows in a Weaving Area

In a typical weaving area, there are four component flows that may exist. By definition, the two that cross each other's path are called *weaving flows*, while those that do not are called *non-weaving flows*. Figure 13.3 illustrates.

Vehicles entering on leg A and exiting on leg D cross the path of vehicles entering on leg B and exiting on leg C. These are the weaving flows. Movements A-C

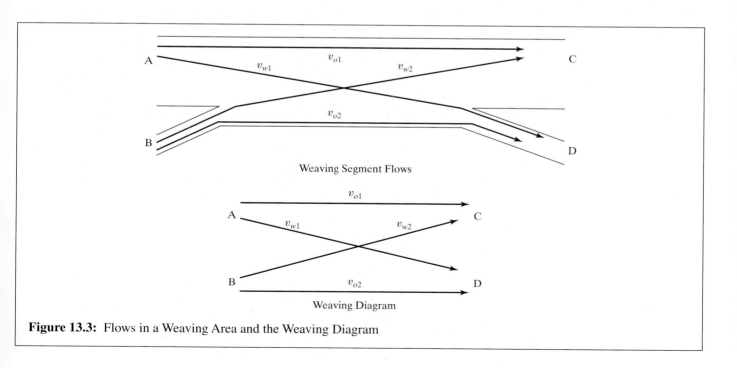

Figure 13.3: Flows in a Weaving Area and the Weaving Diagram

and B-D do not have to cross the path of any other movement, even thought they may share lanes, and are referred to as non-weaving or outer flows.

By convention, weaving flows use the subscript "*w*," while outer or non-weaving flows use the subscript "*o*." The larger of the two outer or weaving flows is given the second subscript "1," while the smaller uses the subscript "2." Thus:

v_{o1} = larger outer flow, pc/h, equivalent base conditions

v_{o2} = smaller outer flow, pc/h, equivalent base conditions

v_{w1} = larger weaving flow, pc/h, equivalent base conditions

v_{w2} = smaller weaving flow, pc/h, equivalent base conditions

The schematic line drawing of Figure 13.3 is called the *weaving diagram*. In block form, it shows the weaving and non-weaving flows and their relative positions on the roadway. By convention, it is always drawn with traffic moving from left to right. It is a convenient form to illustrate the component flows in a consistent way for analysis.

Other critical variables, used in analysis algorithms, may be computed from these:

v_w = total weaving flow, pc/h = $v_{w1} + v_{w2}$

v_{nw} = total non-weaving or outer flow, pc/h = $v_{o1} + v_{o2}$

v = total demand flow, pc/h = $v_w + v_{nw}$

VR = volume ratio = v_w/v

R = weaving ratio = v_{w2}/v_w

13.4.2   Critical Geometric Variables

Three geometric variables have a significant effect on the quality of weaving operations:

- Lane configuration
- Length of the weaving area, ft
- Width (number of lanes) in the weaving area

Each of these has an impact on the amount of lane-changing that must or may occur, and the intensity of that lane-changing.

Lane Configuration

Lane configuration refers to the manner in which entry and exit legs "connect." Seven different configurations have been identified, but many of these are rare, and little is known concerning their characteristics. The seven have been grouped into three major categories of configuration, each with unique features that significantly affect operations. These configurations are illustrated in Figure 13.4.

Type A Configurations   The most common Type A weaving configuration is often referred to as a *ramp-weave* (Figure 13.4 1a). It is formed by a one-lane on-ramp followed by a one-lane off-ramp connected by a continuous auxiliary lane. The unique feature of a Type A weaving configuration is that each and every weaving vehicle must make at least one lane-change. Further, all of these required lane-changes are across a single lane line, which is referred to as a *crown line*. The crown line is a lane line that joins the nose of the merge junction directly to the nose of the diverge junction. While one lane-change is required of all weaving vehicles, they may make additional lane-changes on a discretionary basis. The importance of the required lane-changes is that they must take place within the limits of the weaving section.

The second form of Type A weaving configuration (Figure 13.4 1b) has similar characteristics, in that a crown line exists and all weaving vehicles must make one lane-change across it. It is formed, however, by a major merge junction and a major diverge junction (i.e., both legs have multiple lanes). While the lane-changing patterns are similar to those of ramp-weaves, there is a difference in operating characteristics. As weaving vehicles in a ramp-weave either originate from or are destined for a ramp roadway with restrictive geometry, they are generally accelerating or decelerating through the weaving area. When formed by a major merge and diverge junction, vehicles are more likely to be entering and leaving the section at freeway speeds. The analysis methodology of HCM 2000 was calibrated primarily for

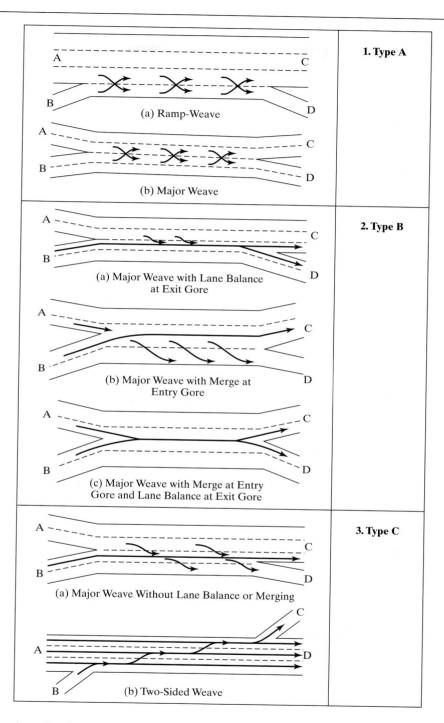

Figure 13.4: Weaving Area Configurations Illustrated (Used with permission of Transportation Research Board, National Research Council, *Highway Capacity Manual*, 2000, Exhibits 13-8, 13-9, and 13-10, pgs. 13-15, 13-16, and 13-17.)

ramp-weaves and is more approximate for other Type A configurations. Type A configurations of the type shown in Figure 13.4 1b, however, are relatively rare.

Type B Configurations   In Type B weaving configurations, one of the two weaving movements can be completed without making a lane-change, as illustrated in Figure 13.4 2a, b, and c.

Thus, a Type B weaving configuration must feature a continuous lane that enters on the left leg of the section and leaves on the right leg, or vice versa. A second and critical characteristic of a Type B weaving configuration is that while one weaving movement can take place without a lane-change, the other weaving movement can be completed with only *one* lane-change.

The configuration shown in Figure 13.4 2a is the most common form of Type B section. The "through" lane for weaving vehicles is created by having *lane balance* at the exit gore area. "Lane balance" is a diverge characteristic in which there is one more lane leaving the diverge than there is entering it. By providing lane balance, a lane is created in which vehicles can take either the left or right exit leg without changing lanes.

Type B sections are very effective in carrying large weaving flows. Because of the "through" lane for weaving vehicles, required lane-changing is minimized and overall turbulence is reduced compared with a Type A or (as will be seen) a Type C configuration. As a practical issue, the "through" lane for weaving vehicles should be provided for the larger of the two weaving demand flows.

Figure 13.4 2b shows another way of creating a similar lane-changing scenario. Instead of lane balance at the exit gore, two lanes from entry legs A and B merge to form a single lane. While this also creates a "through" lane for weaving vehicles, merging lanes usually create more turbulence than diverging lanes. In Figure 13.4 2c, a configuration is shown in which *both* weaving movements can be made without a lane-change. This occurs when there are both merging lanes at the entry gore and lane balance at the exit gore. This type of configuration occurs most often on collector-distributor roadways as part of interchanges. The HCM 2000 analysis methodology is calibrated primarily for sections of the type shown in Figure 13.4 2a. Its application to the other configurations is more approximate.

Type C Configurations   In a Type C configuration, there is also a "through" lane for one of the weaving movements. Again, this weaving maneuver may be completed without a lane-change. The other weaving movement, however, requires at least *two* lane-changes to completed the weaving maneuver. Thus, this type of configuration is most efficient when one weaving flow is considerably larger than the other. Any significant secondary weaving flow (v_{w2}), forced to make two or more lane-changes, will create operational problems.

The most frequently occurring Type C configuration is shown in Figure 13.4 3a. Figure 13.4 3b shows a *two-sided* weaving configuration. These are formed when a left-hand on-ramp is followed by a right-hand off-ramp, or vice versa. In such cases, the entire through freeway flow becomes a weaving flow. Ramp-to-ramp vehicles must cross all freeway lanes as well as the entire through freeway flow to successfully execute the weaving maneuver. The methodologies of the HCM 2000 were calibrated for the former. Their application to the analysis of two-sided weaving areas is extremely approximate at best. Fortunately, two-sided weaving areas are relatively rare. As a general design guideline, they should be avoided unless they are absolutely necessary based on local conditions and constraints.

The Importance of Configuration   Table 13.2 summarizes the number of lane-changes required to execute a successful weaving maneuver in each type of configuration.

Consider a case in which the primary weaving movement has a flow rate of 1,000 veh/h and the secondary weaving movement has a flow rate of 500 veh/h. If a Type A configuration is provided, the total number

Table 13.2: Required Lane-Changes in Weaving Areas

Configuration	No. of Required Lane-Changes for:	
	Primary Weaving Movement	Secondary Weaving Movement
Type A	1	1
Type B	0	1
Type C	0	≥ 2

of required lane-changes for weaving vehicles is $(1*1,000) + (1*500)$ or 1,500 lane-changes. If a Type B configuration is provided, the total number of lane-changes is $(0*1,000) + (1*500)$ or 500 lane-changes. In a Type C configuration, the number of required lane-changes is $(0*1,000) + (2*500) = 1,000$ lane-changes. Thus, configuration would have a significant impact on lane-changing behavior, which is the primary source of turbulence within the weaving area.

It should be noted that drivers may *choose* to make additional discretionary lane-changes within the weaving area boundaries. Table 13.2 only accounts for those lane-changes that *must* be executed to successfully complete a weaving maneuver. These are the only lane-changes that are confined to the limits of the weaving area.

Length of the Weaving Area

While configuration has a tremendous impact on the number of lane-changes that must be made within the confines of the weaving area, the length of the section is a critical determinant of the *intensity* of lane-changing within the section. As all of the *required* lane-changes must take place between the entry and exit gores of the weaving area, the length of the section controls the intensity of lane-changing. If 1,000 lane-changes must be made within the weaving area, then the intensity of those lane-changes will be half as high if the section length is 1,000 ft as compared with 500 ft.

Figure 13.5 shows how the length of a weaving area is measured. The length is measured from a point at the merge gore area where the right edge of the right freeway lane is 2 ft away from the left edge of the ramp lane edge to a point at the diverge gore area where these two points are separated by 12 ft. The original logic for this interpretation is not certain. It was the convention used in a large data base assembled by the then Bureau of Public Roads (now the Federal Highway Administration) in 1963. Subsequent studies have stuck to this convention for consistency. It is likely that this practice relates to the geometry of early ramp-weave sections involving the loop ramps of a cloverleaf interchange. In such sections, the angle of departure of the off-ramp is generally greater than the merge angle of the on-ramp. The definition may also reflect the behavior of 1960s-era drivers with respect to paint markings in gore areas.

Width of a Weaving Area and Type of Operation

The total width of the weaving area is measured as the total number of lanes available for all flows, N. The width of the section has an impact on the total number of lane-changes that drivers can choose to make.

More important, however, is the proportional use of lanes by weaving and non-weaving vehicles in the section. Under normal conditions, vehicles on a highway "compete" for space, and operations across all lanes tend to reach an equilibrium in which all drivers are experiencing relatively similar conditions. In weaving areas, there is always some segregation of weaving and non-weaving flows. Non-weaving drivers tend to stay in outside lanes to avoid the turbulence caused by the intense lane-changing of weaving vehicles. Weaving vehicles, by necessity, tend to occupy lanes appropriate to their particular maneuver. Nevertheless, many lanes are shared by weaving and non-weaving vehicles. Barring any external constraints, weaving vehicles and non-weaving vehicles will share lanes in a manner that

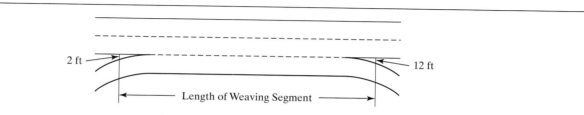

Figure 13.5: Length of a Weaving Area Illustrated (Used with permission of Transportation Research Board, National Research Council, *Highway Capacity Manual*, 2000, Exhibit 13-11, pg. 13-18.)

provides them with similar operating quality. Because of this, the variable N_w is defined as the number of lanes weaving vehicles must occupy to achieve balanced operation with non-weaving vehicles in a weaving area.

Unfortunately, the lane configuration limits the total number of lanes that can be occupied by weaving vehicles because of the specific lane-changes weaving vehicles must make. The following general observations apply:

- Weaving vehicles may occupy all of a lane in which weaving is accomplished without a lane-change.
- Weaving vehicles may occupy most of a lane from which a weaving maneuver can be accomplished with a single lane-change.
- Weaving vehicles may occupy a small portion of a lane from which a weaving maneuver can be completed by making two lane-changes.
- Weaving vehicles do not usually occupy a measurable portion of any lane from which a weaving maneuver would require three or more lane-changes.

These guidelines translate into limitations on the maximum number of lanes that weaving can occupy in any given configuration, as illustrated in Figure 13.6.

In a typical Type A configuration, almost all ramp vehicles are weaving. Thus, the auxiliary lane of a ramp-weave is virtually 100% utilized by weaving vehicles. Weaving vehicles, however, are somewhat restricted to the right lane of the freeway, as they must all execute a lane change from the auxiliary lane to the right freeway lane or vice versa. On the other hand, the shoulder lane of the freeway is always shared with some non-weaving through freeway vehicles. Thus, weaving vehicles rarely occupy more than 1.4 lanes in a Type A configuration.

Type B configurations are more flexible. There is always one "through" lane that can be fully occupied by weaving vehicles. Further, the two lanes adjacent to this through lane can also be heavily occupied by weaving vehicles, as they are only one lane-change away from successfully completing the weaving maneuver. Lanes that are further from the through lane

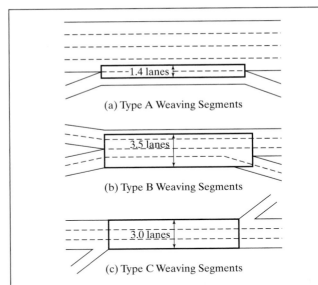

(a) Type A Weaving Segments

(b) Type B Weaving Segments

(c) Type C Weaving Segments

Figure 13.6: Limitations on Lane Use by Weaving Vehicles (Used by permission of Transportation Research Board, National Research Council, *Highway Capacity Manual*, 2000, Exhibit 13-12, pg. 13-19.)

can be only partially occupied by weaving vehicles. Weaving vehicles can occupy up to 3.5 lanes in a Type B configuration.

Type C configurations are similar to Type B configurations in that they have a "through" lane for the major weaving movement. Again, this can be fully occupied by weaving vehicles. Immediately adjacent lanes may also be used by weaving vehicles. Because Type C configurations require the smaller weaving movement to execute two lane-changes, there is somewhat less use of outer lanes by weaving vehicles than in a Type B configuration. Weaving vehicles can occupy no more than 3.0 lanes of a Type C configuration.

While logic supports these limitations on lane use, several studies have provided some validation from field observations, and they may be used with reasonable confidence.

Given that the configuration of the weaving area can effectively "constrain" the number of lanes occupied by weaving vehicles, two distinctly different types of operation may result, based upon the following two defined parameters:

N_w = number of lanes weaving vehicles must occupy to achieve balanced equilibrium operation with non-weaving vehicles, lanes

N_w (max) = maximum number of lanes that can be occupied by weaving vehicles in a given configuration, lanes

Unconstrained operation occurs when $N_w \leq N_w$ (max). In such cases, weaving vehicles can occupy the number of lanes needed to achieve balanced or equilibrium operation with non-weaving vehicles. *Constrained operation* occurs when $N_w > N_w$ (max). In this situation, weaving vehicles are limited, or constrained, by the configuration to N_w (max) lanes, which is not sufficient for them to achieve balance or equilibrium with non-weaving vehicles. When this happens, weaving vehicles get less space than they need for balanced operation. Non-weaving vehicles get correspondingly more space than needed for balance. Under constrained operation, weaving vehicles often experience significantly lower service quality than non-weaving vehicles, even though an average level of service is still applied to the entire weaving area. Constrained operation is not a desirable condition and is often a cue that improvements are needed, even when the overall level of service is acceptable.

13.4.3 Computational Procedures for Weaving Area Analysis

The computational procedures for weaving areas are used only in the operational analysis mode (i.e., all geometric and traffic conditions are specified), and the analysis results in a determination of level of service and weaving area capacity (a new feature in the HCM 2000). The steps in the procedure are illustrated by the flow chart in Figure 13.7 and are summarized as follows:

1. Specify all traffic and geometric conditions for the site.

2. Convert all component demand volumes to peak flow rates in pc/h under equivalent base conditions, using Equation 13-1.

3. Assume that operations are *unconstrained*, and estimate the resulting speed of weaving and non-weaving vehicles in the weaving area.

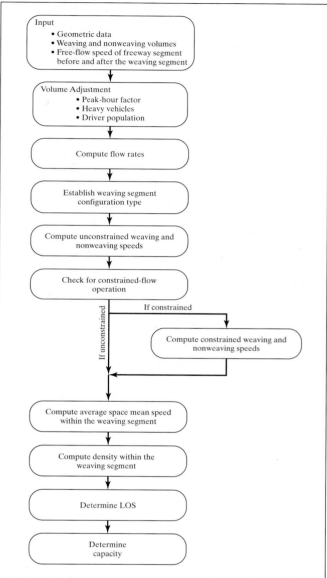

Figure 13.7: Flow Chart Illustrating Weaving-Area Analysis Procedures (Used with permission of Transportation Research Board, National Research Council, *Highway Capacity Manual*, 2000, Exhibit 24-1, pg. 24-2.)

4. Using the results of Step 3, determine whether actual operations are *unconstrained* or *constrained*. If they are constrained, re-estimate the speed of weaving and non-weaving vehicles assuming the constrained result.

5. Compute the weighted average speed and density for all vehicles in the weaving area.

6. Determine level of service from the estimated density in the weaving area using Table 13.1.

7. Check input variables against limitations of the methodology.

8. Determine the capacity of the weaving section.

Once the input information is summarized and the component flows (weaving and non-weaving) have been converted to flow rates in equivalent pc/h under base conditions (using Equation 13-1), the remaining steps of the computational procedure may be implemented.

Estimating the Average Speed of Weaving and Non-weaving Vehicles

The general form of the algorithm used to estimate average speeds of weaving and non-weaving vehicles was originally developed in 1983 as part of an NCHRP-sponsored study of weaving area operations [7]. Because weaving-area operations can sometimes result in markedly different speeds for weaving and non-weaving vehicles, the two speeds are separately estimated, using an equation of the following form:

$$S_i = S_{min} + \frac{S_{max} - S_{min}}{1 + W_i} \qquad (13-2)$$

where: S_i = average speed of weaving (w) or non-weaving (nw) vehicles, mi/h

S_{min} = minimum expected average speed in a weaving area, mi/h

S_{max} = maximum expected average speed in a weaving area, mi/h

W_i = weaving intensity factor for weaving (w) or non-weaving (nw) vehicles

In the HCM 2000, S_{min} is set at 15 mi/h and S_{max} is set as the free-flow speed (FFS) of the freeway plus 5 mi/h. Five mi/h is added to the free-flow speed to account for the nature of the algorithm, which tends to

underpredict high speeds. Thus, the speed-estimation algorithm becomes:

$$S_i = 15 + \frac{FFS - 10}{1 + W_i} \qquad (13-3)$$

The weaving intensity factor (W), as its name suggests, reflects the intensity of weaving within the defined weaving area. As weaving becomes more intense, the value of "W" increases and average speeds decrease. The weaving intensity factor is computed as:

$$W = \frac{a(1 + VR)^b (v/N)^c}{L^d} \qquad (13-4)$$

where: VR = volume ratio (v_w/v)

v = total demand flow rate in the weaving area, equivalent pc/h under base conditions

L = length of the weaving area, ft

a,b,c,d = constants of calibration

The algorithm has logical sensitivities. As the volume ratio increases, the intensity of weaving increases. As the average demand flow per lane increases, the intensity of weaving increases. As the length of the section increases, the intensity of weaving decreases.

The term $(1 + VR)$ is used because values of VR are always less than 1.00. Applying a power to a term that is less than 1.00 results in mathematically complex trends. Using $(1 + VR)$ assures that the value is always greater than 1.00.

Constants of calibration a, b, c, and d vary with three conditions:

• Whether the speed of weaving or non-weaving vehicles is being estimated

• Configuration of the weaving area (Type A, B, or C)

• Whether the operation of the section is *constrained* or *unconstrained*

Thus, there are 12 sets of calibration constants for the weaving-intensity factor. As indicated in the procedural

Table 13.3: Constants of Calibration for Weaving-Intensity Factors

General Form

$$W = \frac{a(1 + VR)^b \left(\dfrac{v}{N}\right)^c}{L^d}$$

	Constants for Weaving Speed, S_w				Constants for Non-weaving Speed, S_{nw}			
	a	b	c	d	a	b	c	d
Type A Configuration								
Unconstrained	0.15	2.2	0.97	0.80	0.0035	4.0	1.3	0.75
Constrained	0.35	2.2	0.97	0.80	0.0020	4.0	1.3	0.75
Type B Configuration								
Unconstrained	0.08	2.2	0.70	0.50	0.0020	6.0	1.0	0.50
Constrained	0.15	2.2	0.70	0.50	0.0010	6.0	1.0	0.50
Type C Configuration								
Unconstrained	0.08	2.3	0.80	0.60	0.0020	6.0	1.1	0.60
Constrained	0.14	2.3	0.80	0.60	0.0010	6.0	1.1	0.60

(Used with permission of Transportation Research Board, National Research Council, *Highway Capacity Manual*, 2000, Exhibit 24-6, pg. 24-6.)

steps, unconstrained operation is always assumed for the initial computation and checked in a subsequent step. Two weaving-intensity factors—one for the prediction of the average speed of weaving vehicles and the other for prediction of the average speed of non-weaving vehicles—are needed. The type of configuration is known.

Table 13.3 gives the constants of calibration for use in computing weaving intensity factors.

Note that the impact of constrained operation is always to increase the weaving-intensity factor for weaving vehicles and to decrease it for non-weaving vehicles. Thus, in constrained operations, weaving vehicles have lower speeds than in corresponding situations under unconstrained operations, while non-weaving vehicles have higher speeds.

Determining the Type of Operation

Initial speed estimates are computed assuming that unconstrained operations prevail. This must be confirmed, and speeds must be recomputed if it is found that constrained operations exist. As noted previously, unconstrained operations are expected when $N_w \leq N_w$ (max) and constrained operations are expected when $N_w > N_w$ (max). Equations for the estimation of N_w values of N_w (max) are shown in Table 13.4.

The algorithms for estimating N_w are based on speed estimates for unconstrained operations. In effect, these algorithms indicate the number of lanes that weaving vehicles would have to occupy in order for all vehicles to achieve these unconstrained speeds. If operations are constrained, then the average speeds would be recomputed using Equations 13-3 and 13-4, using the constants of calibration for constrained operations.

Determining Level of Service in the Weaving Area

Level-of-service criteria were shown previously in Table 13.1. To apply these criteria, an average density for the weaving area must be determined. To determine

Table 13.4: Criteria for Unconstrained vs. Constrained Operation

Configuration	Equation for N_w (lanes)	N_w (max) (lanes)
Type A	$$\dfrac{0.74 * N * VR^{0.571} * L^{0.234}}{S_w^{0.438}}$$	1.4
Type B	$$N * \left[0.085 + 0.703\,VR + \left(\dfrac{234.8}{L} \right) - 0.018(S_{nw} - S_w) \right]$$	3.5
Type C	$$N * [0.761 + 0.047\,VR - 0.00011L - 0.005(S_{nw} - S_w)]$$	3.0 *

* Except for two-sided weaving areas, where N_w (max) = N.
(Used with permission of Transportation Research Board, National Research Council, *Highway Capacity Manual*, 2000, Exhibit 24-7, pg. 24-7.)

average density, a weighted average space mean speed for all vehicles in the weaving area must be computed:

$$S = \frac{v_{nw} + v_w}{\left(\dfrac{v_{nw}}{S_{nw}} \right) + \left(\dfrac{v_w}{S_w} \right)} \qquad (13\text{-}5)$$

where: S = space mean speed of all vehicles in weaving area, mi/h

All other variables are as previously defined.

This computation results in a space mean speed, which is the relevant value for traffic analysis. Using this result, the average density in the section is computed as:

$$D = \frac{(v/N)}{S} \qquad (13\text{-}6)$$

where: D = density, pc/mi/ln

All other variables as previously defined.

This result may be compared with the criteria of Table 13.1 to determine the overall level of service for the weaving area.

While the HCM 2000 does not permit it, it is possible to use density criteria to create a separate level-of-service measure for weaving and non-weaving vehicles. This is an attractive option, as weaving and non-weaving vehicles may be experiencing very different conditions, particularly if operation is constrained. To do so requires

that the actual number of lanes occupied by weaving and non-weaving vehicles be estimated.

For weaving areas operating under unconstrained conditions:

$$\begin{aligned} N_{wA} &= N_w \\ N_{nwA} &= N - N_w \end{aligned} \qquad (13\text{-}7)$$

For weaving areas operating under constrained conditions:

$$\begin{aligned} N_{wA} &= N_w\,(\text{max}) \\ N_{nwA} &= N - N_w\,(\text{max}) \end{aligned} \qquad (13\text{-}8)$$

where: N_{wA} = number of lanes actually occupied by weaving vehicles

N_{nwA} = number of lanes actually occupied by non-weaving vehicles

All other variables are as previously defined.

Using these estimates, separate densities for weaving and non-weaving vehicles may be computed as:

$$D_w = \frac{\left(v_w / N_{wA} \right)}{S_w}$$

$$D_{nw} = \frac{\left(v_{nw} / N_{nwA} \right)}{S_{nw}} \qquad (13\text{-}9)$$

where: D_w = average density of weaving vehicles, pc/mi/ln

D_{nw} = average density of non-weaving vehicles, pc/mi/ln

All other variables are as previously defined.

In the HCM editions of 1985 and 1994, separate levels of service for weaving and non-weaving vehicles were prescribed. Levels of service, however, were determined directly on the basis of speeds. With the use of density as a measure of effectiveness in 1997 and 2000, it was thought that the meaning of separate densities would be confusing, particularly as weaving and non-weaving vehicles share lanes. The concept is presented here to allow analysts to at least consider the very different qualities of service that may be experienced by weaving and non-weaving vehicles in some cases.

Additional Limitations on Weaving-Area Operations

Because Equation 13-3 (for estimating speeds) constrains results to a reasonable range (by using a minimum and maximum value as part of the algorithm), it is possible to get a "reasonable" answer even when input values are quite unreasonable. Thus, even if the average flow rate per lane were entered as 6,000 pc/h/ln, the resulting speed estimate would still be higher than the minimum value set by the algorithm.

Because of this, it is necessary to apply some limitations to input values independently. Table 13.5 summarizes these limitations. The footnotes to the table are important, as they provide an explanation of likely outcomes should any of these limitations be exceeded.

It is important to recognize that only two of these limitations result in failure of the section when exceeded. These involve the weaving capacity and maximum value of v/N. Other limitations, when exceeded, will generally result in poorer operating conditions than those predicted by the weaving-area analysis methodology, but would not necessarily cause a failure of the section.

Capacity of Weaving Areas

The capacity of a weaving area is a complex phenomenon. From Table 13.1, capacity occurs when the resulting average density in the section reaches 43 pc/mi/ln

Table 13.5: Additional Limitations on Weaving-Area Operations

Configuration	Weaving Capacity v_w (max)[1] (pc/h)	Maximum v/N[2] (pc/h/ln)	Maximum VR[3]		Maximum R[4]	Maximum Weaving Length, L[5] (ft)
A	2,800	c[6]	$\frac{N}{2}$	$\frac{VR}{1.00}$	0.50	2,500
			3	0.45		
			4	0.35		
			5	0.20		
B	4,000	c[6]	0.80		0.50	2,500
C	3,500	c[6]	0.50		0.40[7]	2,500

1. Section is likely to fail at higher weaving flow rates.
2. Section is likely to fail at higher total demand flow rates per lane.
3. Section likely to operate at lower speeds and higher densities than predicted by this methodology at higher VR values.
4. Section likely to operate at lower speeds and higher densities than predicted by this methodology at higher R values.
5. Section should be analyzed as isolated merge and diverge junctions when this length is exceeded.
6. Basic freeway or multilane highway section capacity per lane for the specified free-flow speed.
7. Larger weaving flow must be in the direction of the through weaving lane; if not, the section is likely to operate at lower speeds and higher densities than predicted by this algorithm.

(for freeways) or 40 pc/mi/ln (for multilane highways). From Table 13.5, it may also occur when v/N reaches the capacity of a basic freeway or multilane highway segment, or when the total weaving flow rate reaches the maximum values shown in the table. Further, ultimate capacity may occur under unconstrained or constrained operating conditions. Other limiting values from Table 13.5 must also be taken into account when estimating the capacity of a weaving area.

Finding the total flow rate for a weaving area that produces a density of 43 pc/mi/ln (or 40 pc/mi/ln for multilane highways) is a complex process requiring multiple trial-and-error computations. Given the methodology for estimating density in weaving areas, the result will vary with the type of weaving area, the length of the weaving area, the number of lanes in the weaving area, and the volume ratio (VR). Using fixed values for each of the foregoing variables, the total demand flow rate is incrementally increased until the limiting density value is reached. The demand flow rate at which this occurs defines capacity, at least as limited by density. Other limiting conditions must be superimposed on this determination.

Fortunately, spreadsheets can be programmed to perform such iterative computations. Thus, for the HCM 2000, a series of tables was prepared showing capacity for various combinations of underlying variables. The capacities shown in these tables are stated as flow rates under base conditions. They may be converted to prevailing conditions and to full-hour capacities as follows:

$$c = c_b * f_{HV} * f_p$$
$$c_h = c * PHF \qquad (13\text{-}10)$$

where: c = capacity of the weaving area, veh/h

c_b = capacity of the weaving area in equivalent pc/h under base conditions

c_h = capacity of the weaving area expressed as a full-hour volume under prevailing conditions

All other variables are as previously defined.

As the capacity tables for weaving areas consume a number of pages, they are included in Appendix I at the end of this chapter. Values intermediate to those shown in the tables can be estimated using straight-line interpolation.

13.4.4 Multiple Weaving Areas

Through the 1994 edition, the HCM specified a procedure for analyzing two-segment multiple weaving areas. These sections were formed either by a single merge point being followed by two diverge points, or two merge points followed by a single diverge. While the procedures were logical extensions of simple weaving analysis, they were not well supported by field observations, with only one such section being fully studied since the early 1970s. Further, no procedures were specified for three- or more-segment weaving areas.

For the HCM 2000, all references to multiple weaving areas were deleted. Such sections must be broken up into individual simple weaving areas (one merge followed by one diverge) and merge and/or diverge junctions for analysis.

13.4.5 Weaving on Collector-Distributor Roadways and Other Types of Facilities

A common design practice for freeways, sometimes also used on multilane highways, often results in weaving sections located on collector-distributor roadways that are part of an interchange. The weaving analysis methodology presented herein may be used with caution as a very rough estimate of operating conditions on such roadways. Appropriate free-flow speeds, often far lower than those on the main freeway or multilane highway, must be used. Level-of-service criteria from Table 13.1 may not be appropriate in many cases. Many such segments operate at low speeds with correspondingly high densities, and stable operations may exist beyond the maximum densities specified herein.

While there is a great deal of interest in weaving on interrupted flow facilities, such as arterials, the procedures of this chapter should not be applied, even for a very rough estimate. These procedures do not account for the most unique aspect of arterial weaving areas—platoon flow and the effects of signalization.

This subject, unfortunately, has not been extensively researched, and there are no generally accepted techniques for their analysis at present.

13.5 Analysis of Merge and Diverge Areas

As illustrated in Figure 13.2, analysis procedures for merge and diverge areas focus on the merge or diverge influence area that encompasses lanes 1 and 2 (shoulder and adjacent) freeway lanes and the acceleration or deceleration lane for a distance of 1,500 ft upstream of a diverge or 1,500 downstream of a merge area.

Analysis procedures provide algorithms for estimating the density in these influence areas. Estimated densities are compared to the criteria of Table 13.1 to establish the level of service.

13.5.1 Structure of the Methodology for Analysis of Merge and Diverge Areas

Because the analysis of merge and diverge areas focuses on influence areas including only the two right-most lanes of the freeway, a critical step in the methodology is the estimation of the lane distribution of traffic immediately upstream of the merge or diverge. Specifically, a determination of the approaching demand flow remaining in lanes 1 and 2 immediately upstream of the merge or diverge is required. Figure 13.8 shows the key variables involved in the analysis.

The variables included in Figure 13.8 are defined as follows:

v_F = freeway demand flow rate immediately upstream of merge or diverge junction, in pc/h under equivalent base conditions

v_{12} = freeway demand flow rate in lanes 1 and 2 of the freeway immediately upstream of the merge or diverge junctions, in pc/h under equivalent base conditions

v_R = ramp demand flow rate, in pc/h under equivalent base conditions

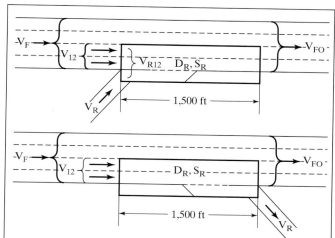

Figure 13.8: Critical Variables in Merge and Diverge Analysis (Used with permission of Transportation Research Board, National Research Council, *Highway Capacity Manual*, 2000, Exhibit 25-2, pg. 25-2.)

v_{R12} = total demand flow rate entering a merge influence area, $v_R + v_{12}$, in pc/h under equivalent base conditions

v_{FO} = total outbound demand flow continuing downstream on the freeway, pc/h under equivalent base conditions

D_R = average density in the ramp influence area, pc/mi/ln

S_R = space mean speed of all vehicles in the ramp influence area, mi/h

Other than the standard geometric characteristics of the freeway that are used to determine its free-flow speed and adjustments to convert demand volumes in veh/h to pc/h under equivalent base conditions (Equation 13-1), there are two specific geometric variables of importance in merge and diverge analysis:

$L_{a \text{ or } d}$ = length of the acceleration or deceleration lane, ft

$RFFS$ = free-flow speed of the ramp, mi/h

The length of the acceleration or deceleration lane is measured from the point at which the ramp lane and lane 1 of the main facility touch to the point at which the

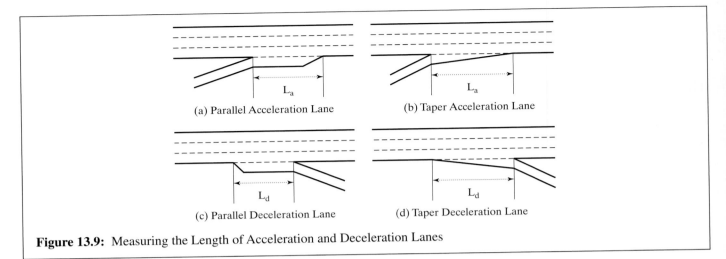

Figure 13.9: Measuring the Length of Acceleration and Deceleration Lanes

acceleration or deceleration lane begins or ends. This definition includes the taper portion of the acceleration or deceleration lane and is the same for both parallel and tapered lanes. Figure 13.9 illustrates the measurement of length of acceleration and deceleration lanes.

The free-flow speed of the ramp is best observed in the field but may be estimated as the design speed of the most restrictive element of the ramp. Many ramps include compound horizontal curves or a number of separate horizontal or vertical curves. The free-flow speed is generally controlled by the design speed (or maximum safe operating speed) of the most severe of these.

Figure 13.10 is a flow chart of the analysis methodology for merge or diverge junctions. It illustrates the following analysis steps:

1. Specify all traffic and roadway data for the junction to be analyzed: peak-hour demands, PHF, traffic composition, driver population, and geometric details of the site, including the free-flow speed for the facility and for the ramp.

2. Convert all demand volumes to flow rates in pc/h under equivalent base conditions using Equation 13-1.

3. Determine the demand flow in lanes 1 and 2 of the facility immediately upstream of the merge or diverge junction using the appropriate algorithm as specified.

4. Determine whether the demand flow exceeds the capacity of any critical element of the junction.

Where demand exceeds capacity, level of service F is assigned and the analysis is complete.

5. If operation is determined to be stable, determine the density and speed of all vehicles within the ramp influence area. Table 13.1 is used to determine level of service based on the density in the ramp influence area.

Once all input characteristics of the merge or diverge junction are specified and all demand volumes have been converted to flow rates in pc/h under equivalent base conditions, remaining parts of the methodology may be completed.

13.5.2 Estimating Demand Flow Rates in Lanes 1 and 2

The starting point for analysis is the determination of demand flow rates in lanes 1 and 2 (the two right-most lanes) of the facility immediately upstream of the merge or diverge junction. This is done using a series of regression-based algorithms developed as part of a nationwide study of ramp-freeway junctions [15]. Different algorithms are used for merge and diverge areas.

Merge Areas

For merge areas, the flow rate remaining in lanes 1 and 2 immediately upstream of the junction is computed

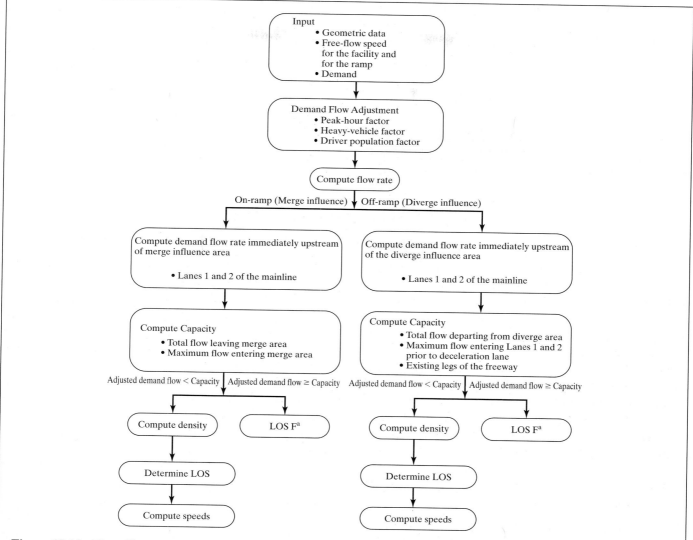

Figure 13.10: Flow Chart of Analysis Methodology for Merge and Diverge Junctions (Used with permission of Transportation Research Board, National Research Council, *Highway Capacity Manual*, 2000, Exhibit 25-1, pg. 25-2.)

simply as a proportion of the total approaching facility flow:

$$v_{12} = v_F * P_{FM} \qquad (13\text{-}11)$$

where: P_{FM} = proportion of approaching vehicles remaining in lanes 1 and 2 immediately upstream of the merge junction, in decimal form

All other variables are as previously defined.

The value of P_{FM} varies with the number of lanes on the facility, demand flow levels, the proximity of adjacent ramps (in some cases), the length of the acceleration lane (in some cases), and the free-flow speed on the ramp (in some cases.) Table 13.6 summarizes algorithms for estimating P_{FM}, and contains a matrix for determining which algorithm is appropriate for various analysis scenarios.

Table 13.6: Algorithms for Estimation of P_{FM} for Use in Equation 13-11

Facility Size	Algorithm for Estimation of P_{FM}	Equation No.
4-lane (2 lanes ea. dir.)	1.00	N/A
6-lane (3 lanes ea. dir.)	$0.5775 + 0.000028L_a$	(13-12)
	$0.7289 - 0.0000135(v_F + v_R) - 0.003296RFFS + 0.000063L_{up}$	(13-13)
	$0.5487 + 0.2628\left(v_d/L_{dn}\right)$	(13-14)
8-lane (4 lanes ea. dir.)	$0.2178 - 0.000125v_R + 0.01115\left(L_a/RFFS\right)$	(13-15)

Selection Matrix for Six-Lane Facilities

Equation for Use	Adjacent Upstream Ramp	Subject Ramp	Adjacent Downstream Ramp
13-12	None	ON	None
13-12	None	ON	On
13-14 or 13-12	None	ON	Off
13-12	On	ON	None
13-13 or 13-12	Off	ON	None
13-12	On	ON	On
13-14 or 13-12	On	ON	Off
13-13 or 13-12	Off	ON	On
13-13, 13-14, or 13-12	Off	ON	Off

(Used by permission of Transportation Research Board, National Research Council, *Highway Capacity Manual*, 2000, compiled from Exhibits 25-5 and 25-6, pg. 25-6.)

For four-lane facilities (two lanes in each direction), the value is trivial, as the entire flow is in lanes 1 and 2. For such cases, P_{FM} is 1.00. For six- and eight-lane facilities, the values are established using the appropriate algorithm shown in Table 13.6.

In Table 13.6, all variables are as previously defined. L_{up} is the distance to the adjacent upstream ramp in feet, and L_{dn} is the distance to the adjacent downstream ramp.

For six- and eight-lane facilities, it is believed that the flow remaining in lanes 1 and 2 is dependent upon the distance to and flow rate on adjacent upstream and downstream ramps. A driver entering the facility on a nearby upstream on-ramp is more likely to remain in lanes 1 and 2 if the distance between ramps is insufficient to allow the driver to make two lane-changes to reach outer lanes. Likewise, a driver knowing he or she has to exit at a nearby downstream off-ramp is more likely to move into lanes 1 and 2 than a driver proceeding downstream on the main facility. While these are logical expectations, the database on ramp junctions was sufficient to establish these relationships for only a few scenarios on six-lane freeways. Thus, Equation 13-13 considers the impact of adjacent upstream off-ramps, while Equation 13-14 considers the impact of adjacent downstream off-ramps on lane distribution at merge areas on six-lane facilities. There are no relationships considering the impact of adjacent upstream or downstream on-ramps on six-lane facilities, or for any adjacent ramps on eight-lane facilities. Equation 13-12 is used for isolated merge areas on six-lane freeways, and is used as the default algorithm for all merge areas on six-lane freeways that cannot be addressed using Equations 13-13 or 13-14.

As shown in the selection matrix of Table 13.6, there are situations in which more than one algorithm may be appropriate. In such cases, it must be determined whether or not the subject ramp is far enough away from the adjacent ramp to be considered "isolated."

Equation 13-13 or 13-12 may be used to analyze merge areas on six-lane facilities with an adjacent upstream off-ramp. The selection of the appropriate equation for use is based on the distance at which the two equations yield the same result, or:

$$L_{EQ} = 0.214(v_F + v_R) + 0.444L_a$$
$$+ 52.32RFFS - 2,403 \qquad (13\text{-}16)$$

where: L_{EQ} = equivalence distance, ft

All other variables are as previously defined.

When the actual distance to the upstream off-ramp is greater than or equal to the equivalence distance ($L_{up} \geq L_{EQ}$), the subject ramp may be considered to be "isolated." Equation 13-12 is used. Where the reverse is true ($L_{up} < L_{EQ}$), the effect of the adjacent upstream off-ramp must be considered. Equation 13-13 is used.

Similarly, Equation 13-14 or 13-12 may be used to analyze merge areas on six-lane facilities with an adjacent downstream off-ramp. Again, the equivalence distance is established:

$$L_{EQ} = \frac{v_d}{0.1096 + 0.000107L_a} \qquad (13\text{-}17)$$

where: L_{EQ} = equivalence distance, ft

v_d = demand flow rate on the adjacent downstream off-ramp, pc/h under equivalent base conditions

All other variables are as previously defined.

As previously, if the actual distance to the adjacent downstream off-ramp (L_{dn}) is greater than or equal to the equivalence distance ($L_{dn} \geq L_{EQ}$), the subject merge may be considered to be "isolated." Equation 13-12 would be used. If the reverse is true ($L_{dn} < L_{EQ}$), the

effect of the adjacent downstream off-ramp must be considered, and Equation 13-14 is used.

It is also possible that two equations will legitimately apply to a given situation. Where a merge area has both an upstream and a downstream adjacent off-ramp, it would be considered first in combination with the adjacent downstream ramp, then with the adjacent upstream ramp. Two different answers may result. In such situations, the approach yielding the worst prediction of operating conditions is used.

Diverge Areas

The general approach to estimating the demand flow rate in lanes 1 and 2 immediately upstream of a diverge is somewhat different from the one used for merge areas. This is because all of the off-ramp traffic is assumed to be in lanes 1 and 2 at this point. Thus, the flow in lanes 1 and 2 is taken as the off-ramp flow plus a proportion of the through traffic on the facility.

$$v_{12} = v_R + (v_F - v_R)P_{FD} \qquad (13\text{-}18)$$

where: P_{FD} = proportion of approaching vehicles remaining in lanes 1 and 2 immediately upstream of the diverge junction, in decimal form

All other variables are as previously defined.

Table 13.7 shows the algorithms used to estimate P_{FD}, as well as a matrix that may be consulted when selecting the appropriate algorithm. L_{up} is the distance to the adjacent upstream ramp, and L_{dn} is the distance to the adjacent downstream ramp.

Again, the data base for ramp junctions was sufficient to establish relationships for the impact of adjacent ramp activity for only two cases, restricted to six-lane facilities: Equation 13-20 considers the impact of an adjacent upstream on-ramp and Equation 13-21 considers the impact of an adjacent downstream off-ramp. There are no algorithms for the effects of adjacent upstream off-ramps or adjacent downstream on-ramps on six-lane facilities. For eight-lane facilities, the value of P_{FD} is taken to be a constant (0.436) primarily due to a sparse database of off-ramps on an eight-lane facility.

Table 13.7: Algorithms for Estimation of P_{FD} for Use in Equation 13-18

Facility Size	Algorithm for Estimation of P_{FD}	Equation No.
4-lane (2 lanes ea. dir.)	1.00	N/A
6-lane (3 lanes ea. dir.)	$0.760 - 0.000025v_F - 0.000046v_R$	(13-19)
	$0.717 - 0.000039v_F + 0.604\left[\dfrac{v_u}{L_{up}}\right]$	(13-20)
	$0.616 - 0.000021v_F + 0.1248\left[\dfrac{v_d}{L_{dn}}\right]$	(13-21)
8-lane (4 lanes ea. dir.)	0.436	N/A

Selection Matrix for 6-Lane Facilities

Equation for Use	Adjacent Upstream Ramp	Subject Ramp	Adjacent Downstream Ramp
13-19	None	OFF	None
13-19	None	OFF	On
13-21 or 13-19	None	OFF	Off
13-20 or 13-19	On	OFF	None
13-19	Off	OFF	None
13-20 or 13-19	On	OFF	On
13-20, 13-21, or 13-19	On	OFF	Off
13-19	Off	OFF	On
13-21 or 13-19	Off	OFF	Off

(Used by permission of Transportation Research Board, National Research Council, *Highway Capacity Manual*, 2000, compiled from Exhibits 25-12 and 25-13, pgs. 25-12 and 25-13.)

When an adjacent upstream on-ramp exists, an equilibrium distance must be established beyond which the subject off-ramp is considered to be isolated:

$$L_{EQ} = \frac{v_u}{0.071 + 0.000023v_F - 0.000076v_R} \quad (13\text{-}22)$$

where: L_{EQ} = equivalence distance, ft

v_u = demand flow rate on the adjacent upstream on-ramp pc/h under equivalent base conditions

All other variables are as previously defined.

If $L_{up} \geq L_{EQ}$, the subject ramp may be considered to be isolated, and Equation 13-19 would be used. If $L_{up} < L_{EQ}$, the effect of the adjacent upstream on-ramp is taken into account by using Equation 13-20.

When an adjacent downstream off-ramp exists, an equivalence distance must also be computed:

$$L_{EQ} = \frac{v_d}{1.15 - 0.000032v_F - 0.000369v_R} \quad (13\text{-}23)$$

where all variables have been previously defined. If $L_{dn} \geq L_{EQ}$, the subject off-ramp may be considered to be isolated, and Equation 13-19 is applied. If $L_{dn} < L_{EQ}$, the effect of the downstream off-ramp is considered by using Equation 13-21.

Special Cases

It must be noted that the procedures for estimating v_{12} illustrated in Tables 13.6 and 13.7 apply to subject merge and diverge areas consisting of one-lane, right-hand ramps. A series of special modifications to these procedures is presented in a later section to deal with a variety of other situations.

13.5.3 Capacity Considerations

The analysis procedure for merge and diverge areas determines whether the segment in question has failed (LOS = F) based upon a comparison of demand flow rates to critical capacity values.

In general, the basic capacity of the facility is not affected by merging or diverging activities. Because of this, the basic facility capacity must be checked immediately upstream and/or downstream of the merge or diverge. Ramp roadway capacities must also be examined for adequacy. When demand flows exceed any of these capacities, a failure is expected, and the level of service is determined to be F.

The total flow entering the ramp influence area is also checked. While a maximum desirable value is set for this flow, exceeding it does not imply level of service F if no other capacity value is exceeded. In cases where only this maximum is violated, expectations are that service quality will be less than that predicted by the methodology.

Capacity values are given in Table 13.8.

Table 13.8: Capacity Values for Merge and Diverge Checkpoints

Freeway FFS (mi/h)	Maximum Freeway Flow Upstream/Downstream of Merge or Diverge (pc/h)				Maximum Desirable Flow Entering Merge Influence Area (pc/h)	Maximum Desirable Flow Entering Diverge Influence Area (pc/h)
	Number of Lanes in One Direction					
	2	3	4	≥5		
≥70	4,800	7,200	9,600	2,400/ln	4,600	4,400
65	4,700	7,050	9,400	2,350/ln	4,600	4,400
60	4,600	6,900	9,200	2,300/ln	4,600	4,400
55	4,500	6,750	9,000	2,250/ln	4,600	4,400

Ramp Free-Flow Speed RFFS (mi/h)	Capacity of Ramp Roadway (pc/h)	
	Single-Lane Ramps	Two-Lane Ramps
>50	2,200	4,400
>40−50	2,100	4,100
>30−40	2,000	3,800
≥20−30	1,900	3,500
<20	1,800	3,200

(Used with permission of Transportation Research Board, National Research Council, *Highway Capacity Manual*, 2000, compiled from Exhibits 25-3, pg. 25-4, 25-7, pg. 25-9, and 25-14, pg. 25-14.)

The freeway capacity values shown are the same as those for basic freeway sections used in Chapter 12. When applying these procedures to merging or diverging segments on multilane highways, use the values indicated in Chapter 12 for basic sections directly. Other values shown in Table 13.8 may be approximately applied to merging or diverging multilane highway segments.

The specific checkpoints that should be compared to the capacity criteria of Table 13.8 may be summarized as follows:

- For merge areas, the maximum facility flow occurs downstream of the merge. Thus, the facility capacity is compared with the downstream facility flow ($v_{FO} = v_F + v_R$).

- For diverge areas, the maximum facility flow occurs upstream of the diverge. Thus, the facility capacity is compared to the approaching upstream facility flow, v_F.

- Where lanes are added or dropped at a merge or diverge, both the upstream (v_F) and downstream (v_{FO}) facility flows must be compared to capacity criteria.

- For merge areas, the flow entering the ramp influence area is $v_{R12} = v_{12} + v_R$. This sum is compared to the maximum desirable flow indicated in Table 13.8.

- For diverge areas, the flow entering the ramp influence area is v_{12}, as the off-ramp flow is already included. It is compared directly with the maximum desirable flow indicated in Table 13.8.

- All ramp flows, v_R, must be checked against the ramp capacities given in Table 13.8.

The ramp capacity check is most important for diverge areas. Diverge segments rarely fail unless the capacity of one of the diverging legs is exceeded by the demand flow. This is most likely to happen on the off-ramp. It should also be noted that the capacities shown in Table 13.8 for two-lane ramps may be quite misleading. They refer to the ramp roadway itself, not to the junction with the main facility. There is no evidence, for example, that a two-lane on-ramp junction can accommodate any greater flow than a one-lane junction. It is unlikely that a two-lane on-ramp can handle more than 2,250 to 2,400 pc/h through the merge area. For higher on-ramp demands, a two-lane on-ramp would have to be combined with a lane addition at the facility junction.

13.5.4 Determining Density and Level of Service in the Ramp Influence Area

If all facility and ramp capacity checks indicate that stable flow prevails in the merge or diverge area, the density in the ramp influence area may be estimated using Equation 13-24 for merge areas and Equation 13-25 for diverge areas:

$$D_R = 5.475 + 0.00734 v_R$$
$$+ 0.0078 v_{12} - 0.00627 L_a \quad (13\text{-}24)$$

$$D_R = 4.252 + 0.0086 v_{12} - 0.009 L_d \quad (13\text{-}25)$$

where all variables have been previously defined. In both cases, the density in the ramp influence area is dependent upon the flow entering it (v_R and v_{12} for merge areas, and v_{12} for diverge areas), and the length of the acceleration or deceleration lane. The density computed by Equation 13-24 or 13-25 is directly compared to the criteria of Table 13.1 to determine the expected level of service.

13.5.5 Determining Expected Speed Measures

Although it is not a measure of effectiveness, and the determination of an expected speed is not required to estimate density (as was the case for weaving areas), it is often convenient to have an average speed as an additional measure or as an input to system analyses. Because speed behavior in the vicinity of ramps (vicinity = 1,500-ft segment encompassing the ramp influence area) is different from basic sections, three algorithms are provided for merge areas and three for diverge areas as follows:

- Estimation algorithm for average speed within the ramp influence area.
- Estimation algorithm for average speed in outer lanes (where they exist) within the 1,500-ft boundaries of the ramp influence area.
- Algorithm for combining the above into an average space mean speed across all lanes within the 1,500-ft boundaries of the ramp influence area.

Table 13.9 summarizes these algorithms for merge areas, and Table 13.10 summarizes them for diverge areas.

The variables in Tables 13.9 and 13.10 are defined as follows:

S_R = space mean speed of vehicles within the ramp influence area; v_{R12} for merge areas; v_{12} for diverge areas; mi/h

S_o = space mean speed of vehicles traveling in outer lanes (lanes 3 and 4 where they exist) within the 1,500 ft length range of the ramp influence area, mi/h

S = space mean speed of vehicles in all lanes within the 1,500-ft range of the ramp influence area, mi/h

M_s = speed proportioning factor for merge areas

D_s = speed proportioning factor for diverge areas

v_{oa} = average demand flow rate in outer lanes, computed as $(v_F - v_{12})/N_o$, pc/h/ln

N_o = number of outer lanes (one for three-lane segments, two for four-lane segments)

All other variables are as previously defined.

13.5.6 Special Cases

As noted previously, the methodology for merge and diverge analysis presented herein applies directly to one-lane, right-hand on-ramps and off-ramps. A number of special modifications can be used to apply them to a variety of other merge and diverge situations. These special cases are described in Appendix II of this chapter.

Table 13.9: Estimating Average Speeds in Merge Areas

Avg Spd In	Estimation Algorithm
Ramp Influence Area	$S_R = FFS - (FFS - 42)M_S$ $M_S = 0.321 + 0.0039e^{(v_{R12}/1000)} - 0.002(L_a * RFFS/1,000)$
Outer Lanes	$S_o = FFS \qquad v_{oa} < 500$ pc/h $S_o = FFS - 0.0036(v_{oa} - 500) \quad v_{oa} = 500 - 2,300$ pc/h $S_o = FFS - 6.53 - 0.006(v_{oa} - 2300) \quad v_{oa} > 2,300$ pc/h
All Lanes	$S = \dfrac{v_{R12} + v_{oa}N_o}{\left(\dfrac{v_{R12}}{S_R}\right) + \left(\dfrac{v_{oa}N_o}{S_o}\right)}$

Table 13.10: Estimating Average Speeds in Diverging Areas

Avg Spd In _____	Estimation Algorithm
Ramp Influence Area	$S_R = FFS - (FFS - 42)D_S$ $D_S = 0.883 + 0.00009v_{12} - 0.013RFFS$
Outer Lanes	$S_o = 1.097FFS \qquad v_{oa} < 1,000$ pc/h $S_o = 1.097FFS - 0.0039(v_{oa} - 1,000) \quad v_{oa} \geq 1000$ pc/h
All Lanes	$S = \dfrac{v_{12} + v_{oa}N_o}{\left(\dfrac{v_{12}}{S_R}\right) + \left(\dfrac{v_{oa}N_o}{S_o}\right)}$

13.6   Sample Problems in Weaving, Merging, and Diverging Analysis

Example 13-1 Analysis of a Ramp-Weave Area

Figure 13.11 illustrates a typical ramp-weave section on a six-lane freeway (three lanes in each direction). The analysis is to determine the expected level of service and capacity for the prevailing conditions shown.

Solution:

    Step 1:   Convert All Demand Volumes to Flow Rates in pc/h Under Equivalent Base Conditions

Each of the component demand volumes is converted to a demand flow rate in pc/h under equivalent base conditions using Equation 13-1:

$$v_p = \frac{V}{PHF * f_{HV} * f_p}$$

where: $PHF = 0.9$ (given)
            $f_p = 1.00$ (assume drivers are familiar with the site)

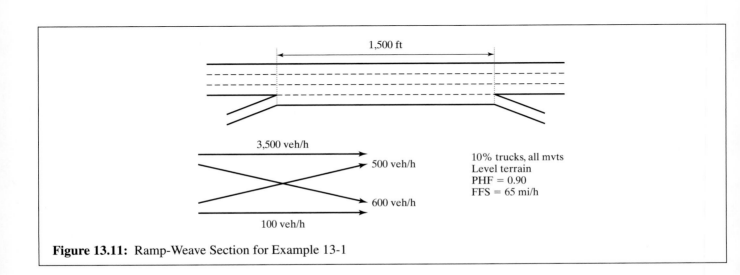

Figure 13.11: Ramp-Weave Section for Example 13-1

The heavy-vehicle factor, f_{HV}, is computed using Equation 12-9 and a value of E_T selected from Table 12.13 for trucks on level terrain ($E_T = 1.5$). Then:

$$f_{HV} = \frac{1}{1 + P_T (E_T - 1)} = \frac{1}{1 + 0.10(1.5 - 1)}$$
$$= 0.952$$

and:

$$v_{o1} = \frac{3,500}{0.90 * 0.952 * 1.00} = 4,085 \text{ pc/h}$$

$$v_{o2} = \frac{100}{0.90 * 0.952 * 1.00} = 117 \text{ pc/h}$$

$$v_{w1} = \frac{600}{0.90 * 0.952 * 1.00} = 700 \text{ pc/h}$$

$$v_{w2} = \frac{500}{0.90 * 0.952 * 1.00} = 584 \text{ pc/h}$$

Other critical variables used in the analysis may now be computed and/or summarized:

$v_w = 700 + 584 = 1,284$ pc/h

$v_{nw} = 4,085 + 117 = 4,202$ pc/h

$v = 1,284 + 4,202 = 5,486$ pc/h

$v/N = 5,486/4 = 1,372$ pc/h/ln

$VR = 1,284/5,485 = 0.23$

$L = 1,500$ ft

Step 2: Estimate the Average (Space Mean) Speed of Weaving and Non-weaving Vehicles in the Section

Speeds are estimated using Equation 13-3:

$$S_i = 15 + \left[\frac{FFS - 10}{1 + W_i} \right]$$

The free-flow speed of the freeway is given as 65 mi/h. Weaving intensity factors, W, must be computed using Equation 13-4 and the constants of calibration included in Table 13.3. The section under study is a Type A configuration (ramp-weave) in which all weaving vehicles must make one lane-change across the lane line separating the shoulder lane of the freeway and the auxiliary lane. The initial computation of speeds assumes that *unconstrained* operations exist, an assumption that will be tested in the next step of the procedure.

Then:

$$W_i = \frac{a(1 + VR)^b (v/N)^c}{L^d}$$

The constants of calibration are drawn from Table 13.3 for Type A configurations operating under unconstrained conditions. They are:

a (weaving)   = 0.15
a (non-weaving) = 0.0035
b (weaving)   = 2.2
b (non-weaving) = 4.0
c (weaving)   = 0.97
c (non-weaving) = 1.3
d (weaving)   = 0.80
d (non-weaving) = 0.75

The site variables were determined in Step 1. Then:

$$W_w = \frac{0.15(1 + 0.23)^{2.2} (1,372)^{0.97}}{1,500^{0.80}} = 0.752$$

$$S_w = 15 + \left(\frac{65 - 10}{1 + 0.752} \right) = 46.4 \text{ mi/h}$$

$$W_{nw} = \frac{0.0035(1 + 0.23)^{4.0} (1,372)^{1.3}}{1500^{0.75}} = 1.372$$

$$S_{nw} = 15 + \left(\frac{65 - 10}{1 + 0.398} \right) = 54.3 \text{ mi/h}$$

Step 3: Determine the Type of Operations

As Step 2 assumes unconstrained operation, it is now necessary to confirm that assumption. The equations and criteria of Table 13.4 are used. The number of lanes weaving vehicles must occupy to achieve unconstrained operation is given by:

$$N_w = \frac{0.74 * N * VR^{0.571} * L^{0.234}}{S_w^{0.438}}$$

$$= \frac{0.74 * 4 * 0.23^{0.571} * 1,500^{0.234}}{46.4^{0.438}} = 1.32 \text{ lanes}$$

From Table 13.4, N_w (max) for Type A configurations is 1.4 lanes. As $1.32 < 1.4$, the section is operating under unconstrained conditions and the speeds estimated in Step 2 are correct.

Step 4: Determine the Weighted Average Speed for the Weaving Area

The weighted average space mean speed for the weaving area is computed using Equation 13-5:

$$S = \frac{v_w + v_{nw}}{\left(\dfrac{v_w}{S_w}\right) + \left(\dfrac{v_{nw}}{S_{nw}}\right)} = \frac{1,284 + 4,202}{\left(\dfrac{1,284}{46.4}\right) + \left(\dfrac{4,202}{54.3}\right)}$$

$$= 52.2 \text{ mi/h}$$

Step 5: Determine Average Density and Level of Service for the Weaving Area

The average density for the weaving area is given by Equation 13-6:

$$D = \frac{(v/N)}{S} = \frac{1,372}{52.2} = 26.3 \text{ pc/min/ln}$$

Comparing this with the criteria of Table 13.1 indicates that the prevailing level of service is C.

Although not recommended by HCM 2000, separate densities may be computed for weaving and non-weaving vehicles using Equations 13-7 and 13-9:

$$N_{wA} = 1.32 \text{ lanes}$$

$$N_{nwA} = 4 - 1.32 = 2.68 \text{ lanes}$$

$$D_w = \frac{(v_w/N_{wA})}{S_w} = \frac{(1,284/1.32)}{46.4} = 20.9 \text{ pc/mi/ln}$$

$$D_{nw} = \frac{(v_{nw}/N_{nwA})}{S_{nw}} = \frac{(4,202/2.68)}{54.3} = 28.9 \text{ pc/mi/ln}$$

If the criteria of Table 13.1 are applied to these densities, weaving vehicles are experiencing level of service C, while non-weaving vehicles are experiencing level of service D.

Step 6: Check Other Limitations on Weaving-Area Operations

The limitations of Table 13.5 should be checked to ensure that none of the input parameters exceed maximum values for Type A configurations. These comparisons

are shown below; they indicate that none of the maximums are violated:

Parameter	Maximum Value (Table 13.5)	Actual Value
Weaving capacity	2,800 pc/h	1,283 pc/h
Maximum v/N	2,350 pc/h/ln	1,372 pc/h/ln
Maximum VR	0.35	0.23
Maximum L	2,500 ft	1,500 ft

Step 7: Determine the Capacity of the Weaving Area

The capacity of this weaving area (Type A, four lanes, 1,500 ft, VR = 0.23, FFS = 65 mi/h) is found from the tables in Appendix I of this chapter. Part B of these tables is used; interpolation between values of $VR = 0.20$ and $VR = 0.30$ is required:

VR	Capacity
0.20	8,170 pc/h
0.23	c_b
0.30	7,470 pc/h

Using straight-line interpolation:

$$c_b = 7,470 + \left(\frac{0.30 - 0.23}{0.30 - 0.20}\right) * (8,170 - 7,470)$$

$$= 7,960 \text{ pc/h}$$

This result is a maximum flow rate in terms of pc/h under equivalent base conditions. This may be converted to a capacity (maximum flow rate) in veh/h using Equation 13-10:

$$c = c_b * f_{HV} * f_p = 7,960 * 0.952 * 1.00 = 7,578 \text{ veh/h}$$

and may be further converted to a full-hour maximum volume as follows:

$$c_h = c * PHF = 7,578 * 0.90 = 6,820 \text{ veh/h}$$

Analysis The analysis results indicate that weaving vehicles are traveling somewhat slower than non-weaving vehicles, but at a lower density than non-weaving vehicles. This is typical of ramp-weave sections, as ramp vehicles enter and leave the traffic stream at reduced speeds due to often restrictive ramp geometries. Density in the auxiliary lane is generally lower than for mainline freeway lanes, resulting in lower densities for weaving vehicles. As the operation is unconstrained, this should not be viewed as a problem.

Example 13-2 Analysis of a Major Weaving Area

The freeway weaving area shown in Figure 13.12 is to be analyzed to determine the expected level of service for the conditions shown and the capacity of the weaving area. For convenience, all demand volumes have already been converted to flow rates in pc/h under equivalent base conditions. For information purposes, the following values were used to make these conversions: $PHF = 0.95$, $f_{HV} = 0.93$, $f_p = 1.00$.

Step 1:   Determine All Required Input Variables

As all demands are specified as flow rates in pc/h under equivalent base conditions, no further conversion of these is necessary. Key analysis variables are summarized below:

$v_w = 800 + 1,700 = 2,500$ pc/h

$v_{nw} = 1,700 + 1,500 = 3,200$ pc/h

$v = 2,500 + 3,200 = 5,700$ pc/h

$v/N = 5,700/3 = 1,900$ pc/h/ln

$VR = 2,500/5,700 = 0.439$

$L = 2,000$ ft

Note that this is a Type B configuration. The weave from left to right can be made with no lane-changes, while the weave from right to left requires one lane-change.

Step 2: Determine the Average (Space Mean) Speed of Weaving and Non-weaving Vehicles in the Section

Equations 13-3 and 13-4 are used to determine the average speeds of weaving and non-weaving vehicles in the section, using the constants of calibration given in Table 13.3 (for Type B configurations). Unconstrained operation is assumed for initial speed estimates.

$$W_w = \frac{0.08(1 + 0.439)^{2.2}\,(1,900)^{0.70}}{2,000^{0.5}} = 0.786$$

$$S_w = 15 + \left(\frac{70 - 10}{1 + 0.786}\right) = 48.6 \text{ mi/h}$$

$$W_{nw} = \frac{0.002(1 + 0.439)^{6.0}\,(1,900)^{1.0}}{2,000^{0.5}} = 0.754$$

$$S_{nw} = 15 + \left(\frac{70 - 10}{1 + 0.754}\right) = 49.2 \text{ mi/h}$$

Step 3:   Determine the Type of Operation

The assumption of unconstrained operation is checked using the equations and criteria of Table 13.4 for Type B configurations:

$$N_w = N * \left[0.085 + 0.703VR + \left(\frac{234.8}{L}\right)\right.$$
$$\left. - 0.018(S_{nw} - S_w)\right]$$

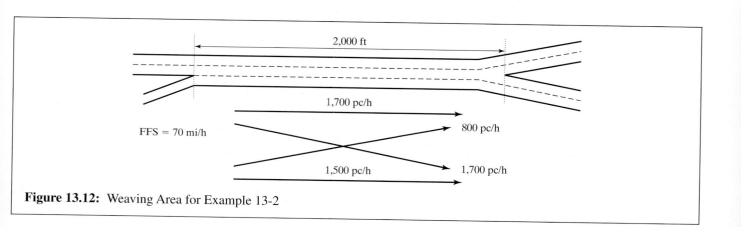

Figure 13.12: Weaving Area for Example 13-2

2,000 ft

1,700 pc/h

FFS = 70 mi/h

800 pc/h

1,500 pc/h

1,700 pc/h

$$N_w = 3 * \left[0.085 + (0.703 * 0.439) + \left(\frac{234.8}{2,000} \right) \right.$$
$$\left. - 0.018(49.2 - 48.6) \right]$$

$N_w = 1.5$ lanes

As this is less than the maximum allowable use of lanes for weaving vehicles in a Type B configuration (3.5), the operation is unconstrained and the speeds determined in Step 2 are correct.

Step 4: Determine the Average (Space Mean) Speed for All Vehicles in the Weaving Area

The space mean speed for all vehicles in the weaving area is given by Equation 13-5:

$$S = \frac{2,500 + 3,200}{\left(\frac{2,500}{48.6} \right) + \left(\frac{3,200}{49.2} \right)} = 48.9 \text{ mi/h}$$

Step 5: Determine the Density and Level of Service in the Weaving Area

The density in the weaving area is computed using Equation 13-6:

$$D = \frac{(v/N)}{S} = \frac{1,900}{48.9} = 38.9 \text{ pc/mi/ln}$$

From the criteria of Table 13.1, this is level of service E. Again, while not recommended by the HCM 2000, separate densities for weaving and non-weaving vehicles can be estimated:

$N_{wA} = 1.5$ lanes

$N_{nwA} = 3 - 1.5 = 1.5$ lanes

$$D_w = \frac{(v_w/N_{wA})}{S_w} = \frac{(2,500/1.5)}{48.6} = 34.3 \text{ pc/mi/ln}$$

$$D_{nw} = \frac{(v_{nw}/N_{nwA})}{S_{nw}} = \frac{(3,200/1.5)}{49.2} = 43.4 \text{ pc/mi/h}$$

If the criteria of Table 13.1 are applied, the weaving vehicles are experiencing level of service D, while the non-weaving vehicles are experiencing level of service F. As the density for the entire section is less than that specified for the level-of-service E/F boundary (43.0 pc/mi/ln), a failure of the section is not expected. Some localized queuing might occur, however, and operations should be considered to be marginal in this section.

Step 6: Check Other Limitations on Weaving Area Operation

The limitations on weaving areas of Table 13.5 should be checked to ensure that input variables do not exceed any of the maximums stated. This comparison is summarized below:

Parameter	Maximum Value	Actual Value
Weaving capacity	4,000 pc/h	2,500 pc/h
Maximum v/N	2,400 pc/h	1,900 pc/h
Maximum VR	0.80	0.439
Maximum L	2,500 ft	2,000 ft

None of these limits are exceeded.

Step 7: Determine the Capacity of the Weaving Area

The capacity of the section is found using the tables in Appendix I of this chapter. Table E is used for this three-lane, Type B section of 2,000 ft, with a VR of 0.439 and a freeway FFS of 70 mi/h. Interpolation will be used between VR values of 0.40 and 0.50.

VR	Capacity
0.40	6,650 pc/h
0.439	c_b
0.50	6,170 pc/h

Using straight-line interpolation:

$$c_b = 6,170 + \left(\frac{0.50 - 0.439}{0.50 - 0.40} \right) * (6,650 - 6,170)$$
$$= 6,643 \text{ pc/h}$$

This is a maximum flow rate in pc/h under equivalent base conditions. It could be converted to a flow rate under prevailing conditions and/or a full-hour maximum volume under prevailing conditions. As the demands are given in terms of base flow rates, such conversions would not be useful in this case. The base capacity of 6,643 pc/h can be directly compared to the converted total demand flow of 5,700 pc/h.

Example 13-3: Analysis of an Isolated On-Ramp

An on-ramp to a busy eight-lane urban freeway is illustrated in Figure 13.13. An analysis of this merge area is to determine the likely level of service under the prevailing conditions shown.

Solution:

Step 1: Convert All Demand Volumes to Flow Rates in pc/h Under Equivalent Base Conditions

The freeway and ramp flows approaching the merge area must be converted to flow rates in pc/h under equivalent base conditions using Equation 13-1. In this case, note that the truck percentages and PHF are different for the two. Table 13.8 is used to obtain a value of E_T for rolling terrain (2.5), and Equation 12-9 is used to compute the heavy-vehicle factor. It is assumed that drivers are familiar users, and that $f_p = 1.00$.
For the ramp demand flow:

$$f_{HV} = \frac{1}{1 + 0.10(2.5 - 1)} = 0.870$$

$$v_R = \frac{900}{0.89 * 0.870 * 1.00} = 1,162 \text{ pc/h}$$

For the freeway demand flow:

$$f_{HV} = \frac{1}{1 + 0.05(2.5 - 1)} = 0.930$$

$$v_F = \frac{5200}{0.92 * 0.930 * 1.00} = 6,078 \text{ pc/h}$$

Step 2: Determine the Demand Flow Remaining in Lanes 1 and 2 Immediately Upstream of the Merge

Table 13.6 gives values of P_{FM}, the proportion of freeway vehicles remaining in lanes 1 and 2 immediately upstream of a merge. For an eight-lane freeway (four lanes in each direction), Equation 13-15 is used to estimate P_{FM}. This value is then used in Equation 13-11 to determine v_{12}.

$$P_{FM} = 0.2178 - 0.000125v_R + 0.01115(^{L_a}/_{RFFS})$$

$$P_{FM} = 0.2178 - (0.000125 * 1,162)$$
$$+ 0.01115 * (^{1,000}/_{40}) = 0.3513$$

$$v_{12} = v_F * P_{FM} = 6,078 * 0.3513 = 2,135 \text{ pc/h}$$

Step 3: Check Capacity Values

To determine whether the section will fail (LOS F), the capacity values of Table 13.8 must be consulted. For a merge section, the critical capacity check is on the downstream freeway section.

$$v_{FO} = v_F + v_R = 6,078 + 1,162 = 7,240 \text{ pc/h}$$

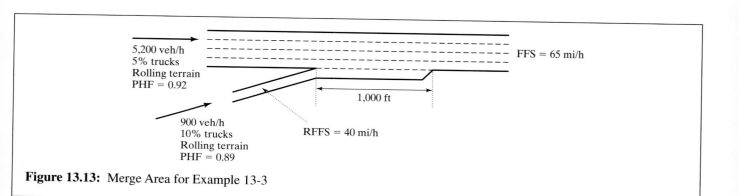

Figure 13.13: Merge Area for Example 13-3

From Table 13.8, the capacity of a four-lane freeway section is 9,400 pc/h when the FFS is 65 mi/h. As 9,400 > 7,240, no failure is expected due to total downstream flow.

The capacity of a one-lane ramp with a free-flow speed of 40 mi/h must also be checked. From Table 13.8, such a ramp has a capacity of 2,000 pc/h. As this is greater than the ramp demand flow of 1,162 pc/h, this element will not fail either.

Total flow entering the merge influence area is:

$$v_{R12} = v_R + v_{12} = 1,162 + 2,135 = 3,297 \text{ pc/h}$$

As the maximum desirable entering flow for single-lane merge area is 4,600 pc/h, this element is also acceptable. No failure (LOS F) is, therefore, expected.

Step 4: Estimate Density and Level of Service in the Ramp Influence Area

As stable operations are expected, Equation 13-24 may be used to estimate the density in the ramp influence area:

$$D_R = 5.475 + 0.00743v_R + 0.0078v_{12} - 0.00627L_a$$

$$D_R = 5.475 + (0.00743 * 1162) + (0.0078 * 2135) \\ - (0.00627 * 1,000)$$

$$D_R = 24.4 \text{ pc/mi/ln}$$

From the criteria in Table 13.1, this is level of service C.

Step 5: Estimate Speed Parameters for Information

While not used to determine level of service, the algorithms of Table 13.9 may be used to estimate speed parameters of interest:

$$M_S = 0.321 + 0.0039e^{\left(v_{R12}/1,000\right)} \\ - 0.002 * \left(L_a * RFFS/1,000\right)$$

$$M_S = 0.321 + 0.0039e^{3.297} - 0.002 * \left(\frac{1,000 * 40}{1,000}\right)$$

$$= 0.346$$

$$S_R = FFS - (FFS - 42)M_S \\ = 65 - (65 - 42) * 0.346 = 57.0 \text{ mi/h}$$

$$S_o = FFS - 0.0036 * (v_{oa} - 500) \\ = 65 - 0.0036 * \left[\left(\frac{6,078 - 2,135}{2}\right) - 500\right] \\ = 59.7 \text{ mi/h}$$

$$S = \frac{v_{R12} + v_{oa}N_o}{\left(\frac{v_{R12}}{S_R}\right) + \left(\frac{v_{oa}N_o}{S_o}\right)} = \frac{3,297 + (3,943)}{\left(\frac{3,297}{57.0}\right) + \left(\frac{3,943}{59.7}\right)} \\ = 58.4 \text{ mi/h}$$

S_R is the space mean speed of vehicles entering the ramp influence area (i.e., v_{R12}). S_o is the space mean speed of vehicles in the outer lanes. There are two outer lanes in a four-lane freeway, and the total flow in those outer lanes is $v_F - v_{12}$ or $6,078 - 2,135 = 3,943$ pc/h. As this outer lane flow is divided into two lanes, the average flow (v_{oa}) is $3,943/2 = 1,972.5$ pc/h/ln. In this rare case, the 0.5 pc/h/ln has meaning, so that when multiplied by the number of outer lanes, the total outer flow is correctly stated. S is the space mean speed of all vehicles. In all cases, the speeds apply to the 1,500-ft length defining the ramp influence area.

Analysis   Several additional items may be of interest. The lane distribution of the incoming freeway flow (v_F) should be checked for reasonableness. In this case 2,135 pc/h use lanes 1 and 2, while 3,943 pc/h use lanes 3 and 4. This is not unexpected, given the large ramp flow (1,162 pc/h) entering at the on-ramp.

It is also useful to check the level of service on the downstream basic freeway section. It carries a total of 7,240 pc/h in four lanes, or 1,810 pc/h/ln. Entering Figure 12.3 with 1,810 pc/h/ln and a 65-mi/h free-flow speed, it can be seen that the level of service on the downstream freeway is D.

What does this mean, considering that the LOS for the ramp influence area is determined to be C? It means that the total freeway flow is the determining element in overall level of service. This is as it should be, as it is always undesirable to have minor movements (in this case, the on-ramp), controlling the overall operation of the facility.

Example 13-4: Analysis of a Sequence of Freeway Ramps

Figure 13.14 shows a series of three ramps on a six-lane freeway (three lanes in each direction). All three ramps are to be analyzed to determine the level of service expected under the prevailing conditions shown.

Solution:

> Step 1: Convert All Demand Volumes to Flow Rates in pc/h Under Equivalent Base Conditions

Before applying any of the models for ramp analysis, all demand volumes must be converted to flow rates in pc/h under equivalent base conditions. This is done using Equation 13-1. Peak-hour factors for each movement are given, as are truck percentages. The heavy-vehicle factor is computed using Equation 12-9 and values of E_T drawn from Table 12.13. For level terrain, $E_T = 1.5$ for all movements. It is assumed that the driver population consists primarily of familiar users and that f_p is, therefore, 1.00. The conversion computations are shown below:

$$f_{HV}(\text{freeway}) = \frac{1}{1 + 0.10(1.5 - 1)} = 0.952$$

$$v_F = \frac{4,000}{0.90 * 0.952 * 1.00} = 4,669 \text{ pc/h}$$

$$f_{HV}(\text{Ramp 1}) = \frac{1}{1 + 0.15(1.5 - 1)} = 0.930$$

$$v_{R1} = \frac{500}{0.95 * 0.930 * 1.00} = 566 \text{ pc/h}$$

$$f_{HV}(\text{Ramp 2}) = \frac{1}{1 + 0.05(1.5 - 1)} = 0.976$$

$$v_{R2} = \frac{600}{0.92 * 0.976 * 1.00} = 668 \text{ pc/h}$$

$$f_{HV}(\text{Ramp 3}) = \frac{1}{1 + 0.12(1.5 - 1)} = 0.943$$

$$v_{R3} = \frac{400}{0.91 * 0.943 * 1.00} = 466 \text{ pc/h}$$

Step 2: Determine the Flow in Lanes 1 and 2 Immediately Upstream of Each Ramp in the Sequence

Each of the ramps in the analysis section must now be considered for the potential impact of the adjacent ramp(s) on lane distribution.

Ramp 1: The first ramp is part of a three-ramp sequence that can be described as None-OFF-On (no upstream adjacent ramp; an adjacent downstream on-ramp). Using Table 13.7, for a six-lane freeway and the sequence indicated, Equation 13-19 should be used to determine v_{12}.

$$v_{12(1)} = v_{R1} + (v_F - v_{R1})P_{FD}$$

$$P_{FD} = 0.760 - 0.000025v_F - 0.000046v_{R1}$$

$$P_{FD} = 0.760 - (0.000025 * 4669) \\ - (0.000046 * 566) = 0.617$$

$$v_{12(1)} = 566 + (4,669 - 566) * 0.617 = 3,098 \text{ pc/h}$$

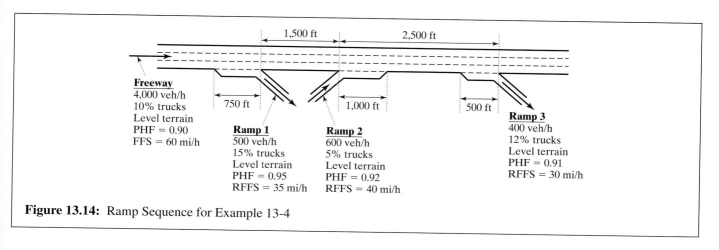

Freeway
4,000 veh/h
10% trucks
Level terrain
PHF = 0.90
FFS = 60 mi/h

750 ft

1,000 ft

500 ft

1,500 ft

2,500 ft

Ramp 1
500 veh/h
15% trucks
Level terrain
PHF = 0.95
RFFS = 35 mi/h

Ramp 2
600 veh/h
5% trucks
Level terrain
PHF = 0.92
RFFS = 40 mi/h

Ramp 3
400 veh/h
12% trucks
Level terrain
PHF = 0.91
RFFS = 30 mi/h

Figure 13.14: Ramp Sequence for Example 13-4

Ramp 2: The second ramp is an on-ramp that can be described as part of an Off-ON-Off sequence. From Table 13.6, there are three potential equations that might apply: Equation 13-13, which considers the effect of the upstream off-ramp; Equation 13-14, which considers the effect of the downstream off-ramp; or Equation 13-12, which treats the ramp as if it were isolated. It is even possible that two of these apply, in which case the equation yielding the larger v_{12} estimate is used. To determine which of these apply, however, requires the use of Equations 13-16 and 13-17 to determine equivalence distances.

In considering whether the impact of the upstream off-ramp must be considered, Equation 13-16 is used:

$$L_{EQ} = 0.214(v_R + v_F) + 0.444L_a \\ + 52.32RFFS - 2,403$$

Note that for Ramp 2, the approaching freeway flow is the beginning freeway flow minus the off-ramp flow at Ramp 1:

$$v_{F2} = v_F - v_{R1} = 4,669 - 566 = 4,103 \text{ pc/h}$$

Thus:

$$L_{EQ} = 0.214(668 + 4,103) + (0.444*1,000) \\ + (52.32*40) - 2,403 = 1,155 \text{ ft}$$

As the actual distance to the upstream ramp is 1,500 ft ($>1,155$ ft), the impact of the upstream off-ramp need not be considered, and Equation 13-12 is used.

To determine whether or not the effect of the downstream off-ramp must be considered, Equation 13-17 is used to compute L_{EQ}:

$$L_{EQ} = \frac{v_d}{0.1096 + 0.000107L_a} \\ = \frac{466}{0.1096 + (0.000107*1,000)} = 2,151 \text{ ft}$$

The actual distance to the downstream off-ramp is 2,500 ft ($>2,151$ ft). Thus, the impact of the downstream off-ramp is also not considered, and Equation 13-12 is used. Through the determination of these equivalence distances, it is seen that Ramp 2 may be considered to be an isolated ramp. Only one—Equation 13-12—applies to the estimation of $v_{12(2)}$.

$$v_{12(2)} = v_{F2}*P_{FM}$$
$$P_{FM} = 0.5775 + 0.000028L_a = 0.5775 \\ + (0.000028*1,000) = 0.6055$$
$$v_{12(2)} = 4,103*0.6055 = 2,484 \text{ pc/h}$$

Ramp 3: The third ramp is now considered as part of an On-OFF-None sequence. From Table 13.7, Equation 13-20 or 13-19 is used. To determine which is the appropriate one for application, Equation 13-22 is used to compute L_{EQ}. In applying this equation, note that v_{F3} includes the on-ramp flow from Ramp 2. Thus:

$$v_{F3} = v_{F2} + v_{R2} = 4,103 + 668 = 4,771 \text{ pc/h}$$

$$L_{EQ} = \frac{v_u}{0.071 + 0.000023v_F - 0.000076v_R} \\ = \frac{668}{0.071 + (0.000023*4771) - (0.000076*466)}$$

$$L_{EQ} = 4,597 \text{ ft}$$

As the actual distance to the upstream on-ramp is only 2,500 ft ($<4,597$ ft), Equation 13-20 is used to consider the impact of Ramp 2 on lane distribution at Ramp 3:

$$v_{12(3)} = v_{R3} + (v_{F3} - v_{R3})*P_{FD}$$

$$P_{FD} = 0.717 - 0.000039v_F + 0.604\left(\frac{v_u}{L_{up}}\right)$$

$$P_{FD} = 0.717 - (0.000039*4771) \\ + 0.604\left(\frac{668}{2500}\right) = 0.6923$$

$$v_{12(3)} = 466 + (4,771 - 466)*(0.6923) \\ = 3,446 \text{ pc/h}$$

Summarizing the results for v_{12} immediately upstream of each of the three ramps:

$$v_{12(1)} = 3,098 \text{ pc/h}$$
$$v_{12(2)} = 2,484 \text{ pc/h}$$
$$v_{12(3)} = 3,446 \text{ pc/h}$$

Step 3: Check Capacities

The capacities and limiting values of Table 13.8 must now be checked to see whether operations are stable or whether level of service F exists. The freeway flow check is made between Ramps 2 and 3, as this is the point where total freeway flow is greatest (v_{F3}). These checks are performed below. Remember that the freeway FFS is 60 mi/h.

Item	Demand Flow	Capacity (Table 13.8)
v_{F3}	4,771 pc/h	6,900 pc/h
$v_{12(1)}$	3,098 pc/h	4,400 pc/h
$v_{R12(2)}$	2,484 + 668 = 3,152 pc/h	4,600 pc/h
$v_{12(3)}$	3,446 pc/h	4,400 pc/h
v_{R1}	566 pc/h	2,000 pc/h (RFFS = 35 mi/h)
v_{R2}	668 pc/h	2,000 pc/h (RFFS = 40 mi/h)
v_{R3}	466 pc/h	1,900 pc/h (RFFS = 30 mi/h)

None of the demand flows exceed the capacities or limiting values of Table 13.8. Thus, stable operation is expected throughout the section.

Step 4: Determine Densities and Levels of Service in Each Ramp Influence Area

The density in the ramp influence area is estimated using Equations 13-24 for on-ramps and 13-25 for off-ramps:

$$D_{R1} = 4.252 + 0.0086v_{12(1)} - 0.009L_d$$

$$D_{R1} = 4.252 + (0.0086*3098) - (0.009*750)$$
$$= 24.1 \text{ pc/mi/ln}$$

$$D_{R2} = 5.475 + 0.00734v_R + 0.0078v_{12(2)} - 0.00627L_a$$

$$D_{R2} = 5.475 + (0.00734*668) + (0.0078*2,484) - (0.00627*1,000) = 23.5 \text{ pc/mi/ln}$$

$$D_{R3} = 4.252 + (0.0086*3446) - (0.009*500)$$
$$= 29.4 \text{ pc/mi/ln}$$

From Table 13.1, Ramp 1 operates at LOS C, Ramp 2 at LOS C, and Ramp 3 at LOS D.

Step 5: Determine Speeds for Each Ramp

As was done in Example 13-3, the algorithms of Tables 13.9 and 13.10 may be used to estimate space mean speeds within each ramp influence area and across all freeway lanes within the 1,500-ft range of each ramp influence area. Because of the length of these computations, they are not shown here. Each would follow the sequence illustrated in Example 13-3.

Analysis Once again, the reasonableness of the predicted lane distributions in the vicinity of each ramp should be checked.

At Ramp 1, 3,098 pc/h are in lanes 1 and 2 immediately upstream of the junction, leaving 4,669 − 3,098 = 1,571 pc/h in lane 3 (the single outer lane in this case). This appears to be a reasonable distribution. At Ramp 2, 2,484 pc/h are in lanes 1 and 2, leaving 4,103 (v_{F2}) − 2,484 = 1,619 pc/h in lane 3. This also appears to be reasonable. At Ramp 3, 3,446 pc/h are in lanes 1 and 2, while 4,771 (v_{F3}) − 3,446 = 1,323 pc/h are in lane 3. This is also a reasonable distribution.

Note that the ramp influence areas of Ramps 2 and 3 overlap for a distance of 500 ft (1,500 + 1,500 − 2,500). For this overlapping segment, the influence area having the highest density and lowest LOS would be used. In this case, Ramp 3, with LOS D, controls this overlap area.

Again, it is interesting to check the basic freeway level of service associated with the controlling (or largest) total freeway flow, which occurs between Ramps 2 and 3. The demand flow per lane for this segment is 4,771/3 = 1,591 pc/h/ln. Entering Table 12.3 with this value, and a FFS of 60 mi/h, the level of service is found to be almost exactly on the border between levels of service C and D. This is consistent with the Ramp 3 LOS, which is D but is only barely above the LOS C/D boundary. Thus, the operation of the freeway as a whole and ramp sequence are somewhat in balance, a desirable condition.

13.7 Analysis of Freeway Facilities

The HCM 2000 contains a new methodology for the analysis of long stretches of freeway facilities, containing many basic, weaving, merge, and/or diverge sections. The methodology is reasonably straightforward for cases in which no sections fail (i.e., LOS F) but is extremely complex in cases that encompass section failures. The models involved cannot be easily implemented by hand, so the overall procedure is outlined here without detail. User-friendly software is not yet available for the freeway facility methodology (as of January 2002), but the HCS software package contains an undocumented spreadsheet program that implements the model.

13.7.1 Segmenting the Freeway

The analysis of a freeway facility must begin by breaking the facility into component sections. Sections are fairly easily established using the definitions of basic, weaving, merge, and diverge areas as defined in Chapters 12 and 13. All weaving, merge, and diverge areas are isolated as sections. All other sections are basic freeway sections.

For analysis, however, sections must be further divided into segments. In addition to section boundaries, segment boundaries must be established at all points where a change in geometric or traffic conditions occurs. As the influence areas of ramp junctions extend 1,500 ft downstream of an on-ramp and 1,500 ft upstream of an off-ramp, longer acceleration or deceleration lanes will be treated as basic freeway segments outside the influence area; a separate basic freeway segment would then have to be established at the point where the acceleration lane ends (or the deceleration lane begins). A basic freeway section might have to be divided into several segments if there are changes in geometry, such as a change in terrain or specific grades.

The general process of segmenting the facility for analysis is illustrated in Figure 13.15.

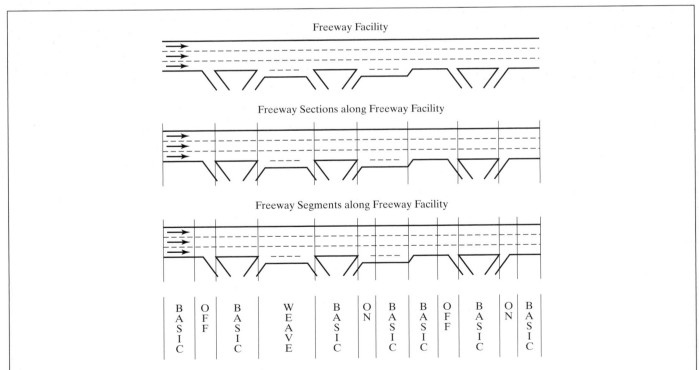

Figure 13.15: Conversion of Freeway Sections into Freeway Segments for Analysis (Used with permission of Transportation Research Board, *Highway Capacity Manual*, 4th Edition, Washington DC, 2000, Exhibit 22-3, pg. 22-5.)

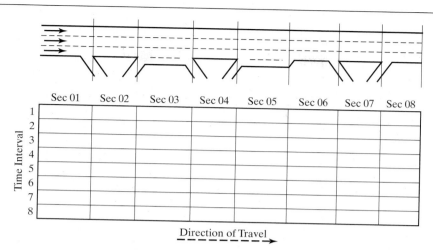

Figure 13.16: Time-Space Domain of a Freeway Facility (Used with permission of Transportation Research Board, *Highway Capacity Manual*, 4th Edition, Washington DC, 2000, Exhibit 22-2, pg. 22-4.)

13.7.2 Analysis Models

The analysis procedure treats the freeway facility as a time-space domain, as illustrated in Figure 13.16. Input demands are established for each section during each time interval. Each time interval is usually 15 minutes, but longer periods may be used. Each time-section cell is then analyzed, establishing appropriate segments within each and obtaining level-of-service and speed estimates. If any cell breaks down within the analysis period, new models track the spread of queuing to upstream segments, and shift the demand in both space and time that results from such queuing. The time-space matrix should be set up so that the beginning and ending time periods have no segment failures or residual queues.

Using this procedure, the models produce level-of-service and speed predictions for all segments and time periods and show both the build-up and dissipation of queues due to segment failures.

References

1. *Highway Capacity Manual*, Transportation Research Board, Washington DC, 2000.

2. *Highway Capacity Manual*, "Special Report 87," Transportation Research Board, Washington DC, 1965.

3. Pignataro, L., *et al.*, "Weaving Areas—Design and Analysis," *NCHRP Report 159*, Transportation Research Board, Washington DC, 1975.

4. Roess, R., *et al.*, "Freeway Capacity Analysis Procedures," *Final Report*, Project No. DOT-FH-11-9336, Polytechnic University, Brooklyn, NY, 1978.

5. "Interim Procedures on Highway Capacity," *Circular 212*, Transportation Research Board, Washington DC, 1980.

6. Leisch, J, "Completion of Procedures for Analysis and Design of Traffic Weaving Areas," *Final Report*, Vols 1 and 2, U.S. Department of Transportation, Federal Highway Administration, Washington DC, 1983.

7. Reilly W., *et al.*, "Weaving Analysis Procedures for the New Highway Capacity Manual," *Technical Report*, JHK & Associates, Tucson, AZ, 1983.

8. Roess, R., "Development of Weaving Area Analysis Procedures for the 1985 Highway Capacity Manual," *Transportation Research Record 1112*, Transportation Research Board, Washington DC, 1987.

9. Fazio, J., "Development and Testing of a Weaving Operational Design and Analysis Procedure," *Master's Thesis*, University of Illinois at Chicago, Chicago IL, 1985.

10. Cassidy, M. and May, A., Jr., "Proposed Analytic Technique for Estimating Capacity and Level of Service of Major Freeway Weaving Sections," *Transportation Research Record 1320*, Transportation Research Board, Washington DC, 1992.

11. Ostram, B., *et al.*, "Suggested Procedures for Analyzing Freeway Weaving Sections, *Transportation Research Record 1398*, Transportation Research Board, Washington DC, 1993.

12. Windover, J. and May, A., Jr., "Revisions to Level D Methodology of Analyzing Freeway Ramp-Weaving Sections, *Transportation Research Record 1457*, Transportation Research Board, Washington DC, 1995.

13. "Weaving Zones," *Draft Report*, NCHRP Project 3-55(5), Viggen Corporation, Sterling, VA, 1998.

14. Roess, R. and Ulerio, J., "Weaving Area Analysis in the HCM 2000," *Transportation Research Record*, Transportation Research Board, Washington DC, 2000.

15. Roess, R. and Ulerio, J., "Capacity of Ramp-Freeway Junctions," *Final Report*, Polytechnic University, Brooklyn, NY, 1993.

16. Leisch, J., *Capacity Analysis Techniques for Design and Operation of Freeway Facilities*, Federal Highway Administration, U.S. Department of Transportation, Washington DC, 1974.

Problems

13-1. Consider the pair of ramps shown in Figure 13.17 below. It may be assumed that there is no ramp-to-ramp flow.

(a) Given the existing demand volumes and other prevailing conditions, at what level of service is this section expected to operate? If problems exist, which elements appear to be causing the difficulty?

(b) It is proposed that the acceleration and deceleration lanes be joined to form a continuous auxiliary lane. How will this affect the operation? What level of service would be expected? Would you recommend this change?

(c) What is the capacity of the section under the two scenarios described in parts (a) and (b)?

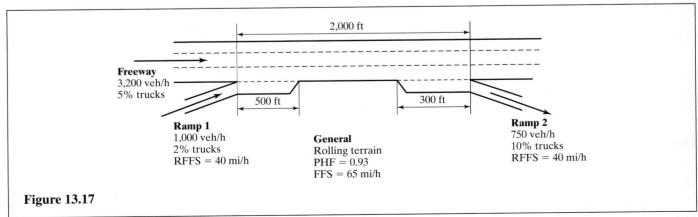

Figure 13.17

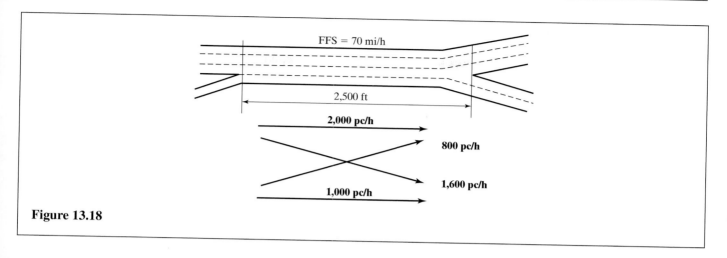

Figure 13.18

13-2. Consider the weaving area in Figure 13.18. All demands are shown as flow rates in pc/h under equivalent base conditions.

(a) What type of configuration is shown?

(b) What is the expected level of service for these conditions?

(c) What is the capacity of the weaving section?

13-3. Consider the on-ramp shown in Figure 13.19.

(a) At what level of service would the merge area be expected to operate?

(b) A new development nearby opens and increases the on-ramp volume to 1,100 veh/h. How does this affect the level of service?

13-4. Figure 13.20 illustrates two consecutive ramps on an older freeway. It may be assumed that there is a ramp-to-ramp flow of 200 veh/h.

(a) What is the expected level of service for the conditions shown?

(b) Several improvement plans are under consideration:

i. Connect the two ramps with a continuous auxiliary lane.

ii. Add a third lane to the freeway and extend the length of acceleration and deceleration lanes to 300 ft.

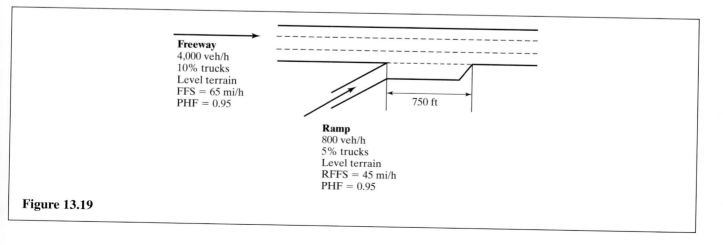

Figure 13.19

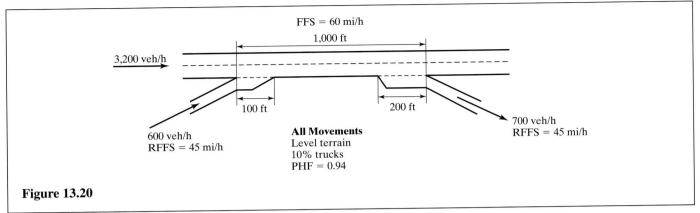

Figure 13.20

iii. Provide a lane addition at the on-ramp that continues past the off-ramp on the downstream freeway section. The off-ramp deceleration lane remains 200 ft long.

Which of these three improvements would you recommend? Why? Justify your answer.

13-5. What is the expected level of service and capacity for the weaving area shown in Figure 13.21. All demands are stated in terms of flow rates in pc/h under equivalent base conditions.

Comment on the results—are they acceptable? If not, suggest any solutions that might mitigate some of the problems. You do not have to analyze the suggested improvements but should provide a verbal description of why they would improve on current operations.

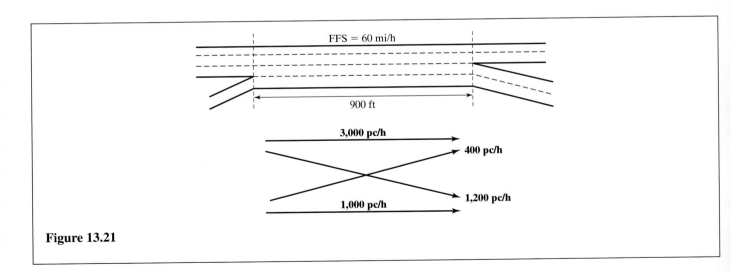

Figure 13.21

Appendix I: Capacity Tables for Freeway Weaving Sections

(A) Type A Weaving Segments—75-mi/h Free-Flow Speed and 70 mi/h FFS

Volume Ratio, VR	Length of Weaving Segment (ft)				
	500	1,000	1,500	2,000	2,500[a]
Three-Lane Segments					
0.10	6,030	6,800	7,200[d]	7,200[d]	7,200[d]
0.20	5,460	6,230	6,680	7,010	7,200[d]
0.30	4,990	5,740	6,210	6,530	6,790
0.40	4,620	5,340	5,480[c]	5,790[c]	6,040[c]
0.45[b]	4,460	4,840[c]	5,240[c]	5,540[c]	5,780[c]
Four-Lane Segments					
0.10	8,040	9,070	9,600[d]	9,600[d]	9,600[d]
0.20	7,280	8,300	8,910	9,350	9,600[d]
0.30	6,660	7,520[c]	8,090[c]	8,520[c]	8,830[c]
0.35[e]	6,250[c]	7,120[c]	7,690[c]	8,000[h]	8,000[h]
Five-Lane Segments					
0.10	10,050	11,340	12,000[d]	12,000[d]	12,000[d]
0.20[f]	9,100	10,540[c]	11,270[c]	11,790[c]	12,000[c]

(B) Type A Weaving Segments—65-mi/h Free-Flow Speed

Volume Ratio, VR	Length of Weaving Segment (ft)				
	500	1,000	1,500	2,000	2,500[a]
Three-Lane Segments					
0.10	5,570	6,230	6,620	6,890	7,050[d]
0.20	5,070	5,740	6,130	6,410	6,620
0.30	4,670	5,320	5,720	6,000	6,220
0.40	4,330	4,950	5,090[c]	5,360[c]	5,570[c]
0.45[b]	4,190	4,520[c]	4,870[c]	5,140[c]	5,340[c]
Four-Lane Segments					
0.10	7,430	8,310	8,830	9,190	9,400[d]
0.20	6,760	7,660	8,170	8,550	8,830
0.30	6,180[c]	6,970[c]	7,470[c]	7,830[c]	8,110[c]
0.35[e]	5,870[c]	6,620[c]	7,120[c]	7,470[c]	7,760[c]
Five-Lane Segments					
0.10	9,290	10,390	11,040	11,490	11,750[d]
0.20[f]	8,450	9,700[c]	10,320[c]	10,760[c]	11,090[c]

(C) Type A Weaving Segments—60-mi/h Free-Flow Speed

Volume Ratio, VR	Length of Weaving Segment (ft)				
	500	1,000	1,500	2,000	2,500[a]
Three-Lane Segments					
0.10	5,330	5,940	6,280	6,520	6,700
0.20	4,870	5,480	5,850	6,100	6,280
0.30	4,490	5,100	5,460	5,720	5,920
0.40	4,180	4,570[c]	4,880[c]	5,140[c]	5,330[c]
0.45[b]	4,040	4,360[c]	4,680[c]	4,920[c]	5,120[c]
Four-Lane Segments					
0.10	7,110	7,920	8,380	8,690	8,930
0.20	6,500	7,310	7,800	8,140	8,420[c]
0.30	5,960[c]	6,680[c]	7,140[c]	7,470[c]	7,710[c]
0.35[e]	5,660[c]	6,370[c]	6,810[c]	7,140[c]	7,400[c]
Five-Lane Segments					
0.10	8,880	9,900	10,480	10,870	11,500[d]
0.20[f]	8,120	9,270[c]	9,830[c]	10,220[c]	10,530[c]

(D) Type A Weaving Segments—55-mi/h Free-Flow Speed

Volume Ratio, VR	Length of Weaving Segment (ft)				
	500	1,000	1,500	2,000	2,500[a]
Three-Lane Segments					
0.10	5,080	5,630	5,940	6,140	6,300
0.20	4,660	5,210	5,550	5,770	5,940
0.30	4,300	4,850	5,190	5,430	5,610
0.40	4,010	4,360[c]	4,670[c]	4,890[c]	5,070[c]
0.45[b]	3,880	4,180[c]	4,480[c]	4,700[c]	4,880[c]
Four-Lane Segments					
0.10	6,780	7,500	7,920	8,190	8,400
0.20	6,210	6,950	7,400	7,690	7,940[c]
0.30	5,710[c]	6,380[c]	6,780[c]	7,090[c]	7,310[c]
0.35[e]	5,440[c]	6,090[c]	6,490[c]	6,800[c]	7,030[c]
Five-Lane Segments					
0.10	8,480	9,380	9,900	10,240	10,510
0.20[f]	7,960[c]	8,800[c]	9,320[c]	9,660[c]	9,930[c]

(E) Type B Weaving Segments—75-mi/h Free-Flow Speed and 70 mi/h FFS

Volume Ratio, VR	Length of Weaving Segment (ft)				
	500	1,000	1,500	2,000	2,500[a]
Three-Lane Segments					
0.10	7,200[d]	7,200[d]	7,200[d]	7,200[d]	7,200[d]
0.20	6,810	7,200[d]	7,200[d]	7,200[d]	7,200[d]
0.30	6,120	6,670	7,000	7,230	7,200[d]
0.40	5,540	6,090	6,430	6,650	6,830
0.50	5,080	5,620	5,940	6,170	6,360
0.60	4,750	5,250	5,560	5,790	5,970
0.70	4,180	4,980	5,290	5,510	5,690
0.80[g]	3,890	4,810	5,000[h]	5,000[h]	5,000[h]
Four-Lane Segments					
0.10	9,600[d]	9,600[d]	9,600[d]	9,600[d]	9,600[d]
0.20	9,090	9,600[d]	9,600[d]	9,600[d]	9,600[d]
0.30	8,160	8,900	9,330	9,600[d]	9,600[d]
0.40	7,390	8,120	8,570	8,860	9,110
0.50	6,660[c]	7,490	7,920	8,000[h]	8,000[h]
0.60	6,060[c]	6,670[h]	6,670[h]	6,670[h]	6,670[h]
0.70	5,570[c]	5,760[h]	5,760[h]	5,760[h]	5,760[h]
0.80[g]	5,000[h]	5,000[h]	5,000[h]	5,000[h]	5,000[h]
Five-Lane Segments					
0.10	12,000[d]	12,000[d]	12,000[d]	12,000[d]	12,000[d]
0.20	11,360	12,000[d]	12,000[d]	12,000[d]	12,000[d]
0.30	10,200	11,120	11,670	12,000[d]	12,000[d]
0.40	9,250[c]	10,000[h]	10,000[h]	10,000[h]	10,000[h]
0.50	8,000[h]	8,000[h]	8,000[h]	8,000[h]	8,000[h]
0.60	6,670[h]	6,670[h]	6,670[h]	6,670[h]	6,670[h]
0.70	5,760[h]	5,760[h]	5,760[h]	5,760[h]	5,760[h]
0.80[g]	5,000[h]	5,000[h]	5,000[h]	5,000[h]	5,000[h]

(F) Type B Weaving Segments—65-mi/h Free-Flow Speed

Volume Ratio, VR	Length of Weaving Segment (ft)				
	500	1,000	1,500	2,000	2,500[a]
Three-Lane Segments					
0.10	6,930	7,050[d]	7,050[d]	7,050[d]	7,050[d]
0.20	6,220	6,670	6,930	7,050[d]	7,050[d]
0.30	5,610	6,090	6,360	6,560	6,700
0.40	5,110	5,590	5,870	6,070	6,220
0.50	4,710	5,170	5,460	5,650	5,810
0.60	4,410	4,850	5,120	5,320	5,470
0.70	4,190	4,620	4,880	5,070	5,230
0.80[g]	3,650[c]	4,460	4,720	4,920	5,000[h]
Four-Lane Segments					
0.10	9,240	9,400[d]	9,400[d]	9,400[d]	9,400[d]
0.20	8,300	8,900	9,240	9,400[d]	9,400[d]
0.30	7,490	8,120	8,480	8,740	8,930
0.40	6,810	7,450	7,830	8,090	8,300
0.50	6,180[c]	6,900	7,280	7,540	7,740
0.60	5,640[c]	6,470	6,670[h]	6,670[h]	6,670[h]
0.70	5,210[c]	5,730	5,760[h]	5,760[h]	5,760[h]
0.80[g]	4,870[c]	5,000[h]	5,000[h]	5,000[h]	5,000[h]
Five-Lane Segments					
0.10	11,550	11,750[d]	11,750[d]	11,750[d]	11,750[d]
0.20	10,370	11,120	11,550	11,750[d]	11,750[d]
0.30	9,360	10,150	10,610	10,930	11,170
0.40	8,540[c]	9,320	9,790	10,000[h]	10,000[h]
0.50	7,720[c]	8,000[h]	8,000[h]	8,000[h]	8,000[h]
0.60	6,670[h]	6,670[h]	6,670[h]	6,670[h]	6,670[h]
0.70	5,760[h]	5,760[h]	5,760[h]	5,760[h]	5,760[h]
0.80[g]	5,000[h]	5,000[h]	5,000[h]	5,000[h]	5,000[h]

(G) Type B Weaving Segments—60-mi/h Free-Flow Speed

Volume Ratio, VR	Length of Weaving Segment (ft)				
	500	1,000	1,500	2,000	2,500[a]
Three-Lane Segments					
0.10	6,540	6,890	6,900[d]	6,900[d]	6,900[d]
0.20	5,900	6,320	6,540	6,700	6,810
0.30	5,340	5,780	6,040	6,210	6,340
0.40	4,880	5,320	5,570	5,760	5,900
0.50	4,520	4,940	5,200	5,380	5,520
0.60	4,230	4,640	4,890	5,080	5,220
0.70	4,020	4,420	4,670	4,850	4,990
0.80[g]	3,520[c]	4,280	4,520	4,700	4,840
Four-Lane Segments					
0.10	8,720	9,190	9,200[d]	9,200[d]	9,200[d]
0.20	7,860	8,420	8,730	8,930	9,090
0.30	7,120	7,710	8,050	8,280	8,450
0.40	6,510	7,090	7,430	7,690	7,860
0.50	5,920[c]	6,590	6,930	7,180	7,370
0.60	5,420[c]	6,190	6,520	6,670[h]	6,670[h]
0.70	5,020[c]	5,520[c]	5,760[h]	5,760[h]	5,760[h]
0.80[g]	4,700[c]	5,000[h]	5,000[h]	5,000[h]	5,000[h]
Five-Lane Segments					
0.10	10,900	11,490	11,500[d]	11,500[d]	11,500[d]
0.20	9,830	10,530	10,910	11,170	11,360
0.30	8,910	9,640	10,070	10,350	10,560
0.40	8,170[c]	8,860	9,290	9,610	9,830
0.50	7,400[c]	8,000[h]	8,000[h]	8,000[h]	8,000[h]
0.60	6,670[h]	6,670[h]	6,670[h]	6,670[h]	6,670[h]
0.70	5,760[h]	5,760[h]	5,760[h]	5,760[h]	5,760[h]
0.80[g]	5,000[h]	5,000[h]	5,000[h]	5,000[h]	5,000[h]

(H) Type B Weaving Segments—55-mi/h Free-Flow Speed

Volume Ratio, VR	Length of Weaving Segment (ft)				
	500	**1,000**	**1,500**	**2,000**	**2,500**[a]
Three-Lane Segments					
0.10	6,160	6,470	6,630	6,750[d]	5,750[d]
0.20	5,570	5,950	6,160	6,300	5,400
0.30	5,070	5,460	5,690	5,850	5,960
0.40	4,640	5,040	5,280	5,450	5,570
0.50	4,310	4,700	4,940	5,100	5,240
0.60	4,050	4,420	4,660	4,830	4,950
0.70	3,870	4,230	4,450	4,620	5,750
0.80[g]	3,390[c]	4,090	4,310	4,480	4,610
Four-Lane Segments					
0.10	8,210	8,620	8,850	9,000[d]	9,000[d]
0.20	7,430	7,930	8,210	8,400	8,540
0.30	6,760	7,290	7,590	7,800	7,950
0.40	6,190	6,730	7,040	7,260	7,430
0.50	5,660[c]	6,260	6,590	6,800	6,990
0.60	5,210[c]	5,900	6,210	6,440	6,610
0.70	4,830[c]	5,280[c]	5,760[h]	5,760[h]	5,760[h]
0.80[g]	4,520[c]	4,950[c]	5,000[h]	5,000[h]	5,000[h]
Five-Lane Segments					
0.10	10,260	10,780	11,060	11,250[d]	11,250[d]
0.20	9,290	9,920	10,260	10,500	10,670
0.30	8,450	9,110	9,490	9,750	9,940
0.40	7,770[c]	8,410	8,810	9,080	9,280
0.50	7,080[c]	7,830	8,000[h]	8,000[h]	8,000[h]
0.60	6,520[c]	6,670[h]	6,670[h]	6,670[h]	6,670[h]
0.70	5,760[h]	5,760[h]	5,760[h]	5,760[h]	5,760[h]
0.80[g]	5,000[h]	5,000[h]	5,000[h]	5,000[h]	5,000[h]

(I) Type C Weaving Segments—75-mi/h Free-Flow Speed and 70 mi/h FFS

Volume Ratio, VR	Length of Weaving Segment (ft)				
	500	1,000	1,500	2,000	2,500[a]
Three-Lane Segments					
0.10	7,200[d]	7,200[d]	7,200[d]	7,200[d]	7,200[d]
0.20	6,580	7,200[d]	7,200[d]	7,200[d]	7,200[d]
0.30	5,880	6,530	6,920	7,190	7,200[d]
0.40	5,330	5,960	6,340	6,610	6,830
0.50[i]	4,890	5,490	5,870	6,140	6,350
Four-Lane Segments					
0.10	9,600[d]	9,600[d]	9,600[d]	9,600[d]	9,600[d]
0.20	8,780	9,600[d]	9,600[d]	9,600[d]	9,600[d]
0.30	7,850	8,710	9,220	9,590	9,600[d]
0.40	7,110	7,950	8,450	8,750	8,750
0.50[i]	6,520	7,000[h]	7,000[h]	7,000[h]	7,000[h]
Five-Lane Segments					
0.10	12,000[d]	12,000[d]	12,000[d]	12,000[d]	12,000[d]
0.20	11,510[c]	12,000[d]	12,000[d]	12,000[d]	12,000[d]
0.30	10,120[c]	11,160[c]	11,530	11,670[h]	11,670[h]
0.40	8,750[h]	8,750[h]	8,750[h]	8,750[h]	8,750[h]
0.50[i]	7,000[h]	7,000[h]	7,000[h]	7,000[h]	7,000[h]

(J) Type C Weaving Segments—65-mi/h Free-Flow Speed

Volume Ratio, VR	Length of Weaving Segment (ft)				
	500	1,000	1,500	2,000	2,500[a]
Three-Lane Segments					
0.10	6,740	7,050[d]	7,050[d]	7,050[d]	7,050[d]
0.20	6,010	6,570	6,870	7,050[d]	7,050[d]
0.30	5,420	5,970	6,310	6,530	6,700
0.40	4,930	5,480	5,810	6,040	6,230
0.50[i]	4,540	5,070	5,400	5,640	5,820
Four-Lane Segments					
0.10	8,980	9,400[d]	9,400[d]	9,400[d]	9,400[d]
0.20	8,020	8,760	9,160	9,400[d]	9,400[d]
0.30	7,230	7,970	8,410	8,710	8,930
0.40	6,570	7,310	7,750	8,060	8,310
0.50[i]	6,060	6,760	7,000[h]	7,000[h]	7,000[h]

(continued)

(J) Type C Weaving Segments—65-mi/h Free-Flow Speed

Volume Ratio, VR	Length of Weaving Segment (ft)				
	500	1,000	1,500	2,000	2,500[a]
Five-Lane Segments					
0.10	11,500[d]	11,500[d]	11,500[d]	11,500[d]	11,500[d]
0.20	10,500[c]	11,320[c]	11,460	11,500[d]	11,500[d]
0.30	9,320[c]	10,180[c]	10,520	10,890	11,170
0.40	8,330[c]	8,750[h]	8,750[h]	8,750[h]	8,750[h]
0.50[i]	7,000[h]	7,000[h]	7,000[h]	7,000[h]	7,000[h]

(K) Type C Weaving Segments—60-mi/h Free-Flow Speed

Volume Ratio, VR	Length of Weaving Segment (ft)				
	500	1,000	1,500	2,000	2,500[a]
Three-Lane Segments					
0.10	6,380	6,830	6,900[d]	6,900[d]	6,900[d]
0.20	5,730	6,230	6,490	6,680	6,830
0.30	5,170	5,690	5,990	6,190	6,350
0.40	5,720	5,240	5,540	5,740	5,910
0.50[i]	4,360	4,850	5,160	5,370	5,540
Four-Lane Segments					
0.10	8,500	9,100	9,200[d]	9,200[d]	9,200[d]
0.20	7,640	8,310	8,660	8,910	9,100
0.30	6,900	7,590	7,990	8,260	8,470
0.40	6,300	6,990	7,380	7,660	7,880
0.50[i]	5,820	6,470	6,880	7,000[h]	7,000[h]
Five-Lane Segments					
0.10	11,250	11,500[d]	11,500[d]	11,500[d]	11,500[d]
0.20	9,980[c]	10,720[c]	10,820	11,140	11,380
0.30	8,880[c]	9,680[c]	9,980	10,330	10,590
0.40	7,980[c]	8,750[h]	8,750[h]	8,750[h]	8,750[h]
0.50[i]	7,000[h]	7,000[h]	7,000[h]	7,000[h]	7,000[h]

(L) Type C Weaving Segments—55-mi/h Free-Flow Speed

Volume Ratio, VR	Length of Weaving Segment (ft)				
	500	1,000	1,500	2,000	2,500[a]
Three-Lane Segments					
0.10	6,010	6,400	6,610	6,740	6,750[d]
0.20	5,420	5,870	6,120	6,270	6,400
0.30	4,930	5,380	5,650	5,850	5,970
0.40	4,510	4,980	5,250	5,450	5,590
0.50[i]	4,180	4,630	4,900	5,100	5,250
Four-Lane Segments					
0.10	8,020	8,540	8,810	8,980	9,000[d]
0.20	7,230	7,830	8,160	8,360	8,540
0.30	6,570	7,180	7,540	7,800	7,970
0.40	6,020	6,640	7,000	7,260	7,450
0.50[i]	5,570	6,180	6,540	6,800	7,000[h]
Five-Lane Segments					
0.10	10,560[c]	11,100[c]	11,020	11,230	11,250[d]
0.20	9,420[c]	10,090[c]	10,200	10,460	10,670
0.30	8,430[c]	9,160[c]	9,420	9,750	9,960
0.40	7,610[c]	8,350[c]	8,750[h]	8,750[h]	8,750[h]
0.50[i]	6,930[c]	7,000[h]	7,000[h]	7,000[h]	7,000[h]

Notes For Appendix I

Used with permission of Transportation Research Board, National Research Council, *Highway Capacity Manual*, 2000, Exhibit 24-8, pgs. 24-10 through 24-18.

For FFS = 70 mi/h, use tables labeled 75 mi/h.

Footnotes:

(a) Weaving segments longer than 2,500 ft are treated as isolated merge and diverge areas using the procedures for ramp junctions outlined in this chapter.

(b) Capacity constrained by basic freeway section capacity.

(c) Capacity occurs under constrained conditions.

(d) Three-lane Type A sections do not operate well at volume ratios greater than 0.45. Poor operations and some local queuing are expected in such situations.

(e) Four-lane Type A sections do not operate well at volume ratios greater than 0.35. Poor operations and some local queuing are expected in such situations.

(f) Capacity constrained by maximum allowable weaving flow rate: 2,800 pc/h (Type A); 4,000 pc/h (Type B); 3,500 pc/h (Type C).

(g) Five-lane Type A sections do not operate well at volume ratios greater than 0.20. Poor operations and some local queuing are expected in such situations.

(h) Type B weaving sections do not operate well at volume ratios greater than 0.80. Poor operations and some local queuing are expected in such situations.

(i) Type C weaving sections do not operate well at volume ratios greater than 0.50. Poor operations and some local queuing are expected in such situations.

Appendix II: Special Cases in Merge and Diverge Analysis

As noted in the body of the chapter, merge and diverge analysis procedures were calibrated primarily for single-lane, right-hand on- and off-ramps. Modifications have been developed so that a broad range of merge and diverge geometries can be analyzed using these procedures. These "special cases" include:

- Two-lane, right-hand on- and off-ramps
- On- and off-ramps on five-lane (one direction) freeway sections
- One-lane, left-hand on- and off-ramps

- Major merge and diverge areas
- Lane drops and lane additions

Each of these special cases is addressed in the sections that follow.

Two-Lane on-Ramps

Figure 13.A.II.1 illustrates the typical geometry of a two-lane on-ramp. Two lanes join the freeway at the merge point. There are, in effect, two acceleration lanes. First, the right ramp lane merges into the left ramp lane; subsequently, the left ramp lane merges into the right freeway lane. The lengths of these two acceleration lanes are as shown in the figure below.

The general procedure for on-ramps is modified in two ways. When estimating the demand flow in lanes 1 and 2 immediately upstream of the on-ramp (v_{12}), the standard equation is used:

$$v_{12} = v_F * P_{FM}$$

Instead of using the standard equations to find P_{FM}, the following values are used:

$$P_{FM} = 1.000 \text{ four-lane freeways}$$

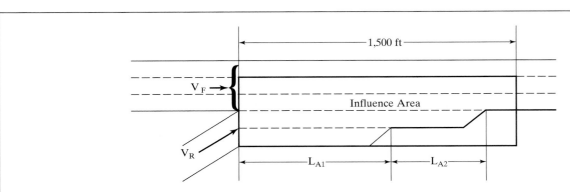

Figure 13.A.II.1: Typical Two-Lane On-Ramp (Used with permission of Transportation Research Board, National Research Council, *Highway Capacity Manual*, 2000, Exhibit 25-8, pg. 25-9.)

$P_{FM} = 0.555$ six-lane freeways

$P_{FM} = 0.209$ eight-lane freeways

In addition, in the density equation, the length of the acceleration lane is replaced by an effective length which considers both lanes of the two-lane merge area:

$$L_{aEFF} = 2L_{A1} + L_{A2}$$

where L_{A1} and L_{A2} are defined in Figure 13.A.II.1.

Occasionally, a two-lane on-ramp will be used at a location where one or two lanes are being added to the downstream freeway section. Depending on the details of such merge areas, they could be treated as lane additions or as major merge areas.

Two-Lane Off-Ramps

Figure 13.A.II.2 illustrates two common geometries used with two-lane off-ramps. The first is a mirror image of a typical two-lane on-ramp junction, with two deceleration lanes provided. The second provides a single deceleration lane, with the left-hand ramp lane originating at the diverge point without a separate deceleration lane.

As was the case with two-lane on-ramps, the standard procedures are applied to the analysis of two-lane off-ramps with two modifications. In the standard equation,

$$v_{12} = v_R + (v_F - v_R)P_{FD}$$

the following values are used for P_{FD}:

$P_{FD} = 1.000$ four-lane freeways

$P_{FD} = 0.450$ six-lane freeways

$P_{FD} = 0.260$ eight-lane freeways

Also, the length of the acceleration lane in the density equation is replaced with an effective length, computed as follows:

$$L_{dEFF} = 2L_{D1} + L_{D2}$$

where L_{D1} and L_{D2} are defined in Figure 13.A.II.2. This modification is applied only in the case of the geometry shown in the first part of Figure 13.A.II.2. Where there is only one deceleration lane, it is used without modification.

On- and Off-Ramps On Five-Lane Freeway Sections

In some areas of the country, freeway sections with five lanes in a single direction are not uncommon. The procedure for analyzing right-hand ramps on such sections is relatively simple: an estimate of the demand flow in lane 5 (the left-most lane) of the section is made. This is deducted from the total approaching freeway flow; the remaining flow is in the right four lanes of the section. Once this deduction is made, the section can be analyzed as if it were a ramp on an eight-lane freeway (four

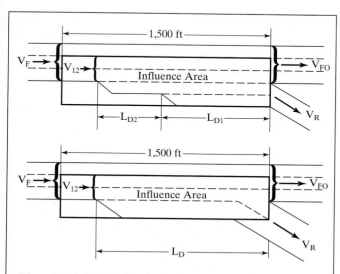

Figure 13.A.II.2: Typical Geometries for Two-Lane Off-Ramps (Used with permission of Transportation Research Board, National Research Council, *Highway Capacity Manual*, 2000, Exhibit 25-15, pg. 25-15.)

Table 13.A.II.1: Estimating Demand Flow in Lane 5 of a Five-Lane Freeway Section

On-Ramps		Off-Ramps	
v_F (pc/h)	v_5 (pc/h)	v_F (pc/h)	v_5 (pc/h)
≥8,500	2,500	≥7,000	$0.200\ v_F$
7,500–8,499	$0.295\ v_F$	5,500–6,999	$0.150\ v_F$
6,500–7,499	$0.270\ v_F$	4,000–5,499	$0.100\ v_F$
5,500–6,499	$0.240\ v_F$	<4,000	0
<5,500	$0.220\ v_F$		

(Used with permission of Transportation Research Board, National Research Council, *Highway Capacity Manual*, 2000, compiled from Exhibits 25-11 and 15-18, pgs. 25-11 and 25-17.)

lanes in one direction). Table 13.A.II.1 gives simple algorithms for determining the flow in lane 5 (v_5). Then:

$$v_{4EFF} = v_F - v_5$$

and the remainder of the problem is analyzed using v_{4EFF} as the approaching freeway flow on a four-lane (one direction) freeway section.

Left-Hand On- and Off-Ramps

Left-hand on- and off-ramps are found, with varying frequency, in most parts of the nation. A technique for modifying analysis procedures for application to left-hand ramps was developed in the 1970s by Leisch (16). The technique follows the following steps:

- Estimate v_{12} for the prevailing conditions as if the ramp were on the right-hand side of the freeway.

- To estimate the traffic remaining in the two left-most lanes of the freeway (v_{12} for a four-lane freeway, v_{23} for a six-lane freeway, v_{34} for an eight-lane freeway), multiply the result by the appropriate factor selected from Table 13.A.II.2.

- Using the demand flow in the two left-most freeway lanes instead of v_{12}, check capacities and estimate density in the ramp influence area without further modification to the methodology.

Table 13.A.II.2: Conversion of v_{12} Estimates for Left-Hand Ramps

$v_{xy} = v_{12} * f_{LH}$		
Adjustment Factor, f_{LH}		
To Estimate:	**For On-Ramps**	**For Off-Ramps**
v_{12} on four-lane freeways (2 lanes ea. dir.)	1.00	1.00
v_{23} on six-lane freeways (3 lanes ea. dir.)	1.12	1.05
v_{34} on eight-lane freeways (4 lanes ea. dir.)	1.20	1.10

- Speed algorithms should be viewed as only very rough estimates for left-hand ramps. Speed predictions for "outer lanes" may not be applied.

Lane Additions and Lane Drops

Many merge and diverge junctions involve the addition of a lane (at a merge area) or the deletion of a lane (at a diverge area). In general, these areas are relatively straightforward to analyze, applying the following general principles:

- Where a single-lane ramp adds a lane (at a merge) or deletes a lane (at a diverge), the capacity of the ramp is determined by its free-flow speed, and it is analyzed as a ramp roadway using the criteria of Table 13.8. Level-of-service criteria for basic freeway sections are applied to upstream and downstream freeway segments, which will have a different number of lanes.
- Where a two-lane ramp results in a lane addition or a lane deletion, it is treated as a major merge or diverge area. The techniques described in the next section are applied.
- Where a lane drop occurs within 2,500 ft of an adjacent upstream lane addition, the entire section is treated as a weaving area.

Major Merge and Diverge Areas

A major merge area is formed when two multilane roadways join to form a single freeway or multilane highway segment. A major diverge area occurs when a freeway or multilane highway segment splits into two multilane downstream roadways. These multilane merge and diverge situations may be part of major freeway interchanges or may involve significant multilane ramp connections to surface streets. The typical characteristic of these roadways is that they are often designed to accommodate relatively high speeds, which somewhat changes the dynamics of merge and diverge operations.

At a major merge area, a lane may be dropped, or the number of lanes in the downstream section may be the same as the total approaching the merge. Similarly, at a diverge area, a lane may be added, or the total lanes leaving the diverge area may be equal to the number on the approaching facility segment. Figure 13.A.II.3 illustrates these configurations.

The analysis of major merge and diverge areas is generally limited to an examination of the demand-capacity balance of approaching and departing facility segments. No level-of-service criteria are applied.

For major diverge areas, an algorithm has been developed to roughly estimate the density across all approaching freeway lanes for a segment 1,500 ft upstream of the diverge:

$$D = 0.0109 \left(\frac{v_F}{N} \right)$$

where: D = density across all freeway lanes, from diverge to a point 1,500 ft upstream of the diverge, pc/mi/ln

v_F = approaching freeway demand flow, pc/h

N = number of freeway lanes approaching the diverge

This is an approximation at best and is not used to assign a level of service to the diverge area.

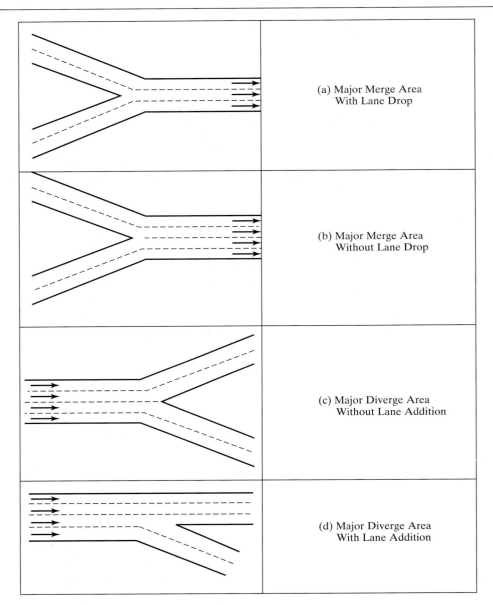

Figure 13.A.II.3: Major Merge and Diverge Areas (Used with permission of Transportation Research Board, National Research Council, *Highway Capacity Manual*, 2000, Exhibits 25-9 and 25-10, pg. 25-10, and Exhibits 15-16 and 15-17, pg. 25-16.)

CHAPTER
14

Two-Lane, Two-Way Rural Highways

14.1 Introduction

A significant portion of the nation's almost four million miles of paved highway are considered to be "rural." Of these, a very large percentage are two-lane highways, (i.e., one lane for traffic in each direction). On such highways, passing takes place in the opposing lane when the sight distance and opposing traffic conditions permit. Because of these passing operations, the two-lane, two-way rural highway is the only type of highway link on which traffic in one direction has a distinct operational impact on traffic in the other direction.

These roadways range from heavily traveled intercity routes to sparsely traveled links to isolated areas. They provide a vast network connecting the fringes of urban areas, agricultural regions, resource development areas, and remote outposts. Because of the varied functions they serve, two-lane rural highways are built to widely varying geometric standards and display a wide range of operating characteristics.

Rural two-lane highways serve two primary functions in the nation's highway network:

- Mobility
- Accessibility

As part of state and county primary highway systems, they serve a critical mobility function. Large numbers of road users rely on these highways for regular trips of significant length. Design standards for this type of two-lane highway generally reflect their use in serving higher demand volumes. Higher design speeds reflect the primary mobility service provided.

Many two-lane rural highways, however, serve low volumes, sometimes under 100 vehs/day. The primary function of such highways is to provide for basic all-weather access to remote or sparsely-developed areas. Because such highways are not used by large numbers of people or vehicles, their design speeds and related geometric features are often not a major concern.

Because of the broad diversity of use on these highways, the 2000 edition of the *Highway Capacity Manual* (HCM 2000) [1] has created two distinct classes of rural two-lane, two-way highways:

- *Class I.* These are highways on which motorists expect to travel at relatively high speeds, including major intercity routes, primary arterials, and daily commuter routes.
- *Class II.* These are highways on which motorists do not necessarily expect to travel at high speeds, including access routes, scenic and recreational routes that are not primary arterials, and routes through rugged terrain.

Class I two-lane highways serve primarily mobility needs, while Class II two-lane highways serve primarily access needs.

Even this categorization does not completely describe the diversity in the "look and feel" of such highways. Routes through rugged terrain are classified as Class II, primarily because the terrain limits the geometry of the roadway, forcing low-speed operation and providing few or no passing opportunities. Nevertheless, some of these roads must serve mobility needs where demand is sufficient.

The American Association of State Highway and Transportation Officials (AASHTO) classifies two-lane rural highways as "rural local roads," "rural collectors," or "rural arterials." Unfortunately, these categories overlap the HCM classifications. Virtually all rural arterials would be Class I facilities. Rural collectors, however, could fall into either HCM class depending upon the specifics of terrain and geometry. Virtually all rural local roads would fall into Class II. Some judgment, therefore, is needed to properly classify rural highways.

Figure 14.1 contains four illustrations of rural two-lane highways, together with their likely classifications according to the HCM and AASHTO.

Class I highways generally feature gentle geometries allowing for higher speed operation, as illustrated in Figures 14.1 (a) and (b), usable paved shoulders and/or stabilized roadside recovery areas, and full pavement markings and signage. Figure 14.1 (c) illustrates a class II rural collector. The rural local road depicted in Figure 14.1 (d) is clearly a Class II facility. It lacks pavement markings and usable shoulders, and low-type pavements are evident.

Because of the wide diversity in the function and physical characteristics of rural roadways, both design standards and level of service criteria must be flexible and must address the full range of situations in which rural two-lane, two-way highways exist.

This chapter provides an overview of the following subjects related to two-lane, two-way rural highways:

- Design standards
- Passing sight distance requirements and the impact of "No Passing" zones
- Capacity and level of service analysis of two-lane highways

Each of these is treated in the major sections that follow.

14.2 Design Standards

Design standards for urban, suburban, and rural highways are set by AASHTO in the current version of *A Policy on Geometric Design of Highways and Streets* [2], often referred to as the "Green Book," because of the color of its cover in recent editions. The latest version of the Policy is the fourth edition, released in early 2001.

The single most important design factor controlling the specifics of geometry is the design speed. Every element of a highway (horizontal alignment, vertical alignment, cross-section) must be designed to allow safe operation at the "design speed." Thus, a design speed of 60 mi/h, for example, requires that every element and segment of the facility must allow for safe operation at 60 mi/h. In Chapter 3, the impacts of speed on horizontal and vertical alignment is discussed in detail.

Table 14.1 shows recommended design speeds for rural two-lane highways based on function, classification, terrain, and ADT demand volumes.

Figure 14.2 shows reasonable recommended design criteria for maximum grades on two-lane, two-way rural highways. Maximum grade relates to the function of the facility, its design speed, and the terrain in which

(a) A Class I Rural Arterial

(b) A Class I Rural Collector

(c) A Class II Rural Collector

(d) A Class II Rural Local Road

Figure 14.1: Rural Two-Lane Highways Illustrated

Table 14.1: Recommended Minimum Design Speeds for Rural Two-Lane Highways (mi/h)

Type of Facility	ADT (veh/day)	Minimum Design Speed in		
		Level Terrain	Rolling Terrain	Mountainous Terrain
Rural Local Roads	**<50**	30	20	20
	50–249	30	30	20
	250–399	40	30	20
	400–1499	50	40	30
	1500–1999	50	40	30
	≥2000	50	40	30
Rural Collectors	**<400**	40	30	20
	400–2000	50	40	30
	>2000	60	50	40
Rural Arterials	**All**	60	50	40

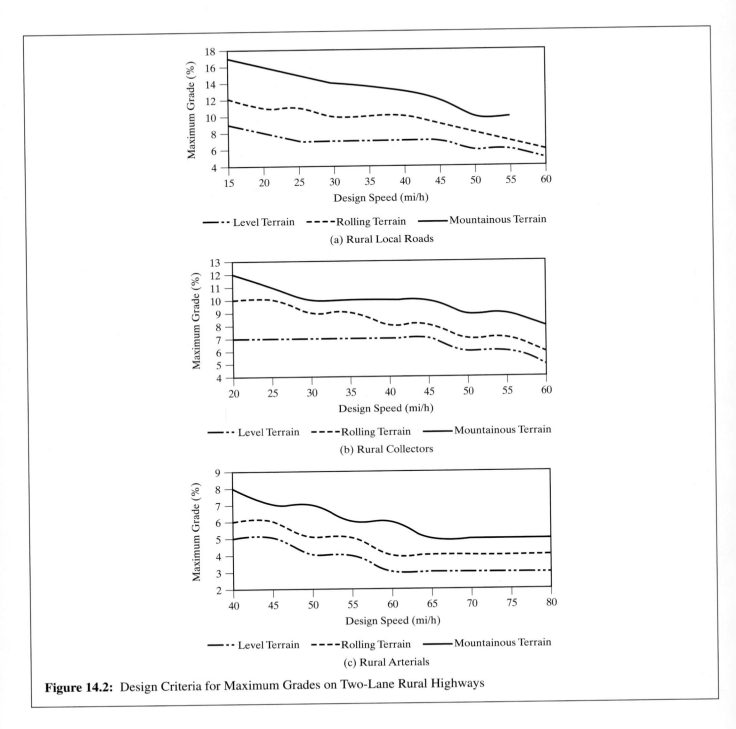

Figure 14.2: Design Criteria for Maximum Grades on Two-Lane Rural Highways

it is located. The design speed, once selected, limits horizontal and vertical alignment. Consult Chapter 3 for specific relationships between design speed and horizontal and vertical curvature.

It should be noted that the criteria of Table 14.1 and Figure 14.2 represent *minimum* recommended design speeds and *maximum* grades. When conditions permit, higher design speeds and less severe grades should be used.

14.3   Passing Sight Distance on Rural Two-Lane Highways

In Chapters 2 and 3, the importance of safe stopping sight distance was discussed and illustrated. The safe stopping sight distance *must* be provided at every point on every roadway for the selected design speed. No driver must ever be confronted with a sudden obstacle on the roadway and insufficient distance (and time) to stop before colliding with it.

On two-lane highways, another critical safety feature is the passing sight distance. Because vehicles pass by using the opposing traffic lane, passing maneuvers on two-lane rural highways are particularly dangerous. Passing sight distance is the minimum sight distance required to safely begin and complete a passing maneuver under the assumed conditions for the highway.

Passing sight distance need not be provided at every point along a two-lane rural highway. Passing should not, however, be permitted where passing sight distance is not available. Roadway markings and "No Passing" zone signs are used to prohibit passing where sight distance is insufficient to do so safely. It should be noted that on unmarked rural two-lane highways, drivers *may not* assume that passing sight distance is available. On unmarked facilities, driver judgment is the only control on passing maneuvers.

AASHTO criteria for minimum passing sight distances are based on a number of assumptions concerning driver behavior. These assumptions accommodate most drivers and are not based on averages.

The minimum passing sight distance is determined as the sum of four component distances, as illustrated in Figure 14.3. The component distances are defined as:

d_1 = distance traversed during perception and reaction time and during the initial acceleration to the point of encroachment on the left lane.

d_2 = distance traveled while the passing vehicle occupies the left lane.

d_3 = distance between the passing vehicle at the end of its maneuver and the opposing vehicle.

d_4 = distance traversed by the opposing vehicle for two-thirds of the time the passing vehicle occupies the left lane, or two-thirds of d_2 above.

Criteria for the speeds and distances used in determining passing sight distances are based on extensive field studies [3] and validations [4, 5]. The component distances are computed as follows.

$$d_1 = 1.47t_1\left(S - m + \frac{at_1}{2}\right) \qquad (14\text{-}1)$$

where: t_1 = reaction time, s

   S = speed of passing vehicle, mi/h

   m = difference between speed of passing vehicle and passed vehicle, mi/h

   a = average acceleration of passing vehicle, mi/h/s

$$d_2 = 1.47\, St_2 \qquad (14\text{-}2)$$

where: t_2 = time passing vehicle occupies the left lane, s

   S = speed of the passing vehicle, mi/h

$$d_3 = 100 - 300 \text{ ft} \qquad (14\text{-}3)$$

$$d_4 = \left(\frac{2}{3}\right)d_2 \qquad (14\text{-}4)$$

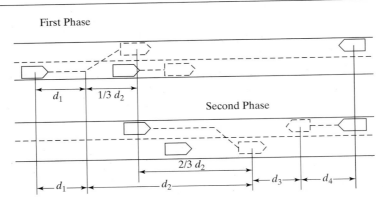

Figure 14.3: Elements of Passing Sight Distance (Used with permission of the American Association of State Highway and Transportation Officials, *A Policy on Geometric Design of Highways and Streets*, 4th Edition, Washington DC, 2001, Exhibit 3-4, pg. 119.)

In using these equations, AASHTO assumes that the speed of the passing vehicle is 10 mi/h (m) greater than the speed of the passed vehicle. Acceleration rates between 1.4 and 1.5 mi/h/s are used. Reaction times range between 3.6 and 4.5 s (t_1). The time that the left lane is occupied is based upon speed parameters, and ranges between 9.3 and 11.3 s (t_2). The clearance distance range is as shown, with the lower end of the scale used for slower speeds.

To be useful in design, safe passing sight distances must be related to the design speed of the facility. Table 14.2 shows AASHTO passing sight distance criteria based on design speed. The assumed passed and passing vehicle speeds are such that they accommodate the vast majority of potential passing maneuvers occurring on a two-lane rural highway.

It is always desirable to have passing sections provided as frequently as possible. The effects of "No Passing" zones on operations are made clear by the capacity and level-of-service models presented in the following section. Where passing is permitted, it is also desirable to provide as much sight distance as is practical, using the values of Table 14.2 as minimums.

The design criteria of Table 14.2 are not the same as the warrants for signing and marking "No Passing" zones presented in the MUTCD [6]. The MUTCD requirements, shown in Table 14.3, are substantially less than the design criteria. The assumptions used in establishing the warrants are different from those described for design standards. When a centerline is used, and where sight distances are less than the criteria of Table 14.3, "No Passing" zone markings and signs must be installed. A more conservative, but safer, practice would be to post "No Passing" signs and markings wherever the design criteria for passing sight distance are not met.

14.4   Capacity and Level-of-Service Analysis of Two-Lane Rural Highways

The HCM 2000 methodology for the analysis of two-lane, two-way rural highways is a new procedure based on an extensive research study conducted at the Midwest Research Institute [7]. It builds conceptually on the methodology in use since 1985 but introduces a number of new elements.

One of the principal difficulties in studying the behavior of two-lane rural highways is the fact that few operate under conditions that are even near capacity. Thus, research into two-lane highway operations relies heavily on simulation modeling in addition to limited field observations.

Among the unique features of this methodology are the definition of two classes of rural two-lane highways

Table 14.2: Design Values for Passing Sight Distance on Two-Lane Highways

Design Speed (mi/h)	Assumed Speeds (mi/h)		Passing Sight Distance (ft)	
	Passed Veh	Passing Veh	Exact	Rounded for Design
20	18	28	706	710
25	22	32	897	900
30	26	36	1,088	1,090
35	30	40	1,279	1,280
40	34	44	1,470	1,470
45	37	47	1,625	1,625
50	41	51	1,832	1,835
55	44	54	1,984	1,985
60	47	57	2,133	2,135
65	50	60	2,281	2,285
70	54	64	2,479	2,480
75	56	66	2,578	2,580
80	58	68	2,677	2,680

(Used with permission of the American Association of State Highway and Transportation Officials, *A Policy on Geometric Design of Highways and Streets,* 4th Edition, Washington DC, 2001, Exhibit 3-7, pg. 124.)

Table 14.3: Minimum Passing Sight Distance Criteria for Placement of "No Passing" Markings and Signs

85th Percentile Speed or Statutory Speed Limit (mi/h)	Minimum Passing Sight Distance (ft)
25	450
30	500
35	550
40	600
45	700
50	800
55	900
60	1,000
65	1,100
70	1,200

(Used with Permission of Federal Highway Administration, U.S. Department of Transportation, *Manual of Uniform Traffic Control Devices,* Millennium Edition, Washington DC, 2000, Table 3B-1, pg. 3B-9.)

(described previously), the use of two measures of effectiveness to define levels of service, and the ability to analyze two-way or one-way sections under a variety of conditions.

14.4.1 Capacity

The capacity of a two-lane highway has been significantly increased from previous values to accommodate frequent, higher observations throughout North America. The capacity of a two-lane highway under base conditions is now established as 3,200 pc/h in both directions, with a maximum of 1,700 pc/h in one direction. The base conditions for which this capacity is defined include:

- 12-foot (or greater) lanes
- 6-foot (or greater) usable shoulders
- Level terrain
- No heavy vehicles
- 100% passing sight distance available (no "No Passing" zones)
- 50/50 directional split of traffic
- No traffic interruptions

As with all capacity values, these standards reflect "reasonable expectancy" (i.e., most two-lane highway segments operating under base conditions should be able to achieve such capacities most of the time). Isolated observations of higher volumes do not negate the standard. In new procedures for analysis of two-lane highways, level of service F exists only when adjusted demand flow rates exceed the capacity of a segment.

14.4.2 Level of Service

Level of service for two-lane rural highways is defined in terms of two measures of effectiveness:

- Average travel speed (ATS)
- Percent time spent following (PTSF)

Average travel speed is the average speed of all vehicles traversing the defined analysis segment for the specified time period, which is usually the peak 15-minutes of a peak hour. When analysis of both directions is used, the average travel speed includes vehicles in both directions. When analysis of single direction is used, the average travel speed includes those vehicles in the analysis direction only.

Percent time spent following is similar to "percent time delay," which was used in the 1985 through 1997 *Highway Capacity Manuals*. It is the aggregate percentage of time that all drivers spend in queues, unable to pass, with the speed restricted by the queue leader. A surrogate measure for PTSF is the percentage of vehicles following others at headways of 3.0 s or less.

Level of service criteria for two-lane rural highways are shown in Table 14.4. The criteria vary for Class I and Class II highways. Class II highways, where mobility is not a principal function, use only the PTSF criteria for determination of level of service. For Class I highways, the LOS is determined by the measure yielding the poorest result.

Figure 14.4 illustrates the relationships between ATS, PTSF, and two-way flow rate on a two-lane highway with base conditions. Figure 14.4 (b) clearly illustrates the unique nature of operations on a two-lane highway. For multilane highways and freeways, operational deterioration does not occur until *v/c* ratios are quite high. Drivers on such facilities maintain high speeds in the vicinity of free-flow speed for *v/c* ratios in excess of 0.75. On a two-lane highway, however, operational deterioration, particularly with respect to PTSF, occurs at relatively low v/c ratios.

As illustrated in Figure 14.4 (b), at a demand flow of 1,500 pc/h (a *v/c* ratio of 1500/3200 = 0.47), PTSF is already at 64%. This is for a highway with base, or nearly ideal, conditions. As the analysis methodology makes clear, the value would be considerably higher where conditions are worse than those defined for the base.

Table 14.4: Level-of-Service Criteria for Two-Lane Rural Highways

Level of Service	Class I Facilities		Class II Facilities
	Average Travel Speed (mi/h)	**Percent Time Spent Following (%)**	**Percent Time Spent Following (%)**
A	>55	≤35	≤40
B	>50–55	>35–50	>40–55
C	>45–50	>50–65	>55–70
D	>40–45	>65–80	>70–85
E	≤40	>80	>85

Note: Level of service F occurs whenever the demand flow rate exceeds the segment capacity.
(Used with permission of Transportation Research Board, National Research Council, *Highway Capacity Manual*, 4th Edition, Washington DC, 2000, Exhibits 20-2 and 20-4, pgs. 20-3 and 20-4.)

This characteristic explains why few two-lane highway segments are observed operating at flows near capacity. Long before demands approach capacity, operational quality has seriously deteriorated. On two-lane highways, this can lead to safety problems, as drivers attempt unsafe passing maneuvers to avoid these delays. Most two-lane highways, therefore, are reconfigured or re-constructed because of such problems, which occur long before demand levels approach the capacity of the facility.

As will be seen as a result of the analysis methodology, poor levels of service on two-lane highways can exist at extremely low demand flow rates. No other type of traffic facility exhibits this characteristic.

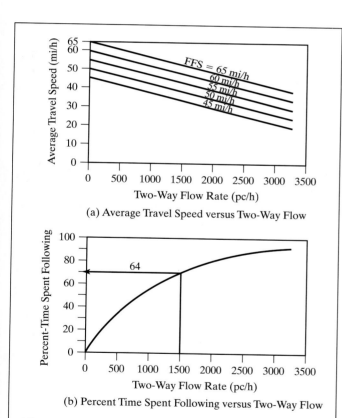

(a) Average Travel Speed versus Two-Way Flow

(b) Percent Time Spent Following versus Two-Way Flow

Figure 14.4: Flow Versus Average Travel Speed and Percent Time Spent Following for Two-Lane Highways (Used with permission of Transportation Research Board, National Research Council, *Highway Capacity Manual*, 4th Edition, Washington DC, 2000, Exhibit 12-6, pg. 12-14.)

14.4.3 Types of Analysis

Previous editions of the HCM required that analysis of two-lane rural highways operations be limited to a composite analysis of both directions. This requirement reflected the operational interaction between directional flows on a two-lane highway. Despite this, there are many occasions where operational characteristics in the two directions of a two-lane highway varied considerably, particularly where significant grades were involved. Because of this, the HCM 2000 allows for both two-direction and single-direction analysis. Three distinct methodologies are provided:

- Two-directional analysis of general extended sections (≥ 2.0 mi) in level or rolling terrain
- Single-directional analysis of general extended sections (≥ 2.0 mi) in level or rolling terrain
- Single-direction analysis of specific grades

For specific grades, only single-direction analysis of the upgrade and downgrade is permitted, as these tend to differ significantly. In what is usually referred to as "mountainous" terrain, all analysis is on the basis of specific grades comprising that terrain. Any grade of 3% or more and at least 0.6 mi long must be addressed using specific grade procedures.

14.4.4 Free-Flow Speed

As was the case for multilane highways and freeways, the free-flow speed of a two-lane highway is a significant variable used in estimating expected operating conditions. HCM 2000 recommends that free-flow speeds be measured in the field where practical, but also offers a methodology for their estimation where measurement is not practical.

Field Measurement of Free-Flow Speed

The free-flow speed of a two-lane rural highway may be measured directly in the field. The speed study should be conducted at a representative site within the study section. Free-flow speeds may be directly measured as follows:

- A representative speed sample of 100 or more vehicles should be obtained.

- Total two-way traffic flow should be 200 pc/h or less.

- All vehicle speeds should be observed during the study period, or a systematic sampling (such as 1 vehicle out of every 10) should be applied.

- When two-direction analysis is contemplated, the speed sample should be selected from both directions of flow; when a one-direction analysis is contemplated, the speed sample should be selected only from the direction under study.

If field measurements must be made at total flow levels higher than 200 pc/h, the free-flow speed may be estimated as:

$$FFS = S_m + 0.00776\left(\frac{v_f}{f_{HV}}\right) \qquad (14\text{-}5)$$

where: FFS = free-flow speed for the facility, mi/h

S_m = mean speed of the measured sample (where total flow > 200 pc/h), mi/h

v_f = observed flow rate for the period of the speed sample, veh/h

f_{HV} = heavy vehicle adjustment factor

Chapter 9 details procedures for measuring speeds in the field.

Estimating Free-Flow Speeds

If field observation of free-flow speed is not practical, free-flow speed on a two-way rural highway may be estimated as follows:

$$FFS = BFFS - f_{LS} - f_A \qquad (14\text{-}6)$$

where: FFS = free-flow speed for the facility, mi/h

$BFFS$ = base free-flow speed for the facility, mi/h

f_{LS} = adjustment for lane and shoulder width, mi/h

f_A = adjustment for access point density, mi/h

Unfortunately, the HCM 2000 does not provide any detailed criteria for the base free-flow speed, $BFFS$. It is limited to a range of 45–65 mi/h, with Class I highways usually in the 55–65 mi/h range and Class II highways usually in the 45–50 mi/h range. Design speed and statutory speed limits may be used as inputs to establishing an appropriate value for $BFFS$. The design speed, which represents the maximum safe speed for the horizontal and vertical alignment of the highway, is a reasonable surrogate for the $BFFS$. Adjustment factors account for the impact of lane and shoulder width and access points, none of which are accounted for in the basic design speed of the facility. Speed limits are not as good a guide, as various jurisdictions apply different philosophies and policies in setting such limits.

Adjustment factors for lane and shoulder width are shown in Table 14.5; adjustment factors for access point density are shown in Table 14.6.

Access point density is computed by dividing the total number of driveways and intersections on both sides of the highway by the total length of the segment in miles.

A Sample Problem

Find the free-flow speed of a two-lane rural highway segment in rolling terrain. The lane width is 11.0 ft with 2.0-ft shoulders. Access point density is 30 per mile, and the base free-flow speed may be taken to be 60 mi/h.

The estimation technique is used. The free-flow speed is determined as follows:

$$FFS = BFFS - f_{LS} - f_A$$

where: $BFFS$ = 60 mi/h (given)

f_{LS} = 3.0 mi/h (Table 14.5)

f_A = 7.5 mi/h (Table 14.6)

$$FFS = 60.0 - 3.0 - 7.5 = 49.5 \text{ mi/h}$$

Table 14.5: Free-Flow Speed Adjustments for Lane and Shoulder Width

Lane Width (ft)	Reduction in FFS (mi/h), f_{LS}			
	Shoulder Width (ft)			
	≥0 < 2	≥2 < 4	≥4 < 6	≥6
≥9 < 10	6.4	4.8	3.5	2.2
≥10 < 11	5.3	3.7	2.4	1.1
≥11 < 12	4.7	3.0	1.7	0.4
≥12	4.2	2.6	1.3	0.0

(Used with permission of Transportation Research Board, National Science Council, *Highway Capacity Manual*, 4th Edition, Washington DC, 2000, Exhibit 20-5, pg. 20-6.)

Table 14.6: Free-Flow Speed Adjustments for Access Point Density

Access Points per Mile	Reduction in *FFS* (mi/h) f_A
0	0.0
10	2.5
20	5.0
30	7.5
40	10.0

(Used with permission of Transportation Research Board, National Science Council, *Highway Capacity Manual*, 4th Edition, Washington DC, 2000, Exhibit 20-6, pg. 20-6.)

Note that in selecting the lane and shoulder width adjustment, f_{LS}, care must be taken to observe the boundary conditions in Table 14.5. The lane width of 11.0 ft falls within the "≥11 < 12" category, while the 2.0 ft shoulders fall within the "≥2 < 4" category.

14.4.5 Estimating Demand Flow Rate

As for most HCM 2000 methodologies, a critical computational step is the determination of a demand flow rate reflecting the base conditions for the facility type being analyzed. This requires that an hourly volume reflecting prevailing conditions be adjusted to reflect peak flow rates within the hour and base conditions.

For two-lane rural highways, this adjustment is made as follows:

$$v = \frac{V}{PHF * f_{HV} * f_G} \qquad (14\text{-}7)$$

where: v = demand flow rate, pc/h

V = hourly demand volume under prevailing conditions, veh/h

PHF = peak hour factor

f_{HV} = adjustment for heavy vehicle presence

f_G = adjustment for grades

The HCM 2000 methodology, however, introduces several levels of complexity to this seemingly simple determination:

1. Because there are *two* measures of effectiveness to be determined (ATS, PTSF), there are *two* such conversions to be made based on *two* different sets of adjustment factors (f_{HV} and f_G) for each measure.

2. When doing two-direction analysis, the total volume is converted to a total demand flow rate. When conducting a one-direction analysis, the subject direction volume and the opposing direction volume must be separately converted.

3. The determination of demand flow rate is iterative, as adjustment factors f_{HV} and f_G vary with demand flow rate.

In the most complex case of a one-direction analysis, there are four different determinations of demand flow rate: two demand flow rates in *each direction*, one for ATS determination, and one for PTSF determination. Each of the four determinations is also iterative. Fortunately, all iterations are limited to a single recomputation, as will be seen.

Determining Grade Adjustment Factors

For every computation, two grade adjustment factors will be required: one for the ATS determination and one for the PTSF determination. Selection of appropriate adjustment factors also depends upon the type of analysis being conducted. Grade adjustment factors are found as follows:

- Two-direction analysis of general terrain segments for both ATS and PTSF determinations: Table 14.7.
- One-direction analysis of general terrain segments for both ATS and PTSF determinations: Table 14.7.
- One-direction analysis of specific upgrades for ATS determination: Table 14.8.
- One-direction analysis of specific upgrades for PTSF determination: Table 14.9.
- One-direction analysis of specific downgrades for both ATS and PTSF determination: Table 14.7.

Iteration is necessary in all of these exhibits as the selection of adjustment factors requires an assumption of the demand flow rate to be determined. Because of the step-function nature of the adjustments, it is possible to obtain mutually exclusive estimates (i.e., assumption of demand flow rate category A results in a demand flow rate in category B, while assumption of category B results in a demand flow rate in category A). In such cases, the following iteration rule is applied: *the computation resulting in a demand flow rate that is below the highest value of the assumed range is used.*

For two-direction analysis, the flow ranges in the first column are used; for one-direction analysis, the flow ranges in the second column are used. The iteration mode is better illustrated by the following example:

A two-direction analysis is underway. An initial estimate that the demand flow rate will wind up in the "0–600 pc/h" category is made. The result, however, is a demand flow rate of 850 pc/h, which is not in the assumed category. The next iteration starts with an assumption that the resulting demand flow rate will be in the ">600–1,200 pc/h"

Table 14.7: Grade Adjustment Factor (f_G) for General Terrain Segments and Specific Downgrades (ATS and PTSF Determinations)

Range of Two-Way Flow (pc/h)	Range of One-Way Flow (pc/h)	For ATS Determination		For PTSF Determination	
		Level Terrain or Specific Downgrades	Rolling Terrain	Level Terrain or Specific Downgrades	Rolling Terrain
0–600	0–300	1.00	0.71	1.00	0.77
>600–1200	>300–600	1.00	0.93	1.00	0.94
>1200	>600	1.00	0.99	1.00	1.00

(Used with permission of Transportation Research Board, National Science Council, *Highway Capacity Manual*, 4th Edition, Washington DC, 2000, Exhibits 20-7 and 20-8, pg. 20-7.)

Table 14.8:   Grade Adjustment Factor (f_G) for Specific Upgrades: ATS Determinations

Grade (%)	Length of Grade (mi)	Range of Direction Flow Rate (pc/h)		
		0–300	>300–600	>600
≥3.0 < 3.5	0.25	0.81	1.00	1.00
	0.50	0.79	1.00	1.00
	0.75	0.77	1.00	1.00
	1.00	0.75	1.00	1.00
	1.50	0.75	0.99	1.00
	2.00	0.75	0.97	1.00
	3.00	0.75	0.95	0.97
	≥4.00	0.75	0.94	0.95
≥3.5 < 4.5	0.25	0.79	1.00	1.00
	0.50	0.76	1.00	1.00
	0.75	0.72	1.00	1.00
	1.00	0.69	0.93	1.00
	1.50	0.68	0.92	1.00
	2.00	0.66	0.91	1.00
	3.00	0.65	0.91	0.96
	≥4.00	0.65	0.90	0.94
≥4.5 < 5.5	0.25	0.75	1.00	1.00
	0.50	0.65	0.93	1.00
	0.75	0.60	0.89	1.00
	1.00	0.59	0.89	1.00
	1.50	0.57	0.86	0.99
	2.00	0.56	0.85	0.98
	3.00	0.56	0.84	0.97
	≥4.00	0.56	0.82	0.93
≥5.5 < 6.5	0.25	0.63	0.91	1.00
	0.50	0.57	0.85	0.99
	0.75	0.52	0.83	0.97
	1.00	0.51	0.79	0.97
	1.50	0.49	0.78	0.95
	2.00	0.48	0.78	0.94
	3.00	0.46	0.76	0.93
	≥4.00	0.45	0.76	0.93
≥6.5	0.25	0.59	0.86	0.98
	0.50	0.48	0.76	0.94
	0.75	0.44	0.74	0.91
	1.00	0.41	0.70	0.91
	1.50	0.40	0.67	0.91
	2.00	0.39	0.67	0.89
	3.00	0.39	0.66	0.88
	≥4.00	0.38	0.66	0.87

Note: Straight line interpolation of f_G for length of grade permitted to nearest 0.01.
(Used with permission of Transportation Research Board, National Science Council, *Highway Capacity Manual*, 4th Edition, Washington DC, 2000, Exhibit 20-13, pg. 20-15.)

Table 14.9: Grade Adjustment Factor (f_G) for Specific Upgrades: PTSF Determinations

Grade (%)	Length of Grade (mi)	Range of Direction Flow Rate (pc/h)		
		0–300	>300–600	>600
≥3.0 < 3.5	0.25	1.00	0.92	0.92
	0.50	1.00	0.93	0.93
	0.75	1.00	0.93	0.93
	1.00	1.00	0.93	0.93
	1.50	1.00	0.94	0.94
	2.00	1.00	0.95	0.95
	3.00	1.00	0.97	0.96
	≥4.00	1.00	1.00	0.97
≥3.5 < 4.5	0.25	1.00	0.94	0.92
	0.50	1.00	0.97	0.96
	0.75	1.00	0.97	0.96
	1.00	1.00	0.97	0.97
	1.50	1.00	0.97	0.97
	2.00	1.00	0.98	0.98
	3.00	1.00	1.00	1.00
	≥4.00	1.00	1.00	1.00
≥4.5 < 5.5	0.25	1.00	1.00	0.97
	≥0.50	1.00	1.00	1.00
≥5.5 < 6.5	All	1.00	1.00	1.00
≥6.5	All	1.00	1.00	1.00

Note: Straight-line interpolation of f_G for length of grade permitted to nearest 0.01.
(Used with permission of Transportation Research Board, National Science Council, *Highway Capacity Manual*, 4th Edition, Washington DC, 2000, Exhibit 20-14, pg. 20-16, *format modified*.)

category. It, however, results in a demand flow rate of 550 pc/h. By the rules of iteration for this procedure, the second computation is accepted, as 550 pc/h is less than the maximum of 1,200 pc/h in the assumed range.

Mutually exclusive estimates of demand flow rate usually occur when conducting analysis of rolling terrain segments, where the step-function difference between adjustment factors in the "0–600" and ">600–1,200" flow categories is substantial. Such situations will almost never arise when analyzing level terrain segments, and rarely arise when dealing with specific grade segments.

Determining the Heavy-Vehicle Adjustment Factor

The heavy-vehicle adjustment factors for ATS and PTSF determinations are found from passenger-car equivalents as follows:

$$f_{HV} = \frac{1}{1 + P_T(E_T - 1) + P_R(E_R - 1)} \quad (14\text{-}8)$$

where: f_{HV} = heavy-vehicle adjustment factor

P_T = proportion of trucks and buses in the traffic stream

P_R = proportion of recreational vehicles in the traffic stream

E_T = passenger-car equivalent for trucks and buses

E_R = passenger-car equivalent for recreational vehicles

As in multilane methodologies, the *passenger-car equivalent* is the number of passenger cars displaced by one truck (or RV) under the prevailing conditions on the analysis segment.

As in the determination of the grade-adjustment factor, values of E_T and E_R depend upon initial estimates of the demand flow rate and are therefore iterative. Iteration rules are the same as described for the grade-adjustment factor. Passenger-car equivalents also depend upon which measure of effectiveness is being predicted (ATS or PTSF), and the type of analysis being applied. Passenger-car equivalents are found from the following tables:

- Two-direction analysis of general terrain segments for both ATS and PTSF determination: Table 14.10.

- One-direction analysis of general terrain segments for both ATS and PTSF determination: Table 14.10.

- One-direction analysis of specific upgrades for ATS determination: *Trucks:* Table 14.11; *RV's:* Table 14.12.

- One-direction analysis of specific upgrades for PTSF determination: Table 14.13.

- One-direction analysis of specific downgrades: Table 14.10.

Some specific downgrades are steep enough to require some trucks to shift into low gear and travel at crawl speeds to avoid loss of control. In such situations, the effect of trucks traveling at crawl speed may be taken into account by replacing Equation 14-8 with the following when computing the heavy vehicle adjustment factor, f_{HV}, for ATS determination:

$$f_{HV} = \frac{1}{1 + P_{TC}*P_T(E_{TC} - 1) + (1 - P_{TC})*P_T(E_T - 1) + P_R(E_R - 1)} \quad (14\text{-}9)$$

where: P_{TC} = proportion of heavy vehicles forced to travel at crawl speeds

Table 14.10: Passenger-Car Equivalents for General Terrain Segments: ATS and PTSF Determinations

| Vehicle Type | Range of Two-Way Flows (pc/h) | Range of One-Way Flows (pc/h) | For ATS Determination | | For PTSF Determination | |
			Level Terrain or Specific Downgrades	Rolling Terrain	Level Terrain or Specific Downgrades	Rolling Terrain
Trucks and Buses	0–600	0–300	1.7	2.5	1.1	1.8
E_T	≥600–<1,200	≥300–<600	1.2	1.9	1.1	1.5
	≥1200	≥600	1.1	1.5	1.0	1.0
Recreational Vehicles	0–600	0–300	1.0	1.1	1.0	1.0
E_R	≥600–<1,200	≥300–<600	1.0	1.1	1.0	1.0
	≥1,200	≥600	1.0	1.1	1.0	1.0

(Used with permission of Transportation Research Board, National Science Council, *Highway Capacity Manual*, 4th Edition, Washington DC, 2000, Exhibits 20-9 and 20-10, pg. 20-8, *format modified.*)

Table 14.11:   Passenger-Car Equivalents of Trucks for Specific Upgrades: ATS Determination

Grade (%)	Length of Grade (mi)	Range of Direction Flow Rate (pc/h)		
		0–300	>300–600	>600
≥3.0 < 3.5	0.25	2.5	1.9	1.5
	0.50	3.5	2.8	2.3
	0.75	4.5	3.9	2.9
	1.00	5.1	4.6	3.5
	1.50	6.1	5.5	4.1
	2.00	7.1	5.9	4.7
	3.00	8.2	6.7	5.3
	≥4.00	9.1	7.5	5.7
≥3.5 < 4.5	0.25	3.6	2.4	1.9
	0.50	5.4	4.6	3.4
	0.75	6.4	6.6	4.6
	1.00	7.7	6.9	5.9
	1.50	9.4	8.3	7.1
	2.00	10.2	9.6	8.1
	3.00	11.3	11.0	8.9
	≥4.00	12.3	11.9	9.7
≥4.5 < 5.5	0.25	4.2	2.7	2.6
	0.50	6.0	6.0	5.1
	0.75	7.5	7.5	7.5
	1.00	9.2	9.0	8.9
	1.50	10.6	10.5	10.3
	2.00	11.8	11.7	11.3
	3.00	13.7	13.5	12.4
	≥4.00	15.3	15.0	12.5
≥5.5 < 6.5	0.25	4.7	4.1	3.5
	0.50	7.2	7.2	7.2
	0.75	9.1	9.1	9.1
	1.00	10.3	10.3	10.2
	1.50	11.9	11.8	11.7
	2.00	12.8	12.7	12.6
	3.00	14.4	14.3	14.2
	≥4.00	15.4	15.2	15.0
≥6.5	0.25	5.1	4.8	4.6
	0.50	7.8	7.8	7.8
	0.75	9.8	9.8	9.8
	1.00	10.4	10.4	10.3
	1.50	12.0	11.9	11.8
	2.00	12.9	12.8	12.7
	3.00	14.5	14.4	14.3
	≥4.00	15.4	15.3	15.2

Note: Straight-line interpolation of E_T for length of grade permitted to nearest 0.1.
(Used with permission of Transportation Research Board, National Science Council, *Highway Capacity Manual*, 4th Edition, Washington DC, 2000, Exhibit 20-15, pg. 20-17.)

Table 14.12: Passenger-Car Equivalents of RVs for Specific Upgrades: ATS Determination

Grade (%)	Length of Grade (mi)	Range of Direction Flow Rate (pc/h)		
		0–300	>300–600	>600
≥3.0 < 3.5	0.25	1.1	1.0	1.0
	0.50	1.2	1.0	1.0
	0.75	1.2	1.0	1.0
	1.00	1.3	1.0	1.0
	1.50	1.4	1.0	1.0
	2.00	1.4	1.0	1.0
	3.00	1.5	1.0	1.0
	≥4.00	1.5	1.0	1.0
≥3.5 < 4.5	0.25–0.75	1.3	1.0	1.0
	1.00–3.00	1.4	1.0	1.0
	≥4.00	1.5	1.0	1.0
≥4.5 < 5.5	0.25–2.00	1.5	1.0	1.0
	≥3.00	1.6	1.0	1.0
≥5.5 < 6.5	0.25–0.75	1.5	1.0	1.0
	1.00–2.00	1.6	1.0	1.0
	3.00	1.6	1.2	1.0
	≥4.00	1.6	1.5	1.2
≥6.5	0.25–2.00	1.6	1.0	1.0
	3.00	1.6	1.3	1.3
	≥4.00	1.6	1.5	1.4

Note: Straight-line interpolation of E_R for length of grade permitted to nearest 0.1.
(Used with permission of Transportation Research Board, National Science Council, *Highway Capacity Manual*, 4th Edition, Washington DC, 2000, Exhibit 20-17, pg. 20-19, *format modified.*)

E_{TC} = passenger care equivalents for trucks at crawl speed Table 14.14

All other variables are as previously defined.

In applying Equation 14-19, note that P_{TC} is stated as a proportion of the *truck* population, not of the entire traffic stream. Thus, a P_{TC} of 0.50, means that 50% of the *trucks* are operating down the grade at crawl speeds.

Note that for two-lane highways, all composite grades are treated using the *average grade* of the analysis section. The average grade for any segment is the total change in elevation (ft) divided by the length of the segment (ft).

14.4.6 Estimating Average Travel Speed

Once the appropriate demand flow rate(s) are computed, the average travel speed in the section is estimated using Equation 14-10 for two-direction analysis and Equation 14-11 for single-direction analysis:

$$ATS = FFS - 0.00776v - f_{np} \qquad (14\text{-}10)$$
$$ATS_d = FFS_d - 0.00776(v_d + v_o) - f_{np} \qquad (14\text{-}11)$$

where: ATS = average travel speed, both directions, mi/h

ATS_d = average travel speed in the direction of analysis, mi/h

Table 14.13: Passenger-Car Equivalents for Trucks and RV's on Specific Upgrades: PTSF Determination

Grade (%)	Length of Grade (mi)	Passenger Car Equivalent for Trucks, E_T Range of Directional Flows (pc/h)			Passenger-Car Equivalent for RVs, E_R
		0–300	>300–600	>600	
≥3.0 < 3.5	0.25–2.00	1.0	1.0	1.0	1.0
	3.00	1.4	1.0	1.0	1.0
	≥4.00	1.5	1.0	1.0	1.0
≥3.5 < 4.5	0.25–1.00	1.0	1.0	1.0	1.0
	1.50	1.1	1.0	1.0	1.0
	2.00	1.6	1.0	1.0	1.0
	3.00	1.7	1.1	1.2	1.0
	≥4.00	2.0	1.5	1.4	1.0
≥4.5 < 5.5	0.25–1.00	1.0	1.0	1.0	1.0
	1.50	1.1	1.2	1.2	1.0
	2.00	1.6	1.3	1.5	1.0
	3.00	2.3	1.9	1.7	1.0
	≥4.00	3.3	2.1	1.8	1.0
≥5.5 < 6.5	0.25–0.75	1.0	1.0	1.0	1.0
	1.00	1.0	1.2	1.2	1.0
	1.50	1.5	1.6	1.6	1.0
	2.00	1.9	1.9	1.8	1.0
	3.00	3.3	2.5	2.0	1.0
	≥4.00	4.3	3.1	2.0	1.0
≥6.5	0.25–0.50	1.0	1.0	1.0	1.0
	0.75	1.0	1.0	1.3	1.0
	1.00	1.3	1.4	1.6	1.0
	1.50	2.1	2.0	2.0	1.0
	2.00	2.8	2.5	2.1	1.0
	3.00	4.0	3.1	2.2	1.0
	≥4.00	4.8	3.5	2.3	1.0

Note: Straight-line interpolation of E_R for length of grade permitted to nearest 0.1.
(Used with permission of Transportation Research Board, National Science Council, *Highway Capacity Manual,* 4th Edition, Washington DC, 2000, Exhibit 20-16, pg. 20-18, format modified.)

FFS = free-flow speed, both directions, mi/h

FFS_d = free-flow speed in the direction of analysis, mi/h

v = demand flow rate, both directions, pc/h

v_d = demand flow rate in the direction of analysis, pc/h

v_o = demand flow rate in the opposing direction, pc/h

f_{np} = adjustment for the existence of "No Passing" zones in the study segment

Values of the adjustment factor, f_{np}, are given in Table 14.15 for two-direction analyses, and in Table 14.16 for single-direction analyses. The adjustment is based on flow rates, the percentage of the analyses segment for which passing is prohibited, and (for single-direction analyses) the free-flow speed of the facility.

Table 14.14: Passenger-Car Equivalents for Trucks Operating at Crawl Speeds on Specific Downgrades: ATS Determination

Difference Between FFS and Crawl Speed of Truck (mi/h)	Range of Directional Flows (pc/h)		
	0–300	300–600	>600
≤15	4.4	2.8	1.4
25	14.3	9.6	5.7
≥40	34.1	23.1	13.0

Note: Straight-line interpolation of E_R for length of grade permitted to nearest 0.1.

(Used by Permission of Transportation Research Board, National Science Council, *Highway Capacity Manual*, 4th Edition, Washington DC, 2000, Exhibit 20-18, pg. 20-20, *format modified.*)

Table 14.15: Adjustment for Effect of "No Passing" Zones (f_{np}) on ATS: Two-Direction Segments

Two-Way Demand Flow Rate (pc/h)	Reduction in Average Travel Speed, f_{np} (mi/h) Percent "No Passing" Zones in Segment					
	0	20	40	60	80	100
0	0.0	0.0	0.0	0.0	0.0	0.0
200	0.0	0.6	1.4	2.4	2.6	3.5
400	0.0	1.7	2.7	3.5	3.9	4.5
600	0.0	1.6	2.4	3.0	3.4	3.9
800	0.0	1.4	1.9	2.4	2.7	3.0
1,000	0.0	1.1	1.6	2.0	2.2	2.6
1,200	0.0	0.8	1.2	1.6	1.9	2.1
1,400	0.0	0.6	0.9	1.2	1.4	1.7
1,600	0.0	0.6	0.8	1.1	1.3	1.5
1,800	0.0	0.5	0.7	1.0	1.1	1.3
2,000	0.0	0.5	0.6	0.9	1.0	1.1
2,200	0.0	0.5	0.6	0.9	0.9	1.1
2,400	0.0	0.5	0.6	0.8	0.9	1.1
2,600	0.0	0.5	0.6	0.8	0.9	1.0
2,800	0.0	0.5	0.6	0.7	0.8	0.9
3,000	0.0	0.5	0.6	0.7	0.7	0.8
3,200	0.0	0.5	0.6	0.6	0.6	0.7

Note: Straight-line interpolation of f_{np} for % "No Passing" zones and demand flow rate permitted to nearest 0.1.

(Used with permission of Transportation Research Board, National Science Council, *Highway Capacity Manual*, 4th Edition, Washington DC, 2000, Exhibit 20-11, pg. 20-10.)

Table 14.16: Adjustment for Effect of "No Passing" Zones (f_{np}) on ATS Single-Direction Segments

Opposing Demand Flow Rate v_0 (pc/h)	Percent "No Passing" Zones (%)				
	≤20	40	60	80	100
FFS = 65 mi/h					
≤100	1.1	2.2	2.8	3.0	3.1
200	2.2	3.3	3.9	4.0	4.2
400	1.6	2.3	2.7	2.8	2.9
600	1.4	1.5	1.7	1.9	2.0
800	0.7	1.0	1.2	1.4	1.5
1,000	0.6	0.8	1.1	1.1	1.2
1,200	0.6	0.8	0.9	1.0	1.1
1,400	0.6	0.7	0.9	0.9	0.9
≥1,600	0.6	0.7	0.7	0.7	0.8
FFS = 60 mi/h					
≤100	0.7	1.7	2.5	2.8	2.9
200	1.9	2.9	3.7	4.0	4.2
400	1.4	2.0	2.5	2.7	3.9
600	1.1	1.3	1.6	1.9	2.0
800	0.6	0.9	1.1	1.3	1.4
1,000	0.6	0.7	0.9	1.1	1.2
1,200	0.5	0.7	0.9	0.9	1.1
1,400	0.5	0.6	0.8	0.8	0.9
≥1,600	0.5	0.6	0.7	0.7	0.7
FFS = 55 mi/h					
≤100	0.5	1.2	2.2	2.6	2.7
200	1.5	2.4	3.5	3.9	4.1
400	1.3	1.9	2.4	2.7	2.8
600	0.9	1.1	1.6	1.8	1.9
800	0.5	0.7	1.1	1.2	1.4
1,000	0.5	0.6	0.8	0.9	1.1
1,200	0.5	0.6	0.7	0.9	1.0
1,400	0.5	0.6	0.7	0.7	0.9
≥1,600	0.5	0.6	0.6	0.6	0.7
FFS = 50 mi/h					
≤100	0.2	0.7	1.9	2.4	2.5
200	1.2	2.0	3.3	3.9	4.0
400	1.1	1.6	2.2	2.6	2.7
600	0.6	0.9	1.4	1.7	1.9
800	0.4	0.6	0.9	1.2	1.3

(Continued)

Table 14.16: Adjustment for Effect of "No Passing" Zones (f_{np}) on ATS Single-Direction Segments (*Continued*)

Opposing Demand Flow Rate v_0 (pc/h)	Percent "No Passing" Zones (%)				
	≤20	40	60	80	100
FFS = 50 mi/h					
1,000	0.4	0.4	0.7	0.9	1.1
1,200	0.4	0.4	0.7	0.8	1.0
1,400	0.4	0.4	0.6	0.7	0.8
≥1,600	0.4	0.4	0.5	0.5	0.5
FFS = 45 mi/h					
≤100	0.1	0.4	1.7	2.2	2.4
200	0.9	1.6	3.1	3.8	4.0
400	0.9	0.5	2.0	2.5	2.7
600	0.4	0.3	1.3	1.7	1.8
800	0.3	0.3	0.8	1.1	1.2
1,000	0.3	0.3	0.6	0.8	1.1
1,200	0.3	0.3	0.6	0.7	1.0
1,400	0.3	0.3	0.6	0.6	0.7
≥1,600	0.3	0.3	0.4	0.4	0.6

Note: Straight-line interpolation of f_{np} for % "No Passing" zones, demand flow rate and FFS permitted to nearest 0.1.
(Used with permission of Transportation Research Board, National Science Council, *Highway Capacity Manual*, 4th Edition, Washington DC, 2000, Exhibit 20-20, pg. 20-23.)

14.4.7 Determining Percent Time Spent Following

Percent time spent following (PTSF) is determined using Equation 14-12 for two-direction analyses, and Equation 14-13 for single-direction analyses:

$$PTSF = BPTSF + f_{d/np}$$
$$BPTSF = 100(1 - e^{-0.000879\,v}) \quad (14\text{-}12)$$
$$PTSF_d = BPTSF_d + f_{np}$$
$$BPTSF_d = 100\left(1 - e^{av_d^b}\right) \quad (14\text{-}13)$$

where: $PTSF$ = percent time spent following, two directions,%

$PTSF_d$ = percent time spent following, single direction,%

$BPTSF$ = base percent time spent following, two directions,%

v = demand flow rate, pc/h, both directions

v_d = demand flow rate in analysis direction, pc/h

$f_{d/np}$ = adjustment to PTSF for the combined effect of directional distribution and percent "No Passing" zones on two-way analysis segments, %

f_{np} = adjustment to PTSF for the effect of percent "No Passing" zones on single-direction analysis segments, %

a, b = calibration constants based on opposing flow rate in single-direction analysis

Adjustment factor $f_{d/np}$ is found in Table 14.17. Adjustment factor f_{np} is found in Table 14.18 and calibration constants "a" and "b" are found in Table 14.19.

Table 14.17: Adjustment ($f_{d/np}$) for the Combined Effect of Directional Distribution and Percent "No Passing" Zones on PTSF on Two-Way Segments

Two-Way Flow Rate (pc/h)	Increase in Percent Time Spent Following, %					
	Percent "No Passing" Zones					
	0	20	40	60	80	100
Directional Split = 50/50						
≤200	0.0	10.1	17.2	20.2	21.0	21.8
400	0.0	12.4	19.0	22.7	23.8	24.8
600	0.0	11.2	16.0	18.7	19.7	20.5
800	0.0	9.0	12.3	14.1	14.5	15.4
1,400	0.0	3.6	5.5	6.7	7.3	7.9
2,000	0.0	1.8	2.9	3.7	4.1	4.4
2,600	0.0	1.1	1.6	2.0	2.3	2.4
3,200	0.0	0.7	0.9	1.1	1.2	1.4
Directional Split = 60/40						
≤200	1.6	11.8	17.2	22.5	23.1	23.7
400	0.5	11.7	16.2	20.7	21.5	22.2
600	0.0	11.5	15.2	18.9	19.8	20.7
800	0.0	7.6	10.3	13.0	13.7	14.4
1,400	0.0	3.7	5.4	7.1	7.6	8.1
2,000	0.0	2.3	3.4	3.6	4.0	4.3
≥2,600	0.0	0.9	1.4	1.9	2.1	2.2
Directional Split = 70/30						
≤200	2.8	13.4	19.1	24.8	25.2	25.5
400	1.1	12.5	17.3	22.0	22.6	23.2
600	0.0	11.6	15.4	19.1	20.0	20.9
800	0.0	7.7	10.5	13.3	14.0	14.7
1,400	0.0	3.8	5.6	7.4	7.9	8.3
≥2,000	0.0	1.4	4.9	3.5	3.9	4.2
Directional Split = 80/20						
≤200	5.1	17.5	24.3	31.0	31.3	31.6
400	2.5	15.8	21.5	27.1	27.6	28.0
600	0.0	14.0	18.6	23.2	23.9	24.5
800	0.0	9.3	12.7	16.0	16.5	17.0
1,400	0.0	4.6	6.7	8.7	9.1	9.5
≥2,000	0.0	2.4	3.4	4.5	4.7	4.9

(Continued)

Table 14.17: Adjustment ($f_{d/np}$) for the Combined Effect of Directional Distribution and Percent "No Passing" Zones on PTSF on Two-Way Segments (*Continued*)

Two-Way Flow Rate	Increase in Percent Time Spent Following, %					
	Percent "No Passing" Zones					
(pc/h)	0	20	40	60	80	100
Directional Split = 90/10						
≤200	5.6	21.6	29.4	37.2	37.4	37.6
400	2.4	19.0	25.6	32.2	32.5	32.8
600	0.0	16.3	21.8	27.2	27.6	28.0
800	0.0	10.9	14.8	18.6	19.0	19.4
≥1400	0.0	5.5	7.8	10.0	10.4	10.7

Note: Straight-line interpolation of $f_{d/np}$ for % "No Passing" zones, demand flow rate and directional distribution permitted to nearest 0.1.
(Used with permission of Transportation Research Board, National Science Council, *Highway Capacity Manual*, 4th Edition, Washington DC, 2000, Exhibit 20-12, pg. 20-11.)

Table 14.18: Adjustment (f_{np}) to PTSF for Percent "No Passing" Zones in Single-Direction Segments

Opposing Demand Flow Rate v_o (pc/h)	Percent "No Passing Zones" (%)				
	≤20	40	60	80	100
FFS = 65 mi/h					
≤100	10.1	17.2	20.2	21.0	21.8
200	12.4	19.0	22.7	23.8	24.8
400	9.0	12.3	14.1	14.4	15.4
600	5.3	7.7	9.2	9.7	10.4
800	3.0	4.6	5.7	6.2	6.7
1,000	1.8	2.9	3.7	4.1	4.4
1,200	1.3	2.0	2.6	2.9	3.1
1,400	0.9	1.4	1.7	1.9	2.1
≥1,600	0.7	0.9	0.9	1.2	1.4
FFS = 60 mi/h					
≤100	8.4	14.9	20.9	22.8	26.6
200	11.5	18.2	24.1	26.2	29.7
400	8.6	12.1	14.8	15.9	18.1
600	5.1	7.5	9.6	10.6	12.1
800	2.8	4.5	5.9	6.7	7.7
1,000	1.6	2.8	3.7	4.3	4.9
1,200	1.2	1.9	2.6	3.0	3.4
1,400	0.8	1.3	1.7	2.0	2.3
≥1,600	0.6	0.9	1.1	1.2	1.5

(Continued)

Table 14.18: Adjustment (f_{np}) to PTSF for Percent "No Passing" Zones in Single-Direction Segments (*Continued*)

Opposing Demand Flow Rate v_o (pc/h)	Percent "No Passing Zones" (%)				
	≤20	40	60	80	100
FFS = 55 mi/h					
≤100	6.7	12.7	21.7	24.5	31.3
200	10.5	17.5	25.4	28.6	34.7
400	8.3	11.8	15.5	17.5	20.7
600	4.9	7.3	10.0	11.5	13.9
800	2.7	4.3	6.1	7.2	8.8
1,000	1.5	2.7	3.8	4.5	5.4
1,200	1.0	1.8	2.6	3.1	3.8
1,400	0.7	1.2	1.7	2.0	2.4
≥1,600	0.6	0.9	1.2	1.3	1.5
FFS = 50 mi/h					
≤100	5.0	10.4	22.2	26.3	36.1
200	9.6	16.7	26.8	31.0	39.6
400	7.9	11.6	16.2	19.0	23.4
600	4.7	7.1	10.4	12.4	15.6
800	2.5	4.2	6.3	7.7	9.8
1,000	1.3	2.6	3.8	4.7	5.9
1,200	0.9	1.7	2.6	3.2	4.1
1,400	0.6	1.1	1.7	2.1	2.6
≥1,600	0.5	0.9	1.2	1.3	1.6
FFS = 45 mi/h					
≤100	3.7	8.5	23.2	28.2	41.6
200	8.7	16.0	28.2	33.6	45.2
400	7.5	11.4	16.9	20.7	26.4
600	4.5	6.9	10.8	13.4	17.6
800	2.3	4.1	6.5	8.2	11.0
1,000	1.2	2.5	3.8	4.9	6.4
1,200	0.8	1.6	2.6	3.3	4.5
1,400	0.5	1.0	1.7	2.2	2.8
≥1,600	0.4	0.9	1.2	1.3	1.7

Note: Straight-line interpolation of f_{np} for % "No Passing" zones, demand flow rate and FFS permitted to nearest 0.1.

(Used with permission of Transportation Research Board, National Science Council, *Highway Capacity Manual*, 4th Edition, Washington DC, 2000, Exhibit 20-20, pg. 20-23.)

Table 14.19: Coefficients "a" and "b" for Use in Equation 14-13

Opposing Demand Flow Rate, pc/h	Coefficient "a"	Coefficient "b"
≤200	−0.013	0.668
400	−0.057	0.479
600	−0.100	0.413
800	−0.173	0.349
1,000	−0.320	0.276
1,200	−0.430	0.242
1,400	−0.522	0.225
≥1,600	−0.665	0.199

Note: Straight-line interpolation of "a" and "b" for opposing demand flow rate permitted to nearest 0.001.
(Used with permission of Transportation Research Board, National Science Council, *Highway Capacity Manual*, 4th Edition, Washington DC, 2000, Exhibit 20-21, pg. 20–24.)

Example 14-1: A Two-Way Analysis in General Terrain

A Class I two-lane highway in rolling terrain has a peak demand volume of 500 veh/h, with 15% trucks and 5% RVs. The highway serves as a main link to a popular recreation area. The directional split of traffic is 60/40 during peak periods, and the peak-hour factor is 0.88. The 15-mile section under study has 40% "No Passing" zones. The base free-flow speed of the facility may be taken to be 60 mph. Lane widths are 12 ft, and shoulder widths are 2 ft. There are 10 access points per mile along this three-mile section.

Solution:

Step 1: Estimate the Free-Flow Speed.
The free-flow speed (*FFS*) is estimated from the base free-flow speed (*BFFS*) and applicable adjustment factors f_{LS} (Table 14.5) and f_A (Table 14.6). Then:

$$FFS = BFFS - f_{LS} - f_A$$
$$FFS = 60.0 - 2.6 - 2.5 = 54.9 \text{ mi/h}$$

where $f_{LS} = 2.6$ (for 12-ft lanes and 2-ft shoulders) and $f_A = 2.5$ (for 10 access points per mile)

Step 2: Estimate Demand Flow Rates for ATS and PTSF Determinations

Equivalent demand flow rate is computed as:

$$v = \frac{V}{PHF * f_G * f_{HV}}$$

$$f_{HV} = \frac{1}{1 + P_T (E_T - 1) + P_R (E_R - 1)}$$

where: $V = 500$ vph (given)
$PHF = 0.88$ (given)
$P_T = 0.15$ (given)
$P_R = 0.05$ (given)

All other values must be selected from the appropriate adjustment tables. To do so requires an iterative procedure, beginning with an estimated value of $v = 500/0.88 = 568$ veh/h. As this is close to a boundary of 600 veh/h in Tables 14.7 and 14.10, an assumption will be made that the final values of v will be in the 600–1200 pc/h range. All adjustment factors are, therefore, chosen accordingly. Two values of v are estimated, one for ATS determination and one for PTSF determination.

For determining average travel speed (ATS):

$$f_G = 0.93 \text{ (Table 14.7, rolling terrain)}$$
$$E_T = 1.9 \text{ (Table 14.10, rolling terrain)}$$
$$E_R = 1.1 \text{ (Table 14.10, rolling terrain)}$$

Then:

$$f_{HV} = \frac{1}{1 + 0.15(1.9 - 1) + 0.05(1.1 - 1)} = 0.877$$
$$v = \frac{500}{0.88 * 0.93 * 0.877} = 697 \text{ pc/h}$$

As the result is within the assumed 600–1200 pc/h range, this result is complete.

For determining percent time spent following (PTSF):

$$f_G = 0.94 \text{ (Table 14.7, rolling terrain)}$$
$$E_T = 1.5 \text{ (Table 14.10, rolling terrain)}$$
$$E_R = 1.0 \text{ (Table 14.10, rolling terrain)}$$

Then:

$$f_{HV} = \frac{1}{1 + 0.15(1.5 - 1) + 0.05(1.0 - 1)} = 0.93$$
$$v = \frac{500}{0.88 * 0.94 * 0.93} = 650 \text{ pc/h}$$

This estimate is also within the assumed range of 600–1200 pc/h. Thus, the computation is complete.

Step 3:   Determine ATS.

The estimation of average travel speed (ATS) uses the free-flow speed (54.9 mi/h), the computed value of v (697 pc/h), and the adjustment factor for "No Passing" zones f_{np}. The latter is found from Table 14.15 for 40% "No Passing" zones and a demand flow of 697 pc/h (which can be approximately interpolated using 700 pc/h). The value found in this manner is 2.15 mi/h. Then:

$$ATS = FFS - 0.00776v - f_{np}$$
$$ATS = 54.9 - 0.00776(697) - 2.15$$
$$= 47.341, \text{ say } 47.3 \text{ mi/h}$$

Step 4:   Determine PTSF.

The estimation of percent time spent following (PTSF) uses the computed value of v (650 pc/h) and the adjustment factor for directional distribution and "No Passing" zones $f_{d/np}$. The latter is found in Table 14.17 for 40% "No Passing" zones, a 60/40 directional split, and a demand flow of 650 pc/h. Interpolating for 650 pc/h, this adjustment is 14.0%. Then:

$$PTSF = 100(1 - e^{-0.000879\,v}) + f_{d/np}$$
$$PTSF = 100(1 - e^{-0.000879 * 650}) + 14.0 = 57.5\%$$

Step 5:   Determine the Level of Service. From Table 14.4, the LOS based on ATS = 47.3 mi/h is C, while the LOS based upon PTSF = 57.5% is also C. The prevailing level of service is therefore C.

Example 14-2: A Directional Analysis in General Terrain

Conduct an analysis of the highway section described for Sample Problem 1 for each direction of flow.

Solution:

Two single-direction analyses will be conducted for the section described in Sample Problem 1. This requires that the demand of 500 veh/h be separated by direction. Note that the two directional analyses may be done concurrently, as the directional demand in one case is the opposing demand in the other. Given the specified 60/40 split:

$$V_1 = 500 * 0.60 = 300 \text{ veh/h}$$
$$V_2 = 500 * 0.40 = 200 \text{ veh/h}$$

Both of these values have to be converted to base passenger-car flow rates. All other information concerning the section remains the same, as does the estimated free-flow speed of 54.9 mph.

Step 1:   Determine the Demand Flow Rates for ATS and PTSF Determinations

Four demand flows will be computed. Both the directional and opposing volumes must be separately converted for ATS determination and for PTSF determination. Both could be iterative. The initial selection of adjustment factors would be based on $v_1 = 300/0.88 = 341$ veh/h and $v_2 = 200/0.88 = 227$ veh/h. The former is in the

300–600 veh/h range, while the latter is in the 0–300 veh/h range. Adjustments will be selected accordingly, as follows:

For the directional volume, V_1:

$$f_G (ATS) = 0.93 \text{ (Table 14.7, rolling terrain)}$$
$$E_T (ATS) = 1.9 \quad \text{(Table 14.10, rolling terrain)}$$
$$E_R (ATS) = 1.1 \quad \text{(Table 14.10, rolling terrain)}$$
$$f_G (PTSF) = 0.94 \text{ (Table 14.7, rolling terrain)}$$
$$E_T (PTSF) = 1.5 \quad \text{(Table 14.10, rolling terrain)}$$
$$E_R (PTSF) = 1.0 \quad \text{(Table 14.10, rolling terrain)}$$

Then:

$$f_{HV} (ATS) = \frac{1}{1 + 0.15(1.9 - 1) + 0.05(1.1 - 1)}$$
$$= 0.877$$

$$v_1 (ATS) = \frac{300}{0.88 * 0.93 * 0.877} = 418 \text{ pc/h}$$

$$f_{HV} (PTSF) = \frac{1}{1 + 0.15(1.5 - 1) + 0.05(1.0 - 1)}$$
$$= 0.93$$

$$v_1 (PTSF) = \frac{300}{0.88 * 0.94 * 0.93} = 390 \text{ pc/h}$$

As both of these results are in the assumed range of 300–600 pc/h, no iteration is required.

For the directional volume, V_2:

$$f_G (ATS) = 0.71 \quad \text{(Table 14.7, rolling terrain)}$$
$$E_T (ATS) = 2.5 \quad \text{(Table 14.10, rolling terrain)}$$
$$E_R (ATS) = 1.1 \quad \text{(Table 14.10, rolling terrain)}$$
$$f_G (PTSF) = 0.77 \text{ (Table 14.7, rolling terrain)}$$
$$E_T (PTSF) = 1.8 \quad \text{(Table 14.10, rolling terrain)}$$
$$E_R (PTSF) = 1.0 \quad \text{(Table 14.10, rolling terrain)}$$

Then:

$$f_{HV} (ATS) = \frac{1}{1 + 0.15(2.5 - 1) + 0.05(1.1 - 1)}$$
$$= 0.813$$

$$v_2 (ATS) = \frac{200}{0.88 * 0.71 * 0.813} = 394 \text{ pc/h}$$

$$f_{HV} (PTSF) = \frac{1}{1 + 0.15(1.8 - 1) + 0.05(1.0 - 1)}$$
$$= 0.893$$

$$v_2 (PTSF) = \frac{200}{0.88 * 0.77 * 0.893} = 331 \text{ pc/h}$$

Both of these results lie outside the assumed range of 0–300 pc/h. They must now be iterated, using the 300–600 pc/h range as a base. In this case, the factors will be the same as those found for V_1 previously. Then:

$$v_2(ATS) = \frac{200}{0.88 * 0.93 * 0.877} = 279 \text{ pc/h}$$

$$v_2(PTSF) = \frac{200}{0.88 * 0.94 * 0.93} = 260 \text{ pc/h}$$

These results, however, are outside the assumed range 300–600 pc/h. Thus, when a range of 0–300 pc/h is assumed, the results were higher than 300 pc/h. When the higher range is assumed, the result is less than 300 pc/h. The operative selection rule for this iteration is: If the computed value of v is less than the upper limit of the selected flow-range for which f_G, E_T, and E_R were determined, then the computed value of v should be used. [HCM 2000, pg. 20-9]. As these results are less than 600 pc/h (the upper limit of the assumed range), they are used in subsequent computations.

Step 2: Determine ATS Values for Each Direction

These values are estimated as:

$$ATS_d = FFS_d - 0.00776(v_d + v_o) - f_{np}$$

where: FFS_d = 54.9 mph (directions 1 and 2, previously computed)

$v_1 (ATS)$ = 418 pc/h (previously computed)

$v_2 (ATS)$ = 279 pc/h (previously computed)

f_{np1} = 2.20 mi/h (Table 14.16, FFS = 55 mi/h, v_o = 279 pc/h, 40% no passing zones, interpolated)

f_{np2} = 1.83 (Table 14.16, FFS = 55 mi/h, v_o = 418 pc/h, 40% no passing zones, interpolated)

Then:

$$ATS_1 = 54.9 - 0.00776(418 + 279) - 2.20$$
$$= 47.3 \text{ mi/h}$$
$$ATS_2 = 54.9 - 0.00776(279 + 418) - 1.83$$
$$= 47.7 \text{ mi/h}$$

Step 3:   Determine PTSF Values for Each Direction

These values are determined as:

$$PTSF_d = BPTSF_d + f_{np}$$

$$BPTSF_d = 100(1 - e^{av_d^b})$$

where: $v_1 = 390$ pc/h (computed previously)

$v_2 = 260$ pc/h (computed previously)

$a_1 = -0.026$ (Table 14.19, $v_o = 260$ pc/h, interpolated)

$a_2 = -0.055$ (Table 14.19, $v_o = 390$ pc/h, interpolated)

$b_1 = 0.611$ (Table 14.19, $v_o = 260$ pc/h, interpolated)

$b_2 = 0.488$ (Table 14.19, $v_o = 390$ pc/h, interpolated)

$f_{np1} = 15.8$ (Table 14.18, FFS = 55 mi/h, 40% no passing zones, $v_o = 260$ pc/h)

$f_{np2} = 12.1$ (Table 14.18, FFS = 55 mi/h, 40% no passing zones, $v_o = 390$ pc/h)

Then:

$$BPTSF_1 = 100(1 - e^{-0.026*390^{0.611}}) = 63.1\%$$

$$BPTSF_2 = 100(1 - e^{-0.055*260^{0.488}}) = 56.4\%$$

$$PTSF_1 = 63.1 + 15.8 = 78.9\%$$

$$PTSF_2 = 56.4 + 12.1 = 68.5\%$$

Step 4:   Determine Directional Levels of Service

From Table 14.4, Direction 1 has an LOS of C based on an ATS of 47.3 mi/h and an LOS of D based on a PTSF of 78.9%. The latter controls, and the overall LOS is D. Direction 2 has an LOS of C based on an ATS of 47.7 mi/h and an LOS of D based on a PTSF of 68.5%. Again, the latter controls, and LOS D prevails.

Table 14.20: Comparison of Examples 14-1 and 14-2

Analysis Section	ATS (mi/h)	PTSF (%)	LOS
Both Directions	47.3	57.5	C
Direction 1	47.3	78.9	D
Direction 2	47.7	68.5	D

Discussion of Examples 14-1 and 14-2

The results of Sample Problems 1 and 2, which consider the same general section of the same facility, reveal some anomalies in the HCM 2000 analysis methodologies. Table 14.20 summarizes the results of the two analyses.

The worst inconsistency is in the PTSF results. If the combined directions result in 57.5% time spent following, it is illogical that *both* directions, analyzed separately, result in substantially higher values (78.9% and 68.5%). A logical result would have the two-directional value represent some form of average (not necessarily a simple one) of the two directions, considered alone. The average travel speeds are similarly inconsistent, but less so. The resulting levels of service are illogical as well. If the entire facility is at LOS C, how can *both* directions, studied separately, be operating at LOS D, especially when the same LOS criteria (Table 14.4) are used in both cases.

This inconsistency was related to members of the Highway Capacity and Quality of Service Committee of the Transportation Research Board shortly after the publication of the HCM 2000, and it is currently under review (as of Spring 2001). The directional methodology was developed for specific grades and was later adapted to the analysis of general terrain segments. It is, therefore, likely that for general terrain segments, the two-direction procedure yields the more reliable results.

Example 14-3: Single-Direction Analysis of a Specific Grade

Class II two-lane highway serving a rural logging area has a 2-mi grade of 4%. Peak demand on the grade is 250 veh/h, with a 70/30 directional split and a PHF of 0.82. Because of active logging operations in the area, demand contains 20% trucks. The grade has 100% "No Passing" zones. The free-flow speed of the facility has been measured to be 45 mph. At what level of service does the facility operate, assuming that 70% of the traffic is on the upgrade?

Solution:

Both the upgrade and downgrade must be analyzed in this case. Because this is a Class II facility, percent time spent following (PTSF) is the only parameter that need be established. It is necessary to divide the demand volume into the upgrade volume, V_u ($250 * 0.70 = 175$ veh/h), and downgrade volume, V_d ($250 * 0.30 = 75$ veh/h). As the upgrade and downgrade are analyzed, these will serve as directional and opposing volumes.

Step 1: Determine Demand Flow Rate

Base passenger-car equivalent demand flows are computed as:

$$v = \frac{V}{PHF * f_G * f_{HV}}$$

Selecting values of f_G, E_T, and E_R from the appropriate tables for PTSF determination only results in the values shown in Table 14.21. Note that for the downgrade, Equation 14-9 must be used with an appropriate value of E_{TC}, as tractor-trailers are traveling at crawl speeds downgrade.

Then:

$$f_{HV} \text{ (up)} = \frac{1}{1 + 0.20(1.6 - 1) + 0.00(1.0 - 1)} = 0.893$$

$$v_u = \frac{175}{0.82 * 1.00 * 0.893} = 239 \text{ pc/h}$$

$$f_{HV} \text{ (down)} = \frac{1}{1 + 0.20(1.1 - 1) + 0.00(1.0 - 1)} = 0.980$$

$$v_d = \frac{75}{0.82 * 1.00 * 0.980} = 93 \text{ pc/h}$$

As both values are within the assumed flow range of 0–300 pc/h, no iteration of these values is necessary.

Step 2: Estimate the PTSF for the Specific Upgrade and Specific Downgrade

The percent time spent following is estimated as:

$$PTSF_d = BPTSF + f_{np}$$
$$BPTSF_d = 100(1 - e^{av_d^b})$$

where:

f_{np} (upgrade) = 41.6% (Table 14.18, FFS = 45 mi/h, 100% "No Passing" zones, and $v_o = 93$ pc/h)

a (upgrade) = -0.013 (Table 14.19, $v_o < 200$ pc/h)

b (upgrade) = 0.668 (Table 14.19, $v_o < 200$ pc/h)

f_{np} (downgrade) = 41.5% (Table 14.18, FFS = 45 mph, 100% "No Passing" zones, $v_o = 239$ pc/h, interpolated)

Table 14.21: Factors for PTSF Determination in Example 14-4

Direction	f_G	E_T	E_R
Upgrade	1.00	1.6	1.0
Downgrade	1.00	1.1	1.0

Notes:

f_G (upgrade) from Table 14.9 for 4% grade, 2 mi long, assumed flow range 0–300 pc/h
f_G (downgrade) from Table 14.7
E_T (upgrade) from Table 14.13 for 4% grade, 2 mi long, assumed flow range 0–300 pc/h
E_T (downgrade) from Table 14.10, Specific Downgrades, assumed flow range 0–300 pc/h
E_R (upgrade) from Table 14.13 for 4% grade, 2 mi long
E_R (downgrade) from Table 14.10, Specific Downgrades, assumed flow range 0–300 pc/h

a (downgrade) $= -0.022$ (Table 14.19,
$\qquad v_o = 239$ pc/h, interpolated)

b (downgrade) $= 0.631$ (Table 14.19,
$\qquad v_o = 239$ pc/h, interpolated)

Then:

$$BPTSF_u = 100\left(1 - e^{-0.013*239^{0.668}}\right) = 39.6\%$$

$$BPTSF_d = 100\left(1 - e^{-0.022 \times 123^{0.631}}\right) = 36.8\%$$

$$PTSF_u = 39.6 + 41.6 = 81.2\%$$

$$PTSF_d = 36.8 + 41.5 = 78.3\%$$

Step 3:   Determine Level of Service

From Table 14.4, this is level of service D for the up-grade and D for the downgrade. This problem illustrates how small demands can cause poor operations on two-lane highways, particularly in severe terrain.

14.4.8   Impacts of Passing Lanes

The single operational aspect that makes two-lane highways unique is the need to pass in the opposing lane of traffic. Both PTSF and ATS are affected when passing lanes are periodically provided. This allows platoons (in the direction of the passing lane) to break up through unrestricted passing for the length of the passing lane. If such lanes are provided periodically, the formation of long platoons behind a single slow-moving vehicle can be avoided.

The HCM 2000 provides a methodology for estimating the impact of a passing lane on a single-direction segment of a two-lane highway in level or rolling terrain. The procedure is intended for use on a directional segment containing only *one* passing lane of appropriate length, based on the criteria of Table 14.22.

The first step in the procedure is to complete a single-direction analysis for the cross-section as if the passing lane did not exist. This will result in an estimation of ATS_d and $PTSF_d$ for the section without a passing lane. The analysis segment is now broken into four subsegments, as follows:

Table 14.22: Optimal Lengths of Passing Lanes on Two-Lane Highways

Directional Flow Rate (pc/h)	Optimal Length of Passing Lane (mi)
100	≤ 0.50
200	$>0.50–0.75$
400	$>0.75–1.00$
≥ 700	$>1.00–2.00$

(Used with permission of Transportation Research Board, National Research Council, *Highway Capacity Manual*, 4th Edition, Washington DC, 2000, Exhibit 12-12, pg. 12-18.)

- The subsection upstream of the passing lane, length L_u (mi)
- The passing lane, including tapers, length L_{pl} (mi)
- The effective downstream length of the passing lane, length L_{de} (mi)
- The subsection downstream of the effective length of the passing lane, length L_d (mi)

The sum of the four lengths must be equal to the total length of the directional segment.

The "effective downstream length of the passing lane" reflects observations indicating that the passing lane improves both ATS and PTSF for a distance downstream of the actual passing lane itself. These "effective" distances differ depending upon which measure of effectiveness is involved as well as by the demand flow rate. The "effective" distance *does not* include the passing lane itself, and is measured from the end of the passing lane taper. Effective lengths, L_{de}, are shown in Table 14.23.

Lengths L_u and L_{pl} are known from existing conditions or plans. Length L_{de} is obtained from Table 14.23. Then:

$$L_d = L - (L_u + L_{pl} + L_{de}) \qquad (14\text{-}14)$$

where: L = total length of the single-direction
$\qquad\qquad$ analysis segment, mi

All other variables are as previously defined

Note that both L_{de} and L_d will differ for ATS and PTSF determinations.

Effect of Passing Lanes on PTSF

It is assumed that PTSF for the upstream subsection, L_u, and downstream subsection, L_d, are the same as if the

Table 14.23: Length of Downstream Roadway Affected by Passing Lanes on Single-Direction Segments in Level and Rolling Terrain

Direction Flow Rate, pc/h	Downstream Length of Roadway Affected, L_{de} (mi)	
	ATS	PTSF
≤200	1.7	13.0
400	1.7	8.1
700	1.7	5.7
≥1,000	1.7	3.6

Note: Straight-line interpolation for L_{de} for directional flow rate is permitted to the nearest 0.1 mi.
(Used with permission of Transportation Research Board, National Research Council, *Highway Capacity Manual*, 4th Edition, Washington DC, 2000, Exhibit 20-23, pg. 20-24)

where: $PTSF_{pl}$ = percent time spent following, adjusted to account for the impact of a passing lane in a directional segment (level or rolling terrain only), %

$PTSF_d$ = percent time spent following, assuming no passing lane in the directional segment, %

f_{pl} = adjustment factor for the effect of the passing lane on PTSF (Table 14.24)

All lengths are as previously defined

Effect of Passing Lane on ATS

Again, it is assumed that the ATS upstream (L_u) and downstream (L_d) of the passing lane and its effective length is the same as if no passing lane existed, (i.e., ATS_d). Within the passing lane, speeds are generally 8% to 11% higher than the upstream value. Within the effective length, L_{de}, ATS is assumed to move linearly from the passing lane value to the downstream value. Thus, the adjusted value of ATS is found as follows:

$$ATS_{pl} = \frac{ATS_d * L}{L_u + L_d + \left(\dfrac{L_{pl}}{f_{pl}}\right) + \left(\dfrac{2L_{de}}{1 + f_{pl}}\right)} \quad (14\text{-}16)$$

where: ATS_{pl} = average travel speed, adjusted to account for the effect of a passing lane, mi/h

passing lane did not exist (i.e., $PTSF_d$). Within the passing lane segment, L_{pl}, PTSF is usually between 58% and 62% of the upstream value. With the effective length segment, L_{de}, PTSF is assumed to rise linearly from its passing lane value to the downstream segment value. Thus, the adjusted value of PTSF, accounting for the impact of the passing lane on the single-direction segment found as:

$$PTSF_{pl} = \frac{PTSF_d\left[L_u + L_d + f_{pl}L_{pl} + \left(\dfrac{1 + f_{pl}}{2}\right)L_{de}\right]}{L}$$

$$(14\text{-}15)$$

Table 14.24: Adjustment Factors for Passing Lanes in Single-Direction Segments of Two-Lane Rural Highways

Directional Demand Flow Rate (pc/h)	Adjustment Factor f_{pl} for ATS	Adjustment Factor f_{pl} for PTSF
0–300	1.08	0.58
>300–600	1.10	0.61
>600	1.11	0.62

(Used with permission of Transportation Research Board, National Research Council, *Highway Capacity Manual*, 4th Edition, Washington DC, 2000, Exhibit 20-24, pg. 20-26.)

ATS_d = average travel speed for the directional segment without a passing lane, mi/h

f_{pl} = adjustment factor for the effect of the passing lane on ATS (Table 14.24)

All lengths as previously defined

Adjustment factors for use with Equations 14-15 and 14-16 are shown in Table 14.24.

A Sample Problem

Consider the 15.0-mile segment of two-lane rural highway described in Examples 14-1 and 14-2 of the previous section. Demand flow rates were found to be 418 pc/h for ATS determination and 390 pc/h for PTSF determination. In the heaviest direction of travel, measures of effectiveness were determined to be:

$$ATS_d = 47.3 \text{ mi/h}$$
$$PTSF_d = 78.9\%$$

A passing lane of 0.75 miles is to be added to this directional segment as illustrated in Figure 14.5. Determine the impact of the passing lane on ATS and PTSF for the section.

From Table 14.23, L_{de} is 1.7 mi for ATS and 8.3 mi for PTSF. The latter is interpolated based upon a demand flow rate of 390 pc/h (obtained from Example 14-2, previous section). From Table 14.24, f_{pl} is 1.10 for ATS and 0.61 for PTSF. Then:

$$L_{d,ATS} = 15.0 - (2.00 + 0.75 + 1.70) = 10.55 \text{ mi}$$

$$L_{d,PTSF} = 15.0 - (2.00 + 0.75 + 8.3) = 3.95 \text{ mi}$$

and:

$$ATS_{pl} = \frac{47.3 * 15.0}{2.00 + 10.55 + \left(\dfrac{0.75}{1.10}\right) + \left(\dfrac{2*1.7}{1+1.10}\right)}$$
$$= 47.8 \text{ mi/h}$$

$$PTSF_{pl} = \frac{78.9 \left[\begin{array}{l} 2.00 + 3.95 + 0.61 * 0.75 \\ + \left(\dfrac{1 + 0.61}{2}\right) 8.3 \end{array} \right]}{15.0}$$
$$= 68.8\%$$

Thus, the placement of a single 0.75-mi passing lane in this 15-mile directional segment increases the ATS slightly (from 47.3 mi/h to 47.8 mi/h) and decreases the PTSF by 10%, from 78.9% to 68.8%.

While the single passing lane as described has a small impact on operations over the 15-mile analysis segment, increasing the length of the passing lane would have a larger impact. Table 14.25 illustrates the impact of increasing the length of the passing lane by 0.50-mile increments to a total of 4.0 miles on ATS and PTSF. Increasing the length of the passing lane fourfold adds only about 1 mi/h to ATS and reduces PTSF another 6.6% to 62.2%. In this case, where PTSF controls the LOS, the passing lane must be at least 3.0 mi long to reduce PTSF to below 65%, the threshold for LOS C.

This example illustrates that the impact of having a single passing lane in the 15-mile directional segment is more important than its length. Simply having a passing lane (of minimum length 0.75 mi) reduces PTSF by 10%. Increasing its length to 4.0 miles provides additional reductions of only 6.6%. ATS is not significantly affected in this case. It should be noted, however, that this lane is inserted in a section with 40% "No Passing" zones, indicating that there are a number of passing opportunities in the

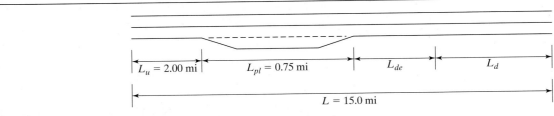

Figure 14.5: Directional Segment for Passing Lane Example

Table 14.25: Impact of Length of Passing Lane on Example

L_{pl}	1.00	1.50	2.00	2.50	3.00	3.50	4.00
L	15.00	15.00	15.00	15.00	15.00	15.00	15.00
L_u	2.00	2.00	2.00	2.00	2.00	2.00	2.00
$L_{de,ATS}$	1.70	1.70	1.70	1.70	1.70	1.70	1.70
$L_{d,ATS}$	10.30	9.80	9.30	8.80	8.30	7.80	7.30
$L_{de,PTSF}$	8.30	8.30	8.30	8.30	8.30	8.30	8.30
$L_{d,PTSF}$	3.70	3.20	2.70	2.20	1.70	1.20	0.70
$f_{pl,ATS}$	1.10	1.10	1.10	1.10	1.10	1.10	1.10
$f_{pl,PTSF}$	0.61	0.61	0.61	0.61	0.61	0.61	0.61
ATS_d	47.30	47.30	47.30	47.30	47.30	47.30	47.30
$PTSF_d$	78.90	78.90	78.90	78.90	78.90	78.90	78.90
ATS_{pl}	47.85	48.00	48.14	48.29	48.44	48.59	48.74
$PTSF_{pl}$	68.34	67.31	66.28	65.26	64.23	63.21	62.18

Note: All L in mi; ATS in mi/h; $PTSF$ in %.

segment even without a passing lane. The impact of adding a passing lane would be greater in a section with more restricted passing in the normal two-lane cross-section.

The HCM 2000 provides a modified procedure for segments in which L_{de} would extend beyond the boundaries of the analysis segment. Consult the manual [1] directly for this detail.

14.4.9   Impact of Climbing Lanes

On specific upgrades, climbing lanes are often added to avoid development of long queues behind slow-moving vehicles, particularly heavy vehicles unable to maintain speed on upgrades. According to AASHTO criteria [2], climbing lanes are warranted on two-lane rural highways when:

- The directional flow rate on the upgrade exceeds 200 veh/h.

- The directional flow rate for trucks on the upgrade exceeds 20 veh/h.

- Any of the following conditions apply: (1) a speed reduction of 10 mi/h for a typical heavy truck, (2) LOS E or F exists on the upgrade, or

(3) the LOS on the upgrade is two or more levels below that existing on the approach to the upgrade.

Values of ATS and PTSF for the upgrade may be modified to take into account the existence of a climbing lane using Equations 14-15 and 14-16 from the previous section, except that:

- L_u, L_{de}, and L_d are generally equal to 0.0 mi

- Adjustment factors f_{pl} are selected from Table 14.26 instead of Table 14.24.

14.5   Summary

Two-lane rural highways continue to form a vital part of the nation's intercity and rural roadway networks. Their unique operating characteristics require careful consideration, as poor operations will often be a harbinger of safety problems, particularly on Class I roadways. Passing and climbing lanes are often used to improve the operation of two-lane highways, but serious accident or operational problems generally require more comprehensive solutions. Such solutions may involve creating a continuous three-lane alignment with alternating passing

Table 14.26: Adjustment Factors for Climbing Lanes in Single-Direction Segments of Two-Lane Rural Highways

Directional Flow Rate (pc/h)	Adjustment Factor f_{pl} for PTSF	Adjustment Factor f_{pl} for ATS
0–300	0.20	1.02
>300–600	0.21	1.07
>600	0.23	1.14

(Used with permission of Transportation Research Board, National Research Council, *Highway Capacity Manual*, 4th Edition, Washington DC, 2000, Exhibit 20-27, pg. 20-29.)

zones for each direction of flow, or reconstruction as a multilane or limited-access facility.

Operational quality is not as serious a concern on Class II highways, where local service or all-weather access to remote areas is the primary function served.

References

1. *Highway Capacity Manual*, 4th Edition, Transportation Research Board, National Research Council, Washington DC, 2000.

2. *A Policy on Geometric Design of Highways and Streets*, 4th Edition, American Association of State Highway and Transportation Officials, Washington DC, 2001.

3. Prisk, C.W., "Passing Practices on Two-Lane Highways," *Proceedings of the Highway Research Board*, Transportation Research Board, National Research Council, Washington DC, 1941.

4. Weaver, G.D. and Glennon, J.C., *Passing Performance Measurements Related to Sight Distance Design*, Report No. 134-6, Texas Transportation Institute, Texas A&M University, College Station TX, July 1971.

5. Weaver, G.D. and Woods, D.L., *Passing and No-Passing Signs, Markings, and Warrants*, Report No. FHWA-RD-79-5, Federal Highway Administration, U.S. Department of Transportation, Washington DC, Sept 1978.

6. *Manual of Uniform Traffic Control Devices*, Millennium Edition, Federal Highway Administration, U.S. Department of Transportation, Washington DC, 2000.

7. Harwood, D.W. et al., *Capacity and Quality of Service of Two-Lane Highways*, Final Report, NCHRP Project 3-55(3), Midwest Research Institute, Kansas City MO, 2000.

Problems

14-1. A Class I rural two-lane highway segment of 20 miles in rolling terrain has a base free-flow speed (BFFS) of 60 mi/h. It consists of 12-ft lanes, with 4-ft clear shoulders on each side of the roadway. There are an average of 10 access points per mile along the segment in question, and 50% of the segment consists of "no-passing" zones. Current traffic on the facility is 350 veh/h (total, both ways), including 12% trucks and 6% RVs. The directional distribution of traffic is 60/40, and the PHF is 0.82.

(a) For the current conditions described, at what level of service would the facility be expected to operate? Use a two-way analysis.

(b) If traffic is growing at a rate of 8% per year, how many years will it be before the capacity of this facility is reached?

14-2. A Class I rural highway segment of 10 miles in level terrain has a base free-flow speed (BFFS) of 65 mi/h. The cross-section has 12-ft lanes and 6-ft clear shoulders on both sides of the road. There are an average of 20 access points per mile along the segment, and 20% of the segment consists of "no-passing" zones. Current traffic on the facility is 800 veh/h, with 10% trucks and no RVs. The directional distribution of traffic is 70/30, and the PHF is 0.87. Using a directional analysis, what is the level of service expected in each direction on this segment?

14-3. A Class I rural two-lane highway segment contains a significant grade of 4%, 2 miles in length. The base free-flow speed (BFFS) of the facility is 55 mi/h, and the cross-section is 11-ft lanes with 2-ft clear shoulders on both sides. There are 5 access points per mile along the segment, and 80% of the segment consists of "no-passing" zones. Current traffic on the facility is 250 veh/h, including 20% trucks and 10% RVs. Directional distribution is 70/30 (70% traveling upgrade), and the PHF is 0.88. What level of service is expected on the upgrade and downgrade of this segment?

14-4. A Class II rural two-lane highway segment in rolling terrain has a base free-flow speed (BFFS) of 50 mi/h. The highway has 11-ft lanes and 4-ft clear shoulders on both sides. There are 10 access points/mile, and 70% of the segment is in "no-passing" zones. Current traffic is 400 veh/h, with 10% trucks, 3% RVs, and a PHF of 0.81. The directional distribution is 60/40. Using a two-way analysis, what level of service is expected?

14-5. Reconsider Problem 14-1 by doing a single-direction analysis of the peak (60%) direction of flow. What are the resulting ATS, PTSF, and LOS? How would this be affected by adding a single 3-mile passing lane beginning 5.0 miles from the beginning of the segment?

14-6. What would be the impact of adding a truck climbing lane to the upgrade described in Problem 14-3?

Signing and Marking for Freeways and Rural Highways

The principal forms of traffic control implemented on freeways and rural highways involve road markings and signing. At-grade intersections occur on most rural highways (multilane and two-lane) and in some cases signalized intersections. The conventions and norms for signalization are the same as for urban and suburban signalization, with the exception that higher approach speeds are generally present in such cases. This, however, is taken into account as part of the normal design and analysis procedures for signalized intersections.

This chapter presents some of the special conventions for application of traffic signs and markings on freeways and rural highways.

15.1 Traffic Markings on Freeways and Rural Highways

Traffic markings on freeways and rural highways involve lane lines, edge markings, gore area and other specialized markings at and in the vicinity of ramps or interchanges. On two-lane rural highways, centerline markings are used in conjunction with signs to designate passing and no-passing zones.

15.1.1 Freeway Markings

Figure 15.1 illustrates typical mainline markings on a freeway. Standard lane lines are provided to delineate lanes for travel. Right-edge markings consist of a solid white line, and left-edge markings consist of a solid yellow line. On freeways, lane markings and edge markings are mandated by the MUTCD [1].

15.1.2 Rural Highway Markings

General highway segment marking conventions for rural highways vary according to the specific configuration that is present. MUTCD criteria for placement of centerline and edge markings on rural highways are:

- Centerline markings *shall* (standard) be placed on all paved two-way streets or highways that have three or more traffic lanes.
- Centerline markings *should* (guidance) be placed on all rural arterials and collectors that have a travelled width of more than 18 ft *and* an ADT of 3,000 veh/day or greater.

Figure 15.1: Typical Markings on a Freeway

- Edge lane markings *shall* (standard) be placed on paved streets or highways with the following characteristics: freeways, expressways, or rural highways with a traveled way of 20 ft or more in width and an ADT of 6,000 veh/day or greater.
- Edge lines *should* (guidance) be placed on rural highways with a traveled way of 20 ft or more in width and an ADT of 3,000 veh/day.

From these standards and guidelines, two-lane rural highways with an ADT of less than 3,000 veh/day are the only types of highways for which centerline and edge-line markings need not be used. In many of these cases, it would be wise to provide centerlines, if only to clarify whether or not passing is permitted. On an unmarked, two-lane rural highway, it is the driver's responsibility to recognize situations in which a passing maneuver would be unsafe.

Typical two-lane, two-way rural highway markings are illustrated in Figure 15.2. Both edge markings are white, while the centerline markings are yellow. Standard centerline markings for two-lane highways are as follows:

- A single dashed yellow line signifies that passing is permitted from either lane.

- A double solid yellow line signifies that passing is prohibited from either lane.
- A double yellow line, one solid, one dashed signifies that passing is permitted from the lane adjacent to the dashed yellow line, and that passing is prohibited from the lane adjacent to the solid yellow line.

The "No Passing Zone" sign is classified as a warning sign, although it has a unique pennant shape. It is placed on the left side of the road at the beginning of a marked "No Passing" zone.

Most rural highways of four or six lanes are marked similarly to the freeway example shown in Figure 15.1. Right- and left-edge markings are provided, as well as lane lines and centerlines. While centerline markings are mandated for four- and six-lane highways, edge markings are required only where ADTs exceed 6,000 veh/day. There is no requirement for use of lane lines on surface rural four- or six-lane highways. Nevertheless, usual practice calls for use of both centerlines and lane lines as a minimum on four-lane and six-lane rural highways. In addition, such highways would not normally be provided for low ADTs, and edge markings are therefore often installed as well.

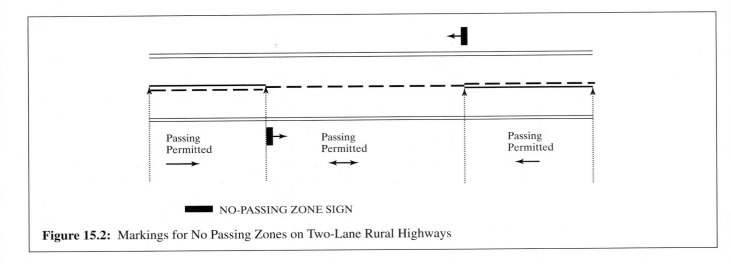

Figure 15.2: Markings for No Passing Zones on Two-Lane Rural Highways

One rural highway alignment calls for special markings. In the 1940s and 1950s, three-lane rural highway alignments were provided in areas where two-lane highways were becoming congested or where additional capacity beyond what a two-lane highway could provide was considered necessary. They were initially viewed as a low-cost alternative to improve two-lane highway operations without creating a full multilane highway of four to six lanes. Initially, they were marked to provide for one lane in each direction and a center passing lane shared by both directions of flow. As there were not restrictions to the use of the passing lane, these roadways eventually produced high accident and fatality rates. Modern practice is to assign two lanes to one direction and one lane to the other, alternating the assignment at appropriate intervals and locations. The marking conventions for this type of operation are shown in Figure 15.3.

Where passing from the single-lane direction is permitted, it is necessary that passing sight distance be available to drivers in that direction. These sight distances are the same as those defined for two-lane highways, as discussed in Chapter 14. Where the additional lane assignment is alternated at frequent intervals (less than every two miles or so), passing from the single-lane direction is usually not permitted.

As the direction of the center lane in a three-lane alignment must be periodically alternated, special transition markings must be provided at these locations, as illustrated in Figure 15.4.

The advance warning distance, d, is found from Chapter 4, where advance warning sign distances are prescribed. A warning sign indicating the merging of two lanes into one would be posted at this point, and the lane line would be discontinued as shown.

Distance L is the length of the taper markings. These are related to the posted or statutory speed limit as indicated in Equations 15-1:

For speed limits of 70 mi/h or more:

$$L = WS \qquad (15\text{-}1a)$$

For speed limits less than 70 mi/h:

$$L = \frac{WS^2}{60} \qquad (15\text{-}1b)$$

where L =   length of taper, ft

        W =   width of the center lane, ft

        S =   85th percentile speed, or speed limit, mi/h

Distance B is the buffer distance between the two transition markings. It should be a minimum of 50 ft long. If there is a no passing restriction in both directions at the location of the transition zone, the buffer zone is the distance between the beginning of the no-passing zones in each direction. This is an issue only in cases marked to allow passing from the single-lane direction.

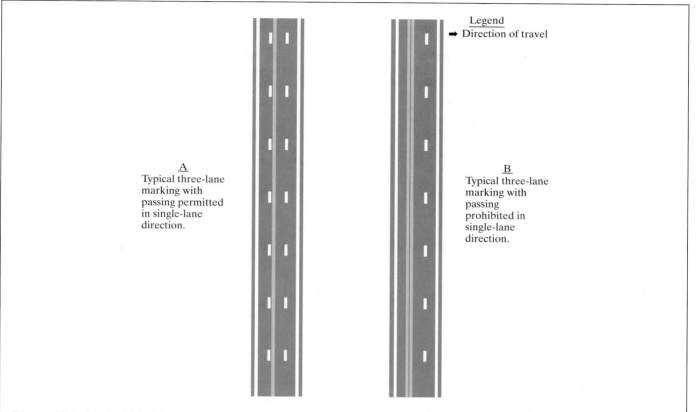

Figure 15.3: Typical Markings on Three-Lane Rural Highways (Used with permission of Federal Highway Administration, *Manual on Uniform Traffic Control Devices*, Millennium Edition, Washington DC, 2000, Figure 3B-3, pg. 3B-5.)

15.1.3 Ramp Junction Markings

Ramp junctions occur at all freeway interchanges and on all types of rural highways at locations where grade separated interchanges are provided. Figure 15.5 illustrates typical markings for off-ramps.

Two basic design types are used: parallel deceleration lanes and tapered deceleration lanes. The former are the more common type. Typical edge and lane markings are carried through the off-ramp junction; the right-edge marking, however, follows the right edge of the deceleration lane and ramp. In the case of a parallel deceleration lane, the lane line between the deceleration lane and lane 1 (right lane) of the freeway or highway is carried one-half the distance between the beginning of the deceleration lane (taper included) and the gore area. A dotted line extension of this lane line is optional, and is

often included. In the case of a tapered deceleration lane, a dotted line is used between the deceleration lane and the right lane of the freeway or highway.

The gore area itself is delineated with channelizing lines. The theoretical gore point is the beginning of the channelizing line of the gore area. The interior of the gore area is often marked with chevron markings, as shown. Note the orientation of the chevron markings. They are positioned to visually guide a driver who encroaches onto the gore area back into the appropriate lane (either the ramp or right lane of the freeway or highway).

On-ramp junctions are similarly marked, except that chevron markings are not used in the interior of the gore area. This is because vehicles rarely encroach on this area at an on-ramp. Tapered on-ramp markings involve an extended lane line between the ramp lane and the right lane

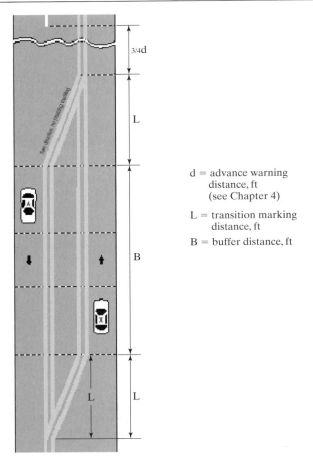

d = advance warning
       distance, ft
       (see Chapter 4)

L = transition marking
       distance, ft

B = buffer distance, ft

Figure 15.4: Transition Markings on a Three-Lane Rural Highway (Modified from Federal Highway Administration, *Manual on Uniform Traffic Control Devices*, Millennium Edition, Washington DC, 2000, Figure 3B-4, pg. 3B-7.)

of the freeway or highway. It is extended to a point where the width of the ramp lane is equal to the width of the right-most lane. This is illustrated in Figure 15.6.

15.2 Establishing and Posting of Speed Limits

A brief discussion of speed limit signs was included in Chapter 4. On freeways and rural highways, speed limits are generally of the linear type (i.e., applying to a

specified linear section of a designated highway). The MUTCD requires the posting of speed limit signs at:

- Points of change from one speed limit to another
- Beyond major intersections and at other locations where it is necessary to remind drivers of the speed limit that is applicable.

In practical terms, this is generally interpreted to mean that speed-limit signs should be located at points of change and within 1,000 ft of major entry locations. "Major" entry locations would include all on-ramps on freeways or other rural highways and significant at-grade intersections on rural highways.

Where the state statutory speed limit is in effect, signs should be periodically posted reminding drivers of this fact. Placement of such signs along freeways and rural highways follows the same pattern as for the posting of other speed limits. At borders between states, signs indicating the statutory speed limit of the state being entered should also be placed.

The setting of an appropriate speed limit for a freeway or rural highway calls for much judgment to be exercised. While the national 55-mi/h speed limit was in effect, most freeways and high-type rural highways were set at this level. With this restriction no longer in effect, the selection of an appropriate speed limit is once again a significant control decision for freeways and rural highways.

The general philosophy applied to setting speed limits is that the majority of drivers are not suicidal. They will, with no controls imposed, tend to select a range of speeds that is safe for the conditions that exist. Using this approach, speed limits should be set at the 85th percentile speed of free-flowing traffic on the facility, rounded up to the nearest 5 mi/h (or 10 km/h).

The traffic engineer, however, must also take into account other factors that might make it prudent to establish a speed limit that is slower than the 85th percentile speed of free-flowing traffic. Some of these factors include:

- Design speed of the facility section
- Details of the roadway geometry, including sight distances
- Roadside development intensity and roadside environment

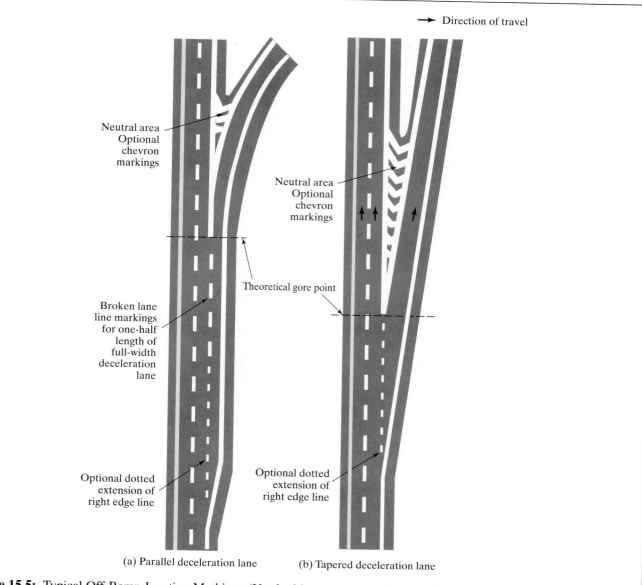

Direction of travel

Neutral area
Optional
chevron
markings

Neutral area
Optional
chevron
markings

Theoretical gore point

Broken lane
line markings
for one-half
length of
full-width
deceleration
lane

Optional dotted
extension of
right edge line

Optional dotted
extension of
right edge line

(a) Parallel deceleration lane   (b) Tapered deceleration lane

Figure 15.5: Typical Off-Ramp Junction Markings (Used with permission of Federal Highway Administration, *Manual on Uniform Traffic Control Devices*, Millennium Edition, Washington DC, 2000, Figure 3B-8a,b, pg. 3B-16.)

- Accident experience
- Observed pace speeds (10 mi/h increment with the highest percentage of drivers)

A reduction of the speed limit below the 85th percentile speed is usually required in traffic environments that contain elements that are difficult for drivers to perceive. Such limits, however, require vigilant enforcement to deter drivers from following their own judgments on appropriate speeds.

A number of different types of speed limits may be established. In addition to the primary speed limit,

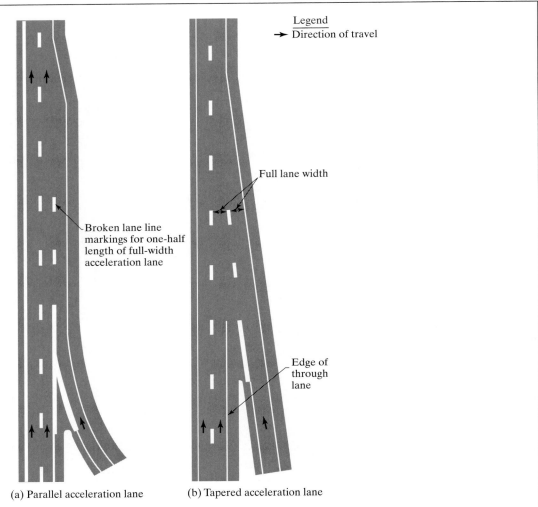

Figure 15.6: Typical On-Ramp Junction Markings (Used with permission of Federal Highway Administration, *Manual on Uniform Traffic Control Devices*, Millennium Edition, Washington DC, 2000, Figure 3B-9a,b, pg. 3B-18.)

which applies to all vehicles, additional speed limits may be set as follows:

- Truck speed limits
- Night speed limits
- Minimum speed limits

Truck speed limits only apply to trucks (as defined in each state's vehicle and traffic code or law). They are generally introduced in situations where operation of trucks at the general speed limits involves safety issues. Night speed limits are frequently used in harsh terrain, where reduced night visibility would make it unsafe to drive at the general speed limit. The night speed limit sign has white reflectorized lettering on a black background.

Minimum speed limits are employed to reduce the variability of individual vehicle speeds within the traffic stream. They are generally applied to freeways, and are infrequently used on other types of rural highways.

All applicable speed limits should be posted at the same locations, with the general caveat that no more than three speed limits should be posted at any given location.

Additional speed signs can be used to inform drivers of forthcoming changes in the speed limit. These include signs indicating "Reduced Speed Ahead" or "Speed Zone Ahead."

15.3 Guide Signing of Freeways and Rural Highways

One of the most important elements of traffic control on freeways and rural highways is proper guide signing. These types of highways often serve significant numbers of drivers who are not regular users of the facility. These drivers are the primary target of guide signing. A confused driver is a dangerous driver. Thus, it is important that unfamiliar users be properly informed in terms of routing to key destinations that they may be seeking. There are many elements involved in doing this effectively, as discussed in the sections that follow.

15.3.1 Reference Posts

Reference posts provide a location system along highways on which they are installed. Formerly (and still often) referred to as "mileposts," reference posts indicate the number of miles (or kilometers in states using a metric system) along a highway from a designated terminus. They provide a location system for accidents and emergencies and may be used as the basis for exit numbering.

The numbering system must be continuous within a state. By convention, mile (or kilometer) "0" is located:

- At the southern state boundary or southernmost point on a route beginning within a state, on north-south routes.

- At the western state boundary or westernmost point on a route beginning within a state, on east-west routes.

Cardinal directions used in highway designations recognize only two axes: North-South or East-West. Each highway is designated based on the general direction of the route within the state. Thus, a north-south route may have individual sections that are aligned east-west and vice versa. In general, if a straight line between the beginning and end of a route within a given state makes an angle of less than 45° *or* more than 135° with the horizontal, it is designated as an east-west route. If such a line makes an angle of between 45° and 135° with the horizontal, it is designated as a north-south route.

The MUTCD requires that reference posts be placed on all freeway facilities and on expressway facilities that are "located on a route where there is reference post continuity." Other rural highways may also use reference posts, but this is entirely optional.

When used, reference posts are placed every mile (or kilometer) along the route. They are located in line with delineators, with the bottom of each post at the same height as the delineator. On interstate facilities, additional reference posts are generally placed for each 1/10th mile. Typical reference posts are shown in Figure 15.7.

15.3.2 Numbered Highway Systems

There are four types of highway systems that are numbered in the United States. The two national systems are the National System of Interstate and Defense Highways

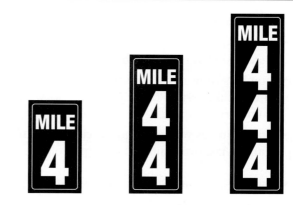

Figure 15.7: Reference Posts for Freeways and Expressways (Used with permission of Federal Highway Administration, *Manual on Uniform Traffic Control Devices*, Millennium Edition, Washington DC, 2000, pg. 2D-42.)

(the Interstate System), and the U.S. system. These are numbered by the American Association of State Highway and Transportation Officials (AASHTO) based on recommendations from individual state highway departments in accordance with published policies [2, 3]. State and county roadway systems are numbered by the appropriate agencies in accordance with standards and criteria established by each state.

The oldest system of numbered highways is the U.S. system, developed in an often-heated series of meetings of the then American Association of State Highway Officials (AASHO) and representatives from state highway agencies between 1923 and 1927 [4]. A loose system of named national routes existed at the time, with most of these "named" routes (such as the Lincoln Highway) sponsored by private organizations and motorists' clubs. This was to be replaced by a national system of numbered highways. Highways that afforded significant travel over more than one state were eligible to be considered for inclusion. The U.S. system was initially envisioned to include approximately 50,000 miles of highway, but when it was formally established on November 11, 1926, it included close to 75,000 miles. The numbering system loosely followed a convention:

- Principal north-south routes were given numbers (of one or two digits) ending in "1."
- North-south routes of secondary importance were given numbers (of one or two digits) ending in "5."
- Transcontinental and principal east-west routes were assigned route numbers in multiples of 10.
- Numbers of principal and secondary routes were to be in numerical order from east to west and from north to south.
- Branch routes were assigned three-digit numbers, with the last two being the principal route.

Many of these conventions were adapted for the Interstate System route designations:

- All primary east-west routes have two-digit even numbers.
- All primary north-south routes have two-digit odd numbers.

- All branch routes have three-digit numbers, with the last two representing the primary route.

Figure 15.8 illustrates this system, depicting the primary interstate routes emanating from New York City. The principal north-south route serving the entire East Coast is I-95. The principal east-west routes serving New York are I-80 and I-78. Branch routes connecting with I-95, such as the Baltimore and Washington Beltways, have 3-digit numbers ending in "95." Note that three-digit route numbers need not be unique. There are several routes 495, 695, and 895 along the eastern coast, all providing major connections to I-95. Principal interstate route numbers are, however, unique.

Numbered routes are identified by the appropriate shield bearing the route number and an auxiliary panel indicating the cardinal direction of the route. Standard shield designs are illustrated in Figure 15.9. The U.S. and Interstate shields each have a standard design. The shape of each state shield is designated by the state transportation agency. County shields are uniform throughout the nation; the name of the county appears as part of the shield design.

When numbered routes converge, both route numbers are signed using the appropriate shields. All route shields are posted at common locations. As cardinal directions indicate the general direction of the route, as described previously, it is possible to have a given section of highway with multiple route numbers signed with different cardinal directions. For example, a section of the New York State Thruway (a north-south route) that is convergent with the Cross-Westchester Expressway (an east-west route) is signed as "I-87 North" and "I-278 West." The reverse direction bears "south" and "east" cardinal directions.

15.3.3 Exit Numbering Systems

On freeways and some expressways, interchanges are numbered using one of two systems:

- *Milepost numbering.* The exit number is the milepost number closest to the interchange.
- *Sequential numbering.* Exits are sequentially numbered; Exit 1 begins at the westernmost or southernmost interchange within the state.

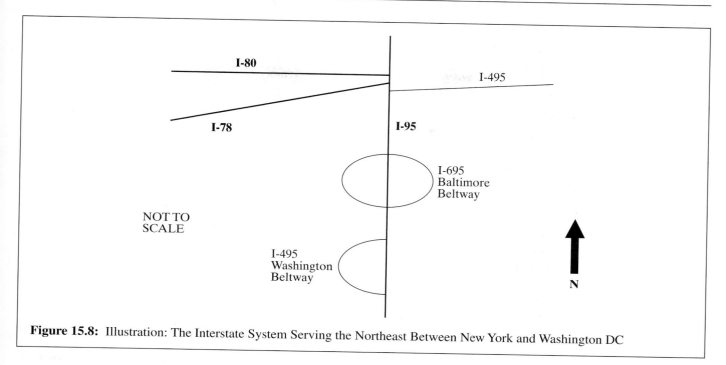

Figure 15.8: Illustration: The Interstate System Serving the Northeast Between New York and Washington DC

Milepost numbering is the preferred system according to the MUTCD, and many states have been converting from sequential numbering to this system. While milepost numbers are not sequential, drivers can use the exit number in conjunction with mileposts to determine the distance to the desired interchange. Another advantage is that interchanges may be added without disrupting the numbering sequence. In sequential systems, addition of an interchange forces the use of "A" and "B" supplemental designations, which are also used for interchanges with separate ramps for each direction of the intersecting route. This dual use of supplemental designations can cause confusion for unfamiliar motorists.

The most significant argument in favor of sequential numbering is that it is a system with which drivers have become familiar and comfortable. Historically, all freeways were initially signed using sequential numbering, with milepost numbering introduced primarily over the past 20 years.

Where routes converge, mileposts and interchange numbers are continuous for only one route. In terms of hierarchy, Interstate routes take precedence over all other systems, followed (in order) by U.S. routes, state routes, and county routes. Where two routes of similar precedence converge, primary routes take precedence over branch routes; where two routes have exactly the same precedence, the state transportation agencies determines which one is continuously numbered.

Mileposts, numbered routes, and interchange numbers are important elements in route guidance for both freeways and conventional rural highways. All can be used in guide signing to present information in a clear and consistent fashion that will minimize confusion to motorists.

15.3.4 Route Sign Assemblies

A route sign assembly is any posting of single or multiple route number signs. Where numbered routes converge or divide or intersect with other numbered routes, the proper design of route sign assemblies is a critical element in providing directional guidance to motorists. The MUTCD defines a number of different route sign assemblies, as follows:

- *Junction assembly*. Used to indicate an upcoming intersection with another numbered route(s).

Figure 15.9: Route Markers Illustrated (Used with permission of Federal Highway Administration, *Manual on Uniform Traffic Control Devices*, Millennium Edition, Washington DC, 2000, pg. 2D-7.)

- *Advance route turn assembly*. Used to indicate that a turn must be made at an upcoming intersection to remain on the indicated route.

- *Directional assembly*. Used to indicate required turn movements for route continuity at an intersection of numbered routes as well as the beginning and end of numbered routes.

- *Confirming or reassurance assemblies*. Used after motorists have passed through an intersection of numbered routes. Within a short distance, such an assembly assures motorists that they are on the intended route.

- *Trailblazer assemblies*. Used on non-numbered routes that lead to a numbered route; "To" auxiliary panel is used in conjunction with the route shield of the numbered route.

The MUTCD gives relatively precise guidelines on the exact placement and arrangement of these assemblies. A junction assembly, for example, must be placed in the block preceding the intersection in urban areas or at least 400 ft in advance of the intersection in rural areas. Further, it must be at least 200 ft in advance of the directional guide sign for the intersection and 200 ft in advance of the advance route turn assembly.

The directional assembly is an arrangement of route shields and supplementary directional arrows for all intersection routes. Routes approaching the intersection from the left are posted at the top and/or left of the assembly. Routes approaching the intersection from the right are posted at the right and/or bottom of the assembly. Routes passing straight through the intersection are posted in the center of a horizontal or vertical display.

Figure 15.10 shows two typical examples of the use of route marker assemblies. Both show signing in only one direction; each approach to the intersection would have similar signing.

In both illustrations, signing is for a driver approaching from the south. The first assembly in both cases is the junction assembly. Note that the two illustrations show two different styles of sign that can be used for this purpose. The junction assembly is followed by a directional guide sign (form for conventional highways). Neither example includes an advance turn assembly, as drivers approaching from the south do not have to execute a turn to stay on the same route. The

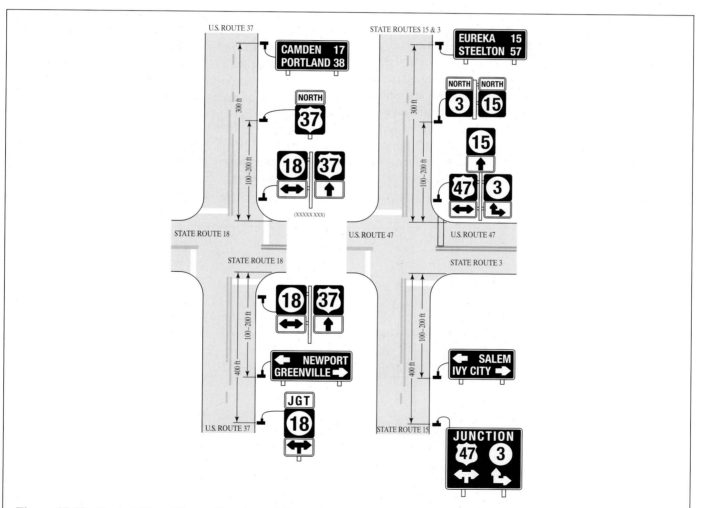

Figure 15.10: Typical Use of Route Sign Assemblies (Used with permission of Federal Highway Administration, *Manual of Uniform Traffic Control Devices*, Millennium Edition, Washington DC, Figure 2D-2, pg. 2D-18.)

next placement is the directional assembly. The standard location for this is on the far side of the intersection. In one of the illustrations, a duplicate is provided on the near side of the intersection. Once the intersection is crossed, a confirming assembly is posted within 200 ft of the intersection, so that drivers may be assured that they are indeed on their desired route. These are followed by optional destination distance guide signs.

Route sign assemblies provide critical information to motorists on numbered routes. Figure 15.11 illustrates the typical situation that such assemblies must clarify.

Before the age of the Interstate System, numbered U.S. and state routes served most interstate or long-distance intercity travel by vehicle. Unlike freeways, however, most surface routes go through every significant town and city along the way. Thus, Figure 15.11 illustrates a typical condition: many numbered routes converging on a town or city, diverging again on the other side of town. These routes, however, are generally not continuous, and they often converge and diverge several times within the town or city. The only way for an unfamiliar driver to negotiate the proper path through the town to either stay on a particular route or turn onto another is to follow appropriate route sign assemblies.

15.3.5 Freeway and Expressway Guide Signing

Freeways and most expressways have numbered exits and reference posts, and guide signing is keyed to these features. As noted in Chapter 4, guide signing provides route or directional guidance, information on services, and information on historical and natural locations. Signs are rectangular, with the long dimension horizontal, and are color coded: white on green for directional guidance, white on blue for service information, and white on brown for information on historical destinations.

Directional guide signs, however, make up the majority of these signs and are the most important in assuring safe and unconfused operation of vehicles. A typical freeway guide sign is illustrated in Figure 15.12. In this case, it is a guide sign located just at the exit. It contains a number of informative features:

- The ramp leads to U.S. Route 56, West (U.S. shield used).
- The primary destination reached by selecting this ramp is the city or town of Utopia.

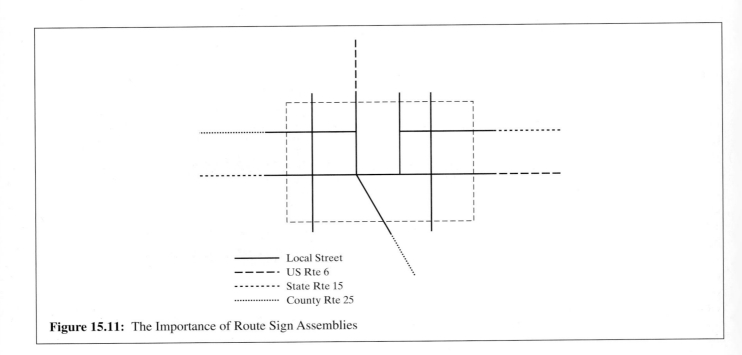

Local Street
US Rte 6
State Rte 15
County Rte 25

Figure 15.11: The Importance of Route Sign Assemblies

Figure 15.12: A Typical Freeway Directional Guide Sign (Used with permission of Federal Highway Administration, *Manual on Uniform Traffic Control Devices*, Millennium Edition, Washington DC, Figure 2E-18, pg. 2E-33.)

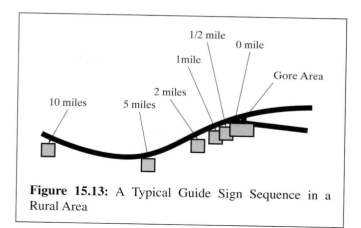

Figure 15.13: A Typical Guide Sign Sequence in a Rural Area

- The exit number is 211A. Assuming that this is a milepost-numbered sign, the exit is located at milepost 211. Exit 211A refers to the ramp leading to Route 56 WB. Exit 211B would lead to Route 56 EB.

- The exit number tab is located on the right side of the sign, indicating that this is a right-hand exit. For a left-hand exit, the tab would be located on the left side of the sign.

In general, drivers should be given as much advance warning of interchanges and destinations as possible. This leads to very different signing approaches in urban and rural areas. In rural areas, advance signing is much easier to accomplish, as there are long distances between interchanges. Figure 15.13 shows a typical signing sequence for a rural area.

The first advance directional guide sign can be as far as 10 miles away, assuming that there are no other exits between the sign and the subject interchange. If a 10-mile sign is placed, then the usual sequence would be to repeat advance signs at 5 miles, 2 miles, 1 mile, and $1/2$ mile from the interchange. Where distance between interchanges permits, the first advance sign should be *at least* 2 miles from the exit. The $1/2$-mile sign is optional, but often helpful. At the point of the exit, a large directional sign of the type shown in Figure 15.12 is placed. Another small exit number sign, of the

type illustrated in Figure 15.14, is located in the gore area, as shown in Figure 15.13.

The location of a final directional guide sign in the gore area of the off-ramp is desirable from an information point of view, but it is a problem from the point of view of safety. The gore area is an area that unsure or confused drivers often encroach upon. Thus, signs located in this area are subject to being hit in accidents. Such signs should be mounted on break-away sign supports to avoid injury to motorists and extensive damage to vehicles that enter the gore area. Where warranted, the gore-area exit panel can be replaced by the final destination

Figure 15.14: Gore Area Exit Sign (Used with permission of Federal Highway Administration, *Manual on Uniform Traffic Control Devices*, Millennium Edition, Washington DC, Figure 2E-19b, pg. 2E-50.)

sign mounted on an overhead cantilevered sign support such that the sign is situated over a portion of the gore area. The sign support would be located on the right side of the exit ramp, protected by guardrail.

In urban areas, advance signing is more difficult to achieve. To avoid confusion, advance signs for a specific interchange should not be placed in advance of the previous interchange. Thus, for example, the first advance interchange sign for Exit 2 should not be placed before Exit 1.

In urban situations, signing must be done very carefully to avoid confusing unfamiliar drivers. "Sign spreading" is a technique employed on urban freeways and expressways to avoid confusing sign sequences and the appearance of too many signs at one location. Figure 15.15 illustrates the concept. The sign for Exit 7 is placed on an overhead cantilevered support over the gore area. The first advance sign for Exit 8 is *not* placed at the same location, but rather is on another overhead support (in this case an overpass) a short distance beyond Exit 7. This is preferable to an older practice in which the advance sign for Exit 8 was placed at the same location as the Exit 7 gore area sign on a single overhead sign structure.

The signing of freeway and expressway interchanges is complex, and virtually every situation contains unique features that must be carefully addressed. Several examples from the MUTCD (Millennium Edition) are included and discussed for illustration. Additional examples exist in the MUTCD and should be consulted directly.

Figure 15.16 illustrates guide signing for a series of closely-spaced interchanges on an urban freeway. Distances are such that only one advance sign is provided for each exit, using the sign-spreading technique previously discussed. Note also that sequential exit numbers are used, with Exits 22A and 22B representing separate interchanges, doubtless due to the addition of one after the original exit numbering had been assigned. The unique feature here is the use of "interchange sequence signs" to provide additional advance information on upcoming destinations. The signs list three upcoming interchanges each, but do not list exit numbers. This avoids overlapping advance exit number sequences. Street names are given for each interchange with the mileage to each one. This effectively supplements the single advance interchange sign for each exit without presenting a confusing numbering sequence.

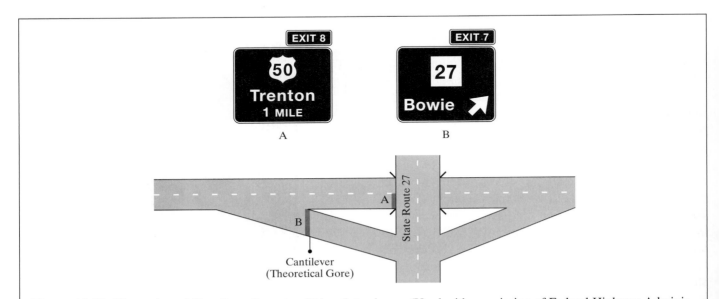

Figure 15.15: Illustration of Sign Spreading at an Urban Interchange (Used with permission of Federal Highway Administration, *Manual on Uniform Traffic Control Devices*, Millennium Edition, Washington DC, Figure 2E-1, pg. 2E-7.)

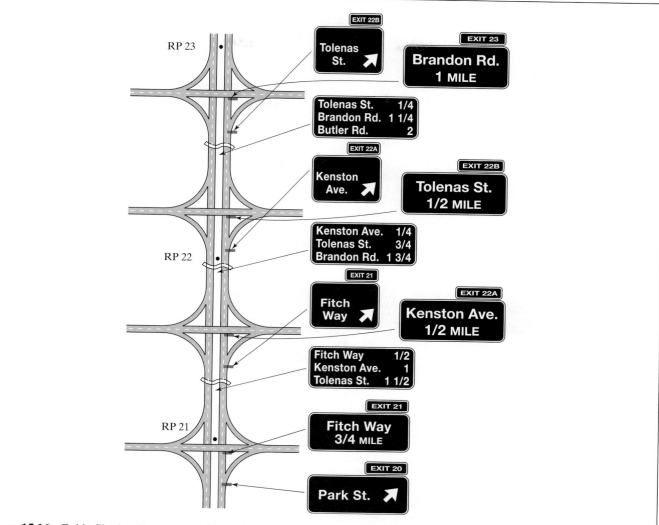

Figure 15.16: Guide Signing for a Series of Closely-Spaced Urban Interchanges (Used with permission of Federal Highway Administration, *Manual on Uniform Traffic Control Devices*, Millennium Edition, Washington DC, Figure 2E-21, pg. 2E-53.)

Figure 15.17 illustrates a complex interchange between two interstate highways, I-42 and I-17. Signing is shown for only two of the four approach directions. In the EB direction, the critical feature is that a single off-ramp serves both directions of I-17. The single ramp, however, splits shortly after leaving I-42. Thus, two closely spaced sequential diverges must be negotiated by motorists. Advance exit signs are placed at 2 miles, 1 mile, and $1/2$ mile from the exit. Arrows indicate that the exit has two lanes, and the exit tab indicates that it is

a right-side ramp. Pull-through signs for Springfield are located to the left of exit signs at the 1-mile, $1/2$ mile, and exit location. All of the advance exit signs indicate both Miami and Portland as destinations. Because of the distances involved, a single set of overhead signs indicates I-17 North and I-17 South at the location where the ramp divides. The signing of the EB approach to the interchange involves nine signs and five sign support structures.

In the NB direction, separate ramps are provided for the EB and WB directions on I-42. "A" and "B"

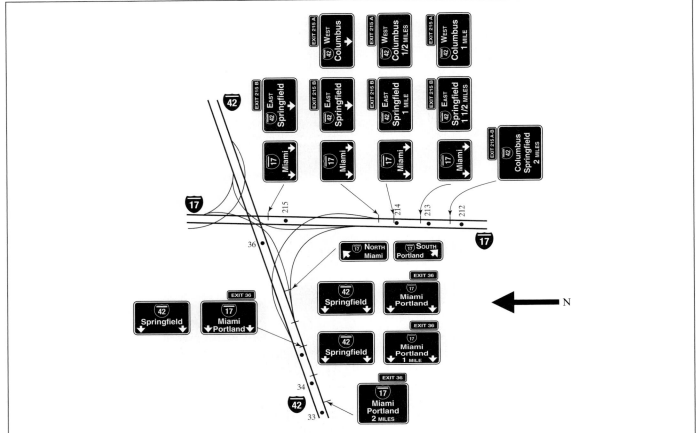

Figure 15.17: A Complex Interchange Between Two Interstate Facilities (Used with permission of Federal Highway Administration, *Manual on Uniform Traffic Control Devices*, Millennium Edition, Washington DC, Figure 2E-25, pg. 2E-58.)

exits are used to differentiate between the two, but advance signs for both are placed at the same locations, except for the first advance sign at two miles, which lists both ramps on the same sign. Pull-through signs for Miami are used at each advance location, except at two miles. For the NB approach, there are 12 signs and 5 sign-support structures.

Proper signing of a major interchange can be a considerable expense. In the case shown in Figure 15.17, 21 signs and 10 support structures are needed for two of the four approach directions. The expense is necessary, however, to provide drivers with the required information to safely navigate the interchange.

Diagrammatic guide signs may be of great utility when approaching major diverge junctions on freeways and expressways. MUTCD, however, *does not* permit their use in signing for cloverleaf interchanges, as studies have shown these to be more confusing than helpful.

Figure 15.18 illustrates the signing of a major diverge of two interstate routes, I-50 West and I-79 South. Advance signs at 2 miles, 1 mile, and $^1/_2$ mile are diagrammatic. The diagram shows how the three approach lanes split at the diverge, and allows drivers to move into a lane appropriate to their desired destinations well in advance of the gore area. The last sign at the diverge uses arrows to indicate which two lanes can be used for each route.

The guiding principle for destination guide signing is to keep it as simple as possible. Consult Chapter 4 for guidelines on how to keep the content of each sign understandable. Chapter 4 also contains guidelines and illustrations of the application of service information and historical/cultural/recreational information signing.

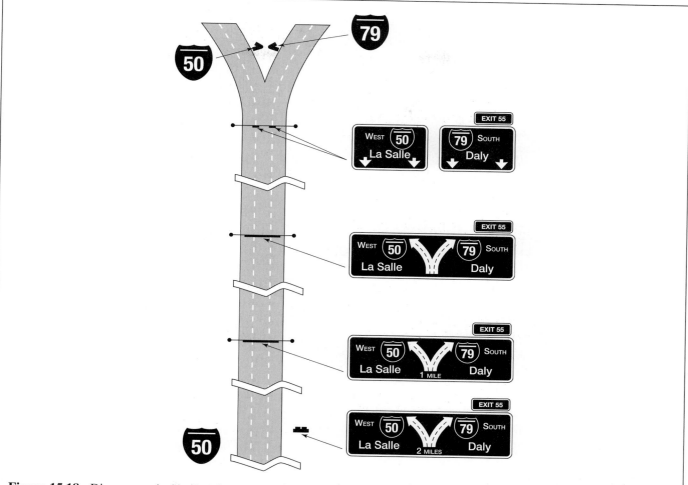

Figure 15.18: Diagrammatic Signing for a Major Diverge of Interstate Routes (Used with permission of Federal Highway Administration, *Manual on Uniform Traffic Control Devices*, Millennium Edition, Washington DC, Figure 2E-25, pg. 2E-58.)

15.3.6 Guide Signing for Conventional Roads

Guide signing for conventional roadways primarily consists of route sign assemblies, as previously discussed, and destination signing. As numbered routes are not involved in most cases, destination names become the primary means of conveying information. Advance destination signs are generally placed at least 200 ft from an intersection, with confirming destination signs located after passing through an intersection with route marker assemblies at the intersection. A sample of guide signs for conventional roads is shown in Figure 15.19.

15.4 Other Signs on Freeways and Rural Highways

Other than regulatory and guide signs, freeways and expressways do not have extensive additional signing. Warning signs are relatively rare on freeways and expressways, but would be used to provide advance warning of at-grade intersections on expressways, and would

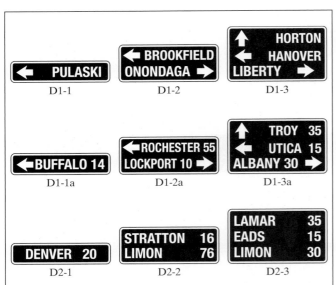

Figure 15.19: Illustrations of Conventional Road Guide Signs (Used with permission of Federal Highway Administration, *Manual on Uniform Traffic Control Devices*, Millennium Edition, Washington DC, pg. 2D-26.)

be used to warn of extended downgrades (primarily for trucks). Rarely are other elements warranting warning signs present in these types of facilities, except for animal-crossing warnings, particularly in deer country.

On conventional rural highways, warning signs are frequently warranted to warn of various hazards, including unexpected restricted geometric elements, blind driveways and intersections, crossings of various types, and advance warning of control devices, such as STOP signs and signals. Consult Chapter 4 for a detailed discussion of the application of warning signs on all types of highways.

References

1. *Manual of Uniform Traffic Control Devices*, Millennium Edition, Federal Highway Administration, U.S. Department of Transportation, Washington DC, 2000.

2. "Purpose and Policy in the Establishment and Development of United States Numbered Highways," American Association of State Highway and Transportation Officials, Washington DC, revised September 15, 1970.

3. "In the Establishment of a Marking System of Routes Comprising the National System of Interstate and Defense Highways," American Association of State Highway and Transportation Officials, Washington DC, adopted August 14, 1954, revised August 10, 1973.

4. Weingroff, R. F., "From Names to Numbers: The Origins of the U.S. Numbered Highway System," *AASHTO Quarterly*, American Association of State Highway and Transportation Officials, Washington DC, Spring 1997.

Problems

15-1. A three-lane rural highway has 12-foot lanes and a speed limit of 55 mi/h. There is no passing permitted in the direction with a single lane. What is the minimum length of the transition and buffer markings at a location where the center-lane direction is to be switched?

15-2. An expressway in a suburban area has a design speed of 65 mi/h and an 85th percentile speed of 72 mi/h. It is experiencing a high accident rate compared to similar highways in the same jurisdiction. What speed limit would you recommend? What additional information would you like to have before making such a recommendation?

Class Project

15-3. This class project should be assigned to groups with a minimum of two persons in each group.

Select a 5-mile section of freeway or rural highway in your area. Survey the section in both directions, making note of all traffic signs and markings that exist. Evaluate the effectiveness of these signs and markings, and suggest improvements that might result in better communication with motorists. Write a report on your findings, including photographs where appropriate to illustrate your comments.

15-4. Figure 15.20 illustrates a diamond interchange between a state-numbered freeway and a county road. The diamond–county road intersections are STOP-controlled. Indicate what guide signs and route signs you would place, specifying their location. Prepare a rough sketch of each sign to indicate its precise content.

Note that there are no other exits on State Route 50 within 25 miles of this location.

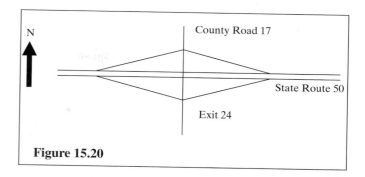

Figure 15.20

PART 4

Applications to Urban and Suburban Street Systems

CHAPTER

16

Introduction
to Intersection Control

The most complex individual locations within any street and highway system are at-grade intersections. At a typical intersection of two two-way streets, there are 12 legal vehicular movements (left turn, through, and right-turn from four approaches) and four legal pedestrian crossing movements. As indicated in Figure 16.1, these movements create many potential conflicts where vehicles and/or pedestrian paths may try to occupy the same physical space at the same time.

As illustrated in Figure 16.1, there are a total of 16 potential vehicular crossing conflicts: four between through movements from the two streets; four between left-turning movements from the two streets, and eight between left-turning movements and through movements from the two streets. In addition, there are eight vehicular merge conflicts, as right- and left-turning vehicles merge into a through flow at the completion of their desired maneuver. Pedestrians add additional potential conflicts to the mix.

The critical task of the traffic engineer is to control and manage these conflicts in a manner that ensures safety and provides for efficient movement through the intersection for both motorists and pedestrians.

16.1 The Hierarchy of Intersection Control

There are three basic levels of control that can be implemented at an intersection:

- *Level I*—Basic rules of the road
- *Level II*—Direct assignment of right-of-way using YIELD or STOP signs
- *Level III*—Traffic signalization

There are variations within each level of control as well. The selection of an appropriate level of control involves a determination of which (and how many) conflicts a driver should be able to perceive and avoid through the exercise of judgment. Where it is not reasonable to expect a driver to perceive and avoid a particular conflict, traffic controls must be imposed to assist.

Two factors affect a driver's ability to avoid conflicts: (1) a driver must be able to see a potentially conflicting vehicle or pedestrian in time to implement an avoidance maneuver and (2) the volume levels that

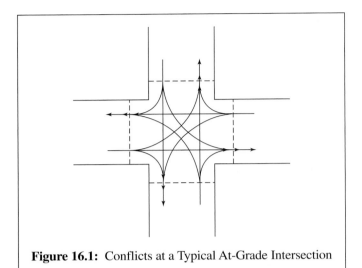

Figure 16.1: Conflicts at a Typical At-Grade Intersection

exist must present reasonable opportunities for a safe maneuver to take place. The first involves considerations of sight distance and avoidance maneuvers, while the second involves an assessment of demand intensity and the complexity of potential conflicts that exist at a given intersection.

A rural intersection of two farm roads contains all of the potential conflicts illustrated in Figure 16.1. However, pedestrians are rare, and vehicular flows may be extremely low. There is a low probability of any two vehicles and/or pedestrians attempting to use a common physical point simultaneously. At the junction between two major urban arterials, the probability of vehicles or pedestrians on conflicting paths arriving simultaneously is quite high. The sections that follow discuss how a determination of an appropriate form of intersection control can be made, highlighting the important factors to consider in making such critical decisions.

16.2 Level I Control: Basic Rules of the Road

Basic rules of the road apply at any intersection where right-of-way is not explicitly assigned through the use of traffic signals, STOP, or YIELD signs. These rules are spelled out in each state's vehicle and traffic law, and drivers are expected to know them. At intersections, all states follow a similar format. In the absence of control devices, the driver on the left must yield to the driver on the right when the vehicle on the right is approaching in a manner that may create an impending hazard. In essence, the responsibility for avoiding a potential conflict is assigned to the vehicle on the left. Most state codes also specify that through vehicles have the right of way over turning vehicles at uncontrolled intersections.

Operating under basic rules of the road does not imply that no control devices are in place at or in advance of the intersection, although that could be the case. Use of street-name signs, other guide signs, or advance intersection warning signs do not change the application of the basic rules. They may, however, be able to contribute to the safety of the operation by calling the driver's attention to the existence and location of the intersection.

In order to safely operate under basic rules of the road, drivers on conflicting approaches must be able to see each other in time to assess whether an "impending hazard" is imposed, and to take appropriate action to avoid an accident. Figure 16.2 illustrates a visibility triangle at a typical intersection. Sight distances must be analyzed to ensure that they are sufficient for drivers to judge and avoid conflicts.

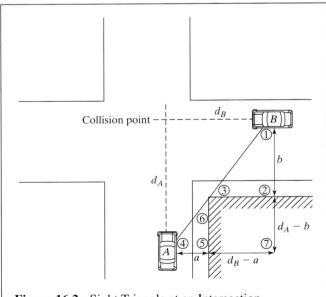

Figure 16.2: Sight Triangle at an Intersection

At intersections, sight distances are normally limited by buildings or other sight-line obstructions located on or near the corners. There are, of course, four sight triangles at every intersection with four approaches. At the point where the drivers of both approaching vehicles first see each other, Vehicle A is located a distance of d_A from the collision or conflict point, and Vehicle B is located a distance d_B from the collision point. The sight triangle must be sufficiently large to ensure that at no time could two vehicles be on conflicting paths at distances and speeds that might lead to an accident, without sufficient time and distance being available for either driver to take evasive action.

Note that the sight line forms three similar triangles with sides of the sight obstruction: $\Delta 123$, $\Delta 147$, and $\Delta 645$. From the similarity of the triangles, a relationship between the critical distances in Figure 16.2 can be established:

$$\frac{b}{d_B - a} = \frac{d_A - b}{a}$$

$$d_B = \frac{a\, d_A}{d_A - b} \qquad (16\text{-}1)$$

where: d_A = distance from Vehicle A to the collision point, ft

   d_B = distance from Vehicle B to the collision point, ft

   a = distance from driver position in Vehicle A to the sight obstruction, measured parallel to the path of Vehicle B, ft

   b = distance from driver position in Vehicle B to the sight obstruction, measured parallel to the path of Vehicle A, ft

Thus, when the position of one vehicle is known, the position of the other when they first become visible to each other can be computed. The triangle is dynamic, and the position of one vehicle affects the position of the other when visibility is achieved.

The American Association of State Highway and Transportation Officials (AASHTO) suggests that to ensure safe operation with no control, both drivers should be able to stop before reaching the collision point when they first see each other. In other words, both d_A and d_B should be equal to or greater than the safe stopping distance at the points where visibility is established. AASHTO standards [1] suggest that a driver reaction time of 2.5 s be used in estimating safe stopping distance and that the 85th percentile speed of immediately approaching vehicles be used. AASHTO does suggest, however, that drivers slow from their midblock speeds when approaching uncontrolled intersections, and use an immediate approach speed that is assumed to be lower than the design speed of the facility. From Chapter 2, the safe stopping distance is given by:

$$d_s = 1.47\, S_i t + \frac{S_i^2}{30(0.348 \pm 0.01G)} \qquad (16\text{-}2)$$

where: d_s = safe stopping distance, ft

   S_i = initial speed of vehicle, mi/h

   G = grade, %

   t = reaction time, s

   0.348 = standard friction factor for stopping maneuvers

Using this equation, the following analysis steps may be used to test whether an intersection sight triangle meets these sight distance requirements:

1. Assume that Vehicle A is located one safe stopping distance from the collision point (i.e., $d_A = d_s$), using Equation 16-2. By convention, Vehicle A is generally selected as the vehicle on the *minor* street.

2. Using Equation 16-1, determine the location of Vehicle B when the drivers first see each other. This becomes the actual position of Vehicle B when visibility is established, d_{Bact}.

3. Since the avoidance rule requires that both vehicles have one safe stopping distance available, the minimum requirement for d_B is the safe stopping distance for Vehicle B, computed using Equation 16-2. This becomes d_{Bmin}.

4. For the intersection to be safely operated under basic rules of the road (i.e., with no control), $d_{Bact} \geq d_{Bmin}$.

Historically, another approach to ensuring safe operation with no control has also been used. In this case, to avoid collision from the point at which visibility is established, *Vehicle A must travel 18 ft past the collision point in the same time that Vehicle B travels to a point 12 ft before the collision point.* This can be expressed as:

$$\frac{d_A + 18}{1.47\,S_A} = \frac{d_B - 12}{1.47 S_B}$$

$$d_B = (d_A + 18)\frac{S_B}{S_A} + 12 \qquad (16\text{-}3)$$

where all variables are as previously defined. This, in effect, provides another means of estimating the minimum required distance, d_{Bmin}. In conjunction with the four-step analysis process outlined previously, it can also be used as a criterion to ensure safe operation.

At any intersection, all of the sight triangles must be checked and must be safe in order to implement basic rules of the road. If, for any of the sight triangles, $d_{Bact} < d_{Bmin}$, then operation with no control cannot be permitted. When this is the case, there are three potential remedies:

- Implement intersection control, using STOP- or YIELD-control, or traffic signals.
- Lower the speed limit on the major street to a point where sight distances are adequate.
- Remove or reduce sight obstructions to provide adequate sight distances.

The first is the most common result. The exact form of control implemented would require consideration of warrants and other conditions, as discussed in subsequent portions of this chapter. The second approach is viable where sight distances at series of uncontrolled intersections can be remedied by a reduced, but still reasonable speed limit. The latter depends upon the type of obstruction and ownership rights.

Consider the intersection illustrated in Figure 16.3. It shows an intersection of a one-way minor street and a two-way major street. In this case, there are two sight triangles that must be analyzed. The 85th percentile immediate approach speeds are shown.

First, it is assumed that Vehicle A is one safe stopping distance from the collision point:

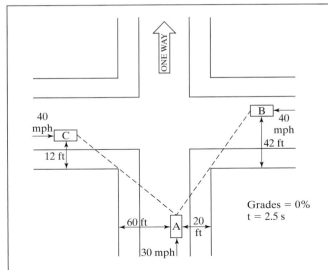

Figure 16.3: Intersection for Case Study of Sight Distance

$$d_A = 1.47*30*2.5 + \frac{30^2}{30(0.348 + 0)}$$
$$= 110.3 + 86.2 = 196.5 \text{ ft}$$

where 2.5 s is the standard driver reaction time used in safe stopping sight distance computations. Using Equation 16-1, the actual position of Vehicle B when it is first visible to the driver of Vehicle A is found:

$$d_{Bact} = \frac{a\,d_A}{d_A - b} = \frac{20*196.5}{196.5 - 42} = \frac{3,930}{154.5} = 25.4 \text{ ft}$$

This must be compared with the minimum requirement for d_B, estimated as either one safe stopping distance (Equation 16-2), or using Equation 16-3:

$$d_{Bmin} = 1.47*40*2.5 + \frac{40^2}{30(0.348 + 0)}$$
$$= 147.0 + 153.3 = 300.3 \text{ ft}$$

or:

$$d_{Bmin} = (196.5 + 18)\frac{40}{30} + 12 = 298.0 \text{ ft}$$

In this case, both of the minimum requirements are similar, and both are far larger than the actual distance of

25.4 ft. Thus, the sight triangle between Vehicles A and B fails to meet the criteria for safe operation under basic rules of the road.

Consider the actual meaning of this result. Clearly, if Vehicle A is 196.5 ft away from the collision point when Vehicle B is only 25.4 ft away from it, they will not collide. Why, then, is this condition termed "unsafe?" It is unsafe because there could be a Vehicle B, further away than 25.4 ft, on a collision path with Vehicle A and the drivers would not be able to see each other.

Since the sight triangle between Vehicles A and B did not meet the sight-distance criteria, it is not necessary to check the sight triangle between vehicles A and C. Basic rules of the road may not be permitted at this intersection without reducing major street speeds or removing sight obstructions. This implies that, in many cases, YIELD or STOP control should be imposed on the minor street as a minimum form of control.

Even if the intersection met the sight distance criteria, this does not mean that basic rules of the road should be applied to the intersection. Adequate sight distance is a necessary, but not sufficient, condition for adopting a "no-control" option. Traffic volumes or other conditions may make a higher level of control desirable or necessary.

16.3   Level II Control: YIELD and STOP Control

If a check of the intersection sight triangle indicates that it would not be safe to apply the basic rules of the road, then as a minimum, some form of level II control must be imposed. Even if sight distances are safe for operating under no control, there may be other reasons to implement a higher level of control as well. Usually, these would involve the intensity of traffic demand and the general complexity of the intersection environment.

The *Manual of Uniform Traffic Control Devices* (MUTCD) [2] gives some guidance as to conditions for which imposition of STOP or YIELD control is justified. It is not very specific, and it requires the exercise of engineering judgment.

16.3.1   Two-Way Stop Control

The most common form of Level II control is the two-way STOP sign. In fact, such control may involve one or two STOP signs, depending upon the number of intersection approaches. It is not all-way STOP control, which is discussed later in this chapter.

Under the heading of "guidance," the MUTCD suggests several conditions under which the use of STOP signs would be justified. Table 16.1 shows these warrants.

Warrant A is little more than a statement that adequate sight distances for uncontrolled operation are not available. Warrant B allows control of all local streets entering a major through facility, a measure that helps to reinforce the through character of the facility. Warrant C addresses a case in which most intersections along an arterial are signalized. If only an occasional intersection is not signalized, it is often wise to install STOP signs on the minor streets of other intersections. Leaving an intersection uncontrolled, when the arterial right-of-way is reinforced with coordinated, signals at most intersections, might give drivers (on the minor approach) a false sense of security when looking for conflicting vehicles. Warrant D addresses almost every other condition in

Table 16.1: Warrants for Two-Way STOP Control

STOP signs should not be used unless engineering judgment indicates that one or more of the following conditions are met:

a. Intersection of a less important road with a main road where application of the normal right-of-way rule would not be expected to provide reasonably safe operation.

b. Street entering a through highway or street.

c. Unsignalized intersection in a signalized area.

d. High speeds, restricted view, or crash records indicate a need for control by the STOP sign.

(Used by permission of Federal Highway Administration, U.S. Dept. of Transportation, *Manual on Uniform Traffic Control Devices*, Millennium Edition, Washington DC, 2000, pg. 2B-8.)

which the engineer judges that a STOP sign should be installed.

The MUTCD is somewhat more explicit in dealing with inappropriate uses of the STOP sign. Under the heading of a "standard" (i.e., a mandatory condition), STOP signs *shall not* be installed at intersections where traffic control signals are installed and operating. This disallows a past practice in which some jurisdictions turned signals off at night, leaving STOP signs in place for the evening hours. During the day, however, an unfamiliar driver approaching a green signal with a STOP sign could become significantly confused. The manual also disallows the use of portable or part-time STOP signs except for emergency and temporary traffic control.

Under the heading of "guidance," STOP signs *should not* be used for speed control, although this is frequently done on local streets designed in a straight grid pattern. In modern designs, street layout and geometric design would be used to discourage excessive speeds on local streets.

In general, STOP signs should be installed in a manner that minimizes the number of vehicles affected, which generally means installing them on the minor street.

AASHTO [1] also provides sight distance criteria for STOP-controlled intersections. A methodology based upon observed gap acceptance behavior of drivers at STOP-controlled intersections is used. A standard stop location is assumed for the minor street vehicle (Vehicle A in Figure 16.2). The distance to the collision point (d_A) has three components:

- Distance from the driver's eye to the front of the vehicle (assumed to be 8 ft)
- Distance from the front of the vehicle to the curb line (assumed to be 10 ft)
- Distance from the curb line to the center of the right-most travel lane approaching from the left, or from the curb line to the left-most travel lane approaching from the right

Thus:

$$d_{A\text{-}STOP} = 18 + d_{cl} \qquad (16\text{-}4)$$

where: $d_{A\text{-}STOP}$ = distance of Vehicle A on a STOP-controlled approach from the collision point, ft

d_{cl} = distance from the curb line to the center of the closest travel lane from the direction under consideration, ft

The required sight distances for Vehicle B, on the major street for STOP-controlled intersections is found as follows:

$$d_{Bmin} = 1.47 * S_{maj} * t_g \qquad (16\text{-}5)$$

where: d_{Bmin} = minimum sight distance for Vehicle B approaching on major (uncontrolled) street, ft

S_{maj} = design speed of major street, mi/h

t_g = average gap accepted by minor street driver to enter the major road, s

Average gaps accepted are best observed in the field for the situation under study. In general, they range from 6.5 s to 12.5 s depending upon the minor street movement and vehicle type, as well as some of the specific geometric conditions that exist.

For most STOP-controlled intersections, the design vehicle is the passenger car, and the criteria for left-turns are used, as they are the most restrictive. Trucks or combination vehicles are considered only when they make up a substantial proportion of the total traffic on the approach. Values for right-turn and through movements are used when no left-turn movement is present. For these typical conditions, AASHTO recommends the use of $t_g = 7.5$ s.

Consider the case of a STOP-controlled approach at an intersection with a two-lane arterial with a design speed of 40 mi/h, as shown in Figure 16.4.

Using Equation 16-4, the position of the stopped vehicle on the minor approach can be determined.

$$d_{A\text{-}STOP}(\text{from left}) = 18.0 + 6.0 = 24.0 \text{ ft}$$

$$d_{A\text{-}STOP}(\text{from right}) = 18.0 + 18.0 = 36.0 \text{ ft}$$

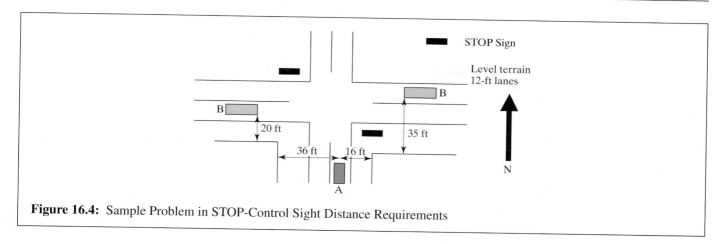

Figure 16.4: Sample Problem in STOP-Control Sight Distance Requirements

The minimum sight distance requirement for Vehicle B is determined from Equation 16-5, using a time gap (t_g) of 7.5 s for typical conditions.

$$d_{Bmin} = 1.47 * 40 * 7.5 = 441 \text{ ft}$$

Now the actual distance of Vehicle B from the collision point when visibility is established is determined using Equation 16-1:

$$d_{Bact}(\text{from left}) = \frac{36 * 24}{24 - 20} = 216 \text{ ft} < 441 \text{ ft}$$

$$d_{Bact}(\text{from right}) = \frac{16 * 36}{36 - 35} = 576 \text{ ft} > 441 \text{ ft}$$

In the case of a major street Vehicle B approaching from the left, there is not sufficient sight distance to meet the criteria. The sight distance for Vehicle B approaching from the right meets the criteria. Note that it is possible for d_{Bact} to be negative. This would indicate that there was no sight obstruction from the direction analyzed.

Where the STOP-sign sight-distance criterion is not met, it is recommended that speed limits be reduced (with signs posted) to a level that would allow appropriate sight distance to the minor street. Removal or cutting back of sight obstructions is also a potential solution, but this is often impossible in developed areas, where buildings are the principal obstructions.

16.3.2 Yield Control

A YIELD sign assigns right-of-way to the major uncontrolled street. It requires vehicles on the minor approach(es) to slow and yield the right-of-way to any major street vehicle approaching at a distance and speed that would present an impending hazard to the minor street vehicle if it entered the major street. Most state laws require that drivers on YIELD-controlled approaches slow to 8–10 mi/h before entering the major street.

Warrants for YIELD control in the MUTCD are hardly definitive, and are given only under the heading of "options." The warrants are summarized in Table 16.2.

The first condition represents a significant policy change from previous editions of the MUTCD. In essence, to use a YIELD sign, sight distances should be at least as good as those required for application of basic rules of the road. Previously, a sight distance analysis using the reduced approach speed of minor street vehicles required by the YIELD sign (8–10 mi/h) was performed. Now, the same approach speed used in the consideration of basic rules of the road must be accommodated. This requirement will probably lead to diminished use of the YIELD sign for normal intersection control in the future.

The principal uses of the YIELD sign emanate from Warrants B and C. Warrant B allows use of the YIELD sign to control channelized right turns at signalized and

Table 16.2: Warrants for YIELD Control

YIELD signs may be installed:

a. When the ability to see all potentially conflicting traffic is sufficient to allow a road use traveling at the posted speed, 85th percentile speed, or the statutory speed to pass through the intersection or stop in a safe manner.

b. If controlling a merge-type movement on the entering roadway where acceleration geometry or sight distance is not adequate for merging traffic operations.

c. At a second crossroad of a divided highway, where the median width is 30 ft or greater. A STOP sign may be installed at the entrance to the first roadway of a divided highway, and a YIELD sign may be installed at the entrance to the second roadway.

d. At an intersection where a special problem exists and where engineering judgment indicates that the problem is susceptible to correction by use of a YIELD sign.

(Used by permission of Federal Highway Administration, U.S. Dept. of Transportation, *Manual on Uniform Traffic Control Devices*, Millennium Edition, Washington DC, 2001, pg. 2B-12.)

unsignalized intersections, and on-ramp or other merge situations. Both are frequent uses in which adequate sight distance or geometry (i.e., inadequate length of the acceleration lane) make an uncontrolled merge potentially unsafe.

Warrant C indicates another common use of the YIELD sign to control entries to the second roadway of a major street with a wide median (>30 ft). In this case, however, there should be no sight distance restrictions in the median. If such obstructions exist, then an analysis of the sight triangle formed, using the same criteria as for basic rules of the road, must be done to ensure safety.

AASHTO provides a relatively complex procedure for analysis of YIELD sign sight triangles and requirements at intersections. Unfortunately, they consider the reduced approach speeds of drivers in response to YIELD control, which contradicts the latest MUTCD warrants. In general, if the sight triangles are insufficient for basic rules

of the road, then a YIELD sign should not be used on a minor intersection approach. In such cases, if Level II control is desired, a STOP sign should be placed.

16.3.3 Multiway Stop Control

Multiway STOP control, where all intersection approaches are controlled using STOP signs, remains a controversial form of control. Some agencies find it attractive, primarily as a safety measure. Others believe that the confusion that drivers often exhibit under this form of control negates any of the benefits it might provide.

MUTCD warrants and provisions with regard to multiway STOP control reflect this ongoing controversy. Multiway STOP control is most often used where there are significant conflicts between vehicles and pedestrians and/or bicyclists in all directions, and where vehicular demands on the intersecting roadways are approximately equal. Table 16.3 shows the warrants for multiway STOP control.

Despite continuing controversy over the use of multiway STOP control, these warrants are considerably more specific, particularly with regard to volume requirements, than in previous editions of the MUTCD. It should be noted that such control is generally implemented as a safety measure, as operations at such locations are often not very efficient. The fourth edition of the *Highway Capacity Manual*, [3] includes a methodology for analysis of the capacity and level of service provided by multiway STOP control. This method is discussed in Chapter 25.

16.4   Level III Control: Traffic Control Signals

The ultimate form of intersection control is the traffic signal. Because it alternately assigns right-of-way to specific movements, it can substantially reduce the number and nature of intersection conflicts as no other form of control can.

If drivers obey the signal, then driver judgment is not needed to avoid some of the most critical intersection conflicts. Imposition of traffic signal control does not, however, remove all conflicts from the realm of

Table 16.3: Warrants for Multiway STOP Control

The following criteria should be considered in the engineering study for a multiway STOP sign:

a. Where traffic control signals are justified, the multiway STOP is an interim measure that can be installed quickly to control traffic while arrangements are being made for the installation of the traffic control signal.

b. A crash problem, as indicated by five or more reported crashes in a 12-month period that are susceptible to correction by a multiway STOP installation. Such crashes include right- and left-turn collisions as well as right-angle collisions.

c. Minimum volumes:

   1. The vehicular volume entering the intersection from the major street approaches (total of both approaches) averages at least 300 veh/h for any eight hours of an average day.
   2. The combined vehicular, pedestrian, and bicycle volume entering the intersection from the minor street approaches (total of both approaches) averages at least 200 units/h for the same eight hours, with an average delay to minor-street vehicular traffic of at least 30 s/veh during the highest hour.
   3. If the 85th percentile approach speed of the major highway exceeds 40 mi/h, the minimum vehicular volume warrants are 70% of the above values.

d. Where no single criterion is satisfied, but where criteria B, C1, and C2 are all satisfied to 80% of the minimum values. Criterion C3 is excluded from this condition.

(Used by permission of Federal Highway Administration, U.S. Dept. of Transportation, *Manual on Uniform Traffic Control Devices*, Millennium Edition, Washington DC, 2001, pg. 2B-10.)

driver judgment. At two-phase signals, where all left-turns are made against an opposing vehicular flow, drivers must still evaluate and select gaps in opposing traffic through which to safely turn. At virtually all signals, some pedestrian-vehicle and bicycle-vehicle conflicts remain between legal movements, and driver vigilance and judgment are still required to avoid accidents. Nevertheless, drivers at signalized intersections do not have to negotiate the critical conflicts between crossing vehicle streams, and where exclusive left-turn phases are provided, critical conflicts between left turns and opposing through vehicles are also eliminated through signal control. This chapter deals with the issue of whether or not signal control is warranted or needed. Given that it is needed, Chapter 18 deals with the design of a specific phasing plan and the timing of the signal.

While warrants and other criteria for STOP and YIELD signs are somewhat general in the MUTCD, warrants for signals are quite detailed. The cost involved in installation of traffic signals (e.g., power supply, signal controller, detectors, signal heads, and support structures, and other items) is considerably higher than for STOP or YIELD signs and can run into the hundreds of thousands of dollars for complex intersections. Because of this, and because traffic signals introduce a fixed source of delay into the system, it is important that they not be overused; they should be installed only where no other solution or form of control would be effective in assuring safety and efficiency at the intersection.

16.4.1   Advantages of Traffic Signal Control

The MUTCD lists the following advantages of traffic control signals that are "properly designed, located, operated, and maintained" [MUTCD, Millennium Edition, pg. 4b-2]. These advantages include:

1. They provide for the orderly movement of traffic.
2. They increase the traffic-handling capacity of the intersection if proper physical layouts and control measures are used and if the signal timing is reviewed and updated on a regular basis (every two years) to ensure that it satisfies the current traffic demands.
3. They reduce the frequency and severity of certain types of crashes, especially right-angle collisions.
4. They are coordinated to provide for continuous or nearly continuous movement at a definite speed along a given route under favorable conditions.

5. They are used to interrupt heavy traffic at intervals to permit other traffic, vehicular or pedestrian, to cross.

These specific advantages address the primary reasons why a traffic signal would be installed: to increase capacity (thereby improving level of service), to improve safety, and to provide for orderly movement through a complex situation. Coordination of signals provides other benefits, but not all signals are necessarily coordinated.

16.4.2 Disadvantages of Traffic Signal Control

The description of the second advantage in the foregoing list indicates that capacity is increased by a well-designed signal at a well-designed intersection. Poor design of either the signalization or the geometry of the intersection can significantly reduce the benefits achieved or negate them entirely. Improperly designed traffic signals, or the placement of a signal where it is not justified, can lead to some of the following disadvantages [MUTCD, Millennium Edition, pg. 4B-3].

1. Excessive delay
2. Excessive disobedience of the signal indications
3. Increased use of less adequate routes as road users attempt to avoid the traffic control signal
4. Significant increases in the frequency of collisions (especially rear-end collisions)

Item 4 is of some interest. Even when they are properly installed and well-designed, traffic signal controls can lead to increases in rear-end accidents because of the cyclical stopping of the traffic stream. Where safety is concerned, signals can reduce the number of right-angle, turning, and pedestrian/bicycle accidents; they might cause an increase in rear-end collisions (which tend to be less severe); they will have almost no impact on head-on or sideswipe accidents, or on single-vehicle accidents involving fixed objects.

Excessive delay can result from an improperly installed signal, but it can also occur if the signal timing is inappropriate. In general, excessive delay results from cycle lengths that are either too long or too short for the existing demands at the intersection. Further, drivers will tend to assume that a signal is broken if they experience an excessive wait, particularly when there is little or no demand occurring on the cross street.

16.4.3 Warrants for Traffic Signals

The MUTCD specifies eight different warrants that justify the installation of a traffic signal. The manual *requires*, however, that a comprehensive engineering study be conducted to determine whether or not installation of a signal is justified. The study *must* include applicable factors reflected in the specified warrants, but could extend to other factors as well. The manual also goes on to indicate that "satisfaction of a traffic signal warrant or warrants shall not in itself require the installation of a traffic signal" [MUTCD, Millennium Edition, pg. 4C-1]. On the other hand, traffic signal control *should not* be implemented if none of the warrants are met. The warrants, therefore, still require the exercise of engineering judgment. In the final analysis, if engineering studies and/or judgment indicate that signal installation *will not* improve the overall safety or operational efficiency at a candidate location, it should not be installed.

While offered only under the heading of an option, the MUTCD suggests that the following data be included in an engineering study of the need for a traffic signal:

1. Traffic volumes for each approach for at least 12 hours of a typical day. It would be preferable to have volumes for 24 hours, and 12-hour studies must be sure that peak periods have been observed as part of the study.

2. Vehicular volumes by movement and vehicle class during each 15 minutes of the two hours of the AM and two hours of the PM during which total traffic entering the intersection is greatest.

3. Pedestrian volume counts on each crosswalk for the same periods as noted in item 2. Where necessary, crossing times should be measured, particularly where many pedestrians are young children, elderly, or disabled.

4. Information on nearby facilities and centers serving the young, elderly, or disabled, as pedestrian

observations might not include these because there is no signal at the time the engineering study is conducted.

5. Posted or statutory speed limit and/or the 85th percentile approach speed on all uncontrolled approaches.

6. A condition diagram showing all physical and geometric features of the intersection, including channelization, grades, parking locations, driveways, roadside appurtenances, bus stops, and others.

7. A collision diagram showing accident experience by type, location, direction, time of day, weather, and the like, including all accidents for at least one (most recent) year.

8. For the two peak AM and two peak PM hours (the same as those referred to in item 2 previously):

   (a) Vehicle-hours of stopped delay for each approach

   (b) Number and distribution of gaps in vehicular traffic on the major street

   (c) Posted or statutory speed limits and/or the 85th percentile approach speed on controlled approaches—measured at a point near the intersection, but where speed is unaffected by control

   (d) Pedestrian delays for at least two 30-minute peak pedestrian delay periods of an average weekday, or like periods on a Saturday or Sunday

   (e) Queue lengths on STOP-controlled approaches

This data will allow the engineer to fully evaluate whether or not the intersection satisfies the requirements of one or more of the following warrants:

- *Warrant 1*: Eight-Hour Vehicular Volume
- *Warrant 2*: Four-Hour Vehicular Volume
- *Warrant 3*: Peak Hour
- *Warrant 4*: Pedestrian Volume
- *Warrant 5*: School Crossing
- *Warrant 6*: Coordinated Signal System
- *Warrant 7*: Crash Experience
- *Warrant 8*: Roadway Network

It also provides a sufficient base for the exercise of engineering judgment in determining whether a traffic signal should be installed at the study location. Each of these warrants is presented and discussed in the sections that follow.

Warrant 1: Eight-Hour Vehicular Volume

The eight-hour vehicular volume warrant represents a merging of three different warrants in the pre-2000 MUTCD (old Warrants 1, 2, and 8). It addresses the need for signalization for conditions that exist over extended periods of the day (a minimum of eight hours). Two of the most fundamental reasons for signalization are addressed:

- Heavy volumes on conflicting cross-movements that make it impractical for drivers to select gaps in an uninterrupted traffic stream through which to safely pass. This requirement is often referred to as the "minimum vehicular volume" condition (Condition A).

- Vehicular volumes on the major street are so heavy that no minor-street vehicle can safely pass through the major-street traffic stream without the aid of signals. This requirement is often referred to as the "interruption of continuous traffic" condition (Condition B).

Details of this warrant are shown in Table 16.4. The warrant is met when:

- Either Condition A or Condition B is met to the 100% level.

- Either Condition A or Condition B is met to the 70% level, where the intersection is located in an isolated community of population 10,000 or less, or where the major-street approach speed is 40 mi/h or higher.

- Both Conditions A and B are met to the 80% level.

Table 16.4: Warrant 1: Eight-Hour Vehicular Volume

Condition A: Minimum Vehicular Volume							
Number of lanes for moving traffic on each approach		Vehicles per hour on major street (total, both approaches)			Vehicles per hour on higher-volume minor-street approach (one direction only)		
Major Street	Minor Street	100%	80%	70%	100%	80%	70%
1	1	500	400	350	150	120	105
2 or more	1	600	480	420	150	120	105
2 or more	2 or more	600	480	420	200	160	140
1	2 or more	500	400	350	200	160	140

Condition B: Interruption of Continuous Traffic							
Number of lanes for moving traffic on each approach		Vehicles per hour on major street (total, both approaches)			Vehicles per hour on higher-volume minor street approach (one direction only)		
Major Street	Minor Street	100%	80%	70%	100%	80%	70%
1	1	750	600	525	75	60	53
2 or more	1	900	720	630	75	60	53
2 or more	2 or more	900	720	630	100	80	70
1	2 or more	750	600	525	100	80	70

(Used with permission of Federal Highway Administration, U.S. Department of Transportation, *Manual on Uniform Traffic Control Devices*, Millennium Edition, Washington DC, 2000, Table 4C-1, pg. 4C-5.)

Note that in applying these warrants, the major-street volume criteria are related to the total volume in both directions, while the minor-street volume criteria are applied to the highest volume in one direction. The volume criteria in Table 16.4 must be met for a minimum of eight hours on a typical day. The eight hours do not have to be consecutive, and often involve four hours around the morning peak and four hours around the evening peak. Major- and minor-street volumes used must, however, be for the same eight hours.

Either of the intersecting streets may be treated as the "major" approach, but the designation must be consistent for a given application. If the designation of the "major" street is not obvious, a warrant analysis can be conducted considering each as the "major" street in turn. While the designation of the major street may not be changed within any one analysis, the direction of peak one-way volume for the minor street need not be consistent.

The 70% reduction allowed for rural communities of population 10,000 or less reflects the fact that drivers in small communities have little experience in driving under congested situations. They will require the guidance of traffic signal control at volume levels lower than those for drivers more used to driving in congested situations. The same reduction applies where the major-street speed limit is 40 mi/h or greater. As gap selection is more difficult through a higher-speed major-street flow, signals are justified at lower volumes.

The various elements of the eight-hour vehicular volume warrant are historically the oldest of the warrants, having been initially formulated and disseminated in the 1930s.

Warrant 2: Four-Hour Vehicular Volume

The four-hour vehicular volume warrant was introduced in the 1970s to assist in the evaluation of situations where volume levels requiring signal control might exist for periods shorter than eight hours. Prior to the MUTCD Millennium Edition, this was old Warrant 9.

Figure 16.5 shows the warrant, which is in the form of a continuous graph. Because this warrant is expressed as a continuous relationship between major and minor street volumes, it addresses a wide variety of conditions. Indeed,

Conditions A and B of the eight-hour warrant represent two points in such a continuum for each configuration, but the older eight-hour warrant did not investigate or create criteria for the full range of potential conditions.

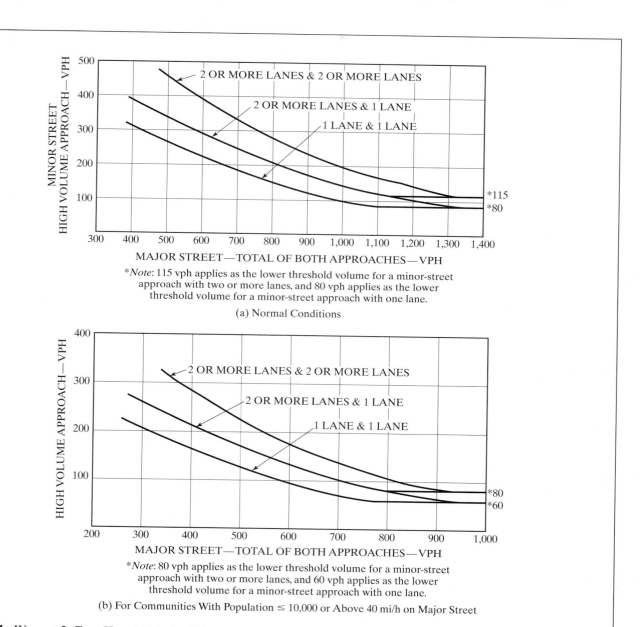

Note: 115 vph applies as the lower threshold volume for a minor-street approach with two or more lanes, and 80 vph applies as the lower threshold volume for a minor-street approach with one lane.

(a) Normal Conditions

Note: 80 vph applies as the lower threshold volume for a minor-street approach with two or more lanes, and 60 vph applies as the lower threshold volume for a minor-street approach with one lane.

(b) For Communities With Population ≤ 10,000 or Above 40 mi/h on Major Street

Figure 16.5: Warrant 2: Four-Hour Vehicular Volume Warrant (Used with permission of Federal Highway Administration, U.S. Department of Transportation, *Manual on Uniform Traffic Control Devices*, Millennium Edition, Washington DC, 2000, Figures 4C-1, 4C-2, pg. 4C-7.)

Figure 16.5(a) is the warrant for normal conditions, while Figure 16.5(b) reflects the 70% reduction applied to isolated small communities (with population less than 10,000) or where the major-street speed limit is above 40 mi/h. Because the four-hour warrant represents a continuous set of conditions, there is no need to include an 80% reduction for two discrete conditions within the relationship.

To test the warrant, the two-way major-street volume is plotted against the highest one-way volume on the minor street for each hour of the study period. To meet the warrant, at least four hours must plot *above* the appropriate decision curve. The three curves represent intersections of (1) two streets with one lane in each direction, (2) one street with one lane in each direction with another having two or more lanes in each direction, and (3) two streets with more than one lane in each direction. In Case (2), the distinction between which intersecting street has one lane in each direction (major or minor) is no longer relevant, except for the footnotes.

Warrant 3: Peak Hour

Warrant 3 addresses two critical situations that might exist for only one hour of a typical day. The first is a volume condition, similar in form to Warrant 2, and shown in Figure 16.6 (old Warrant 11). The second is a delay warrant (old Warrant 10). If either condition is satisfied, the peak-hour warrant is met.

The volume portion of the warrant is implemented in the same manner as the four-hour warrant. For each hour of the study, the two-way major street volume is plotted against the high single-direction volume on the minor street. For the Peak-Hour Volume Warrant, however, only one hour must plot above the appropriate decision line to meet the criteria. Criteria are given for normal conditions in Figure 16.6(a), and the 70% criteria for small isolated communities and high major-street speeds are shown in Figure 16.6(b). The Peak-Hour Delay Warrant is summarized in Table 16.5.

It is important to recognize that the delay portion of Warrant 3 applies only to cases in which STOP control is already in effect for the minor street. Thus, delay during the peak hour is not a criterion that allows going from no control or YIELD control to signalization directly.

Table 16.5: Warrant 3B: Peak-Hour Delay Warrant

The need for traffic signal control shall be considered if an engineering study finds that the following three conditions exist for the same one hour (any four consecutive 15-minute periods) of an average day:

1. The total stopped-time delay experienced by traffic on one minor street approach (one direction only) controlled by a STOP sign equals or exceeds four veh-hours for a one-lane approach or five veh-hours for a two-lane approach.

2. The volume on the same minor street approach (one direction only) equals or exceeds 100 veh/h for one moving lane of traffic or 150 veh/h for two moving lanes.

3. The total entering volume serviced during the hour equals or exceeds 650 veh/h for intersections with three approaches, or 800 veh/h for intersections with four or more approaches.

(Used by permission of Federal Highway Administration, U.S. Dept. of Transportation, *Manual on Uniform Traffic Control Devices*, Millennium Edition, Washington DC, 2001, pg. 4C-8.)

The MUTCD also emphasizes that the Peak-Hour Warrant should be applied only in special cases, given the following advice under the heading of a mandatory standard: this signal warrant shall be applied only in unusual cases. Such cases include, but are not limited to, office complexes, manufacturing plants, industrial complexes, or high-occupancy vehicle facilities that attract or discharge large numbers of vehicles over a short time [MUTCD, Millennium Edition, pg. 4C-6].

Warrant 4: Pedestrians

The Pedestrian Warrant addresses situations in which the need for signalization is the frequency of vehicle-pedestrian conflicts and the inability of pedestrians to avoid such conflicts due to the volume of traffic present. Signals may be placed under this warrant at mid-block locations, as well as at intersections. The criteria for the warrant are summarized in Table 16.6.

If the traffic signal is justified at an intersection by this warrant only, it will usually be at least a semi-actuated

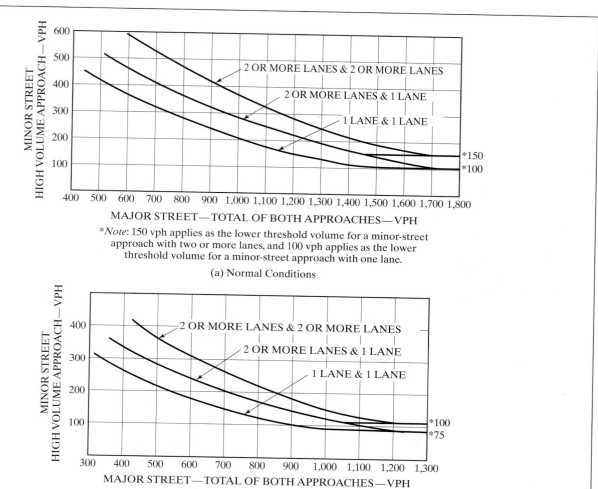

Note: 150 vph applies as the lower threshold volume for a minor-street approach with two or more lanes, and 100 vph applies as the lower threshold volume for a minor-street approach with one lane.

(a) Normal Conditions

Note: 100 vph applies as the lower threshold volume for a minor-street approach with two or more lanes, and 75 vph applies as the lower threshold volume for a minor-street approach with one lane.

(b) For Communities With Population ≤ 10,000 or Above 40 mi/h on Major Street

Figure 16.6: Warrant 3A: Peak-Hour Vehicular Volume Warrant (Used with permission of Federal Highway Administration, U.S. Department of Transportation, *Manual on Uniform Traffic Control Devices*, Millennium Edition, Washington DC, 2000, Figures 4C-3, 4C-4, pg. 4C-9.)

signal (a full actuated signal is also a possibility at an isolated intersection) with pedestrian push-buttons and signal heads for pedestrians crossing the major street. If it is within a coordinated signal system, it would also be coordinated into the system. If such a signal is located in mid-block, it will always be pedestrian-actuated, and parking and other sight restrictions should be eliminated within 20 ft of both sides of the crosswalk. Standard reinforcing markings and signs should also be provided.

If the intersection meets this warrant but also meets other vehicular warrants, any type of signal could be installed as appropriate to other conditions. Pedestrian signal heads would be required for major-street crossings.

Table 16.6: Warrant 4: Pedestrian Warrant

The need for a traffic control signal at an intersection or mid-block crossing shall be considered if an engineering study finds that both of the following criteria are met:

1. The pedestrian volume crossing the major street at an intersection or mid-block location during an average day is 100 or more for each of any four hours, or 190 or more during any one hour.

2. There are fewer than 60 gaps per hour in the traffic stream of adequate length to allow pedestrians to cross the street during the same period when the pedestrian volume criteria are satisfied. Where there is a divided street having a median of sufficient width for pedestrians to wait, the requirement applies separately to each direction of vehicular traffic.

The Pedestrian Signal Warrant shall not be applied at locations where the distance to the nearest traffic control signal along the major street is less than 300 ft, unless the proposed traffic control signal will not restrict the progressive movement of traffic.

The criterion for pedestrian volume crossing the major roadway may be reduced as much as 50% if the average crossing speed of pedestrians is less than 4 ft/s.

If a traffic control signal is justified by both this signal warrant and a traffic engineering study, the traffic control signal shall be equipped with pedestrian signal heads conforming to requirements set forth in Chapter 4E (of the MUTCD).

(Used by permission of Federal Highway Administration, U.S. Dept. of Transportation, *Manual on Uniform Traffic Control Devices*, Millennium Edition, Washington DC, 2001, pgs. 4C-10, 4C-11.)

Pedestrian push-buttons would be installed unless the vehicular signal timing safely accommodates pedestrians in every signal cycle.

Figure 16.7 illustrates the concept of an "acceptable gap." A gap is defined as the time interval between two vehicles in any lane or direction encroaching on the crosswalk. An "acceptable" gap is one that is large enough to allow a pedestrian crossing at 4 ft/s to completely cross the street, plus a buffer of 2–4 s. The gap criteria indicate that there must be at least 60 acceptable gaps per hour for safe pedestrian crossings without a signal. This is an average of one gap per minute and an average time between gaps of one minute. When pedestrians have to wait too long to safely cross, they will begin to select unsafe gaps to quickly walk or run through, exposing themselves to a dangerous hazard.

The MUTCD also indicates that a signal may not be needed where adjacent coordinated signals consistently provide gaps of adequate length to cross the street, even if their frequency is less than one per minute.

Warrant 5: School Crossing

This warrant is similar to the pedestrian warrant but is limited to application at designated school crossing locations, either at intersections or at mid-block locations. Unlike Warrant 4, "adequacy" of gaps must be related to the numbers of school children crossing the major street in groups (i.e., through a single gap). Thus, an acceptable gap would include the crossing time, the buffer time, and an allowance for groups of children to start crossing the street.

The minimum number of children crossing the major street is 20 during the highest crossing hour. The frequency of acceptable gaps must be no less than one for each minute during which school children are crossing.

Traffic signals are rarely implemented under this warrant. Children do not usually observe and obey signals regularly, particularly if they are very young. Thus, traffic signals would have to be augmented by crossing guards in most cases. Except in unusual circumstances involving a very heavily traveled major street, the crossing guard, perhaps augmented with STOP signs, would suffice under most circumstances without signalization. Where extremely high volumes of school children cross a very wide and heavily traveled major street, overpasses or underpasses should be provided with barriers preventing entry onto the street.

Warrant 6: Coordinated Signal System

Chapter 24 of this text addresses signal coordination and progression systems for arterials and networks. Critical

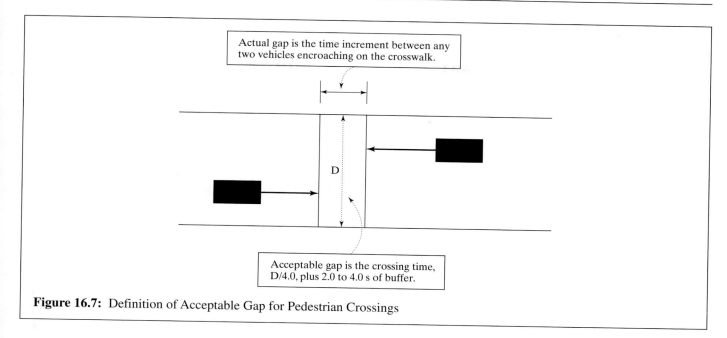

Figure 16.7: Definition of Acceptable Gap for Pedestrian Crossings

to such systems is the maintenance of platoons of vehicles moving together through a "green wave" as they progress along an arterial. If the distance between two adjacent coordinated signals is too large, platoons begin to dissipate and the positive impact of the progression is sharply reduced. In such cases, the traffic engineer may place a signal at an intermediate intersection where it would not otherwise be warranted to reinforce the coordination scheme and to help maintain platoon coherence. The application of this warrant, shown in Table 16.7, should not result in signal spacing of less than 1,000 ft. Such signals, when placed, are often referred to as "spacer signals."

The two criteria are similar, but not exactly the same. Inserting a signal in a one-way progression is always possible without damaging the progression. On a two-way street, it is not always possible to place a signal that will maintain the progression in both directions acceptably. This issue is discussed in greater detail in Chapter 24.

Warrant 7: Crash Experience

The Crash Experience Warrant addresses cases in which a traffic control signal would be installed to alleviate an observed high-accident occurrence at the intersection. The criteria are summarized in Table 16.8.

Table 16.7: Warrant 6: Coordinated Signal System

The need for a traffic control signal shall be considered if an engineering study finds that one of the following criteria is met:

1. On a one-way street or a street that has traffic predominantly in one direction, the adjacent traffic control signals are so far apart that they do not provide the necessary degree of vehicular platooning.

2. On a two-way street, adjacent traffic control signals do not provide the necessary degree of platooning and the proposed and adjacent traffic control signals will collectively provide a progressive operation.

(Used by permission of Federal Highway Administration, U.S. Dept. of Transportation, *Manual on Uniform Traffic Control Devices*, Millennium Edition, Washington DC, 2001, pg. 4C-12.)

The requirement for an adequate trial of alternative methods means that either YIELD or STOP control is already in place and properly enforced. These types of control can also address many of the same accident problems as signalization. Thus, a signal is justified

Table 16.8: Warrant 7: Crash Experience

The need for a traffic control signal shall be considered if an engineering study finds that all of the following criteria are met:

1. Adequate trial of alternatives with satisfactory observance and enforcement has failed to reduce the crash frequency.

2. Five or more reported crashes of types susceptible to correction by a traffic control signal have occurred within a 12-month period, each involving a personal injury or property damage apparently exceeding the applicable requirements for a reportable crash.

3. For each of any eight hours of the day, vehicular volumes meet either Warrant 1A or Warrant 1B at the 80% level.

(Used by permission of Federal Highway Administration, U.S. Dept. of Transportation, *Manual on Uniform Traffic Control Devices*, Millennium Edition, Washington DC, 2001, pg. 4C-13, item 3 paraphrased for simplicity.)

only when these lesser measures have failed to adequately address the situation.

Accidents that are susceptible to correction by signalization include right-angle accidents, accidents involving turning vehicles from the two streets, and accidents between vehicles and pedestrians crossing the street on which the vehicle is traveling. Rear-end accidents are often increased with imposition of traffic signals (or STOP/YIELD signs), as some drivers may be induced to stop quickly or suddenly. Head-on and sideswipe collisions are not addressed by signalization; accidents between vehicles and fixed objects at corners are also not correctable through signalization.

Warrant 8: Roadway Network

The last warrant is the only one that addresses a developing situation (i.e., a case in which present volumes would not justify signalization but where new development is expected to generate substantial traffic that would justify signalization).

Large traffic generators, such as regional shopping centers, sports stadiums and arenas, and similar facilities are often built in areas that are sparsely populated and where existing roadways have light traffic. Such projects often require substantial roadway improvements that change the physical layout of the roadway network and create new or substantially enlarged intersections that will require signalization. Since no volume studies can be made to check other warrants and since it is not practical to open such intersections without proper control, a warrant addressing future conditions is necessary. The specific criteria for this warrant are summarized in Table 16.9.

"Immediately projected" generally refers to the traffic expected on day one of the opening of new facilities and/or traffic generators that create the need for signalization.

While this warrant does NOT address the need for signalization of entrances and exits to major new developments, many local agencies apply this criteria to such locations (or apply other warrants on the basis of forecasted traffic). This is a critical part of development control. In many areas, the developers must pay to install signals at entry and exit points where they are needed, and may even be asked to pay for signals required at other intersections as a result of traffic generated by the development.

16.4.4 Summary

It is important to reiterate the basic meaning of these warrants. No signal should be placed without an engineering study showing that the criteria of at least one of the warrants are met. On the other hand, meeting one or more of these warrants *does not* necessitate signalization. Note that every warrant uses the language "The need for a traffic control signal *shall be considered* . . . ." (emphasis added). While the "shall" is a mandatory standard, it calls only for consideration, not placement, of a traffic signal. The engineering study must also convince the traffic engineer that installation of a signal will improve the safety of the intersection, increase the capacity of the intersection, or improve the efficiency of operation at the intersection before the signal is installed. That is why the recommended information to be collected during an "engineering study" exceeds that

Table 16.9: Warrant 8: Roadway Network

The need for a traffic control signal shall be considered if an engineering study finds that the common intersection of two or more major routes meets one or both of the following criteria:

1. The intersection has a total existing, or immediately projected, entering volume of at least 1,000 veh/h during the peak hour of a typical weekday, and has five-year projected traffic volumes, based upon an engineering study, that meet one or more of Warrants 1, 2 and 3 during an average weekday.

2. The intersection has a total existing, or immediately projected, entering volume of at least 1,000 veh/h for each of any five hours of a non-normal business day (Saturday or Sunday).

A major route as used in this warrant shall have one or more of the following characteristics:

1. It is part of the street or highway system that serves as the principal roadway network for through traffic flow.

2. It includes rural or suburban highways outside, entering, or traversing a city.

3. It appears as a major route on an official plan, such as a major street plan in an urban area traffic and transportation study.

(Used by permission of Federal Highway Administration, U.S. Dept. of Transportation, *Manual on Uniform Traffic Control Devices*, Millennium Edition, Washington DC, 2001, pgs. 4C-13, 4C-14.)

needed to simply apply the eight warrants of the MUTCD. In the end, engineering judgment is called for, as is appropriate in any professional practice.

16.4.5 A Sample Problem in Application of Signal Warrants

Consider the intersection and related data shown in Figure 16.8.

Note that the data is formatted in a way that is conducive to comparing with warrant criteria. Thus, a column adding the traffic in each direction on the major street is included, and a column listing the "high volume" in one direction on the minor street is also included.

Pedestrian volumes are summarized for those crossing the major street, as this is the criterion used in the pedestrian warrant. The number of acceptable gaps in the major traffic stream is also listed for convenience. As will be seen, not every warrant applies to every intersection, and data for some warrants is not provided. The following analysis is applied:

- *Warrant 1:* There is no indication that the 70% reduction factor applies, so it is assumed that either Condition A or Condition B must be met at 100%, or both must be met at 80%. Condition A requires 600 veh/h in both directions on the multilane major street and 150 veh/h in the high-volume direction on the one-lane minor street. While all 12 hours shown in Figure 16.8 are greater than 600 veh/h, none have a one-way volume equal to or higher than 150 veh/h on the minor street. Condition A is not met. Condition B requires 900 veh/h on the major street (both directions) and 75 veh/h on the minor street (one direction). The 10 hours between 12:00 Noon and 10:00 PM meet the major-street criterion. The same 10 hours meet the minor-street criterion as well. Therefore, Condition B is met. As one condition is met at 100%, the consideration of whether both conditions are met at 80% is not necessary. Warrant 1 is satisfied.

- *Warrant 2:* Figure 16.9 shows the hourly volume data plotted against the four-hour warrant graph. The center decision curve (one street with multilane approaches, one with one-lane approaches) is used. Only one of the 12 hours of data is above the criterion. To meet the warrant, four are required. The warrant is not met.

- *Warrant 3:* Figure 16.10 shows the hourly volume data plotted against the peak-hour volume warrant graph. Again, the center decision curve is used. None of the 12 hours of data is above the criterion. The volume portion of this warrant is not met.

The delay portion of the peak-hour warrant requires 4 veh-hours of delay in the high-volume direction on a STOP-controlled approach. The intersection data indicates that each vehicle experiences 30 s of delay. The peak one-direction

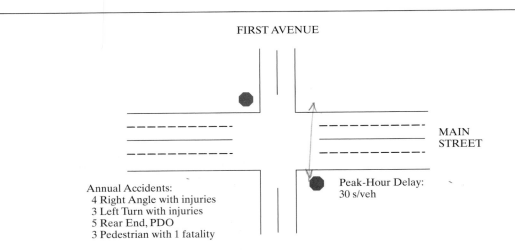

FIRST AVENUE

MAIN STREET

Annual Accidents:
4 Right Angle with injuries
3 Left Turn with injuries
5 Rear End, PDO
3 Pedestrian with 1 fatality

Peak-Hour Delay:
30 s/veh

Time	Main Street Volume (veh/h)			First Ave Volume (veh/h)			Ped Volume (ped/h) Crossing Main	No. of Acceptable Gaps
	EB	WB	TOT	NB	SB	High Vol		
11 AM–12	400	425	825	75	80	80	115	65
12–1 PM	450	465	915	85	85	85	120	71
1–2 PM	485	500	985	90	100	100	125	68
2–3 PM	525	525	1,050	110	115	115	130	61
3–4 PM	515	525	1,040	100	95	100	135	55
4–5 PM	540	550	1,090	90	100	100	140	50
5–6 PM	550	580	1,130	110	125	125	120	52
6–7 PM	545	525	1,070	96	103	103	108	58
7–8 PM	505	506	1,011	90	95	95	100	63
8–9 PM	485	490	975	85	75	85	90	69
9–10 PM	475	475	950	75	60	75	50	75
10–11 PM	400	410	810	50	55	55	25	85

Figure 16.8: Intersection and Data for Sample Problem in Signal Warrants

volume is 125 veh/h, resulting in 125 * 30 = 3,750 veh-secs of aggregate delay, or 3,750/3,600 = 1.04 veh-hrs of delay. This is less than that required by the warrant. The delay portion of this warrant is not met.

- *Warrant 4:* The warrant requires 190 ped/h crossing the major street for one hour, or 100 ped/h crossing for four hours. In the same hours, the number of acceptable gaps must be less than 60 per hour. The data indicates that there are no hours with more than 190 pedestrians crossing the major street, but that the nine hours between 11:00 AM and 8:00 PM have 100 pedestrians or more cross-

ing the major street. In four of those hours from 3:00 PM to 7:00 PM, the number of acceptable gaps is less than 60. The warrant is met.

- *Warrant 5:* The school-crossing warrant does not apply. This is not a school crossing.

- *Warrant 6:* No information on signal progression is given, so this warrant cannot be applied.

- *Warrant 7:* The crash experience warrant has several criteria: Have lesser measures been tried? Yes, as the minor street is already STOP-controlled. Have five accidents susceptible to correction by signalization occurred in a 12-month

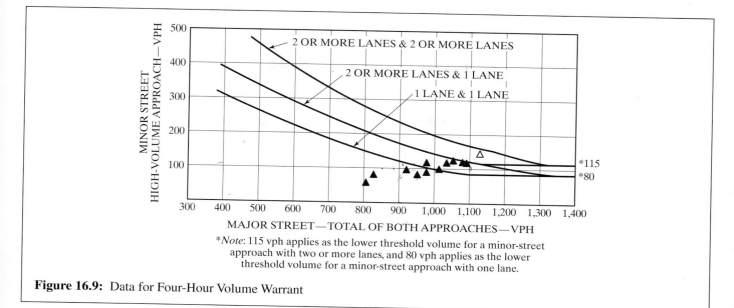

Figure 16.9: Data for Four-Hour Volume Warrant

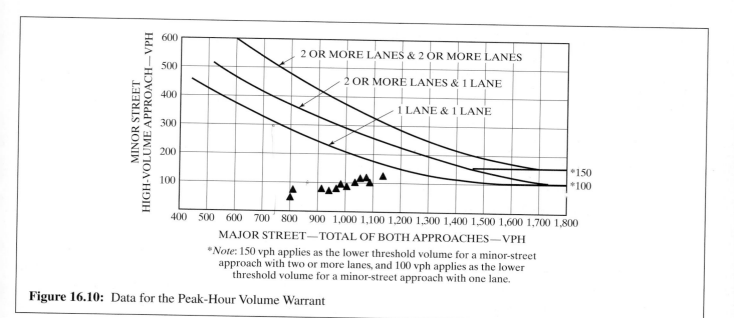

Figure 16.10: Data for the Peak-Hour Volume Warrant

period? Yes—four right-angle, three left-turn, and three pedestrian. Are the criteria for Warrants 1A or 1B met to the extent of 80%? Yes, Warrant 1B is met at 100%. Therefore, the crash experience warrant is met.

- *Warrant 8:* There is no information given concerning the roadway network, and the data reflects

an existing situation. This warrant is not applicable in this case.

In summary, a signal should be considered at this location, as the criteria for Warrants 1B (Interruption of Continuous Traffic), 4 (Pedestrians), and 7 (Crash Experience) are all met. Unless unusual circumstances are present, it would be reasonable to expect that the accident experience will improve with signalization, and it is, therefore, likely that one would be placed.

The fact that Warrant 1B is satisfied may suggest that a semi-actuated signal be considered. In addition, Warrant 4 requires the use of pedestrian signals, at least for pedestrians crossing the major street. If a semi-actuated signal is installed, it must have a pedestrian push-button (for pedestrians crossing the major street). The number of left turning accidents may also suggest consideration of protected left-turn phasing.

16.5   Closing Comments

In selecting an appropriate type of control for an intersection, the traffic engineer has many things to consider, including sight distances and warrants. In most cases, the objective is to provide the minimum level of control that will assure safety and efficient operations. In general, providing unneeded or excessive control leads to additional delay to drivers and passengers. With all of the analysis procedures and guidelines, however, engineering judgment is still required to make intelligent decisions. It is always useful to view the operation of existing intersections in the field in addition to reviewing study results before making recommendations on the best form of control.

References

1. *A Policy on Geometric Design of Highways and Streets*, 4th Edition, American Association of State Highway and Transportation Officials, Washington DC, 2001.

2. *Manual on Uniform Traffic Control Devices*, Millennium Edition, Federal Highway Administration, U.S. Department of Transportation, Washington DC, 2000.

3. *Highway Capacity Manual*, 4th Edition, Transportation Research Board, National Research Council, Washington DC, 2000.

Problems

16-1. For the intersection of two rural roads shown in Figure 16.11, determine whether or not operation under basic rules of the road would be safe. If not, what type of control would you recommend, assuming that traffic signals are not warranted?

16-2. Determine whether the intersection shown in Figure 16.12 can be safely operated under basic rules of the road. If not, what form of control would you recommend, assuming that signalization is not warranted?

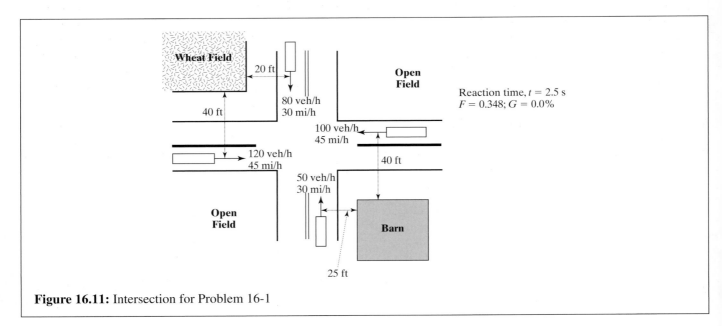

Figure 16.11: Intersection for Problem 16-1

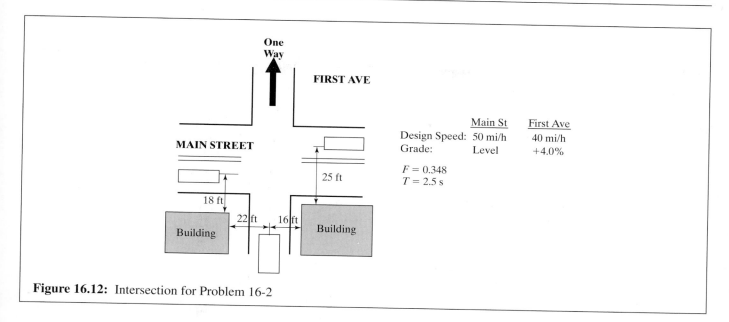

Figure 16.12: Intersection for Problem 16-2

16-3. Determine whether the sight distances for the STOP-controlled intersection shown in Figure 16.13 are adequate. If not, what measures would you recommend to ensure safety?

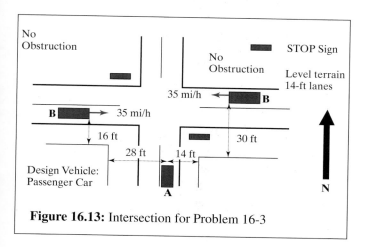

Figure 16.13: Intersection for Problem 16-3

16-4–16-7. For each of the intersections shown in the following figures, determine whether the data supports each of the eight signal warrants. For each problem, and each warrant, indicate whether the warrant is: (a) met, (b) not met, (c) not applicable, or (d) insufficient information given to assess.

For each problem, indicate: (a) whether a signal is warranted, (b) the type of signalization that should be considered, and (c) whether pedestrian signals and/or push-buttons are recommended.

In all cases, assume that no warrants are met for the hours that are not included in the study data.

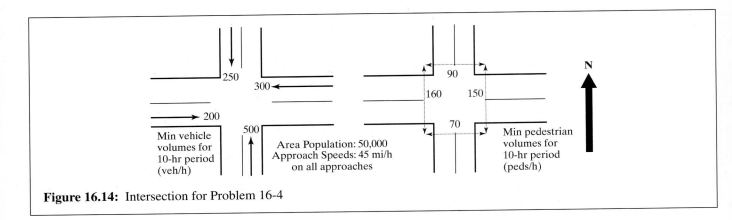

Figure 16.14: Intersection for Problem 16-4

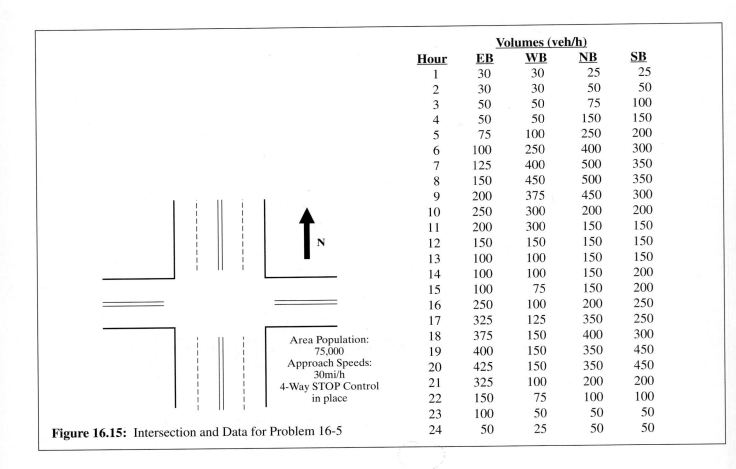

Figure 16.15: Intersection and Data for Problem 16-5

		Volumes (veh/h)		
Hour	**EB**	**WB**	**NB**	**SB**
1	30	30	25	25
2	30	30	50	50
3	50	50	75	100
4	50	50	150	150
5	75	100	250	200
6	100	250	400	300
7	125	400	500	350
8	150	450	500	350
9	200	375	450	300
10	250	300	200	200
11	200	300	150	150
12	150	150	150	150
13	100	100	150	150
14	100	100	150	200
15	100	75	150	200
16	250	100	200	250
17	325	125	350	250
18	375	150	400	300
19	400	150	350	450
20	425	150	350	450
21	325	100	200	200
22	150	75	100	100
23	100	50	50	50
24	50	25	50	50

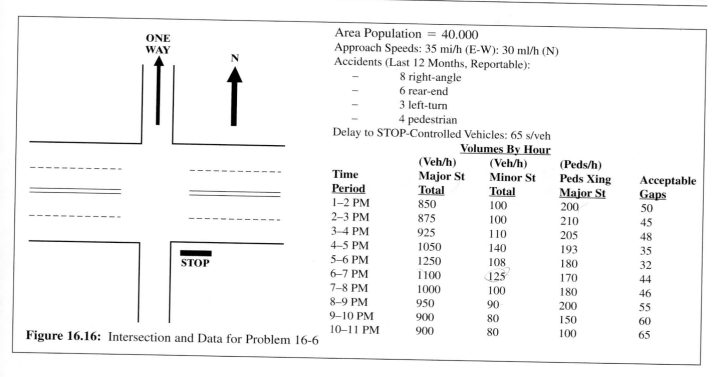

Figure 16.16: Intersection and Data for Problem 16-6

Area Population = 40.000
Approach Speeds: 35 mi/h (E-W): 30 ml/h (N)
Accidents (Last 12 Months, Reportable):
- – 8 right-angle
- – 6 rear-end
- – 3 left-turn
- – 4 pedestrian
Delay to STOP-Controlled Vehicles: 65 s/veh

Volumes By Hour

Time Period	(Veh/h) Major St Total	(Veh/h) Minor St Total	(Peds/h) Peds Xing Major St	Acceptable Gaps
1–2 PM	850	100	200	50
2–3 PM	875	100	210	45
3–4 PM	925	110	205	48
4–5 PM	1050	140	193	35
5–6 PM	1250	108	180	32
6–7 PM	1100	125	170	44
7–8 PM	1000	100	180	46
8–9 PM	950	90	200	55
9–10 PM	900	80	150	60
10–11 PM	900	80	100	65

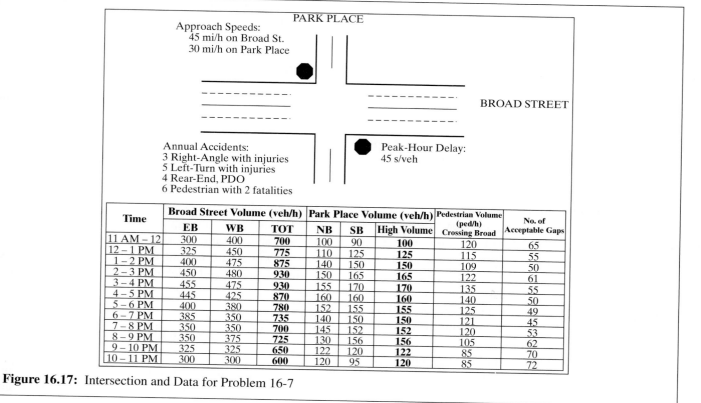

Figure 16.17: Intersection and Data for Problem 16-7

PARK PLACE

Approach Speeds:
45 mi/h on Broad St.
30 mi/h on Park Place

BROAD STREET

Annual Accidents:
3 Right-Angle with injuries
5 Left-Turn with injuries
4 Rear-End, PDO
6 Pedestrian with 2 fatalities

Peak-Hour Delay:
45 s/veh

Time	Broad Street Volume (veh/h)			Park Place Volume (veh/h)			Pedestrian Volume (ped/h) Crossing Broad	No. of Acceptable Gaps
	EB	WB	TOT	NB	SB	High Volume		
11 AM – 12	300	400	700	100	90	100	120	65
12 – 1 PM	325	450	775	110	125	125	115	55
1 – 2 PM	400	475	875	140	150	150	109	50
2 – 3 PM	450	480	930	150	165	165	122	61
3 – 4 PM	455	475	930	155	170	170	135	55
4 – 5 PM	445	425	870	160	160	160	140	50
5 – 6 PM	400	380	780	152	155	155	125	49
6 – 7 PM	385	350	735	140	150	150	121	45
7 – 8 PM	350	350	700	145	152	152	120	53
8 – 9 PM	350	375	725	130	156	156	105	62
9 – 10 PM	325	325	650	122	120	122	85	70
10 – 11 PM	300	300	600	120	95	120	85	72

Basic Principles of Intersection Signalization

In Chapter 16, various options for intersection control were presented and discussed. Warrants for implementation of traffic control signals at an intersection, presented in the *Manual on Uniform Traffic Control Devices* [*1*], provide general and specific criteria for selection of an appropriate form of intersection control. At many intersections, the combination of traffic volumes, potential conflicts, overall safety of operation, efficiency of operation, and driver convenience will lead to a decision to install traffic control signals.

The operation of signalized intersections is often complex, involving competing vehicular and pedestrian movements. Appropriate methodologies for design and timing of signals and for the operational analysis of signalized intersections require that the behavior of drivers and pedestrians at a signalized intersection be modeled in a form that can be easily manipulated and optimized. This chapter discusses some of the fundamental operational characteristics at a signalized intersection and the ways in which they may be effectively modeled.

In Chapter 18, these principles are applied to a signalized intersection design and timing process. In Chapter 21, they are augmented and combined into an overall model of signalized intersection operations. The particular model presented is that of the *Highway Capacity Manual* [*2*].

This chapter focuses on four critical aspects of signalized intersection operation:

1. Discharge headways, saturation flow rates, and lost times
2. Allocation of time and the critical lane concept
3. The concept of left-turn equivalency
4. Delay as a measure of service quality

There are other aspects of signalized intersection operation that are also important, and the *Highway Capacity Manual* analysis model addresses many of them. These four, however, are central to understanding traffic behavior at signalized intersections and are highlighted here.

17.1 Terms and Definitions

Traffic signals are complex devices that can operate in a variety of different modes. A number of key terms and definitions should be understood before pursuing a more substantive discussion.

17.1.1   Components of a Signal Cycle

The following terms describe portions and subportions of a signal cycle. The most fundamental unit in signal design and timing is the *cycle*, as defined below.

1. *Cycle.* A signal cycle is one complete rotation through all of the indications provided. In general, every legal vehicular movement receives a "green" indication during each cycle, although there are some exceptions to this rule.

2. *Cycle length.* The cycle length is the time (in seconds) that it takes to complete one full cycle of indications. It is given the symbol "C."

3. *Interval.* The interval is a period of time during which no signal indication changes. It is the smallest unit of time described within a signal cycle. There are several types of intervals within a signal cycle:

   (a) *Change interval.* The change interval is the "yellow" indication for a given movement. It is part of the transition from "green" to "red," in which movements about to lose "green" are given a "yellow" signal, while all other movements have a "red" signal. It is timed to allow a vehicle that cannot safely stop when the "green" is withdrawn to enter the intersection legally. The change interval is given the symbol "y_i" for movement(s) i.

   (b) *Clearance interval.* The clearance interval is also part of the transition from "green" to "red" for a given set of movements. During the clearance interval, all movements have a "red" signal. It is timed to allow a vehicle that legally enters the intersection on "yellow" to safely cross the intersection before conflicting flows are released. The clearance interval is given the symbol "ar_i" (for "all red") for movement(s) i.

   (c) *Green interval.* Each movement has one green interval during the signal cycle. During

a green interval, the movements permitted have a "green" light, while all other movements have a "red" light. The green interval is given the symbol "G_i" for movement(s) i.

   (d) *Red interval.* Each movement has a red interval during the signal cycle. All movements not permitted have a "red" light, while those permitted to move have a "green" light. In general, the red interval overlaps the green intervals for all other movements in the intersection. The red interval is given the symbol "R_i" for movement(s) i.

4. *Phase.* A signal phase consists of a green interval, plus the change and clearance intervals that follow it. It is a set of intervals that allows a designated movement or set of movements to flow and to be safely halted before release of a conflicting set of movements.

17.1.2   Types of Signal Operation

Traffic signals can operate on a pretimed basis or may be partially or fully actuated by arriving vehicles sensed by detectors. In networks, or on arterials, signals may be coordinated through computer control.

1. *Pretimed operation.* In pretimed operation, the cycle length, phase sequence, and timing of each interval are constant. Each cycle of the signal follows the same predetermined plan. "Multi-dial" controllers will allow different pretimed settings to be established. An internal clock is used to activate the appropriate timing. In such cases, it is typical to have at least an AM peak, a PM peak, and an off-peak signal timing.

2. *Semi-actuated operation.* In semi-actuated operation, detectors are placed on the minor approach(es) to the intersection; there are no detectors on the major street. The light is green for the major street at all times except when a "call" or actuation is noted on one of the minor approaches. Then, subject to limitations such as a minimum major-street green, the green is transferred to the minor street. The green returns

to the major street when the maximum minor-street green is reached or when the detector senses that there is no further demand on the minor street. Semi-actuated operation is often used where the primary reason for signalization is "interruption of continuous traffic," as discussed in Chapter 16.

3. *Full actuated operation.* In full actuated operation, every lane of every approach must be monitored by a detector. Green time is allocated in accordance with information from detectors and programmed "rules" established in the controller for capturing and retaining the green. In full actuated operation, the cycle length, sequence of phases, and green time split may vary from cycle to cycle. Chapter 20 presents more detailed descriptions of actuated signal operation, along with a methodology for timing such signals.

4. *Computer control.* Computer control is a system term. No individual signal is "computer controlled," unless the signal controller is considered to be a computer. In a computer-controlled system, the computer acts as a master controller, coordinating the timings of a large number (hundreds) of signals. The computer selects or calculates an optimal coordination plan based on input from detectors placed throughout the system. In general, such selections are made only once in advance of an AM or PM peak period. The nature of a system transition from one timing plan to another is sufficiently disruptive to be avoided during peak-demand periods. Individual signals in a computer-controlled system generally operate in the pretimed mode. For coordination to be effective, all signals in the network must use the same cycle length (or an even multiple thereof), and it is therefore difficult to maintain a progressive pattern where cycle length or phase splits are allowed to vary.

17.1.3   Treatment of Left Turns

The modeling of signalized intersection operation would be straightforward if left turns did not exist. Left turns at a signalized intersection can be handled in one of three ways:

1. *Permitted left turns.* A "permitted" left turn movement is one that is made across an opposing flow of vehicles. The driver is permitted to cross through the opposing flow, but must select an appropriate gap in the opposing traffic stream through which to turn. This is the most common form of left-turn phasing at signalized intersections, used where left-turn volumes are reasonable and where gaps in the opposing flow are adequate to accommodate left turns safely.

2. *Protected left turns.* A "protected" left turn movement is made without an opposing vehicular flow. The signal plan protects left-turning vehicles by stopping the opposing through movement. This requires that the left turns and the opposing through flow be accommodated in separate signal phases and leads to multiphase (more than two) signalization. In some cases, left turns are "protected" by geometry or regulation. Left turns from the stem of a T-intersection, for example, face no opposing flow, as there is no opposing approach to the intersection. Left turns from a one-way street similarly do not face an opposing flow.

3. *Compound left turns.* More complicated signal timing can be designed in which left turns are protected for a portion of the signal cycle and are permitted in another portion of the cycle. Protected and permitted portions of the cycle can be provided in any order. Such phasing is also referred to as *protected plus permitted* or *permitted plus protected*, depending upon the order of the sequence.

The permitted left turn movement is very complex. It involves the conflict between a left turn and an opposing through movement. The operation is affected by the left-turn flow rate and the opposing flow rate, the number of opposing lanes, whether left turns flow from an exclusive left-turn lane or from a shared lane, and the details of the signal timing. Modeling the interaction among these elements is a complicated process, one that often involves iterative elements.

The terms *protected* and *permitted* may also be applied to right turns. In this case, however, the conflict is between the right-turn vehicular movement and the pedestrian movement in the conflicting crosswalk. The vast majority of right turns at signalized intersections are handled on a permitted basis. Protected right turns generally occur at locations where there are overpasses or underpasses provided for pedestrians. At these locations, pedestrians are prohibited from making surface crossings; barriers are often required to enforce such a prohibition.

17.2 Discharge Headways, Saturation Flow, Lost Times, and Capacity

The fundamental element of a signalized intersection is the periodic stopping and restarting of the traffic stream. Figure 17.1 illustrates this process. When the light turns GREEN, there is a queue of stored vehicles that were stopped during the preceding RED phase, waiting to be discharged. As the queue of vehicles moves, headway measurements are taken as follows:

- The first headway is the time lapse between the initiation of the GREEN signal and the time that the front wheels of the first vehicle cross the stop line.
- The second headway is the time lapse between the time that the first vehicle's front wheels cross the stop line and the time that the second vehicle's front wheels cross the stop line.
- Subsequent headways are similarly measured.
- Only headways through the last vehicle in queue (at the initiation of the GREEN light) are considered to be operating under "saturated" conditions.

If many queues of vehicles are observed at a given location and the average headway is plotted vs. the queue position of the vehicle, a trend similar to that shown in Figure 17.1 (b) emerges.

The first headway is relatively long. The first driver must go through the full perception-reaction sequence, move his or her foot from the brake to the accelerator, and accelerate through the intersection. The second headway is shorter, because the second driver can overlap the perception-reaction and acceleration process of the first driver. Each successive headway is a little bit smaller. Eventually, the headways tend to level out. This generally occurs when queued vehicles have fully accelerated by the time they cross the stop line. At this point, a stable moving queue has been established.

17.2.1 Saturation Headway and Saturation Flow Rate

As noted, average headways will tend towards a constant value. In general, this occurs from the fourth or fifth headway position. The constant headway achieved is referred to as the *saturation headway*, as it is the average headway that can be achieved by a saturated, stable moving queue of vehicles passing through the signal. It is given the symbol "*h*," in units of seconds/vehicle.

It is convenient to model behavior at a signalized intersection by assuming that every vehicle (in a given lane) consumes an average of "*h*" seconds of green time to enter the intersection. If every vehicle consumes "*h*" seconds of green time and if the signal were *always* green, then "*s*" vehicles per hour could enter the intersection. This is referred to as the *saturation flow rate*:

$$s = \frac{3,600}{h} \qquad (17-1)$$

where: s = saturation flow rate, vehicles per hour of green per lane (veh/hg/ln)

h = saturation headway, seconds/vehicle (s/veh)

Saturation flow rate can be multiplied by the number of lanes provided for a given set of movements to obtain a saturation flow rate for a lane group or approach.

The saturation flow rate is, in effect, the capacity of the approach lane or lanes if they were available for use all of the time (i.e., if the signal were always GREEN). The signal, of course, is not always GREEN for any given movement. Thus, some mechanism (or

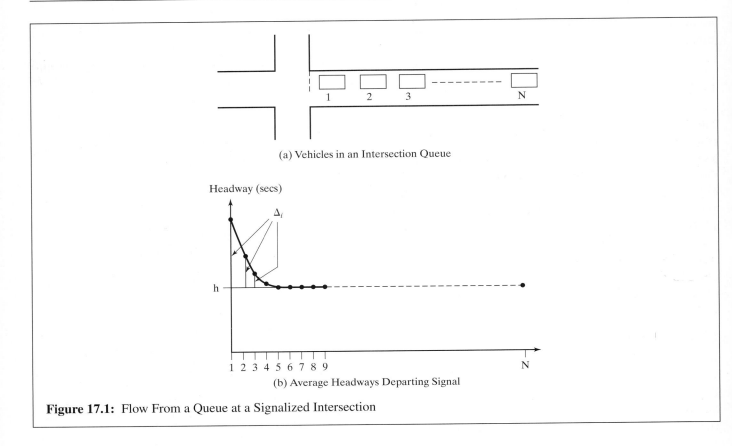

(a) Vehicles in an Intersection Queue

(b) Average Headways Departing Signal

Figure 17.1: Flow From a Queue at a Signalized Intersection

model) for dealing with the cyclic starting and stopping of movements must be developed.

17.2.2   Start-Up Lost Time

The average headway per vehicle is actually greater than "h" seconds. The first several headways are, in fact, larger than "h" seconds, as illustrated in Exhibit 17-1 (b). The first three or four headways involve additional time as drivers react to the GREEN signal and accelerate. The additional time involved in each of these initial headways (above and beyond "h" seconds) is noted by the symbol Δ_i (for headway i). These additional times are added, and are referred to as the *start-up lost time*:

$$\ell_1 = \sum_i \Delta_i \qquad (17\text{-}2)$$

where: ℓ_1 = start-up lost time, s/phase
  Δ_i = incremental headway (above "h" seconds) for vehicle i, s

Thus, it is possible to model the amount of GREEN time required to discharge a queue of "n" vehicles as:

$$T_n = \ell_1 + nh \qquad (17\text{-}3)$$

where: T_n = GREEN time required to move queue of "n" vehicles through a signalized intersection, s

  ℓ_1 = start-up lost time, s/phase

  n = number of vehicles in queue

  h = saturation headway, s/veh

While this particular model is not of great use, it does illustrate the basic concepts of saturation headway and start-up lost times. The start-up lost time is thought of as a period of time that is "lost" to vehicle use. Remaining

GREEN time, however, may be assumed to be usable at a rate of h s/veh.

17.2.3   Clearance Lost Time

The start-up lost time occurs every time a queue of vehicles starts moving on a GREEN signal. There is also a lost time associated with stopping the queue at the end of the GREEN signal. This time is more difficult to observe in the field, as it requires that the standing queue of vehicles be large enough to consume all of the GREEN time provided. In such a situation, the clearance lost time, ℓ_2, is defined as the time interval between the last vehicle's front wheels crossing the stop line, and the initiation of the GREEN for the *next* phase. The clearance lost time occurs each time a flow of vehicles is stopped.

17.2.4   Total Lost Time and the Concept of Effective GREEN Time

If the start-up lost time occurs each time a queue starts to move and the clearance lost time occurs each time the flow of vehicles stops, then for each GREEN phase:

$$t_L = \ell_1 + \ell_2 \qquad (17\text{-}4)$$

where: t_L = total lost time per phase, s/phase

All other variables are as previously defined.

The concept of lost times leads to the concept of *effective green time*. The actual signal goes through a sequence of intervals for each signal phase:

- Green
- Yellow
- All-red
- Red

The "yellow" and "all-red" intervals are a transition between GREEN and RED. This must be provided because vehicles cannot stop instantaneously when the light changes. The "all-red" is a period of time during which all lights in all directions are red. During the RED interval for one set of movements, another set of movements goes

through the green, yellow, and all-red intervals. These intervals are defined more precisely in Chapter 18.

In terms of modeling, there are really only two time periods of interest: *effective green time* and *effective red time*. For any given set of movements, *effective green time* is the amount of time that vehicles are moving (at a rate of one vehicle every h seconds). The *effective red time* is the amount of time that they are not moving. Effective green time is related to actual green time as follows:

$$g_i = G_i + Y_i - t_{Li} \qquad (17\text{-}5)$$

where: g_i = effective green time for movement(s) i, s
  G_i = actual green time for movement(s) i, s
  Y_i = sum of yellow and all red intervals for movement(s) i, s, ($Y_i = y_i + ar_i$)
  y_i = yellow interval for movement(s) i, s
  ar_i = all-red interval for movement(s) i, s
  t_{Li} = total lost time for movement(s) i, s

This model results in an effective green time that may be fully utilized by vehicles at the saturation flow rate (i.e., at an average headway of h s/veh).

17.2.5   Capacity of an Intersection Lane or Lane Group

The saturation flow rate(s) represents the capacity of an intersection lane or lane group assuming that the light is always GREEN. The portion of real time that is effective green is defined by the "green ratio," the ratio of the effective green time to the cycle length of the signal (g/C). The capacity of an intersection lane or lane group may then be computed as:

$$c_i = s_i * \left({g_i}/{C} \right) \qquad (17\text{-}6)$$

where: c_i = capacity of lane or lane group i, veh/h
  s_i = saturation flow rate for lane or lane group i, veh/hg
  g_i = effective green time for lane or lane group i, s
  C = signal cycle length, s

A Sample Problem

These concepts are best illustrated using a sample problem. Consider a given movement at a signalized intersection with the following known characteristics:

- Cycle length, $C = 60$ s
- Green time, $G = 27$ s
- Yellow plus all-red time, $Y = 3$ s
- Saturation headway, $h = 2.4$ s/veh
- Start-up lost time, $\ell_1 = 2.0$ s
- Clearance lost time, $\ell_2 = 1.0$ s

For these characteristics, what is the capacity (per lane) for this movement?

The problem will be approached in two different ways. In the first, a ledger of time within the hour is created. Once the amount of time per hour used by vehicles at the saturation flow rate is established, capacity can be found by assuming that this time is used at a rate of one vehicle every h seconds. Since the characteristics stated are given on a *per phase* basis, these would have to be converted to a *per hour* basis. This is easily done knowing the number of signal cycles that occur within an hour. For a 60-s cycle, there are 3,600/60 = 60 cycles within the hour. The subject movements will have one GREEN phase in each of these cycles. Then:

- Time in hour: 3,600 s
- RED time in hour: $(60 - 27 - 3) \times 60$ = 1,800 s
- Lost time in hour: $(2 + 1) \times 60 = 180$ s
- Remaining time in hour: $3,600 - 1,800 - 180$ = 1,620 s

The 1,620 remaining seconds of time in the hour represent the amount of time that can be used at a rate of one vehicle every h seconds, where $h = 2.4$ s/veh in this case. This number was calculated by deducting the periods during which no vehicles (in the subject movements) are effectively moving. These periods include the RED time as well as the start-up and clearance lost times in each signal cycle. The capacity of this movement may then be computed as:

$$c = \frac{1620}{2.4} = 675 \text{ veh/h/ln}$$

A second approach to this problem utilizes Equation 17-6, with the following values:

$$s = \frac{3,600}{h} = \frac{3,600}{2.4} = 1,500 \text{ veh/hg/ln}$$
$$g = G + Y - t_L = 27 + 3 - 3 = 27 \text{ s}$$
$$c = s * \left(\frac{g}{C}\right) = 1,500 * \left(\frac{27}{60}\right) = 675 \text{ veh/h/ln}$$

The two results are, as expected, the same. Capacity is found by isolating the effective green time available to the subject movements and by assuming that this time is used at the saturation flow rate (or headway).

17.2.6 Notable Studies on Saturation Headways, Flow Rates, and Lost Times

For purposes of illustrating basic concepts, subsequent sections of this chapter will assume that the value of saturation flow rate (or headway) is known. In reality, the saturation flow rate varies widely with a variety of prevailing conditions, including lane widths, heavy-vehicle presence, approach grades, parking conditions near the intersection, transit bus presence, vehicular and pedestrian flow rates, and other conditions.

The first significant studies of saturation flow were conducted by Bruce Greenshields in the 1940s [3]. His studies resulted in an average saturation flow rate of 1,714 veh/hg/ln and a start-up lost time of 3.7 s. The study, however, covered a variety of intersections with varying underlying characteristics. A later study in 1978 [4] reexamined the Greenshields hypothesis; it resulted in the same saturation flow rate (1,714 veh/hg/ln) but a lower start-up lost time of 1.1 s. The latter study had data from 175 intersections, covering a wide range of underlying characteristics.

A comprehensive study of saturation flow rates at intersections in five cities was conducted in 1987–1988 [5] to determine the effect of opposed left turns. It also produced, however, a good deal of data on saturation flow rates in general. Some of the results are summarized in Table 17.1.

These results show generally lower saturation flow rates (and higher saturation headways) than previous studies. The data, however, reflect the impact of opposed

Table 17.1: Saturation Flow Rates from a Nationwide Survey

Item	Single-Lane Approaches	Two-Lane Approaches
Number of Approaches	14	26
Number of 15-Minute Periods	101	156
Saturation Flow Rates		
Average	1,280 veh/hg/ln	1,337 veh/hg/ln
Minimum	636 veh/hg/ln	748 veh/hg/ln
Maximum	1,705 veh/hg/ln	1,969 veh/hg/ln
Saturation Headways		
Average	2.81 s/veh	2.69 s/veh
Minimum	2.11 s/veh	1.83 s/veh
Maximum	5.66 s/veh	4.81 s/veh

left turns, truck presence, and a number of other "non-standard" conditions, all of which have a significant impeding effect. The most remarkable result of this study, however, was the wide variation in measured saturation flow rates, both over time at the same site and from location to location. Even when underlying conditions remained fairly constant, the variation in observed saturation flow rates at a given location was as large as 20%–25%. In a doctoral dissertation using the same data, Prassas demonstrated that saturation headways and flow rates have a significant stochastic component, making calibration of stable values difficult [6].

The study also isolated saturation flow rates for "ideal" conditions, which include all passenger cars, no turns, level grade, and 12-ft lanes. Even under these conditions, saturation flow rates varied from 1,240 pc/hg/ln to 2,092 pc/hg/ln for single-lane approaches, and from 1,668 pc/hg/ln to 2,361 pc/hg/ln for multilane approaches. The difference between observed saturation flow rates at single and multilane approaches is also interesting. Single-lane approaches have a number of unique characteristics that are addressed in the *Highway Capacity Manual* model for analysis of signalized intersections (see Chapters 21 and 22).

Current standards in the *Highway Capacity Manual* [1] use an ideal saturation flow rate of 1,900 pc/hg/ln for both single and multilane approaches. This ideal rate is then adjusted for a variety of prevailing conditions. The manual also provides default values for lost times. The default value for start-up lost time (ℓ_1) is 2.0 s. For the clearance lost time (ℓ_2), the default value varies with the "yellow" and "all-red" timings of the signal:

$$\ell_2 = y + ar - e \qquad (17\text{-}7)$$

where: ℓ_2 = clearance lost time, s

y = length of yellow interval, s

ar = length of all-red interval, s

e = encroachment of vehicles into yellow and all-red, s

A default value of 2.0 s is used for e.

17.3 The Critical-Lane and Time-Budget Concepts

In signal analysis and design, the "critical-lane" and "time budget" concepts are closely related. The time budget, in its simplest form, is the allocation of time to various vehicular and pedestrian movements at an intersection through signal control. Time is a constant: there are always 3,600 seconds in an hour, and all of them must be allocated. In any given hour, time is "budgeted" to legal vehicular and pedestrian movements and to lost times.

The "critical-lane" concept involves the identification of specific lane movements that will control the timing of a given signal phase. Consider the situation illustrated in Figure 17.2. A simple two-phase signal

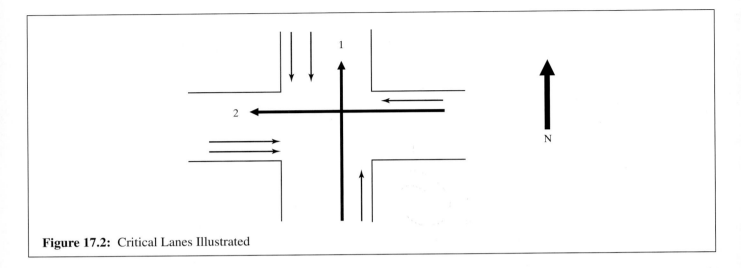

Figure 17.2: Critical Lanes Illustrated

controls the intersection. Thus, all E-W movements are permitted during one phase, and all N-S movements are permitted in another phase. During each of these phases, there are four lanes of traffic (two in each direction) moving simultaneously. Demand is not evenly distributed among them; one of these lanes will have the most intense traffic demand. The signal must be timed to accommodate traffic in this lane—the "critical lane" for the phase.

In the illustration of Figure 17.2, the signal timing and design must accommodate the total demand flows in lanes 1 and 2. As these lanes have the most intense demand, if the signal accommodates them, all other lanes will be accommodated as well. Note that the critical lane is identified as the lane with the most *intense traffic demand*, not the lane with the highest volume. This is because there are many variables affecting traffic flow. A lane with many left-turning vehicles, for example, may require more time than an adjacent lane with no turning vehicles, but a higher volume. Determining the intensity of traffic demand in a lane involves accounting for prevailing conditions that may affect flow in that particular lane.

In establishing a time budget for the intersection of Figure 17.2, time would have to be allocated to four elements:

- Movement of vehicles in critical lane 1
- Movement of vehicles in critical lane 2

- Start-up and clearance lost times for vehicles in critical lane 1
- Start-up and clearance lost times for vehicles in critical lane 2

This can be thought of in the following way: lost times are not used by any vehicle. When deducted from total time, remaining time is effective green time and is allocated to critical-lane demands—in this case, in lanes 1 and 2. The total amount of effective green time, therefore, must be sufficient to accommodate the total demand in lanes 1 and 2 (the critical lanes). These critical demands must be accommodated one vehicle at a time, as they cannot move simultaneously.

The example of Figure 17.2 is a relatively simple case. In general, the following rules apply to the identification of critical lanes:

(a) There is a critical lane and a critical-lane flow for each discrete signal phase provided.

(b) Except for lost times, when no vehicles move, there must be *one* and *only one* critical lane moving during every second of effective green time in the signal cycle.

(c) Where there are overlapping phases, the potential combination of lane flows yielding the highest sum of critical lane flows while preserving the requirement of item (b) identifies critical lanes.

Chapter 20 contains a detailed discussion of how to identify critical lanes for any signal timing and design.

17.3.1   The Maximum Sum of Critical-Lane Volumes: One View of Signalized Intersection Capacity

It is possible to consider the maximum possible sum of critical-lane volumes to be a general measure of the "capacity" of the intersection. This is not the same as the traditional view of capacity presented in the *Highway Capacity Manual*, but it is a useful concept to pursue.

By definition, each signal phase has one and only one critical lane. Except for lost times in the cycle, one critical lane is always moving. Lost times occur for each signal phase and represent time during which *no* vehicles in any lane are moving. The maximum sum of critical lane volumes may, therefore, be found by determining how much total lost time exists in the hour. The remaining time (total effective green time) may then be divided by the saturation headway.

To simplify this derivation, it is assumed that the total lost time per phase (t_L) is a constant for all phases. Then, the total lost time per signal cycle is:

$$L = N * t_L \qquad (17\text{-}8)$$

where: L = lost time per cycle, s/cycle

t_L = total lost time per phase (sum of $\ell_1 + \ell_2$), s/phase

N = number of phases in the cycle

The total lost time in an hour depends upon the number of cycles occurring in the hour:

$$L_H = L * \left(\frac{3,600}{C}\right) \qquad (17\text{-}9)$$

where: L_H = lost time per hour, s/hr

L = lost time per cycle, s/cycle

C = cycle length, s

The remaining time within the hour is devoted to effective green time for critical lane movements:

$$T_G = 3,600 - L_H \qquad (17\text{-}10)$$

where: T_G = total effective green time in the hour, s

This time may be used at a rate of one vehicle every h seconds, where h is the saturation headway:

$$V_c = \left(\frac{T_G}{h}\right) \qquad (17\text{-}11)$$

where: V_c = maximum sum of critical lane volumes, veh/h

h = saturation headway, s/veh

Merging Equations 17-8 through 17-11, the following relationship emerges:

$$V_c = \frac{1}{h}\left[3,600 - Nt_L\left(\frac{3,600}{C}\right)\right] \qquad (17\text{-}12)$$

where all variables are as previously defined.

Consider the example of Figure 17.2 again. If the signal at this location has two phases, a cycle length of 60 seconds, total lost times of 4 s/phase, and a saturation headway of 2.5 s/veh, the maximum sum of critical lane flows (the sum of flows in lanes 1 and 2) is:

$$V_c = \frac{1}{2.5}\left[3,600 - 2 * 4 * \left(\frac{3,600}{60}\right)\right] = 1,248 \text{ veh/h}$$

The equation indicates that there are $3,600/60 = 60$ cycles in an hour. For each of these, $2 * 4 = 8$ s of lost time is experienced, for a total of $8 * 60 = 480$ s in the hour. The remaining $3,600 - 480 = 3,120$ s may be used at a rate of one vehicle every 2.5 s.

If Equation 17-12 is plotted, an interesting relationship between the maximum sum of critical lane volumes (V_c), cycle length (C), and number of phases (N) may be observed, as illustrated in Figure 17.3.

As the cycle length increases, the "capacity" of the intersection also increases. This is because of lost times, which are constant per cycle. The longer the cycle length, the fewer cycles there are in an hour. This leads to less lost time in the hour, more effective green time in the hour, and a higher sum of critical-lane volumes. Note, however, that the relationship gets flatter as cycle length increases. As a general rule, increasing the cycle length may result in small increases in capacity. On the other hand, capacity can rarely be increased significantly by only increasing the cycle length. Other measures, such as adding lanes, are often also necessary.

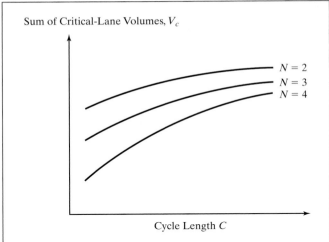

Sum of Critical-Lane Volumes, V_c

$N = 2$
$N = 3$
$N = 4$

Cycle Length C

Figure 17.3: Maximum Sum of Critical-Lane Volumes Plotted

Capacity also decreases as the number of phases increases. This is because for each phase, there is one full set of lost times in the cycle. Thus, a two-phase signal has only two sets of lost times in the cycle, while a three-phase signal has three.

These trends provide insight, but also raise an interesting question: Given these trends, it appears that all signals should have two phases and that the maximum practical cycle length should be used in all cases. After all, this combination would, apparently, yield the highest "capacity" for the intersection.

Using the maximum cycle length is not practical unless truly needed. Having a cycle length that is considerably longer than what is desirable causes increases in delay to drivers and passengers. The increase in delay is because there will be times when vehicles on one approach are waiting for the green while there is no demand on conflicting approaches. Shorter cycle lengths yield less delay. Further, there is no incentive to maximize the cycle length. There will always be 3,600 seconds in the hour, and increasing the cycle length to accommodate increasing demand over time is quite simple, requiring only a resetting of the local signal controller. The shortest cycle length consistent with a v/c ratio in the range of 0.80–0.95 is generally used to produce optimal delays. Thus, the view of signal capacity is quite different from that of pavement capacity. When

deciding on the number of lanes on a freeway (or on an intersection approach), it is desirable to build excess capacity (i.e., achieve a low v/c ratio). This is because once built, it is unlikely that engineers will get an opportunity to expand the facility for 20 or more years, and adjacent land development may make such expansion impossible. The 3,600 seconds in an hour, however, are immutable, and retiming the signal to allocate more of them to effective green time is a simple task requiring no field construction.

17.3.2 Finding an Appropriate Cycle Length

If it is assumed that the demands on an intersection are known and that the critical lanes can be identified, then Equation 17-12 could be solved using a known value of V_c to find a minimum acceptable cycle length:

$$C_{min} = \frac{Nt_L}{1 - \left(\dfrac{V_c}{3,600/h}\right)} \quad (17\text{-}13)$$

Thus, if in the example of Figure 17.2, the actual sum of critical-lane volumes was determined to be 1,000 veh/h, the minimum feasible cycle length would be:

$$C_{min} = \frac{2*4}{1 - \left(\dfrac{1,000}{3,600/2.5}\right)} = \frac{8}{0.3056} = 26.2 \ s$$

The cycle length could be reduced, in this case, from the given 60 s to 30 s (the effective minimum cycle length used). This computation, however, assumes that the demand (V_c) is uniformly distributed throughout the hour and that every second of effective green time will be used. Neither of these assumptions is very practical. In general, signals would be timed for the flow rates occurring in the peak 15 minutes of the hour. Equation 17-13 could be modified by dividing V_c by a known peak-hour factor (PHF) to estimate the flow rate in the worst 15-minute period of the hour. Similarly, most signals would be timed to have somewhere between 80% and 95% of the available capacity actually used. Due to the normal stochastic variations in demand on a cycle-by-cycle and daily basis, some excess capacity must be provided to avoid failure of individual cycles or peak periods on a specific day. If demand, V_c, is also divided by the expected utilization of capacity (expressed in decimal form), then this is also accommodated. Introducing these changes transforms Equation 17-13 to:

$$C_{des} = \frac{Nt_L}{1 - \left[\dfrac{V_c}{(3,600/h) * PHF * (v/c)}\right]} \quad (17\text{-}14)$$

where: C_{des} = desirable cycle length, s
PHF = peak hour factor
v/c = desired volume to capacity ratio

All other variables are as previously defined.
Returning to the example, if the PHF is 0.95 and it is desired to use no more than 90% of available capacity during the peak 15-minute period of the hour, then:

$$C_{des} = \frac{2 * 4}{1 - \left[\dfrac{1,000}{(3,600/2.5) * 0.95 * 0.90}\right]}$$

$$= \frac{8}{0.188} = 42.6 \text{ s}$$

In practical terms, this would lead to the use of a 45-second cycle length.
The relationship between a desirable cycle length, the sum of critical-lane volumes, and the target v/c ratio is quite interesting and is illustrated in Figure 17.4.
Figure 17.4 illustrates a typical relationship for a specified number of phases, saturation headway, lost times, and peak-hour factor. If a vertical is drawn at any specified value of V_c (sum of critical lane volumes), it is clear that the resulting cycle length is very sensitive to the target v/c ratio. As the curves for each v/c ratio are eventually asymptotic to the vertical, it is not always possible to achieve a specified v/c ratio.
Consider the case of a three-phase signal, with $t_L = 4$ s/phase, a saturation headway of 2.2 s/veh, a PHF of 0.90 and $V_c = 1,200$ veh/h. Desirable cycle lengths will be computed for a range of target v/c ratios varying from 1.00 to 0.80.

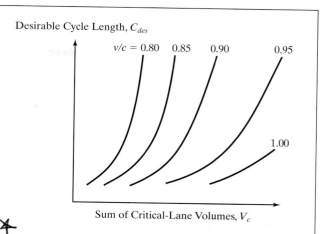

Figure 17.4: Desirable Cycle Length vs. Sum of Critical-Lane Volumes

$$C_{des} = \frac{3 * 4}{1 - \left[\dfrac{1,200}{(3,600/2.2) * 0.90 * 1.00}\right]}$$

$$= \frac{12}{0.1852} = 64.8 \Rightarrow 65 \text{ s}$$

$$C_{des} = \frac{3 * 4}{1 - \left[\dfrac{1,200}{(3,600/2.2) * 0.90 * 0.95}\right]}$$

$$= \frac{12}{0.1423} = 84.3 \Rightarrow 85 \text{ s}$$

$$C_{des} = \frac{3 * 4}{1 - \left[\dfrac{1,200}{(3,600/2.2) * 0.90 * 0.90}\right]}$$

$$= \frac{12}{0.0947} = 126.7 \Rightarrow 130 \text{ s}$$

$$C_{des} = \frac{3 * 4}{1 - \left[\dfrac{1,200}{(3,600/2.2) * 0.90 * 0.85}\right]}$$

$$= \frac{12}{0.0414} = 289.9 \Rightarrow 290 \text{ s}$$

$$C_{des} = \cfrac{3*4}{1 - \left[\cfrac{1,200}{(3,600/2.2)*0.90*0.80} \right]}$$

$$= \frac{12}{-0.0185} = -648.6 \text{ s}$$

For this case, reasonable cycle lengths can provide target v/c ratios of 1.00 or 0.95. Achieving v/c ratios of 0.90 or 0.85 would require long cycle lengths beyond the practical limit of 120 s for pretimed signals. The 130 s cycle needed to achieve a v/c ratio of 0.90 might be acceptable for an actuated signal location. However, a v/c ratio of 0.80 cannot be achieved under any circumstances. The negative cycle length that results signifies that there is not enough time within the hour to accommodate the demand with the required green time plus the 12 s of lost time per cycle. In effect, more than 3,600 s would have to be available to accomplish this.

A Sample Problem

Consider the intersection shown in Figure 17.5. The critical directional demands for this two-phase signal are shown with other key variables. Using the time-budget and critical-lane concepts, determine the number of lanes required for each of the critical movements and the minimum desirable cycle length that could be used. Note that an initial cycle length is specified, but will be modified as part of the analysis.

Assuming that the initial specification of a 60-s cycle is correct and given the other specified conditions, the maximum sum of critical lanes that can be accommodated is computed using Equation 17-12:

$$V_c = \frac{1}{2.3}\left[3,600 - 2*4*\left(\frac{3,600}{60} \right) \right] = 1,357 \text{ veh/h}$$

The critical SB volume is 1,200 veh/h, and the critical EB volume is 1,800 veh/h. The number of lanes each must be divided into is now to be determined. Whatever combination is used, the sum of the critical-lane volumes for these two approaches must be below 1,357 veh/h. Figure 17.6 shows a number of possible lane combinations and the resulting sum of critical lane volumes. As can be seen from the scenarios of Figure 17.6, in order to have a sum of critical-lane volumes less than 1,357 veh/h, the SB approach must have at least two lanes, and the EB approach must have three lanes. Realizing that these demands probably reverse in the other peak hour (AM or PM), the N–S artery would probably require four lanes, and the E–W artery six lanes.

This is a very basic analysis, and it would have to be modified based on more specific information regarding individual movements, pedestrians, parking needs, and other factors.

If the final scenario is provided, V_c is only 1,200 veh/h. It is possible that the original cycle length of 60 s

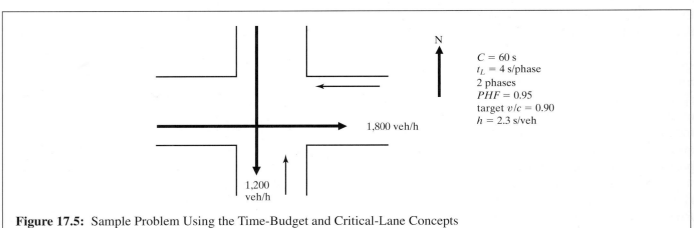

Figure 17.5: Sample Problem Using the Time-Budget and Critical-Lane Concepts

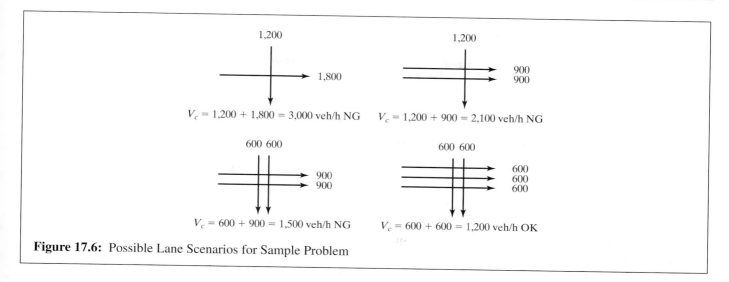

Figure 17.6: Possible Lane Scenarios for Sample Problem

could be reduced. A minimum desirable cycle length may be computed from Equation 17-14:

$$C_{des} = \frac{2*4}{1 - \left[\dfrac{1,200}{(3,600/2.3)*0.95*0.90}\right]}$$

$$= \frac{8}{0.103} = 77.7 \Rightarrow 80 \text{ s}$$

The resulting cycle length is larger than the original 60 s because the equation takes both the *PHF* and target v/c ratios into account. Equation 17-12 for computing the maximum value of V_c does not and assumes full use of capacity ($v/c = 1.00$) and no peaking within the hour. In essence, the 2 × 3 lane design proposal should be combined with an 80 s cycle length to achieve the desired results.

This problem illustrates the critical relationship between number of lanes and cycle lengths. Clearly, there are other scenarios that would produce desirable results. Additional lanes could be provided in either direction, which would allow the use of a shorter cycle length. Unfortunately, for many cases, signal timing is considered with a fixed design already in place. Only where right-of-way is available or a new intersection is being constructed can major changes in the number of lanes be considered. Allocation of lanes to various movements is also a consideration. Optimal solutions

are generally found more easily when the physical design and signalization can be treated in tandem.

If, in the problem of Figure 17.5, space limited both the EB and SB approaches to two lanes, the resulting V_c would be 1,500 veh/h. Would it be possible to accommodate this demand by lengthening the cycle length? Again, Equation 17-14 is used:

$$C_{des} = \frac{2*4}{1 - \left[\dfrac{1,500}{(3,600/2.3)*0.95*0.90}\right]}$$

$$= \frac{8}{-0.121} = -66.1 \text{ s NG}$$

The negative result indicates that there is no cycle length that can accommodate a V_c of 1,500 veh/h at this location.

17.4 The Concept of Left-Turn Equivalency

The most difficult process to model at a signalized intersection is the left turn. Left turns are made in several different modes using different design elements. Left turns may be made from a lane shared with through vehicles (shared-lane operation) or from a lane dedicated to left-turning vehicles (exclusive-lane operation). Traffic signals

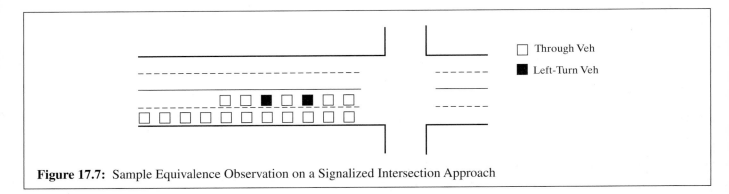

Figure 17.7: Sample Equivalence Observation on a Signalized Intersection Approach

may allow for permitted or protected left turns, or some combination of the two.

Whatever the case, however, a left-turning vehicle will consume more effective green time traversing the intersection than will a similar through vehicle. The most complex case is that of a permitted left turn made across an opposing vehicular flow from a shared lane. A left-turning vehicle in the shared lane must wait for an acceptable gap in the opposing flow. While waiting, the vehicle blocks the shared lane, and other vehicles (including through vehicles) in the lane are delayed behind it. Some vehicles will change lanes to avoid the delay, while others are unable to and must wait until the left-turner successfully completes the turn.

Many models of the signalized intersection account for this in terms of "through vehicle equivalents" (i.e., how many through vehicles would consume the same amount of effective green time traversing the stop-line as *one* left-turning vehicle?). Consider the situation depicted in Figure 17.7. If both the left lane and the right lane were observed, an equivalence similar to the following statement could be determined:

> In the same amount of time, the left lane discharges five through vehicles and two left-turning vehicles, while the right lane discharges eleven through vehicles.

In terms of effective green time consumed, this observation means that eleven through vehicles are equivalent to five through vehicles plus two left turning vehicles. If the left-turn equivalent is defined as E_{LT}:

$$11 = 5 + 2E_{LT}$$
$$E_{LT} = \frac{11 - 5}{2} = 3.0$$

It should be noted that this computation holds only for the prevailing characteristics of the approach during the observation period. The left-turn equivalent depends upon a number of factors, including how left turns are made (protected, permitted, compound), the opposing traffic flow, and the number of opposing lanes. Figure 17.8 illustrates the general form of the relationship for through vehicle equivalents of *permitted* left turns.

The left-turn equivalent, E_{LT}, increases as the opposing flow increases. For any given opposing flow, however, the equivalent decreases as the number of opposing lanes is increased from one to three. This latter relationship is not linear, as the task of selecting a gap through multilane opposing traffic is more difficult than selecting a gap through single-lane opposing traffic. Further, in a multilane traffic stream, vehicles do not pace each other side-by-side, and the gap distribution does not improve as much as the per-lane opposing flow decreases.

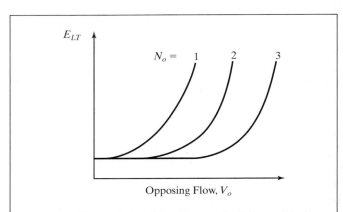

Figure 17.8: Relationship Among Left-Turn Equivalents, Opposing Flow, and Number of Opposing Lanes

To illustrate the use of left-turn equivalents in modeling, consider the following problem:

> An approach to a signalized intersection has two lanes, permitted left-turn phasing, 10% left-turning vehicles, and a left-turn equivalent of 5.0. The saturation headway for through vehicles is 2.0 s/veh. Determine the equivalent saturation flow rate and headway for all vehicles on this approach.

The first way to interpret the left-turn equivalent is that each left-turning vehicle consumes 5.0 times the effective green time as a through vehicle. Thus, for the situation described, 10% of the traffic stream has a saturation headway of $2.0 \times 5.0 = 10.0$ s/veh, while the remainder (90%) have a saturation headway of 2.0 s/veh. The average saturation headway for all vehicles is, therefore:

$$h = (0.10 * 10.0) + (0.90 * 2.0) = 2.80 \text{ s/veh}$$

This corresponds to a saturation flow rate of:

$$s = \frac{3,600}{2.80} = 1,286 \text{ veh/hg/ln}$$

A number of models, including the *Highway Capacity Manual* approach, calibrate a multiplicative adjustment factor that converts an ideal (or through) saturation flow rate to a saturation flow rate for prevailing conditions:

$$s_{prev} = s_{ideal} * f_{LT}$$

$$f_{LT} = s_{prev}/s_{ideal} = (3,600/h_{prev})/(3,600/h_{ideal})$$

$$= h_{ideal}/h_{prev} \qquad (17\text{-}15)$$

where: s_{prev} = saturation flow rate under prevailing conditions, veh/hg/ln

s_{ideal} = saturation flow rate under ideal conditions, veh/hg/ln

f_{LT} = left-turn adjustment factor

h_{ideal} = saturation headway under ideal conditions, s/veh

h_{prev} = saturation headway under prevailing conditions, s/veh

In effect, in the first solution, the prevailing headway, h_{prev}, was computed as follows:

$$h_{prev} = (P_{LT} E_{LT} h_{ideal}) + [(1 - P_{LT})h_{ideal}] \quad (17\text{-}16)$$

Combining Equations 17-15 and 17-16:

$$f_{LT} = \frac{h_{ideal}}{(P_{LT} E_{LT} h_{ideal}) + [(1 - P_{LT})h_{ideal}]}$$

$$f_{LT} = \frac{1}{P_{LT} E_{LT} + (1 - P_{LT})} = \frac{1}{1 + P_{LT}(E_{LT} - 1)} \quad (17\text{-}17)$$

The problem posed may now be solved using a left-turn adjustment factor. Note that the saturation headway under ideal conditions is $3,600/2.0 = 1,800$ veh/hg/ln. Then:

$$f_{LT} = \frac{1}{1 + 0.10(5 - 1)} = 0.714$$

$$s_{prev} = 1800 * 0.714 = 1,286 \text{ veh/hg/ln}$$

This, of course, is the same result.

It is important that the concept of left-turn equivalence be understood. Its use in multiplicative adjustment factors often obscures its intent and meaning. The fundamental concept, however, is unchanged—the equivalence is based on the fact that the effective green time consumed by a left-turning vehicle is E_{LT} times the effective green time consumed by a similar through vehicle.

Signalized intersection and other traffic models use other types of equivalents that are similar. Heavy-vehicle, local-bus, and right-turn equivalents have similar meanings and result in similar equations. Some of these have been discussed in previous chapters, and others will be discussed in subsequent chapters.

17.5   Delay as a Measure of Effectiveness

Signalized intersections represent point locations within a surface street network. As point locations, the measures of operational quality or effectiveness used for highway sections are not relevant. Speed has no meaning at a point, and density requires a section of some length for measurement. A number of measures have been used to characterized the operational quality of a signalized intersection, the most common of which are:

- Delay
- Queuing
- Stops

These are all related. Delay refers to the amount of time consumed in traversing the intersection—the difference between the arrival time and the departure time, where these may be defined in a number of different ways. Queuing refers to the number of vehicles forced to queue behind the stop-line during a RED signal phase; common measures include the average queue length or a percentile queue length. Stops refer to the percentage or number of vehicles that must stop at the signal.

17.5.1   Types of Delay

The most common measure used is delay, with queuing often used as a secondary measure of operational quality. While it is possible to measure delay in the field, it is a difficult process, and different observers may make judgments that could yield different results. For many purposes, it is, therefore, convenient to have a predictive model for the estimate of delay. Delay, however, can be quantified in many different ways. The most frequently used forms of delay are defined as follows:

1. *Stopped-time delay*. Stopped-time delay is defined as the time a vehicle is stopped in queue while waiting to pass through the intersection; average stopped-time delay is the average for all vehicles during a specified time period.

2. *Approach delay*. Approach delay includes stopped-time delay but adds the time loss due to deceleration from the approach speed to a stop and the time loss due to reacceleration back to the desired speed. Average approach delay is the average for all vehicles during a specified time period.

3. *Time-in-queue delay*. Time-in-queue delay is the total time from a vehicle joining an intersection queue to its discharge across the STOP line on departure. Again, average time-in-queue delay is the average for all vehicles during a specified time period.

4. *Travel time delay*. This is a more conceptual value. It is the difference between the driver's expected travel time through the intersection (or any roadway segment) and the actual time taken. Given the difficulty in establishing a "desired" travel time to traverse an intersection, this value is rarely used, other than as a philosophic concept.

5. *Control delay*. The concept of control delay was developed in the 1994 *Highway Capacity Manual*, and is included in the HCM 2000. It is the delay caused by a control device, either a traffic signal or a STOP-sign. It is approximately equal to time-in-queue delay plus the acceleration-deceleration delay component.

Figure 17.9 illustrates three of these delay types for a single vehicle approaching a RED signal.

Stopped-time delay for this vehicle includes only the time spent stopped at the signal. It begins when the vehicle is fully stopped and ends when the vehicle begins to accelerate. Approach delay includes additional time losses due to deceleration and acceleration. It is

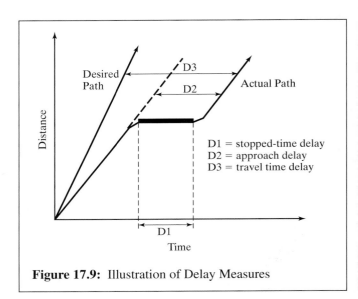

Figure 17.9: Illustration of Delay Measures

found by extending the velocity slope of the approaching vehicle as if no signal existed; the approach delay is the horizontal (time) difference between the hypothetical extension of the approaching velocity slope and the departure slope after full acceleration is achieved. Travel time delay is the difference in time between a hypothetical desired velocity line and the actual vehicle path. Time-in-queue delay cannot be effectively shown using one vehicle, as it involves joining and departing a queue of several vehicles.

Delay measures can be stated for a single vehicle, as an average for all vehicles over a specified time period, or as an aggregate total value for all vehicles over a specified time period. Aggregate delay is measured in total *vehicle-seconds, vehicle-minutes,* or *vehicle-hours* for all vehicles in the specified time interval. Average individual delay is generally stated in terms of s/veh for a specified time interval.

17.5.2   Basic Theoretical Models of Delay

Virtually all analytic models of delay begin with a plot of cumulative vehicles arriving and departing vs. time at a given signal location. The time axis is divided into periods of effective green and effective red as illustrated in Figure 17.10.

Vehicles are assumed to arrive at a uniform rate of flow of v vehicles per unit time, seconds in this case. This is shown by the constant slope of the arrival curve. Uniform arrivals assume that the inter-vehicle arrival time between vehicles is a constant. Thus, if the arrival flow rate, v, is 1,800 vehs/h, then one vehicle arrives every $3,600/1,800 = 2.0$ s.

Assuming no preexisting queue, vehicles arriving when the light is GREEN continue through the intersection, (i.e., the departure curve is the same as the arrival curve). When the light turns RED, however, vehicles continue to arrive, but none depart. Thus, the departure curve is parallel to the x-axis during the RED interval. When the next effective GREEN begins, vehicles queued during the RED interval depart from the intersection, now at the saturation flow rate, s, in veh/s. For stable operations, depicted here, the departure curve "catches up" with the arrival curve before the next RED interval begins (i.e., there is no residual or unserved queue left at the end of the effective GREEN).

This simple depiction of arrivals and departures at a signal allows the estimation of three critical parameters:

- The total time that any vehicle i spends waiting in the queue, $W(i)$, is given by the horizontal time-scale difference between the time of arrival and the time of departure.

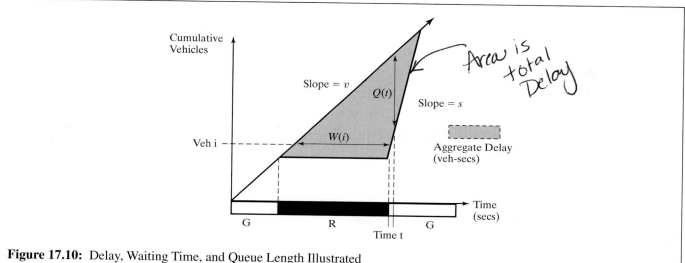

Figure 17.10: Delay, Waiting Time, and Queue Length Illustrated

- The total number of vehicles queued at any time t, $Q(t)$, is the vertical vehicle-scale difference between the number of vehicles that have arrived and the number of vehicles that have departed.

- The aggregate delay for all vehicles passing through the signal is the area between the arrival and departure curves (vehicles × time).

Note that since the plot illustrates vehicles arriving in queue and departing from queue, this model most closely represents what has been defined as *time-in-queue delay*. There are many simplifications that have been assumed, however, in constructing this simple depiction of delay. It is important to understand the two major simplifications:

- The assumption of a uniform arrival rate is a simplification. Even at a completely isolated location, actual arrivals would be *random* (i.e., would have an average rate over time), but inter-vehicle arrival times would vary around an average rather than being constant. Within coordinated signal systems, however, vehicle arrivals are in platoons.

- It is assumed that the queue is building at a point location (as if vehicles were stacked on top of one another). In reality, as the queue grows, the rate at which vehicles arrive at its end is the arrival rate of vehicles (at a point), plus a component representing the backward growth of the queue in space.

Both of these can have a significant effect on actual results. Modern models account for the former in ways that will be discussed subsequently. The assumption of a "point queue," however is imbedded in many modern applications.

Figure 17.11 expands the range of Figure 17.10 to show a series of GREEN phases and depicts three different types of operation. It also allows for an arrival function, $a(t)$, that varies, while maintaining the departure function, $d(t)$, described previously.

Figure 17.11 (a) shows stable flow throughout the period depicted. No signal cycle "fails" (i.e., ends with some vehicles queued during the preceding RED unserved). During every GREEN phase, the departure

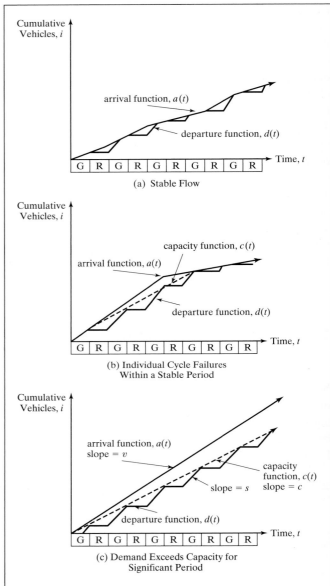

Figure 17.11: Three Delay Scenarios (Adapted with permission of Transportation Research Board, National Research Council, Washington DC, from V.F. Hurdle, "Signalized Intersection Delay Model: A Primer for the Uninitiated," *Transportation Research Record 971*, pgs. 97, 98, 1984.)

function "catches up" with the arrival function. Total aggregate delay during this period is the total of all the triangular areas between the arrival and departure curves. This type of delay is often referred to as "uniform delay."

In Figure 17.11 (b), some of the signal phases "fail." At the end of the second and third GREEN intervals, some vehicles are not served (i.e., they must wait for a second GREEN interval to depart the intersection). By the time the entire period ends, however, the departure function has "caught up" with the arrival function and there is no residual queue left unserved. This case represents a situation in which the overall period of analysis is stable (i.e., total demand does not exceed total capacity). Individual cycle failures within the period, however, have occurred. For these periods, there is a second component of delay in addition to uniform delay. It consists of the area between the arrival function and the dashed line, which represents the capacity of the intersection to discharge vehicles, and has the slope c. This type of delay is referred to as "overflow delay."

Figure 17.11 (c) shows the worst possible case: Every GREEN interval "fails" for a significant period of time, and the residual, or unserved, queue of vehicles continues to grow throughout the analysis period. In this case, the overflow delay component grows over time, quickly dwarfing the uniform delay component.

The latter case illustrates an important practical operational characteristic. When demand exceeds capacity ($v/c > 1.00$), the delay depends upon *the length of time* that the condition exists. In Figure 17.11 (b), the condition exists for only two phases. Thus, the queue and the resulting overflow delay is limited. In Figure 17.11 (c), the condition exists for a long time, and the delay continues to grow throughout the oversaturated period.

Components of Delay ✗

In analytic models for predicting delay, there are three distinct components of delay that may be identified:

- *Uniform delay* is the delay based on an assumption of uniform arrivals and stable flow with no individual cycle failures.

- *Random delay* is the additional delay, above and beyond uniform delay, because flow is randomly distributed rather than uniform at isolated intersections.

- *Overflow delay* is the additional delay that occurs when the capacity of an individual phase or

series of phases is less than the demand or arrival flow rate.

In addition, the delay impacts of platoon flow (rather than uniform or random) is treated as an adjustment to uniform delay. Many modern models combine the random and overflow delays into a single function, which is referred to as "overflow delay," even though it contains both components.

The differences between uniform, random, and platooned arrivals are illustrated in Figure 17.12. As noted, the analytic basis for most delay models is the assumption of uniform arrivals, which are depicted in Figure 17.12 (a). Even at isolated intersections, however, arrivals would be random, as shown in Figure 17.12 (b). With random arrivals, the underlying rate of arrivals is a constant, but the inter-arrival times are exponentially distributed around an average. In most urban and suburban cases, where a signalized intersection is likely to be part of a coordinated signal system, arrivals will be in organized platoons that move down the arterial in a cohesive group. The exact time that a platoon arrives at a downstream signal has an enormous potential effect on delay. A platoon of vehicles arriving at the beginning of the RED forces most vehicles to stop for the entire length of the RED phase. The same platoon of vehicles arriving at the beginning of the GREEN phase may flow through the intersection without any vehicles stopping. In both cases, the arrival flow, v, and the capacity of the intersection, c, are the same. The resulting delay, however, would vary significantly. The existence of platoon

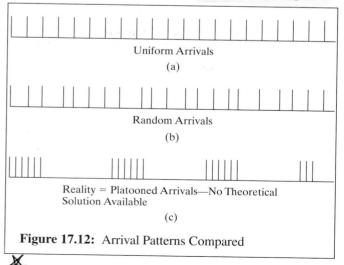

Uniform Arrivals

(a)

Random Arrivals

(b)

Reality = Platooned Arrivals—No Theoretical Solution Available

(c)

Figure 17.12: Arrival Patterns Compared

arrivals, therefore, necessitates a significant adjustment to models based on theoretically uniform or random flow.

Webster's Uniform Delay Model

Virtually every model of delay starts with Webster's model of uniform delay. Initially published in 1958 [7], this model begins with the simple illustration of delay depicted in Figure 17.10, with its assumptions of stable flow and a simple uniform arrival function. As noted previously, aggregate delay can be estimated as the area between the arrival and departure curves in the figure. Thus, Webster's model for uniform delay is the area of the triangle formed by the arrival and departure functions. For clarity, this triangle is shown again in Figure 17.13.

The area of the aggregate delay triangle is simply one-half the base times the height, or:

$$UD_a = \frac{1}{2}RV$$

where: UD_a = aggregate uniform delay, veh-secs

R = length of the RED phase, s

V = total vehicles in queue, vehs

By convention, traffic models are not developed in terms of RED time. Rather, they focus on GREEN time.

Thus, Webster substitutes the following equivalence for the length of the RED phase:

$$R = C\left[1 - \left(\frac{g}{C}\right)\right]$$

where: C = cycle length, s

g = effective green time, s

In words, the RED time is the portion of the cycle length that is not effectively green.

The height of the triangle, V, is the total number of vehicles in the queue. In effect, it includes vehicles arriving during the RED phase, R, plus those that join the end of the queue while it is moving out of the intersection (i.e., during time t_c in Figure 17.13). Thus, determining the time it takes for the queue to clear, t_c, is an important part of the model. This is done by setting the number of vehicles arriving during the period $R + t_c$ equal to the number of vehicles departing during the period t_c, or:

$$v(R + t_c) = st_c$$

$$R + t_c = \left(\frac{s}{v}\right)t_c$$

$$R = t_c\left(\frac{s}{v} - 1\right)$$

$$t_c = \frac{R}{\left(\frac{s}{v} - 1\right)}$$

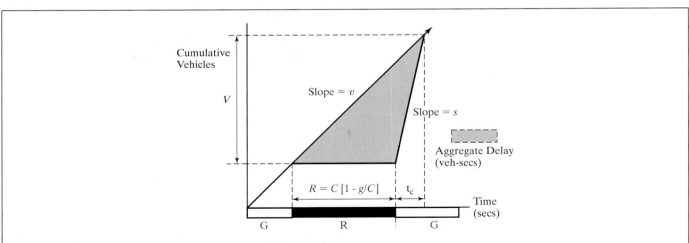

Figure 17.13: Webster's Uniform Delay Model Illustrated

Then, substituting for t_c:

$$V = v(R + t_c) = v\left[R + \frac{R}{\left(\frac{s}{v} - 1\right)}\right] = R\left[\frac{vs}{s - v}\right]$$

and for R:

$$V = C\left[1 - \left(\frac{g}{C}\right)\right]\left[\frac{vs}{s - v}\right]$$

Then, aggregate delay can be stated as:

$$UD_a = \frac{1}{2}RV = \frac{1}{2}C^2\left[1 - \frac{g}{C}\right]^2\left[\frac{vs}{s - v}\right] \quad (17\text{-}18)$$

where all variables are as previously defined.

Equation 17-18 estimates aggregate uniform delay in vehicle-seconds for one signal cycle. To get an estimate of average uniform delay per vehicle, the aggregate is divided by the number of vehicles arriving during the cycle, vC. Then:

$$UD = \frac{1}{2}C\frac{\left[1 - \frac{g}{C}\right]^2}{\left[1 - \frac{v}{s}\right]} \quad (17\text{-}19)$$

Another form of the equation uses the capacity, c, rather than the saturation flow rate, s. Noting that $s = c/(g/C)$, the following form emerges:

$$UD = \frac{1}{2}C\frac{\left[1 - \frac{g}{C}\right]^2}{\left[1 - \left(\frac{g}{C}\right)\left(\frac{v}{c}\right)\right]} = \frac{0.50C\left[1 - \frac{g}{C}\right]^2}{1 - \left(\frac{g}{C}\right)X}$$

$$(17\text{-}20)$$

where: UD = average uniform delay per vehicles, s/veh

C = cycle length, s

g = effective green time, s

v = arrival flow rate, veh/h

c = capacity of intersection approach, veh/h

X = v/c ratio, or degree of saturation

This average includes the vehicles that arrive and depart on green, accruing no delay. This is appropriate. One of the objectives in signalizations is to minimize the number or proportion of vehicles that must stop. Any meaningful quality measure would have to include the positive impact of vehicles that are not delayed.

In Equation 17-20, it must be noted that the maximum value of X (the v/c ratio) is 1.00. As the uniform delay model assumes no overflow, the v/c ratio cannot be more than 1.00.

Modeling Random Delay

The uniform delay model assumes that arrivals are uniform and that no signal phases fail (i.e., that arrival flow is less than capacity during every signal cycle of the analysis period).

At isolated intersections, vehicle arrivals are more likely to be random. A number of stochastic models have been developed for this case, including those by Newall [8], Miller [9,10], and Webster [7]. Such models assume that inter-vehicle arrival times are distributed according to the Poisson distribution, with an underlying average arrival rate of v vehicles/unit time. The models account for both the underlying randomness of arrivals and the fact that some individual cycles within a demand period with v/c < 1.00 could fail due to this randomness. This additional delay is often referred to as "overflow delay," but it does not address situations in which v/c > 1.00 for the entire analysis period. This text refers to additional delay due to randomness as "random delay," RD, to distinguish it from true overflow delay when v/c > 1.00. The most frequently used model for random delay is Webster's formulation:

$$RD = \frac{X^2}{2v(1 - X)} \quad (17\text{-}21)$$

where: RD = average random delay per vehicle, s/veh

X = v/c ratio

This formulation was found to somewhat overestimate delay, and Webster proposed that total delay (the sum of uniform and random delay) be estimated as:

$$D = 0.90(UD + RD) \quad (17\text{-}22)$$

where: D = sum of uniform and random delay

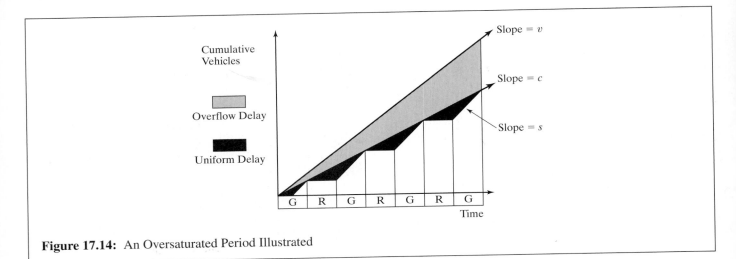

Figure 17.14: An Oversaturated Period Illustrated

Modeling Overflow Delay

"Oversaturation" is used to describe extended time periods during which arriving vehicles exceed the capacity of the intersection approach to discharge vehicles. In such cases, queues grow, and overflow delay, in addition to uniform delay, accrues. As overflow delay accounts for the failure of an extended series of phases, it encompasses a portion of random delay as well.

Figure 17.14 illustrates a time period for which $v/c > 1.00$. Again, as in the uniform delay model, it is assumed that the arrival function is uniform.

During the period of oversaturation, delay consists of both uniform delay (in the triangles between the capacity and departure curves) and overflow delay (in the growing triangle between the arrival and capacity curves). The formula for the uniform delay component may be simplified in this case, as the v/c ratio (X) is the maximum value of 1.00 for the uniform delay component. Then:

$$UD_o = \frac{0.50C\left[1 - \frac{g}{C}\right]^2}{1 - \left(\frac{g}{C}\right)X} = \frac{0.50C\left[1 - \frac{g}{C}\right]^2}{1 - \left(\frac{g}{C}\right)1.00}$$

$$= 0.50C\left[1 - \frac{g}{C}\right] \qquad (17\text{-}23)$$

To this, the overflow delay must be added. Figure 17.15 illustrates how the overflow delay is estimated. The aggregate and average overflow delay can be estimated as:

$$OD_a = \frac{1}{2}T(vT - cT) = \frac{T^2}{2}(v - c)$$

$$OD = \frac{T}{2}[X - 1] \qquad (17\text{-}24)$$

where: OD_a = aggregate overflow delay, veh-secs

OD = average overflow delay per vehicle, s/veh

Other parameters as previously defined

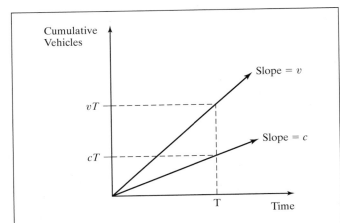

Figure 17.15: Derivation of the Overflow Delay Formula

In Equations 17-24, the average overflow delay is obtained by dividing the aggregate delay by *the number of vehicles discharged within time T, cT.* Unlike the formulation for uniform delay, where the number of vehicles arriving and the number of vehicles discharged during a cycle were the same, the overflow delay triangle includes vehicles that arrive within time T, but are not discharged within time T. The delay triangle, therefore, includes only the delay accrued by vehicles through time T, and excludes additional delay that vehicles still "stuck" in the queue will experience after time T.

Equations 17-24 may use any unit of time for "T." The resulting overflow delay, OD, will have the same units as specified for T, on a per-vehicle basis.

Equations 17-24 are time-dependent (i.e., the longer the period of oversaturation exists, the larger delay becomes). The predicted delay per vehicle is averaged over the entire period of oversaturation, T. This masks, however, a significant issue: vehicles arriving early during time T experience far less delay than vehicles arriving later during time T. A model for average overflow delay during a time period T_1 through T_2 may be developed, as illustrated in Figure 17.16. Note that the delay area formed is a trapezoid, not a triangle.

The resulting model for average delay per vehicle during the time period T_1 through T_2 is:

$$OD = \frac{T_1 + T_2}{2}(X - 1) \qquad (17\text{-}25)$$

where all terms are as previously defined. Note that the trapezoidal shape of the delay area results in the $T_1 + T_2$ formulation, emphasizing the growth of delay as the oversaturated condition continues over time. Also, this formulation predicts the average delay per vehicle that occurs during the specified interval, T_1 through T_2. Thus, delays to vehicles arriving before time T_1 but discharging after T_1 are included only to the extent of their delay within the specified times, not any delay they may have experienced in queue before T_1. Similarly, vehicles discharging after T_2 do have a delay component after T_2 that is not included in the formulation.

The three varieties of delay—uniform, random, and overflow delay—can be modeled in relatively simple terms as long as simplifying assumptions are made in terms of arrival and discharge flows, and in the nature of the queuing that occurs, particularly during periods of oversaturation. The next section begins to consider some of the complications that arise from the direct use of these simplified models.

17.5.3   Inconsistencies in Random and Overflow Delay

Figure 17.17 illustrates a basic inconsistency in the random and overflow delay models previously discussed.

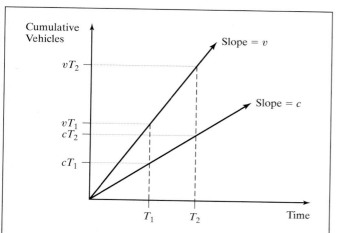

Figure 17.16: A Model for Overflow Delay Between Times T_1 and T_2

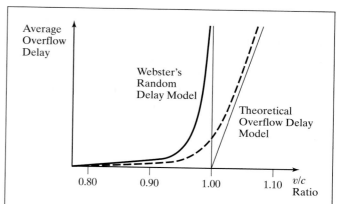

Figure 17.17: Random and Overflow Delay Models Compared (Adapted with permission of Transportation Research Board, National Research Council, Washington DC, from Hurdle, V.F. "Signalized Intersection Delay Model: A Primer for the Uninitiated, *Transportation Research Record 971*, pg. 101, 1984.)

The inconsistency occurs when the v/c ratio (X) is in the vicinity of 1.00. When the v/c ratio is below 1.00, a random delay model is used, as there is no "overflow" delay in this case. Webster's random delay model (Equation 17-22), however, contains the term $(1-X)$ in the denominator. Thus, as X approaches a value of 1.00, random delay increases asymptotically to an infinite value. When the v/c ratio (X) is greater than 1.00, an overflow delay model is applied. The overflow delay model of Equation 17-24, however, has an overflow delay of 0 when $X = 1.00$, and increases uniformly with increasing values of X thereafter.

Neither model is accurate in the immediate vicinity of $v/c = 1.00$. Delay does not become infinite at $v/c = 1.00$. There is no true "overflow" at $v/c = 1.00$, although individual cycle failures due to random arrivals do occur. Similarly, the overflow model, with overflow delay $= 0.0$ s/veh at $v/c = 1.00$ is also unrealistic. The additional delay of individual cycle failures due to the randomness of arrivals is not reflected in this model.

In practical terms, most studies confirm that the uniform delay model is a sufficiently predictive tool (except for the issue of platooned arrivals) when the v/c ratio is 0.85 or less. In this range, the true value of random delay is minuscule, and there is no overflow delay. Similarly, the simple theoretical overflow delay model (when added to uniform delay) is a reasonable predictor when $v/c \geq 1.15$ or so. The problem is that the most interesting cases fall in the intermediate range ($0.85 < v/c < 1.15$), for which neither model is adequate. Much of the more recent work in delay modeling involves attempts to bridge this gap, creating a model that closely follows the uniform delay model at low v/c ratios, and approaches the theoretical overflow delay model at high v/c ratios (≥ 1.15), producing "reasonable" delay estimates in between. Figure 17.17 illustrates this as the dashed line.

The most commonly used model for bridging this gap was developed by Akcelik for the Australian Road Research Board's signalized intersection analysis procedure [11, 12]:

$$OD = \frac{cT}{4}\left[(X - 1)\right.$$

$$\left. + \sqrt{(X - 1)^2 + \left(\frac{12(X - X_o)}{cT}\right)}\right]$$

$$X_o = 0.67 + \left(\frac{sg}{600}\right)$$

$$OD = 0.0 \text{ s/veh for } X \leq X_o \qquad (17\text{-}26)$$

where: T = analysis period, h
$\quad\quad X$ = v/c ratio
$\quad\quad c$ = capacity, veh/h
$\quad\quad s$ = saturation flow rate, veh/sg, (vehs per second of green)
$\quad\quad g$ = effective green time, s

The only relatively recent study resulting in large amounts of delay measurements in the field was conducted by Reilly, *et al.* [13] in the early 1980s to calibrate a model for use in the 1985 edition of the *Highway Capacity Manual*. The study concluded that Equation 17-26 substantially overestimated field-measured values of delay and recommended that a factor of 0.50 be included in the model to adjust for this. The version of the delay equation that was included in the 1985 *Highway Capacity Manual* ultimately did not follow this recommendation, and included other empiric adjustments to the theoretical equation.

17.5.4   Delay Models in the HCM 2000

The delay model incorporated into the HCM 2000 [2] includes the uniform delay model, a version of Akcelik's overflow delay model, and a term covering delay from an existing or residual queue at the beginning of the analysis period. The model is:

$$d = d_1 PF + d_2 + d_3$$

$$d_1 = \left(\frac{C}{2}\right) * \left\{\frac{\left(1 - g/c\right)^2}{1 - \left[\min(1, X) * g/c\right]}\right\}$$

$$d_2 = 900\, T\left[(X - 1) + \sqrt{(X - 1)^2 + \left(\frac{8klX}{cT}\right)}\right]$$

$$(17\text{-}27)$$

where: d = control delay, s/veh
$\quad\quad d_1$ = uniform delay component, s/veh
$\quad\quad PF$ = progression adjustment factor

d_2 = overflow delay component, s/veh

d_3 = delay due to pre-existing queue, s/veh

T = analysis period, h

X = v/c ratio

C = cycle length, s

k = incremental delay factor for actuated controller settings; 0.50 for all pre-timed controllers

I = upstream filtering/metering adjustment factor; 1.00 for all individual intersection analyses

c = capacity, veh/h

The progression factor is an empirically calibrated adjustment to uniform delay that accounts for the effect of platooned arrival patterns. This adjustment is detailed in Chapter 21. The delay due to preexisting queues, d_3, is found using a relatively complex model. (see Chapter 21).

In the final analysis, all delay modeling is based on the determination of the area between an arrival curve and a departure curve on a plot of cumulative vehicles vs. time. As the arrival and departure functions are permitted to become more complex and as rates are permitted to vary for various sub-parts of the signal cycle, the models become more complex as well.

17.5.5   Examples in Delay Estimation

Example 17-1:

Consider the following situation: An intersection approach has an approach flow rate of 1,000 veh/h, a saturation flow rate of 2,800 veh/hg, a cycle length of 90 s, and a g/C ratio of 0.55. What average delay per vehicle is expected under these conditions?

Solution:

To begin, the capacity and v/c ratio for the intersection approach must be computed. This will determine what model(s) are most appropriate for application in this case:

$$c = s*\left(g/C\right) = 2,800*0.55 = 1,540 \text{ veh/h}$$

$$v/c = X = \frac{1,000}{1,540} = 0.649$$

As this is a relatively low value, the uniform delay equation (Equation 17-19) may be applied directly. There is little random delay at such a v/c ratio and no overflow delay to consider. Thus:

$$d = \left(\frac{C}{2}\right)*\left[\frac{\left(1 - g/C\right)^2}{1 - v/s}\right]$$

$$= \left(\frac{90}{2}\right)*\left[\frac{(1 - 0.55)^2}{\left(1 - 1,000/2,800\right)}\right] = 14.2 \text{ s/veh}$$

Note that this solution assumes that arrivals at the subject intersection approach are random. Platooning effects are not taken into account.

Example 17-2:

How would the above result change if the demand flow rate increased to 1,600 veh/h?

Solution:

In this case, the v/c ratio now changes to 1,600/1,540 = 1.039. This is in the difficult range of 0.85–1.15 for which neither the simple random flow model nor the simple overflow delay model are accurate. The Akcelik model of Equation 17-26 will be used. Total delay, however, includes

both uniform delay and overflow delay. The uniform delay component when $v/c > 1.00$ is given by Equation 17-23:

$$UD = 0.50\, C\left(1 - g/C\right) = 0.50*90*(1 - 0.55)$$

$$= 20.3 \text{ s/veh}$$

Use of Akcelik's overflow delay model requires that the analysis period be selected or arbitrarily set. If a one-hour time period is used (this assumes that the demand flows and capacities exist over a one-hour period), then:

$$OD = \frac{1{,}540*1}{4}\left[(1.039 - 1)\right.$$

$$\left. + \sqrt{(1.039 - 1)^2 + \left(\frac{12(1.039 - 0.734)}{1{,}540*1}\right)}\,\right]$$

$$= 39.1 \text{ s/veh}$$

$$X_o = 0.67 + \left[\frac{0.778*49.5}{600}\right] = 0.734$$

where: $g = 0.55*90 = 49.5$ s

$$s = 2{,}800/3{,}600 = 0.778 \text{ veh/sg}$$

In this case, even with the "overflow" quite small (approximately 4% of the demand flow), the additional average delay due to this overflow is considerable. The total expected delay in this situation is the sum of the uniform and overflow delay terms, or:

$$d = 20.3 + 39.1 = 59.4 \text{ s/veh}$$

Note that this computation, as in Sample Problem 1, assumes random arrivals on this intersection approach.

Example 17-3:

How would the result change if the demand flow rate increased to 1,900 veh/h over a two-hour period.?

Solution:
The *v/c* ratio in this case is now $1{,}900/1{,}540 = 1.23$. In this range, the simple theoretical overflow model is an adequate predictor. As in Example 17-2, the Uniform Delay component must also be included:

$$UD = 0.50\,C\left(1 - \frac{g}{c}\right)$$

$$= 0.50*90*(1 - 0.55) = 20.3 \text{ s/veh}$$

The overflow delay component may be estimated using the simple theoretical Equation 17-24:

$$OD = \frac{T}{2}(X - 1) = \frac{7{,}200}{2}(1.23 - 1) = 828.0 \text{ s/veh}$$

As the period of oversaturation is given as two hours, and a result in seconds is desired, T is entered as $2*3{,}600 = 7{,}200$ s.

The total delay experienced by the average motorist is the sum of uniform and overflow delay, or:

$$d = 20.3 + 828.0 = 848.3 \text{ s/veh}$$

This is a very large value but represents an average over the full two-hour period of oversaturation. Equation 17-25 may be used to examine the average delay to vehicles arriving in the first 15 minutes of oversaturation to those arriving in the last 15 minutes of oversaturation:

$$OD_{first\ 15} = \frac{T_1 + T_2}{2}(X - 1)$$

$$= \frac{0 + 900}{2}(1.23 - 1) = 103.5 \text{ s/veh}$$

$$OD_{last\ 15} = \frac{6{,}300 + 7{,}200}{2}(1.23 - 1) = 1{,}552.5 \text{ s/veh}$$

As previously noted, the delay experienced during periods of oversaturation is very much influenced by the length of time that oversaturated operations have prevailed. Total delay for each case would also include the 20.3 s/veh of uniform delay. As in Examples 17-1 and 17-2, random arrivals are assumed.

17.6 Overview

This chapter has reviewed four key concepts necessary to understand the operation of signalized intersections:

1. Saturation flow rate and lost times
2. The time budget and critical lanes
3. Left-turn equivalency
4. Delay as a measure of effectiveness

These fundamental concepts are also the critical components of models of signalized intersection analysis. In Chapter 18, some of these concepts are implemented in a simple methodology for signal timing. In Chapters 21 and 22, all are used as parts of the HCM 2000 analysis procedure for signalized intersections.

References

1. *Manual of Uniform Traffic Control Devices*, Millennium Edition, Federal Highway Administration, U.S. Department of Transportation, Washington DC, 2000.

2. *Highway Capacity Manual*, 4th Edition, Transportation Research Board, National Research Council, Washington DC, 2000.

3. Greenshields, B., "Traffic Performance at Intersections," *Yale Bureau Technical Report No. 1*, Yale University, New Haven CT, 1947.

4. Kunzman, W., "Another Look at Signalized Intersection Capacity," *ITE Journal*, Institute of Transportation Engineers, Washington DC, August 1978.

5. Roess, R. et al., "Level of Service in Shared, Permissive Left-Turn Lane Groups," *Final Report*, FHWA Contract DTFH-87-C-0012, Transportation Training and Research Center, Polytechnic University, Brooklyn NY, September 1989.

6. Prassas, E., "Modeling the Effects of Permissive Left Turns on Intersection Capacity," *Doctoral Dissertation*, Polytechnic University, Brooklyn NY, December 1994.

7. Webster, F., "Traffic Signal Settings," *Road Research Paper No. 39*, Road Research Laboratory, Her Majesty's Stationery Office, London UK, 1958.

8. Newall, G., "Approximation Methods for Queues with Application to the Fixed-Cycle Traffic Light," *SIAM Review*, Vol. 7, 1965.

9. Miller, A., "Settings for Fixed-Cycle Traffic Signals," *ARRB Bulletin 3*, Australian Road Research Board, Victoria, Australia, 1968.

10. Miller, A., "The Capacity of Signalized Intersections in Australia," *ARRB Bulletin 3*, Australian Road Research Board, Victoria, Australia, 1968.

11. Akcelik, R., "Time-Dependent Expressions for Delay, Stop Rate, and Queue Lengths at Traffic Signals," *Report No. AIR 367-1*, Australian Road Research Board, Victoria, Australia, 1980.

12. Akcelik, R., "Traffic Signals: Capacity and Timing Analysis," *ARRB Report 123*, Australian Road Research Board, Victoria, Australia, March 1981.

13. Reilly, W. and Gardner, C., "Technique for Measuring Delay at Intersections," *Transportation Research Record 644*, Transportation Research Board, National Research Council, Washington DC, 1977.

Problems

17-1. Consider the headway data shown in Table 17.2. Data was taken from the center lane of a three-lane intersection approach for a total of 10 signal cycles. For the purposes of this analysis, the data may be considered to have been collected under ideal conditions.

(a) Plot the headways vs. position in queue for the data shown. Sketch an approximate best-fit curve through the data.

(b) Using the approximate best-fit curve constructed in (a), determine the saturation headway and the start-up lost time for the data.

(c) What is the saturation flow rate for this data?

17-2. A signalized intersection approach has three lanes with no exclusive left- or right-turning lanes. The approach has a 35-second green out of a 75-s cycle. The yellow plus all-red intervals for the

Table 17.2: Data for Problem 17-1

Q	Headways (s) for Cycle No.									
Pos	1	2	3	4	5	6	7	8	9	10
1	3.5	3.4	3.0	3.6	3.5	3.1	3.2	3.5	3.0	3.5
2	2.9	2.6	2.9	2.7	2.5	2.6	2.9	3.1	2.6	3.0
3	2.3	2.5	2.3	2.0	2.7	2.7	2.4	2.5	2.8	2.3
4	2.1	2.4	2.3	2.4	2.6	2.1	2.0	2.6	2.2	2.2
5	2.2	1.9	2.0	2.3	2.1	2.0	2.1	1.8	1.9	1.8
6	1.9	2.0	2.1	2.2	1.8	2.3	2.2	1.8	2.1	1.7
7	1.9	2.1	2.1	2.3	1.9	2.3	2.1	2.0	2.2	2.0
8	X	2.2	2.1	1.9	2.0	2.1	2.3	2.1	X	2.1
9	X	1.9	X	2.2	X	2.2	1.9	X	X	1.8
10	X	1.9	X	2.1	X	X	2.1	X	X	2.0

phase total 3.0 s. If the start-up lost time is 1.5 s/phase, the clearance lost time is 1.0 s/phase, and the saturation headway is 2.87 s/veh under prevailing conditions, what is the capacity of the intersection approach?

17-3. An equation has been calibrated for the amount of time required to clear N vehicles through a given signal phase:

$$T = 1.35 + 2.35N$$

(a) What start-up lost time does this equation suggest exists?

(b) What saturation headway and saturation flow rate is implied by the equation?

(c) A simple two-phase signal is to be placed at this location, and the two critical volumes are 550 veh/h/ln and 485 veh/h/ln. Using the equation, estimate an appropriate signal timing for the two phases. Is iteration necessary? Why? How does the equation close to a solution?

17-4. What is the maximum sum of critical-lane volumes that may be served by an intersection having three phases, a cycle length of 75 s, a saturation headway of 2.75 s/veh, and a total lost time per phase of 3.1 s?

17-5–17-6. For the two intersections illustrated in Figure 17.5 and 17.6, find the appropriate number of lanes for each lane group needed. Assume that all volumes shown have been converted to compatible "through-car equivalent" values for the conditions shown. Assume that critical volumes reverse in the other daily peak hour.

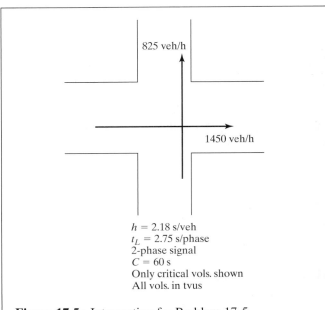

825 veh/h

1450 veh/h

$h = 2.18$ s/veh
$t_L = 2.75$ s/phase
2-phase signal
$C = 60$ s
Only critical vols. shown
All vols. in tvus

Figure 17.5: Intersection for Problem 17-5

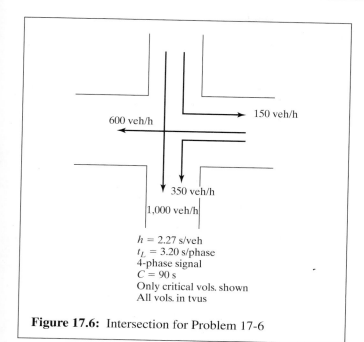

600 veh/h

150 veh/h

350 veh/h

1,000 veh/h

$h = 2.27$ s/veh
$t_L = 3.20$ s/phase
4-phase signal
$C = 90$ s
Only critical vols. shown
All vols. in tvus

Figure 17.6: Intersection for Problem 17-6

17-7. For the intersection of Problem 17-5, consider a case in which the E–W arterial has two lanes in each direction and the N–S arterial has only one lane in each direction. For this case:

(a) What is the absolute minimum cycle length that could be used?

(b) What cycle length would be required to provide for a v/c ratio of 0.90 during the worst 15-minutes of the hour if the PHF is 0.92?

17-8. At a signalized intersection, one lane is observed to discharge 18 through vehicles in the same time as the left lane discharges 9 through vehicles and 3 left-turning vehicles. For this case:

(a) What is the through-vehicle equivalent, E_{LT}, for left-turning vehicles?

(b) What is the left-turn adjustment factor, f_{LT}, for the case described?

(c) What variables can be expected to affect the observed value of E_T?

17-9. An intersection approach volume is 1,500 veh/h and includes 7.5% left turns with a through-vehicle equivalent of 3.0 tvus/left turn. What is the total equivalent through volume on the approach?

17-10. An intersection approach has three lanes, permitted left turns, and 8% left-turning volume with a through vehicle equivalent of 4.50 tvus/left turn. The saturation flow rate for *through vehicles* under prevailing conditions is 1,750 veh/hg/ln.

(a) What is the left-turn adjustment factor for the case described?

(b) Determine the saturation flow rate and saturation headway for the approach, including the impact of left-turning vehicles.

(c) If the approach in question has an effective green time of 45 s in a 75-s cycle, what is the capacity of the approach in veh/h?

17-11. An intersection approach has a demand volume of 500 veh/h, a saturation flow rate of 1,250 veh/hg, a cycle length of 60 s, and 35 s of effective green time. What average delay per vehicle is expected under these conditions?

17-12. A signalized intersection approach operates at an effective v/c ratio of 1.15 for a peak 30-minute period each evening. If the approach has a g/C ratio of 0.50 and the cycle length is 60 s:

(a) What is the average control delay for the entire 30-minute period?

(b) What is the average control delay during the last five minutes of the peak period?

(c) What is the average control delay per vehicle during the first 15 minutes of the peak period? Why is this period significant?

17-13. A signalized intersection approach experiences chronic oversaturation for a one-hour period each day. During this time, vehicles arrive at a rate of 2,000 veh/h. The saturation flow rate for the approach is 3,250 veh/hg, with a 100-s cycle length, and 55 s of effective green.

(a) What is the average control delay per vehicle for the full hour?

(b) What is the average control delay per vehicle for the first 15 minutes of the peak period? Compare the delay predicted by theoretical equations to those in the 2000 *Highway Capacity Manual*.

(c) What is the average control delay per vehicle for the last 15 minutes of the peak hour?

CHAPTER

18

Fundamentals of Signal Timing and Design

Signal timing and design involve several important components, including the physical design and layout of the intersection itself. Physical design is treated in some detail in Chapter 19. This chapter focuses on the design and timing of traffic control signals.

The key elements in signal design and timing are:

1. Development of a safe and effective phase plan and sequence

2. Determination of vehicular signal needs:

   (a) Timing of "yellow" (change) and "all-red" (clearance) intervals for each signal phase

   (b) Determination of the sum of critical lane volumes (V_c)

   (c) Determination of lost times per phase (t_L) and per cycle (L)

   (d) Determination of an appropriate cycle length (C)

   (e) Allocation of effective green time to the various phases defined in the phase plan—often referred to as "splitting" the green

3. Determination of pedestrian signal needs:

   (a) Determine minimum pedestrian "green" times

   (b) Check to see if vehicular greens meet minimum pedestrian needs

   (c) If pedestrian needs are unmet by the vehicular signal timing, adjust timing or add pedestrian actuators to ensure pedestrian safety

While most signal timings are developed for vehicles and checked for pedestrian needs, it is critical that signal timings provide safety and relative efficiency for both. Approaches vary with relative vehicular and pedestrian flows, but every signal timing must consider and provide for the requirements of both groups.

Many aspects of signal timing are tied to the principles discussed in Chapter 17 and elsewhere in this text.

The process, however, is not exact, nor is there often a single "right" design and timing for a traffic control signal. Thus, signal timing does involve judgmental elements and represents true engineering design in a most fundamental way.

All of the key elements of signal timing are discussed in some detail in this chapter, and various illustrations are offered. It must be noted, however, that it is virtually impossible to develop a complete and final signal timing that will not be subject to subsequent fine-tuning when the proposed design is analyzed using the HCM 2000 analysis model or some other analysis model or simulation. This is because no straightforward signal design and timing process can hope to include and fully address all of the potential complexities that may exist in any given situation. Thus, initial design and timing is often a starting point for analysis using a more complex model. Chapters 21 and 22 of this text discuss the HCM 2000 analysis model and show a number of sample applications. While there are a number of other models available, the HCM 2000 is widely used and accepted in the United States and a number of other countries.

18.1 Development of Signal Phase Plans

The most critical aspect of signal design and timing is the development of an appropriate phase plan. Once this is done, many other aspects of the signal timing can be analytically treated in a deterministic fashion. The phase plan and sequence involves the application of engineering judgment while applying a number of commonly used guidelines. In any given situation, there may be a number of feasible approaches that will work effectively.

18.1.1   Treatment of Left Turns

The single most important feature that drives the development of a phase plan is the treatment of left turns. As discussed in Chapter 17, left turns may be handled as permitted movements (with an opposing through flow), as protected movements (with the opposing vehicular through movement stopped), or as a combination of the

two (compound phasing). The simplest signal phase plan has two phases, one for each of the crossing streets. In this plan, all left turns are permitted. Additional phases may be added to provide protection for some or all left turns, but additional phases add lost time to the cycle. Thus, the consideration of protection for left turns must weigh the inefficiency of adding phases and lost time to the cycle against the improved efficiency in operation of left-turning and other vehicles gained from that protection.

There are two general guidelines that provide initial insight into whether or not a particular left-turn movement requires a protected or a partially protected phase. Such phasing should be considered whenever there is an opposed left turn that satisfies one of the two following criteria:

$$v_{LT} \geq 200 \text{ veh/h} \qquad (18\text{-}1a)$$

$$v_{LT} * \left({v_o}/{N_o} \right) \geq 50{,}000 \qquad (18\text{-}1b)$$

where: v_{LT} = left-turn flow rate, veh/h

v_o = opposing through movement flow rate, veh/h

N_o = number of lanes for opposing through movement

Equation 18-1b is often referred to as the "cross-product" rule. Various agencies may use different forms of this particular guideline. These criteria, however, are not absolute. They provide a starting point for considering whether or not left-turn protection is needed for a particular left-turn movement.

There are other considerations. Left-turn protection, for example, is rarely provided when left-turn flows are less than two vehicles per cycle. It is generally assumed that in the worst case, where opposing flows are so high that *no* left turns may filter through it, an average of two vehicles each cycle will wait in the intersection until the opposing flow is stopped and then complete their turns. Such vehicles are usually referred to as "sneakers." Where a protected phase is needed for one left-turning movement, it is often convenient to provide one for the opposing left turn, even if it does not meet any of the normal guidelines. Sometimes, a left turn that does not meet any of the guidelines will present a particular problem that is revealed during the signal timing or in later analysis, and protection will be added.

The *Traffic Engineering Handbook* [1] provides some additional criteria for consideration of protected or partially protected left-turn movements, based on a research study [2]. Permitted phasing should be provided when the following conditions exist:

1. The left-turn demand flow within the peak hour, as plotted on Figure 18.1 against the speed limit for opposing traffic, falls within the "permitted" portion of the exhibit.

2. The sight distance for left-turning vehicles is not restricted.

3. Fewer than eight left-turn accidents have occurred within the last three years at any one approach with permitted-only phasing.

The third criterion may be applied only to intersections that are already signalized using permitted left-turn phasing.

Guidance is also provided concerning the choice between fully protected and compound phasing. Fully protected phasing is recommended when any *two* of the following criteria are met:

1. Left-turn flow rate is greater than 320 veh/h.

2. Opposing flow rate is greater than 1,100 veh/h.

3. Opposing speed limit is greater than or equal to 45 mi/h.

4. There are two or more left-turn lanes.

Fully protected phasing is also recommended when any *one* of the following conditions exist:

1. There are three opposing traffic lanes, and the opposing speed is 45 mi/h or greater.

2. Left-turn flow rate is greater than 320 veh/h, and the percent of heavy vehicles exceeds 2.5%.

3. The opposing flow rate exceeds 1,100 veh/h, and the percent of left turns exceeds 2.5%.

4. Seven or more left-turn accidents have occurred within three years under compound phasing.

5. The average stopped delay to left-turning traffic is acceptable for fully protected phasing, and the engineer judges that additional left-turn accidents would occur under the compound phasing option.

A complex criterion based upon observed left-turn conflict rates is also provided. The conflict criteria as well as criteria 4 and 5 above apply only where an existing intersection is being operated with compound phasing.

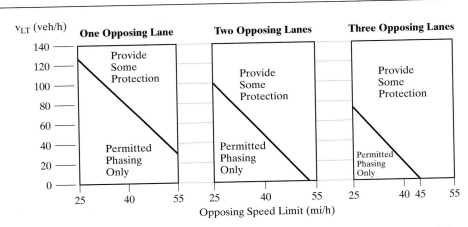

Figure 18.1: Recommended Selection Criteria for Left-Turn Protection (Used with permission of Transportation Research Board, National Research Council, Asante, S., Ardekani, S, and Williams, J, "Selection Criteria for Left-Turn Phasing and Indication Sequence," *Transportation Research Record 1421*, Washington DC, 1993, pg. 11.)

These criteria essentially indicate when compound phasing *should not* be implemented. They suggest that compound phasing may be considered when left-turn protection is needed but none of these criteria are met.

It should be noted, however, that use of compound phasing is also subject to many local agency policies that may supercede or supplement the more general guidelines presented herein. Use of compound phasing varies widely from jurisdiction to jurisdiction. Some agencies use it only as a last resort for the reasons stated previously. Others provide compound phasing wherever a left-turn phase is needed because of the delay reductions usually achieved by compound vs. fully protected left-turn phasing.

In extreme cases, it may be necessary to ban left turns entirely. This must be done, however, with the utmost care. It is essential that alternative paths for vehicles wishing to turn left are available and that they do not unduly inconvenience the affected motorists. Further, the additional demands on alternative routes should not cause worse problems at nearby intersections. Special design treatments for left turns are also discussed in Chapter 19.

18.1.2   General Considerations in Signal Phasing

There are several important considerations that should be kept in mind when establishing a phase plan:

1. Phasing can be used to minimize accident risks by separating competing movements. A traffic signal always eliminates the basic crossing conflicts present at intersections. Addition of left-turn protection can also eliminate some or all of the conflicts between left-turning movements and their opposing through movements. Additional phases generally lead to additional delay, which must be weighed against the safety and improved efficiency of protected left turns.

2. While increasing the number of phases increases the total lost time in the cycle, the offsetting benefit is an increase in affected left-turn saturation flow rates.

3. All phase plans must be implemented in accordance with the standards and criteria of the MUTCD [3], and must be accompanied by the necessary signs, markings, and signal hardware needed to identify appropriate lane usage.

4. The phase plan must be consistent with the intersection geometry, lane-use assignments, volumes and speeds, and pedestrian crossing requirements.

For example, it is not practical to provide a fully protected left-turn phase where there is no exclusive left-turn lane. If such phasing were implemented with a shared lane, the first vehicle in queue may be a through vehicle. When the protected left-turn green is initiated, the through vehicle blocks all left-turning vehicles from using the phase. Thus, protected left-turn phases *require* exclusive left-turn lanes.

18.1.3   Phase and Ring Diagrams

A number of typical and a few not-so-typical phase plans are presented and discussed herein. Signal phase plans are generally illustrated using *phase diagrams* and *ring diagrams*. In both cases, movements allowed during each phase are shown using arrows. Herein, only those movements allowed in each phase are shown; in some of the literature, movements not allowed are also shown with a straight line at the head of the arrow, indicating that the movement is stopped during the subject phase. Figure 18.2 illustrates some of the basic conventions used in these diagrams.

A more complete definition and discussion of the use and interpretation of these symbols follows:

1. A solid arrow denotes a movement without opposition. All through movements are unopposed by definition. An unopposed left turn has no opposing through vehicular flow. An unopposed right turn has no opposing pedestrian movement in the crosswalk through which the right turn is made.

2. Opposed left- and or right-turn movements are shown as a dashed line.

3. Turning movements made from a shared lane(s) are shown as arrows connected to the through movement that shares the lane(s).

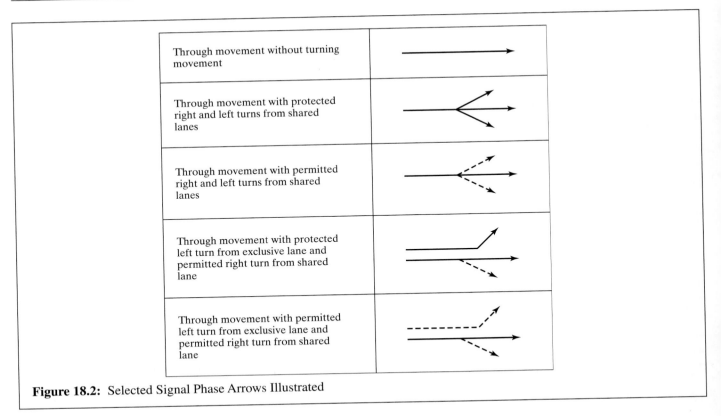

Through movement without turning movement	
Through movement with protected right and left turns from shared lanes	
Through movement with permitted right and left turns from shared lanes	
Through movement with protected left turn from exclusive lane and permitted right turn from shared lane	
Through movement with permitted left turn from exclusive lane and permitted right turn from shared lane	

Figure 18.2: Selected Signal Phase Arrows Illustrated

4. Turning movements from an exclusive lane(s) are shown as separate arrows, not connected to any through movement.

Although not shown in Figure 18.2, pedestrian paths may also be shown on phase or ring diagrams. They are generally shown as dotted lines with a double arrowhead, denoting movement in both directions in the crosswalk.

A *phase diagram* shows all movements being made in a given phase within a single block of the diagram. A *ring diagram* shows which movements are controlled by which "ring" on a signal controller. A "ring" of a controller generally controls one set of signal faces. Thus, while a phase involving two opposing through movements would be shown in one block of a phase diagram, each movement would be separately shown in a ring diagram. As will be seen in the next section, the ring diagram is more informative, particularly where overlapping phase sequences are involved.

Chapter 17 describes signal hardware and the operation of signal controllers in more detail.

18.1.4   Common Phase Plans and Their Use

Simple two-phase signalization is the most common plan in use. If guidelines or professional judgment indicate the need to fully or partially protect one or more left-turn movements, there are a variety of options available for doing so. The following sections illustrate and discuss the most common phase plans in general use.

Basic Two-Phase Signalization

Figure 18.3 illustrates basic two-phase signalization. Each street receives one signal phase, and all left and right turns are made on a permitted basis. Exclusive lanes for left- and/or right-turning movements may be used but are not required for two-phase signalization.

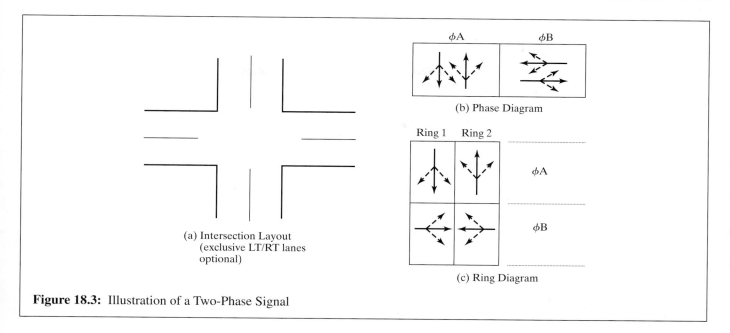

Figure 18.3: Illustration of a Two-Phase Signal

This form of signalization is appropriate where the mix of left turns and opposing through flows is such that no unreasonable delays or unsafe conditions are created by and/or for left-turners.

In this case, the phase diagram shows all N–S movements occurring in Phase A and all E–W movements occurring in Phase B. The ring diagram shows that in each phase, each set of directional movements is controlled by a separate ring of the signal controller. As the basic signalization is relatively simple, both the phase and ring diagrams are quite similar, and both are relatively easy to interpret. As will be seen, this is not the case for more complex signal phase plans.

Note that all phase boundaries cut across both rings of the controllers, meaning that all transitions occur at the same times in both rings. Also, it would make little difference which movements appear in which rings. The combination shown could be easily reversed without affecting the operation of the signal.

Exclusive Left-Turn Phasing

When a need for left-turn protection is indicated by guidelines or professional judgment, the simplest way to provide it is through the use of an exclusive left-turn phase(s). Two opposing left-turn movements are provided with a simultaneous and exclusive left-turn green, during which the two through movements on the subject street are stopped. An exclusive left-turn phase may be provided either before or after the through/right-turn phase for the subject street, although the most common practice is to provide it *before* the through phase. Because this is the most often used sequence, drivers have become more comfortable with left-turn phases placed before the corresponding through phase.

As noted previously, when an exclusive left-turn phase is used, an exclusive left-turn lane of sufficient length to accommodate expected queues must be provided. If an exclusive left-turn phase is implemented on one street and not the other, a three-phase signal plan emerges. Where an exclusive left-turn phase is implemented on both intersecting streets, a four-phase signal plan is formed. Figure 18.4 illustrates the use of an exclusive left-turn phase on the E–W street, but not on the N–S street, where left turns are made on a permitted basis.

The phase plan of Figure 18.4 can be modified to provide for protected plus permitted left turns on the E-W street. This is done by adding a permitted left-turn movement to Phase B. In general, such compound phasing is used where the combination of left turns and

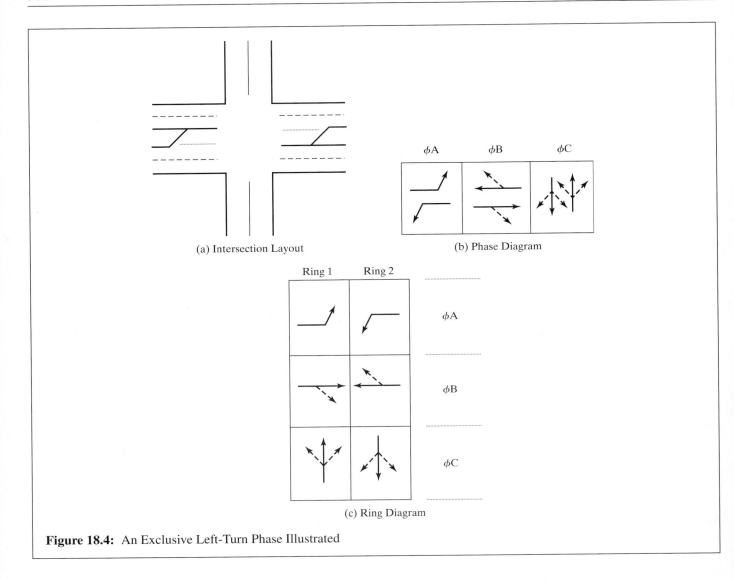

(a) Intersection Layout

(b) Phase Diagram

(c) Ring Diagram

Figure 18.4: An Exclusive Left-Turn Phase Illustrated

opposing flows is so heavy that provision of fully protected phasing leads to undesirably long or unfeasible cycle lengths. Compound phasing is more difficult for drivers to comprehend and is more difficult to display.

Most exclusive or unopposed left-turn movements are indicated by use of a green arrow. The arrow indication may be used only when there is no opposing through movement. In the case of a protected-plus-permitted compound phase, the green arrow is followed by a yellow arrow; the yellow arrow is then followed by a green ball indication during the permitted portion of the phase.

Leading and Lagging Green Phases

When exclusive left-turn phases are used, a potential inefficiency exists. If the two left-turning movements have very different demand flow rates (on a per-lane basis), then providing them with protected left-turn phases of equal length assures that the smaller of the two left-turn movements will have excess green time that cannot be used. Where this inefficiency leads to excessive or unfeasible cycle lengths and/or excessive delays, a phase plan in which opposing protected left-turn phases are separated should be considered. If a NB protected left-turn phase is separated from the SB protected left-turn phase, the two can be assigned different green times in accordance with their individual demand flow rates.

The traditional approach to accomplishing this is referred to as "leading and lagging" green phases. A leading and lagging green sequence for a given street has three components:

1. *The leading green.* Vehicles in one direction get the green, while vehicles in the opposing direction are stopped. Thus, the left-turning movement in the direction of the "green" is protected.

2. *The overlapping through green.* Left-turning vehicles in the initial green direction are stopped, while through (and right-turning) vehicles in both directions are released. As an option, left turns may be allowed on a permitted basis in both directions during this portion of the phase, creating a compound phase plan.

3. *The lagging green.* Vehicles in the initial direction (all movements) are stopped, while vehicles in the opposing direction continue to have the green. As the opposing flow is stopped, left turns made during this part of the phase are protected.

The leading and lagging green sequence is no longer a standard phasing supported by the National Electronics Manufacturing Association (NEMA), which creates standards for signal controllers and other electronic devices. Such controllers, however, are still available, and this sequence is still used in many jurisdictions.

Figure 18.5 illustrates a leading and lagging green sequence in the E-W direction. A similar sequence can be used in the N-S direction as well. Again, an exclusive left-turn lane must be provided when a leading and lagging green is implemented.

The leading and lagging green phase plan involves "overlapping" phases. The EB through is moving in Phases A1 and A2, while the WB through is moving in Phases A2 and A3. One critical question arises in this case: How many phases are there in this plan? It might be argued that there are *four* distinct phases: A1, A2, A3,

and B. It might also be argued that Phases A1, A2, and A3 form a single overlapping phase and that the plan therefore involves only *two* phases. In fact, both analyses are incorrect.

The ring diagram is critical in the analysis of overlapping phase plans. At the end of Phase A1, only Ring 1 goes though a transition, transferring the green from the EB left turn to the WB through and right-turn movements. At the end of Phase A2, only Ring 2 goes through a transition, transferring the green from the EB through and right-turn movements to the WB left turn. Each ring, therefore, goes through *three* transitions in a cycle. In effect, this is a *three*-phase signal plan. The ring diagram makes the difference between partial and full phase boundaries clear, while the phase diagram can easily mask this important feature.

This distinction is critical to subsequent signal-timing computations. For each phase transition, a set of lost times (start-up plus clearance) is experienced. If the sum of the lost times per phase (t_L) were 4.0 seconds per phase, then a two-phase signal would have 8.0 seconds of lost time per cycle (L), a three-phase signal would have 12.0 seconds of lost time per cycle, and a four-phase signal would have 16.0 seconds of lost time per cycle. As will be seen, the lost time per cycle has a dramatic affect on the required cycle length.

There are a number of interesting options to the leading and lagging phase plan that could be implemented:

1. A leading green may be used *without* a lagging green or vice versa. This is usually done where a one-way street or T-intersection creates a case in which there is only one left turn from the major street.

2. A compound phasing can be created by allowing permitted left turns in Phase A2. This would create a protected-plus-permitted phase for the EB left turn, and a permitted-plus-protected phase for the WB left turn.

3. A leading and/or lagging green may be added to the N–S street, assuming that an exclusive left-turn lane could be provided within the right-of-way.

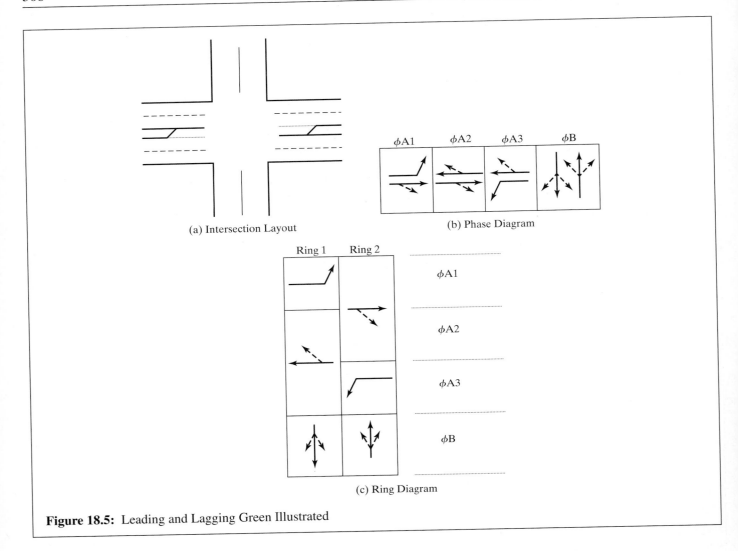

Figure 18.5: Leading and Lagging Green Illustrated

Exclusive Left-Turn Phase With Leading Green

It was previously noted that NEMA does not have a set of controller specifications to implement a leading and lagging green phase plan. The NEMA standard phase sequence for providing unequal protected left-turn phases employs an exclusive left-turn phase followed by a leading green phase in the direction of the heaviest left-turn demand flow. In effect, this sequence provides the same benefits as the leading and lagging green, but allows that all protected left-turn movements are made before the opposing through movements are released. Figure 18.6 illustrates such a phase plan for the E–W street.

In the case illustrated, the EB left-turn movement receives the leading green as the heavier of the two left-turn demand flows. If the WB left turn required the leading green, this is easily accomplished by reversing the positions of the partial boundaries between Phases A1 and A2 and between A2 and A3.

Note that there is a similarity between the leading and lagging green phase plan and the exclusive left-turn plus leading green phase plan. In both cases, the partial transition in each ring is between a protected left turn and the opposing through and right-turn movements, or vice versa. Virtually all overlapping phase sequences involve such transfers.

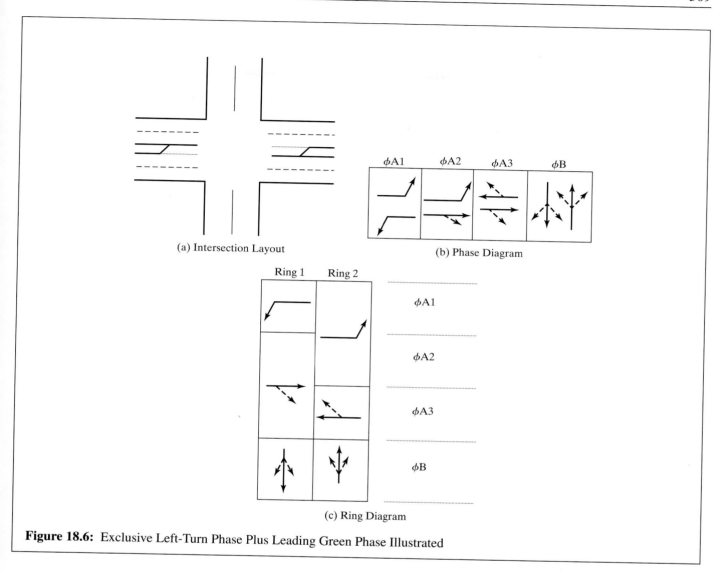

(a) Intersection Layout

(b) Phase Diagram

(c) Ring Diagram

Figure 18.6: Exclusive Left-Turn Phase Plus Leading Green Phase Illustrated

A compound phase can be implemented by allowing EB and WB permitted left-turn movements in Phase A2. In both cases, this creates a protected-plus-permitted phase sequence. This phasing can also be implemented for the N–S street if needed, as long as an exclusive left-turn lane is provided.

The issue of number of phases is also critical in this phase plan. The phase plan of Figure 18.5 involves *three* discrete phases and *three* phase transitions on each ring.

Eight-Phase Actuated Control

Any of the previous phase plans may be implemented using a pretimed or an actuated controller with detectors.

However, actuated controllers offer the additional flexibility of skipping phases when no demand is detected. This is most often done for left-turn movements. Protected left-turn phases may be skipped in any cycle where detectors indicate no left-turn demand. The most flexible controller follows the phase sequence of an exclusive left-turn phase plus a leading green. Figure 18.7 shows the actuated phase plan for such a controller.

In this case, exclusive left-turn phases and leading greens are provided for both streets, and both streets have exclusive left-turn lanes as shown.

This type of actuated signalization provides for complete flexibility in both the phase sequence and in

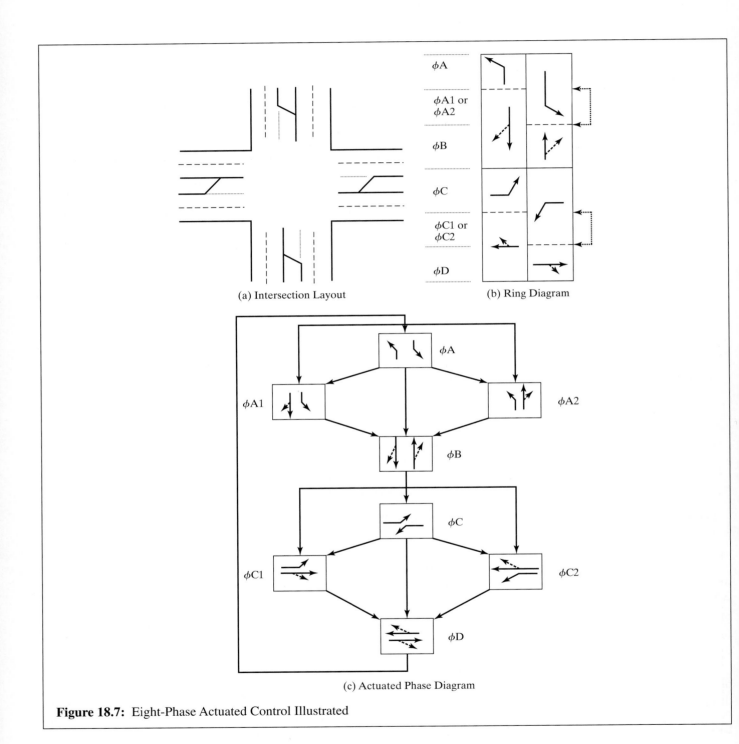

(a) Intersection Layout

(b) Ring Diagram

(c) Actuated Phase Diagram

Figure 18.7: Eight-Phase Actuated Control Illustrated

the timing of each phase. Each street may start its green phases in one of three ways, depending upon demand:

- An exclusive left-turn phase in both directions if left-turn demand is present in both directions

- A leading green phase (in the appropriate direction) if only one left-turn demand is present

- A combined through and right-turn phase in both directions if no left-turn demand is present in either direction

If the first option is selected, the next phase may be a leading green if one direction still has left-turn demand when the other has none, or a combined through and right-turn phase if both left-turn demands are simultaneously satisfied during the exclusive left-turn phase.

The ring diagram assumes that a full sequence requiring both the exclusive left-turn phase and the leading green phase (for one direction) are needed. The partial phase boundaries are shown as dashed lines, as the relative position of these may switch from cycle to cycle depending upon which left-turn demand flow is greater. If the entire sequence is needed, there are *four* phase transitions in either ring, making this (as a maximum) a *four-phase* signal plan. Thus, even though the controller defines eight potential phases, during any given cycle, a maximum of four phases may be activated.

As has been noted previously, actuated control is generally used where signalized intersections are relatively isolated. The type of flexibility provided by eight-phase actuated control is most effective where left-turn demands vary significantly over the course of the day.

The Exclusive Pedestrian Phase

The exclusive pedestrian phase was a unique approach to the control of situations in which pedestrian flows are a significant, or even dominant, movement at a traffic signal. Originally developed by New York City traffic engineer Henry Barnes in the 1960s for Manhattan, this type of phasing is often referred to as the "Barnes Dance."

Figure 18.8 illustrates this phasing. During the exclusive pedestrian phase, pedestrians are permitted to cross the intersection in any direction, including diagonally. All vehicular movements are stopped during the exclusive pedestrian phase. The exclusive pedestrian phase is virtually never used where more than two vehicular phases are needed.

The exclusive pedestrian phase has several drawbacks. The primary problem is that the entire pedestrian phase must be treated as lost time in terms of the vehicular signalization. Delays to vehicles are substantially increased because of this, and vehicular capacity is significantly reduced.

The exclusive pedestrian phase never worked well in the city of its birth. Where extremely heavy pedestrian flows exist, such as in Manhattan, the issue of clearing them out of the intersection at the close of the

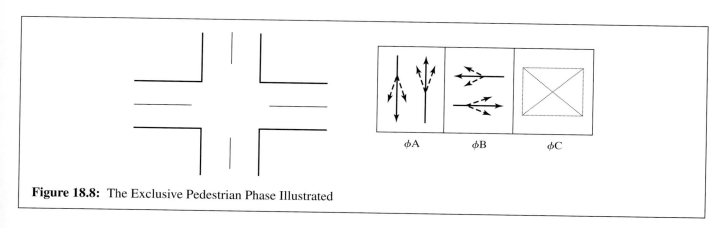

Figure 18.8: The Exclusive Pedestrian Phase Illustrated

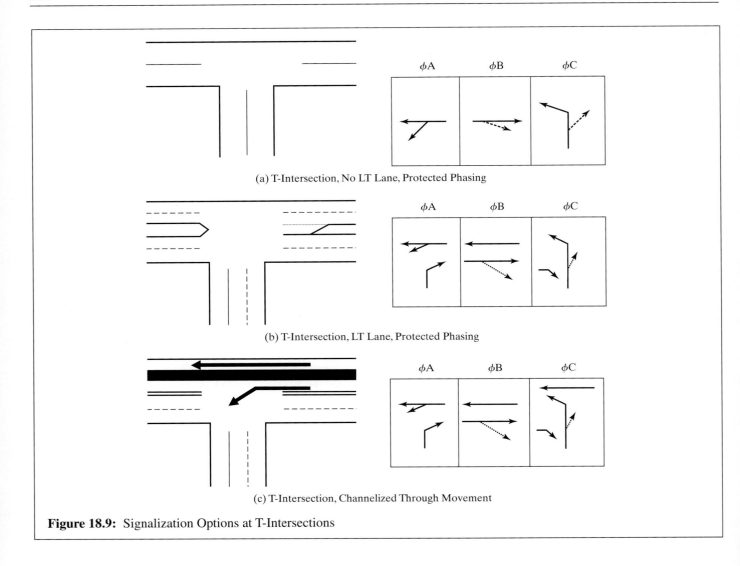

(a) T-Intersection, No LT Lane, Protected Phasing

(b) T-Intersection, LT Lane, Protected Phasing

(c) T-Intersection, Channelized Through Movement

Figure 18.9: Signalization Options at T-Intersections

pedestrian phase is a major enforcement problem. In New York, pedestrians occupied intersections for far longer periods than intended, and the negative impacts on vehicular movement were intolerable.

The exclusive pedestrian phase works best in small rural or suburban centers, where vehicular flows are not extremely high and where the volume of pedestrians is not likely to present a clearance problem at the end of the pedestrian phase. In such cases, it can provide additional safety for pedestrians in environments where drivers are not used to negotiating conflicts between vehicles and pedestrians.

Signalization of T-Intersections

T-intersections present unique problems along with the opportunity for unique solutions using the combination of geometric design and imaginative signal phasing. This is particularly true where the one opposed left turn that exists at a T-intersection requires a protected phase.

Figure 18.9 illustrates such a situation along with several candidate solutions. In Figure 18.9(a), there are no turning lanes provided. In such a case, providing the WB left turn with a protected phase requires that each of the three approach legs have its own signal phase. While achieving the required protected phasing for the op-posed left turn, such phasing is not very efficient in that each movement uses only one of three phases. Delays to all vehicles tend to be longer than they would be if more efficient phasing could be implemented.

If an exclusive left-turn lane is provided for the WB left-turn movement and if separate lanes for left and right turns are provided on the stem of the T, a more efficient phasing can be implemented. In this plan, the intersection geometry is used to allow several vehicular movements to use two of the three phases, including some overlaps be-tween right turns from one street and selected movements from the other. This is illustrated in Figure 18.9(b).

If a left-turn lane for the WB left turn can be com-bined with a channelizing island separating the WB through movement from all other vehicle paths, a signal-ization can be adopted in which the WB through movement is never stopped. Figure 18.9(c) illustrates this approach. Note that this particular approach can be used only where

there are no pedestrians present or where an overpass or underpass is provided for those crossing the E-W artery.

In each of the cases shown in Figure 18.9, a three-phase signal plan is used. Using geometry, however, ad-ditional movements can be added to each of the signal phases, improving the overall efficiency of the signal-ization. As the signal plan becomes more efficient, de-lays to drivers and passengers will be reduced, and the capacity for each movement will be increased.

Note that this example starts with the assumption that the WB left turn needs to be protected. If this were not true, a simple two-phase signal could have been used.

Five- and Six-Leg Intersections

Five- and six-leg intersections are a traffic engineer's worst nightmare. While somewhat rare, these intersections do occur with sufficient frequency to present major prob-lems in signal networks. Figure 18.10 illustrates a five-leg

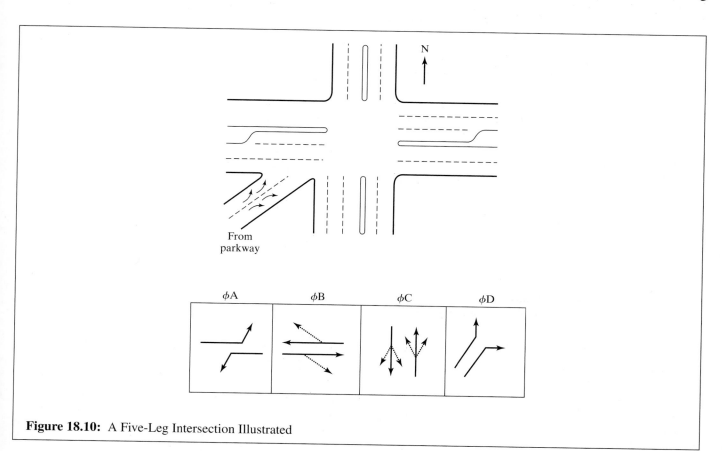

Figure 18.10: A Five-Leg Intersection Illustrated

intersection formed when an off-ramp from a limited access facility is fed directly into a signalized intersection.

In the example shown, a four-phase signal phase plan is needed to provide a protected left-turn phase for the E–W artery. Had the N–S artery required a protected left-turn phase as well (an exclusive LT lane would have to be provided), then a five-phase signalization could have resulted. Addition of a fifth, and, potentially, even a sixth phase creates inordinate amounts of lost time, increases delay, and reduces capacity to critical approaches and lane groups.

Wherever possible, design alternatives should be considered to eliminate five- and six-leg intersections. In the case illustrated in Figure 18.10, for example, redesign of the ramp to create another separate intersection should be considered. The ramp could be connected to either of the intersecting arteries in a T-intersection. The distance from the new intersection to the main intersection would be a critical feature, and should be arranged to avoid queuing that would block egress from the ramp. It may be necessary to signalize the new intersection as well.

In Manhattan (New York City), Broadway created a major problem in traffic control. The street system in most of Manhattan is a perfect grid, with the distance between N–S avenues (uptown/downtown) a uniform 800 ft, and the distance between E–W streets (crosstown) a uniform 400 ft. Such a regular grid, particularly when combined with a one-way street system (initiated in the late 1950s), is relatively easy to signalize. Broadway, however, runs diagonally through the grid, creating a series of major multileg intersections involving three major intersecting arteries. Some of these "major" intersections include Times Square and Herald Square, and all involve major vehicular and major pedestrian flows. To take advantage of the signalization benefits of a one-way, uniform grid street system, through flow on Broadway is banned at most of these intersections. This has effectively turned Broadway into a local street, with little through traffic. Through traffic is forced back onto the grid. Channelization is provided that forces vehicles approaching on Broadway to join either the avenue or the street, eliminating the need for multi-phase signals. In addition, channelizing islands have been used to create unique pedestrian environments at these intersections.

Right-Turn Phasing

While the use of protected left-turn phasing is common, the overwhelming majority of signalized intersections handle right turns on a permitted basis, mostly from shared lanes. Protected right-turn phasing is used only where the number of pedestrians is extremely high. Modern studies show that a pedestrian flow of 1,700 peds/h in a crosswalk can effectively block all right turns on green. Such a pedestrian flow is, however, extremely rare, and exists only in major urban city centers. While use of a protected right-turn phase in such circumstances may help motorists, it may worsen pedestrian congestion on the street corner and on approaching sidewalks. In extreme cases, it is often useful to examine the feasibility of pedestrian overpasses or underpasses. These would generally be coordinated with barriers preventing pedestrians from entering the street at the corner. It should be noted, however, that pedestrian overpasses and/or underpasses are inconvenient for pedestrians and may pose security risks, primarily at night.

Compound right-turn phasing is usually implemented only in conjunction with an exclusive left-turn phase on the intersecting street. For example, NB and SB right turns may be without pedestrian interference during an EB and WB exclusive left-turn phase. Permitted right turns would then continue during the NB and SB through phase.

Exclusive right-turn lanes are useful where heavy right-turn movements exist, particularly where right-turn-on-red is permitted. Such lanes can be easily created on streets where curb parking is permitted. Parking may be prohibited within several hundred feet of the STOP line; the curb lane may then be used as an exclusive right-turn lane. Channelized right turns may also be provided. Channelized right turns are generally controlled by a YIELD sign and need not be included in the signalization plan. Chapter 19 contains a more detailed discussion of exclusive right-turn lanes and channelized right-turn treatments.

Right-Turn-on-Red

"Right-turn-on-red" (RTOR) was first permitted in California in 1937 only in conjunction with a sign authorizing the movement [4]. In recent years, virtually all states allow RTOR unless it is specifically prohibited by a sign. The federal government encouraged this approach in the 1970s by linking implementation of RTOR to receipt of federal-aid highway funds. In some urban areas, like New York City, right-turn-on-red is still generally prohibited. Signs indicating this general prohibition must be posted on all roadways entering the area. All RTOR laws require that the motorist stop before executing the right-turn movement on red.

When implemented using a shared right-turn through lane, the utility of RTOR is affected by the proportion of through vehicles using the lane. When a through vehicle reaches the STOP line, it blocks subsequent right-turners from using RTOR. Thus, provision of an exclusive right-turn lane greatly enhances the effectiveness of RTOR.

The major issues regarding RTOR continue to be (1) the delay savings to right-turning vehicles and (2) the increased accident risk such movements cause. An early study by McGee et al. [5] concluded that the delay to the average right-turning vehicle was reduced by 9% in central business districts (CBDs), 31% in other urban areas, and 39% in rural areas. Another early study on the safety of RTOR [6] found that only 0.61% of all intersection accidents involved RTOR vehicles and that these accidents tended to be less severe than other intersection accidents.

As there are potential safety issues involving RTOR, its use and application should be carefully considered. The primary reasons for prohibiting RTOR include:

1. Restricted sight-distance for right-turning motorist

2. High speed of conflicting through vehicles

3. High flow rates of conflicting through vehicles

4. High pedestrian flows in crosswalk directly in front of right-turning vehicles

Any of these conditions would make it difficult for drivers to discern and avoid conflicts during the RTOR maneuver.

18.1.5 Summary and Conclusion

The subject of phasing along with the selection of an appropriate phase plan is a critical part of effective intersection signalization. While general criteria have been presented to assist in the design process, there are few firm standards. The traffic engineer must apply a knowledge

and understanding of the various phasing options and how they affect other critical aspects of signalization, such as capacity and delay.

Phasing decisions are made for each approach on each of the intersection streets. It is possible, for example, for the E–W street to use an exclusive left-turn phase while the N–S street uses leading and lagging greens and compound phasing. The number of potential combinations for the intersection as a whole is, therefore, large.

The final signalization should also be analyzed using a comprehensive signalized intersection model (such as the 2000 *Highway Capacity Manual*) or simulation (such as CORSIM/NETSIM). This will allow for fine-tuning of the signalization on a trial-and-error basis and for a wider range of alternatives to be quickly assessed.

18.2 Determining Vehicular Signal Requirements

Once a candidate phase plan has been established, it is possible to establish the "timing" of the signal that would most effectively accommodate the vehicular demands present.

18.2.1 Change and Clearance Intervals

The terms "change" and "clearance" interval are used in a variety of ways in the literature. They refer to the *yellow* and *all-red* indications respectively that mark the transition from GREEN to RED in each signal phase. The *all-red* interval is a period during which all signal faces show a RED indication. The MUTCD specifically prohibits the use of a *yellow* indication to mark the transition from RED to GREEN, a practice common in many European countries.

While the MUTCD does not strictly require *yellow* and/or *all-red* intervals, the Institute of Transportation Engineers (ITE) recommends that both be used at all signals. In most states, it is legal to *enter* an intersection on *yellow*. Therefore, the function of these critical intervals is as follows:

- *Change interval (yellow)*. This interval allows a vehicle that is one safe stopping distance away from the STOP line when the GREEN is withdrawn to continue at the approach speed and enter the intersection legally on *yellow*. "Entering the intersection" is interpreted to be the front wheels crossing over the STOP line.

- *Clearance interval (all-red).* Assuming that a vehicle has just entered the intersection legally on *yellow*, the *all-red* must provide sufficient time for the vehicle to cross the intersection and clear its back bumper past the far curb line (or crosswalk line) before conflicting vehicles are given the GREEN.

The ITE recommends the following methodology for determining the length of the *yellow* or change interval [7]:

$$y = t + \frac{1.47 \, S_{85}}{2a + (64.4 * 0.01G)} \qquad (18\text{-}2)$$

where: y = length of the *yellow* interval, s

t = driver reaction time, s

S_{85} = 85th percentile speed of approaching vehicles, or speed limit, as appropriate, mi/h

a = deceleration rate of vehicles, ft/s^2

G = grade of approach, %

64.4 = twice the acceleration rate due to gravity, which is 32.2 ft/s^2

This equation was derived as the time required for a vehicle to traverse one safe stopping distance at its approach speed. Commonly used values for key parameters include a deceleration rate of 10.0 ft/s^2 and a driver reaction time of 1.0 s.

The ITE also recommends the following policy for determining the length of *all-red* clearance intervals [7]: For cases in which there is no pedestrian traffic:

$$ar = \frac{w + L}{1.47 \, S_{15}} \qquad (18\text{-}3a)$$

For cases in which significant pedestrian traffic exists:

$$ar = \frac{P + L}{1.47 \, S_{15}} \qquad (18\text{-}3b)$$

For cases in which some pedestrian traffic exists:

$$ar = \max\left[\left(\frac{w + L}{1.47 \, S_{15}}\right), \left(\frac{P}{1.47 \, S_{15}}\right)\right] \qquad (18\text{-}3c)$$

where: ar = length of the all-red phase, s

w = distance from the departure STOP line to the far side of the farthest conflicting traffic lane, ft

P = distance from the departure STOP line to the far side of the farthest conflicting crosswalk, ft

L = length of a standard vehicle, usually taken to be 18–20 ft

S_{15} = 15th percentile speed of approaching traffic, or speed limit, as appropriate, mi/h

The difference between the three equations involves pedestrian activity levels and the decision to clear vehicles beyond the line of potential conflicting vehicle paths and/or conflicting pedestrian paths before releasing the conflicting flows. Equation 18-3c, which addresses the most frequently occurring situations—some, but not significant pedestrian flows—is a compromise. If the pedestrian clearance distance, P, is used, the length of the vehicle is not added (i.e., the timing would provide for the *front* bumper of a vehicle to reach the far crosswalk line before releasing the conflicting vehicular and pedestrian flows).

To provide for optimal safety, the equations for *yellow* and *all-red* intervals use different speeds: the 85th percentile and the 15th percentile, respectively. Because speed appears in the numerator of the *yellow* determination and in the denominator of the *all-red* determination, accommodating the majority of motorists safely requires the use of different percentiles. If only the average approach speed is known, the percentile speeds may be estimated as:

$$\begin{aligned} S_{15} &= S - 5 \\ S_{85} &= S + 5 \end{aligned} \qquad (18\text{-}4)$$

where: S_{15} = 15th percentile speed, mi/h

   S_{85} = 85th percentile speed, mi/h

   S = average speed, mi/h

Where approach speeds are not measured and the speed limit is used, both the *yellow* and *all-red* intervals will be determined using the same value of speed.

Use of these ITE policies to determine *yellow* and *all-red* intervals assures that drivers will not be presented with a "dilemma zone." A dilemma zone occurs when the combined length of the change and clearance intervals is not sufficient to allow a motorist who cannot safely stop when the *yellow* is initiated to cross through the intersection and out of conflicting vehicular and/or pedestrian paths before those flows are released. Where *yellow* and *all-red* phases are mistimed and a dilemma zone is created, agencies face possible liability for accidents that occur as a result.

Consider the following example: Compute the appropriate change and clearance intervals for a signalized intersection approach with the following characteristics:

- Average approach speed = 35 mi/h

- Grade = −2.5%

- Distance from STOP line to far side of the most distant lane = 48 ft

- Distance from STOP line to far side of the most distant cross-walk = 60 ft

- Standard vehicle length = 20 ft

- Reaction time = 1.0 s

- Deceleration rate = 10 ft/s^2

- Some pedestrians present

To apply Equations 18-2 and 18-3, estimates of the 15th and 85th percentile speeds are needed. Using Equation 18-4:

$$S_{85} = 35 + 5 = 40 \text{ mi/h}$$

$$S_{15} = 35 - 5 = 30 \text{ mi/h}$$

Using Equation 18-2, the length of the change or *yellow* interval should be:

$$y = 1.0 + \frac{1.47 * 40}{[2 * 10] + [64.4 * 0.01 * (-2.5)]}$$

$$= 1.0 + \frac{58.8}{20 - 1.61} = 4.2\text{s}$$

Equation 18-3c is used to compute the length of the clearance or *all-red* phase, as there are some, but not significant, pedestrian flows present. The length of the clearance interval is the maximum of:

$$ar = \frac{48 + 20}{1.47 * 30} = \frac{68}{44.1} = 1.5 \text{ s}$$

$$ar = \frac{60}{1.47 * 30} = \frac{60}{44.1} = 1.4 \text{ s}$$

In this case, 1.5 s would be applied.

18.2.2   Determining Lost Times

The 2000 edition of the *Highway Capacity Manual* [8] indicates that lost times vary with the length of the *yellow* and *all-red* phases in the signal timing. Thus, it is no longer appropriate to use a constant default value for lost times, as has historically been done in most signal timing methodologies. The HCM now recommends the use of the following default values for this determination:

- Start-up lost time, ℓ_1 = 2.0 s/phase

- Motorist use of *yellow* and *all-red*, e = 2.0 s/phase

Using these default values, lost time per phase and lost time per cycle may be estimated as follows:

$$\ell_2 = Y - e \qquad (18\text{-}5)$$

$$Y = y + ar \qquad (18\text{-}6)$$

$$t_L = \ell_1 + \ell_2 \qquad (18\text{-}7)$$

where: ℓ_1 = start-up lost time, s/phase

   ℓ_2 = clearance lost time, s/phase

t_L = total lost time, s/phase

y = length of *yellow* change interval, s

ar = length of *all red* clearance interval, s

Y = total length of change and clearance interval, s

In the example of the previous section, the *yellow* interval was computed as 4.2 s, and the *all-red* interval was found to be 1.5 s. Using the recommended default values for ℓ_1 and e, respectively, lost times would be computed as:

$$Y = 4.2 + 1.5 = 5.7 \text{ s}$$
$$\ell_2 = 5.7 - 2.0 = 3.7 \text{ s}$$
$$t_L = 2.0 + 3.7 = 5.7 \text{ s}$$

Note that when the HCM-recommended default values for ℓ_1 and e (both 2.0 s) are used, the lost time per phase, t_L, is always equal to the total of the *yellow* and *all red* intervals, Y. As the lost time for each phase may differ, based on different *yellow* and *all-red* intervals, the total lost time per cycle is merely the sum of lost times in each phase, or:

$$L = \sum_i^n t_{Li} \qquad (18\text{-}8)$$

where: L = total lost time per cycle, s

t_{Li} = total lost time for phase i, s

n = number of discrete phases in cycle

18.2.3 Determining the Sum of Critical-Lane Volumes

To estimate an appropriate cycle length and to split the cycle into appropriate green times for each phase, it is necessary to find the *critical-lane volume* for each discrete phase or portion of the cycle.

As discussed in Chapter 17, the *critical-lane volume* is the per-lane volume that controls the required length of a particular phase. For example, in the case of a simple two-phase signal, on a given phase the EB and WB flows move simultaneously. One of these per-lane

volumes represents the most intense demand, and that is the one that will determine the appropriate length of the phase.

Making this determination is complicated by two factors:

- Simple volumes cannot be simply compared. Trucks require more time than passenger cars, left- and right-turns require more time than through vehicles, vehicles on a downgrade approach require less time than vehicles on a level or upgrade approach. Thus, *intensity* of demand is not measured accurately by simple volume.

- Where phase plans involve overlapping elements, the ring diagram must be carefully examined to determine which flows constitute *critical-lane volumes*.

Ideally, demand volumes would be converted to equivalents based on all of the traffic and roadway factors that might affect intensity. For initial signal timing, however, this is too complex a process. Demand volumes can, however, be converted to reflect the influence of the most significant factor affecting intensity: left and right turns. This is accomplished by converting all demand volumes to *equivalent through vehicle units* (tvus). Through vehicle equivalents for left and right turns are shown in Tables 18.1 and 18.2, respectively.

Table 18.1: Through Vehicle Equivalents for Left-Turning Vehicles, E_{LT}

Opposing Flow	Number of Opposing Lanes, N_o		
V_o (veh/h)	1	2	3
0	1.1	1.1	1.1
200	2.5	2.0	1.8
400	5.0	3.0	2.5
600	10.0[*]	5.0	4.0
800	13.0[*]	8.0	6.0
1,000	15.0[*]	13.0[*]	10.0[*]
$\geq$1,200	15.0[*]	15.0[*]	15.0[*]

E_{LT} for all *protected* left turns = 1.05

[*]indicates that the LT capacity is only available through "sneakers."

Table 18.2: Through Vehicle Equivalent for Right-Turning Vehicles, E_{RT}

Pedestrian Volume In Conflicting Crosswalk (peds/h)	Equivalent
None (0)	1.18
Low (50)	1.21
Moderate (200)	1.32
High (400)	1.52
Extreme (800)	2.14

These values are actually a simplification of a more complex approach in the *Highway Capacity Manual* analysis model for signalized intersections, and they form an appropriate basis for signal timing and design. In using these tables, the following should be noted:

- Opposing volume, V_o, includes only the through volume on the opposing approach, in veh/h.

- Interpolation in Table 18.1 for opposing volume is appropriate, but values should be rounded to the nearest tenth.

- For right turns, the "conflicting crosswalk" is the crosswalk through which right-turning vehicles must pass.

- Pedestrian volumes indicated in Table 18.1 represent typical situations in moderate-sized communities. Pedestrian volumes in large cities, like New York, Chicago, or Boston, may be much higher, and the relative terms used (low, moderate, high, extreme) are not well correlated to such situations.

- Interpolation in Table 18.2 is not recommended.

Once appropriate values for E_{LT} and E_{RT} have been selected, all right- and left-turn volumes must be converted to units of "through-vehicle equivalents." Subsequently, the demand intensity *per lane* is found for each approach or lane group.

$$V_{LTE} = V_{LT} * E_{LT}$$
$$V_{RTE} = V_{RT} * E_{RT} \qquad (18\text{-}9)$$

where: V_{LTE} = left-turn volume in through-vehicle equivalents, tvu/h

V_{RTE} = right-turn volume in through-vehicle equivalents, tvu/h

Other variables are as previously defined.

These equivalents are added to through vehicles that may be present in a given approach or lane group to find the total equivalent volume and equivalent volume per lane in each approach or lane group:

$$V_{EQ} = V_{LTE} + V_{TH} + V_{RTE}$$
$$V_{EQL} = {}^{V_{EQ}}\!/_{N} \qquad (18\text{-}10)$$

where: V_{EQ} = total volume in a lane group or approach, tvu/h

V_{EQL} = total volume per lane in a lane group or approach, tvu/h/ln

N = number of lanes

Finding the critical-lane volumes for the signal phase plan requires determining the critical path through the plan (i.e., the path that controls the signal timing). This is done by finding the path through the signal phase plan that results in the highest possible sum of critical-lane volumes. Since most signal plans involve two "rings," alternative paths must deal with two potential rings for each discrete portion of the phase plan. It must also be noted that the critical path may "switch" rings at any full phase boundary (i.e., a phase boundary that cuts through both rings). This process is best understood through examples. Figure 18.11 shows a ring diagram for a signalization with overlapping phases. Lane volumes, V_{EQL}, are shown for each movement in the phase diagram.

To find the critical path, the controlling (maximum) equivalent volumes must be found for each portion of the cycle, working between full phase transition boundaries. For the combined Phase A in Figure 18.11, the volumes that control the total length of A1, A2, and A3 are on Ring 1 or Ring 2. As shown, the maximum total comes from Ring 2 and yields a total critical-lane volume of 800 tvu's. For Phase B, the choice is much simpler, as there are no overlapping phases. Thus, the Ring 1 total of 300 tvu's is identified as critical. The critical path through the cycle is now indicated by asterisks,

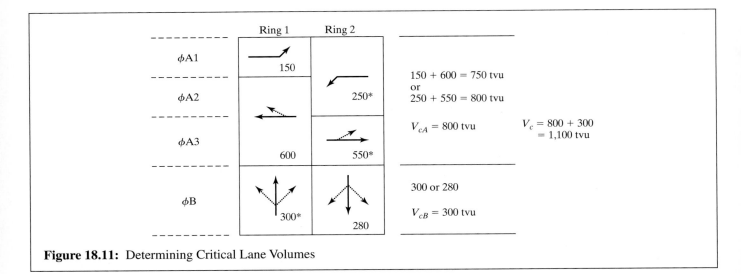

Figure 18.11: Determining Critical Lane Volumes

and the sum of critical-lane volumes is $800 + 300 =$ 1,100 tvus. In essence, if the intersection is thought of in terms of a number of vehicles in single lanes seeking to move through a single common conflicting point, the signal, in this case, must have a timing that is sufficient to handle 1,100 tvu's through this point. The determination of critical-lane volumes is further illustrated in the complete signal-timing examples included in the last section of this chapter.

18.2.4 Determining the Desired Cycle Length

In Chapter 17, an equation describing the maximum sum of critical lane volumes that could be handled by a signal was manipulated to find a desirable cycle length. That equation is used to find the desired cycle length, based on tvu volumes, and a default value for saturation flow rate. The default saturation flow rate, 1,615 tvu's per hour of green, assumes typical conditions of lane width, heavy-vehicle presence, grades, parking, pedestrian volumes, local buses, area type, and lane utilization. Common default values for saturation flow rate range from 1,500 to 1,700 in the literature, but these commonly also account for typical left-turn and right-turn percentages as well. The method presented here makes these adjustments by converting demand to equivalent through-vehicle units.

When the default value for saturation flow rate is inserted into the relationship, the desired cycle length is computed as:

but level of service has not been checked

$$C_{des} = \frac{L}{1 - \left[\dfrac{V_c}{1615 * PHF * (v/c)} \right]} \qquad (18\text{-}11)$$

where: C_{des} = desirable cycle length, s

L = total lost time per cycle, s/cycle

PHF = peak-hour factor

v/c = target v/c ratio for the critical movements in the intersection

Use of the peak-hour factor ensures that the signal timing is appropriate for the peak 15-minutes of the design hour. Target v/c ratios are generally in the range of 0.85 to 0.95. Very low values of v/c increase delays, as vehicles are forced to wait while an unused green phase times out. Values of $v/c > 0.95$ indicate conditions in which frequent individual phase or cycle failures are possible, thereby increasing delay.

Consider the example illustrated previously in Figure 18.11. The sum of the critical-lane volumes for this case was shown to be 1,100 veh/h. What is the desirable cycle length for this three-phase signal if the total lost time per cycle is 4 s/phase $\times$ 3 phases/cycle = 12 s/cycle, the peak-hour factor is 0.92, and the target v/c ratio is 0.90. Using Equation 18-11:

$$C_{des} = \frac{12}{1 - \left[\dfrac{1,100}{1,615*0.92*0.90}\right]} = 67.6 \text{ s}$$

If this were a pretimed signal, timing dials (or modules) are available in 5-second increments between cycle lengths of 30 and 90 s, and in 10-second increments between 90 and 120 s. Thus, a 70-second cycle would be adopted in this case. For actuated controllers, cycle lengths vary, although this equation might be used to obtain a very rough estimate of the expected average cycle length.

18.2.5 Splitting the Green

Once the cycle length is determined, the available *effective green time* in the cycle must be divided amongst the various signal phases. The available effective green time in the cycle is found by deducting the lost time per cycle from the cycle length:

$$g_{TOT} = C - L \qquad (18\text{-}12)$$

where: g_{TOT} = total effective green time in the cycle, s
C, L as previously defined

The total effective green time is then allocated to the various phases or sub-phases of the signal plan in proportion to the critical lane volumes for each phase or sub-phase:

$$g_i = g_{TOT}*\left(\frac{V_{ci}}{V_c}\right) \qquad (18\text{-}13)$$

where: g_i = effective green time for Phase i, s

g_{TOT} = total effective green time in the cycle, s

V_{ci} = critical lane volume for Phase or Sub-phase i, veh/h

V_c = sum of the critical-lane volumes, veh/h

Returning to the example of Figure 18.11, the situation is complicated somewhat by the presence of overlapping phases. For the critical path, the following critical-lane volumes were obtained:

- 250 veh/h/ln for the *sum* of Phases A1 and A2
- 550 veh/h/ln for Phase A3
- 300 veh/h/ln for Phase B

Remembering that the desired cycle length of 70 s contains 12 s of lost time, the total effective green time may be computed using Equation 18-12:

$$g_{TOT} = 70 - 12 = 58 \text{ s}$$

Using Equation 18-13 and the critical-lane volumes noted above, the effective green times for the signal are estimated as:

$$g_{A1+A2} = 58*\left(\frac{250}{1,100}\right) = 13.2 \text{ s}$$

$$g_{A3} = 58*\left(\frac{550}{1,100}\right) = 29.0 \text{ s}$$

$$g_B = 58*\left(\frac{300}{1,100}\right) = 15.8 \text{ s}$$

The sum of these times (13.2 + 29.0 + 15.8) must equal 58.0 seconds, and it does. Together with the 12.0 seconds of lost time in the cycle, the 70-second cycle length is now fully allocated.

Because of the overlapping phases illustrated in this example, the signal timing is still not complete. The split between phases A1 and A2 must still be addressed. This can be done only by considering the noncritical Ring 1 for Phase A, as this ring contains the transition between these two sub-phases. The total length of Phase A is 13.2 + 29.0 = 42.2 s. On the noncritical ring (Ring 1), critical-lane volumes are 150 for Phase A1 and 600 for the sum of Phases A2 and A3. Using these critical-lane volumes:

$$g_{A1} = 42.2*\left(\frac{150}{150 + 600}\right) = 8.4 \text{ s}$$

By implication, g_{A2} is now computed as the total length of Phase A, 42.2 s, minus the effective green times for

Phases A1 and A3, both of which have now been determined (8.4 s and 29.0 s respectively). Thus:

$$g_{A2} = 42.2 - 8.4 - 29.0 = 4.8 \ s$$

The signal timing is now complete except for the conversion of effective green times to actual green times:

$$G_i = g_i - Y_i + t_{Li} \qquad (18\text{-}14)$$

where: G_i = actual green time for Phase i, s

g_i = effective green time for Phase i, s

Y_i = total of yellow and all-red intervals for Phase i, s

t_{Li} = total lost time for Phase i, s

As information on the timing of yellow and all-red phases was not provided for the example, this step cannot be completed. Full signal-timing examples in the last section of this chapter will fully illustrate determination of actual green times.

As a general rule, very short phases should be avoided. In this case, the overlapping Phase A2 has an effective green time of only 4.8 s. When converted to actual green, this value would either stay the same or decrease. In either case, the short overlap period may not provide sufficient efficiency to warrant the potential confusion of drivers. The short Phase A2, in this case, may be one argument in favor of simplifying the phase plan by using a common exclusive left-turn phase.

18.3 Determining Pedestrian Signal Requirements

To this point in the process, the signal design has considered vehicular requirements. Pedestrians, however, must also be accommodated by the signal timing. Problems arise because pedestrian requirements and vehicular requirements are often quite different. Consider the intersection of a wide major arterial and a small local collector. Vehicle demand on the major arterial is more intense than on the small collector, and the green split for vehicles would generally result in the arterial receiving a long green and the collector a relatively short green.

This, unfortunately, is exactly the opposite of what pedestrians would require. During the short collector green, pedestrians are crossing the wide arterial. During the long arterial green, pedestrians are crossing the narrower collector. In summary, pedestrians require a longer green during the shorter vehicular green, and a shorter green during the longer vehicular green.

The 2000 edition of the *Highway Capacity Manual* [8] suggests the following minimum green-time requirements for pedestrians:

$$G_p = 3.2 + \left(\frac{L}{S_p}\right) + \left(2.7 * \frac{N_{ped}}{W_E}\right)$$

$$for \ W_E > 10 \text{ ft}$$

$$G_p = 3.2 + \left(\frac{L}{S_p}\right) + (0.27 * N_{ped})$$

$$for \ W_E \leq 10 \text{ ft} \qquad (18\text{-}15)$$

where: G_p = minimum pedestrian crossing time, s

L = length of the crosswalk, ft

S_p = average walking speed of pedestrians, ft/s

N_{ped} = number of pedestrians crossing per phase in a single crosswalk, peds

W_E = width of crosswalk, ft

In Equation 18-15, 3.2 seconds is allocated as a minimal start-up time for pedestrians. A pedestrian just starting to cross the street at the end of 3.2 seconds requires an additional (L/S_p) seconds to safely cross. The last term of the equation allocates additional start-up time based on the volume of pedestrians that need to cross the street. In effect, this equation provides that the minimum pedestrian green (or WALK) indication (where pedestrian signals are employed) would be the sum of the first and third terms of the equation or:

$$WALK_{min} = 3.2 + \left(2.7 * \frac{N_{ped}}{W_E}\right) \ for \ W_E > 10 \text{ ft}$$

$$WALK_{min} = 3.2 + (0.27 * N_{ped})$$

$$for \ W_E \leq 10 \text{ ft} \qquad (18\text{-}16)$$

The flashing "up-raised hand" (DON'T WALK) signal (which is the pedestrian clearance interval) is always (L/S_p) measured from the end of the vehicular all-red phase.

The WALK interval may be longer than the minimum required by pedestrians, if the vehicular green is longer than needed. The total length of the WALK + Flashing DON'T WALK intervals must equal the sum of the vehicular green, plus the yellow and all-red transitions intervals that follow it.

For a signal timing to be viable for pedestrians, the minimum pedestrian crossing requirement, G_p, in each phase must be compared with the total length of the vehicular green, yellow, and all-red intervals:

$$G_p \leq G + Y \qquad (18\text{-}17)$$

If this condition is not met, pedestrians are not safely accommodated, and changes must be made to provide for their needs.

Where the minimum pedestrian condition is not met in a given phase, two approaches may be taken:

1. A pedestrian actuator may be provided. In this case, when pushed, the *next* green phase is lengthened to provide $G_p = G + Y$. The additional green time is subtracted from other phases (in a pretimed signal) to maintain the cycle length. When pedestrian actuators are provided, pedestrian signals *must* be used.

2. Retime the signal to provide the minimum need in all cycles. This must be done in a manner that also maintains the vehicular balance of green times and results in a longer cycle length.

The first approach has limited utility. Where pedestrians are present in most cycles, it is reasonable to assume that the actuator will always be pushed, thus destroying the planned vehicular signal timing. In such cases, the approach should be to retime the signal to satisfy both vehicular and pedestrian needs in every cycle. Pedestrian actuators are useful in cases where pedestrians are relatively rare or where actuated signal controllers are used.

In the second case, the task is to provide the minimum pedestrian crossing time while maintaining the balance of effective green needed to accommodate vehicles.

Consider the case of the vehicular signal timing for a two-phase signal shown in Table 18.3. Minimum pedestrian needs are also shown for comparison.

In this case, Phase A serves a major arterial and thus has the longer vehicular green but the shorter pedestrian requirement. Phase B serves a minor cross-street, but has the longer pedestrian requirement. Pedestrian requirements must be compared with the vehicular signal timing, using Equation 18-17:

$$G_{pA} = 20.0 \geq G_A + Y_A = 40.0 + 5.0 = 45.0 \text{ s OK}$$

$$G_{pB} = 30.0 \geq G_B + Y_B = 15.0 + 5.0 = 20.0 \text{ s NG}$$

The effective green times for Phases A and B may be computed as:

$$g = G + Y - t_L$$
$$g_A = 40.0 + 5.0 - 4.0 = 41.0 \text{ s}$$
$$g_B = 15.0 + 5.0 - 4.0 = 16.0 \text{ s}$$

The signal must be retimed to result in a $G + Y$ for Phase B of at least 30.0 seconds, while maintaining the relative balance of effective green time needed by vehicles in both phases (i.e., a ratio of 41.0 to 16.0). For Phase B to have a $G + Y$ of 30.0 seconds, the effective green time would have to be increased to:

$$g_B = 30.0 - 4.0 = 26.0 \text{ s}$$

Table 18.3: Sample Signal Timing

Phase	Green Time G (s)	Yellow + All-Red Y (s)	Lost Time t_L (s)	Pedestrian Requirement G_p (s)
A	40.0	5.0	4.0	20.0
B	15.0	5.0	4.0	30.0

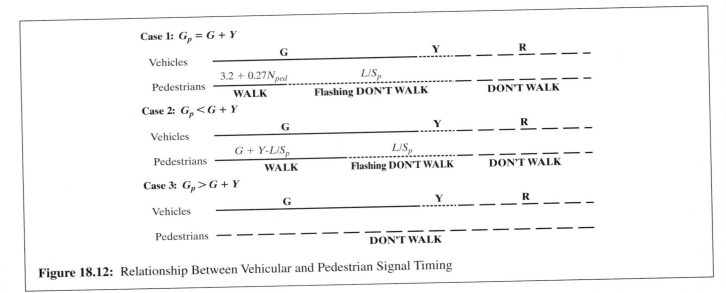

Figure 18.12: Relationship Between Vehicular and Pedestrian Signal Timing

To maintain the original ratio of vehicular green time, the effective green time for Phase A must also be increased:

$$\frac{g_A}{26.0} = \frac{41.0}{16.0}$$

$$g_A = \frac{41.0 * 26.0}{16.0} = 66.6 \text{ s}$$

The actual green times would become:

$$G_A = 66.6 - 5.0 + 4.0 = 65.6 \text{ s}$$

$$G_B = 26.0 - 5.0 + 4.0 = 25.0 \text{ s}$$

This yields a cycle length of 65.6 + 5.0 + 25.0 + 5.0 = 100.6 s. If this intersection were under pre-timed control, a 110-second cycle would be needed, and would be re-split to maintain the original proportion of effective green times. The provision of a timing that safely accommodates both pedestrians and vehicles results in an increase in the cycle length from 65 seconds to 110 seconds. The downside of this retiming would be an increase in delay to motorists and passengers.

Figure 18.12 further illustrates the relationship between vehicular and pedestrian phases for three basic cases: a) $G_p = G + Y$, b) $G_p < G + Y$, and c) $G_p > G + Y$.

In Case 1, where $G_p = G + Y$, the minimum WALK period is given with a pedestrian clearance interval of L/S_p. In this case, as in all cases, the vehicular RED indication coincides with the pedestrian DON'T WALK interval. In Case 2, where the vehicular signal is more than adequate for pedestrians, the WALK interval is longer than the minimum, essentially whatever time can be given after providing the pedestrian clearance interval of L/S_p. In Case 3, the vehicular green is not sufficient for pedestrians, so a permanent DON'T WALK is present. In this case, pedestrian push-button actuators *must* be provided. When pushed, the next vehicular green phase will be lengthened to provide for $G_p = G + Y$ (Case 1). It should be noted that pedestrian signals are not required in all cases. They should, however, be provided whenever pedestrian safety might be compromised without them.

18.4 Sample Signal Timing Applications

The procedures presented in Sections 18.1 through 18.3 will be illustrated in a series of signal-timing applications. The following steps should be followed:

1. Develop a reasonable phase plan in accordance with the principles discussed in Section 18.1. Use Equation 18-1 or local agency guidelines

to make an initial determination of whether left-turn movements need to be protected. Do not include compound phasing in preliminary signal timing; this may be tried as part of a more comprehensive intersection analysis later.

2. Convert all left-turn and right-turn movements to equivalent through vehicle units (tvu's) using the equivalents of Tables 18.1 and 18.2, respectively.

3. Draw a ring diagram of the proposed phase plan, inserting lane volumes (in tvu's) for each set of movements. Determine the critical path through the signal phasing as well as the sum of the critical-lane volumes (V_c) for the critical path.

4. Determine *yellow* and *all-red* intervals for each signal phase.

5. Determine lost times per cycle using Equations 18-5 through 18-7.

6. Determine the desirable cycle length, C, using Equation 18-11. For pretimed signals, round up to reflect available controller cycle lengths. An appropriate *PHF* and reasonable target *v/c* ratio should be used.

7. Allocate the available effective green time within the cycle in proportion to the critical lane volumes for each portion of the phase plan.

8. Check pedestrian requirements and adjust signal timing as needed.

Example 18-1: Signal-Timing Case 1: A Simple Two-Phase Signal

Consider the intersection layout and demand volumes shown in Figure 18.13. It shows the intersection of two streets with one lane in each direction and relatively low turning volumes. Moderate pedestrian activity is present, and the *PHF* and target *v/c* ratio is specified.

Solution:

Step 1:   Develop a Phase Plan

Given that there is only one lane for each approach, it is not possible to even consider including protected left turns in the phase plan. However, a check of the criteria of Equation 18-1 shows that no protected left turns are required for this case:

- EB: $V_{LT} = 10 < 200$
  xprod $= 10*315/1 = 3,150 < 50,000$
- WB: $V_{LT} = 12 < 200$
  xprod $= 12*420/1 = 5,040 < 50,000$
- NB: $V_{LT} = 10 < 200$
  xprod $= 10*400/1 = 4,000 < 50,000$
- SB: $V_{LT} = 10 < 200$
  xprod $< 10*375/1 = 3,750 < 50,000$

A simple two-phase signal, therefore, will be adopted for this intersection.

Step 2:   Convert Volumes to Through-Vehicle Equivalents

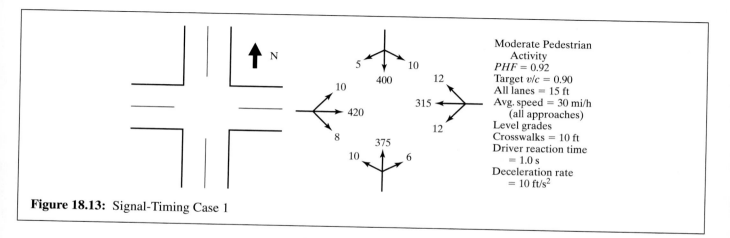

Figure 18.13: Signal-Timing Case 1

Table 18.4: Computation of Through Vehicle Equivalent Volumes for Signal Timing Case 1

Approach	Movement	Volume (Veh/h)	Equivalent Tables 18.1, 18.2	Volume (tvu/h)	Lane Group Vol (tvu/h)	Vol/Lane (tvu/h/ln)
EB	L	10	3.94	39	470	470
	T	420	1.00	420		
	R	8	1.32	11		
WB	L	12	5.50	66	397	397
	T	315	1.00	315		
	R	12	1.32	16		
NB	L	10	5.00	50	433	433
	T	375	1.00	375		
	R	6	1.32	8		
SB	L	10	4.69	47	454	454
	T	400	1.00	400		
	R	5	1.32	7		

The conversion of volumes to tvus is illustrated in Table 18.4. Equivalent values are taken from Tables 18.1 and 18.2, and are interpolated for intermediate values of opposing volume. Note that all through vehicles are equivalent to 1.0 tvu.

Step 3:   Determine Critical-Lane Volumes

The critical path through the signal phase plan is illustrated in Figure 18.14. As a two-phase signal, this is a relatively simple determination. For Phase A, either the EB or WB approach is critical. As the EB approach has the higher lane volume, 470 tvu/h, this is the critical movement for Phase A. For Phase B, either the NB or SB approach is critical; SB has the higher lane volume

(454 tvu/h), so this is the critical movement for Phase B. The sum of the critical-lane volumes is, therefore, 470 + 454 = 924 tvu/h.

Step 4:   Determine Yellow and All-Red Intervals

Yellow and *all-red* intervals are found using Equations 18-2 and 18-3. The average approach speed for all approaches is 30 mi/h. Thus, the $S_{85} = 30 + 5 = 35$ mi/h, and the $S_{15} = 30 - 5 = 25$ mi/h. As there are moderate numbers of pedestrians present, the all-red interval will be computed using Equation 18-3b, which allows vehicles to clear beyond the far crosswalk line. The distance to be crossed during the *all-red* clearance interval

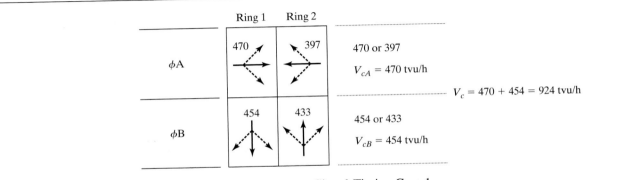

Figure 18.14: Determination of Critical Lane Volumes—Signal-Timing Case 1

is the sum of two 15-ft lanes and a 10-ft crosswalk, or $P = 15 + 15 + 10 = 40$ ft. Then:

$$y = t + \frac{1.47\,S_{85}}{2a + (64.4*0.01G)}$$

$$= 1.0 + \frac{1.47*35}{(2*10) + (0)}$$

$$= 3.6 \text{ s}$$

$$ar = \frac{P + L}{1.47 S_{15}} = \frac{40 + 20}{1.47*25} = 1.6 \text{ s}$$

Because both streets have the same width, crosswalk width, and approach speed, the values of y and ar are the same for both Phases A and B of the signal.

Step 5: Determination of Lost Times

Lost times are found using Equations 18-5 through 18-7. In this case, the recommended 2.0-s default values for start-up lost time (ℓ_1) and extension of effective green into yellow and all-red (e) are used:

$$Y = y + ar = 3.6 + 1.6 = 5.2 \text{ s}$$

$$\ell_2 = Y - e = 5.2 - 2.0 = 3.2 \text{ s}$$

$$t_L = \ell_1 + \ell_2 = 2.0 + 3.2 = 5.2 \text{ s}$$

As both phases have the same value, the total lost time per cycle, L, is $5.2 + 5.2 = 10.4$ s. Note that in all cases where the recommended default values for ℓ_1 (2.0 s) and e (2.0 s) are used, lost time per phase (t_L) is the same numerical value as the sum of the yellow and all red intervals (Y).

Step 6: Determine the Desirable Cycle Length

Equation 18-11 is used to determine the desirable cycle length:

$$C_{des} = \frac{L}{1 - \left(\dfrac{V_c}{1,615 * PHF * v/c} \right)}$$

$$= \frac{10.4}{1 - \left(\dfrac{924}{1,615 * 0.92 * 0.90} \right)}$$

$$= \frac{10.4}{0.31} = 33.5 \text{ s}$$

Assuming that this is a pretimed controller, a desirable cycle length of 35 s or a 40 s would be used. For the purposes of this signal timing case, the minimum value of 35 s will be used.

Step 7: Allocate Effective Green to Each Phase

Given a 35-second cycle length with 10.4 s of lost time per cycle, the amount of effective green time to be allocated is $35.0 - 10.4 = 24.6$ s. The allocation is done using Equation 18-13:

$$g_A = g_{TOT} * \left(\frac{V_{cA}}{V_c} \right) = 24.6 * \left(\frac{470}{924} \right) = 12.5 \text{ s}$$

$$g_B = g_{TOT} * \left(\frac{V_{cB}}{V_c} \right) = 24.6 * \left(\frac{454}{924} \right) = 12.1 \text{ s}$$

The cycle length may be checked as the total of effective green times plus the lost time per cycle, or $12.5 + 12.1 + 10.4 = 35.0$ s. Effective green times may be converted to actual green times using Equation 18-14:

$$G_A = g_A - Y_A + t_{LA} = 12.5 - 5.2 + 5.2 = 12.5 \text{ s}$$

$$G_B = g_B - Y_B + t_{LB} = 12.1 - 5.2 + 5.2 = 12.1 \text{ s}$$

Again, note that when default values for start-up lost time (2.0 s) and extension of effective green into yellow and all-red (2.0 s) are used, the actual green time is numerically the same as effective green time.

Step 8: Check Pedestrian Requirements

Equation 18-15 is used to compute the minimum pedestrian green requirement for each phase. Because both streets have equal width and equal crosswalk widths and because pedestrian traffic is "moderate" in all crosswalks, the requirements will be the same for each phase in this case. From Table 18.2, the default pedestrian volume for "moderate" activity is 200 peds/h. The number of pedestrians per cycle (N_{ped}) is based on the number of cycles per hour ($3,600/35 = 102.9$, say 103 cycles/h). The number of pedestrians per cycle is then $200/103 = 1.94$, say 2 peds/cycle. Then:

$$G_{pA,B} = 3.2 + \left(\frac{L}{S_p} \right) + (0.27 N_{ped})$$

$$= 3.2 + \left(\frac{30}{4.0} \right) + (0.27*2) = 11.2 \text{ s}$$

For this signal to be safe for pedestrians:

$$G_p < G + Y$$
$$G_{pA} = 11.2 < 12.5 + 5.2 = 17.7 \text{ s OK}$$
$$G_{pB} = 11.2 < 12.1 + 5.2 = 17.3 \text{ s OK}$$

The signal safely accommodates all pedestrians. No changes in the signal timing for vehicular needs is required.

Example 18-2: Signal Timing Case 2: Intersection of Major Arterials

Figure 18.15 illustrates the intersection of two four-lane arterials with significant demand volumes and exclusive left-turn lanes provided on each approach.

Step 1:   Develop a Phase Plan

Each left-turn movement should be checked against the criteria of Equation 18-1 to determine whether or not it needs to be protected:

- EB: $V_{LT} = 35 < 200$
  xprod $= 35 * (500/2) = 8,750 < 50,000$
  No protection needed.

- WB: $V_{LT} = 25 < 200$
  xprod $= 25 * (610/2) = 22,875 < 50,000$
  No protection needed.

- NB: $V_{LT} = 250 > 200$
  Protection needed.

- SB: $V_{LT} = 220 > 200$
  Protection needed.

Given that the NB and SB left turns require a protected phase, the next issue is how to provide it. The two opposing left-turn volumes, 220 veh/h (NB) and 250 veh/h (SB), are not numerically very different. Therefore, there appears to be little reason to separate the NB and SB protected phases. An exclusive left-turn phase will be used on the N–S arterial. A single phase using permitted left turns will be used on the E–W arterial.

Step 2:   Convert Volumes to Through Vehicle Equivalents

Through-vehicle equivalents are obtained from Tables 18.1 and 18.2 for left and right turns, respectively. The computations are illustrated in Table 18.5.

Note that exclusive LT lanes must be established as separate lane groups, with their demand volumes separately computed, as shown in Table 18.5. The equivalent for all protected left turns (Table 18.1) is 1.05.

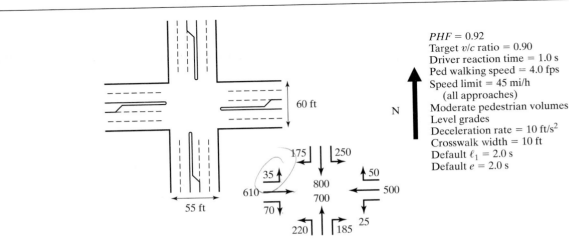

Figure 18.15: Signal-Timing Case 2

Table 18.5: Computation of Through Vehicle Equivalent Volumes for Signal-Timing Case 2

Approach	Movement	Volume (Veh/h)	Equivalent Tables 18.1, 18.2	Volume (tvu/h)	Lane Group Vol (tvu/h)	Vol/Lane (tvu/h/ln)
EB	L	35	4.00*	140	140	140
	T	610	1.00	610	702	351
	R	70	1.32	92		
WB	L	25	5.15*	129	129	129
	T	500	1.00	500	566	283
	R	50	1.32	66		
NB	L	220	1.05	231	231	231
	T	700	1.00	700	944	472
	R	185	1.32	244		
SB	L	250	1.05	263	263	263
	T	800	1.00	800	1,031	516
	R	175	1.32	231		

*Interpolated by opposing volume.

Step 3:   Determine Critical Lane Volumes

As noted in Step 1, the signal phase plan includes an exclusive LT phase for the N–S artery and a single phase with permitted left turns for the E–W artery. Figure 18.16 illustrates this and the determination of critical lane volumes.

Phase A is the exclusive N–S LT phase. The heaviest movement in the phase is 263 tvu/h for the SB left turn. In Phase B, the heavier movement is the SB through and right turn, with 516 tvu/h. In Phase C, both E–W left-turn lane groups and through/right-turn lane groups move at the same time. The heaviest movement is the EB TH/RT lanes, with 351 tvu/h. The sum of critical-lane volumes, V_C, is, therefore, $263 + 516 + 351 = 1,130$ tvu/h.

Note that each "ring" handles two sets of movements in Phase C. This is possible, of course, because it is the same signal face that controls all movements in a given direction. The left-turn lane volume cannot be averaged with the through/right-turn movement as there are lane-use restrictions involved. All left turns must be in the left-turn lane; none may be in the through/right-turn lanes.

Step 4:   Determine Yellow and All-Red Intervals

Equation 18-2 is used to determine the length of the *yellow* interval; Equation 18-3b is used to determine the length of the *all-red* interval. As a speed limit—45 mi/h—is given rather than a measured average approach speed, there will be no differentiation between the S_{85} and S_{15}.

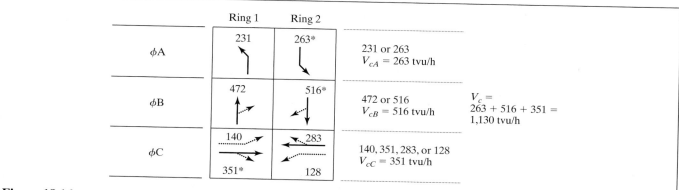

Figure 18.16: Determination of Critical-Lane Volumes—Signal-Timing Case 2

As the speed limits on both arteries are the same, the *yellow* intervals for all three phases will also be the same:

$$y_{A,B,C} = 1.0 + \frac{1.47*45}{(2*10) + (0)} = 4.3 \text{ s}$$

The *all-red* intervals will reflect the need to clear the full width of the street plus the width of the far crosswalk. The width of the N–S street is 55 ft, while the width of the E–W street is 60 ft. The width of a crosswalk is 10 ft. During the N–S left-turn phase, it will be assumed that a vehicle must clear the entire width of the E–W artery. Thus, for Phase A, the width to be cleared (P) is 60 +10 = 70 ft; for Phase B, it is also 60 +10 = 70 ft; for Phase C, the distance to be cleared is 55 +10 = 65 ft. Thus:

$$ar_{A,B} = \frac{70 + 20}{1.47*45} = 1.4 \text{ s}$$

$$ar_C = \frac{65 + 20}{1.47*45} = 1.3 \text{ s}$$

where 20 ft is the assumed length of a typical vehicle.

Step 5:   Determination of Lost Times

Remembering that where the default values for ℓ_1 and e are both 2.0 s, that the lost time per phase, t_L, is the same as the sum of the yellow plus all-red intervals, Y:

$$Y_{A,B} = t_{LA,B} = 4.3 + 1.4 = 5.7 \text{ s}$$

$$Y_C = t_{LC} = 4.3 + 1.3 = 5.6 \text{ s}$$

Based on this, the total lost time per cycle, L, is 5.7 + 5.7 + 5.6 = 17.0 s.

Step 6:   Determine the Desirable Cycle Length

The desirable cycle length is found using Equation 18-11:

$$C_{des} = \frac{17}{1 - \left(\dfrac{1,130}{1,615*0.92*0.90}\right)} = \frac{17}{0.155} = 109.7 \text{ s}$$

Assuming that this is a pretimed signal controller, a cycle length of 110 s would be selected.

Step 7:   Allocate Effective Green to Each Phase

In a cycle length of 110 s, with 17 s of lost time per cycle, the amount of effective green time that must be allocated to the three phases is 110 − 17 = 93 s.

Using Equation 18-13, the effective green time is allocated in proportion to the phase critical lane volumes:

$$g_A = 93*\left(\frac{263}{1,130}\right) = 21.6 \text{ s}$$

$$g_B = 93*\left(\frac{516}{1,130}\right) = 42.5 \text{ s}$$

$$g_C = 93*\left(\frac{351}{1,130}\right) = 28.9 \text{ s}$$

The cycle length is now checked to ensure that the sum of all effective green times and the lost time equals 110 s: 21.6 + 42.5 + 28.9 + 17.0 = 110 OK. Note that when the default values for ℓ_1 and e (both 2.0 s) are used, actual green times, G, equal effective green times, g.

Step 8:   Check Pedestrian Requirements

Pedestrian requirements are estimated using Equation 18-15. In this case, note that pedestrians will be permitted to cross the E–W artery only during Phase B. Pedestrians will cross the N–S artery during Phase C. The number of pedestrians per cycle for all crosswalks is the default pedestrian volume for "moderate" activity, 200 peds/h, divided by the number of cycles in an hour (3600/110 = 32.7 cycles/h). Thus, N_{ped} = 200/32.7 = 6.1 peds/cycle. Required pedestrian green times are:

$$G_{pB} = 3.2 + \left(\frac{60}{4.0}\right) + (0.27*6.1)$$

$$= 3.2 + 15.0 + 1.6 = 19.8 \text{ s}$$

$$G_{pC} = 3.2 + \left(\frac{55}{4.0}\right) + (0.27*6.1)$$

$$= 3.2 + 13.8 + 1.6 = 18.6 \text{ s}$$

The minimum requirements are compared to the sum of the green, yellow, and all-red times provided for vehicles:

$$G_{pB} = 19.8 \text{ s} < G_B + Y_B = 42.5 + 5.7$$

$$= 48.2 \text{ s}   OK$$

$$G_{pC} = 18.6 \text{ s} < G_C + Y_C = 28.9 + 5.6$$

$$= 34.5 \text{ s}   OK$$

Therefore, no changes to the vehicular signal timing are required to accommodate pedestrians safely.

For major arterial crossings, pedestrian signals would normally be provided. During Phase A, all pedestrian signals would indicate "DON'T WALK." During

Phase B, the pedestrian clearance interval (the flashing DON'T WALK) would be L/S_p or $60/4.0 = 15.0$ s. The WALK interval is whatever time is left in $G + Y$,

counting from the end of Y: $48.2 - 15.0 = 33.2$ s. During Phase C, L/S_p is $55/4.0 = 13.8$ s, and the WALK interval would be $34.5 - 13.8 = 20.7$ s.

Example 18-3: Signal-Timing Case 3: Another Junction of Major Arterials

Figure 18.17 illustrates another junction of major arterials. In this case, the E–W artery has three through lanes, plus an exclusive LT lane and an exclusive RT lane in each direction. In effect, each movement on the E–W artery has its own lane group. The N–S artery has two lanes in each direction, with no exclusive LT or RT lanes. There are no pedestrians present at this intersection.

Step 1:   Develop a Phase Plan
Phasing is determined by the need for left-turn protection. Using the criteria of Equation 18-1, each left turn movement is examined.

- EB: $V_{LT} = 300$ veh/h > 200 veh/h
  Protected phase needed.

- WB: $V_{LT} = 150$ veh/h < 200 veh/h
  $$xprod = 150*(1200/3)$$
  $$= 60,000 > 50,000$$
  Protected phase needed.

- NB: $V_{LT} = 50$ veh/h < 200 veh/h
  $$xprod = 50*(400/2)$$
  $$= 10,000 < 50,000$$
  Protected phase not needed.

- SB: $V_{LT} = 30$ veh/h < 200 veh/h
  $$xprod = 30*(500/2)$$
  $$= 7,500 < 50,000$$
  Protected phase not needed.

The results are fortunate. Had protected phasing been required for the NB and SB approaches, the lack of an exclusive LT lane on these approaches would have caused a problem.

The E–W approaches have LT lanes, and protected left-turns are needed on both approaches. As the LT volumes EB and WB are very different (300 veh/h vs. 150 veh/h), a phase plan that splits the protected LT phases would be advisable. A NEMA phase plan, utilizing an exclusive LT phase followed by a leading green for the EB direction, will be employed for the E–W artery.

Step 2:   Convert Volumes to Through-Vehicle Equivalents

Tables 18.1 and 18.2 are used to find through-vehicle equivalents for left- and right-turn volumes respectively. Conversion computations are illustrated in Table 18.6.

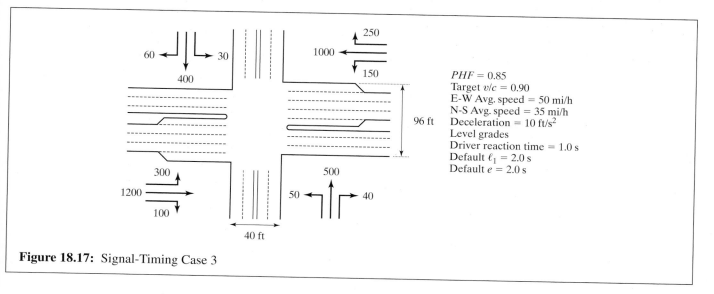

Figure 18.17: Signal-Timing Case 3

PHF = 0.85
Target v/c = 0.90
E-W Avg. speed = 50 mi/h
N-S Avg. speed = 35 mi/h
Deceleration = 10 ft/s²
Level grades
Driver reaction time = 1.0 s
Default ℓ_1 = 2.0 s
Default e = 2.0 s

Table 18.6: Computation of Through Vehicle Equivalent Volumes for Signal-Timing Case 3

Approach	Movement	Volume (Veh/h)	Equivalent Tables 18.1, 18.2	Volume (tvu/h)	Lane Group Vol. (tvu/h)	Vol./Lane (tvu/h/ln)
EB	L	300	1.05	315	315	315
	T	1,200	1.00	1,200	1,200	400
	R	100	1.18	118	118	118
WB	L	150	1.05	158	158	158
	T	1,000	1.00	1,000	1,000	334
	R	250	1.18	295	295	295
NB	L	50	3.00	150		
	T	500	1.00	500	697	349
	R	40	1.18	47		
SB	L	30	4.00[*]	120		
	T	400	1.00	400	591	296
	R	60	1.18	71		

[*] Interpolated by opposing volume.

Note that the EB and WB approaches have a separate lane group for each movement, while the NB and SB approaches have a single lane group serving all movements from shared lanes.

Step 3:   Determine Critical-Lane Volumes

Figure 18.18 shows a ring diagram for the phase plan discussed in Step 1 and illustrates the selection of the critical-lane volumes.

The phasing involves overlaps. For the combined Phase A, the critical path is down Ring 1, which has a sum of critical-lane volumes of 649 tvu/h. For Phase B,

the choice is simpler, as there are no overlapping phases. Ring 2, serving the NB approach, has the critical-lane volume of 349 tvu/h. The sum of all critical-lane volumes (V_c) is $649 + 349 = 998$ tvu/h.

Note also that overlapping phases have a unique characteristic. In this example, for overlapping Phase A, the largest left-turn movement is EB and the largest through movement is EB as well. Because of this, the overlapping phase plan will yield a smaller sum of critical lane volumes than one using an exclusive left-turn phase for both left-turn movements. Had the largest left-turn and through movements been from opposing

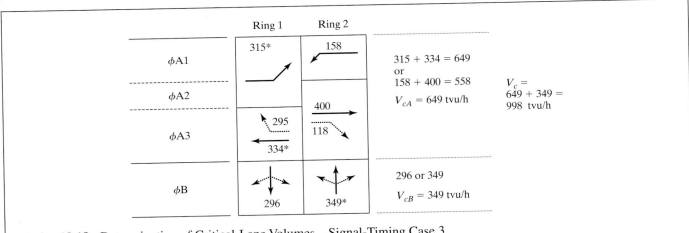

Figure 18.18: Determination of Critical-Lane Volumes—Signal-Timing Case 3

approaches, the sum of critical-lane volumes would be the same for the overlapping sequence and for a single exclusive LT phase. In other words, little is gained by using overlapping phases where a left turn and its opposing through (through plus right turn) movement are the larger movements.

Step 4:   Determine Yellow and All-Red Intervals

Equation 18-2 is used to determine the appropriate length of the *yellow* change intervals. Note that the signal design is a *three*-phase signal and that there are three transitions in the cycle. Because of the overlapping sequence, the transition at the end of the protected EB/WB left turns occur at different times on Ring 1 and Ring 2. For simplicity, it is assumed that left-turning vehicles from the EB and WB approaches cross the entire width of the N–S artery. *All-red* intervals are determined using Equation 18-3a, as there are no pedestrians present.

Percentile speeds are estimated from the measured average approach speeds given:

$$S_{85EW} = 50 + 5 = 55 \text{ mi/h}$$
$$S_{15EW} = 50 - 5 = 45 \text{ mi/h}$$
$$S_{85NS} = 35 + 5 = 40 \text{ mi/h}$$
$$S_{15NS} = 35 - 5 = 30 \text{ mi/h}$$

Then:

$$y_{A1,A2,A3} = 1.0 + \frac{1.47 * 55}{(2 * 10) + (0)} = 5.0 \text{ s}$$

$$y_B = 1.0 + \frac{1.47 * 40}{(2 * 10) + (0)} = 3.9 \text{ s}$$

$$ar_{A1,A2,A3} = \frac{40 + 20}{1.47 * 45} = 0.9 \text{ s}$$

$$ar_B = \frac{96 + 20}{1.47 * 30} = 2.6 \text{ s}$$

where 20 ft is the assumed average length of a typical vehicle.

Step 5:   Determination of Lost Times

As the problem statement specifies the default values of 2.0 s each for start-up lost time and extension of effective green into yellow and all-red intervals, the total lost time in each phase, t_L, is equal to the sum of the yellow and all red intervals, Y. Thus:

$$t_{LA1/A2} = Y_{A1/A2} = 5.0 + 0.9 = 5.9 \text{ s}$$
$$t_{LA3} = Y_{A3} = 5.0 + 0.9 = 5.9 \text{ s}$$
$$t_{LB} = Y_B = 3.9 + 2.6 = 6.5 \text{ s}$$

Note from Figure 18.18 that the first phase transition occurs at the end of Phase A1, but only on Ring 2. A similar transition occurs at the end of Phase A2, but only on Ring 1. The two other transitions, at the end of Phases A3 and B, occur on both rings. Thus, the total lost time per cycle, L is $5.9 + 5.9 + 6.5 = 18.3$ s, and the phase plan represents a three-phase signal.

Step 6:   Determine the Desirable Cycle Length

The desirable cycle length is found using Equation 18-11:

$$C_{des} = \frac{18.3}{1 - \left(\dfrac{998}{1,615 * 0.85 * 0.90} \right)}$$

$$= \frac{18.3}{0.192} = 95.3 \text{ s}$$

Assuming that this is a pretimed controller, a cycle length of 100 s would be selected.

Step 7:   Allocate Effective Green to Each Phase

A signal cycle of 100 s with 18.3 s of lost time has $100.0 - 18.3 = 81.7$ s of effective green time to allocate in accordance with Equation 18-13. Note that in allocating green to the critical path, Phases A1 and A2 are treated as a single segment. Subsequently, the location of the Ring 2 transition between Phases A1 and A2 will have to be established.

$$g_{A1+A2} = 81.7 * \left(\frac{315}{998} \right) = 25.8 \text{ s}$$

$$g_{A3} = 81.7 * \left(\frac{334}{998} \right) = 27.3 \text{ s}$$

$$g_B = 81.7 * \left(\frac{349}{998} \right) = 28.6 \text{ s}$$

The specific lengths of Phases A1 and A2 are determined by fixing the Ring 2 transition between them. This requires consideration of the noncritical path through combined Phase A, which occurs on Ring 2. The total length of combined Phase A is the sum of g_{A1+A2} and g_{A3}, or $25.8 + 27.3 = 53.1$ s. The Ring 2 transition is based upon the relative values of the lane volumes for Phase A1 and the combined Phase A2/A3, or:

$$g_{A1} = 53.1 * \left(\frac{158}{158 + 400}\right) = 15.0 \text{ s}$$

By implication, Phase A2 is the total length of combined Phase A minus the length of Phase A1 and Phase A3, or:

$$g_{A2} = 53.1 - 15.0 - 27.3 = 10.8 \text{ s}$$

Now, the signal has been completely timed for vehicular needs. With the assumption of default values for ℓ_1(2.0 s) and e (2.0 s), actual green times are equal to effective green times (numerically, although they do not occur simultaneously):

$$G_{A1} = 15.0 \text{ s}$$
$$G_{A2} = 10.8 \text{ s}$$
$$Y_{A1/A2} = 5.9 \text{ s}$$
$$G_{A3} = 27.3 \text{ s}$$
$$Y_{A3} = 5.9 \text{ s}$$
$$G_B = 28.6 \text{ s}$$
$$Y_B = 6.5 \text{ s}$$
$$C = 100.0 \text{ s}$$

There is no Step 8 in this case, as there are no pedestrians at this intersection and, therefore, no pedestrian requirements to be checked.

Example 18-4: Signal-Timing Case 4: A T-Intersection

Figure 18.19 illustrates a typical T-intersection, with exclusive lanes for various movements as shown. Note that there is only one opposed left turn in the WB direction.

Step 1:   Develop a Phase Plan

In this case, there is only one opposed left turn to check for the need of a protected phase. As the WB left turn > 200 veh/h, it should be provided with a protected left-turn phase. There is no EB or SB left turn, and the NB left turn is unopposed. The standard way of providing for the necessary phasing would be to utilize a leading WB green with no lagging EB green.

Step 2:   Convert Volumes to Through-Vehicle Equivalents

Table 18.7 shows the conversion of volumes to through vehicle equivalents, using the equivalent values given in Tables 18.1 and 18.2 for left and right turns respectively.

Note that the NB left turn is treated as an opposed turn with $V_o = 0$ veh/h. There are different approaches that have been used to address left turns that are unopposed due to one-way streets and T-intersections, reasons other than the presence of a protected left-turn

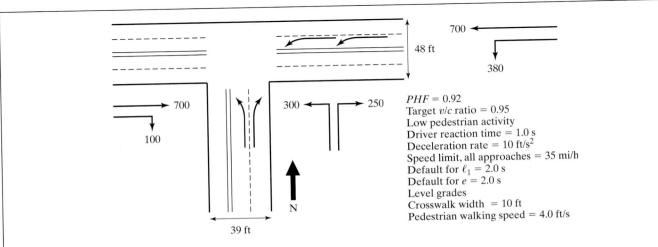

PHF = 0.92
Target v/c ratio = 0.95
Low pedestrian activity
Driver reaction time = 1.0 s
Deceleration rate = 10 ft/s²
Speed limit, all approaches = 35 mi/h
Default for ℓ_1 = 2.0 s
Default for e = 2.0 s
Level grades
Crosswalk width = 10 ft
Pedestrian walking speed = 4.0 ft/s

Figure 18.19: Signal-Timing Case 4

Table 18.7: Computation of Through-Vehicle Equivalent Volumes for Signal-Timing Case 3

Approach	Movement	Volume (veh/h)	Equivalent Tables 18.1, 18.2	Volume (tvu/h)	Lane Group Vol. (tvu/h)	Vol./Lane (tvu/h/ln)
EB	T	700	1.00	700	821	411
	R	100	1.21	121		
WB	L	380	1.05	399	399	399
	T	700	1.00	700	700	700
NB	L	300	1.10	330	330	330
	R	250	1.21	303	303	303

phase. Such a movement could also be treated as any protected left turn and an equivalent of 1.05 applied. In some cases, particularly unopposed left turns from a one-way street, the movement is treated as a right turn, using the appropriate factor based on pedestrian interference.

Step 3:   Determine Critical-Lane Volumes

Figure 18.20 shows the ring diagram for the phasing described in Step 1 and illustrates the determination of the sum of critical-lane volumes.

In this case, the selection of the critical path through combined Phase A is interesting. Ring 1 goes through two phases, while Ring 2 goes through only one. In this case, the critical path goes through Ring 1 and has a total of three phases. Had the Phase A critical path been through Ring 2, the signal would have only two critical phases. In such cases, the highest critical-lane volume total does not alone determine the critical

path. Because one path has an additional phase and, therefore, an additional set of lost times, it could possibly be critical even if it has the lower total critical-lane volume. In such a case, the cycle length would be computed using *either* path, and the one yielding the largest desirable cycle length would be critical. In this case, the path yielding three phases has the highest sum of critical-lane volumes, so only one cycle length will have to be computed.

Step 4:   Determine Yellow and All-Red Intervals

Both *yellow* and *all-red* intervals for both streets will be computed using Equations 18-2 and 18-3a (low pedestrian activity) and the speed limit of 35 mi/h for both streets. As a measured average speed was not given, the 85th and 15th percentile speeds cannot be differentiated. For Phases A1 and A2, it will be assumed that both the left-turn and through movements from the E–W street cross the entire 39-ft width of the N–S street. Similarly,

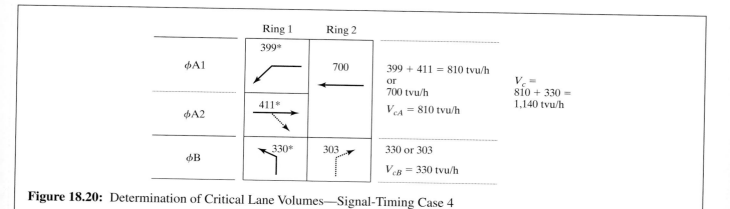

Figure 18.20: Determination of Critical Lane Volumes—Signal-Timing Case 4

in Phase B, it will be assumed that both movements cross the entire 48-ft width of the E–W street. Then:

$$y_{A1,A2,B} = 1.0 + \frac{1.47 * 35}{(2 * 10) + (0)} = 3.6 \text{ s}$$

$$ar_{A1,A2} = \frac{39 + 20}{1.47 * 35} = 1.1 \text{ s}$$

$$ar_B = \frac{48 + 20}{1.47 * 35} = 1.3 \text{ s}$$

Step 5:   Determination of Lost Times

Once again, 2.0-s default values are used for start-up lost time (ℓ_1) and extension of effective green into yellow and all-red (e), so that the total lost time for each phase is equal to the sum of the yellow plus all-red intervals:

$$Y_{A1} = t_{LA1} = 3.6 + 1.1 = 4.7 \text{ s}$$

$$Y_{A2} = t_{LA2} = 3.6 + 1.1 = 4.7 \text{ s}$$

$$Y_B = t_{LB} = 3.6 + 1.3 = 4.9 \text{ s}$$

The total lost time per cycle is, therefore, 4.7 + 4.7 + 4.9 = 14.3 s.

Step 6:   Determine the Desirable Cycle Length

Equation 18-11 is once again used to determine the desirable cycle length, using the sum of critical-lane volumes, 1,140 tvu/h:

$$C_{des} = \frac{14.3}{1 - \left(\dfrac{1,140}{1,615 * 0.92 * 0.95} \right)} = \frac{14.3}{0.192} = 74.5 \text{ s}$$

For a pretimed controller, a cycle length of 75 s would be implemented.

Step 7:   Allocate Effective Green to Each Phase

The available effective green time for this signal is 75.0 − 14.3 = 60.7 s. It is allocated in proportion to the critical-lane volumes for each phase:

$$g_{A1} = 60.7 * \left(\frac{399}{1140} \right) = 21.2 \text{ s}$$

$$g_{A2} = 60.7 * \left(\frac{411}{1140} \right) = 21.9 \text{ s}$$

$$g_B = 60.7 * \left(\frac{330}{1140} \right) = 17.6 \text{ s}$$

As the usual defaults for ℓ_1 and e are used, actual green times are numerically equal to effective green times.

Step 8:   Check Pedestrian Requirements

While there is low pedestrian activity at this intersection, pedestrians must still be safely accommodated by the signal phasing. It will be assumed that pedestrians cross the N–S street only during Phase A2 and that pedestrians crossing the E–W street will use Phase B. The number of pedestrians per cycle in each crosswalk is based on the default volume for "low" activity—50 peds/h (Table 18.2)—and the number of cycles per hour—3,600/75 = 48. Then, N_{ped} in each crosswalk would be 50/48 = 1.0 ped/cycle. Equation 18-15 is used to compute minimum pedestrian requirements:

$$G_{pA2} = 3.2 + \left(\frac{39}{4.0} \right) + (0.27 * 1.0) = 13.2 \text{ s}$$

$$G_{pB} = 3.2 + \left(\frac{48}{4.0} \right) + (0.27 * 1.0) = 15.5 \text{ s}$$

These requirements must be checked against the vehicular green, yellow, and all-red intervals:

$$G_{pA2} = 13.2 \text{ s} < G_{A2} + Y_{A2}$$
$$= 21.9 + 4.7 = 26.6 \text{ s OK}$$

$$G_{pB} = 15.5 \text{ s} < G_B + Y_B = 17.6 + 4.9$$
$$= 22.5 \text{ s OK}$$

Pedestrians are safely accommodated by the vehicular signalization, and no changes are required.

References

1.   Pusey, R. and Butzer, G., "Traffic Control Signals," *Traffic Engineering Handbook*, 5th Edition

(Pline, J. ed.), Institute of Transportation Engineers, Washington DC, 2000.

2. Asante, S., Ardekani, S., and Williams, J., "Selection Criteria for Left-Turn Phasing and Indication Sequence," *Transportation Research Record 1421*, Transportation Research Board, Washington DC, 1993.

3. *Manual of Uniform Traffic Control Devices*, Millennium Edition, Federal Highway Administration, U.S. Department of Transportation, Washington DC, 2000.

4. McGee, H. and Warren, D., "Right Turn on Red," *Public Roads*, Federal Highway Administration, U.S. Department of Transportation, Washington DC, June 1976.

5. "Driver Behavior at RTOR Locations," ITE Technical Committee 4M-20, *ITE Journal*, Institute of Transportation Engineers, Washington DC, April 1992.

6. McGee, H., "Accident Experience with Right Turn on Red," *Transportation Research Record 644*, Transportation Research Board, Washington DC, 1977.

7. "Recommended Practice: Determining Vehicle Change Intervals," ITE Technical Committee 4A-16, *ITE Journal*, Institute of Transportation Engineers, Washington DC, May 1985.

8. *Highway Capacity Manual*, 4th Edition, Transportation Research Board, Washington DC, 2000.

Problems

18-1. What change and clearance intervals are recommended for an intersection with an average approach speed of 42 mi/h, a grade of +3%, a cross-street width of 40 ft, and 10-ft crosswalks? Assume a standard vehicle length of 20 ft, a driver reaction time of 1.0 s, and significant pedestrian movements.

18-2. An analysis of pedestrian needs at a signalized intersection is undertaken. Important parameters concerning pedestrian needs and the existing vehicular signal timing are given in the table below. Are pedestrians safely accommodated by this signal timing? If not, what signal timing should be implemented? Assume that the standard default values for start-up lost time and extension of effective green into yellow and all-red (2.0 s each) are in effect.

Phase	G (s)	Y (s)	G_p (s)
A	20.0	4.5	28.0
B	70.0	5.0	13.0

18-3–18-7. Develop a signal design and timing for the intersections shown in Figures 18-21–18-25. In each case, accommodate both vehicular and pedestrian needs. Where necessary to make assumptions on key values, state these explicitly. If a successful signal timing *requires* geometric changes, indicate these with an appropriate drawing.

In general, the following values should be used for all problems:

- All volumes are in veh/h
- Pedestrian walking speed = 4.0 ft/s
- Vehicle deceleration speed = 10.0 ft/s^2
- Driver reaction time = 1.0 s
- Length of typical vehicle = 18 ft
- Level grades unless otherwise indicated

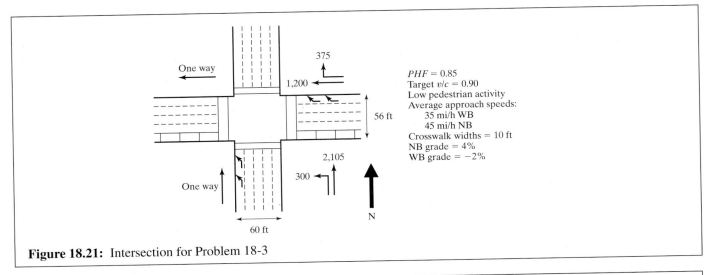

Figure 18.21: Intersection for Problem 18-3

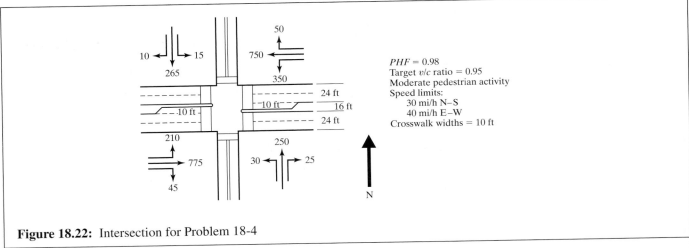

Figure 18.22: Intersection for Problem 18-4

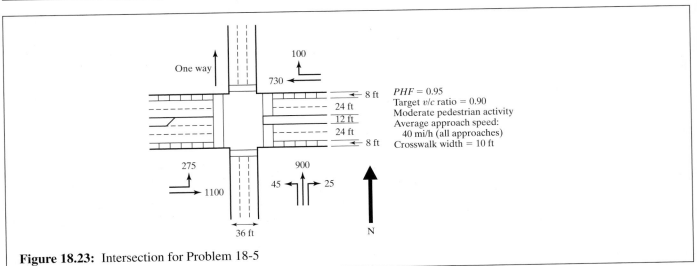

Figure 18.23: Intersection for Problem 18-5

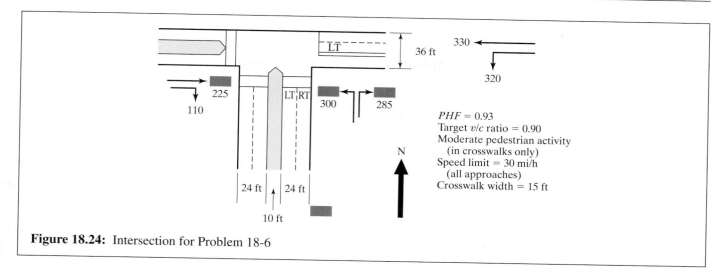

Figure 18.24: Intersection for Problem 18-6

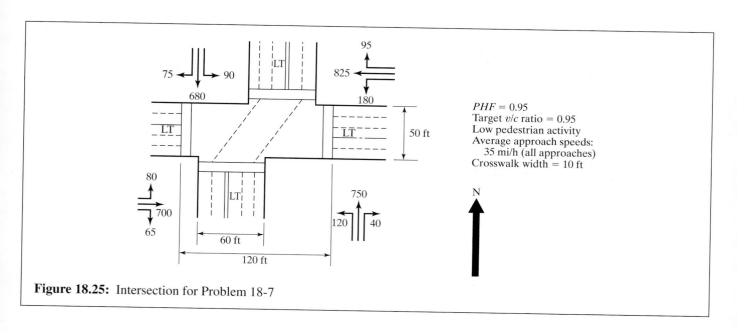

Figure 18.25: Intersection for Problem 18-7

CHAPTER
19

Elements of Intersection Design and Layout

In Chapters 16 through 18, the selection of appropriate control measures for intersections was addressed, along with the detailed characteristics of signalized intersection operation and signal-timing techniques. Whether signalized or unsignalized, the control measures implemented at an intersection must be synergistic with the design and layout of the intersection. In this chapter, an overview of several important intersection design features is provided. It is emphasized that this is only an overview, as the details of intersection design could be the subject of a textbook on its own.

The elements treated here include techniques for determining the appropriate number and use of lanes at an intersection approach, channelization, right- and left-turn treatments, special safety issues at intersections, and location of intersection signs and signal displays. There are a number of standard references for more detail on these and related subject areas, including the AASHTO *Policy on Geometric Design of Highways and Streets* [*1*], the *Manual on Uniform Traffic Control Devices* [*2*], the *Manual of Traffic Signal Design* [*3*], the *Traffic Detector Handbook* [*4*], and the *Highway Capacity Manual* [*5*].

19.1 Intersection Design Objectives and Considerations

As in all aspects of traffic engineering, intersection design has two primary objectives: (1) to ensure safety for all users, including drivers, passengers, pedestrians, bicyclists, and others and (2) to promote efficient movement of all users (motorists, pedestrians, bicyclists, etc.) through the intersection. Achievement of both is not an easy task, as safety and efficiency are often competing rather than mutually reinforcing goals.

In developing an intersection design, AASHTO [*1*] recommends that the following elements be considered:

- Human factors
- Traffic considerations
- Physical elements
- Economic factors
- Functional intersection area

Human factors must be taken into account. Thus, intersection designs should accommodate reasonable approach speeds, user expectancy, decision and reaction times, and other user characteristics. Design should, for example, reinforce natural movement paths and trajectories, unless doing so presents a particular hazard.

Traffic considerations include provision of appropriate capacity for all user demands; the distribution of vehicle types and turning movements; approach speeds; and special requirements for transit vehicles, pedestrians, and bicyclists.

Physical elements include the nature of abutting properties, particularly traffic movements generated by these properties (parking, pedestrians, driveway movements, etc.). They also include the intersection angle, existence and location of traffic control devices, sight distances, and specific geometric characteristics, such as curb radii.

Economic factors include the cost of improvements (construction, operation, maintenance), the effects of improvements on the value of abutting properties (whether used by the expanded right-of-way or not), and the effect of improvements on energy consumption.

Finally, intersection design must encompass the full functional intersection area. The operational intersection area includes approach areas that fully encompass deceleration and acceleration zones as well as queuing areas. The latter are particularly critical at signalized intersections.

19.2   A Basic Starting Point: Sizing the Intersection

One of the most critical aspects of intersection design is the determination of the number of lanes needed on each approach. This is not an exact science, as the result is affected by the type of control at the intersection, parking conditions and needs, availability of right-of-way, and a number of other factors that are not always directly under the control of the traffic engineer. Further, considerations of capacity, safety, and efficiency all influence the desirable number of lanes. As is the case in most design exercises, there is no one correct answer, and many alternatives may be available that provide for acceptable safety and operation.

19.2.1   Unsignalized Intersections

Unsignalized intersections may be operated under basic rules of the road (no control devices other than warning and guide signs), or under STOP or YIELD control.

When totally uncontrolled, intersection traffic volumes are generally light, and there is rarely a clear "major" street with significant volumes involved. In such cases, intersection areas do not often require more lanes than on the approaching roadway. Additional turning lanes are rarely provided. Where high speeds and/or visibility problems exist, channelization may be used in conjunction with warning signs to improve safety.

The conditions under which two-way (or one-way at a T-intersection or intersection of one-way roadways) STOP or YIELD control are appropriate are treated in Chapter 16. The existence of STOP- or YIELD-controlled approach(es), however, adds some new considerations into the design process:

- Should left-turn lanes be provided on the major street?
- Should right-turn lanes be provided on the major street?
- Should a right-turn lane be provided on minor approaches?
- How many basic lanes does each minor approach require?

Most of these issues involve capacity considerations (see Chapter 21). For convenience, however, some general guidelines are presented herein.

When left turns are made from a mixed lane on the major street, there is the potential for unnecessary delay to through vehicles that must wait while left-turners find a gap in the opposing major-street traffic. The impact of major-street left turns on delay to all major-street approach traffic becomes noticeable when left turns exceed 150 vehs/h. This may be used as a general guideline indicating the probable need for a major-street left-turn lane, although a value as low as 100 vehs/h could be justified.

Right-turning vehicles from the major street do not have a major impact on the operation of STOP- or

YIELD-controlled intersections. While they do not technically conflict with minor-street movements when they are made from shared lanes, they may impede some minor-street movements when drivers do not clearly signal that they are turning or approach the intersection at high speed. When major-street right turns are made from an exclusive lane, their intent to turn is more obvious to minor-street drivers. Right-turn lanes for major-street vehicles can be easily provided where on-street parking is permitted. In such situations, parking may be prohibited for 100–200 ft from the STOP line, thus creating a short right-turn lane.

Most STOP-controlled approaches have a single lane shared by all minor-street movements. Occasionally, two lanes are provided. Any approach with sufficient demand to require three lanes is probably inappropriate for STOP control. Approximate guidelines for the number of lanes required may be developed from the unsignalized intersection analysis methodology of the *Highway Capacity Manual*. Table 19.1 shows various combinations of minor-approach demand vs. total crossing traffic on the major street, along with guidelines as to whether one or two lanes would be needed. They are based on assumptions that (1) all major-street traffic is through traffic, (2) all minor- approach traffic is through traffic, and (3) various impedances and other non-ideal characteristics reduce the capacity of a lane to about 80% of its original value.

The other issue for consideration on minor STOP-controlled approaches is whether or not a right-turning lane should be provided. Because the right-turn movement at a STOP-controlled approach is much more efficient than crossing and left-turn movements, better operation can usually be accomplished by providing a right-turn lane. This is often as simple as banning parking within 200 ft of the STOP line, and it prevents right-turning drivers from being stuck in a queue when they could easily be executing their movements. Where a significant proportion of the minor-approach traffic is turning right ($>20\%$), provision of a right-turning lane should always be considered.

Note that the lane criteria of Table 19.1 are approximate. Any finalized design should be subjected to detailed analysis using the appropriate procedures of the HCM 2000.

Consider the following example: a two-lane major roadway carries a volume of 800 veh/h, of which 10% turn left and 5% turn right at a local street. Both approaches on the local street are STOP-controlled and carry 150 veh/h, with 50 turning left and 50 turning right. Suggest an appropriate design for the intersection.

Given the relatively low volume of left turns (80/h) and right turns (40/h) on the major street, neither left- nor right-turn lanes would be required, although they could be provided if space is available. From Table 19.1, it appears that one lane would be sufficient for each of

Table 19.1: Guidelines for Number of Lanes on STOP-Controlled Approaches[1]

Total Volume on Minor Approach (veh/h)	Total Volume on Major Street (veh/h)			
	500	1,000	1,500	2,000
100	1 lane	1 lane	1 lane	2 lanes
200	1 lane	1 lane	2 lanes	NA
300	1 lane	2 lanes	2 lanes	NA
400	1 lane	2 lanes	NA	NA
500	2 lanes	NA	NA	NA
600	2 lanes	NA	NA	NA
700	2 lanes	NA	NA	NA
800	2 lanes	NA	NA	NA

[1]Not including multiway STOP-controlled intersections.

NA = STOP control probably not appropriate for these volumes.

the minor-street approaches. The relatively heavy percentage of right turns (33%), however, suggests that a right-turn lane on each minor approach would be useful.

19.2.2   Signalized Intersections

Approximating the required size and layout of a signalized intersection involves many factors, including the demands on each lane group, the number of signal phases, and the signal cycle length.

Determining the appropriate number of lanes for each approach and lane group is not a simple design task. Like so many design tasks, there is no absolutely unique result, and many different combinations of physical design and signal timing can provide for a safe and efficient intersection.

The primary control on number of lanes is the *maximum sum of critical-lane volumes* that the intersection can support. This was discussed and illustrated in Chapter 17. The equation governing the maximum sum of critical lane volumes is repeated here for convenience:

$$V_c = \frac{1}{h}\left[3{,}600 - Nt_L\left(\frac{3{,}600}{C}\right)\right] \qquad (19\text{-}1)$$

where: V_c = maximum sum of critical-lane volumes, veh/h

   h = average headway for prevailing conditions on the lane group or approach, s/veh

   N = number of phases in the cycle

   t_L = lost time per phase, s/phase

   C = cycle length, s

Table 19.2 gives approximate maximum sums of critical lane volumes for typical prevailing conditions. An average headway of 2.6 s/veh is used, along with a typical lost time per phase of 4.0 s (t_L). Maximum sums are tabulated for a number of combinations of N and C.

Consider the case of an intersection between two major arterials. Arterial 1 has a peak directional volume of 900 veh/h; Arterial 2 has a peak directional volume of 1,100 veh/h. Turning volumes are light, and a two-phase signal is anticipated. As a preliminary estimate, what number of lanes is needed to accommodate these volumes, and what range of cycle lengths might be appropriate?

Table 19.2: Maximum Sums of Critical Lane Volumes for a Typical Signalized Intersection

Cycle Length (s)	No. of Phases		
	2	3	4
30	1,015	831	646
40	1,108	969	831
50	1,163	1,052	942
60	1,200	1,108	1,015
70	1,226	1,147	1,068
80	1,246	1,177	1,108
90	1,262	1,200	1,138
100	1,274	1,218	1,163
110	1,284	1,234	1,183
120	1,292	1,246	1,200

From Table 19.2, the range of maximum sums of critical lane volumes is between 1,015 veh/h for a 30-s cycle length to 1,292 veh/h for a 120-s cycle length. The two critical volumes are given as 900 veh/h and 1,100 veh/h. If only one lane is provided for each, then the sum of critical-lane volumes is 900 + 1,100 = 2,000 veh/h, well outside the range of maximum values for reasonable cycle lengths. Table 19.3 shows a number of reasonable scenarios for the number of lanes on each critical approach along with the resulting sum of critical-lane volumes.

With one lane on Arterial 1 and 3 lanes on Arterial 2, the sum of critical-lane volumes is 1,267 veh/h. From Table 19.2, this would be a workable solution with a cycle length over 100 s. With two lanes on each arterial, the sum of critical-lane volumes is 1,000 veh/h. This situation would be workable at any cycle length between 30 and 120 s. All other potentially workable scenarios in Table 19.3 could accommodate any cycle length between 30 and 120 s as well.

This type of analysis does not yield a final design or cycle length, as it is approximate. But it does give the traffic engineer a basic idea of where to start. In this case, providing two lanes on each arterial in the peak direction appears to be a reasonable solution. As peaks tend to be reciprocal (what goes one way in the morning comes back the opposite way in the evening), two lanes

Table 19.3: Sum of Critical-Lane Volumes (veh/h) for Various Scenarios: Sample Problem

No. of Lanes on Arterial 2	Critical-Lane Volume for Arterial (veh/h)	No. of Lanes on Arterial 1		
		1	2	3
		900/1 = 900	900/2 = 450	900/3 = 300
1	1,100/1 = 1,100	2,000	1,550	1,400
2	1,100/2 = 550	1,450	1,000[1]	850[1]
3	1,100/3 = 367	1,267[1]	817[1]	667[1]

[1] Acceptable lane plan with V_c acceptable at some cycle length.

would also be provided for the off-peak directions on each arterial as well.

The signal timing should then be developed using the methodology of Chapter 18. The final design and timing should then be subjected to analysis using the *Highway Capacity Manual* (see Chapters 21 and 22) or some other appropriate analysis technique.

The number of anticipated phases is, of course, critical to a general analysis of this type. Suggested criteria for determining when protected left-turn phases are needed are given in Chapter 18. Remember that because there is a critical-lane volume for *each* signal phase; a four-phase signal involves four critical-lane volumes, for example.

Exclusive left-turn lanes must be provided whenever a fully protected left-turn phase is used and is highly desirable when compound left-turn phasing (protected + permitted or vice-versa) is used.

19.3   Intersection Channelization

19.3.1   General Principles

Channelization can be provided through the use of painted markings or by installation of raised channelizing islands. The AASHTO *Policy on Geometric Design of Highways and Streets* [1] gives a number of reasons for considering channelization at an intersection:

- Vehicle paths may be confined so that no more than two paths cross at any one point.
- The angles at which merging, diverging, or weaving movements occur may be controlled.

- Pavement area may be reduced, decreasing the tendency to wander and narrowing the area of conflict between vehicle paths.
- Clearer indications of proper vehicle paths may be provided.
- Predominant movements may be given priority.
- Areas for pedestrian refuge may be provided.
- Separate storage lanes may be provided to permit turning vehicles to wait clear of through-traffic lanes.
- Space may be provided for the mounting of traffic control devices in more visible locations.
- Prohibited turns may be physically controlled.
- Vehicle speeds may be somewhat reduced.

The decision to channelize an intersection depends upon a number of factors, including the existence of sufficient right-of-way to accommodate an effective design. Factors such as terrain, visibility, demand, and cost also enter into the decision. Channelization supplements other control measures but can sometimes be used to simplify other elements of control.

19.3.2   Some Examples

It is difficult to discuss channelization in the abstract. A selection of examples illustrates the implementation of the principles noted previously.

Figure 19.1 shows the intersection of a major street (E–W) with a minor crossroad (N–S). A median island is provided on the major street. Partial channelization is provided for the SB right turn, and a left-turn

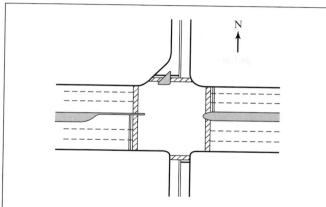

Figure 19.1: A Four-Leg Intersection with Partial Channelization for SB-EB and EB-SB Movements

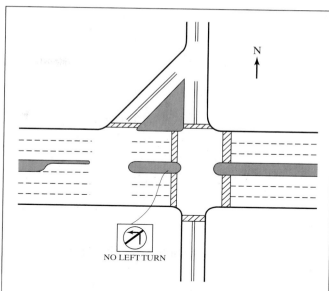

Figure 19.2: A Four-Leg Intersection Channelization for Major SB-EB and EB-SB Movements

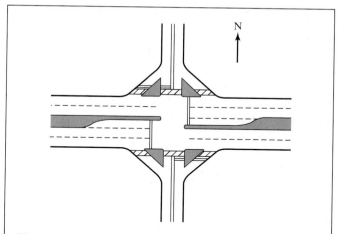

Figure 19.3: A Four-Leg Intersection with Full Channelization of Right Turns

lane is provided for the EB left turn. The two channelized turns are reciprocal, and the design reflects a situation in which these two turning movements are significant. The design illustrated minimizes the conflict between SB right turns and other movements and provides a storage lane for EB left turns, removing the conflict with EB through movements. The lack of any channelization for other turning movements suggests that they have light demand. The design does not provide for a great deal of pedestrian refuge, except for the wide median on the east leg of the intersection. This suggests that pedestrian volumes are relatively low at this location; if this is so, the crosswalk markings are optional. The channelization at this intersection is appropriate for both an unsignalized or a signalized intersection.

Figure 19.2 shows a four-leg intersection with similar turning movements as in Figure 19.1. In this case, however, the SB-EB and EB-SB movements are far heavier, and require a more dramatic treatment. Here channelization is used to create two additional intersections to handle these dominant turns. Conflicts between the various turning movements are minimized in this design.

Figure 19.3 is a similar four-leg intersection with far greater use of channelization. All right turns are channelized, and both major street left-turning movements have an exclusive left-turn lane. This design addresses a situation in which turning movements are more dominant. Pedestrian refuge is provided only on the right-turn channelizing islands and this may be limited

by the physical size of the islands. Again, the channelization scheme is appropriate for either signalized or unsignalized control.

Channelization can also be used at locations with significant traffic volumes to simplify and reduce the number of conflicts and to make traffic control simpler and more effective. Figure 19.4 illustrates such a case.

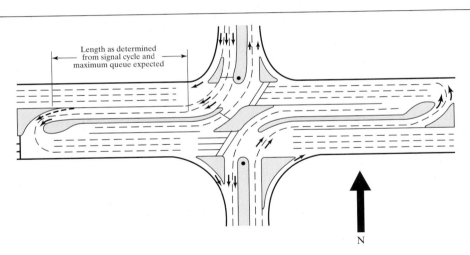

Figure 19.4: Channelization of a Complex Intersection (Used with permission of Institute of Transportation Engineers, R.P. Kramer, "New Combinations of Old Techniques to Rejuvenate Jammed Suburban Arterials," *Strategies to Alleviate Traffic Congestion*, Washington DC, 1988.)

In this case, a major arterial is fed by two major generators, perhaps two large shopping centers, on opposite sides of the roadway. Through movements across the arterial are prevented by the channelization scheme, as are left turns from either generator onto the arterial. The channelization allows only the following movements to take place:

- Through movements on the arterial
- Right-turn movements into either generator
- Left-turn movements into either generator
- Right-turn movements onto the arterial

Double-left-turn lanes on the arterial are provided for storage and processing of left turns entering either generator. A wide median is used to nest a double U-turn lane next to the left-turn lanes. These U-turn lanes allow vehicles to exit either generator and accomplish either a left-turning movement onto the arterial or a through movement into the opposite generator. In this case, it is highly likely that the main intersection and the U-turn locations would be signalized. However, all movements at this complex location could be handled with two-phase signalization, because the channelization design limits the signal to the control of two conflicting movements at

each of the three locations. The distance between the main intersection and the U-turn locations must consider the queuing characteristics in the segments between intersections to avoid spillback and related demand starvation issues.

From these examples, it is seen that channelization of intersections can be a powerful tool to improve both the safety and efficiency of intersection operation.

19.3.3 Channelizing Right Turns

When space is available, it is virtually always desirable to provide a channelized path for right-turning vehicles. This is especially true at signalized intersections where such channelization accomplishes two major benefits:

- Where "right-turn on red" regulations are in effect, channelized right turns minimize the probability of a right-turning vehicle or vehicles being stuck behind a through vehicle in a shared lane.
- Where channelized, right turns can effectively be removed from the signalization design, as they would, in most cases, be controlled by a YIELD sign and would be permitted to move continuously.

The accomplishment of these benefits, however, depends upon some of the details of the channelization design.

Figure 19.5 shows three different schemes for providing channelized right turns at an intersection. In Figure 19.5 (a), a simple channelizing triangle is provided. This design has limited benefits for two reasons: (1) through vehicles in the right lane may queue during the "red" signal phase, blocking access to the channelized right-turn lane and (2) high right-turn volumes may limit the utility of the right-hand lane to through vehicles during "green" phases.

In the second design (shown in Figure 19.5 (b)), acceleration and deceleration lanes are added for the channelized right turn. If the lengths of the acceleration and deceleration lanes are sufficient, this design can avoid the problem of queues blocking access to the channelized right turn.

In the third design (Figure 19.5 (c)), a very heavy right-turn movement can run continuously. A lane drop on the approach leg and a lane addition to the departure leg provide a continuous lane and an unopposed path for right-turning vehicles. This design requires unique

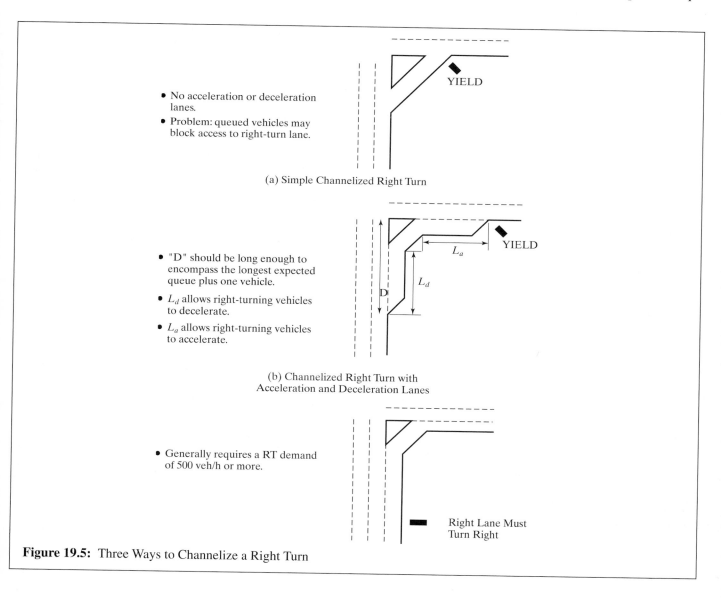

Figure 19.5: Three Ways to Channelize a Right Turn

situations in which the lane drop and lane addition are appropriate for the arterials involved. To be effective, the lane addition on the departure leg cannot be removed too close to the intersection. It should be carried for at least several thousand feet before it is dropped, if necessary.

Right-turn channelization can simplify intersection operations, particularly where the movement is significant. It can also make signalization more efficient, as channelized right turns, controlled by a YIELD sign, do not require green time to be served.

19.4 Special Situations at Intersections

This section deals with four unique intersection situations that require attention: (1) intersections with junction angles less than 60° or more than 120°, (2) T-intersections, (3) offset intersections, and (4) special treatments for heavy left-turn movements.

19.4.1 Intersections at Skewed Angles

Intersections, both signalized and unsignalized, work best when the angle of the intersection is 90°. Sight distances are easier to define, and drivers tend to expect intersections at right angles. Nevertheless, there are many situations in which the intersection angle is not 90°. Such angles may present special challenges to the traffic engineer, particularly when they are less than 60° or

more than 120°. These occur relatively infrequently. Drivers are generally less familiar with their special characteristics, particularly vis-à-vis sight lines and distances.

Skewed-angled intersections are particularly hazardous when uncontrolled and combined with high intersection-approach speeds. Such cases generally occur in rural areas and involve primary state and/or county routes. The situation illustrated in Figure 19.6 provides an example.

The example is a rural junction of two-lane, high-speed arterials, Routes 160 and 190. Given relatively gentle terrain, low volumes, and the rural setting, speed limits of 50 mi/h are in effect on both facilities. Figure 19.6 also illustrates the two movements representing a hazard. The conflict between the WB movement on Route 160 and the EB movement on Route 190 is a significant safety hazard. At the junction shown, both roadways have similar designs. Thus, there is no visual cue to the driver indicating which route has precedence or right-of-way. Given that signalization is rarely justifiable in low-volume rural settings, other means must be considered to improve the safety of operations at the intersection.

The most direct means of improving the situation is to change the alignment of the intersection, making it clear which of the routes has the right-of-way. Figure 19.7 illustrates the two possible realignments. In the first case, Route 190 is given clear preference; vehicles arriving or departing on the east leg of Route 160 must go through a 90° intersection to complete their maneuver. In the second case, Route 160 is dominant, and

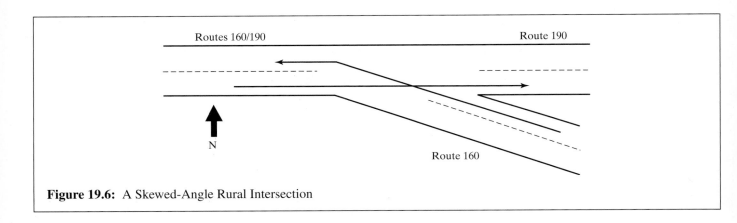

Figure 19.6: A Skewed-Angle Rural Intersection

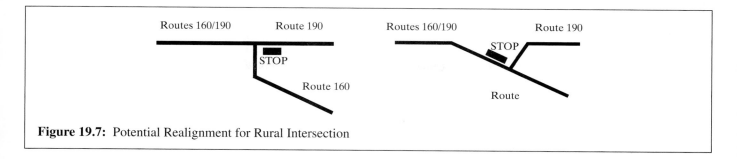

Figure 19.7: Potential Realignment for Rural Intersection

those arriving or departing on the east leg of Route 190 go through the 90° intersection. In either case, the 90° intersection would be controlled using a STOP sign to clearly designate right-of-way.

While basic realignment is the best solution for high-speed odd-angle intersections, it requires that right-of-way be available to implement the change. Even in a rural setting, sufficient right-of-way to realign the intersection may not always be available. Other solutions can also be considered. Channelization can be used to better define the intersection movements, and control devices can be used to designate right-of-way. Figure 19.8 shows another potential design that requires less right-of-way than full realignment.

In this case, only the WB movement on Route 106 was realigned. While this would still require some right-of-way, the amount needed is substantially less than for full realignment. Additional channelization is provided to separate EB movements on Routes 106 and 109. In

addition to the regulatory signs indicated in Figure 19.8, warning and directional guide signs would be placed on all approaches to the intersection. In this solution, the WB left turn from Route 109 must be prohibited; an alternative route would have to be provided and appropriate guide signs designed and placed.

The junction illustrated is, in essence, a three-leg intersection. Skewed-angle four-leg intersections also occur in rural, suburban, and urban settings and present similar problems. Again, total realignment of such intersections is the most desirable solution. Figure 19.9 shows an intersection and the potential realignments that would eliminate the odd-angle junction. Where a four-leg intersection is involved, however, the realignment solution creates two separate intersections.

Depending upon volumes and the general traffic environment of the intersection, the realignments proposed in Figure 19.9 could result in signalized or unsignalized intersections.

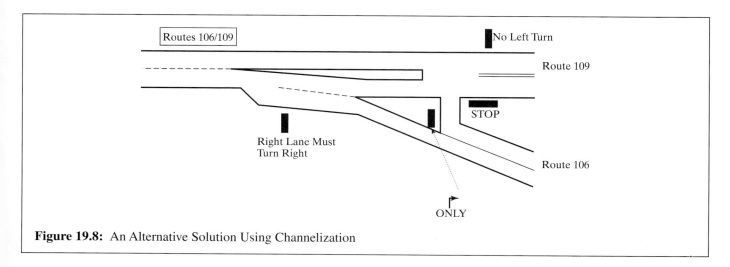

Figure 19.8: An Alternative Solution Using Channelization

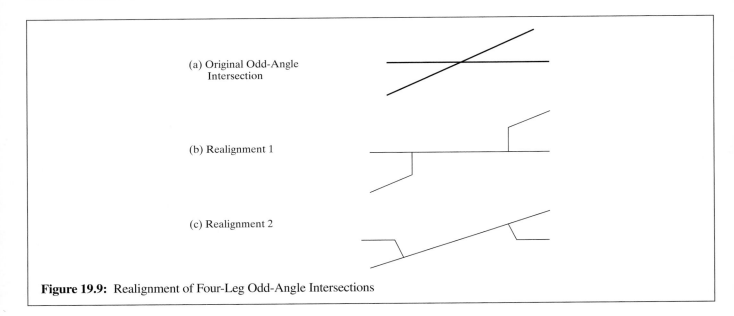

(a) Original Odd-Angle
Intersection

(b) Realignment 1

(c) Realignment 2

Figure 19.9: Realignment of Four-Leg Odd-Angle Intersections

In urban and suburban settings, where right-of-way is a significant impediment to realigning intersection, signalization of the odd-angle intersection can be combined with channelization to achieve safe and efficient operations. Channelized right turns would be provided for acute-angle turns, and left-turn lanes (and signalization) would be provided as needed.

In extreme cases, where volumes and approach speeds present hazards that cannot be ameliorated through normal traffic engineering measures, consideration may be given to providing a full or partial interchange with the two main roadways grade-separated. Providing grade-separation would also involve some expansion of the traveled way, and overpasses in some suburban and urban surroundings may involve visual pollution and/or other negative environmental impacts.

19.4.2 T-Intersections: Opportunities for Creativity

In many ways, T-intersections are far simpler than traditional four-leg intersections. The typical four-leg intersection contains twelve vehicular movements and four crossing pedestrian movements. At a T-intersection, only six vehicular movements exist and there are only three crossing pedestrian movements. These are illustrated in Figure 19.10.

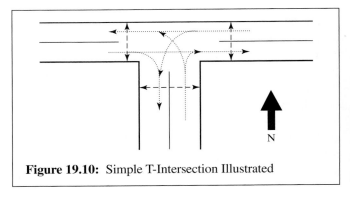

Figure 19.10: Simple T-Intersection Illustrated

Note that in the set of T-intersection vehicular movements, there is only one opposed left turn—the WB left-turn movement in this case. Because of this, conflicts are easier to manage, and signalization, when necessary, is easier to address.

Control options include all generally applicable alternatives for intersection control:

- Uncontrolled (warning and guide signs only)
- STOP or YIELD control
- Signal control

The intersection shown in Figure 19.10 has one lane for each approach. There are no channelized movements or left-turn lanes. If visibility is not appropriate for uncontrolled operation under basic rules of the road, then the options of STOP/YIELD control or signalization must be considered. The normal warrants would apply.

The T-intersection form, however, presents some relatively unique characteristics that influence how control is applied. With the issuance of the Millennium Edition of the MUTCD [2], the use of the YIELD sign for primary intersection control will decline. STOP-control is usually applied to the stem of the T-intersection, although it is possible to apply 2-way STOP control to the cross street if movements into and out of the stem dominate.

If needed, the form of signalization applied to the intersection of Figure 19.10 depends entirely on the need to protect the (WB) opposed left turn. A protected phase is normally suggested if the left-turn volume exceeds 200 veh/h or the cross-product of the left-turn volume and the opposing volume per lane exceeds 50,000. If left-turn protection is not needed, a simple two-phase signal plan is used. If the opposed left-turn must be protected and there is no left-turn lane available (as in Figure 19.10), a three-phase plan must be used. Figure 19.11 illustrates the possible signal plans for the T-intersection of Figure 19.10. The three-phase plan is relatively inefficient, because a separate phase is needed for each of the three approaches.

Where a protected left-turn phase is desirable, the addition of an exclusive left-turn lane would simplify the signalization. Channelization and some additional right-of-way would be required to do this. Channelization can also be applied in other ways to simplify the overall operation and control of the intersection. Channelizing islands can be used to create separated right-turn paths for vehicles entering and leaving the stem via right turns. Such movements would be YIELD-controlled, regardless of the primary form of intersection control.

Figure 19.12 shows a T-intersection in which a left-turn lane is provided for the opposed left turn. Right turns are also channelized. Assuming that a signal with a protected left turn is needed at this location, the signal plan shown could be implemented. This plan is far more efficient than that of Figure 19.11, as EB and WB through flows can move simultaneously. Right turns move more or less continuously through the YIELD-controlled channelized turning roadways. The potential for queues to block access to the right-turn roadways, however, should be considered in timing the signal.

Right turns can be completely eliminated from the signal plan if volumes are sufficient to allow lane drops or additions for the right-turning movements, as illustrated in Figure 19.13. Right turns into and out of the stem of the T-intersection become continuous movements.

19.4.3 Offset Intersections

One of the traffic engineer's most difficult problems is the safe operation of high-volume offset intersections. Figure 19.14 illustrates such an intersection with a modest right offset. In the case illustrated, the driver needs more sight distance (when compared with a perfectly-aligned 90° intersection) to observe vehicles approaching from the right. The obstruction caused by the building becomes a more serious problem because of this. In addition to sight distance problems, the offset intersection distorts the normal trajectory of all movements, creating accident risks that do not exist at aligned intersections.

Offset intersections are rarely consciously designed. They are necessitated by a variety of situations, generally involving long-standing historic development patterns. Figure 19.15 illustrates a relatively common situation in which offset intersections occur.

In many older urban or suburban developments, zoning and other regulations were (and in some cases, still are) not particularly stringent. Additional development was considered to be an economic benefit because it added to the property tax base of the community involved.

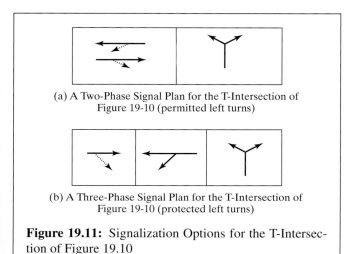

(a) A Two-Phase Signal Plan for the T-Intersection of Figure 19-10 (permitted left turns)

(b) A Three-Phase Signal Plan for the T-Intersection of Figure 19-10 (protected left turns)

Figure 19.11: Signalization Options for the T-Intersection of Figure 19.10

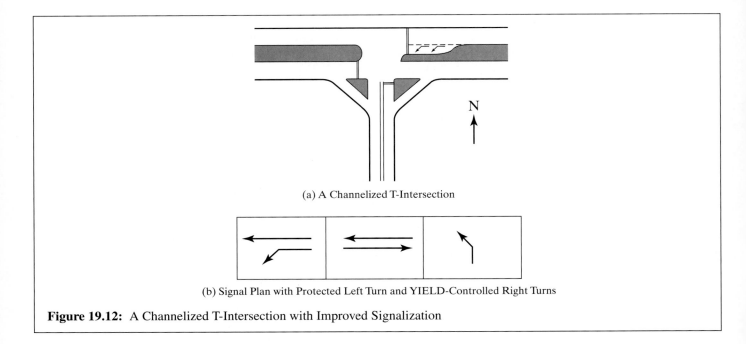

(a) A Channelized T-Intersection

(b) Signal Plan with Protected Left Turn and YIELD-Controlled Right Turns

Figure 19.12: A Channelized T-Intersection with Improved Signalization

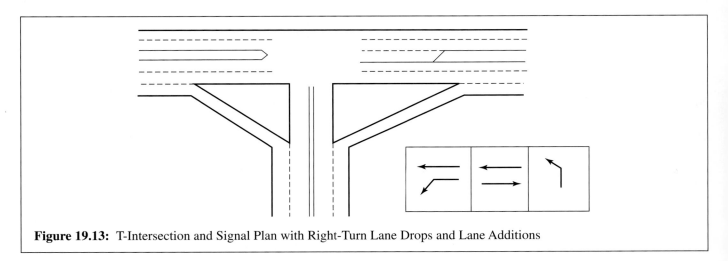

Figure 19.13: T-Intersection and Signal Plan with Right-Turn Lane Drops and Lane Additions

Firm control over the specific design of subdivision developments, therefore, is not always exercised by zoning boards and authorities.

The situation depicted in Figure 19.15 occurs when Developer A obtains the land to the south of a major arterial and lays out a circulation system that will maximize the number of building lots that can be accommodated on the parcel. At a later time, Developer B obtains the rights to land north of the same arterial.

Again, an internal layout that provides the maximum number of development parcels is selected. Without a strong planning board or other oversight group requiring it, there is no guarantee that opposing local streets will "line up." Offsets can and do occur frequently in such circumstances. In urban and suburban environments, it is rarely possible to acquire sufficient right-of-way to realign the intersections; therefore, other approaches to control and operation of such intersections must be considered.

Figure 19.14: Offset Intersection with Sight Distance and Trajectory Problems

There are two major operational problems posed by a right-offset intersection, as illustrated in Figure 19.16.

In Figure 19.16 (a), the left-turn trajectories from the offset legs involve a high level of hazard. Unlike the situation with an aligned intersection, a vehicle turning left from either offset leg is in conflict with the opposing through vehicle almost immediately after crossing the STOP line. To avoid this conflict, left-turning vehicles must bear right as if they were going to go through to the opposite leg, beginning their left turns only when they are approximately halfway through the intersection. This, of course, is not a natural movement, and a high incidence of left-turn accidents often result at such intersections.

In Figure 19.16 (b), the hazard to pedestrians crossing the aligned roadway is highlighted. Two paths are possible, and both are reasonably intuitive for pedestrians:

They can cross from corner to corner, following an angled crossing path, or they can cross perpendicularly. The latter places one end of their crossing away from the street corner. Perpendicular crossings, however, minimize the crossing time and distance. On the other hand, right-turning vehicles encounter the pedestrian conflict at an unexpected location, after they have virtually completed their right turn. Diagonal crossings increase the exposure of pedestrians, but conflicts with right-turning vehicles are closer to the normal location.

Yet another special hazard at offset intersections, not clearly illustrated by Figure 19.16, is the heightened risk of sideswipe accidents as vehicles cross between the offset legs. Since the required angular path is not necessarily obvious, more vehicles will stray from their lane during the crossing.

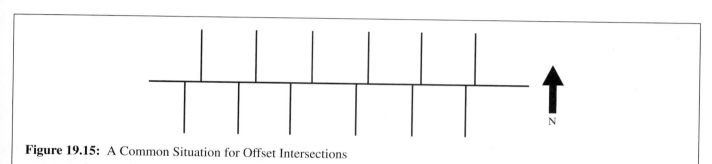

Figure 19.15: A Common Situation for Offset Intersections

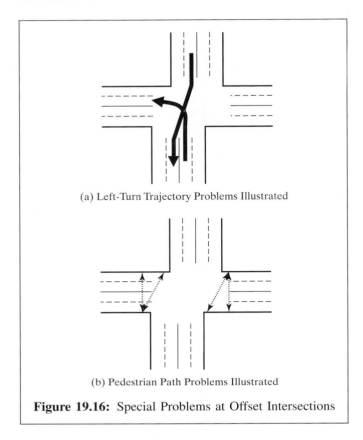

(a) Left-Turn Trajectory Problems Illustrated

(b) Pedestrian Path Problems Illustrated

Figure 19.16: Special Problems at Offset Intersections

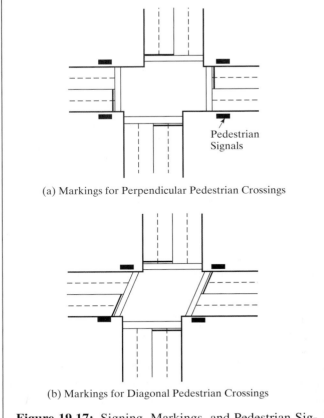

(a) Markings for Perpendicular Pedestrian Crossings

(b) Markings for Diagonal Pedestrian Crossings

Figure 19.17: Signing, Markings, and Pedestrian Signals for a Right-Offset Intersection

There are, however, remedies that will minimize these additional hazards. Where the intersection is signalized, the left-turn conflict can be eliminated through the use of a fully protected left-turn phase in the direction of the offset. In this case, the left-turning vehicles will not be entering the intersection area at the same time as the opposing through vehicles. This requires, however, that one of the existing lanes be designated an exclusive turning lane, or that a left-turn lane can be added to each offset leg. If this is not possible, a more extreme remedy is to provide each of the offset legs with an exclusive signal phase. While this separates the left-turning vehicles from the opposing flows, it is an inefficient signal plan and can lead to four-phase signalization if left-turn phases are needed on the aligned arterial.

For pedestrian safety, it is absolutely necessary that the traffic engineer clearly designate the intended path they are to take. This is done through proper use of markings, signs, and pedestrian signals, as shown in Figure 19.17.

Crosswalk locations influence the location of STOP-lines and the position of pedestrian signals, which must be located in the line of sight (which is the walking path) of pedestrians. Vehicular signal timing is also influenced by the crossing paths implemented. Where perpendicular crossings are used, the distance between STOP-lines on the aligned street can be considerably longer than for diagonal crossings. This increases the length of the all-red interval for the aligned street and adds lost time to the signal cycle.

In extreme cases, where enforcement of perpendicular crossings becomes difficult, barriers can be placed at normal street corner locations, preventing pedestrians from entering the street at an inappropriate or unintended location.

To help vehicles follow appropriate paths through the offset intersection, dashed lane and centerline markings through the intersection may be added, as illustrated

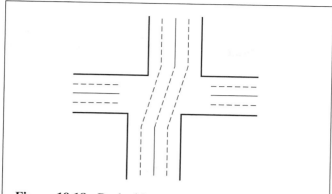

Figure 19.18: Dashed Lane and Centerline Through an Offset Intersection

in Figure 19.18. The extended centerline marking would be yellow, and the lane lines would be white.

Left-offset intersections share some of the same problems as right-offset intersections. The left-turn interaction with the opposing through flow is not as critical, however. The pedestrian–right-turn interaction is different, but potentially just as serious. Figure 19.19 illustrates.

The left-turn trajectory through the offset intersection is still quite different from an aligned intersection, but the left-turn movement does not thrust the vehicle immediately into the path of the oncoming through movement, as in a right-offset intersection. Sideswipe accidents are still a risk, and extended lane markings would be used to minimize this risk.

At a left-offset intersection, the diagonal pedestrian path is more difficult, as it brings the pedestrian into immediate conflict with right-turning vehicles more quickly than at an aligned intersection. For this reason,

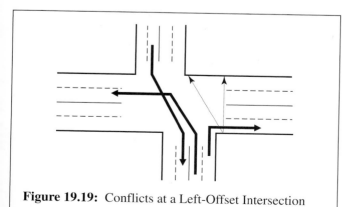

Figure 19.19: Conflicts at a Left-Offset Intersection

diagonal crossings are generally not recommended at left-offset intersections. The signing, marking, and signalization of perpendicular pedestrian crossings is similar to that used at a right-offset intersection.

When at all possible, offset intersections should be avoided. If sufficient right-of-way is available, basic realignment should be seriously considered. When confronted with such a situation however, the traffic engineering approaches discussed here can ameliorate some of the fundamental concerns associated with offset alignments. The traffic engineer should recognize that many of these measures will negatively affect capacity of the approaches due to the additional signal phases and longer lost times often involved. This is, however, a necessary price paid to optimize safety of intersection operation.

19.4.4 Special Treatments for Heavy Left-Turn Movements

Some of the most difficult intersection problems to solve involve heavy left-turn movements on major arterials. Accommodating such turns usually requires the addition of protected left-turn phasing, which often reduces the effective capacity to handle through movements. In some cases, adding an exclusive left-turn phase or phases is not practical, given the associated losses in through capacity.

Alternative treatments must be sought to handle such left-turn movements, with the objective of maintaining two-phase signalization at the intersection. Several design and control treatments are possible, including:

- Prohibition of left turns
- Provision of jug-handles
- Provision of at-grade loops and diamond ramps
- Provision of a continuous-flow intersection
- Provision of U-turn treatments

Prohibition of left turns is rarely a practical option for a heavy left-turn demand. Alternative paths would be needed to accommodate the demand for this movement, and diversion of a heavy flow onto an "around-the-block" or similar path often creates problems elsewhere.

Figure 19.20 illustrates the use of jug-handles for handling left turns. In effect, left-turners enter a surface ramp on the right, executing a left turn onto the cross street. The jug-handle may also handle right-turn movements. The design creates two new intersections. Depending upon volumes, these may require signalization

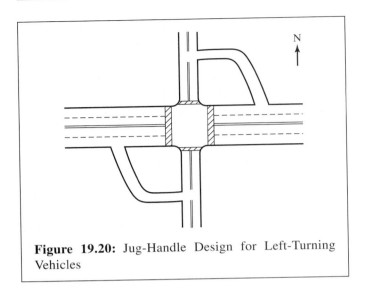

Figure 19.20: Jug-Handle Design for Left-Turning Vehicles

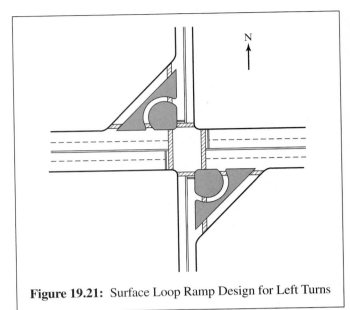

Figure 19.21: Surface Loop Ramp Design for Left Turns

or could be controlled with STOP signs. In either case, queuing between the main intersection and the two new intersections is a critical issue. Queues should not block egress from either of the jug-handle lanes. The provision of jug-handles also requires that there be sufficient right-of-way available to accommodate solution. In some extreme cases, existing local streets may be used to form a jug-handle pattern.

Figure 19.21 illustrates the use of surface loop ramps to handle heavy left-turning movements at an arterial intersection. These are generally combined with surface diamond ramps to handle right turns from the cross street, thus avoiding the conflict between normal right turns and the loop ramp movements on the arterial. Once again, queuing could become a problem if left-turning vehicles back up along the loop ramp far enough to affect the flow of vehicles that can enter the loop ramp. This option also consumes considerable right-of-way and may be difficult to implement in high-density environments.

Figure 19.22 illustrates a continuous-flow intersection, a relatively novel design approach developed during

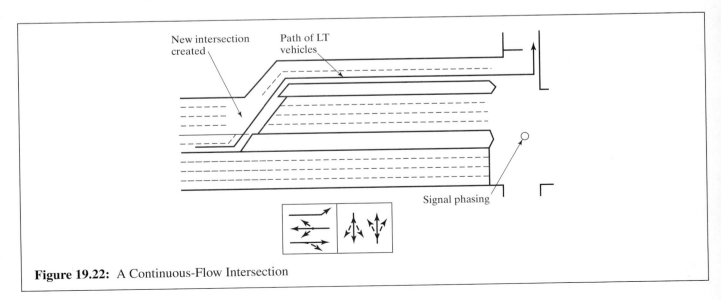

Figure 19.22: A Continuous-Flow Intersection

the late 1980s and early 1990s. The continuous-flow intersection [6] takes a single intersection with complex multiphase signalization and separates it into two intersections, each of which can be operated with a two-phase signal and coordinated. At the new intersection, located upstream of the left-turn location, left-turning vehicles are essentially transferred to a separate roadway on the left side of the arterial. At the main intersection, the left turns can then be made without a protected phase, regardless of the demand level. The design requires sufficient right-of-way on one side of the arterial to create the new left-turn roadway and a median that is wide enough to provide one or two left-turning lanes at the new intersection. Queuing from the main

intersection can become a problem if left-turning vehicles are blocked from entering the left-turn lane(s) at the new upstream intersection.

While a few continuous-flow intersections have been built, they have not seen the widespread use that was originally anticipated. In most cases, right-of-way restrictions will make this solution somewhat impractical.

As a last resort, left turns may be handled in a variety of ways as U-turn movements. Figure 19.23 illustrates four potential designs for doing this. In Figure 19.23 (a), left-turning vehicles go through the intersection and make a U-turn through a wide median downstream. The distance between the U-turn location and the main intersection must be sufficient to avoid blockage by queued

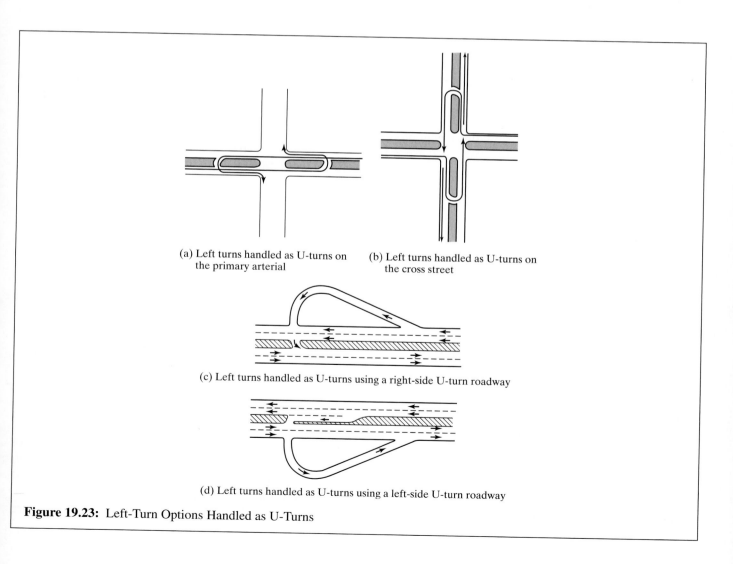

(a) Left turns handled as U-turns on the primary arterial

(b) Left turns handled as U-turns on the cross street

(c) Left turns handled as U-turns using a right-side U-turn roadway

(d) Left turns handled as U-turns using a left-side U-turn roadway

Figure 19.23: Left-Turn Options Handled as U-Turns

vehicles and must provide sufficient distance for drivers to execute the required number of lane changes to get from the median lane to right lane. In Figure 19.23 (b), left-turning vehicles turn right at the main intersection, then execute a U-turn on the cross street. Queuing and lane-changing requirements are similar to those described for Figure 19.23 (a). Where medians are narrow, the U-turn paths of (a) and (b) cannot be provided. Figures 19.23 (c) and (d) utilize U-turn roadways built to the right and left sides of the arterial (respectively) to accommodate left-turning movements. These options require additional right-of-way.

The safe and efficient accommodation of heavy left-turn movements on arterials often requires creative approaches that combine both design and control elements. The examples shown here are intended to be illustrations, not a complete review of all possible alternatives.

19.5 Street Hardware for Signalized Intersections

In Chapter 4, the basic requirements for display of signal faces at a signalized intersection were discussed in detail. The key specifications are:

- A minimum of two signal faces should be visible to each primary movement in the intersection.

- All signal faces should be placed within a horizontal 20° angle around the centerline of the intersection approach (including exclusive left- and/or right-turn lanes).

- All signal faces should be placed at mounting heights in conformance with MUTCD standards, as presented in Figure 4.20 of Chapter 4.

The proper location of signal heads is a key element of intersection design and is critical to maximizing observance of traffic signals.

There are three general types of signal-head mountings that can be used alone or in combination to achieve the appropriate location of signal heads: post mounting, mast-arm mounting, and span-wire mounting.

Figure 19.24 illustrates post mounting. The signal head can be oriented either vertically or horizontally, as

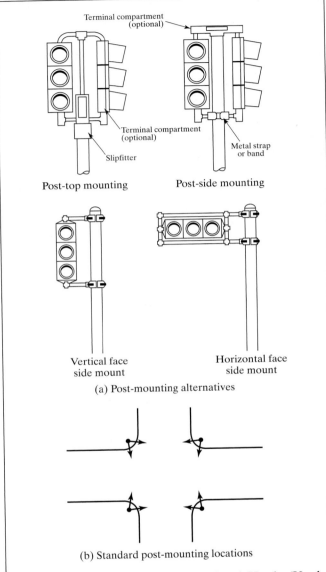

Figure 19.24: Post-Mounting of Signal Heads (Used with permission of Prentice Hall, Inc., Kell, J. and Fullerton, I., *Manual of Traffic Signal Design*, 2nd Edition, 1991, pg. 44.)

shown. Post-mounted signals are located on each street corner. A post-mounted signal head generally has two faces, oriented such that a driver sees two faces located on each of the far intersection corners. Because they are located on street corners, care must be taken to ensure that post-mounted signals fall within the required 20° angle of the approach centerline. Post-mounted signals are often inappropriate for use at intersections with narrow streets, as street corners in such circumstance lie outside of the visibility requirement.

Figure 19.25 illustrates mast-arm mounting of signal heads. Typically, the mast arm is perpendicular to the intersection approach. They are located so that drivers are looking at a signal face or faces on the far side of the intersection. Mast arms can be long enough to accommodate

two signal heads, but they are rarely used for more than two signal heads.

Figure 19.26 shows two typical mast-arm signal installations. The first (a) shows mast-arm signals at a four-leg intersection, with the mast-arm oriented perpendicular to the direction of traffic. Note that the mast-arm signal heads are supplemented by a post-mounted signal in the gore of the four-leg intersection. The second (b) represents a very efficient scheme for mounting signal heads at a simple intersection of two two-lane streets. Two mast arms are

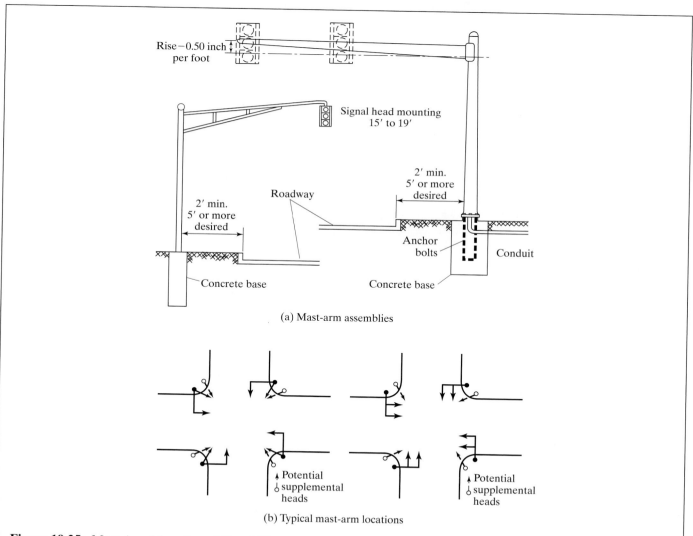

(a) Mast-arm assemblies

(b) Typical mast-arm locations

Figure 19.25: Mast-Arm Mounting of Signal Heads (Used with permission of Prentice Hall, Inc., Kell, J. and Fullerton, I., *Manual of Traffic Signal Design*, 2nd Edition, 1991, pg. 57.)

(a) Mast-Arm Mounted Signals at a Four-Leg Intersection

(b) Mast-Arm Mounted Signals at the Intersection
of Two-Lane Streets

Figure 19.26: Two Examples of Mast-Arm Mounted Signals

used, each extending diagonally across the intersection. Only two signal heads are used, each with a full four faces. In this way, using only two signal heads, all movements have two signal faces displaying the same signal interval.

In the case of both post-mounted and mast-arm–mounted signal heads, power lines are carried to the signal head within the hollow structure of the post or mast arm.

The most common method for mounting signal heads is span wire, as it is the most flexible and can be used in a variety of configurations. Figure 19.27 shows four basic configurations in which span wires can be used.

The first is a single diagonal span wire between two intersection corners. The span wire allows the installation of a number of signal heads, each having between one and three faces, depending upon the exact location. Such installations are generally supplemented by post-mounted signals on the two other intersection corners. The second installation illustrated is a "box" design. Four span wires are installed across each intersection leg. Signal heads are oriented much in the same way as with mast arms. Most signal heads have a single face and are visible from the far side of the intersection. The third example is a "modified box," in which the box is suspended over the middle of the

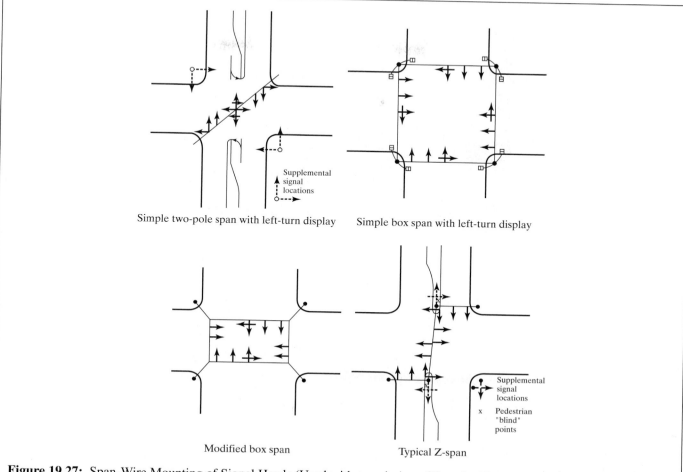

Simple two-pole span with left-turn display   Simple box span with left-turn display

Supplemental
signal
locations

Modified box span   Typical Z-span

Supplemental
signal
locations

x   Pedestrian
"blind"
points

Figure 19.27: Span-Wire Mounting of Signal Heads (Used with permission of Prentice Hall, Inc., Kell, J. and Fullerton, I., *Manual of Traffic Signal Design*, 2nd Edition, 1991, pgs. 51–53.)

intersection. This is done to accomplish signal-face locations that are more visible and more clearly aligned with specific lanes of each intersection approach. The final example of Figure 19.27 is a "lazy Z" pattern in which the primary span wire is anchored on opposing medians. This latter design is possible only where opposite medians exist.

Span wire allows the traffic engineer to place signal faces in almost any desired position and is often used at complex intersections where a signal face for each entering lane is desired.

Figure 19.28 illustrates how signal heads are anchored on span wires. In general, the main cable supports each signal head from above. Signal heads so mounted

can and do sway in the wind. Where wind is excessive or where the exact orientation of the signal face is important, a tether wire may be attached to the bottom of the signal head for restraint. This is most important where Polaroid signal lenses are used. These lenses are visible only when viewed from a designated angle. They are often used at closely-spaced signalized intersections, where the traffic engineer uses them to prevent drivers from reacting to the next downstream signal.

Figure 19.29 illustrates how power is supplied to a span-wire mounted signal head. A shielded power cable is wrapped around the primary support wire and connected to each signal head.

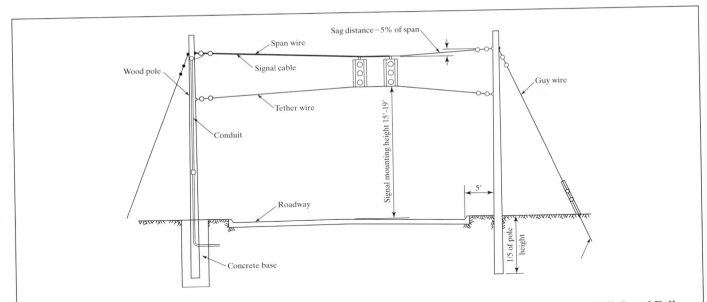

Figure 19.28: Use of Span Wire and Tether Wire Illustrated (Used with permission of Prentice Hall, Inc., Kell, J. and Fullerton, I., *Manual of Traffic Signal Design*, 2nd Edition, 1991.)

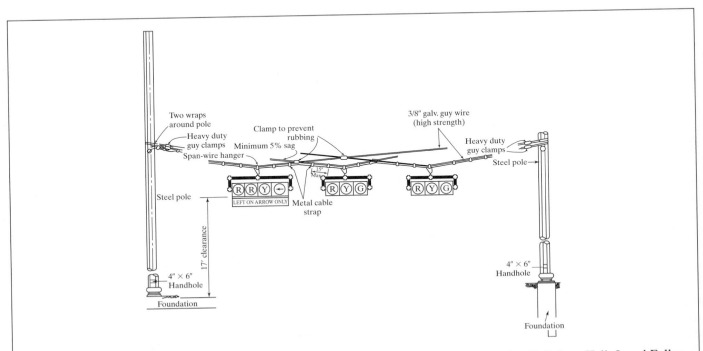

Figure 19.29: Providing Power to Span-Mounted Signals (Used with permission of Prentice Hall, Inc., Kell, J. and Fullerton, I., *Manual of Traffic Signal Design*, 2nd Edition, 1991.)

Figure 19.30: A Typical Span-Wire Signal Installation

Figure 19.30 shows a typical field installation of span-wire mounted signals. In this case, a single span wire supports six signal heads that are sufficient to control all movements, including a left-turn phase on the major street.

Using the three signal mounting options (post mounted, mast-arm mounted, span-wire mounted), either alone or in combination, the traffic engineer can satisfy all of the posting requirements of the MUTCD and present drivers with clear and unambiguous operating instructions. Achieving this goal is critical to ensuring safe and efficient operations at signalized intersections.

19.6 Closing Comments

This chapter has provided an overview of several important elements of intersections design. It is not intended to be exhaustive, and the reader is encouraged to consult standard references for additional relevant topics and detail.

References

1. *A Policy on Geometric Design of Highways and Streets*, 4th Edition, American Association of State Highway and Transportation Officials, Washington DC, 2001.

2. *Manual of Uniform Traffic Control Devices*, Millennium Edition, Federal Highway Administration, U.S. Department of Transportation, 2000.

3. Kell, J. and Fullerton, I., *Manual of Traffic Signal Design*, 2nd Edition, Institute of Transportation Engineers, Washington DC, 1991.

4. *Traffic Detector Handbook*, JHK & Associates, Institute of Transportation Engineers, Washington DC, nd.

5. *Highway Capacity Manual*, 4th Edition, Transportation Research Board, Washington DC, 2000.

6. Hutchinson, T., "The Continuous Flow Intersection: The Greatest New Development Since the Traffic

Signal?", *Traffic Engineering and Control*, Vol 36, No 3, Printhall Ltd., London, England, 1995.

Problems

19-1–19-2. Each of the sets of demands shown in Figures 19.31 and 19.32 represent the forecast flows (already adjusted for PHF) expected at new intersections that are created as a result of large new developments. Assume that each intersection will be signalized. In each case, propose a design for the intersection, including a detailing of where and how signal heads would be located.

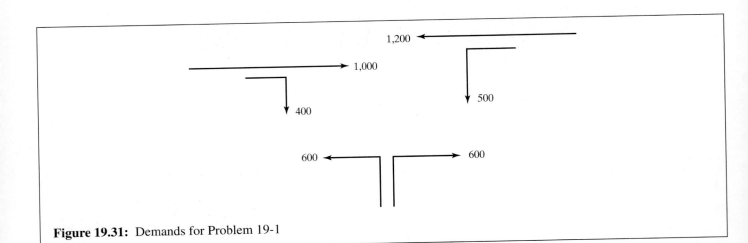

Figure 19.31: Demands for Problem 19-1

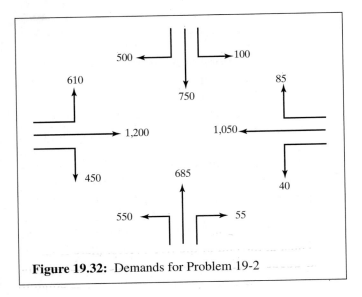

Figure 19.32: Demands for Problem 19-2

CHAPTER
20

Actuated Signal Control and Detection

When pretimed signal controllers are employed, the phase sequence, cycle length, and all interval times are uniform and constant from cycle to cycle. At best, controllers will provide for several predetermined time periods during which different pretimings may be applied. During any one period, however, each signal cycle is an exact replica of every other signal cycle.

Actuated control uses information on current demands and operations, obtained from detectors within the intersection, to alter one or more aspects of the signal timing on a cycle-by-cycle basis. Actuated controllers may be programmed to accommodate:

- Variable phase sequences (e.g., optional protected LT phases)
- Variable green times for each phase
- Variable cycle length, caused by variable green times

Such variability allows the signal to allocate green time based on current demands and operations. Pretimed signals are timed to accommodate average demand flows during a peak 15-min demand period. Even within that period, however, demands vary on a cycle-by-cycle basis. Thus, it is, at least conceptually, more efficient to have signal timing vary in the same way.

Consider the situation illustrated in Figure 20.1. Five consecutive cycles are shown, including the capacity and demand during each. Note that over the five cycles shown, the signal has the capacity to discharge 50 veh and that total demand during the five cycles is also 50 veh. Thus, over the five cycles shown, total demand is equal to total capacity.

Actual operations over the five cycles, however, result in a queue of unserved vehicles with pretimed operation. In the first cycle, 10 vehs arrive and 10 vehs are discharged. In the second, six vehs arrive and six are discharged. In the third cycle, eight vehs arrive and eight are discharged. Note that from the second and third cycles, there is unused capacity for an additional six vehicles. In cycle 4, 12 vehs arrive and only 10 are discharged, leaving a queue of two unserved vehicles. In cycle 5, 14 vehs arrive and only 10 are discharged, leaving an additional four unserved vehicles. Thus, at the end of the five cycles, there is an unserved queue of six vehicles. This occurs despite the fact that over the entire period, the demand is equal to the capacity.

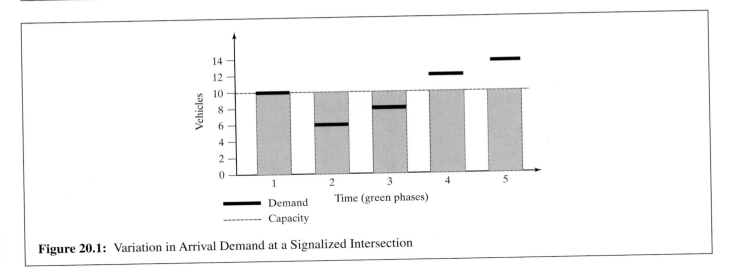

Figure 20.1: Variation in Arrival Demand at a Signalized Intersection

The difficulty with pretimed operation is that the unused capacity of six vehicles in cycles 2 and 3 may not be used by excess vehicles arriving in cycles 4 and 5. If the signal had been a properly timed actuated signal, the green in cycles 2 and 3 could have been terminated when no demand was present and additional green time could have been added to cycles 4 and 5 to accommodate a higher number of vehicles. The ability of the signal timing to respond to short-term variations in arrival demand makes the overall signal operation more efficient. Even if the total amount of green time allocated over the five cycles illustrated did not change, the ability to "save" unused green time from cycles 2 and 3 to increase green time in cycles 4 and 5 would significantly reduce delay and avoid or reduce a residual queue of unserved vehicles at the end of the five-cycle period.

Another major benefit of actuated signal timing is that a single programmed timing pattern can flex to handle varying demand periods throughout the day, including peak and off-peak periods and changes in the balance of movements.

If the advantages of allowing signal timing to vary on a cycle-by-cycle basis are significant, why aren't all signalized intersections actuated? The principal issue is coordination of signal systems. To effectively coordinate a network of signals to provide for progressive movement of vehicles through the system, all signals must operate on a fixed and equal cycle length. Thus, where signals must be interconnected for progressive movement, the cycle length cannot be permitted to vary on a cycle-by-cycle basis. This removes the principal benefit of actuated control in such circumstances, the ability to vary the cycle length.

For the most part, actuated signal control is used at isolated signalized intersections, usually a minimum of 2.0 miles from the nearest adjacent signal. There are some exceptions to this, including semi-actuated controllers (see Section 20.1), which can be included in coordinated signal networks.

20.1 Types of Actuated Control

There are three basic types of actuated control, each using signal controllers that are somewhat different in their design:

1. *Semi-actuated control.* This form of control is used where a small side street intersects with a major arterial or collector. This type of control should be considered whenever Warrant 1B is the principal reason justifying signalization. Semi-actuated signals are always two-phase, with all turns being made on a permitted basis. Detectors are placed only on the side street. The green is on the major street at all times unless a "call" on the side street is noted. The number and duration of side-street greens is

limited by the signal timing and can be restricted to times that do not interfere with progressive signal-timing patterns along the collector or arterial.

2. *Full actuated control.* In full actuated operation, all lanes of all approaches are monitored by detectors. The phase sequence, green allocations, and cycle length are all subject to variation. This form of control is effective for both two-phase and multiphase operations and can accommodate optional phases.

3. *Volume-density control.* Volume-density control is basically the same as full actuated control with additional demand-responsive features, which are discussed later in this chapter.

Computer-controlled signal systems do not constitute actuated control at individual intersections. In such systems, the computer plays the role of a large master controller, establishing and maintaining offsets for progression throughout a network or series of arterials. As such coordination generally requires that individual intersections operate on a common and constant cycle length, most signals within such a system are of the pre-timed variety.

20.2 Detectors and Detection

The hardware for detection of vehicles is advancing rapidly. Pressure-plate detectors, popular in the 1970s and 1980s, are rarely used in modern traffic engineering. Most detectors rely on creating or observing changes in magnetic or electromagnetic fields, which occurs when a metallic object (a vehicle) passes through such a field. The *Traffic Engineering Handbook* [1] contains a useful summary of these detectors, which include:

• *Inductive loop.* A loop assembly is installed in the pavement, usually by saw-cutting through the existing pavement. The loop is laid into the saw cut in a variety of shapes, including square, rectangle, trapezoid, or circle. The saw cut is refilled with an epoxy sealant. The loop is connected to a low-grade electrical source, creating an electromagnetic field that is disturbed whenever a

metallic object (vehicle) moves across it. This is the most common type of detector in use today. Figure 20.2 shows a loop detector installation with the epoxy-covered loop saw cut clearly visible.

• *Microloop.* This is a small cylindrical passive transducer that senses changes in the vertical component of the earth's magnetic field and converts them into electronically discernible signals. The sensor is cylindrical, about 2.5 inches in length and .75 inches in diameter. The probe is placed in a hole drilled in the roadway surface.

• *Magnetic.* These detectors measure changes in the concentration of lines of flux in the earth's magnetic field and convert such changes to an electronically discernible signal. The sensor unit contains a small coil of wire that is placed below the roadway surface.

For all of the magnetic class of detectors, one or more detectors must be used in each lane of each approach. A disadvantage is that all must be placed in or below the pavement. In areas where pavement condition is a serious issue, these detectors could become damaged or inoperable.

Another class of detector uses *sonic* or *ultrasonic* waves that can be emitted from an overhead or elevated roadside location. Such detectors rely on the echoes from reflected waves (ultrasonic) or on the Doppler principle of changes in reflected frequency when waves reflect back from a moving object (sonic). The emitted wave spreads in a cone-like shape and can, therefore, cover more than one lane with a single detector unit, depending upon its exact placement.

Figure 20.2 also illustrates an ultrasonic detector. In the case of an overhead ultrasonic detector, the unit is calibrated for the time it takes a reflected wave to return to the sensor when it is reflected from the pavement. When a vehicle intercepts the emitted wave, it is reflected in a shorter time.

There are other detector types in use, such as radar, optical, and even older pressure-plate systems. The vast majority of detectors are of the types described previously.

A rapidly emerging technology is video imaging, in which real-time video of an intersection approach or

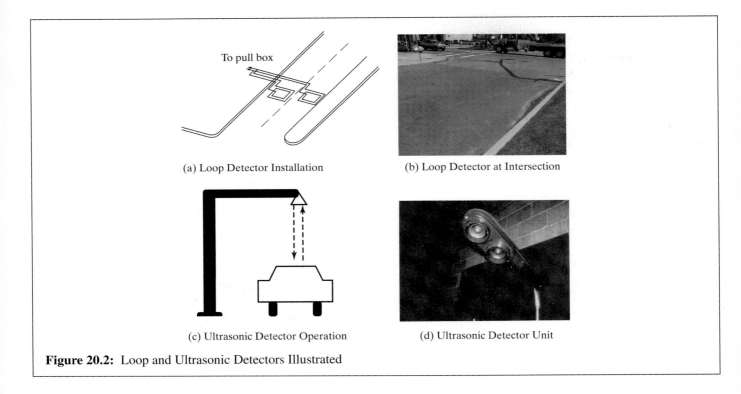

(a) Loop Detector Installation   (b) Loop Detector at Intersection

(c) Ultrasonic Detector Operation   (d) Ultrasonic Detector Unit

Figure 20.2: Loop and Ultrasonic Detectors Illustrated

other traffic location is combined with computerized pattern-recognition software. Virtual detectors are defined within the video screen, and software is programmed to note changes in pixel intensity at the virtual detector location. Now in common use for data and remote observation, such detection systems have not yet been employed to operate signals in real time.

Of greater importance than the specific detection device(s) used is the *type* of detection. There are two types of detection that influence the design and timing of actuated controllers:

- *Passage or point detection.* In this type of detection, only the fact that the detector has been "disturbed" is noted. The detector is installed at a "point," even though the detector unit itself may involve a short length.
- *Presence or area detection.* In this type of detection, a significant length (or area) of an approach lane is included in the detection zone. Entries and exits of vehicles into and out of the detection zone are "remembered." Thus, the number of vehicles

stored in the detection zone is known. Most presence detection is accomplished using a single "long-loop" detector, or using a series of loop detectors covering a significant length of an approach or approach lane.

As will be seen, the timing of an actuated signal is very much influenced by the type of detection in place. Point detection is by far the most frequently used system, as area detection involves significantly higher expense.

20.3 Actuated Control Features and Operation

Actuated signal controllers are manufactured in accordance with one of two standards. The most common is that of the National Electronic Manufacturer's Association (NEMA). NEMA standards specify all features, functions, and timing intervals, and timing software is provided as a built–in feature of the hardware (often referred to as "firmware"). The second set of standards is for the Type 170 class of controllers, originally developed for the New York State Department of Transportation. Type 170 controllers do not come with built-in software, which is generally available through third-party vendors. While NEMA software cannot be modified by an agency, Type 170 software can be modified. U.S. manufacturers of signal controllers include Control Technologies, Eagle, Econolite, Kentronics, Naztec, and others. Most manufacturers maintain current Web sites, and students are urged to consult them for the most up-to-date descriptions of hardware, software, and functions.

20.3.1   Actuated Controller Features

Regardless of the controller type, virtually all actuated controllers offer the same basic functions, although the methodology for implementing them may vary by type and manufacturer. For each actuated phase, the following basic features must be set on the controller:

1. *Minimum green time, G_{min}.* Each actuated phase has a minimum green time, which serves as the smallest amount of green time that may be allocated to a phase when it is initiated, s.

2. *Unit or vehicle extension, U.* This time actually serves three different purposes: (1) It represents the maximum gap between actuations at a single detector required to retain the green. (2) It is the amount of time added to the green phase when an additional actuation is received within the unit extension, *U.* (3) It must be of sufficient length to allow a vehicle to travel from the detector to the STOP line.

3. *Maximum green time, G_{max}.* Each phase has a maximum green time that limits the length of a green phase, even if there are continued actuations that would normally retain the green. The "maximum green time" begins when there is a "call" (or detector actuation) on a competing phase.

4. *Recall switches.* Each actuated phase has a recall switch. The recall switches determine what happens to the signal when there is no demand. Normally, one recall switch is placed in the "on" position, while all others are turned "off."

In this case, when there is no demand present, the green returns to the phase with its recall switch on. If no recall switch is in the "on" position, the green remains on the phase that had the last "call." If all recall switches are "on" and no demand exists, one phase continues to move to the next at the expiration of the minimum green.

5. *Yellow and all-red intervals.* Yellow and all-red intervals provide for safe transition from "green" to "red." They are fixed times and are not subject to variation, even in an actuated controller. They are found in the same manner as for pretimed signals (Chapter 18).

6. *Pedestrian WALK ("Walking Man"), Clearance ("Flashing Up-raised Hand"), and DON'T WALK ("Up-raised Hand") intervals.* Pedestrian intervals must also be set. With actuated signals, however, the total length of the GREEN is not known. Thus, pedestrian intervals are set in accordance with the minimum green time for each phase. Pedestrian push buttons are often, but not always, needed to ensure adequate crossing times.

Volume-density controllers add several other features. They are generally used at intersections with high approach speeds (≥ 45 mi/h), and in conjunction with area detectors (or point detectors set back a considerable length from the STOP line). In addition to the normal features of any actuated controller, the volume-density controller offers two important additions:

1. *Variable minimum green.* Because area detectors are capable of "remembering" the number of queued vehicles, the minimum green time may be varied to reflect the number of queued vehicles that must be served on the next "green" interval.

2. *Gap reduction.* In a standard actuated controller, the unit or vehicle extension is a constant value. Volume density controllers allow the minimum gap required to retain the green to be reduced over time. Doing this makes it more difficult to retain the green on a particular phase as the phase gets longer. Implementing the gap-reduction feature usually involves identifying four different measures:

(a) Initial unit or vehicle extension, U_1 (s) (maximum value)

(b) Final unit or vehicle extension, U_2 (s) (minimum value)

(c) Time into the green that gap reduction begins, t_1 (s)

(d) Time into the green that gap reduction ends, t_2 (s)

Time t_1 begins when a "call" on a competing phase is noted.

Many controllers contain additional features that may be implemented. Those noted here, however, are common to virtually all controllers and controller types.

20.3.2 Actuated Controller Operation

Figure 20.3 illustrates the operation of an actuated phase based on the three critical settings: minimum green, maximum green, and unit or vehicle extension.

When the green is initiated for a phase, it will be *at least* as long as the minimum green period, G_{min}. The controller divides the minimum green into an initial portion and a portion equal to one unit extension. If an additional "call" is received during the initial portion of the minimum green, no time is added to the phase, as there is sufficient time within the minimum green to cross the STOP line (yellow and all-red intervals take care of clearing the intersection). If a "call" is received during the last U seconds of the minimum green, U seconds of green are added to the phase. Thereafter, every time an additional "call" is received during a unit extension of U seconds, an additional period of U seconds is added to the green.

Note that the additional periods of U seconds are added *from the time of the actuation or "call."* They are *not* added to the end of the previous unit extension, as this would accumulate unused green times within each unit extension and include them in the total "green" period.

The "green" is terminated in one of two ways: (1) a unit extension of U seconds expires without an additional actuation, or (2) the maximum green is reached. The maximum green begins timing out when a "call" on a competing phase is noted. During the most congested periods of flow, however, it may be

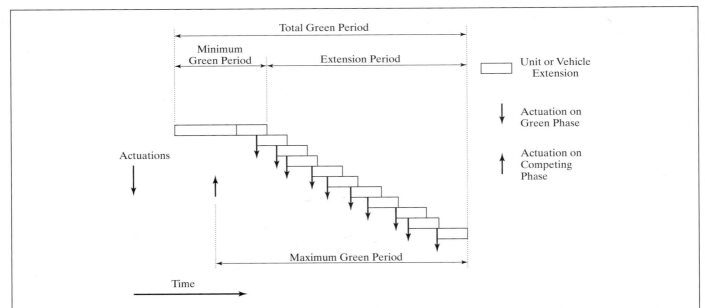

Figure 20.3: Operation of an Actuated Phase (Used with permission of Institute of Transportation Engineers, *Traffic Detector Handbook*, 2nd Edition, JHK & Associates, Tucson AZ, pg. 66.)

assumed that demand exists more or less continuously on all phases. The maximum green, therefore, begins timing out at the beginning of the green period in such a situation.

Assuming that demand exists continuously on all phases, the green period would be limited to a range of G_{min} to G_{max}. During periods of light flow, with no demand on a competing phase, the length of any green period can be unlimited, depending upon the setting of the recall switches.

In most situations, parallel lanes on an approach operate in parallel with each other. For example, in a three-lane approach, there will be three detectors (one for each lane). If *any* of the three lanes receives an additional "call" within U seconds, the green will be extended. Where multiple detectors are connected in series, using a single lead-in cable, gaps may reflect a lead vehicle crossing one detector and a following vehicle crossing another. While this type of operation is less desirable, it is less expensive to install and is, therefore, used frequently.

The gap time measured by a single detector is also somewhat smaller than the actual gap. That is because a detector is activated when the front of the vehicle crosses

the detector and deactivates when the rear of the vehicle clears the detector. The difference between the actual gap and the detected gap is the time it takes the length of the vehicle to cross over the detector.

Figure 20.4 illustrates the operation of the "gap-reduction" feature on volume-density controllers. Note the four critical times that must be set on the controller. Depending upon the manufacturer and model selected, there are a number of different protocols for implementing these four times, including:

- *BY-EVERY option.* Specify the amount of time by which the allowable gap is reduced after a specified amount of time. For example, for every 1.5 seconds of extension (after time x), reduce the allowable gap by 0.2 seconds.

- *EVERY SECOND option.* Specify the amount of time by which the allowable gap is reduced each second (after time x). For example, for every second of extension, reduce the allowable gap by 0.1 s.

- *TIME TO REDUCE option.* Specify a maximum and minimum allowable gap, and specify how long it will take to reduce from the maximum

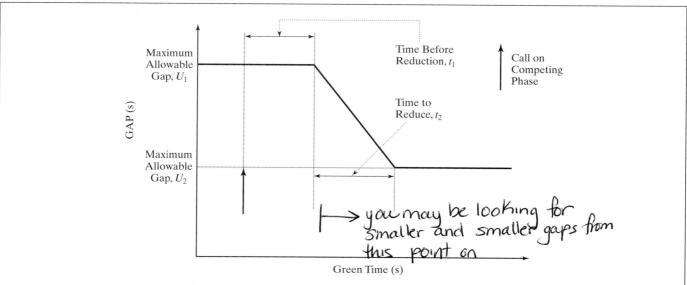

Figure 20.4: Gap-Reduction Feature on Volume-Density Controllers (Used with permission of Institute of Transportation Engineers, *Traffic Detector Handbook*, 2nd Edition, JHK & Associates, Tucson AZ, pg. 68.)

to the minimum (after time *x*). For example, the allowable gap will be linearly reduced from 3.5 s to 1.5 s over a period of 15 s.

20.4 Actuated Signal Timing and Design

In an actuated signal design, the traffic engineer does not provide an exact signal timing. Rather, a phase plan is established, and minima and maxima are set, along with programmed rules for determining the green period between limiting values based on vehicle actuations on detectors.

20.4.1 Phase Plans

phase sequencing may change
if no demand, skip phase

Phase plans are established using the same types of considerations as for pretimed signals (see Chapter 18). The primary difference is the flexibility in phase sequencing offered by actuated controllers.

Protected left-turn phases may be installed at lower left-turn flow rates, as these phases may be skipped during any cycle in which no left-turn demand is present. There are no precise guidelines for minimum left-turn demands and/or cross-products, so the engineer has considerable flexibility in determining an optimum phase plan.

20.4.2 Minimum Green Times

Minimum green times must be set for each phase in an actuated signalization, including the nonactuated phase of a semi-actuated controller. The minimum green timing on an actuated phase is based on the type and location of detectors.

Point or Passage Detectors

Point detectors only provide an indication that a "call" has been received on the subject phase. The number of calls experienced and/or serviced is not retained. Thus, if a point detector is located *d* feet from the STOP line, it must be assumed that a queue of vehicles fully occupies the distance *d*. The minimum green time, therefore, must be long enough to clear a queue of vehicles fully occupying the distance *d*, or:

saturation headway flow rate
2s for each vehicle

$$G_{min} = \ell_1 + 2 * \text{Int}\left(\frac{d}{20}\right) \qquad (20\text{-}1)$$

Int = integer number

where G_{min} = minimum green time, s

ℓ_1 = start-up lost time, s

d = distance between detector and STOP line, ft

20 = assumed head-to-head spacing between vehicles in queue, ft

The integer function requires that the value of $d/20$ be rounded to the next highest integer value. In essence, it requires that a vehicle straddling the detector be serviced within the minimum green period. Various agencies will set the value of ℓ_1 based on local policy. Values between 2.0 and 4.0 s are most often used.

Area or Presence Detectors

Where area detectors are in use, the minimum green time can be variable, based on the number of vehicles sensed in the queue when the green is initiated. In general:

$$G_{min} = \ell_1 + 2n \qquad (20\text{-}2)$$

where ℓ_1 = start-up lost time, s

n = number of vehicles stored in the detection area, vehs

20.4.3 Unit or Vehicle Extension

As noted previously, the unit or vehicle extension serves three different purposes. In terms of signal operation, it serves as both the minimum allowable gap to retain a green signal and as the amount of green time added when an additional actuation is detected within the minimum allowable gap.

The unit extension is selected with two criteria in mind:

• The unit extension should be long enough such that a subsequent vehicle operating in dense traffic at a safe headway will be able to retain a green signal (assuming the maximum green has not yet been reached).

- The unit extension should not be so long that straggling vehicles may retain the green or that excessive time is added to the green (beyond what one vehicle reasonably requires to cross the STOP line on green).

These criteria produce a rather narrow range of feasible values. It must be remembered that even under ideal conditions, saturation headways are in the range of 1.8 to 2.0 s. Thus, it would be impractical to lower the unit extension below these or even to these values. Since the intent of an actuated signal is to maximize the efficient use of available green time, it is illogical to allow for gaps of more than 4.0 to 5.0 s to result in an extension of green when there are calls on competing phases.

The *Traffic Detector Handbook* [2] recommends that a unit extension of 3.0 s be used where approach speeds are equal to or less than 30 mi/h, and that 3.5 s be used at higher approach speeds.

On volume-density controllers with a gap-reduction feature, the initial gap may be about 0.5 s higher then these values, with the minimum gap in the range of 2.0 to 2.5 s.

For all types of controllers, however, the unit extension must be equal to or more than the passage time. Passage time is the time it takes a vehicle to traverse the distance from the detector to the STOP line, or:

$$U \geq P = \frac{d}{1.47 \, S_{15}} \qquad (20\text{-}3)$$

where U = unit extension, s
  P = passage time, s
  d = distance from detector to STOP line, ft
  S_{15} = 15th percentile approach speed, mi/h

20.4.4 Detector Location Strategies

The minimum green time and the detector location are mathematically linked. There are two strategies for detector location that are frequently employed:

- Place the detector to achieve a desired minimum green time.
- Place the detector such that passage time to the STOP line is equal to the unit extension.

Because many actuated signals are at locations where demands are quite low during off-peak periods, there is often the desire to keep minimum green times as low as possible, thus minimizing the waiting period for a vehicle on a competing phase when there is no demand on the subject phase. A practical minimum limit on the minimum green time is the assumed start-up lost time, $\ell_1 + 2.0$ s. This is the amount of time needed to process a single vehicle; it ranges between 4.0 and 6.0 s.

When this strategy is used, Equation 20-1 is used to compute the appropriate detector location for the selected minimum green. Consider the following situation:

A minimum green on an approach to an actuated signal is to be set at 6.0 s, with an assumed start-up lost time of 4.0 s. How far may the detector be located from the STOP line?

From Equation 20-1:

$$G_{min} = 6.0 = 4.0 + 2\text{Int}\left(\frac{d}{20}\right)$$

$$\text{Int}\left(\frac{d}{20}\right) = \frac{6.0 - 4.0}{2} = 1.0$$

Due to the integer function, the detector may be located anywhere between 0.1 and 20.0 ft from the STOP line. Note that where loop detectors are used, the location refers to the *front* of the detector.

The second strategy places the detector to equalize the unit extension and the passage time. Consider the following situation:

At an approach to an actuated signal location, a detector is to be placed such that the unit extension of 3.5 s is equal to the passage time. The 15th percentile approach speed on this approach is 40 mi/h.

Using Equation 20-3:

$$U = 3.5 = \frac{d}{1.47 * 40}$$

$$d = 3.5 * 1.47 * 40 = 205.8 \text{ ft}$$

The advantage of such a setback is that a vehicle arriving when there is no other demand present but the signal is "red," could cross the detector and have the light turn green just as the vehicle arrives at the STOP line. The vehicle would not, therefore, have to stop first and wait for the green.

The disadvantage of such a long setback is that it leads to a very long minimum green time. In the case considered:

$$G_{min} = 4.0 + 2\text{Int}\left(\frac{205.8}{20}\right) = 26.0 \text{ s}$$

In most situations, this would be excessive. During periods of light demand, a competing vehicle might be forced to wait as long as 26.0 s plus the transition intervals (yellow and all-red), while only one vehicle is served. For this reason, longer setbacks in which the unit extension and passage time are equal are generally used only where presence or area detectors are in place, allowing for a variable minimum green assignment.

In any event, there is a practical limitation on the placement of point detectors that must be observed: the detector(s) must be placed such that no vehicle can arrive at the STOP line without having crossed a detector. In practical terms, this means that no detector can be placed where a vehicle can enter the traffic stream from driveway or curb parking space located between the detector and the STOP line. In many urban and suburban settings, this requires that the detector be located quite close to the STOP line.

Area detectors are more flexible, in that they can detect vehicles entering the detection area from the side. Thus, it is only the location of the *front* of the area detector that is limited as described previously.

20.4.5   Yellow and All-Red Intervals → Safety
Same process/thinking

Yellow and all-red intervals are determined in the same fashion as for pretimed signals:

$$y = t + \frac{1.47 * S_{85}}{2a + 64.4 * 0.01G} \tag{20-4}$$

$$ar = \frac{w + L}{1.47 \, S_{15}} \text{ or } \frac{P + L}{1.47 \, S_{15}} \tag{20-5}$$

where
$\quad y =$ yellow interval, s
$\quad ar =$ all red interval, s
$\quad S_{85} =$ 85th percentile speed, mi/h
$\quad S_{15} =$ 15th percentile speed, mi/h
$\quad a =$ deceleration rate (10 ft/s^2)
$\quad G =$ grade $(\%)$
$\quad w =$ width of street being crossed, ft
$\quad P =$ distance from near curb to far side of far crosswalk, ft

As in the case of pretimed signals, yellow and all-red times must be known to determine the total lost time in the cycle, L, which is needed to determine maximum green times. The relationships between yellow and all red times and lost times are repeated here for convenience:

$$L = \sum_i t_{Li}$$
$$t_{Li} = \ell_{1i} + \ell_{2i}$$
$$\ell_{2i} = Y_i - e_i$$
$$Y_i = y_i + ar_i \tag{20-6}$$

where
$\quad L =$ total lost time in the cycle, s/cycle
$\quad t_{Li} =$ total lost time for Phase i, s
$\quad \ell_{1i} =$ start-up lost time for Phase i, s (measured value, or 2.0 s default value)
$\quad \ell_{2i} =$ clearance lost time for Phase i, s
$\quad e_i =$ encroachment of effective green into yellow and all-red periods for Phase i, s (measured value, or 2.0 s default value)
$\quad Y_i =$ sum of yellow and all red intervals for Phase i, s
$\quad y_i =$ yellow interval for Phase i, s
$\quad ar_i =$ all red interval for Phase i, s

Note that when the default values for ℓ_1 and e are used, the total lost time per cycle, L, is equal to the sum of the yellow and all-red phases associated with critical movements in the cycle, and that effective green, g, is equal to actual green, G.

20.4.6 Maximum Green Times and the Critical Cycle

The "critical cycle" for a full actuated signal is one in which each phase reaches its maximum green time. For semi-actuated signals, the "critical cycle" involves the maximum green time for the side street and the minimum green time for the major street, which has no detectors.

Maximum green times for actuated phases and/or the minimum green time for the major street with semi-actuated signalization are found by determining a cycle length and initial green split based on average demands during the peak analysis period. The method is the same as that used for determining cycle lengths and green times for a pretimed signal:

$$C_i = \frac{L}{1 - \left[\dfrac{V_c}{1,615 * PHF * (v/c)} \right]} \tag{20-7}$$

where C_i = initial cycle length, s
   V_c = sum of critical lane volumes, veh/h
   PHF = peak hour factor
   v/c = desired v/c ratio to be achieved

Because the objective in actuated signalization is to have little unused green time during peak periods, the v/c ratio chosen in this determination is taken to be 0.95 or higher in most applications.

Knowing the cycle length, green times are then determined as:

$$g_i = (C - L) * \left(\frac{V_{ci}}{V_c} \right) \tag{20-8}$$

where g_i = effective green time for Phase i, s
   V_{ci} = critical lane volume for Phase i, veh/h

All other variables are as previously defined.

These computations result in a cycle length and green times that would accommodate the average cycle demands in the peak 15 min of the analysis hour. They are not, however, sufficient to handle perturbations occurring during the peak 15-min demand period when individual cycle demands exceed the capacity of the cycle. Thus, to provide enough flexibility in the controller to adequately service peak cycle-by-cycle demands during the analysis period, green times determined from Equation 20-8 are multiplied by a factor of between 1.25 and 1.50. The results would then become the maximum green times for each phase and/or the minimum green time for a major street at a semi-actuated signal.

The "critical cycle length" is then equal to the sum of the actual maximum green times (and/or the minimum green time for a major street at a semi-actuated location) plus yellow and all-red transitions.

$$C_c = \sum_i (G_i + Y_i) \tag{20-9}$$

where C_c = critical cycle length, s
   G_i = actual maximum green time for actuated Phase i, or actual minimum green time for the major street at a semi-actuated signal, s
   Y_i = sum of yellow and all red intervals for Phase i, s

The timing of an actuated signal involves a number of practical considerations that may override the results of the computations as described. Particularly at a semi-actuated signal location with low side-street demands, the maximum green, G_{max}, may compute to a value that is less than the minimum green, G_{min}. Although a rarer occurrence, this could happen on a given phase at full actuated location as well, particularly where protected left-turn phases are involved. In such cases, the G_{max} is judgmentally set as $G_{min} + nU$, where n is the maximum number of vehicles to be served during a single green phase. The value of n is usually approximately set as 1.5 times the average number of vehicles expected per cycle (an iterative concept, as the cycle length would be needed to determine the value of n). However, to maintain an appropriate balance between all phases, values of G_{max} for other phases must then be adjusted to maintain a ratio equal to the balance of critical-lane volumes for each phase.

20.4.7   Pedestrian Requirements for Actuated Signals

As for pretimed signals, pedestrians require the following amount of time to safely cross a street:

$$G_p = 3.2 + \left(\frac{L}{S_p}\right) + 0.27\, N_{peds} \quad (20\text{-}10)$$

where G_p = minimum pedestrian green time, s

L = length of crosswalk, ft

S_p = walking speed of pedestrians, default value 4.0 ft/s

N_{peds} = average number of pedestrians crossing the street in one crosswalk per cycle, peds

As pedestrians are permitted to be in the crosswalk during the vehicular green time plus the yellow and all-red intervals, safe operation occurs only when:

$$G_p \leq G + Y \quad (20\text{-}11)$$

for a given phase.

At actuated signal locations, however, there are several considerations in addition to those at pretimed locations:

1. The value of N_{peds} varies, as the cycle length is variable. For actuated signals, the value of N_{peds} is determined from the critical cycle length. This results in the *maximum feasible* number of pedestrians that might occur during a single cycle.
2. Since the length of green times also varies from cycle to cycle, safety can be assured only when Equation 20-11 is used with $G = G_{min}$ for each cycle.

With pretimed signals, when safe crossing is not assured, either the cycle length must be increased to accommodate both pedestrians and vehicles or pedestrian-actuated push-button and pedestrian signals must be installed.

With actuated signals, safe crossing based on minimum green times is most often *not* provided. Increasing minimum greens to accommodate pedestrians during every cycle is not an option, as that would create inefficiencies for vehicles that the actuated signal was installed to avoid. Thus, whenever the minimum green *does not* provide safe crossing, a pedestrian push button should be installed with pedestrian signals.

In such cases, the pedestrian signal rests on a DON'T WALK indication. When the pedestrian push button is actuated, on the next green phase the minimum green time is increased to:

$$G_{min,\,ped} = 3.2 + \left(\frac{L}{S_p}\right) + 0.27\, N_{peds} - Y \quad (20\text{-}12)$$

When the pedestrian minimum green is called for, the WALK interval (Walking Man) is as follows:

$$WALK = 3.2 + 0.27\, N_{peds} \quad (20\text{-}13)$$

and the pedestrian clearance interval (Flashing Up-raised Hand) is:

$$Up\text{-}raised\ Hand_{Flashing} = \frac{L}{S_p} \quad (20\text{-}14)$$

where all terms are as previously defined.

20.5   Examples in Actuated Signal Design and Timing

Timing of an actuated signal is less definitive than for pretimed signals and calls for more judgment to be exercised by the engineer. In any given instance, it is possible that several different signal timings and designs would work acceptably. Some of the considerations involved are best illustrated by example.

Example 20-1: A Semi-actuated Signal Timing

Figure 20.5 shows an intersection that will be signalized using a semi-actuated controller. For convenience, the demand volumes shown have already been converted to through vehicle units (tvus). This conversion is the same as for pretimed signals (see Chapter 18).

Step 1:   Phasing

As this is a semi-actuated signal, there are only two phases, as follows:

- Phase 1—All First Avenue movements (minor street)
- Phase 2—All Main Street movements (major street)

Step 2:   Minimum Green Time and Detector Location

For a semi-actuated signal, only the side-street phase is actuated and only side-street approaches have detectors. In such cases, the objective is generally to provide only the amount of green time necessary to clear side-street vehicles, with as little unused green time as possible. Therefore, the minimum green time for First Avenue

should be as low as possible. Using a start-up lost time of 2.0 s, the minimum green time that could be allocated would be 4.0 s. If G_{min1} is set at 4.0 s, then the detector placement is determined by solving for d in Equation 20-1:

$$G_{min} = 4.0 = 2.0 + 2 \, \text{Int}\left(\frac{d}{20}\right)$$

$$\text{Int}\left(\frac{d}{20}\right) = \frac{4.0 - 2.0}{2} = 1.0$$

$$d = 0.1 - 20.0 \text{ ft}$$

The detector would be placed anywhere between 0.1 and 20.0 ft from the STOP line. It must be placed such that no vehicle can enter the approach without traversing the detector.

Step 3:   Unit Extension

For the side-street approaches, with an approach speed of 25 mi/h, the recommended unit extension, U, is 3.0 s. This must be greater than the passage time from the

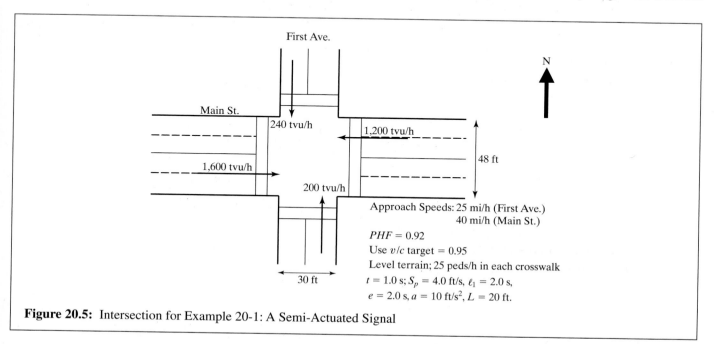

First Ave.

Main St.

N

240 tvu/h

1,200 tvu/h

48 ft

1,600 tvu/h

200 tvu/h

Approach Speeds: 25 mi/h (First Ave.)
40 mi/h (Main St.)

30 ft

PHF = 0.92
Use *v/c* target = 0.95
Level terrain; 25 peds/h in each crosswalk
$t = 1.0$ s; $S_p = 4.0$ ft/s, $\ell_1 = 2.0$ s,
$e = 2.0$ s, $a = 10$ ft/s^2, $L = 20$ ft.

Figure 20.5: Intersection for Example 20-1: A Semi-Actuated Signal

detector to the STOP line, assuming the maximum set-back of 20.0 ft, or:

$$U \geq P = \frac{20}{1.47 * 25} = 0.54 \text{ s}$$

The 3.0-second unit extension is safe and will be implemented.

Step 4: Sum of Critical-Lane Volumes

All of the demand volumes of Figure 20.5 have already been converted into through vehicle equivalents (tvus). The SB movement has a higher volume than the NB movement (both approaches have one lane). Thus, the critical-lane volume for Phase 1 is 240 tvu/h. The EB volume of 1,600 tvu/h is critical for Phase 2, but is divided into two lanes. Thus, the critical-lane volume for Phase 2 is 1,600/2 = 800 tvu/h. The sum of the critical-lane volumes, V_c, is 240 + 800 = 1,040 tvu/h.

Step 5: Yellow and All-Red Times; Lost Time Per Cycle

To determine other signal timing parameters, an initial cycle length must be selected. This requires, however, that all lost times within the cycle be known, which requires that the yellow and all-red intervals be established. Yellow intervals for each phase are estimated using Equation 20-4, while all-red intervals are estimated using Equation 20-5. Average approach speeds are given for Main Street and First Avenue. The 85th percentile speed may be estimated as 5 mi/h more than the average, while the 15th percentile speed is estimated as 5 mi/h less than the average.

$$y = t + \frac{1.47 S_{85}}{2a + 64.4(0.01G)}$$

$$y_1 = 1.0 + \frac{1.47(25 + 5)}{2*10 + 64.4(0.01*0)} = 3.2 \text{ s}$$

$$y_2 = 1.0 + \frac{1.47(40 + 5)}{2*10 + 64.4(0.01*0)} = 4.3 \text{ s}$$

$$ar = \frac{w + L}{1.47 S_{15}}$$

$$ar_1 = \frac{48 + 20}{1.47(25 - 5)} = 2.3 \text{ s}$$

$$ar_2 = \frac{30 + 20}{1.47(40 - 5)} = 1.0 \text{ s}$$

With default values of 2.0 s used for both ℓ_1 and e, the lost time per cycle is equal to the sum of the yellow and all-red times in the cycle, or:

$$L = 3.2 + 2.3 + 4.3 + 1.0 = 10.8 \text{ s/cycle}$$

Step 6: Maximum Green (Phase 1) and Minimum Green (Phase 2)

As a semi-actuated signal, the critical cycle is composed of the maximum green for the side street (First Avenue, Phase 1), the minimum green for the major street (Main Street, Phase 2), and the yellow and all-red intervals from each. The initial cycle length is estimated using Equation 20-7:

$$C_i = \frac{L}{1 - \left[\frac{V_c}{1,615 * PHF * (v/c)} \right]}$$

$$= \frac{10.8}{1 - \left[\frac{1,040}{1,615 * 0.92 * 0.95} \right]}$$

$$= \frac{10.8}{0.263} = 41.1 \text{ s}$$

For a semi-actuated signal, this value does not have to be rounded. Green splits based on this cycle length are determined using Equation 20-8:

$$g_1 = G_1 = (41.1 - 10.8) * \left(\frac{240}{1,040} \right) = 7.0 \text{ s}$$

$$g_2 = G_2 = (41.1 - 10.8) * \left(\frac{800}{1,040} \right) = 23.3 \text{ s}$$

Effective green times and actual green times are equal, given the default values of 2.0 s for both l_1 and e. Standard practice establishes the maximum green for the minor street and the minimum green for the major street as 1.50 times the above values, or:

$$G_{max1} = 1.50 * 7.0 = 10.5 \text{ s}$$
$$G_{min2} = 1.50 * 23.3 = 35.0 \text{ s}$$

The G_{max1} of 10.5 s compares favorably with the G_{min1} of 4.0 s established earlier, so no adjustment of this timing is necessary. The critical cycle length is the sum of $G_{max1} + G_{min2} + Y_1 + Y_2$, or:

$$C_c = 10.5 + 35.0 + 3.2 + 2.3 + 4.3 + 1.0$$

$$= 56.3 \text{ s}$$

Step 7: Pedestrian Requirements

Pedestrians cross the minor street during Phase 2. Thus, the pedestrian crossing requirement must be compared to the minimum green plus yellow and all-red provided in Phase 2. For the purposes of this computation, N_{peds}, which is the same for all crosswalks, will be based on the critical cycle length of 56.3 s. Then:

$$N_{peds} = \frac{25}{(3,600/56.3)} = 0.39 \text{ peds/cycle}$$

$$G_{p2} = 3.2 + \frac{30}{4.0} + 0.27(0.39) = 10.8 \text{ s}$$

$$G_{p2} = 10.8 \leq G_{min2} + Y_2$$

$$= 35.0 + 4.3 + 1.0 = 40.3 \text{ s}$$

The minimum green provides more than enough time for safe crossings of the minor street during Phase 2. No pedestrian push button is needed; pedestrian signals are optional.

Pedestrians cross the major street during Phase 1, which has a minimum green time of 4.0 s. Checking for pedestrian safety:

$$G_{p1} = 3.2 + \frac{48}{4.0} + 0.27(0.39) = 15.3 \text{ s}$$

$$G_{p1} = 15.3 \leq G_{min1} + Y_1$$

$$= 4.0 + 3.2 + 2.3 = 9.5 \text{ s}$$

Pedestrians *are not* safely accommodated by G_{min1}. Thus, for pedestrians crossing the major street, a pedestrian push button must be provided, and pedestrian signals are mandatory. When pushed, the next green phase will provide a minimum green time of:

$$G_{min1,ped} = 15.3 - 3.2 - 2.3 = 9.8 \text{ s}$$

The pedestrian walk and clearance intervals would be as follows:

$$WALK_1 = 3.2 + 0.27(0.38) = 3.3 \text{ s}$$

$$Up\text{-}raised\ Hand_{flashing1} = \frac{48}{4.0} = 12.0 \text{ s}$$

Example 20-2: A Variation on Example 20-1

What would change if the side-street critical-demand volume were only 85 veh/h, instead of the 240 veh/h of Example 20-1? None of the details of detector placement would change, nor would the length of the yellow and all-red phases. Thus, the following values have already been determined:

- $G_{min1} = 4.0$ s
- $Y_1 = 3.2 + 2.3 = 5.5$ s
- $Y_2 = 4.3 + 1.0 = 5.3$ s
- $L = 10.8$ s/cycle

The only significant change is in the critical lane volumes. For the minor street (First Avenue, Phase 1), the critical volume is now 85 veh/h. The major street (Main Street, Phase 2) critical-lane volume is unchanged: $1,600/2 = 800$ veh/h. The sum of critical-lane volumes is, therefore, $800 + 85 = 885$ veh/h. This will change the initial cycle length and the values of G_{max1} and G_{min2} that result from it:

$$C_i = \frac{10.8}{1 - \left[\dfrac{885}{1,615 * 0.92 * 0.95}\right]} = \frac{10.8}{0.373} = 29.0 \text{ s}$$

$$G_1 = (29.0 - 10.8)\left(\frac{85}{885}\right) = 1.7 \text{ s}$$

$$G_2 = (29.0 - 10.8)\left(\frac{800}{885}\right) = 16.5 \text{ s}$$

$$G_{max1} = 1.7 * 1.5 = 2.6 \text{ s}$$

$$G_{min2} = 16.5 * 1.5 = 24.8 \text{ s}$$

This timing is not reasonable, as G_{max1} (2.6 s) is less than G_{min1} (4.0 s). An alternative way of establishing a reasonable G_{max1} must be found. The average number of expected vehicles per cycle arriving on the side street is dependent upon the critical cycle length, which is not yet known. If a 60-s cycle length is assumed as a rough estimate, then $85/60 = 1.4$ veh/cycle would be expected. If this is multiplied by 1.5, the result is 2.1

veh/cycle. This suggests that the maximum green time should accommodate two or three vehicles. The minimum green time of 4.0 s accommodates one vehicle. To guarantee that three vehicles could move through on a single green phase, G_{max1} would be increased by two unit extensions, or $4.0 + 2*3.0 = 10.0$ s. A guarantee of two vehicles would require G_{max1} to be set at $4.0 + 1*3.0 = 7.0$ s. Either choice is defendable, but 7.0 s will be used in the remainder of this illustration. The G_{min2} must now be increased to maintain the original balance of times $(G_{max1} = 2.6$ s; $G_{min2} = 24.8$ s). Then:

$$\frac{24.8}{2.6} = \frac{G_{min2}}{7.0}$$

$$G_{min2} = \frac{24.8*7.0}{2.6} = 66.8 \text{ s}$$

The critical cycle length now becomes:

$$C_c = 7.0 + 5.5 + 66.8 + 5.3 = 84.6 \text{ s}$$

The initial assumption of a 60-s cycle to determine the number of vehicles arriving in one cycle could be iterated, but given the approximate nature of these computations, that would not be necessary. As the larger result would *increase* the number arriving per cycle, it might suggest that G_{max1} be set at 10.0 s rather than 7.0 s, but even this is a judgment call. The pedestrian safety checks could be redone to reflect small changes in N_{peds}, but this value would result in a change in the order of 0.10 and is also not necessary.

With a smaller side-street demand, the balance of green times is significantly changed, with the major street receiving a longer G_{min} and the side street receiving a smaller G_{max} than in the original solution.

Example 20-3: A Full-actuated Signal

An isolated suburban intersection of two major arterials is to be signalized using a full actuated controller. Area detection is to be used, and there are no driveways or other potential entry points for vehicles within 300 ft of the STOP line on all approaches. The intersection is shown in Figure 20.6, and all volumes have already been converted to tvus for convenience. Left-turn slots of 250 ft in length are provided for each approach. The tvu conversions assume that a protected left-turn phase will be provided for all approaches.

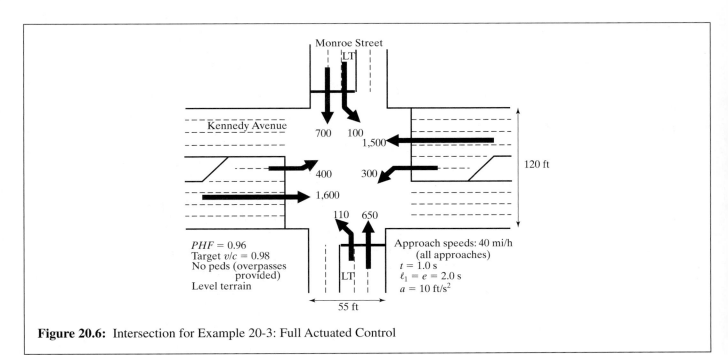

Figure 20.6: Intersection for Example 20-3: Full Actuated Control

Step 1: Phasing

The problem statement indicates that protected left-turn phasing will be implemented on all approaches. Note that Kennedy Avenue has double left-turn lanes in each direction and that Monroe Street has a single left-turn lane in each direction.

At a heavily utilized intersection such as this, quad-eight phasing would be desirable. Each street would have an exclusive LT phase followed by a leading green in the direction of heavier LT flow and a TH/RT phase. As indicated in Chapter 18, such phasing provides much flexibility in that LT phasing is always optional and can be skipped in any cycle in which no LT demand is noted. The resulting signalization has a maximum of four phases in any given cycle and a minimum of two. It is treated as a *four-phase* signal, as this option leads to the maximum lost times.

Quad-eight phasing involves overlaps (again, see Chapter 18) that would be taken into account if this were a pretimed signal. As an actuated signal, the worst-case cycle, however, would occur when there are no overlap periods. This would occur when the LT flow in opposing directions are equal. Thus, the signal timing will be considered as if this were a simple four-phase operation without overlaps. The controller, however, will allow one protected LT to be terminated before the opposing protected LT, creating a leading green phase. The four phases are:

- Phase 1—Protected LT for Kennedy Avenue
- Phase 2—TH/RT for Kennedy Avenue
- Phase 3—Protected LT for Monroe Street
- Phase 4—TH/RT for Monroe Street

Step 2: Unit Extension

For approach speeds of 40 mi/h, the recommended unit extension is 3.5 s.

Step 3: Minimum Green Times and Detector Placement

The problem specifies that area detection shall be employed. For area detection, the far end of the detection zone is placed such that the passage time is equal to unit extension. Since all approaches (including LT approaches) have a 40 mi/h approach speed, the far end of detectors should be located as follows:

$$U = 3.5 = P = \frac{d}{1.47 * 40}$$
$$d = 3.5 * 1.47 * 40 = 205.8, \text{ say } 206 \text{ ft}$$

The near end of the detection zone would be placed within 1.0 ft of the STOP line.

The minimum green time for area detection is variable, based on the number of vehicles sensed within the detection area when the green is initiated. The value can vary from the time needed to service one waiting vehicle to the time needed to service $Int(206/20) = 11$ vehicles. Using Equation 20-2, the range of minimum green times can be established for each approach. In this case, all values will be equal, as the approach speeds are the same for all approaches and the detector location is common to every approach, including the LT lanes, all of which are long enough to accommodate a 206-ft setback.

$$G_{min} = \ell_1 + 2.0n$$
$$G_{min/low} = 2.0 + (2.0 * 1) = 4.0 \text{ s}$$
$$G_{min/high} = 2.0 + (2.0 * 11) = 24.0 \text{ s}$$

Step 4: Critical-Lane Volumes

As the volumes given have already been converted to tvus, critical-lane volumes for each phase are easily identified:

- Phase 1 (Kennedy Ave, LT) 400/2 = 200 tvu/h
- Phase 2 (Kennedy Ave, TH/RT) 1,600/4 = 400 tvu/h
- Phase 3 (Monroe St, LT) 110/1 = 110 tvu/h
- Phase 4 (Monroe St, TH/RT) 700/2 = 350 tvu/h
- $V_c = 1,060$ tvu/h

Step 5: Yellow and All-Red Times: Lost Times

Yellow times are found using Equation 20-4; all-red times are found using Equation 20-5. With a 40-mi/h average approach speed for all movements, the S_{85} may be estimated as $40 + 5 = 45$ mi/h, and the S_{15} may be estimated as $40 - 5 = 35$ mi/h. Then:

$$y_{all} = 1.0 + \frac{1.47 * 45}{2 * 10 + 64.4(0.01 * 0)} = 4.3 \text{ s}$$
$$ar_{1,2} = \frac{55 + 20}{1.47 * 35} = 1.5 \text{ s}$$
$$ar_{3,4} = \frac{120 + 20}{1.47 * 35} = 2.7 \text{ s}$$
$$Y_{1,2} = 4.3 + 1.5 = 5.8 \text{ s}$$
$$Y_{3,4} = 4.3 + 2.7 = 7.0 \text{ s}$$

There are four phases in the worst-case cycle. As the standard defaults are used for ℓ_1 and e, the total lost time is equal to the sum of the yellow and all-red intervals in the cycle:

$$L = 2*5.8 + 2*7.0 = 25.6 \text{ s}$$

Step 6:   Maximum Green Times and the Critical Cycle

The initial cycle length for determining maximum green times is:

$$C_i = \frac{25.6}{1 - \left[\dfrac{1,060}{1,615*0.96*0.98}\right]} = \frac{25.6}{0.302} = 84.8 \text{ s}$$

Green times are found as:

$$G_1 = (84.8 - 25.6)\left(\frac{200}{1,060}\right) = 11.2 \text{ s}$$

$$G_2 = (84.8 - 25.6)\left(\frac{400}{1,060}\right) = 22.3 \text{ s}$$

$$G_3 = (84.8 - 25.6)\left(\frac{110}{1,060}\right) = 6.1 \text{ s}$$

$$G_4 = (84.8 - 25.6)\left(\frac{350}{1,060}\right) = 19.5 \text{ s}$$

$$G_{max1} = 11.2*1.5 = 16.8 \text{ s}$$

$$G_{max2} = 22.3*1.5 = 33.5 \text{ s}$$

$$G_{max3} = 6.1*1.5 = 9.2 \text{ s}$$

$$G_{max4} = 19.5*1.5 = 29.3 \text{ s}$$

With area detection, the minimum green for all lane groups, including LT lanes, can be as high as 24.0 s. This is inconsistent with G_{max} values for the LT Phases 1 and 3. Increasing the maximum greens beyond the computed values, however, will lead to an excessively long critical cycle length.

Thus, it is recommended that the LT lanes use *point* detectors, placed so that the G_{min} for Phases 1 and 3 is a constant 4.0 s. The above G_{max} results will work in this scenario. The G_{max} results for Phases 2 and 4 (through phases) are close to the high value of G_{min} for these phases, but would provide some flexibility even in peak periods. It is, therefore, not recommended that any of these times be arbitrarily increased.

The critical cycle length becomes:

$$C_c = 16.8 + 5.8 + 33.5 + 5.8 + 9.2 \\ + 7.0 + 29.3 + 7.0 = 114.4 \text{ s}$$

As overpasses are provided for pedestrians, there are no at-grade crossings permitted, and no pedestrian checks are required for this signalization.

References

1. Pusey, R.S. and Butzer, G.L, "Traffic Control Signals," *Traffic Engineering Handbook*, 5th Edition, Institute of Transportation Engineers, Washington DC, 1999, pgs. 453–528.
2. *Traffic Detector Handbook*, 2nd Edition, Institute of Transportation Engineers, JHK & Associates, Tucson, AZ, pg. 68.

Problems

Unless otherwise noted, use the following default values for each of the following actuated signal timing problems:

- Driver reaction time $(t) = 1.0$ s
- Vehicle deceleration rate = 10 ft/s^2
- Length of a vehicle = 20 ft
- Level terrain
- Low pedestrian activity at all locations (50 peds/h each cross walk)
- $PHF = 0.90$
- Target v/c ratio for actuated signals = 0.95
- Lane widths = 12 ft
- Crosswalk widths = 10 ft
- Pedestrian crossing speed = 4.0 ft/s
- All volumes in veh/h

If any other assumptions are necessary, specifically indicate them as part of the answer.

20-1. A semi-actuated signal is to be installed and timed for the location shown in Figure 20.7. Because of light side-street demand, a short minimum green of 6.0 s is desired. For the conditions shown:

(a) How far from the STOP line should the side-street detectors be located?

(b) Recommend a unit extension, and ensure that it is larger than the passage time.

(c) Compute yellow and all-red times.

(d) Recommend a maximum side-street green and a minimum main street green.

(e) What is the critical cycle length?

(f) Are pedestrian signals and/or push buttons required for crossing the main street? The side street?

20-2. A full actuated controller must be retimed at the intersection shown in Figure 20.8. Detector locations are fixed from a previous installation and cannot be moved. For the conditions shown:

(a) Recommend a suitable phase plan for the signal.

(b) What minimum greens should be used?

(c) Recommend a unit extension, and ensure that it is larger than the passage time.

(d) Compute yellow and all-red times.

(e) Recommend maximum green times for each phase.

(f) What is the critical cycle length?

(g) Are pedestrian signals and/or push buttons needed for any phases?

20-3. A full actuated signal is to be installed at the major intersection shown in Figure 20.9. For this location and the conditions shown:

(a) Recommend minimum green times and detector placement locations.

(b) Recommend a unit extension, and ensure that it is larger than the passage time.

(c) Compute yellow and all-red times.

(d) Recommend maximum green times for each phase.

(e) What is the critical cycle length?

(f) Are pedestrian signals and/or push buttons needed for any phases?

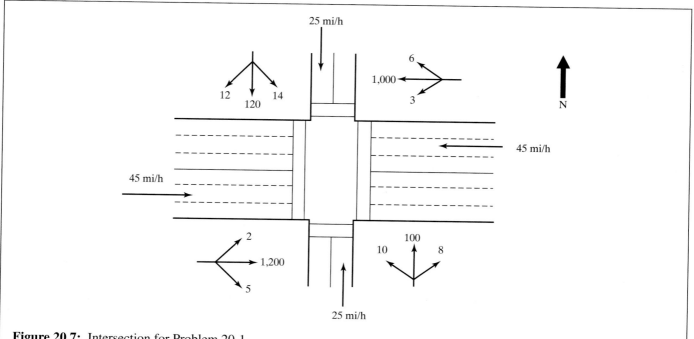

Figure 20.7: Intersection for Problem 20-1

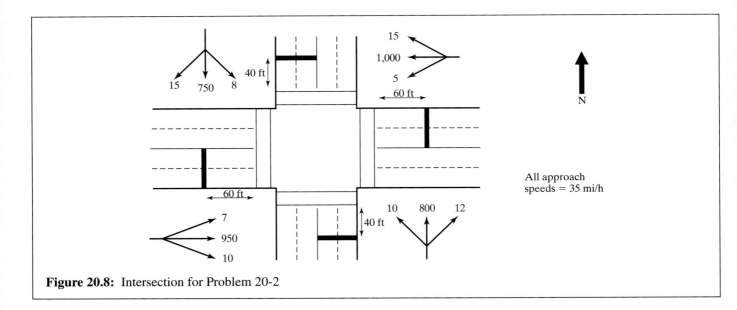

Figure 20.8: Intersection for Problem 20-2

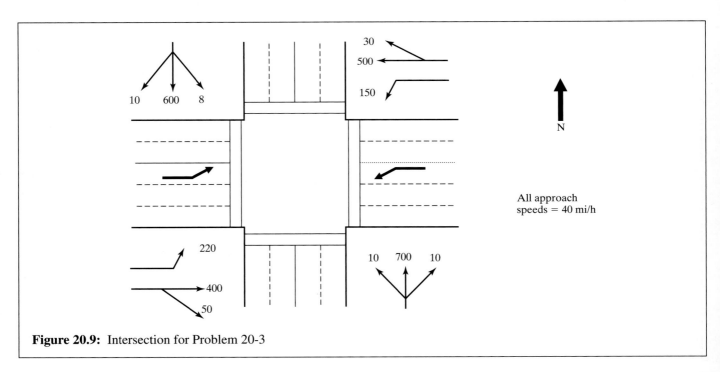

Figure 20.9: Intersection for Problem 20-3

CHAPTER
21

Analysis of Signalized Intersections

21.1 Introduction

The signalized intersection is the most complex location in any traffic system. In Chapter 17, some simple models of critical operational characteristics were presented. In Chapter 18, these were applied to create a simple signal-timing methodology. The signal-timing methodology, however, involved a number of simplifying assumptions, among which was a default value for saturation flow rate that reflected "typical" conditions. A complete analysis of any signalized intersection requires use of a more complex model that addresses all of the many variables affecting intersection operations as well as some of the more intricate interactions among component flows.

The most frequently used model for analysis of signalized intersections in the United States is the model contained in Chapter 16 of the *Highway Capacity Manual* (HCM) [*1*]. This model first appeared in the 1985 edition of the HCM and has been revised and updated in subsequent editions (1994, 1997, 2000). The model has become increasingly complex, involving several iterative elements. It has become difficult, if not impossible, to do complete solutions using this model by hand. Its

implementation, therefore, has been primarily through computer software that replicates the model. Chapter 22 reviews several of the available models and presents a number of complex applications as examples.

There are other models that are in use elsewhere. SIDRA [*2, 3*] is a model and associated computer package developed for use in Australia by the Australian Road Research Board (ARRB). Some of its elements, particularly in delay estimation, have been adapted and applied in the HCM. A Canadian model also exists [*4*] and is the official model used throughout Canada. All of these models are "deterministic" analytic models. In deterministic models, the same input data produces the same result each time the model is applied.

There are also a number of simulation models that may be used to analyze individual intersections, as well as networks. Some of the most frequently used are COR-SIM, TRANSYT 7-F, and PASSER. Simulation often introduces stochastic elements. This means that the same input parameters do not lead to the same results each time the simulation is run. Because of this characteristic, most stochastic simulation models are run multiple times with a given set of input data, and average results of those runs are used in evaluating the intersection.

This chapter attempts to describe the overall concepts and some of the details of the HCM methodology. Some of the complexities are noted without going into full detail, and the student is encouraged to consult the HCM directly for additional information on such subjects.

21.2 Conceptual Framework for the HCM 2000 Methodology

There are five fundamental concepts used in the HCM 2000 signalized intersection analysis methodology that should be understood before considering any of the details of the model. These are:

- The critical lane group concept
- The *v/s* ratio as a measure of demand
- Capacity and saturation flow rate concepts
- Level-of-service criteria and concepts
- Effective green time and lost-time concepts

Some of these are similar to what has been discussed in other chapters; they are repeated here for reinforcement and because of their central importance in understanding the HCM model.

21.2.1 The Critical-Lane Group Concept

The signal-timing methodology of Chapter 18 relied on *critical lanes* and *critical-lane volumes*. In doing so, however, it was generally assumed that movements sharing a set of lanes on an approach would evenly divide among the available lanes, at least in terms of "through vehicle units," (tvu's). In the HCM model, the total demand in a set of lanes is used. Instead of identifying a set of "critical lanes," a set of "critical-lane groups" is identified.

Critical-lane analysis compares actual demand flows in a single lane with the saturation flow rate and capacity of that lane. Critical-lane *group* analysis compares actual flow with the saturation flow rate and capacity of a group of lanes operating in equilibrium. Figure 21.1 illustrates the difference.

Where several lanes operate in equilibrium (i.e., where there are no lane-use restrictions impeding driver selection of lanes), the *lane group* is treated as a single entity.

Not all methodologies do this. Both the Australian and Canadian models focus on individual lanes, taking into account unequal use of lanes. The HCM also accounts for unequal use of lanes through a process of adjustments to saturation flow rates.

The key concept, however, is primarily one of organizing the methodology and approach. Whether individual lanes or lane groups are used, models can and do ensure that the ratio of *v* to *c* based on an individual lane is equal to the ratio of *v* to *c* of the entire lane group.

21.2.2 The *v/s* Ratio as a Measure of Demand

In Chapter 18, a signal timing methodology was based on conversion of demand volumes to "through-vehicle equivalents." This allowed volumes with markedly different percentages of right- and left-turning vehicles to be directly compared in the determination of "critical lanes." It was assumed that all other conditions that might affect the equivalency of volumes (heavy vehicles, grades, parking conditions, etc) were "typical."

In the HCM model, demand flow rates are not converted. They are stated as "veh/h" under prevailing conditions. Without conversion to some common base, flow rates cannot, therefore, be compared directly with critical lanes or lane groups.

The HCM model includes adjustments for a wide variety of prevailing conditions, including the presence of left- and right-turning vehicles. All adjustments, however, are applied to the saturation flow rate, not to demand volumes. As a result, the methodology yields saturation flow rates and capacities that are defined in terms of prevailing conditions. These are then compared with demand volumes that reflect the same prevailing conditions.

To obtain a single parameter that will allow the intensity of demand in each lane group to be compared directly, the demand flow rate, *v*, is divided by the saturation flow rate, *s*, to form the "flow ratio," *v/s*. Since the prevailing conditions in each lane group are reflected in both

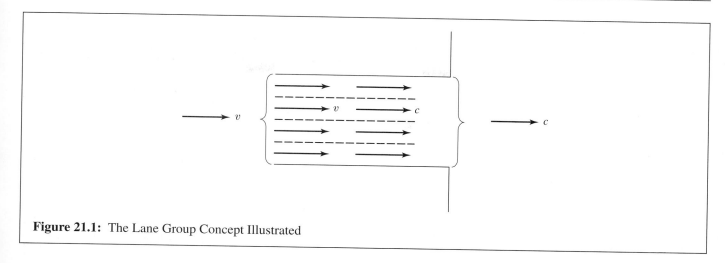

Figure 21.1: The Lane Group Concept Illustrated

the flow rate and the saturation flow rate values, this dimensionless number may be used to represent the magnitude of the demand in each lane group.

21.2.3   Capacity and Saturation Flow Rate Concepts

The HCM model does not produce a value for the capacity of the intersection. Rather, each lane group is considered separately, and a capacity for each is estimated.

Why not simply add all of the lane group capacities to find the capacity of the intersection as a whole? Doing so would ignore the fact that traffic demand does not reach its peak on all approaches at the same time. Unless the demand split on each of the lane groups matched the split of capacities, it would be impossible to successfully accommodate a total demand equal to a capacity so defined. Further, signal timings may change during various periods of the day, yielding significantly different capacities on individual lane groups and, indeed, different sums. In effect, the "capacity" of the intersection as a whole is not a useful or relevant concept. The intent of signalization is to allocate sufficient time to various lane groups and movements to accommodate demand. Capacity is provided to specific movements to accommodate movement demands.

The concept of intersection capacity should not be confused with the "sum of critical-lane volumes" introduced in Chapters 17 and 18. The latter considers only the critical lanes and depends upon a specified cycle length.

The concepts of saturation flow rate, capacity, and v/c ratios are all interrelated in the HCM analysis model.

Saturation Flow Rates

In Chapters 17 and 18, it was assumed that the saturation headway or saturation flow rate reflecting prevailing conditions was known. The key, and most complex, part of the HCM model is a methodology for estimating the saturation flow rate of any lane group based on known prevailing traffic parameters. The algorithm takes the form:

$$s_i = s_o N \prod_i f_i \qquad (21\text{-}1)$$

where: s_i = saturation flow rate of lane group i under prevailing conditions, veh/hg

   s_o = saturation flow rate per lane under base conditions, pc/hg/ln

   N = number of lanes in the lane group

   f_i = multiplicative adjustment factor for each prevailing condition i

The HCM now provides 11 adjustment factors covering a wide variety of potential prevailing conditions. Each adjustment factor involves a separate model, some of which are quite complex. These are described in detail later in the chapter.

Note that the algorithm includes multiplication of the base saturation flow rate by the number of lanes in the lane group, N. This produces a *total* saturation flow rate for the lane group in question.

Capacity of a Lane Group

The relationship between saturation flow rates and capacities is the same as that presented in Chapters 17 and 18. The saturation flow rate is an estimate of the capacity of a lane group if the signal were green 100% of the time. In fact, the signal is only effectively green for a portion of the time. Thus:

$$c_i = s_i\left({g_i}/{C}\right) \qquad (21\text{-}2)$$

where: c_i = capacity of lane group i, veh/h

s_i = saturation flow rate of lane group i, veh/hg

g_i = effective green time for lane group i, s

C = cycle length, s

The *v/c* Ratio

In signal analysis, the *v/c* ratio is often referred to as the "degree of saturation" and given the symbol "*X*." This is convenient, as the term "*v/c*" appears in many equations that can be more simply expressed using a single variable, X.

The *v/c* ratio, or degree of saturation, is a principal output measure from the analysis of a signalized intersection. It is a measure of the sufficiency of available capacity to handle existing or projected demands. Obviously, cases in which *v/c* > 1.00 indicate a shortage of capacity to handle the demand. Care must be taken, however, in analyzing such cases, depending on how the *v/c* value was determined.

Capacity, which is difficult to directly observe in the field, is most often estimated using Equation 21-2. Measured demands are usually a result of counts of *departure flows*. Departure flows are counted because it is easier to classify them by movement as they depart the intersection. True demand, however, must be based on *arrival flows*. There are several different

scenarios in which a *v/c* ratio in excess of 1.00 can be achieved:

1. A departure count is compared to a capacity estimated using Equation 21-2. It is theoretically impossible to count a departure flow that is in excess of the true capacity of the lane group. In this case, obtaining a *v/c* ratio greater than 1.00 (assuming the departure counts are accurate) represents an *underestimate* of the capacity of the lane group. The estimated saturation flow rate resulting from the HCM model is lower than the actual value being achieved.

2. An arrival count is compared to an estimate of capacity using Equation 21-2. In this case, the arrival count (assuming it is accurate) represents existing demand. A *v/c* ratio in excess of 1.00 indicates that queuing is likely to occur. If queues are, in fact, not observed, this is another indication that the capacity has been *underestimated* by the model.

3. A forecast future demand is compared to an estimated capacity using Equation 21-2. In this case, the forecast demand is always an arrival demand flow, and a *v/c* ratio in excess of 1.00 indicates that queuing is likely to occur, based upon the estimated value of capacity.

The key in all cases is that capacity is an *estimate* based upon nationally observed norms and averages. In any given case, the actual capacity can be either higher or lower than the estimate. In fact, actual capacity has stochastic elements and will vary over time at any given location and over space at different locations.

Analytically, the *v/c* ratio for any given lane group is found directly by dividing the demand flow rate by the capacity. Another expression can, however, be derived by inserting Equation 21-2 for capacity:

$$X_i = \frac{v_i}{c_i} = \frac{(v/s)_i}{(g/C)_i} \qquad (21\text{-}3)$$

where: X_i = degree of saturation (*v/c* ratio) for lane group i

v_i = demand flow rate for lane group i, veh/h

c_i = capacity for lane group i, veh/h

$(v/s)_i$ = flow ratio for lane group i

$(g/C)_i$ = green ratio for lane group i

Since demands are eventually expressed as v/s ratios in the HCM model, the latter form of the equation is often convenient for use.

While the HCM does not define a capacity for the entire intersection, it does define a *critical v/c ratio* for the intersection. It is defined as the sum of the demands on critical lane groups divided by the sum of the capacities of critical lane groups, or:

$$X_c = \frac{\sum_i v_{ci}}{\sum_i \left(s_{ci} * \frac{g_{ci}}{C} \right)} = \frac{\sum_i (v/s)_{ci}}{\sum_i \left(g_{ci}/C \right)} \qquad (21\text{-}4)$$

where: X_c = critical v/c ratio for the intersection

v_{ci} = demand flow rate for critical lane group i, veh/h

s_{ci} = saturation flow rate for critical lane group i, veh/hg

g_{ci} = effective green time for critical lane group i, s

C = cycle length, s

The term $\Sigma(g_{ci}/C)$ is the total proportion of the cycle length that is effectively green for all critical-lane groups. Since the definition of a critical-lane group is that one and only one such lane group must be moving during all phases, the only time a critical movement is *not* moving is during the lost times of the cycle. Thus, $\Sigma(g_{ci}/C)$ may also be expressed as:

$$\frac{C - L}{C}$$

where L is the total lost time per cycle. Inserting this into Equation 21-4 yields:

$$X_c = \frac{\sum_i (v/s)_{ci}}{\left(\frac{C - L}{C} \right)} = \sum_i (v/s)_{ci} * \left(\frac{C}{C - L} \right) \qquad (21\text{-}5)$$

This is the form used in the HCM.

As the value of X_c varies with cycle length, it is difficult to apply to future cases in which the exact signal timing may not be known. Thus, for analysis purposes, the 1997 edition of the HCM define a value of X_c based on the maximum feasible cycle length, which results in the minimum feasible value of X_c. For pretimed signals, the maximum feasible cycle length is usually taken to be 120 s, but this is sometimes exceeded in special situations. For actuated signals, longer cycle lengths are not as rare, and 150 s is usually used as a practical maximum. Equation 21-5 then becomes:

$$X_{cmin} = \sum_i (v/s)_{ci} * \left(\frac{C_{max}}{C_{max} - L} \right) \qquad (21\text{-}6)$$

where: X_{cmin} = minimum feasible v/c ratio

C_{max} = maximum feasible cycle length, s

The latter value is more useful in comparing future alternatives, particularly physical design scenarios. The cycle length is assumed to be the maximum and is, in effect, held constant for all cases compared. Use of the maximum cycle length gives a view of the "best" critical v/c ratio achievable through signal timing, given the physical design and the phase plan specified.

The critical v/c ratio, X_c, is an important indicator of capacity sufficiency in analysis. If X_c is ≤ 1.00, then the proposed physical design, cycle length, and phase plan are sufficient to handle all critical demands. This *does not* mean that all lane groups will operate at $X_i \leq 1.00$. It does, however, indicate that all critical lane groups can achieve $X_i \leq 1.00$ by reallocating the green time within the existing cycle and phase plan. When $X_c > 1.00$, then sufficient capacity may be provided only by one or more of the following actions:

• Increasing the cycle length
• Devising a more efficient phase plan
• Adding a lane or lanes to one or more critical lane groups

Increasing cycle length can add small amounts of capacity, as the lost time per hour is diminished. Devising a more efficient phase plan generally means considering additional left-turn protection or making a fully protected left turn a protected plus permitted left turn. It may also mean consideration of more complex phasing such as leading and lagging greens and/or exclusive LT phases followed by a leading green in the direction of heaviest left-turn flow. Chapter 18 contains full discussion of various phasing options.

In many cases, significant capacity shortfalls can be remedied only by adding one or more lanes to critical-lane groups. This increases the saturation flow rate and capacity of these lane groups, while the demand is constant.

21.2.4 Level-of-Service Concepts and Criteria

Level of service is defined in the HCM in terms of *total control delay* per vehicle in a lane group. "Total control delay" is basically *time in queue* delay, as defined in Chapter 17, plus acceleration-deceleration delay. Level-of-service criteria are shown in Table 21.1.

As delay is difficult to measure in the field and because it cannot be measured for future situations, delay is estimated using analytic models, some of which were discussed in Chapter 17. Delay is not a simple measure, however, and varies (in order of importance) with the following measures:

- Quality of progression
- Cycle length
- Green time
- *v/c* ratio

Table 21.1: Level-of-Service Criteria for Signalized Intersections

Level of Service	Control Delay (s/veh)
A	≤ 10
B	$>10-20$
C	$>20-35$
D	$>35-55$
E	$>55-80$
F	>80

(Used with permission of Transportation Research Board, National Research Council, *Highway Capacity Manual*, 4th Edition, Washington DC, 2000, Exhibit 16-2, pgs. 16-2.)

Because of this, level-of-service results must be carefully considered. It is possible, for example, to obtain a result in which delay is greater that 80 s/veh (LOS F) while the *v/c* ratio is less than 1.00. Thus, at a signalized intersection, LOS F does not necessarily imply that there is a capacity deficiency. Such a result is relatively common for short phases (such as LT phases) in a long cycle length, or where the green splits are grossly out of synch with demands.

The reverse is also true, particularly in the case of a short period of oversaturation. If *v/c* > 1.00 for a short time—one 15-minute interval, for example—delay could be less than 80 s/veh. Thus, a situation in which there is a capacity deficiency would be labeled with a level of service of E or better.

Understanding the results of a signalized intersection analysis will require consideration of *both* the level of service and the *v/c* ratio for each lane group. Only then can the results be understood in terms of the sufficiency of the capacity provided and of the acceptability of delays experienced by road users.

21.2.5 Effective Green Times and Lost Times

The relationship between effective green times and lost times is discussed in detail in Chapter 18. In terms of capacity analysis, any given movement has effective green time, g_i, and effective red time, r_i. Table 21.2 illustrates how these values are related to actual green, yellow, and red times in the HCM.

For convenience, the HCM model assumes that the total lost time in a given phase, t_L, occurs at the beginning of the phase. This simplifies some models and does not involve any quantitative errors but rather a shift in when the effective green time, g, is assumed to take place.

Effective green and red times may be found as follows:

$$g_i = G_i + Y_i - t_{Li}$$
$$g_i = G_i - \ell_1 + e$$
$$r_i = C - g_i \qquad (21\text{-}7)$$

Table 21.2: Effective Green Times and Lost Times in the HCM Model

	G			*y*	*ar*		*R*	
ℓ_1				*e*	ℓ_2		*R*	
t_L		*g*					*R*	
r		*g*					*r*	

where:

g_i = effective green time for Phase i, s

G_i = actual green time for Phase i, s

$Y_i = y_i + ar_i$, sum of yellow and all-red time for Phase i, s

t_{Li} = total lost time for Phase i, s

l_1 = start-up lost time, s

e = extension of effective green into yellow and all-red, s

Where there are overlapping phases, care must be taken in the application of lost times. They are applied when movements begin. When a movement continues into a subsequent phase, no additional lost time is incurred. This creates situations in which the lost time for one movement is actually effective green time for another.

The case of a leading and lagging green phase with protected plus permitted left turns is illustrated in Figure 21.2. EB movements begin in Phase 1a and continue in Phase 1b. The lost time is applied only in Phase 1a. WB movements, however, begin in Phase 1b and have their lost times applied there. Thus, in Phase 1b, lost time for WB movements is effective green time for

EB movements. As no movements begin in Phase 1c, no lost times are assessed here. All NB/SB movements flow in Phase 2; their lost times are assessed in this phase. Essentially, three sets of lost times are applied over four sub-phases. As effective green times affect capacity and delay, it is important that a systematic way of properly accounting for lost times be followed.

21.3 The Basic Model

21.3.1 Model Structure

The basic structure of the HCM model for signalized intersections is relatively straightforward and includes many of the conceptual treatments presented in Chapter 17. The model becomes extremely complex, however, when permitted or compound left turns are involved. The complex interactions between permitted left turns and opposing flows must be fully incorporated into models. This results in algorithms with many variables and several iterative aspects.

In this section, the building blocks of the HCM procedure will be described and illustrated for basic

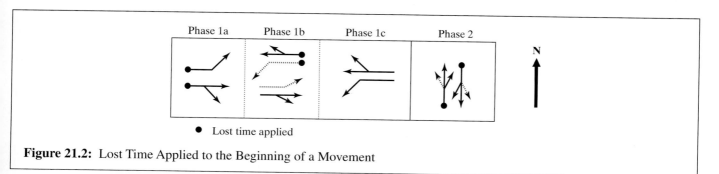

Figure 21.2: Lost Time Applied to the Beginning of a Movement

cases involving only protected left turns. In this way, the fundamental approach of the methodology can be presented without diversions into lengthy detail concerning permitted left turns and other complications. Subsequent sections of the chapter will address these details.

The structure of the HCM model is illustrated in Figure 21.3. The analysis methodology is modular and works with a fully-specified signalized intersection (volumes, geometrics, and signalization fully described) to determine v/c ratios and delays for each lane group in the intersection. When first introduced in the 1985 HCM, each module represented a worksheet. As the methodology has become more complex, there are now many supplementary worksheets,

and some of the iterative portions of the model are difficult to summarize on worksheets.

The model consists of an Input Module, in which all relevant prevailing conditions for all intersection approaches are summarized for the analysis. The Volume Adjustment Module and the Saturation Flow Rate Module are completed in parallel. In the first, peak hour flow rates are estimated and lane groups for analysis are defined. In the second, the saturation flow rate for each lane group is estimated. The results of these computations lead to the Capacity Analysis Module, in which critical lane groups are identified, and lane group v/c ratios, and the critical v/s ratio are computed. The results of the Capacity Analysis Module are used to estimate delays in the Level-of-Service Module.

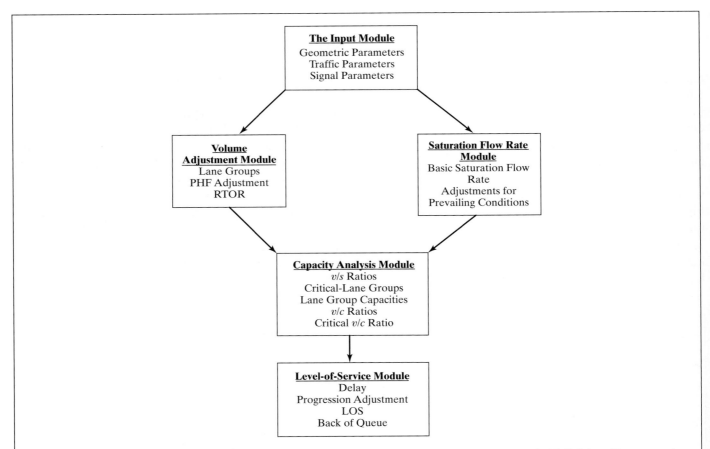

Figure 21.3: Modular Structure of the HCM Model (Modified from *Highway Capacity Manual*, 4th Edition, Transportation Research Board, National Research Council, Washington DC, 2000, Exhibit 16-1, pg. 16-2.)

21.3.2 Analysis Time Periods

The basic time period for analysis recommended by the HCM remains a peak 15-minute period within the analysis hour, which is most often (but need not be) one of the peak hours of the day. The HCM 2000, however, provides for some flexibility in this regard, recognizing that delay is particularly sensitive to the analysis period, especially when oversaturation exists. There are three basic time options for analysis:

1. The peak 15 minutes within the analysis hour
2. The full 60-min analysis hour
3. Sequential 15-min periods for an analysis period of one hour or greater

The first option is appropriate in cases where no oversaturation exists (i.e., no lane groups have $v/c > 1.00$). This focuses attention on the worst period within the analysis hour, where 15 minutes remains the shortest period during which stable flows are thought to exist. The second option allows for an analysis of average conditions over the full analysis hour. It could, however, mask shorter periods during which $v/c > 1.00$, even though the full hour has sufficient capacity.

The third option is the most comprehensive. It requires, however, that demand flows be measured or predicted in 15-minute time increments, which is often difficult. It allows, however, for the most accurate analysis of oversaturation conditions. The initial 15-minute period of analysis would be selected such that all lane groups operate with $v/c \leq 1.00$, and would end in a 15-minute period occurring after all queues have been dissipated. During each 15-minute period, residual queues of unserved vehicles would be estimated and would be used to estimate additional delay due to a queue existing at the start of an analysis period in the subsequent interval. In this way, the impact of residual queues on delay and level of service in each successive period can be estimated.

21.3.3 The Input Module

The Input Module involves a full specification of all prevailing conditions existing on all approaches to the signalized intersection under study. As the model has become more complex in order to address a wider range of conditions, the input information required has also become more comprehensive.

Table 21.3 summarizes all of the input data needed to conduct a full analysis of a signalized intersection.

Most of the variables included in Table 21.3 have been previously defined. Others require some additional definition or discussion. The HCM also provides recommendations for default values that may be used in cases where field data on a particular characteristic is not available. Caution should be exercised in using these, as the accuracy of v/c, delay, and level of service predictions is influenced.

Geometric Conditions

Area type is a somewhat controversial variable. The intersection must be characterized as either being in a central business district (CBD), or not (Other). Calibration studies [5] for the original procedure in the 1985 HCM indicated that intersections located within CBDs have saturation flow rates approximately 10% lower than those located outside the CBD.

This has become a controversial factor, as the definition of a CBD is quite relative. For example, should the business district of a small satellite community be classified as a "CBD" or "Other" location? In general, if drivers are familiar with driving in a big city CBD, then all locations in satellite communities should be classified as "Other." On the other hand, in an isolated rural community, even a small business area would be classified as a "CBD." The general theory is that the busier environment of the CBD causes drivers to be more cautious than in other areas.

Parking conditions must be specified as well. For a typical two-way street, each approach either has curb parking or not. On a one-way street approach, parking may exist on the right and/or left side or on neither. For the purposes of the intersection, curb parking is noted only if it exists within 250 ft of the STOP line of the approach. Most of the other geometric conditions that must be specified are commonly used variables that have been defined elsewhere in this text.

Table 21.3: Data Requirements for Each Lane Group in Signalized Intersection Analysis

Type of Condition	Parameter
Geometric Conditions	Area Type (CBD, Other) Number of Lanes, N Average Lane Width, W (ft) Grade, G (%) Existence of LT or RT Lanes Length of Storage Bay for LT or RT lane (ft) Parking Conditions (Yes/No)
Traffic Conditions	Demand Volume by Movement, V (veh/h) Base Saturation Flow Rate, s_o (pc/hg/ln) Peak Hour Factor, PHF Percent Heavy Vehicles, P_T (%) Pedestrian Flow in Conflicting Crosswalk, v_p (peds/h) Local Buses Stopping at Intersection, N_B (buses/h) Parking Activity, N_m (maneuvers/h) Arrival Type, AT Proportion of Vehicles Arriving on Green, P Approach Speed, S_A (mi/h)
Signalization Conditions	Cycle Length, C (s) Green Time, G (s) Yellow Plus All-Red Interval, Y (s) Type of Operation (Pretimed, Semi-Actuated, Full Actuated) Pedestrian Push Button (Yes/No) Minimum Pedestrian Green, G_p (s) Phase Plan Analysis Period, T (h)

(Modified from *Highway Capacity Manual*, 4th Edition, Transportation Research Board, National Research Council, Washington DC, 2000, Exhibit 16-3, pg. 16-3.)

The HCM provides default values for only two of the geometric variables in Table 21.3, as follows:

- *Lane width*, W: default value = 12 ft
- *Grade*, G: default values = 0% (level); 3% (moderate grades); 6% (steep grades)

Using a default for grade involves at least a general categorization of the grade from field observations.

Where wide lanes exist, some observation of their use should be made. Lanes of 18–20 ft often become two lanes under intense demand, particularly if it is a curb lane that could be used as a through lane plus a narrow RT lane. The analyst should try to characterize lanes as they would be used, not necessarily as they are striped. In most cases, however, these would be the same.

Traffic Conditions

There are a number of interesting variables included in the list of traffic conditions to be specified.

Arrival Type "Arrival type" is used to describe the quality of progression for vehicles arriving on each approach. Arrival type has a major impact on delay predictions but does not have significant influence on other portions of the methodology. There are six defined arrival types, 1 through 6, with AT 1 representing the worst progression quality and AT 6 representing the best progression quality. Definitions are given in Table 21.4.

Arrival types were originally defined only in verbal terms, as in Table 21.4. The intent was that traffic engineers would use their judgment and knowledge of progression to assign a reasonable value based on field observations or study of a time-space diagram for the arterial. This is still an appropriate way to determine the arrival type.

A numerical guideline, the *platoon ratio*, was added to provide for greater clarity. The platoon ratio is defined as the ratio of proportion of vehicles arriving on green to the g/C ratio of the lane group or approach:

$$R_p = \frac{P}{(g/C)} = \frac{P*C}{g} \qquad (21\text{-}8)$$

where: R_p = platoon ratio

P = proportion of vehicles arriving on green

C = cycle length, s

g = effective green time, s

The variable was originally created because the value for random arrivals (AT 3) logically had to be 1.00. If random arrivals occur with equal probability at any time, then it would be expected that vehicles would arrive on green in the same proportion as the green ratio. Subsequent research [6] established the relationship between the platoon ratio and other arrival types, as illustrated in Table 21.5.

If the platoon ratio is observed in the field, then the exact value is used in computations. If the arrival type has been established through exercise of judgment on

Table 21.4: Arrival Types Defined

Arrival Type	Description
1	Dense platoon containing over 80% of the lane group volume, arriving at the start of the red phase. This AT is representative of network links that may experience very poor progression quality as a result of conditions such as overall network optimization.
2	Moderately dense platoon arriving in the middle of the red phase, or dispersed platoon containing 40% to 80% of the lane group volume, arriving throughout the red phase. This AT is representative of unfavorable progression on two-way streets.
3	Random arrivals in which the main platoon contains less than 40% of the lane group volume. This AT is representative of operations at isolated and non-interconnected signalized intersections characterized by highly dispersed platoons. It may also be used to represent uncoordinated operation in which the benefits of progression are minimal.
4	Moderately dense platoon arriving in the middle of the green phase, or dispersed platoon containing 40% to 80% of the lane group volume, arriving throughout the green phase. This AT is representative of favorable progression on a two-way street.
5	Dense to moderately dense platoon containing over 80% of the lane group volume, arriving at the start of the green phase. This AT is representative of highly-favorable progression quality, which may occur on routes with low to moderate side-street entries and which receive high-priority treatment in the signal timing plan.
6	This arrival type is reserved for exceptional progression quality with near-ideal progression characteristics. It is representative of very dense platoons progressing over a number of closely-spaced intersections with minimal or negligible side-street entries.

Table 21.5: Relationship Between Platoon Ratio and Arrival Type

Arrival Type	Range of Platoon Ratio, R_p	Default Value R_p	Progression Quality
1	≤0.50	0.333	Very Poor
2	>0.50−0.85	0.667	Unfavorable
3	>0.85−1.15	1.000	Random Arrivals
4	>1.15−1.50	1.333	Favorable
5	>1.50−2.00	1.667	Highly Favorable
6	2.00	2.000	Exceptional

(Used with Permission of Transportation Research Board, National Research Council, *Highway Capacity Manual*, 4th Edition, Washington DC, 2000, Exhibit 16-11, pg. 16-20.)

the basis of verbal definitions, then the default values for R_p are used in computations.

If the arrival type is not known, the HCM provides the following default recommendations:

- AT-3 for uncoordinated signals
- AT-4 for coordinated signals

Given the significant impact arrival type will have on delay estimates, it is important that a common arrival type be used when comparing different intersection designs and signal timings. High delays should not be simply dismissed or mitigated by assuming an improved progression quality.

Pedestrian Flows Pedestrian flows in conflicting crosswalks must be specified for a signalized intersection analysis. The "conflicting crosswalk" is the crosswalk that right-turning vehicles turn through. Pedestrian flows are best measured in the field, but the HCM provides the following default values:

- *CBD locations*: 400 peds/h per crosswalk
- *Other locations*: 50 peds/h per crosswalk

Parking Activity Parking activity is measured in terms of the number of parking maneuvers per hour into and out of parking spaces (N_m) located within 250 ft of the STOP line of the lane group or approach in question. Movements into and out of parking spaces have additional negative impacts on operations, as the lane adjacent to the parking lane is disrupted for some finite amount of time each time such a maneuver takes place. This is in addition to the frictional impacts of traveling in the lane adjacent to the parking lane. Parking activity should be observed in the field, but it is often not readily available. The HCM recommends the use of the default values shown in Table 21.6 in such cases.

Table 21.6: Recommended Default Values for Parking Activity

Street Type	Number of Parking Spaces Within 250 ft of STOP Line	Parking Time Limit (h)	Maneuvers per Hour (N_m): Recommended Default Value
Two-Way	10	1	16
		2	8
One-Way	20	1	32
		2	16

(Used with Permission of Transportation Research Board, National Research Council, *Highway Capacity Manual*, 4th Edition, Washington DC, 2000, Exhibit 10-20, pg. 10-25.)

Local Buses A local bus is defined as one that stops within the confines of the intersection, either on the near side or far side of the intersection, to pick up and/or discharge passengers. A local bus that does not stop at the intersection is included as a heavy vehicle in the heavy vehicle percentage. Once again, this is a variable best measured in the field. If field measurements and/or bus schedules are not available, the HCM recommends the following default values be used (assuming there is a bus route on the street in question with a bus stop within the confines of the intersection):

- *CBD location*: 12 buses/h
- *Other location*: 2 buses/h

Other Default Values Other traffic conditions are well defined elsewhere in this text. The HCM does recommend default values for some of these variables if they are unavailable from field data or projections:

- *Base saturation flow rate*: 1,900 pc/hg/ln
- *Heavy vehicle presence*: 2.0%
- *Peak hour factor*:

  (a) 0.92 for congested conditions

  (b) 0.88 for uncongested conditions displaying clear variation of flow within the analysis period

As noted previously, use of default values should be avoided whenever local field data or projections are available. For each default value used in lieu of specific data for the intersection, the accuracy of predicted operating conditions becomes less reliable.

Signalization Conditions

Virtually all of the signalization conditions that must be specified for an analysis are available from field observations of signal operation or from a signal design and timing analysis. The HCM provides a procedure for "quick" estimation of signal timing that involves many defaults and assumptions. The "quick" method can become relatively complex, despite the simplifying assumptions. The signal timing approach outlined in Chapter 18 of this text is less complex and can be used to get an initial signal timing for a capacity and

level-of-service analysis. The analysis time period (T) has been discussed earlier in this chapter, and no further discussion is needed.

The analysis methodology uses an algorithm to determine the adequacy of pedestrian crossing times. It is the same as that recommended for use in signal timing in Chapter 18, or:

$$G_p = 3.2 + \left(\frac{L}{S_p}\right) + 0.27 N_{peds} \qquad \text{for } W_E \leq 10 \text{ ft}$$

$$G_p = 3.2 + \left(\frac{L}{S_p}\right) + 2.7\left(\frac{N_{ped}}{W_E}\right) \qquad \text{for } W_E > 10 \text{ ft}$$

(21-9)

where: G_p = minimum green time for safe pedestrian crossings, s

L = length of the crosswalk, ft

S_p = walking speed of pedestrians, ft/s

N_{peds} = number of pedestrians crossing during one green interval, peds/cycle

W_E = width of the crosswalk, ft

The HCM checks green times against these minimum values and issues a warning statement if pedestrian crossings are unsafe under these guidelines. Analysis, however, may continue whether or not the minimum pedestrian green condition is met by the signalization or not.

It should also be noted that many local and state agencies have their own policies on what constitutes safe crossings for pedestrians. These may always be applied as a substitute for Equation 21-9.

21.3.4 The Volume Adjustment Module

There are three analytic functions that take place in the Volume Adjustment Module:

- Conversion of hourly demand volumes to peak 15-min flow rates
- Establishment of lane groups for analysis
- Determination of total lane group demand flow rates

Conversion of hourly demand volumes to flow rates is accomplished using the peak hour factor (*PHF*):

$$v = \frac{V}{PHF} \qquad (21\text{-}10)$$

where: v = demand flow rate, veh/h

V = demand volume, veh/h

PHF = peak-hour factor

Use of this conversion, however, assumes that all component movement flows peak during the *same 15-min interval* of the analysis hour. This is generally not true and is, therefore, a worst-case assumption.

Alternatively, flow rates within 15-min time intervals can be established through field observation of existing conditions or projections of future situations. These flow rates may be used directly in the methodology. To avoid any mistaken adjustment of flow rates, the *PHF* in such cases is set at 1.00.

Vehicles making *right turn on red* may be deducted from the right-turn demand volume (before conversion to flow rates) on the basis of field observations or projection. Channelized right turns that are not controlled by the signal are not included in the signal analysis, although care should be taken to ensure that right-turn queuing is not blocking or partially blocking signal approach lanes.

The rules for establishing lane groups are simple: Lanes (or sets of lanes) serving only designated movements through lane-use controls are considered to be separate lane groups for analysis. In the absence of such lane-use controls, all approach lanes are considered to be one lane group. Separate lane groups on an approach are established only for exclusive left-turn and/or right-turn lanes. Figure 21.4 illustrates the possible lane group designations on a single intersection approach.

It is possible that "de facto" left- and/or right-turn lanes exist where they are not designated. This occurs when the turning demand is so intense that it fully occupies a lane. The methodology contains an internal check for this condition, which requires an iteration when triggered. "De facto" LT or RT lanes are never assumed, unless field observations confirm their existence. In the latter case, the problem is then originally structured to include the exclusive turning lane or lanes.

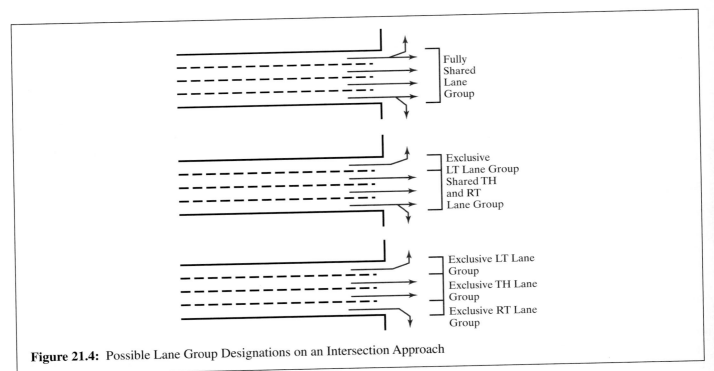

Figure 21.4: Possible Lane Group Designations on an Intersection Approach

A single signalized intersection approach can have up to three separate lane groups when both exclusive RT and LT lanes are provided. Thus, a standard intersection of two two-way streets may have a minimum of four lane groups (one for each approach) and a maximum of 12 lane groups (one for each movement, assuming exclusive lanes for all LT and RT movements).

Once the lane groups are established, the demand flow rates for all movements served by the lane group are added to determine the total lane group demand flow rate:

$$v_{gi} = \sum_j v_j \qquad (21\text{-}11)$$

where: v_{gi} = total demand flow rate for lane group i, veh/h

v_j = demand flow rate for movement j in lane group i, veh/h

21.3.5 The Saturation Flow Rate Module

In the Saturation Flow Rate Module, a base saturation flow rate is modified by a series of multiplicative adjustment factors to determine the total saturation flow rate for each lane group under prevailing demand conditions. It is in this module that most of the complexity of the HCM model occurs, related almost entirely to permitted left-turn movements. The saturation flow rate for a lane group is estimated as:

$$s = s_o N f_w f_{HV} f_g f_p f_{bb} f_a f_{LU} f_{RT} f_{LT} f_{Rpb} f_{Lpb}$$

$$(21\text{-}12)$$

where: s = saturation flow rate for the lane group, veh/h

s_o = base saturation flow rate, pc/hg/ln (1,900 pc/hg/ln unless field data has been used to establish a locally calibrated value)

N = number of lanes in the lane group

f_i = adjustment factor for prevailing condition i (w = lane width; HV = heavy vehicles; g = grade; p = parking; bb = local bus blockage; a = area type; LU = lane use;

RT = right turn; LT = left turn; Rpb = pedestrian/bicycle interference with right turns; and Lpb = pedestrian/bicycle interference with left turns)

Of these 11 adjustment factors, 8 were introduced with the original methodology in the 1985 HCM. The lane-use adjustment was moved from the volume side of the equation to the saturation flow rate side in the 1997 HCM. The pedestrian/bicycle interference factors were added in HCM 2000.

Eight of the adjustments are relatively straightforward. Pedestrian/bicycle adjustments are a little complex, but are still tractable and can be done manually with a little care. The left-turn adjustment is straightforward unless permitted or compound left turns are involved, in which case the model becomes cumbersome and impractical to implement manually. In this section, left-turn adjustments are defined only for protected turns. A subsequent section will discuss the more complex model for permitted left turns.

Adjustment for Lane Width

The lane-width adjustment is computed using Equation 21-13. The standard lane width is 12 ft, and the adjustment can be either negative (for narrower lanes) or positive (for wider lanes). For lane widths of 18 ft–20 ft, consideration should be given to analyzing the situation as two narrower lanes, as drivers will often form two lanes where possible under intense demand pressure. No lane width narrower than 9 ft should be analyzed, and lane widths less than 10 ft are not recommended.

$$f_w = 1 + \left(\frac{W - 12}{30} \right) \qquad (21\text{-}13)$$

where: f_w = lane-width adjustment factor

W = average lane-width for the lane group, ft

Adjustment for Heavy Vehicles

The adjustment for heavy vehicles is done separately from consideration of grade impacts. This separation, different from the approach taken for uninterrupted flow facilities, recognizes that approach grades affect the operation of all vehicles, not just heavy vehicles. A "heavy vehicle" is any vehicle with more than four wheels touching the ground during normal operations.

$$f_{HV} = \frac{1}{1 + P_{HV}(E_{HV} - 1)} \qquad (21\text{-}14)$$

where: f_{HV} = heavy-vehicle adjustment factor

P_{HV} = proportion of heavy vehicles in the lane group demand flow

E_{HV} = passenger-car equivalent for a heavy vehicle

For signalized intersections, the value of E_{HV} is a constant: 2.00.

Heavy vehicles are not segmented into separate classes. Thus, they include trucks, recreational vehicles, and buses not stopping within the confines of the intersection. Buses that do stop within the confines of the intersection are treated as a separate class of vehicles: local buses.

Adjustment for Grade

The adjustment for grade is found as:

$$f_g = 1 - \left(\frac{G}{200}\right) \qquad (21\text{-}15)$$

where: f_g = grade adjustment factor
G = grade %

Remember that downgrades have a negative percentage, resulting in an adjustment in excess of 1.00. Upgrades have a positive percentage, resulting in an adjustment of less than 1.00.

Adjustment for Parking Conditions

The parking adjustment factor involves two variables: parking conditions and movements, and the number of lanes in the lane group. If there is no parking adjacent to the lane group, the factor is, by definition, 1.00. If there is parking adjacent to the lane group, the impact on the

lane directly adjacent to the parking lane is a 10% loss of capacity due to the frictional impact of parked vehicles, plus 18 s of blockage for each movement into or out of a parking space within 250 ft of the STOP line. Thus, the impact on an adjacent lane is:

$$P = 0.90 - \left(\frac{18\,N_m}{3,600}\right)$$

where: P = adjustment factor applied only to the lane adjacent to parking lane

N_m = number of parking movements per hour into and out of parking spaces within 250 ft of the STOP line (mvts/h)

It is then assumed that the adjustment to additional lanes in the lane group is 1.00 (unaffected), or:

$$f_p = \frac{(N - 1) + P}{N}$$

where: N = number of lanes in the lane group

These two expressions are combined to yield the final equation for the parking adjustment factor:

$$f_p = \frac{N - 0.10 - \left(\dfrac{18 N_m}{3,600}\right)}{N} \qquad (21\text{-}16)$$

There are several external limitations on this equation:

- $0 \le N_m \le 180$; if $N_m > 180$, use 180 mvts/h
- $f_p(\text{min}) = 0.05$
- $f_p(\text{no parking}) = 1.00$

On a one-way street with parking on both sides, N_m is the total number of right- and left-parking lane maneuvers.

Adjustment for Local Bus Blockage

The local bus blockage factor accounts for the impact of local buses stopping to pick up and/or discharge passengers at a near-side or far-side bus stop within 250 ft of the near or far STOP line. Again, the primary impact is on the lane in which the bus stops (or the lane adjacent in cases where an off-line bus stop is provided). It assumes that each bus blocks the lane for 14.4 s of green time. Thus:

$$B = 1.0 - \left(\frac{14.4 N_B}{3,600}\right)$$

where: B = adjustment factor applied only to the lane blocked by local buses

N_B = number of local buses per hour stopping

As in the case of the parking adjustment factor, the impact on other lanes in the lane group is assumed to be nil, so that an adjustment of 1.00 would be applied. Then:

$$f_{bb} = \frac{(N - 1) + B}{N}$$

where: N = number of lanes in the lane group

Combining these equations yields:

$$f_{bb} = \frac{N - \left(\dfrac{14.4 N_B}{3,600}\right)}{N} \qquad (21\text{-}17)$$

There are also several limitations on the use of this equation:

- $0 \leq N_B \leq 250$; if $N_B > 250$, use 250 b/h
- $f_{bb}(\min) = 0.05$

If the bus stop involved is a terminal location and/or layover point, field studies may be necessary to determine how much green time each bus blocks. The value of 14.4 s in Equation 21-17 could then be replaced by a field-measured value in such cases.

Adjustment for Type of Area

As noted previously, signalized intersection locations are characterized as "CBD" or "Other," with the adjustment based on this classification:

- *CBD location*: $f_a = 0.90$
- *Other location*: $f_a = 1.00$

This adjustment accounts for the generally more complex driving environment of central business districts and the extra caution that drivers often exercise in such environments.

Adjustment for Lane Utilization

The adjustment for lane utilization accounts for unequal use of lanes by the approaching demand flow in a multi-lane group. Where demand volumes can be observed on a lane-by-lane basis, the adjustment factor may be directly computed as:

$$f_{LU} = \frac{v_g}{v_{g1} N} \qquad (21\text{-}18)$$

where: v_g = demand flow rate for the lane group, veh/h

v_{g1} = demand flow rate for the single lane with the highest volume, veh/h/ln

N = number of lanes in the lane group

When applied in this fashion, the factor adjusts the saturation flow rate downward so that the resulting v/c ratios and delays represent, in effect, conditions in the worst lane of the lane group. While the HCM states that a lane utilization factor of 1.00 can be used "when uniform traffic distribution" can be assumed, the use of this factor is no longer optional, as was the case in the 1985 and 1994 editions of the manual.

Where the lane distribution cannot be measured in the field or predicted reliably, the default values shown in Table 21.7 may be used.

Adjustment for Right Turns

The right-turn adjustment factor accounts for the fact that such vehicles have longer saturation headways than through vehicles, as they are turning on a tight radius requiring reduced speed and greater caution. This factor no longer accounts for pedestrian interference with right-turning vehicles as in previous editions of the HCM. With the introduction of new adjustments for pedestrian/bicycle interference, this element was

Table 21.7: Default Values for the Lane Utilization Adjustment Factor

Lane Group Movements	No. of Lanes In Lane Group	Traffic in Most Heavily Traveled Lane (%)	Lane Utilization Factor f_{LU}
Through or Shared	1	100.0	1.000
	2	52.5	0.952
	3[a]	36.7	0.908
Exclusive Left Turn	1	100.0	1.000
	2[a]	51.5	0.971
Exclusive Right Turn	1	100.0	1.000
	2[a]	56.5	0.885

[a] If lane group has more lanes, field observations are recommended; if not, use smallest value shown in this exhibit.
(Used with Permission of Transportation Research Board, National Research Council, *Highway Capacity Manual*, 4th Edition, Washington DC, 2000, Exhibit 10-23, pg. 10-26.)

removed from the right-turn adjustment factor. Right turns occur under three different scenarios:

- From an exclusive RT lane
- From a shared lane
- From a single-lane approach

Operations in single-lane approaches are somewhat more complex than in multilane groups, as this is the only situation in which left- and right-turning vehicles share a lane and, therefore, interact. In general, a right-turning vehicle has a saturation flow rate that is 15% less than that of a comparable through vehicle. Thus:

$$f_{RT} = 0.85 \qquad \text{for exclusive lanes}$$
$$f_{RT} = 1.0 - 0.15 P_{RT} \qquad \text{for shared lanes}$$
$$f_{RT} = 1.0 - 0.135 P_{RT} \qquad \text{for single lanes}$$

$$(21\text{-}19)$$

where: f_{RT} = right-turn adjustment factor

P_{RT} = proportion of right-turning vehicles in the lane group

Adjustment for Left Turns

If not for the existence of left turns, the HCM model for signalized intersections would be relatively straight-forward. There are six basic situations in which left turns may be made:

- *Case 1*: Exclusive LT lane with protected phasing
- *Case 2*: Exclusive LT lane with permitted phasing
- *Case 3*: Exclusive LT lane with compound phasing
- *Case 4*: Shared lane with protected phasing
- *Case 5*: Shared lane with permitted phasing
- *Case 6*: Shared lane with compound phasing

All of these options are frequently encountered in the field, with the exception of Case 4, which exists primarily on one-way streets with no opposing flows. The modeling of fully protected LT phasing (Cases 1 and 4) is straightforward and is described here. The more complicated cases, all involving permitted or compound left turns, are discussed later in the chapter.

As was the case for right turns, left-turning vehicles have a lower saturation flow rate than through vehicles due to the fact that they are executing a turning maneuver on a restricted radius. Left-turning vehicles are assumed to have a saturation flow rate 5% lower than that of a corresponding through vehicle. The reduction in saturation flow rate for left-turning vehicles is less than that for right-turning vehicles, as the radius of curvature involved in the maneuver is greater. Right-turning vehicles have a sharper turn than left-turning vehicles. The left-turn adjustment factor for fully protected phases is computed as:

$$f_{LT} = 0.95 \qquad \text{for exclusive lanes (Case 1)}$$

$$f_{LT} = \frac{1}{1.0 + 0.05 P_{LT}} \qquad \text{for shared lanes (Case 4)}$$

$$(21\text{-}20)$$

where: f_{LT} = left-turn adjustment factor

P_{LT} = proportion of left-turning vehicles in lane group

Adjustments for Pedestrian and Bicycle Interference with Turning Vehicles

These two new adjustment factors, added in the HCM 2000, account for the interference of both pedestrians and bicycles with right- and left-turning vehicles at a signalized intersection. Their determination is somewhat more complicated than the other adjustments presented in this section:

$$f_{Rpb} = 1.0 - P_{RT} \left(1 - A_{pbT} \right) \left(1 - P_{RTA} \right)$$

$$f_{Lpb} = 1.0 - P_{LT} \left(1 - A_{pbT} \right) \left(1 - P_{LTA} \right)$$

$$(21\text{-}21)$$

where: f_{Rpb} = adjustment factor for ped/bicycle interference with right turns

f_{Lpb} = adjustment factor for ped/bicycle interference with left turns

P_{RT} = proportion of right turns in the lane group

P_{LT} = proportion of left turns in the lane group

A_{pbT} = ped/bicycle adjustment for the permitted portion of the phase

P_{RTA} = proportion of total right-turn green time in a protected phase

P_{LTA} = proportion of total left-turn green time in a protected phase

The interaction of right and left turns with pedestrian and bicycle flows is illustrated in Figure 21.5.

Right-turning vehicles encounter both pedestrian and bicycle interference virtually immediately upon starting their maneuver. Left-turning vehicles will encounter pedestrian interference after moving through a gap in the opposing traffic flow. As bicycles are legally required to be on the right side of the roadway, it is assumed that left turns experience no bicycle interference.

The following steps must be followed to compute the critical variable of Equations 21-21, A_{pbT}. Remember that there are *two* values of this variable, one for interference with right-turning vehicles, and the other for interference with left-turning vehicles.

Step 1: Estimate Pedestrian Flow Rate During Green Phase (Left and Right Turns)

This is the actual pedestrian demand flow rate during the green phase. The rate is adjusted to reflect the fact that pedestrians are moving only during the green phase of the signal.

$$v_{pedg} = v_{ped} \left(C / g_p \right) \qquad (21\text{-}22)$$

where: v_{pedg} = flow rate for pedestrians during the green phase, peds/hg

v_{ped} = demand flow rate for pedestrians during analysis period, peds/h

C = cycle length, s

g_p = green phase for pedestrians, s

The value of g_p is taken to be the sum of the pedestrian walk and clearance intervals where they exist, or the length of the vehicular green where no pedestrian signals exist.

Step 2: Estimate the Average Pedestrian Occupancy in the Crosswalk (Left and Right Turns)

"Occupancy" measures represent the proportion of green time during which pedestrians and/or bicycles are present in a particular area for which the measure is defined. The occupancy of pedestrians

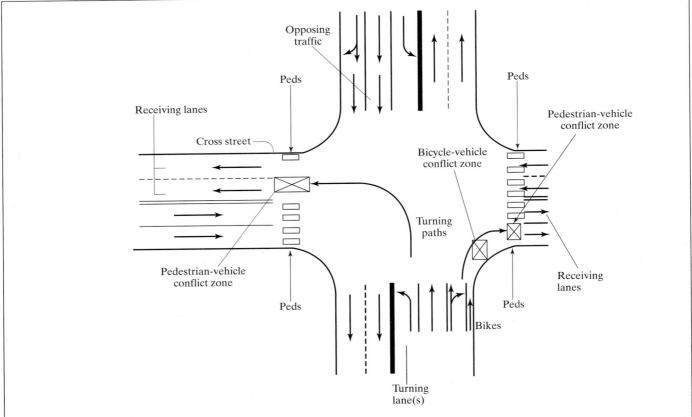

Figure 21.5: Pedestrian and Bicycle Interference with Turning Vehicles (Used with Permission of Transportation Research Board, National Science Foundation, *Highway Capacity Manual*, 4th Edition, Washington DC, 2000, Exhibit D16-1, pg. 16-135.)

in the conflict area of the crosswalk is given by Equation 21-23:

$$OCC_{pedg} = \frac{v_{pedg}}{2,000} \qquad \text{for } v_{pedg} \leq 1,000 \text{ peds/hg}$$

$$OCC_{pedg} = 0.40 + \left(\frac{v_{pedg}}{10,000}\right) \text{ for } 1,000 < v_{pedg} < 5,000$$

$$(21\text{-}23)$$

where: OCC_{pedg} = occupancy of the pedestrian conflict area by pedestrians

All other variables as previously defined

The first equation assumes that each pedestrian blocks the crosswalk conflict area for approximately 1.8 s, considering walking speeds and the likelihood of parallel crossings. At higher demand flows, the likelihood of parallel crossings is much higher, and each additional pedestrian blocks the crosswalk conflict area for another 0.36 s.

Step 3:   Estimate the Bicycle Flow Rate and Occupancy During the Green Phase (Right Turns Only)

The bicycle flow rate during the green phase is found in the same way that the pedestrian flow rate during green was estimated:

$$v_{bicg} = v_{bic}\left(\frac{C}{g}\right) \qquad (21\text{-}24)$$

where: v_{bicg} = bicycle flow rate during the green phase, bic/hg

v_{bic} = demand flow rate for bicycles during the analysis period, vic/h

C = cycle length, s

g = effective green time for vehicular movement, s

Bicycle occupancy may then be estimated as:

$$OCC_{bicg} = 0.02 + \left(v_{bicg} / 2,700 \right) \qquad (21\text{-}25)$$

where: OCC_{bicg} = occupancy of the conflict area by bicycles

All other variables are as previously defined.

Step 4: Estimate the Conflict Zone Occupancy (Left and Right Turns)

The occupancies computed in Steps 2 and 3 treat each element (pedestrians and bicycles) separately. Further, it is assumed that turning vehicles are present to block during all portions of the green phase.

In the simple case of *left turns from a one-way street*, the pedestrian occupancy is the conflict-zone occupancy. There is no bicycle interference with left turns, and there is nothing that limits the access of left-turning vehicles to the conflict zone:

$$OCC_r = OCC_{pedg} \qquad (21\text{-}26)$$

where: OCC_r = occupancy of the conflict zone

Equation 21-26 may also be used when protected left turns are made on a two-way street. In many instances, pedestrians are not permitted to cross during such phases, and no adjustment would be necessary. However, there are cases in which pedestrians crossing move (either legally or illegally) during a protected left-turn phase.

When *permitted left turns from a two-way street* are involved, the situation is more complicated. Left-turning vehicles from a two-way street may be blocked by opposing vehicles from getting to the conflict zone. The desired occupancy rate must indicate the proportion of time *that left turns are actually being made* that is blocked by pedestrians. This is a complex gap-acceptance problem, resulting in the following relationship:

$$OCC_r = 1.0 \qquad \text{for } g_q \geq g_p$$

$$OCC_r = OCC_{pedg}\left[1 - 0.50\left(g_q / g_p \right)\right]\left[e^{-(5/3,600)v_o}\right]$$

$$\text{for } g_q < g_p \qquad (21\text{-}27)$$

where: g_q = portion of green phase blocked by the discharge of an opposing queue of vehicles, s

g_p = WALK plus clearance interval, s

v_o = opposing flow rate, veh/h

The value of g_q has not yet been discussed, as it results from the adjustment process for permitted left turns, covered later in this chapter.

For *right-turning vehicles*, both pedestrian and bicycle occupancies interfere. However, the two interfering flows overlap, and simply adding the two occupancy values would result in too great an adjustment. Allowing for the overlapping impact of pedestrians and bicycles on right-turning vehicles:

$$OCC_r = OCC_{pedg} + OCC_{bicg} - \left(OCC_{pedg} * OCC_{bicg}\right) \qquad (21\text{-}28)$$

where all terms have been previously defined.

Step 5: Estimate the Adjustment(s) for the Permitted Portion of the Phase, A_{pbT} (Right and Left Turns)

Once the occupancies of the conflict zone have been established, the adjustment factors (one for right turns, one for left turns) are computed as follows:

$$A_{pbT} = 1 - OCC_r \qquad \text{if } N_{rec} = N_{turn}$$
$$A_{pbT} = 1 - 0.6OCC_r \qquad \text{if } N_{rec} > N_{turn}$$

$$(21\text{-}29)$$

where: N_{rec} = number of receiving lanes (lanes into which right- or left-turning movement are made)

N_{turn} = number of turning lanes (lanes from which right- or left-turning movement are made)

All other variables are as previously defined.

Occupancies are defined in terms of a conflict zone of one lane. Where the turning vehicle must turn into a designated lane, successful turns may be made only when the conflict zone is unoccupied. Where the turning driver has a choice of exit lanes to turn into, there is some flexibility to avoid a conflict by turning into another lane.

Once values of A_{pbT} are computed, Equation 21-21 may be used to compute the two adjustment factors for pedestrian and bicycle interference, f_{Rpb} and f_{Lpb}.

21.3.6   Capacity Analysis Module

In the Volume Adjustment Module, lane groups were established, as were demand flow rates, v, for each lane group. In the Saturation Flow Rate Module, the saturation flow rate, s, for each lane group has been estimated. Now, both the demand and saturation flow rates for each lane group have been adjusted to reflect the same prevailing conditions. At this point, the ratio of v to s for each lane group can be computed. It is now used as the variable indicating the relative demand intensity on each lane group.

In the Capacity Analysis Module, several important analytic steps are accomplished:

1. The v/s ratio for each lane group is computed.

2. Relative v/s ratios are used to identify the critical lane groups in the phase plan; the sum of critical lane group v/s ratios is computed.

3. Lane group capacities are computed. (Equation 21-2)

4. Lane group v/c ratios are computed. (Equation 21-3)

5. The critical v/c ratio for the intersection is computed. (Equation 21-5)

The first is a simple manipulation of the results of other modules. Steps 3, 4, and 5 conform to the equations indicated, explained in the "concepts" (Section 21.2) portion of this chapter previously.

The remaining critical analysis that must be completed is the identification of critical lane groups (Step 2). In Chapter 18, critical lanes were identified by finding the critical path through the signal ring diagram that resulted in the highest sum of critical-lane volumes, V_c. The procedure here is exactly the same, except that instead of adding critical-lane volumes, v/s ratios are added. This process is best illustrated by an example, shown in Figure 21.6.

The illustration shows a signal with leading and lagging green phases on the E–W arterial and a single phase for the N–S arterial. This involves overlapping phases, which must be carefully considered.

The critical path through Phases A1 through A3 is determined by which ring has the highest sum of v/s ratios. In this case, as shown in Figure 21.6, the left ring has the highest total, yielding a sum of v/s ratios of 0.52. The critical path through Phase B is a straightforward comparison of the two rings, which have concurrent phases. The highest total again is on the left ring, with a v/s of 0.32. In this case, the critical path through the signal is entirely along the left ring, and the sum of critical lane v/s ratios is $0.52 + 0.32 = 0.84$.

The meaning of this sum should be clearly understood. In this case, 84% of real time must be devoted to effective green if all of the critical-lane group flows are to be accommodated. In effect, there must be $0.84 * 3600 = 3,024$ s of effective green time per hour to handle the demands indicated. Conversely, there is at most 16% of real time available to devote to lost times within the cycle. This fact will be used later in the chapter to revise the signal timing if the analysis indicates that improvements are necessary.

Once the critical-lane groups and the sum of critical lane variables are identified, Steps 3, 4, and 5 of the Capacity Analysis Module may be implemented using the equations referenced earlier.

21.3.7   Level-of-Service Module

Levels of service are based on delay, as discussed in Section II of the chapter. Specific criteria were given in

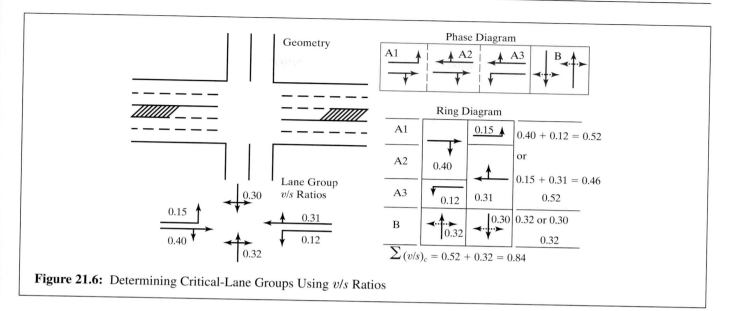

Figure 21.6: Determining Critical-Lane Groups Using v/s Ratios

Table 21.1. In the capacity analysis module, values of the v/c ratio for each lane group will have been established. Using these results, and other signalization information, the delay for each lane group may be computed as:

$$d = d_1 PF + d_2 + d_3 \qquad (21\text{-}30)$$

where:
- d = average control delay per vehicle, s/veh
- d_1 = average uniform delay per vehicle, s/veh
- PF = progression adjustment factor
- d_2 = average incremental delay per vehicle, s/veh
- d_3 = additional delay per vehicle due to a preexisting queue, s/veh

Uniform Delay

Uniform delay is obtained using Webster's uniform delay equation in the following form:

$$d_1 = \frac{0.5C\left[1 - \left(\frac{g}{C}\right)\right]^2}{1 - \left[\min(1, X) * \left(\frac{g}{C}\right)\right]} \qquad (21\text{-}31)$$

where:
- C = cycle length, s
- g = effective green time for lane group, s
- X = v/c ratio for lane group (max value = 1.00)

This equation was discussed in detail in Chapter 17.

Progression Adjustment Factor

Webster's uniform-delay equation assumes that arrivals are uniform over time. In fact, arrivals are at best random and are most often platooned as a result of coordinated signal systems. The quality of signal coordination or progression can have a monumental impact on delay.

Consider the following situation: An approach to a signalized intersection is allocated 30 s of effective green out of a 60 s cycle. A platoon of 15 vehicles at exactly 2.0 s headways is approaching the intersection. Note that the 15 vehicles will exactly consume the 30 s of effective green available ($15 * 2.0 = 30$). Thus, for this signal cycle, the v/c ratio is 1.0.

With perfect progression provided, the platoon arrives at the signal just as the light turns green. The 15 vehicles proceed through the intersection with no delay to any vehicle. In the worst possible case, however, the platoon arrives just as the light turns red. The entire platoon stops for 30 s, with every vehicle experiencing virtually the entire 30 s of delay. When the green is initiated, the platoon fully clears the intersection. In both cases, the v/c ratio is 1.0 for the cycle. The delay, however, could vary from 0 s/veh to almost 30 s/veh, dependent solely upon when the platoon arrives (i.e., the quality of the progression).

The progression adjustment factor, related to the arrival types defined during the Input Module, represents the only way in which the procedure takes this important effect into account. While the HCM does specify an analytic approach to determining the progression adjustment factor from field data, most analysts use the default values shown in Table 21.8. Straight-line interpolation for g/C is appropriate to estimate PF to the nearest 0.001.

The adjustment factor for AT-3 is always 1.000, as it represents random arrivals. Arrival types 1 and 2 are unfavorable; therefore, the adjustment factor is greater than 1.000 for these cases. Arrival types 4, 5, and 6 represent favorable progression; the adjustment factor is, therefore, less than 1.000 for these cases.

When progression is favorable, a larger g/C ratio is beneficial, and the adjustment factor decreases with increasing ratios. When progression is unfavorable, the factor increases with increasing g/C ratios. The latter may, on first glance, appear to be counterintuitive. Even with unfavorable progression, delay should improve with a larger g/C ratio. It must be remembered, however, that the adjustment factor multiplies uniform delay, which decreases rapidly with increasing g/C ratios. What the increasing adjustment factor indicates in these cases is that as g/C increases under unfavorable progression, the delay benefit in d_1 is increasingly understated.

Incremental Delay

The incremental-delay equation is based on Akcelik's equation (see Chapter 17) and includes incremental delay from random arrivals as well as overflow delay when v/c ratio > 1.00. It is estimated as:

$$d_2 = 900\, T\left[(X-1) + \sqrt{(X-1)^2 + \left(\frac{8kIX}{cT}\right)} \right]$$

$$(21\text{-}32)$$

where: T = analysis time period, h

X = v/c ratio for lane group

c = capacity of lane group, veh/h

k = adjustment factor for type of controller

I = upstream filtering/metering adjustment factor

Table 21.8: Progression Adjustment Factors

Green Ratio (g/C)	Arrival Type (AT)					
	AT-1	AT-2	AT-3	AT-4	AT-5	AT-6
0.20	1.167	1.007	1.000	1.000	0.833	0.750
0.30	1.286	1.063	1.000	0.986	0.714	0.571
0.40	1.445	1.136	1.000	0.895	0.555	0.333
0.50	1.667	1.240	1.000	0.757	0.333	0.000
0.60	2.001	1.395	1.000	0.576	0.000	0.000
0.70	2.556	1.653	1.000	0.256	0.000	0.000
Default f_{PA}	1.00	0.93	1.00	1.15	1.00	1.00
Default R_p	0.333	0.667	1.000	1.333	1.667	2.000

Notes: $PF = (1 - P)f_{PA}/(1 - g/C)$
   Tabulation is based upon default values of f_{PA} and R_p
   $P = R_p * (g/C)$; may not exceed 1.00
   PF may not exceed 1.00 for AT-3 through AT-6

(Used with Permission of Transportation Research Board, National Research Council, *Highway Capacity Manual*, 4th Edition, Washington DC 2000, Exhibit 16-12, pg. 16-20.)

Table 21.9: Delay Adjustment (k) for Controller Type

Unit Extension U (s)	Degree of Saturation — v/c Ratio (X)					
	≤ 0.50	0.60	0.70	0.80	0.90	1.00
≤ 2.0	0.04	0.13	0.22	0.32	0.41	0.50
2.5	0.08	0.16	0.25	0.33	0.42	0.50
3.0	0.11	0.19	0.27	0.34	0.42	0.50
3.5	0.13	0.20	0.28	0.35	0.43	0.50
4.0	0.15	0.22	0.29	0.36	0.43	0.50
4.5	0.19	0.25	0.31	0.38	0.44	0.50
5.0[a]	0.23	0.28	0.34	0.39	0.45	0.50
Pretimed or Unactuated Movement	0.50	0.50	0.50	0.50	0.50	0.50

Notes:
[a] if $U > 5.0$ s, extrapolate to find k, with $k_{max} = 0.50$.
Straight-line interpolation for X and U is permissible to find k to the nearest 0.00.
(Used with Permission of Transportation Research Board, National Research Council, *Highway Capacity Manual*, 4th Edition, Washington DC, 2000, Exhibit 16-13, pg. 16-22.)

Adjustment factors for type of controller are shown in Table 21.9. For pretimed controllers, or unactuated movements in a semi-actuated controller, the factor is always 0.50. For actuated movements and controllers, the adjustment is based on the unit extensions, as shown. The upstream filtering/metering adjustment factor is only used in arterial analysis. A value of 1.00 is assumed for all analyses of individual intersections.

Initial Queue Delay

There are five basic delay cases covered in the HCM (excluding special applications for compound phasing):

- *Case 1*: No initial queue, $X \leq 1.00$.
- *Case 2*: No initial queue, $X > 1.00$.
- *Case 3*: Initial queue, $X < 1.00$, no residual queue at end of analysis period.
- *Case 4*: Initial queue, $X < 1.00$, residual queue exists at end of analysis period but is less than the size of the initial queue.
- *Case 5*: Initial queue, $X \geq 1.00$, residual queue exists at end of analysis period and is the same or larger than the size of the initial queue.

These cases can be thought of in a simpler way. For Case 1, there is no queue at beginning or end of the analysis period T. For Case 2, there is no queue at the beginning of analysis period T, but there is a residual queue at the end of period T. For Case 3, there is a queue at the beginning of period T, but there is no queue at the end of period T, as the queue was completely dissipated within the analysis period. For Case 4, there is a queue at the beginning and end of period T, with the end queue smaller than the beginning queue. For Case 5, there is a queue at the beginning and end of period T, with the end queue larger than the beginning queue. A simple comparison between the total demand during the analysis period and the capacity during the analysis period is used to determine whether Case 3, 4, or 5 exists. With an initial queue, the total demand during the analysis period T is given by:

$$N_{dem} = Q_b + vT \qquad (21\text{-}33)$$

where: N_{dem} = total demand during analysis period T, veh

Q_b = initial queue at beginning of analysis period T, veh

v = demand flow rate during analysis period T, veh/h

T = analysis period, h

The capacity during analysis period T is:

$$N_{cap} = cT \qquad (21\text{-}34)$$

where: N_{cap} = total capacity during analysis period T, veh

c = capacity during analysis period T, veh/h

T = analysis period, h

Assuming that an initial queue exists (i.e., $Q_b > 0$):

- Case 3 exists when $N_{dem} \leq N_{cap}$
- Case 4 exists when $N_{dem} > N_{cap}$ and $v < c$
- Case 5 exists when $N_{dem} > N_{cap}$ and $v \geq c$

For Cases 1 and 2, there is no delay due to the existence of an initial queue, and d_3 is, therefore, 0.0 s/veh. For Cases 3, 4, and 5, an existing queue at the beginning of the analysis period results in additional delay beyond adjusted uniform and incremental delay (d_1, PF, and d_2) that must be estimated. Further, the progression adjustment factor is not applied (when initial queues exist) during periods when $X > 1.00$, as the benefits of progression are negligible during period of oversaturation. Technically, this condition also applies to Case 2, after the first signal cycle, as a residual queue will begin to build. The HCM methodology, however, recommends applying PF to the full analysis period where no initial queue exists and $X > 1.00$.

For Cases 3, 4, and 5, the following model is used to predict the additional delay per vehicle caused by the existence of an initial queue at the beginning of the analysis period, T.

$$d_3 = \frac{1{,}800\, Q_b(1 + u)t}{cT} \qquad (21\text{-}35)$$

where: Q_b = size of initial queue at start of analysis period T, veh

c = capacity of lane group, veh/h

T = analysis period, h

t = duration of oversaturation within T, h

u = delay parameter

For Cases 4 and 5, where queues exist for the entire analysis period, T, the equation simplifies, because $t = T$. For Case 3, the queue dissipates at some point during T. Then:

$$t_{CASES4,5} = T$$
$$t_{CASE3} = \frac{Q_b}{c(1 - X)} \qquad (21\text{-}36)$$

where all parameters are as previously defined. Note that for Case 3, X must be less than 1.00.

The "delay parameter," u, is found as:

$$u_{CASE3} = 0.00$$
$$u_{CASE4} = \frac{cT}{Q_b(1 - X)} \qquad (21\text{-}37)$$
$$u_{CASE5} = 1.00$$

where all parameters are as previously defined. Note again that for Case 4, X must be less than 1.00.

In determining the total delay for Cases 4 and 5, the use of the progression adjustment factor on d_1 is eliminated, as noted previously. For Case 3, the progression adjustment factor is applied only during the unsaturated portion of the analysis period T. Thus:

$$d_{CASE3,4,5} = d_1 + d_2 + d_3$$
$$d_{1CASE4,5} = \text{Equation } 21\text{-}31 \qquad (21\text{-}38)$$
$$d_{1CASE3} = \left[d_s\left(\frac{t}{T}\right)\right] + \left[d_u\, PF\left(\frac{T - t}{T}\right)\right]$$

where d_s, saturated delay, is found using Equation 21–31 with $X = 1.0$, and d_u, unsaturated delay, is found using Equation 21–31 with the actual value of X (which, for Case 3, is less than 1.0 for the analysis period).

The ability to predict delay where initial queues exist provides a practical way to consider a series of consecutive analysis periods of arbitrary T (often 0.25 h or 15 m) in sequence. The queue at the beginning of any time period $(i + 1)$ can be related to the queue at the beginning of the previous period (i), as follows:

$$Q_{b(i+1)} = max\left[0, Q_{bi} + cT(X_i - 1)\right] \quad (21\text{-}39)$$

where: $Q_{b(i+1)}$ = initial queue for period $i + 1$, vehs

$\quad\quad\quad Q_{bi}$ = initial queue for period i, vehs

$\quad\quad\quad X_i$ = degree of saturation (v/c ratio) for period i

When oversaturation exists, it is best to analyze consecutive 15-min analysis periods, beginning with the period before a queue appears and ending in the period after dissipation of the queue.

Aggregating Delay

Once appropriate values for d_1, PF, d_2, and d_3 are determined, the total control delay per vehicle for each lane group is known. Levels of service for each lane group are assigned using the criteria of Table 21.1.

The HCM allows for lane group delays to be aggregated to approach delays and an overall intersection delay, as follows:

$$d_A = \frac{\sum_i d_i v_i}{\sum v_i}$$

$$\quad\quad\quad\quad\quad\quad (21\text{-}40)$$

$$d_I = \frac{\sum_A d_A v_A}{\sum_A v_A}$$

where: d_i = total control delay per vehicle, lane group i, s/veh

$\quad\quad\quad d_A$ = total control delay per vehicle, approach A, s/veh

$\quad\quad\quad d_I$ = total control delay per vehicle for the intersection as a whole, s/veh

$\quad\quad\quad v_A$ = demand flow rate, approach A

$\quad\quad\quad v_i$ = demand flow rate, lane group i

Levels of service may be applied to approaches and the intersection as a whole, using the criteria of

Table 21.1. Note that average delays are weighted by the number of vehicles experiencing the delays.

21.3.8 Interpreting the Results of Signalized Intersection Analysis

At the completion of the five-module HCM analysis procedure, the traffic engineer has the following results available for review:

- v/c ratios (X) for every lane group
- Critical v/c ratio (X_c) for the intersection as a whole
- Delays and levels of service for each lane group
- Delays and levels of service for each approach
- Delay and level of service for the overall intersection

All of these results must be considered to obtain a complete overview of predicted operating conditions in the signalized intersection and to get an idea of how to address any problems revealed by the analysis.

As noted earlier, the v/c ratio and delay values are not strongly linked, and a number of interesting combinations can arise. The v/c ratio for any lane group, however, represents an absolute prediction of the sufficiency of the capacity provided to that lane group. Further, the critical v/c ratio represents an absolute prediction of the total sufficiency of capacity in all critical lane groups. The following scenarios may arise:

- *Scenario 1: $X_c \leq 1.00$; all $X_i \leq 1.00$.* These results indicate that there are no capacity deficiencies in any lane group. If there are no initial queues, then there will be no residual queues in any lane group at the end of the analysis period. The analyst may wish to consider the balance of X values among the various lane groups, particularly the critical lane groups. It is often a policy to provide balanced X ratios for all critical lane groups. This is best accomplished when all critical lane groups have $X_i \approx X_c$.
- *Scenario 2: $X_c \leq 1.00$; some $X_i > 1.00$.* As long as $X_c \leq 1.00$, all demands can be handled within the phase plan, cycle length, and physical

design provided. All X_i values may be reduced to values less than 1.00 by reallocation of green time from lane groups with lower X_i values to those with $X_i > 1.00$. A suggested procedure for reallocation of green time is presented later in this chapter.

- *Scenario 3: $X_c > 1.00$; some or all $X_i > 1.00$.* In this case, sufficient capacity can be provided to all critical-lane groups only by changing the phase plan, cycle length, and/or physical design of the intersection. Improving the efficiency of the phase plan involves considering protected left-turn phasing where none exists or protected + permitted phasing where fully protected phasing exists. This may have big benefits, depending on the magnitude of left-turn demands. Increasing the cycle length will add small amounts of capacity. This may not be practical if the cycle length is already long or where the capacity deficiencies are significant. Adding lanes to critical-lane groups will have the biggest impact on capacity and may allow for more effective lane use allocations.

Delays must also be carefully considered, but should be tempered by an understanding of local conditions. Level-of-service designations are based on delay criteria, but acceptability of various delay levels may vary by location. Drivers in a small rural CBD will not accept the delay levels that drivers in a big city will.

As noted earlier, LOS F may exist where v/c ratios are less than 1.00. This situation may imply a poorly timed signal (retiming should be considered), or it may reflect a short protected turning phase in a relatively long cycle length. The latter may not be easily remedied.

Aggregate levels of service for approaches—and particularly for the intersection—may mask problems in one or more lane groups. Individual lane group delays and levels of service should always be reported and must be considered with aggregate measure. This is often a serious problem when consultants or other engineers report only the overall intersection level of service. The Highway Capacity and Quality of Service Committee of the Transportation Research Board is considering whether an aggregate intersection level of service should be permitted in future HCMs because of its potential to mask deficiencies in individual lane groups.

Where lane group delays vary widely, some reallocation of green time may help balance the situation. However, when changing the allocation of green time to achieve better balance in lane group delays, the impact of the reallocation on v/c ratios must be watched carefully.

21.4   Some "Simple" Sample Problems

As long as applications do not involve permitted or compound left turns, a manual implementation of these models can be illustrated. This section provides some "simple" sample problems illustrating the application and interpretation of the HCM 2000 signalized intersection analysis procedure. In subsequent sections, some of the more complex elements of the model will be presented.

21.4.1   Sample Problem 1: Intersection of Two One-Way Streets

The signalized intersection shown in Figure 21.7 involves a simple two-phase signal at the junction of two one-way arterials. Thus, there are no opposed left turns, and all elements of the HCM procedure can be demonstrated in detail.

The problem is an analysis of an existing situation. Improvements should be recommended.

Input Module

All of the input variables needed for this analysis are specified in Figure 21.7. The minimum green times required for safe pedestrian crossings must, however, be computed. These are used to check actual signal timings for pedestrian safety. A warning would be issued if the green times were not sufficient to handle pedestrian safety requirements. Equation 21-9 applies:

$$G_p = 3.2 + \left(\frac{L}{S_p}\right) + 0.27 N_{peds} \le G + Y$$

where: G_p = minimum pedestrian green, s
          L = length of the crosswalk, ft

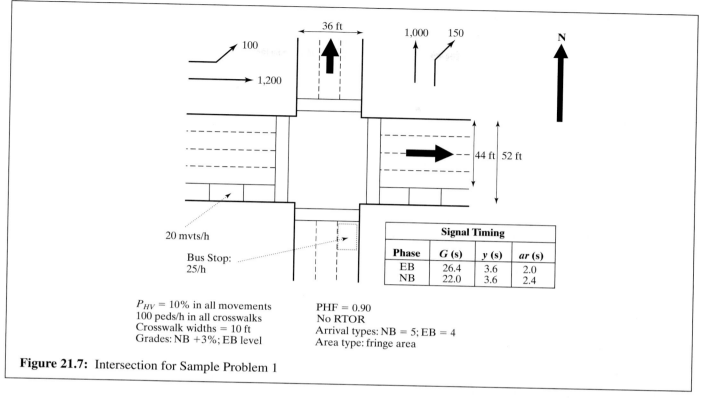

Figure 21.7: Intersection for Sample Problem 1

S_p = crossing speed, default value = 4.0 ft/s

N_{peds} = peds/cycle in crosswalk, peds/cycle = $100/(3,600/60)$ = 1.7 peds/cycle

G = vehicular actual green time, s

Y = sum of yellow plus all-red time, s

Then:

$$G_{p,EB} = 3.2 + \left(\frac{36}{4.0}\right) + (0.27 * 1.7)$$

$$= 12.7 \text{ s} \leq 26.4 + 3.6 + 2.0$$

$$= 32.0 \text{ s} \quad \text{OK}$$

$$G_{p,NB} = 3.2 + \left(\frac{52}{4.0}\right) + (0.27 * 1.7)$$

$$= 16.7 \text{ s} \leq 22.0 + 3.6 + 2.4$$

$$= 28.0 \text{ s} \quad \text{OK}$$

Pedestrians are, therefore, safely accommodated by the signal timing as shown. No warning is necessary.

Volume Adjustment Module

The intersection is fairly simple. There are only two lane groups, EB and NB, as there are no separate turning lanes and only two approaches. In this module, flow rates are estimated using the *PHF*, and proportions of turning movements in each lane group are computed. These computations are summarized in Table 21.10.

Saturation Flow Rate Module

The saturation flow rate for each lane group is computed using Equation 21-12:

$$s = s_o N f_w f_{HV} f_g f_p f_{bb} f_a f_{LU} f_{RT} f_{LT} f_{Rpb} f_{Lpb}$$

where: s = saturation flow rate under prevailing conditions, veh/hg

s_o = saturation flow rate under ideal conditions, default value = 1,900 pc/hg/ln

N = number of lanes in the lane group

f_i = adjustment factor for prevailing condition "i"

Table 21.10: Flow Rate Computations for Sample Problem

Mvt	Vol (veh/h)	PHF	Flow Rate (veh/h)	Lane Group Flow Rate (veh/h)	Proportion of RT's in LG	Proportion of LT's in LG
EB-L	100	0.90	111	1,444	0.000	0.077
EB-T	1,200	0.90	1,333			
NB-T	1,000	0.90	1,111	1,278	0.131	0.000
NB-R	150	0.90	167			

where:
w = lane width
HV = heavy vehicles
g = grade
p = parking
bb = local bus blockage
a = area type
LU = lane utilization
RT = right turn
LT = left turn
Rpb = ped/bike interference with right turns
Lpb = ped/bike interference with left turns

For the simple case posed in this sample problem, the adjustment factors are found in appropriate equations from the chapter. These computations are summarized in Figure 21.8.

The new adjustment factors for pedestrian and bicycle interference with right and left turns require multiple steps, each of which is shown below. Note that there are no bicycles at this location, so only pedestrian interference is evaluated. Also, the EB lane group has only left turns and the NB lane group has only right turns. Thus:

$$f_{Rpb,EB} = 1.000$$
$$f_{Lpb,NB} = 1.000$$

The determination of these factors also requires that g_p, the pedestrian green be determined. This is generally taken to be the length of the pedestrian WALK plus clearance intervals, which (for a pretimed signal) is equal to the green, yellow, and all-red phases. Thus, $g_{pEB} = 26.4 + 3.6 + 2.0 = 32.0$ s, and $g_{pNB} = 22.0 + 3.6 + 2.4 = 28.0$ s.

Equation 21-13
$$f_{wEB} = 1 + \left(\frac{W - 12}{30}\right) = 1 + \left(\frac{11 - 12}{30}\right) = 0.967$$
$$f_{wNB} = 1 + \left(\frac{12 - 12}{30}\right) = 1.000$$

Equation 21-14
$$f_{HV,all} = \frac{1}{1 + P_{HV}(E_{HV} - 1)} = \frac{1}{1 + 0.10\,(2 - 1)} = 0.909$$

Equation 21-15
$$f_{g,EB} = 1 - \frac{G}{200} = 1 - \frac{0}{200} = 1.000$$
$$f_{g,NB} = 1 - \frac{3}{200} = 0.985$$

Equation 21-16
$$f_{p,EB} = \frac{N - 0.10 - \left(\frac{18\,N_m}{3,600}\right)}{N} = \frac{4 - 0.10 - \left(\frac{18 * 20}{3,600}\right)}{4} = 0.950$$
$$f_{p,NB} = 1.000 \text{ (no parking)}$$

Equation 21-17
$$f_{bb,EB} = \frac{N - \left(\frac{14.4 * N_B}{3,600}\right)}{N} = \frac{4 - \left(\frac{14.4 * 0}{3,600}\right)}{4} = 1.000$$
$$f_{bb,NB} = \frac{3 - \left(\frac{14.4 * 25}{3,600}\right)}{3} = 0.967$$
$$f_{a,all} = 1.000 \text{ (fringe area)}$$
$$f_{LU,EB,NB} = 0.908 \text{ (Table 21.7 default value)}$$

Equation 21 - 19
$$f_{RT,EB} = 1.0 - 0.15 P_{RT} = 1.0 - (0.15 * 0.00) = 1.000$$
$$f_{RT,NB} = 1.0 - (0.15 * 0.131) = 0.980$$

Equation 21 - 20
$$f_{LT,EB} = \frac{1}{1.0 + 0.05\,P_{LT}} = \frac{1}{1.0 + (0.05 * 0.077)} = \frac{1}{1.004} = 0.996$$
$$f_{LT,NB} = \frac{1}{1.0 + (0.05 * 0.00)} = 1.000$$

Figure 21.8: Computation of Adjustment Factors for Sample Problem 1

The adjustment factor for right-turn pedestrian interference for the NB lane group is found using Equation 21-21:

$$f_{Rpb,NB} = 1.0 - P_{RT}(1 - A_{pbT})(1 - P_{RTA})$$

where: P_{RT} = proportion of right turns in the lane group (0.131)

A_{pbT} = adjustment factor for permitted portion of the phase

P_{RTA} = proportion of right turns using protected portion of the phase (0.00)

Then, using Equations 21-22, 21-23, 21-28, and 21-29:

$$v_{pedg} = v_{ped} * \left(\frac{C}{g_p}\right) = 100 * \left(\frac{60}{28}\right) = 214.3 \text{ peds/hg}$$

$$OCC_{pedg} = OCC_r = \frac{v_{pedg}}{2,000} = \left(\frac{214.3}{2,000}\right) = 0.107$$

$$A_{pbT} = 1 - 0.6 * OCC_r = 1 - (0.60 * 0.107) = 0.936$$

$$f_{Rpb,NB} = 1.0 - 0.131(1 - 0.936)(1 - 0.000) = 0.992$$

where: v_{pedg} = pedestrian flow rate during WALK interval, peds/hg

v_{ped} = pedestrian demand volume in crosswalk, peds/h

OCC_{pedg} = pedestrian occupancy in the crosswalk

OCC_r = pedestrian occupancy in the conflict zone

The adjustment factor for left-turn pedestrian interference for the EB lane group is also found using Equation 21-21:

$$f_{Lpb,EB} = 1.0 - P_{LT}(1 - A_{pbT})(1 - P_{LTA})$$

where: P_{LT} = proportion of left turns in the lane group (0.077)

A_{pbT} = adjustment factor for permitted portion of the phase

P_{LTA} = proportion of left turns using protected portion of the phase (0.00)

Then, using Equations 21-22, 21-23, 21-28, and 21-29 again:

$$v_{pedg} = v_{ped} * \left(\frac{C}{g_p}\right) = 100 * \left(\frac{60}{32}\right) = 187.5 \text{ peds/hg}$$

$$OCC_{pedg} = OCC_r = \frac{v_{pedg}}{2,000} = \left(\frac{187.5}{2,000}\right) = 0.094$$

$$A_{pnT} = 1 - 0.6 * OCC_r = 1 - (0.60 * 0.094) = 0.944$$

$$f_{Lpb,NB} = 1.0 - 0.077(1 - 0.944)(1 - 0.000) = 0.996$$

The saturation flow rates for each lane group are now estimated as shown in Table 21.11.

Capacity Analysis Module

In this module, the adjusted volumes from the Volume Adjustment Module and the saturation flow rates from the Saturation Flow Rate Module are combined to find v/s ratios and critical-lane groups, lane group capacity, lane group v/c ratios, and the critical v/c ratio (X_c).

In this case, the selection of critical-lane groups is trivial, as there is only one lane group for each signal phase, and it is, by definition, critical for the subject phase.

Capacity of a lane group is computed using Equation 21-2:

$$c_i = s_i * \left(\frac{g_i}{C}\right)$$

Table 21.11: Estimation of Saturation Flow Rates Using Equation 21-12

Lane Group	S_o (pc/hg/ln)	N (lanes)	f_w	f_{HV}	f_g	f_o	f_{bb}	f_a	f_{LU}	f_{RT}	f_{LT}	f_{Rpb}	f_{Lpb}	s (veh/hg)
EB	1,900	4	0.967	0.909	1.000	0.950	1.000	1.000	0.908	1.000	0.996	1.000	0.996	**5,717**
NB	1,900	3	1.000	0.909	0.985	1.000	0.967	1.000	0.908	0.980	1.000	0.992	1.000	**4,356**

Table 21.12: Capacity Analysis Results

Lane Group	v (veh/h)	s (veh/hg)	v/s	g (s)	C (s)	c (veh/h)	X (v/c)
EB	1,444	5,717	0.253	26.4	60	2,515	0.574
NB	1,278	4,356	0.293	22.0	60	1,597	0.800

where: c_i = capacity of lane group i, veh/h

s_i = saturation flow rate for lane group i, veh/hg

g_i = effective green time for lane group i, s

C = cycle length, s

It should be noted that if standard default values for start-up lost time (ℓ_1) − 2.0 s, and for extension of effective green into yellow and all-red (e) −2.0 s are used, the effective green time (g) and the actual green time (G) are numerically equivalent. Capacity analysis computations are summarized in Table 21.12.

The critical v/c ratio is computed using Equation 21-5:

$$X_c = \sum_i \left(\frac{v}{s}\right)_{ci} * \left(\frac{C}{C - L}\right)$$

where $\Sigma(v/s)_{ci}$ = sum of the critical-lane group v/s ratios, $0.253 + 0.293 = 0.546$

L = total lost time per cycle, s

Other terms as previously defined.

The total lost time per cycle is the sum of the start-up and clearance lost times for each phase, using the standard default values for ℓ_1 and e. Thus:

$\ell_{1,EB} = 2.0$ s

$\ell_{2,EB} = y + ar - 2.0 = 3.6 + 2.0 - 2.0 = 3.6$ s

$\ell_{1,NB} = 2.0$ s

$\ell_{2,NB} = 3.6 + 2.4 - 2.0 = 4.0$ s

$L = 2.0 + 3.6 + 2.0 + 4.0 = 11.6$ s

Then:

$$X_c = 0.546 * \left(\frac{60}{60 - 11.6}\right) = 0.677$$

The capacity results indicate that there is no capacity problem at this intersection. Indeed, there may be too much unused green time, indicating that a shorter cycle length might be possible. The difference between the EB and NB v/c ratios also indicates that the green may be somewhat misallocated for the demand flows given. Delay results, however, should be consulted before making this conclusion.

Level-of-Service Module

The average control delay per vehicle for each lane group is computed using Equations 21-30, 21-31, and 21-32:

$$d = d_1 PF + d_2 + d_3$$

$$d_1 = \frac{0.50C\left[1 - \dfrac{g}{C}\right]^2}{1 - \left[\min(X, 1.0) * \dfrac{g}{C}\right]}$$

$$d_2 = 900T\left[(X - 1) + \sqrt{(X - 1)^2 + \frac{8kIX}{cT}}\right]$$

where d = total control delay per vehicle, s/veh

PF = progression adjustment factor

d_1 = uniform delay, s/veh

d_2 = incremental delay (random + overflow), s/veh

d_3 = delay per vehicle due to preexisting queues, s/veh (no preexisting queues in this case, $d_3 = 0.0$ s/veh)

Table 21.13: Delay Computations

Lane Group	C (s)	g/C	X	d_1 (s)	PF	T (h)	K	I	c (veh/h)	d_2 (s)	d_3 (s)	d (s/veh)
EB	60	0.440	0.574	**12.6**	0.840	0.25	0.5	1	2,515	**1.0**	**0.0**	**11.6**
NB	60	0.367	0.800	**17.0**	0.607	0.25	0.5	1	1,597	**4.3**	**0.0**	**14.6**

T = analysis period, h, 0.25 h or 15 m in this case

X = v/c ratio for subject lane group

k = incremental delay factor for controller type 0.50 for all pretimed controllers

I = upstream filtering adjustment factor 1.0 for isolated intersection analysis

Other variables are as previously defined.

The delay for each lane group is computed in Table 21.13. Progression factors are selected from Table 21.8. The g/C ratios are EB 26.4/60 = 0.440; NB 22.0/60 = 0.367. Progression factors are based on the given arrival types (NB = 5; EB = 4) and the g/C ratios. Through interpolation in Table 21.8, the following progression factors are determined:

$$PF_{EB} = 0.757 + (0.895 - 0.757) * \left(\frac{0.50 - 0.44}{0.50 - 0.40}\right)$$

$$= 0.840$$

$$PF_{NB} = 0.555 + (0.714 - 0.555) * \left(\frac{0.40 - 0.367}{0.40 - 0.30}\right)$$

$$= 0.607$$

The aggregate delay for the intersection as a whole is computed using Equation 21-40:

$$d_A = \frac{\sum d_i\, v_i}{\sum v_i} = \frac{(11.6 * 1,444) + (14.6 * 1,278)}{(1,444 + 1,278)}$$

$$= 13.0 \text{ s/veh}$$

From Table 21.1, these delays are in the range of level of service B for both lane groups and the intersection

as a whole, a very acceptable result. In fact, the delays are just barely over the threshold for LOS A.

Analysis

The delay and capacity results both indicate that the intersection is operating acceptably. Some improvement, however, might be possible through a retiming of the signal to achieve better balance between the two phases and to achieve a higher utilization of available green time.

The critical v/c ratio was determined to be 0.677. This is a low value, indicating that 22.3% of green time is unused. Optimal delays usually occur when v/c ratios are in the range of 0.80 to 0.95. The value 0.677 indicates that the cycle length is probably too long for the current demand volumes. Further, the EB lane group has a v/c ratio of 0.574, while the NB lane group has a v/c ratio of 0.800, indicating some imbalance in the allocation of green time. A retiming of the signal might be considered to achieve a shorter cycle length, if there are no system constraints that prevent this.

This was, in effect, a very simple situation with no complicating factors. The solution of the problem, however, is still lengthy and contains many steps and computations. For these reasons, most traffic engineers use a software package to implement this methodology. When more complicated situations arise, there is no choice but to use software.

21.4.2 Sample Problem 2: A Multiphase Signal with No Permitted Left Turns

The intersection to be analyzed is illustrated in Figure 21.9. All necessary data needed to solve the problem are shown, including the specification of all relevant

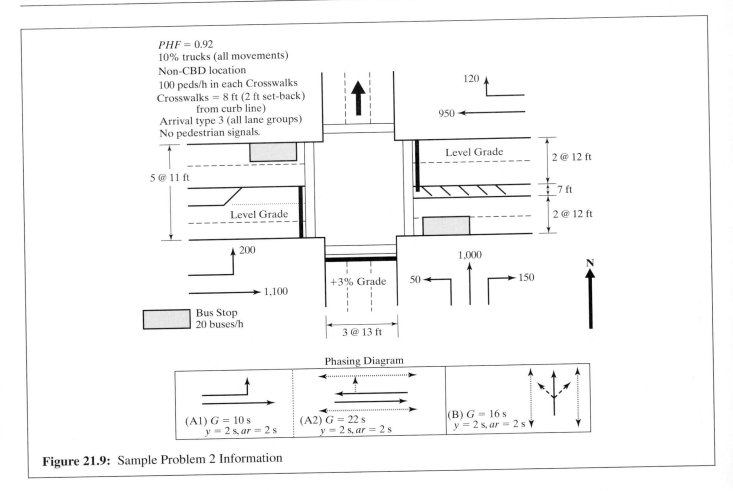

Figure 21.9: Sample Problem 2 Information

geometrics, demand volumes, and signalization details. The signal shown is pretimed, and for simplicity, there are no opposed left turns. This problem is somewhat more complex than Sample Problem 1, but manual computations can still be illustrated. Some of the computational steps are shown in greater detail than in Sample Problem 1.

Input Module

There is one computation that is executed in the Input Module. Where pedestrians are present, a minimum green time to accommodate their safe crossing is used. The minimum green time for a safe pedestrian crossing is estimated using Equation 21-9:

$$G_p = 3.2 + \frac{L}{S_p} + 0.27 N_{ped}$$

where: L = length of the crosswalk, ft

S_p = walking speed (use 4.0 ft/s, default value)

N_{ped} = no. of pedestrians per interval (cycle)

For all crosswalks, with a 60-s cycle (which produces 60 cycles per hour), the value of N_{ped} is $100/60 = 1.67$ peds/cycle. Pedestrians cross the major street during Phase B of the signal and have a crosswalk length of 55 ft. Pedestrians cross the minor street during Phase A2 and have a crosswalk length of 39 ft. Pedestrians are allowed to cross during the green, yellow, and all-red portions of the phase. Thus:

$$G_{pA2} = 3.2 + \frac{39}{4.0} + (0.27 * 1.67)$$

$$= 13.4 \text{ s} < 22.0 + 2 + 2 = 26 \text{ s}$$

$$G_B = 3.2 + \frac{55}{4.0} + (0.27 * 1.67)$$
$$= 17.4 \text{ s} < 16.0 + 2 + 2 = 20 \text{ s}$$

Thus, pedestrians are safely accommodated in all crosswalks, and no warnings need be issued. In practical terms, this means that pedestrian push buttons and an actuated pedestrian phase are not required in this case.

Volume-Adjustment Module

In this module, lane groups are initially identified. In this case, this is relatively simple. There is one LT lane (EB approach) that must be established as a separate lane group. Other lane groups are EB (through), WB (through and right), and NB (left, through, and right). For each lane group, the total demand flow rate must be computed by taking the demand volume shown in the input diagram and adjusting to reflect the peak-hour factor (*PHF*), as follows:

$$v_p = \frac{V}{PHF}$$

where:
v_p = demand flow rate, veh/h
V = hourly demand volume, veh/h
PHF = peak-hour factor

For use later in the procedure, the proportion of left and right turns in each lane group is also computed. Computations are illustrated in Table 21.14.

Note that by using the *PHF* adjustment, it is assumed that all lane groups peak during the same 15-min interval. This is a conservative assumption and represents a worst-case scenario. If flow rates for each 15-min interval during the peak hour were directly observed, the "worst" period would be selected for analysis and the appropriate flow rates entered directly. The PHF would be set at 1.00 to avoid "double-counting" the flow rate adjustment.

Saturation Flow Rate Module

In this module, the actual saturation flow rate for each lane group is computed using Equation 21-12:

$$s = s_o N f_w f_{HV} f_g f_p f_{bb} f_a f_{LU} f_{RT} f_{LT} f_{Rpb} f_{Lpb}$$

where all terms are as previously defined.

For the current problem, appropriate values for all of these variables must be selected for each of the *four* lane groups that have been defined. The process is described in the subsections below.

Ideal Saturation Flow Rate With no other information available, the HCM 2000 recommends the use of 1,900 pc/hg/ln as a default value for ideal saturation flow rate. This is applied to all lane groups in the intersection.

Number of Lanes The number of lanes in each lane group is simply counted from the geometry shown in Figure 21.9, as follows:

Table 21.14: Demand Flow Rate Computations

Lane Group	Demand Flow Rate (veh/h)	Proportion of Left and/or Right Turns
EB LT	200/0.92 = 217	1.00 LT 0.00 RT
EB TH	1,100/0.92 = 1,196	0.00 LT 0.00 RT
WB TH/RT	(950 + 120)/0.92 = 1,163	0.00 LT 120/1,070 = 0.112 RT
NB LT/TH/RT	(50 + 1,000 + 150)/0.92 = 1,304	50/1,200 = 0.042 LT 150/1,200 = 0.125 RT

Lane Group	Number of Lanes
EB LT	1
EB TH	2
WB TH/RT	2
NB LT/TH/LT	3

Adjustment Factor for Lane Width   The lane-width adjustment factor is computed using Equation 21-13:

$$f_w = 1 + \left(\frac{W - 12}{30} \right)$$

where W is the lane width. Based on lane width given in Figure 21.14 for each of the lane groups:

$$f_w \text{ (EB LT)} = 1 + \left(\frac{11 - 12}{30} \right) = 0.967$$

$$f_w \text{ (EB TH)} = 1 + \left(\frac{11 - 12}{30} \right) = 0.967$$

$$f_w \text{ (WB TH/RT)} = 1 + \left(\frac{12 - 12}{30} \right) = 1.000$$

$$f_w \text{ (NB LT/TH/RT)} = 1 + \left(\frac{13 - 12}{30} \right) = 1.033$$

Adjustment Factor for Heavy Vehicles   All of the demand flow rates include 10% trucks. Thus, the same factor is used for all lane groups, based on Equation 21-14:

$$f_{HV} = \frac{1}{1 + P_T (E_T - 1)}$$

where: P_T = proportion of trucks in the lane group

E_T = passenger car equivalent for trucks (two used for all signalized intersection cases)

Then, for all lane groups:

$$f_{HV} = \frac{100}{100 + 10(2 - 1)} = 0.909$$

Adjustment Factor for Grades   The adjustment factor for grades is computed using Equation 21-15:

$$f_g = 1 - \left(\frac{G}{200} \right)$$

where G is the percent grade for each lane group. Only the NB approach has a grade. All other lane groups are on level approaches. Thus $f_g = 1.000$ (EB LT, EB TH, WB TH/RT), and:

$$f_g \text{ (NBLT/TH/RT)} = 1 - \left(\frac{3}{200} \right) = 0.985$$

Adjustment Factor for Parking   There is no parking at the curb in any of the lane groups shown in Figure 21.9. Thus, the adjustment for all lane groups is $f_p = 1.000$.

Adjustment Factor for Bus Blockage   In the problem statement, both the EB TH and WB TH/RT lane groups include a bus stop with an estimated 20 buses per hour stopping to pick up and drop off passengers. Other lane groups (EB LT and NB LT/TH/RT) have no bus stops. From Equation 21-17:

$$f_{bb} = \frac{N - \left(\dfrac{14.4 N_B}{3,600} \right)}{N}$$

where N = number of lanes in the lane group
N_B = number of buses per hour stopping

For both the EB LT and NB LT/TH/RT lane groups, $N_B = 0$, and $f_{bb} = 1.000$. For the EB TH and WB TH/RT lane groups:

$$f_{bb} = \frac{2 - \left(\dfrac{14.4 * 20}{3,600} \right)}{2} = 0.960$$

Adjustment for Area Type   The adjustment for non-CBD areas is $f_a = 1.000$ for all lane groups.

Adjustment for Lane Utilization   Table 21.7, however, gives default values for the lane utilization factor when lane flows cannot be directly observed. The default values are based on the number of lanes in the lane group. Thus:

$$f_{LU} \text{ (EB LT)} = 1.000$$
$$f_{LU} \text{ (EB TH)} = 0.952$$
$$f_{LU} \text{ (WB TH/RT)} = 0.952$$
$$f_{LU} \text{ (NB LT/TH/RT)} = 0.908$$

Adjustment Factor for Right Turns   The adjustment factor for right turns is again obtained using Equation 21-19:

$$f_{RT} = 1.0 - 0.15 P_{RT}$$

where P_{RT} is the decimal proportion of right turns in the lane group. The proportion of right turns in each lane group was computed in Table 21.14. Based on the proportion of right turns in each lane group:

$$f_{RT} \text{ (EB LT)} = 1.0 - (0.15*0.000) = 1.000$$
$$f_{RT} \text{ (EB TH)} = 1.0 - (0.15*0.000) = 1.000$$
$$f_{RT} \text{ (WB TH/RT)} = 1.0 - (0.15*0.112) = 0.983$$
$$f_{RT} \text{ (NB LT/TH/RT)} = 1.0 - (0.15*0.125) = 0.981$$

Lane groups with no right turns have an adjustment factor of 1.00 as expected.

Adjustment Factor for Left Turns   For the EB LT lane group (an exclusive LT lane with protected phasing), Equation 21-20 indicates that $f_{LT} = 0.95$. There are no left turns in the EB TH and WB TH/RT lane groups, for which $f_{LT} = 1.00$.

The NB LT/TH/RT approach is interesting, as it involves a one-way street. Thus, the left turns are not exactly protected, but they are clearly unopposed by vehicles. Thus, the left turn factor for a protected turn from a shared lane group would be applied:

$$f_{LT} = \frac{1}{1 + 0.05 P_{LT}}$$

where P_{LT} is the decimal proportion of left-turning vehicles in the lane group (0.042 in this case—see Table 21.14). Thus:

$$f_{LT} \text{ (NB LT/TH/RT)} = \frac{1}{1 + (0.05*0.042)} = 0.998$$

Adjustment Factor for Pedestrian/Bicycle Interference with Right Turns   This adjustment factor is found using Equation 21-21:

$$f_{Rpb} = 1.0 - P_{RT}(1 - A_{pbT})(1 - P_{RTA})$$

where: P_{RT} = decimal proportion of right turns in lane group

P_{RTA} = decimal proportion of right turns that are protected

A_{pbT} = adjustment factor for permitted phase

For this problem, there are no "protected" right turns (i.e., right turns made without conflicting pedestrians in the crosswalk), so $P_{RTA} = 0.00$ for all lane groups. P_{RT} for each lane group was computed in Table 21.14.

Note that while there are pedestrians in all crosswalks (100 peds/h), there are no bicycles on any approach. To find the factor A_{pbT}, a series of computations is required using Equations 21-22, 21-23, 21-28, and 21-29:

$$v_{pedg} = v_{ped}\left(\frac{C}{g_p}\right)$$

where: v_{ped} = pedestrian volume, peds/h

v_{pedg} = pedestrian flow rate during green phase, peds/hg

C = cycle length, s

g_p = effective green time for pedestrian crossings, s

As there are no pedestrian signals at this intersection, pedestrian green times will be taken to be equal to the vehicular green times. Pedestrians cross the major street during Phase B, with a green time of 20 s. Pedestrians cross the minor street during Phase A2, with a green time of 26 s.

$$v_{pedg} \text{ (Phase B)} = 100*\left(\frac{60}{20}\right) = 300 \text{ peds/hg}$$

$$v_{pedg} \text{ (Phase A2)} = 100*\left(\frac{60}{26}\right) = 231 \text{ peds/hg}$$

Using this result, an average pedestrian occupancy in the crosswalk is computed:

$$OCC_{pedg} = \frac{v_{pedg}}{2,000}$$

$$OCC_{pedg} \text{ (Phase B)} = \frac{300}{2,000} = 0.1500$$

$$OCC_{pedg} \text{ (Phase A2)} = \frac{231}{2,000} = 0.1155$$

For right-turn movements with no bicycle interference, the conflict zone occupancy, OCC_r, is the same as OCC_{pedg}, as computed above. Since the number of receiving lanes for right-turning vehicles exceeds the number of lanes from which right turns are made in all cases, the factor A_{pbT} is computed as follows. Note that Phase B applies to the NB LT/TH/RT lane group, while Phase A2 applies to the WB TH/RT lane group.

$$A_{pbT} = 1 - 0.6 OCC_R$$

$$A_{pbT} \text{ (NB LT/TH/RT)} = 1 - (0.6*0.1500) = 0.910$$

$$A_{pbT} \text{ (WB TH/RT)} = 1 - (0.6*0.1155) = 0.931$$

The adjustment factors may now be computed:

$$f_{Rpb} \text{ (NB LT/TH/RT)} = 1 - 0.125(1 - 0.910)(1 - 0)$$
$$= 0.989$$

$$f_{Rpb} \text{ (WB TH/RT)} = 1 - 0.112(1 - 0.931)(1 - 0)$$
$$= 0.992$$

The other lane groups have no right turns; for them, $f_{Rpb} = 1.00$.

Adjustment Factor for Pedestrian/Bicycle Interference With Left Turns

Equation 21-21 is used again:

$$f_{Lpb} = 1.0 - P_{LT} (1 - A_{pbT})(1 - P_{LTA})$$

where: P_{LT} = decimal proportion of left turns in the lane group

P_{LTA} = decimal proportion of left turns that are protected (from pedestrian interference)

A_{pbT} = adjustment factor for permitted portion of the phase

In the sample problem, there are only two lane groups that have left turns: EB LT and NB LT/TH/RT. The EB LT moves in Phase A1, during which there are no pedestrian movements permitted. Thus, for this movement, $P_{LTA} = 1.00$, and $f_{Lpb} = 1.00$.

For the NB LT/TH/RT lane group, $P_{LTA} = 0.00$. The same basic sequence of equations is used to compute A_{pbT} in this case as was used for right turns in the previous step:

$$v_{pedg} = 100\left(\frac{60}{20}\right) = 300 \text{ peds/hg}$$

$$OCC_{pedg} = \frac{300}{2000} = 0.15$$

$$OCC_r = OCC_{pedg} = 0.15$$

$$A_{pbT} = 1 - (0.6*0.15) = 0.91$$

$$f_{Lpb} = 1 - 0.042(1 - 0.91)(1 - 0) = 0.996$$

None of the other lane groups (EB TH, WB TH/RT) contain left turns; f_{Lpb} for these lane groups is 1.00.

Saturation Flow Rate Computations

As all adjustment factors are now identified, the saturation flow rate for each lane group may now be computed using Equation 21-12:

$$s = s_o N f_w f_{HV} f_g f_p f_{bb} f_a f_{LU} f_{RT} f_{LT} f_{Rpb} f_{Lpb}$$

$$s(\text{EB LT}) = 1,900*1*0.967*0.909*1*1*1*1*$$
$$1*1*0.95*1*1 = 1,587 \text{ veh/hg}$$

$$s(\text{EB TH}) = 1,900*2*0.967*0.909*1*1*0.960*$$
$$1*0.952*1*1*1*1 = 3,053 \text{ veh/hg}$$

$$s(\text{WB TH/RT}) = 1,900*2*1*0.909*1*1*0.960*$$
$$1*0.952*0.983*1*0.992*1$$
$$= 3,078 \text{ veh/hg}$$

$$s(\text{NB L/T/R}) = 1,900*3*1.033*0.909*0.985*1*1*$$
$$1*0.908*0.981*0.998*0.989*0.996$$
$$= 4,617 \text{ veh/hg}$$

Capacity Analysis Module

In this module, three separate computations and determinations are made: (1) computation of v/s ratios and determination of critical lane groups, (2) computation of lane group capacities and v/c ratios, and (3) computation of the critical v/c ratio for the intersection. Each of these is illustrated in the subsections that follow.

Computation of v/s Ratios and Determination of Critical-Lane Groups   The v/s ratio for each lane group is now easily determined. From the volume adjustment module adjusted lane group flow rates, v, have been estimated. In the saturation flow rate module, saturation flow rates for each lane group have also been estimated. Taking the ratio of these two values for each lane group:

Lane Group	v/s Ratio
EB LT	$217/1{,}587 = 0.137$
EB TH	$1{,}196/3{,}053 = 0.392$
WB TH/RT	$1{,}163/3{,}078 = 0.378$
NB LT/TH/RT	$1{,}304/4{,}617 = 0.282$

Using these ratios, it is now possible to determine the critical lane groups. It is useful to draw a ring diagram of the signal phasing, as shown in Figure 21.10, inserting the appropriate v/s ratios, to assist in making this determination.

The critical path through the signal phasing involves critical-lane groups EB LT, EB TH, and NB LT/TH/RT. The sum of critical lane v/s ratios is 0.799, as shown in Figure 21.10.

Computation of Capacities and v/c Ratios for Each Lane Group   The capacity for each lane group is computed using Equation 21-2:

$$c = s\left(\frac{g}{C}\right)$$

where: c = capacity of the lane group, veh/h
   s = saturation flow rate for the lane group, veh/hg
   g = effective green time for the lane group, s
   C = cycle length, s

The effective green time for the lane group is the same as the actual green time, because the standard default values for ℓ_1 and e (both 2.0 s) are in effect. Using these values of effective green time, the capacity of each lane group may now be computed:

$$c(\text{EB LT}) = 1{,}587 * \left(\frac{10}{60}\right) = 265 \text{ veh/h}$$

$$c(\text{EB TH}) = 3{,}053 * \left(\frac{36}{60}\right) = 1{,}832 \text{ veh/h}$$

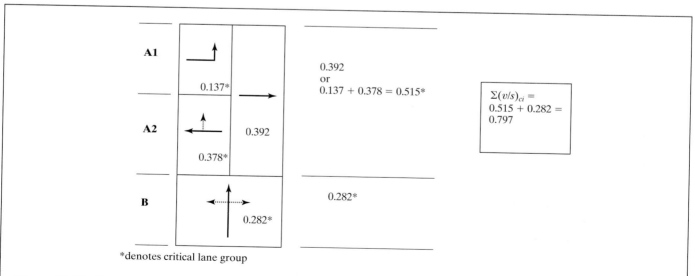

Figure 21.10: Determination of Critical-Lane Groups

*denotes critical lane group

$$c(\text{WB TH/RT}) = 3{,}078 * \left(\frac{22}{60}\right) = 1{,}129 \text{ veh/h}$$

$$c(\text{NB LT/TH/RT}) = 4{,}617 * \left(\frac{16}{60}\right) = 1{,}231 \text{ veh/h}$$

Having determined the capacity of each lane group, the v/c ratio (X) for each lane group may now be computed:

$$X_{EB\ LT} = \frac{217}{265} = 0.819$$

$$X_{EB\ TH} = \frac{1{,}196}{1{,}835} = 0.653$$

$$X_{WB\ TH/RT} = \frac{1{,}163}{1{,}128} = 1.030$$

$$X_{NB\ LT/TH/RT} = \frac{1{,}304}{1{,}231} = 1.059$$

The fact that two of these lane groups are operating at v/c ratios in excess of 1.00 is a significant problem that will have to be dealt with when the analysis is completed.

Computation of the Critical v/c Ratio for the Intersection The critical v/c ratio is computed from Equation 21-5:

$$X_c = \sum_i (v/s)_{ci} * \left(\frac{C}{C - L}\right)$$

where L is the lost time per cycle. In this problem, there is a total of 2 s of start-up lost time and 4 − 2 = 2 s of clearance lost time per phase. With three phases, L = 3 * 4 = 12 s/cycle. Then:

$$X_c = 0.797 * \left(\frac{60}{60 - 12}\right) = 0.996$$

This result indicates that 99.6% of the green time provided within the peak hour must be used to accommodate all of the critical lane group demands. It also indicates that it is at least possible to make the intersection work (i.e., bring all v/c ratios under 1.00, by reallocating green time within the existing cycle length and phase plan. It, however, would also be advisable to reduce this value through increasing the cycle length, creating a more efficient phase plan, and/or adding lanes to critical-lane groups.

Level-of-Service Module

Level of service for signalized intersections is based on *control delay*. Control delay is computed for each lane group and then averaged for approaches and the intersection as a whole. For any lane group, delay is estimated from Equation 21-30:

$$d = d_1 PF + d_2 + d_3$$

where: d = control delay, s/veh

         d_1 = uniform delay, s/veh

         d_2 = overflow delay, s/veh

         d_3 = delay due to preexisting queue, s/veh

       PF = progression adjustment factor

For the current problem, all arrival types are 3. Therefore all progression factors (PF) are 1.00. Further, since we are analyzing only one 15-min period and it is assumed that there were no initial queues existing at the beginning of this period, d_3 for all lane groups is 0.0 s/veh. Thus, to determine level of service, appropriate values for d_1 and d_2 for each lane group must be computed from Equations 21-31 and 21-32:

$$d_1 = \frac{0.5C\left(1 - \dfrac{g}{C}\right)^2}{1 - \left[\min(1,X) * \dfrac{g}{C}\right]}$$

$$d_2 = 900T\left[(X - 1) + \sqrt{(X - 1)^2 + \frac{8kIX}{cT}}\right]$$

where: C = cycle length, s

        g = effective green time for lane group, s

        X = v/c ratio for lane group

        T = length of the analysis period, h

        k = incremental delay factor for controller settings

        I = upstream filtering/metering adjustment factor

        c = lane group capacity, veh/h

For the present problem, T for all lane groups is 0.25 h (15 min); k is 0.50 for all lane groups (pretimed controller); and I = 1.0 for all analyses of individual intersections. Other variables for each lane group have been computed in previous modules or were given as input values. Then:

EB LT Lane Group

$$d_1 = \frac{0.50*60*\left(1 - \frac{10}{60}\right)^2}{1 - \left(0.819*\frac{10}{60}\right)} = 24.1 \text{ s/veh}$$

$$d_2 = 900*0.25*\left[(0.819 - 1) + \sqrt{(0.819 - 1)^2 + \left(\frac{8*0.50*1*0.819}{265*0.25}\right)}\right]$$

$$= 23.8 \text{ s/veh}$$
$$d = (24.1*1.00) + 23.8 + 0.0$$
$$= 47.9 \text{ s/veh (LOS D)}$$

EB TH Lane Group

$$d_1 = \frac{0.50*60*\left(1 - \frac{36}{60}\right)^2}{1 - \left(0.653*\frac{36}{60}\right)} = 7.9 \text{ s/veh}$$

$$d_2 = 900*0.25*\left[(0.653 - 1) + \sqrt{(0.653 - 1)^2 + \left(\frac{8*0.50*1*0.653}{1832*0.25}\right)}\right]$$

$$= 1.8 \text{ s/veh}$$
$$d = (7.9*1.00) + 1.8 + 0.0 = 9.7 \text{ s/veh (LOS A)}$$

WB TH/RT Lane Group

$$d_1 = \frac{0.50*60*\left(1 - \frac{22}{60}\right)^2}{1 - \left(1.000*\frac{22}{60}\right)} = 19.0 \text{ s/veh}$$

$$d_2 = 900*0.25*\left[(1.030 - 1) + \sqrt{(1.030 - 1)^2 + \left(\frac{8*0.50*1*1.030}{1129*0.25}\right)}\right]$$

$$= 34.8 \text{ s/veh}$$
$$d = (19.0*1.00) + 34.8 + 0.0 = 53.8 \text{ s/veh (LOS D)}$$

NB LT/TH/RT Lane Group

$$d_1 = \frac{0.50*60*\left(1 - \frac{16}{60}\right)^2}{1 - \left(1.00*\frac{16}{60}\right)} = 22.0 \text{ s/veh}$$

$$d_2 = 900*0.25*\left[(1.059 - 1) + \sqrt{(1.059 - 1)^2 + \left(\frac{8*0.50*1*1.059}{1,231*0.25}\right)}\right]$$

$$= 42.8 \text{ s/veh}$$
$$d = (22.0*1.00) + 42.8 + 0.0$$
$$= 64.8 \text{ s/veh (LOS E)}$$

Average delays for the EB approach (two lane groups) and the intersection as a whole may now be computed using Equation 21-40. Averages are weighted by the demand flow rates in each respective lane group.

EB Approach

$$d_{EB} = \frac{(217*47.9) + (1,196*9.7)}{(217 + 1,196)}$$
$$= 15.6 \text{ s/veh (LOS B)}$$

Total Intersection

$$d_{INT} = \frac{\begin{array}{c}(1,413*15.6) + (1,163*53.8) \\ + (1,304*64.8)\end{array}}{(1,413 + 1,163 + 1,304)}$$
$$= 43.6 \text{ s/veh (LOS D)}$$

Analysis

Table 21.15 summarizes the v/c ratios, delays, and levels of service predicted for each lane group and for the intersection as a whole for this sample problem.

Table 21.15: Summary Results for the Sample Problem

Lane Group	v/c Ratio (X)	Delay (s/veh)	LOS
EB LT	0.819	47.9	D
EB TH	0.653	9.7	A
WB TH/RT	1.030	53.8	D
NB LT/TH/RT	1.059	64.8	E
Intersection	0.999	43.6	D

As noted previously, two of the lane groups "fail" in that the demand exceeds the capacity of the lane group. At the end of this first 15-min analysis period, there will be a residual queue on the WB TH/RT and NB LT/TH/RT lane groups. Nevertheless, neither lane group is designated as LOS F, as the delays remain under 80.0 s/veh, the maximum for LOS E. No comfort should be taken from this, as the residual queues will rapidly increase the delay per vehicle in subsequent analysis periods. This highlights the need to consider both the v/c ratio and the delay/LOS measure to understand the nature of operations on each lane group.

The intersection as a whole operates at a very high v/c ratio (0.996). This means that if the green time were perfectly distributed, all critical-lane groups would operate at 99.6% of their capacities during the peak 15-min analysis period.

Because of this, it would normally be recommended that capacity for all critical-lane groups be increased and that the green be reallocated to provide for balanced v/c ratios in all critical lane groups.

21.4.3   Sample Problem 3: Dealing with Initial Queues

Consider the results of Sample Problem 2 for the NB lane group, as summarized below:

- $v = 1,304$ veh/h
- $c = 1,231$ veh/h
- $g/C = 16/60 = 0.267$
- $X = 1.059$
- $d = 64.8$ s/veh

- $d_1 = 22.0$ s/veh
- $d_2 = 42.8$ s/veh
- LOS = E

This approach "fails" in that the capacity is insufficient to handle the demand flow. The level of service is not "F" because the analysis of Sample Problem 2 focused on a single 15-min time period (T = 0.25 h), with no initial queue at the start of the period.

What will happen if this condition (i.e., all of the same demands, geometry, and signalization) for four consecutive 15-min periods? As the demand is greater than the capacity in each of these periods, there will be a residual queue at the end of the first period that will increase for each of the three succeeding periods. Thus, for the second, third, and fourth periods, there will be an initial queue, and the delay relating to this (d_3) will have to be considered. Further, there will be changes in the "$d_1 * PF$" term insofar as the progression factor is not applied to some or all of each subsequent analysis period.

This problem will illustrate how the delay computations change to account for these initial queues, while being able to use all of the other results from Sample Problem 2 for the NB approach.

Determining Initial Queues

Equation 21-39 is used to determine the queue remaining at the end of each analysis period. This, of course, becomes the initial queue for the next period.

$$Q_{b(i+1)} = \max[0, Q_{bi} + cT(X_i - 1)]$$

The four 15-min periods will be numbered 1–4 for convenience. Period 1 begins with no residual queue (i.e., $Q_{b1} = 0$). Then:

$$Q_{b2} = 0 + 1{,}231 * 0.25 * (1.059 - 1)$$
$$= 18.2 \Rightarrow 18 \text{ vehs}$$
$$Q_{b3} = 18.2 + 1{,}231 * 0.25 * (1.059 - 1)$$
$$= 36.4 \Rightarrow 36 \text{ vehs}$$
$$Q_{b4} = 36.4 + 1{,}231 * 0.25 * (1.059 - 1)$$
$$= 54.6 \Rightarrow 55 \text{ vehs}$$

Estimating Delay

First, note that periods 2 through 4 represent Case 5—that is, there is an initial queue, and that queue will grow throughout the analysis period. For Case 5, uniform delay, d_1, is not adjusted using the progression adjustment factor. Since the factor was used in Sample Problem 2, NB approach was 1.00 in the first place, there will be no changes to the handling of the d_1 term. For periods 2–4, the additional delay due to initial queues must be estimated and added to uniform and incremental delays (d_1 and d_2). Delay due to an initial queue is estimated using Equation 21-35:

$$d_3 = \frac{1{,}800 Q_b (1 + u) t}{cT}$$

For Case 5, $u = 1$, and $t = T$. Then:

$$d_{3,2} = \frac{1{,}800 * 18 * (1 + 1) * 0.25}{1{,}231 * 0.25} = 52.6 \text{ s/veh}$$

$$d_{3,3} = \frac{1{,}800 * 36 * (1 + 1) * 0.25}{1{,}231 * 0.25} = 105.3 \text{ s/veh}$$

$$d_{3,4} = \frac{1{,}800 * 55 * (1 + 1) * 0.25}{1{,}231 * 0.25} = 160.8 \text{ s/veh}$$

Table 21.16 summarizes the delay results and levels of service for the four consecutive analysis periods.

Analysis

This problem clearly illustrates the impact of consecutive time periods during which arrival demand exceeds capacity. The result is a growing queue of vehicles that adds significant amounts of delay as time progresses. Average control delay to drivers is almost four times as great in period 4 than it was in period 1, when the queue began to develop. This impact occurs even though v/c was 1.059 (i.e., demand exceeds capacity by only 5.9%). The queue on the NB approach grows to 55 vehicles by the *beginning* of the fourth period. Another 18.2 vehicles will accrue during the fourth period, leaving a queue of 73 vehicles at the end of the analysis hour. The NB approach has three lanes. Therefore, if the queued vehicles distribute equally and each takes up approximately 20 ft, the queue will grow to:

$$\left(\frac{73}{3} \right) * 20 = 486.7 \text{ ft}$$

Depending upon the block length, such a queue could begin influencing the discharge from an upstream signal or perhaps create a gridlock situation, if the upstream signal is less than 487 ft away.

Table 21.16: Delay Results for Sample Problem 3 (NB Approach of Sample Problem 2)

Delay Component	Delay (s) in Time Period			
	1	2	3	4
d_1	22.0	22.0	22.0	22.0
d_2	42.8	42.8	42.8	42.8
d_3	0.0	52.6	105.3	160.8
d	64.8	117.4	170.1	225.6
LOS	E	F	F	F

21.5 Complexities

Previous sections of this chapter have dealt with portions of the HCM model for analysis of signalized intersections that are at least somewhat straightforward and that can be reasonably illustrated through manual applications. In this section of the chapter, some of the more intricate portions of the model will be discussed. Some elements will not be completely detailed. The engineer should consult the HCM directly for fuller descriptions.

The following aspects of the model are addressed:

- Left-turn adjustment factor, f_{LT}, for permitted left turns
- Analysis of compound left-turn phasing
- Using analysis parameters to adjust signal timing
- Analysis of actuated signals

21.5.1 Left-Turn Adjustment Factor for Permitted Left Turns

The modeling of permitted left turns must account for the complex interactions between permitted left turns and the opposing flow of vehicles. These interactions involve several discrete time intervals within a green phase that must be separately addressed.

Figure 21.11 illustrates these portions of the green phase. It shows a subject approach with its opposing flow. When the green phase is initiated, vehicles on both approaches begin to move. Vehicles from the standing queue on the opposing approach move through the intersection *with no gaps*. Thus, *no* left turn from the subject approach may proceed during the time it takes this opposing queue of vehicles to clear the intersection. If a left-turning vehicle arrives in the subject approach during this time, it must wait, *blocking the left-most lane*, until the opposing queue has cleared. After the opposing queue has cleared, left turns from the subject approach are made through gaps in the now unsaturated opposing flow. The rate at which they can be made as well as their impact on the operation of the subject approach is dependent on the number of left turns and the magnitude and lane distribution of the opposing flow.

Another fundamental concept is that left-turning vehicles have no impact on the operation of the subject approach *until the first left-turning vehicle arrives*. This is an intuitively obvious point, but it has often been ignored in previous methodologies.

There are three distinct portions of the green phase that may be defined as follows:

g_q = average amount of green time required for the opposing queue of standing vehicles to clear the intersection, s

g_f = average amount of green time before the first left-turning vehicle arrives on the subject approach, s

g_u = average amount of time *after the arrival of the first left-turning vehicle* that is not blocked by the clearance of the opposing queue, s

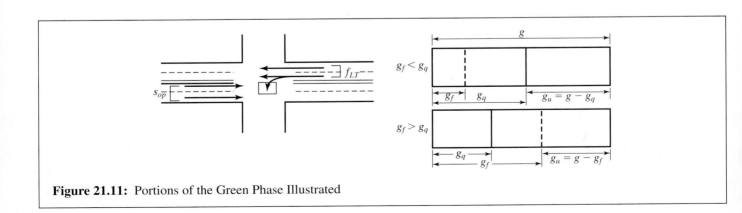

Figure 21.11: Portions of the Green Phase Illustrated

Figure 21.11 also illustrates the relationship between these key variables. The value of g_u depends upon the relative values of g_f and g_q:

$$g_u = g - g_q \qquad \text{for } g_q \geq g_f$$
$$g_u = g - g_f \qquad \text{for } g_q < g_f \qquad \text{(21-41)}$$

where: g = total effective green for phase, s

Other variables as previously defined

When defined in this fashion, g_u represents the actual time (per phase) that left turns actually filter through an unsaturated opposing flow.

Basic Model Structure

Using these definitions of critical portions of the green phase, the model for the left-turn adjustment factor must consider what type of left-turn operations are taking place at various times within a given green phase. The model structure focuses on developing an adjustment that is applied *only to the lane from which left turns are made*. This adjustment factor, f_m, will later be combined with assumed impacts of left turns on other lanes in the lane group.

In deriving an appropriate relationship for f_m, there are three time intervals to consider:

- *Interval 1: g_f.* Before the first left-turning vehicle arrives in the subject approach, left-turning vehicles have no impact on the operation of the left lane. Thus, during this period, an effective left-turn adjustment factor of 1.00 should apply.

- *Interval 2: $(g_q - g_f)$.* If the first left-turning vehicle on the subject approach arrives before the opposing queue clears $(g_q > g_f)$, the vehicle must wait, blocking the left lane during this interval. No vehicle can move in the left lane while the left-turner waits. Therefore, a left-turn adjustment factor of 0.00 applies during this period. Where $g_f \geq g_q$, this time period does not exist. Where the opposing approach is a single-lane approach, some left turns can be made during this period, as will be discussed later.

- *Interval 3: g_u.* This is the period during which left turns from the subject approach filter through

an unsaturated opposing flow. During this period of time, a left-turn adjustment factor between 0.00 and 1.00 applies to reflect the impedance of the opposing flow. The adjustment factor for this period will be referred to as F_1.

Single-lane approaches create some special situations that must be considered. When the *subject approach* has a single lane, then $f_{LT} = f_m$. When there is more than one lane in the lane group, the effect of left turns must be averaged over all lanes. When the *opposing approach* has a single lane, a unique situation for left-turners arises. Left-turning vehicles located within the opposing standing queue will create gaps in the opposing queue as it clears. Left-turners from the subject approach may make use of these gaps to execute their turns. Thus, when the opposing approach has only one lane, some left turns from the subject approach can be made during the time period $(g_q - g_f)$, and the adjustment factor for this period will be not 0.00 but some other value between 0.00 and 1.00, designated as F_2.

If all of these different situations are combined, there are *eight* basic cases under which f_m must be estimated, based on the number of lanes on the subject and opposing approaches $(1, \geq 2)$ and the relative values of g_f and g_q. These eight cases are summarized in Table 21.17.

A general form of the basic model for the factor f_m can be described as follows:

$$f_m = \frac{g_f}{g}(1.0) + \frac{\max\lfloor 0, (g_q - g_f) \rfloor}{g}(F_2) + \frac{g_u}{g}(F_1)$$

$$\text{(21-42)}$$

where f_m = LT adjustment factor applied to the lane from which left turns are made

g_f = effective green time until first left-turning vehicle on subject approach arrives, s

g_q = effective green time until opposing standing queue of vehicles clears the intersection, s

g_u = portion of effective green time during which left turns are made through an unsaturated opposing flow, s

g = effective green time, s

Table 21.17: Cases for Left-Turn Adjustment Factor

Cases	Subject Approach	Opposing Approach	$g_q < g_f$?
Case LT 1	Multilane	Multilane	Yes
Case LT 2	Multilane	Multilane	No
Case LT 3	Single Lane	Single Lane	Yes
Case LT 4	Single Lane	Single Lane	No
Case LT 5	Single Lane	Multilane	Yes
Case LT 6	Single Lane	Multilane	No
Case LT 7	Multilane	Single Lane	Yes
Case LT 8	Multilane	Single Lane	No

F_1 = LT adjustment factor that applies during g_u

F_2 = LT adjustment factor that applies during $(g_q - g_f)$—equals 0.00 when opposing approach is multilane

Based on the cases defined in Table 21.17, this can be resolved into two basic equations. Equation 21-43 applies to all cases with a multilane opposing approach and to all cases where g_q is less than g_f (Cases 1, 2, 3, 5, 6, 7):

$$f_m = \frac{g_f}{g} + \frac{g_u}{g} F_1 \qquad (21\text{-}43)$$

Equation 21-44 applies to all cases with a single-lane opposing approach *and* $g_q > g_f$ (Cases 4, 8):

$$f_m = \frac{g_f}{g} + \left(\frac{g_q - g_f}{g} \right) F_2 + \frac{g_u}{g} F_1 \qquad (21\text{-}44)$$

There are two other important aspects of the general model. Equations 21-43 and 21-44 allow for a result of "0.00," which occurs when the opposing queue consumes the entire effective green phase to clear ($g_q = g$) and there is a left-turning vehicle waiting in the subject approach immediately ($g_f = 0.0$ s). However, even when the entire green is blocked by the opposing queue, one vehicle can move past the STOP line and wait until the opposing queue stops, completing its turn during the yellow or all-red phase. A second vehicle might also do so if it is immediately behind the first left-turner. These vehicles are called *sneakers*, and they represent a practical minimum left-turn flow during a phase. This translates into an effective *minimum value* for the factor f_m:

$$f_m \, (\text{min}) = \frac{2(1 + P_L)}{g} \qquad (21\text{-}45)$$

where: P_L = proportion of left-turning vehicle *in the left lane*

       g = effective green time, s

       2 = assumed headway utilized by a sneaker, s

The remaining issue for the general model is the translation of f_m, which applies only to the lane from which left turns are made, to f_{LT}, which applies to the lane group. When the lane group has only one lane, this is trivial, and $f_m = f_{LT}$. Where the subject lane group has more than one lane, the effect of left turns is averaged over all lanes. With previous factors (bus blockage, parking), it was assumed that lanes adjacent to the primary lane were not affected (i.e., the factor for other lanes in the group was effectively 1.00). This is not true for left turns. As left-turners block the left lane, through vehicles begin to make late lane changes to get around them, causing some disruption to flow in other lanes. Field data suggests that this effect may vary but is insufficient to calibrate a varying relationship. As a default, it is assumed that the effective LT adjustment factor in lanes other than the left lane is 0.91. Then:

$$f_{LT} = \frac{f_m + 0.91(N - 1)}{N} \qquad (21\text{-}46)$$

where N is the number of lanes in the lane group and other variables are as previously defined.

This describes the basic modeling approach. There are several components, however, that must be estimated, including:

- Portions of the effective green phase: g_f, g_q, g_u
- Adjustment factors F_1 and F_2
- Proportion of left turns in the left lane, P_L

Models for finding each of these are discussed below.

Estimating g_f

The methodology for estimating the critical portions of the effective green time (g_f, g_q, and g_u) are based primarily on regression equations developed using a nationwide database of over 50 intersection approaches [6]. The initial portion of the effective green phase, g_f, is the time that elapses in the effective green before the arrival of the first left-turning vehicle. The algorithm depends on whether the subject approach is a multilane or a single-lane group. Departure behavior of queues is somewhat altered when there are left- and right-turning vehicles in the same queue.

$$g_f = 0.0 \qquad \text{Exclusive LT Lanes}$$
$$g_f = Ge^{-(0.882LTC^{0.717})} - t_L \qquad \text{Shared Multilane Groups}$$
$$g_f = Ge^{-(0.860LTC^{0.629})} - t_L \qquad \text{Shared Single-Lane Groups}$$

$$(21\text{-}47)$$

where: g_f = portion of green phase before the arrival of the first left-turning vehicle on the subject approach, s

G = actual green time for lane group, s

t_L = total lost time per phase, s

LTC = left turns per cycle, vehs/cycle [$v_{LT} * C/3{,}600$]

Where permitted left turns are made from an exclusive LT lane, g_f is 0.0, as the first vehicle in queue is, by definition, a left-turning vehicle. As Equations 21-47 are regression-based, they are based on the actual green time, G, rather than the effective green time, g. Numerically, this may not make any difference, as

$g = G$ where standard default values are used for start-up lost time (ℓ_1) and extension of effective green into yellow and all-red (e) are used (2.0 s for each). Because of lost times, however, G and g, even if numerically equal, do not start at the same time. This is why the lost time, t_L, is subtracted in Equations 21-47. The entire lost time is deducted as a result of the simplification of effective green time adopted in the HCM model, which assumes all lost times to occur at the beginning of the phase. As effective green time, therefore, starts t_L seconds after the actual green, and g_f is a measure of the first left-turning vehicle's arrival within g (not G), the t_L term must be deducted.

Estimating g_q

The estimation of g_q depends upon whether the *opposing approach* is a multilane or a single-lane group. When the opposing approach has only a single lane, left turns from that lane within the opposing queue clearance process open gaps for left turns from the subject approach. Thus, queue clearance is somewhat enhanced when the opposing approach has only one lane.

$$g_q = \left[\frac{v_{olc}\, qr_o}{0.5 - [v_{olc}(1 - qr_o)/g_o]} \right] - t_L \qquad \begin{array}{l}\text{Multilane} \\ \text{Opposing} \\ \text{Approach}\end{array}$$

$$g_q = 4.943 v_{olc}^{0.762} qr_o^{1.061} - t_L \qquad \begin{array}{l}\text{Single-Lane Opposing} \\ \text{Approach}\end{array}$$

$$(21\text{-}48)$$

where g_q = average time for opposing standing queue to clear the intersection, s

v_{olc} = opposing flow rate in veh/ln/cycle [$v_o * C/(3{,}600 \times N_o)$]

v_o = opposing flow rate, veh/h (includes only opposing through vehicles)

N_o = opposing lanes (not including exclusive LT or RT lanes)

qr_o = queue ratio for the opposing flow; portion of the opposing queue originating in opposing standing queues, estimated as [$1 - R_{po}(g_o/C)$]

R_{po} = opposing platoon ratio (use default for arrival type unless field measurement is available)

C = cycle length, s

g_o = effective green time for the opposing approach, s

t_L = total lost time per phase for the opposing approach, s/phase

The first Equation 21-48 is theoretically developed, while the second is from the same regression study as Equation 21-47 for g_f. The first equation is based on an assumption that the number of vehicles departing the intersection during g_q must be equal to the number of vehicles arriving during the red interval, plus those that join the end of the queue during g_q.

Estimating g_u

Once g_f and g_q are known, g_u is computed using Equation 21-41, presented previously.

Estimating F_1

Adjustment factor F_1 is the left-turn adjustment applied during g_u, the time in which left turns filter through an unsaturated opposing flow. The factor is based on through-vehicle equivalents for left-turning vehicles filtering through an unsaturated opposing flow, as follows:

$$F_1 = \frac{1}{1 + P_L (E_{L1} - 1)} \qquad (21\text{-}49)$$

where: E_{L1} = through vehicle equivalent of a vehicle executing a left turn during g_u

P_L = proportion of left-turning vehicles in the lane from which left turns are made

Equivalents are based on empiric studies of filtering left turns and are shown in Table 21.18.

Estimating F_2

Adjustment factor F_2 is the adjustment applied during the period $g_q - g_f$ where the opposing approach has a single lane. During this time period, subject left turns can be made through gaps in the opposing queue created by opposing left-turning vehicles. The factor is also based on an equivalence concept, as was F_1.

$$F_2 = \frac{1}{1 + P_L (E_{L2} - 1)} \qquad (21\text{-}50)$$

where: E_{L2} = through vehicle equivalent for a left turn made during period $g_q - g_f$

All other variables are as previously defined.

The value of the equivalent, E_{L2}, is determined using a simple probabilistic model that considers how long a left-turning vehicle in the subject approach would have to wait for a left-turning vehicle in the opposing approach to open a gap in the opposing traffic stream.

$$E_{L2} = \frac{1 - P_{THo}^n}{P_{LTo}} \qquad (21\text{-}51)$$

where: P_{THo} = proportion of through vehicles in the opposing single-lane approach

P_{LTo} = proportion of left turns in the opposing single-lane approach

Table 21.18: Through-Vehicle Equivalents for Permitted Left Turns, E_{L1}

Type of LT Lane	Effective Opposing Flow, $v_{oe} = v_o/f_{Luo}$ (veh/h)						
	1	200	400	600	800	1,000	1,200
Shared	1.4	1.7	2.1	2.5	3.1	3.7	4.5
Exclusive	1.3	1.6	1.9	2.3	2.8	3.3	4.0

Notes: $v_o > 0$ veh/h.

Straight-line interpolation for v_{oe} is appropriate to determine E_{L1} to the nearest 0.1.

(Used with Permission of Transportation Research Board, National Research Council, *Highway Capacity Manual*, 4th Edition, Washington DC, 2000, Exhibit C16-3, pg. 16-124.)

n = number of opposing vehicles in time period $g_q - g_f$, roughly estimated as $(g_q - g_f)/2$

Estimating P_L

The input data specifies a value of P_{LT}, the proportion of left turns in the *lane group*. P_L is the proportion of left turns *in the lane from which left turns are made*. Two cases are trivial:

- Where an exclusive LT lane exists, $P_L = 1.00$
- Where a single-lane approach exists, $P_L = P_{LT}$

It is the remaining case that is difficult: left turns made from a shared lane on a multilane approach. A complex algorithm has been derived to estimate P_L from P_{LT} in such cases:

$$P_L = P_{LT} \left[1 + \frac{(N-1)g}{g_f + \left(\dfrac{g_u}{E_{L1}} \right) + 4.24} \right] \quad (21\text{-}52)$$

where all terms are as previously defined.

The derivation of this equation is somewhat tortuous, and it does not address all relevant factors. By definition, P_L and P_{LT} may be expressed as:

$$P_L = {}^{V_{LT}}\!/_{V_L}$$

$$P_{LT} = {}^{V_{LT}}\!/_{(V_L + V_2)}$$

where V_{LT} = left-turn volume, veh/h

V_L = volume in the left lane of the lane group, veh/h

V_2 = sum of the volumes in all lanes of the lane group *except* the left lane, veh/h (i.e., $V_L + V_2 = V$)

From these definitions, it may be stated that:

$$\frac{P_L}{P_{LT}} = \frac{V_L + V_2}{V_L}$$

$$P_L = P_{LT} \left(1 + \frac{V_2}{V_L} \right)$$

The remaining parts of the derivation concern estimating the value of the ratio V_2/V_L. In essence, it is modeled as the ratio of the green time times the number of lanes available to move the two flows.

There are $N - 1$ lanes available to move flow V_2. It moves throughout the effective green time, g. Thus, this flow moves in $(N - 1)g$ "lane-seconds" of time. The left lane, which services V_L, moves during g_f (before a left-turning vehicle arrives) and during g_u. During g_u, vehicles move less efficiently (as left turns filter through the unsaturated opposing flow), and this time is, therefore, discounted by E_{L1}. This discounting, however, assumes that *all* vehicles moving in the left lane during g_u are left-turners. This is, in fact, a worst-case assumption. The 4.24 term represents additional time added to account for sneakers turning after the end of the effective green. Thus, Equation 21-52 is derived by substituting:

$$\frac{V_2}{V_L} = \frac{(N-1)g}{g_f + \left(\dfrac{g_u}{E_{L1}} \right) + 4.24}$$

There is another critical point that occurs when P_L is computed. If $P_L = 1.00$, a *de facto left turn lane* exists. The entire analysis is then stopped, and it is begun again with the left lane established as an exclusive LT lane.

Example 21-1:

To illustrate the path through these algorithms, consider the following characteristics of a three-lane shared approach at a signalized intersection:

- Demand flow rate, $v = 1,200$ veh/h (adjusted for PHF)
- Left turn flow rate in v, $v_{LT} = 35$ veh/h (also adjusted for PHF)

- Signal timing: $C = 90$ s; $G = g = 50$ s
- Permitted left turns
- Opposing flow rate, $v_o = 900$ veh/h (also adjusted for PHF)
- Arrival Type 3 on subject and opposing approaches
- $t_L = 4.5$ s/phase
- Opposing approach has three lanes

What is the left-turn adjustment factor for this situation?

Solution: Because the opposing approach has more than two lanes, Equation 21-43 defines the basic left-turn adjustment (for the left lane of the subject approach):

$$f_m = \frac{g_f}{g} + \frac{g_u}{g} F_1$$

The initial portion of the green phase, g_f, is found using Equation 21-47 for a multilane subject approach:

$$g_f = Ge^{-(0.882LTC^{0.717})} - t_L$$

where: $g = G = 50$ s

$t_L = 4.5$ s

$LTC = 35 * 90/3,600 = 0.875$

Then:

$$g_f = 50.0e^{-(0.882 * 0.875^{0.717})} - 4.5 = 17.9 \text{ s}$$

The clearance time for the opposing queue of vehicles is given by Equation 21-48 for a multilane opposing approach:

$$g_q = \left[\frac{v_{olc}\, q\, r_o}{0.5 - [v_{olc}\,(1 - q\,r_o)/g_o]} \right] - t_L$$

$$g_o = C - g - Y - Y = C - g - t_L - t_L$$

$$= 90 - 50 - 4.5 - 4.5 = 31$$

where: $v_{olc} = 900 * 90/(3,600 * 3) = 7.5$ veh/cycle/ln

$q\,r_o = 1 - R_{po}\,(g_o/C) = 1 - 1.0(31/90)$

$= 0.656\,(R_{po} = 1.0 \text{ for AT-3})$

$t_L = 4.5$ s

Then:

$$g_q = \left[\frac{7.5 * 0.656}{0.5 - [7.5(1 - 0.656)/31]} \right] - 4.5 = 7.3 \text{ s}$$

and:

$$g_u = g - g_f = 50.0 - 17.9 = 32.1 \text{ s}$$

In computing g_u, Equation 21-41 is used (for cases where $g_f > g_q$).

Equation 21-49 is used to obtain F_1, with Equation 21-52 used to determine P_L, which is a needed input to Equation 21-49. The value of E_{L1} is selected from Table 21-18 for an opposing flow of 900 veh/h and a shared-lane operation. For this condition (interpolating between $v_o = 800$ and $1,000$ veh/h), $E_{L1} = 3.4$. Then:

$$P_L = P_{LT}\left[1 + \frac{(N - 1)g}{g_f + \left(\frac{g_u}{E_{L1}}\right) + 4.24} \right]$$

$$= \left(\frac{35}{1,200}\right) * \left[1 + \frac{(3 - 1)50}{17.9 + \frac{32.1}{3.4} + 4.24} \right] = 0.122$$

$$F_1 = \frac{1}{1 + 0.122(3.4 - 1)} = 0.774$$

Then, using Equations 21-43 and 21-46:

$$f_m = \frac{17.9}{50} + \frac{32.1}{50}(0.774) = 0.358 + 0.497 = 0.855$$

$$f_{LT} = \frac{0.855 + (3 - 1)*0.91}{3} = 0.892$$

21.5.2 Modeling Compound Phasing

Protected plus permitted and/or permitted plus protected phasing is the most complex aspect of signalized intersection operations to model analytically. The approach to estimating saturation flow rates, capacities, and delays must be altered to reflect the unique operating characteristics of such phasing.

This section describes some of these alterations in general, with some illustrations. The engineer is urged to consult the HCM directly for full details of all aspects of compound phasing analysis.

In terms of saturation flow rates and capacities, the general approach taken in the HCM is straightforward. The protected and permitted portions of the phase are separated, with saturation flow rates and capacities computed separately for each. The appropriate green times are associated with each portion of the phase. The protected portion of the phase is analyzed as if it were a fully protected phase (i.e., $f_{LT} = 0.95$), while the permitted portion of the phase is analyzed as if it were a fully permitted phase, using the left-turn model described in the previous section.

In analyzing the permitted portion of the phase, however, the algorithms used to predict g_f, g_q, and g_u

must be modified. Equations 21-47, 21-48, and 21-41 all assume that the permitted phases start at the same time in the subject and opposing directions. In compound phasing, this is not true, and the values of g_f, g_q, and g_u must be altered to reflect this. For example, g_f is the time *within the permitted phase* to the arrival of the first left-turning vehicle in the subject phase. It is indexed to the beginning of the green on the subject approach. If the approach had a protected-plus-permitted phase, however, the predicted value of g_f would be relative to the start of the green—which is the beginning of the *protected* portion of the phase. The value needed must be indexed to the start of the *permitted* portion of the green, which requires an adjustment.

Depending upon the order and type of compound phasing in place, there are many different scenarios requiring different adjustments to the prediction of g_f, g_q, and g_u. All of these are detailed in the HCM. Figure 21.12 illustrates one of the many cases that can arise, a leading green phase in one direction.

The NB permitted phase begins after a NB lead phase. When considering the permitted portion of this phase, several modifications are indicated in Figure 21.12. The effective green time, g^*, for this portion, must be adjusted to remove all lost times, as these are applied at the beginning of the protected portion of the phase. Equation 21-47 for g_f yields a value indexed to the start of the effective green in the protected portion of the phase, when the NB approach begins to move. The impact on the permitted portion of the phase is found by subtracting the portions of g_f occurring during the protected portion of the phase. Even g_q is modified to eliminate the deduction of t_L. This is because the left-turn movement experiences no lost time in the permitted portion of the phase.

Revisions to key variables for the SB permitted phase are similarly altered. In this case, g_q (based on the opposing queue clearance), is indexed to the beginning of the opposing flow, which occurs in the protected portion of the phase. Finding the blockage during the permitted

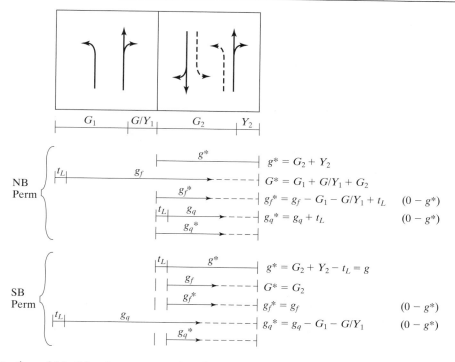

Figure 21.12: Illustration of Modifications to g_f and g_q for Compound Phasing (Leading Green Case) (Used with Permission of Transportation Research Board, National Research Council, *Highway Capacity Manual*, 4th Edition, Washington DC, 2000, Exhibit C16-5, pg. 16-129.)

portion of the phase requires that parts of g_q occurring during the protected portion of the phase be deducted.

These are illustrative. The HCM contains detailed descriptions of five different signalization cases involving compound phasing and shows the specific modifications required in each case.

While the separation of protected and permitted portions of the phase works well in the computation of saturation flow rates and capacities, the determination of v/c ratios is more difficult. In theory, what is needed is an algorithm for assigning the demand flow to the two portions of the phase. Unfortunately, the division of demand between protected and permitted portions of the phase depends upon many factors, including the platoon arrival structure on the approach. The HCM takes a very simplistic view:

- Demand utilizes the full capacity of the *first* portion of the phase, regardless of whether it is protected or permitted.
- All demand unserved by the first portion of the phase is assigned to the second portion of the phase.

Thus, the first portion of the phase has a maximum v/c ratio of 1.00. If all demand can be handled in the first portion of the phase, no demand is assigned to the second. In these cases, it might cause the engineer to question the need for compound phasing.

Where the first portion of the phase cannot accommodate the demand, all remaining flow is assigned to the second. The v/c ratio of the second portion of the phase, therefore, can range from 0.00 (when no demand is assigned) to a value >1.00, when the total demand exceeds the total capacity of the compound phase.

The prediction of delay for compound phasing is even more complex. Appendix E, Chapter 16 of HCM 2000 completely describes the procedure used. Unfortunately, the demand distribution assumed for the compound phasing delay model is not the same as that described previously for the computation of v/c ratios. The approach assumes a uniform arrival pattern throughout the compound phase. This is because it is only the *uniform delay* term, d_1, that is modified for compound phasing.

The HCM identifies five cases for analysis, leading to five different queue accumulation polygons, illustrated in Figure 21.13. The uniform delay model for

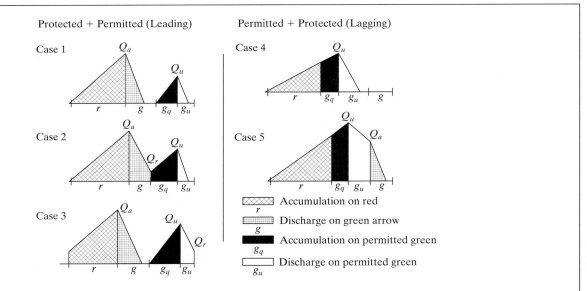

Figure 21.13: Queue Accumulation Polygons for Uniform Delay Estimation: Compound Phasing (Used with Permission of Transportation Research Board, National Research Council, *Highway Capacity Manual*, 4th Edition, Washington DC, 2000, Exhibit E16-1, pg. 16-140.)

each of these cases is the area under the accumulation polygon.

The five cases of Figure 21.13 are defined as follows:

- *Case 1*: Leading LT phase; no queue remains at the end of the protected or permitted phase.
- *Case 2*: Leading LT phase; a queue remains at the end of the protected phase, but not at the end of the permitted phase.
- *Case 3*: Leading LT phase; a queue remains at the end of the permitted phase, but not at the end of the protected phase.
- *Case 4*: Lagging LT phase; no queue remains at the end of the permitted phase.
- *Case 5*: Lagging LT phase; a queue remains at the end of the permitted phase.

The variables shown in Figure 21.13 are:

- Q_a = queue size at the beginning of the green arrow, veh
- Q_u = queue size at the beginning of the unsaturated interval of the permitted green phase, veh
- Q_r = residual queue, veh

The HCM provides algorithms and worksheets to find these important variables. Manual computation is, however, difficult, and most analysts rely on software to implement this portion of the signalized intersection model.

21.5.3 Altering Signal Timings Based on *v/s* Ratios

As discussed earlier, the capacity and/or delay results of a signalized intersection analysis may indicate the need to adjust the cycle length and/or to reallocate green time. While the engineer could return to the methodology of Chapter 18, a retiming of the signal may be accomplished using the results of the analysis as well. The methodology outlined here may *not* be used if the phase plan is to be altered as well, as it assumes that the phase plan on which the initial analysis was conducted does not change.

As a result of an analysis, *v/s* ratios for each lane group have been determined and the critical path through

the phase plan has also been identified. Therefore, the sum of critical-lane group *v/s* ratios is known.

If the cycle length is to be altered, a desired value may be estimated using Equation 21-5, which defines X_c, the critical *v/c* ratio for the intersection. The equation is solved for the cycle length, C:

$$X_c = \sum_i (v/s)_{ci}\left(\frac{C}{C-L}\right)$$

$$C = \frac{LX_c}{X_c - \sum_i (v/s)_{ci}} \tag{21-53}$$

where all terms have been previously defined. In this application, the engineer would choose a target value of X_c. Just as in Chapter 18, the target value is usually somewhere in the range of 0.80 to 0.95. The optimal delays usually occur with *v/c* ratios in this range.

Once a cycle length has been selected, even if it is the same as the initial cycle length used in the analysis, green times may be allocated using Equation 21-3, which is solved for g_i:

$$X_i = \frac{(v/s)_i}{(g/C)_i}$$

$$g_i = (v/s)_i \frac{C}{X_i} \tag{21-54}$$

The most often applied strategy is to allocate the green such that all values of X_i for critical-lane groups are equal to X_c. Other strategies are, however, possible.

Consider the results of Sample Problem 2, discussed earlier. For convenience, the results are summarized in Figure 21.14. Also, instead of demand flow rates, the *v/s* ratios determined for each lane group are shown on the diagram of the intersection.

Recall from the sample problem that two distinct problems were identified. The X_c value of 0.999 indicates that the cycle length is barely adequate to satisfy critical demands. An increase in cycle length appears to be justifiable in this case. Further, two lane groups, the WB and NB lane groups (both critical), have *v/c* ratios in excess of 1.00, indicating that a deficiency of capacity exists in these phases. Thus, some reallocation of green time to these lane groups also appears to be appropriate.

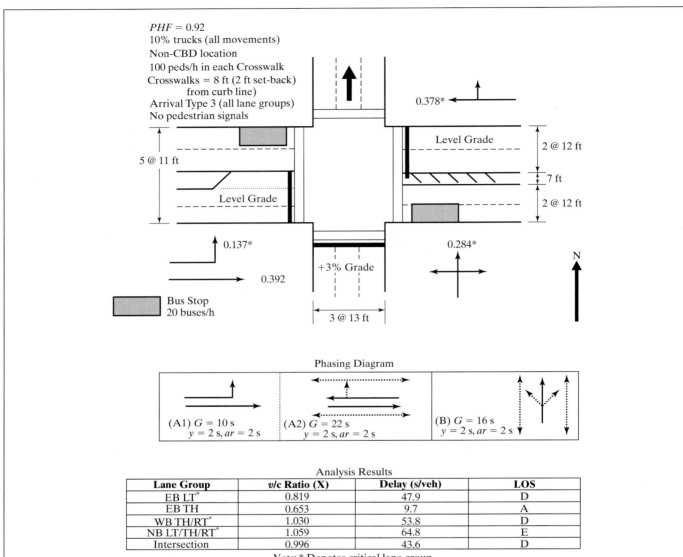

PHF = 0.92
10% trucks (all movements)
Non-CBD location
100 peds/h in each Crosswalk
Crosswalks = 8 ft (2 ft set-back)
        from curb line)
Arrival Type 3 (all lane groups)
No pedestrian signals

Phasing Diagram

(A1) *G* = 10 s
      y = 2 s, *ar* = 2 s

(A2) *G* = 22 s
      y = 2 s, *ar* = 2 s

(B) *G* = 16 s
    y = 2 s, *ar* = 2 s

Analysis Results

Lane Group	*v/c* Ratio (X)	Delay (s/veh)	LOS
EB LT*	0.819	47.9	D
EB TH	0.653	9.7	A
WB TH/RT*	1.030	53.8	D
NB LT/TH/RT*	1.059	64.8	E
Intersection	0.996	43.6	D

Note: * Denotes critical lane group.

$$\sum_i (v/s)_{ci} = 0.137 + 0.378 + 0.282 = 0.797$$

Figure 21.14: Results of Sample Problem 2

Using Equation 21-53, cycle lengths may be computed to provide various target *v/c* ratios. Values of X_c between 0.80 and 0.95 will be tried:

$$C = \frac{12 * 0.95}{0.95 - 0.797} = 74.5 \text{ s}$$

$$C = \frac{12 * 0.90}{0.90 - 0.797} = 104.9 \text{ s}$$

$$C = \frac{12 * 0.85}{0.85 - 0.797} = 192.5 \text{ s}$$

$$C = \frac{12 * 0.80}{0.80 - 0.797} = 3,200 \text{ s}$$

It is impractical to provide for $X_c = 0.80$, or even $X_c = 0.85$, as the resulting cycle lengths required are excessive. Note that this computation *can* result in a negative number, which implies that *no* cycle length can satisfy the target X_c value. In this case, a cycle length between 80 s and 120 s appears to be reasonable. For the remainder of this illustration, a cycle length of 100 s will be used. For a 100-s cycle length:

$$X_c = 0.797 \left(\frac{100}{100 - 12} \right) = 0.906$$

Equation 21-54 will be used to reallocate the green time within the new 100-s cycle length. A target X_i of 0.906 will be used for each critical lane group:

$$g_{A1} = G_{A1} = 0.137 * \left(\frac{100}{0.906} \right) = 15.1 \text{ s}$$

$$g_{A2} = G_{A2} = 0.378 * \left(\frac{100}{0.906} \right) = 41.7 \text{ s}$$

$$g_B = G_B = 0.282 * \left(\frac{100}{0.906} \right) = 31.2 \text{ s}$$

$$15.1 + 41.7 + 31.2 + 12.0 = 100 \text{ OK}$$

The retimed signal will provide for v/c ratios of 0.906 in each critical-lane group, and for a critical v/c ratio of 0.906 for the intersection. Delays could be recalculated based on these results to determine the impact of the retiming on delay and level of service, but these will surely be positive, given that two of the three critical-lane groups "failed" in the initial analysis, with $v/c > 1.00$.

In Sample Problem 2, there were no permitted or compound left turns. Thus, the retiming suggested is exact. Where permitted or compound left turns exist, saturation flow rates include a permitted f_{LT} that is dependent upon signal timing parameters. Thus, the process in such cases is technically iterative. In practical terms, however, retiming the signal as indicated gives a reasonably accurate result that can be used in a revised analysis to determine its exact impact on operations.

21.5.4 Analysis of Actuated Signals

The application of the HCM analysis model requires the specification of signal timing. For actuated signals, the average phase times and cycle length for the period of analysis must be provided. For existing locations, this can be observed in the field. For the analysis of future locations, or consideration of signal retiming, the average phase and cycle lengths must be estimated.

Appendix B, Chapter 16 of HCM 2000 provides a detailed model for the estimation of actuated signal timing, given controller and detector parameters as inputs. It is an algorithm that requires many iterations and is not fully documented for all cases. To date, none of the major software packages include this computational element, so it has, in general, not been applied by users.

Previous editions of the HCM recommended a rough estimation procedure using Equations 21-53 and 21-54 of the previous section with a high target v/c ratio in the range of 0.95 to 1.00. As a rough estimate, this is still a viable approach, pending software to implement the HCM's more detailed approach. It assumes, however, that controller settings and detector locations are optimal, which is not always the case.

The delay model, of course, specifically accounts for the positive impact of actuated control on operations. It too, however, requires that the average signal settings be specified.

21.6 Calibration Issues

The HCM model is based on a default base saturation flow rate of 1,900 pc/hg/ln. This value is adjusted by up to 11 adjustment factors to predict a prevailing saturation flow rate for a lane group. The HCM provides guidance on the measurement of the prevailing saturation flow rate, s. While it allows for substituting a locally calibrated value of the base rate, s_o, it does not provide a means for doing so. It also does not provide a procedure for measuring lost times in the field.

It is also useful to quickly review how the calibration of adjustment factors of various types may be addressed, even if this is impractical in many cases. A study procedure for measuring delays in the field is detailed in Chapter 9.

21.6.1 Measuring Prevailing Saturation Flow Rates

As defined in Chapter 17, saturation flow rate is the maximum average rate at which vehicles in a standing queue may pass through green phase, after start-up lost times have been dissipated. It is measured on a lane-by-lane basis through observations of headways as vehicles pass over the stop line of the intersection approach. The first headway begins when the green is initiated and ends when the first vehicle in queue crosses the stop line (front wheels). The second headway begins when the first vehicle (front wheels) crosses the stop line and ends when the second vehicle in queue (front wheels) crosses the stop line. Subsequent headways are similarly measured.

The HCM suggests that for most cases, the first four headways include an element of lost time and, thus, are not included in saturation flow rate observations. Saturation headways, therefore, begin with the fifth headway in queue and end when the last vehicle in the standing queue crosses the stop line (again, front wheels). Subsequent headways do not necessarily represent saturation flow.

21.6.2 Measuring Base Saturation Flow Rates

The base saturation flow rate assumes a set of "ideal" conditions that include 12-ft lanes, no heavy vehicles, no turning vehicles, no local buses, level terrain, non-CBD location, among others. It is usually impossible to find a location that has all of these conditions.

In calibrating a base saturation flow rate, a location is sought with near ideal physical conditions. An approach with three or more lanes is recommended, as the middle lane can provide for observations without the influence of turning movements. Heavy vehicles cannot be avoided, but sites that have few heavy vehicles provide the best data. Even where data is observed under near ideal physical conditions, *all headways* observed after the first heavy vehicle must be discarded when considering the base rate.

21.6.3 Measuring Start-Up Lost Time

If the first four headways contain a component of start-up lost time, then these headways can be used to measure the start-up lost time. If a saturation headway for the data has been established as h s/veh, then the lost time component in each of the first four headways is $(h_i - h)$, where h_i is the total observed headway for vehicles 1–4

in queue. The start-up lost time is the sum of these increments. Both saturation flow rate and start-up lost time are observed for a given lane during each signal cycle. The calibrated value for use in analysis would be the average of these observations.

Start-up lost time under base conditions can be observed as well by choosing a location and lane that conforms to the base conditions for geometrics with no turning vehicles and by eliminating consideration of any headways observed after the arrival of the first heavy vehicle.

21.6.4 An Example of Measuring Saturation Flow Rates and Start-Up Lost Times

The application of these principles is best illustrated through example. Table 21.19 shows data for six signal cycles of a center lane of a three-lane approach (no turning vehicles) that is geometrically ideal. In general, calibration would involve more cycles and several locations. To keep the illustration to a reasonable size, however, the limited data of Table 21.19 will be used.

Note that saturation conditions are said to exist only between the fifth headway and the headway of the last vehicle present in the standing queue when the signal turns green. Only the headways occurring between these limits can be used to calibrate saturation flow rate. The first four headways in each queue will be used subsequently to establish the start-up lost time.

The saturation headway for the lane in question is the average of all observed headways representing saturated conditions. As seen in Table 21.19, there are 41 observed saturation headways totaling 96.0 s.

From this data, the average saturation headway (under prevailing conditions) at this location is:

$$h = \frac{96.0}{41} = 2.34 \text{ s/veh}$$

From this, the saturation flow rate for this lane may be computed as:

$$s = \frac{3,600}{2.34} = 1,538 \text{ veh/hg/ln}$$

Table 21.19: Sample Data for Measurement of Saturation Flow Rate

Queue Position	Observed Headways (s) in Cycle No.						Sum of Sat. Headways	No. of Sat. Headways
	1	2	3	4	5	6		
1	3.5	2.9	3.9	4.2H	2.9	3.2	0.0	0
2	3.2	3.0	3.3	3.6	3.5H	3.0	0.0	0
3	2.6	2.3	2.4	3.2H	2.7	2.5	0.0	0
4	<u>2.8H</u>	<u>2.2</u>	<u>2.4</u>	<u>2.5</u>	<u>2.1</u>	<u>2.9H</u>	<u>0.0</u>	<u>0</u>
5	2.5	*2.3*	2.1	2.1	2.2	2.5	13.7	6
6	2.3	*2.1*	2.4	2.2	2.0	2.3	13.3	6
7	3.2H	2.0	2.4	2.4	2.2	2.3	14.5	6
8	<u>2.5</u>	*1.9*	2.2	2.3	2.4	2.0	13.3	6
9	4.5	2.9H	2.7H	1.9	2.2	2.4	12.1	5
10	6.0	2.5	<u>2.4</u>	2.3	2.7H	2.1	12.0	5
11		2.8H	<u>4.0</u>	2.2	<u>2.4</u>	2.0	9.4	4
12		<u>2.5</u>	7.0	<u>2.9H</u>	<u>5.0</u>	<u>2.3</u>	7.7	3
13		5.0		4.1		6.0	0.0	0
14		7.5					0.0	0
15							0.0	0
Sum							**96.0**	**41**

Notes: H = heavy vehicle
Single underline: beginning of saturation headways
Double underline: end of standing queue clearance; end of saturation headways
Italics: saturation headway under base conditions

If a lane group had more than one lane, the saturation headways and flow rates would be separately measured for each lane. The saturation flow rate for the lane group is then the sum of the saturation flow rates for each lane.

Measuring the base saturation flow rate for this location involves eliminating the impact of heavy vehicles, assuming that all other features of the lane conform to base conditions. As the heavy vehicles may conceivably influence the behavior of any vehicle in queue behind it, the only headways that can be used for such a calibration are those before the arrival of the first heavy vehicle. Again, saturation headways begin only with the fifth headway. Looking at Table 21.19, there are only eight headways that qualify as saturation headways occurring before the arrival of the first heavy vehicle:

• Headways 5–8 of Cycle 2
• Headways 5–8 of Cycle 3

The sum of these eight headways is 17.4 s, and the base saturation headway and flow rate may be computed as:

$$h_o = \frac{17.4}{8} = 2.175 \text{ s/veh}$$

$$s_o = \frac{3,600}{2.175} = 1,655 \text{ pc/hg/ln}$$

Start-up lost time is evaluated relative to the base saturation headway. It is calibrated using the first four headways in each queue, as these contain a component of start-up lost time in addition to the base saturation headway. As the lost time is relative to base conditions, however, only headways occurring before the arrival of the first heavy vehicle can be used. The average headway for each of the first four positions in queue is determined from the remaining measurements. The component of start-up lost time in each of the first four queue positions is then taken as $(h_i - h_o)$. This computation is shown in

Table 21.20: Calibration of Start-Up Lost Time from Table 21.19 Data

Position In Queue	Observed Headway (s) for Cycle No. _____						Avg. h (s)	h_{avg} − 2.175 (s)
	1	2	3	4	5	6		
1	2.5	2.9	3.9	H	2.9	3.2	3.080	0.905
2	3.2	3.0	3.3	H	H	3.0	3.125	0.950
3	2.6	2.3	2.4	H	H	2.5	2.450	0.275
4	H	2.2	2.4	H	H	H	2.300	0.125
Sum								**2.255**

H = headway occurring after arrival of first heavy vehicle.

Table 21.20, which eliminates all headways occurring after the arrival of the first heavy vehicle. The start-up lost time for this lane is 2.255 s/cycle.

Where more than one lane exists in the lane group, the start-up lost time would be separately calibrated for each lane. The start-up lost time for the lane group would be the *average* of these values.

Clearly, for actual calibration, more data would be needed and should involve a number of different sites. The theory and manipulation of the data to determine actual and base saturation flow rates, however, does not change with the amount of data available.

21.6.5 Calibrating Adjustment Factors

Of the 11 adjustment factors applied to the base saturation flow rate in the HCM model, some are quite complex and would require major research studies for local calibration. Included in this group are the left-turn and right-turn adjustment factors and the pedestrian/bicycle interference adjustment factors. A number of the adjustment factors are relatively straightforward and would not be difficult to calibrate locally, at least on a theoretical basis. It may always be difficult to find appropriate sites with the desired characteristics for calibration. Three adjustment factors involve only a single variable:

- Lane width (12-ft base condition)
- Grade (0% base condition)
- Area type (non-CBD base condition)

Two additional factors involve two variables:

- Parking (no parking base condition)
- Local bus blockage (no buses base condition)

The heavy-vehicle factor involves some very special considerations, and the lane utilization factor should be locally measured in any event and is found using Equation 21-18 or default values.

Calibration of all of these factors involves the controlled observation of saturation headways under conditions in which only one variable does not conform to base conditions. By definition, an adjustment factor converts a base saturation flow rate to one representing a specific prevailing condition, or:

$$s = s_o f_i \qquad (21\text{-}55)$$

where f_i is the adjustment factor for condition i. Thus, by definition, the adjustment factor must be calibrated as:

$$f_i = \frac{s}{s_o} = \frac{(3{,}600/h)}{(3{,}600/h_o)} = \frac{h_o}{h} \qquad (21\text{-}56)$$

where all terms have been previously defined.

For example, to calibrate a set of lane-width adjustment factors, a number of saturation headways would have to be determined at sites representing different lane widths but where all other underlying characteristics conformed to base conditions. For example, if the following data were obtained for various lane widths, i:

$$h_{10} = 2.6 \text{ s/veh}$$
$$h_{11} = 2.4 \text{ s/veh}$$
$$h_{12} = 2.1 \text{ s/veh (base conditions)}$$
$$h_{13} = 2.0 \text{ s/veh}$$
$$h_{14} = 1.9 \text{ s/veh}$$

Adjustment factors for the various observed lane widths could then be calibrated using Equation 21-56:

$$f_{w10} = \frac{2.1}{2.6} = 0.808$$

$$f_{w11} = \frac{2.1}{2.4} = 0.875$$

$$f_{w12} = \frac{2.1}{2.1} = 1.000$$

$$f_{w13} = \frac{2.1}{2.0} = 1.050$$

$$f_{w14} = \frac{2.1}{1.9} = 1.105$$

Adjustment factors for lanes wider than 12 ft are greater than 1.000, indicating that saturation flow rates increase from the base value for wide lanes (>12 ft). For lanes narrower than 12 ft, the adjustment factor is less than 1.000, as expected.

Similar types of calibration can be done for any of the simpler adjustments. If a substantial database of headway measurements can be achieved for any given factor, regression analysis may be used to determine an appropriate relationship that describes the factors.

Calibrating heavy-vehicle factors (or passenger-car equivalents for heavy vehicles) is a bit more complicated and is best explained by example. Refer to the sample problem for calibration of prevailing and base saturation flow rates. In this case, all conditions conformed to the base, except for the presence of heavy vehicles. Of the 41 observed saturation headways, six were heavy vehicles, representing a population of $(6/41)*100 = 14.63\%$.

The actual adjustment factor for this case is easily calibrated. The base saturation headway was calibrated to be 2.175 s/veh, while the prevailing saturation flow rate (representing all base conditions, except for heavy vehicle presence) was 2.34 s/veh. The adjustment factor is:

$$f_{HV} = \frac{2.175}{2.34} = 0.929$$

This calibration, however, is only good for 14.63% heavy vehicles. Additional observations at times and locations with varying heavy-vehicle presence would be required to generate a more complete relationship.

There is another way to look at the situation that produces a more generic calibration. If all 41 headways had been passenger cars, the sum of the headways would have been $41*2.175 = 89.18$ s. In fact, the sum of the 41 headways was 96.0 s. Therefore, the six heavy vehicles caused $96.00 - 89.18 = 6.82$ s of additional time consumption due to their presence. If *all* of the additional time consumed is assigned to the six heavy vehicles, each heavy vehicle accounted for $6.82/6 = 1.137$ s of additional headway time. If the base saturation headway is 2.175 s/veh, the saturation headway for a heavy vehicle would be $2.175 + 1.137 = 3.312$ s/veh. Thus, one heavy vehicle consumes as much headway time as $3.312/2.175 = 1.523$ passenger cars. This is, in effect, the passenger-car equivalent for this case, E_{HV}. This can be converted to an adjustment factor using Equation 21-14:

$$f_{LT} = \frac{1}{1 + 0.146(1.523 - 1)} = 0.929$$

This is the same as the original result. It allows, however, for calibration of adjustment factors for cases with varying heavy-vehicle presence, assuming that the value of E_{HV} is not affected by heavy-vehicle presence.

21.6.6 Normalizing Signalized Intersection Analysis

In many cases, it will be difficult or too expensive to calibrate individual factors involved in signalized intersection analysis. Nevertheless, in some cases, it will be clear that the results of HCM analysis are not correct for local conditions. This occurs when the results of analysis are compared to field measurements and obvious differences arise. It is possible to "normalize" the HCM procedure by observing departure volumes on fully saturated, signalized intersection approaches—conditions that connote capacity operation.

Consider the case of a three-lane intersection approach with a 30-s effective green phase in a 60-s cycle. Assume further that the product of all 11 adjustment factors that apply to the prevailing conditions is 0.80. Then:

$$s = s_o NF = 1,900*3*0.80 = 4,560 \text{ veh/hg}$$
$$c = 4,560*(^{30}/_{60}) = 2,280 \text{ veh/h}$$

This is the predicted capacity of the lane group using the HCM model. Despite this result, field observations measured a peak 15-min departure flow rate from this lane group (under fully saturated conditions) of 2,400 veh/h.

The measured value represents a field calibration of the actual capacity of the lane group, as it was observed under fully saturated conditions. As it is more than the estimated value, the conclusion must be that the estimated value using the HCM model is too low. The difficulty is that it may be too low for many different reasons:

- The base saturation flow rate of 1,900 pc/hg/pl is too low.
- One or more adjustment factors is too low.
- The product of 11 adjustment factors is not an accurate prediction of the *combination* of prevailing conditions existing in the lane group.

All of this assumes that the measured value was accurately observed. The latter point is a significant difficulty with the methodology. Calibration studies for adjustment factors focus on isolated impacts of a single condition. Is the impact of 20% heavy vehicles in an 11-ft lane on a 5% upgrade the same as the product of the three appropriate adjustments, $f_{HV} * f_w * f_g$? This premise, particularly where there are 11 separately calibrated adjustments has never been adequately tested using field data.

The local traffic engineer does not have the resources to check the accuracy of each factor involved in the HCM model, let alone the algorithms used to generate the estimate of capacity. On the other hand, the value of the base saturation flow rate may be adjusted to reflect the field-measured value of capacity. The measured capacity value is first converted to an equivalent value of prevailing saturation flow rate for the lane group:

$$s = \frac{c}{(g/C)} = \frac{2,400}{0.50} = 4,800 \text{ veh/hg}$$

Using Equation 21-12, with the product of all adjustment factors of 0.80, the base saturation flow rate may be normalized:

$$s_o = \frac{s}{NF} = \frac{4,800}{3 * 0.80} = 2,000 \text{ pc/hg/ln}$$

This normalized value may now be used in subsequent analyses concerning the subject intersection. If several such "normalizing" studies at various locations reveal a common area-wide value, it may be more broadly applied.

It must be remembered, however, that this process does *not* mean that the actual base saturation flow rate is 2,000 pc/hg/ln. If this value were observed directly, it might be quite different. It reflects, however, an adjusted value that normalizes the entire model for a number of underlying local conditions that renders some base values used in the model inaccurate.

21.7 Summary

The HCM model for analysis of signalized intersections is complex and incorporates many sub-models and many algorithms. These result from a relatively straightforward model concept to handle the myriad different conditions that could exist at a signalized intersection.

In Chapter 22, an overview of the most frequently used computer tools for implementing this model is presented, along with some more complex sample applications using one of these software packages.

References

1. *Highway Capacity Manual*, 4th Edition, Transportation Research Board, National Research Council, Washington DC, 2000.

2. Akcelik, R., "SIDRA for the Highway Capacity Manual," *Compendium of Papers*, 60th Annual Meeting of the ITE, Institute of Transportation Engineers, Washington DC, 1990.

3. Akcelik, R., *SIDRA 4.1 User's Guide*, Australian Road Research Board, Australia, Aug. 1995.

4. Teply, S., *Canadian Capacity Guide for Signalized Intersections*, 2nd Edition, Institute of Transportation Engineers, District 7—Canada, June 1995.

5. Reilly, W., *et al.*, "Signalized Intersection Capacity Study," *Final Report*, NCHRP Project 3-28 (2), JHK & Associates, Tucson AZ, Dec. 1982.

6. Roess, R., *et al.*, "Levels of Service in Shared-Permissive Left-Turn Lane Groups at Signalized Intersections," *Final Report*, Transportation Research Institute, Polytechnic University, Brooklyn NY, 1989.

Problems

21-1. A one-way shared-lane intersection approach has the following characteristics:

- 60-s effective green time in a 100-s cycle
- Four 11-ft lanes
- 10% heavy vehicles
- 3% upgrade
- Parking on one side with 15 mvts/h within 250 ft of the stop line
- 20 local buses/h stopping to pick up and drop off passengers

- 8% right turns
- 12% left turns
- 100 peds/h in each crosswalk
- No bicycle traffic
- A CBD location
- No opposing approach

Estimate the saturation flow rate and capacity of this approach.

21-2. The intersection shown in Figure 21-15 is to be analyzed using the HCM 2000 methodology. All computations are to be done by hand but may be checked using any appropriate software.

(a) Determine the existing *v/c* ratio and level of service for each lane group in the intersection.

(b) Determine the level of service for each approach and for the intersection as a whole.

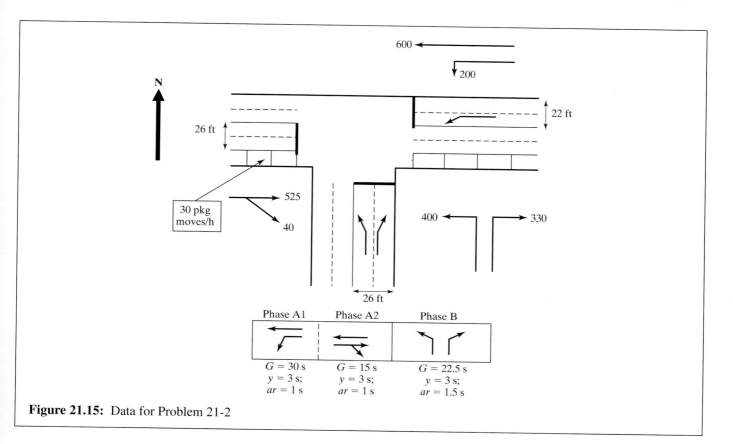

Figure 21.15: Data for Problem 21-2

(c) Make recommendations for improvements in the signal timing, if your results indicate that this is needed.

The following additional information is available concerning the intersection for Problem 21-2: $PHF = 0.92$: Arrival-types: 5 WB, 2 EB, 3 NB; $\%HV = 12\%$ in all movements; no pedestrians—overpasses provided.

21-3. The Input Worksheet for the intersection of Grand Blvd. and Crescent Ave. is shown in Figure 21-16. It is a simple intersection of two one-way arterials in a busy downtown area.

  (a) Determine the delays, levels of service, and v/c ratios for each approach, and for the intersection as a whole.

  (b) If a signal retiming is indicated, propose an appropriate timing that would result in equal v/c ratios for the critical-lane groups.

21-4. Consider the following results of an HCM analysis of a WB lane group at a signalized intersection, as summarized below:

- $v = 800$ veh/h
- $c = 775$ veh/h
- $g/C = 0.40$
- $C = 90$ s
- Initial queue $= 0.0$
- $T = 0.25$ h
- $AT = 3$

  (a) What total control delay is expected during the first 15-min analysis interval that these conditions exist?

  (b) If these conditions exist for two additional successive 15-min periods, what would the total control delay be for those periods?

21-5. A lane group with permitted left turns has the following characteristics:

- Demand flow rate, $v = 750$ veh/h (adjusted for PHF)

- Left turn flow rate in v, $v_{LT} = 10$ veh/h (also adjusted for PHF)
- Signal timing: $C = 60$ s; $G = g = 36$ s
- Permitted left turns
- Opposing flow rate, $v_o = 600$ veh/h (also adjusted for PHF)
- Arrival Type 3 on subject and opposing approaches
- $t_L = 4.0$ s/phase
- Opposing and subject approaches have two lanes

What is the left-turn adjustment factor for this situation?

21-6. For the data shown in Table 21.21:

  (a) Determine the prevailing saturation flow rate for the lane group illustrated in the data.

  (b) Determine the base saturation flow rate for this lane group.

  (c) Determine the start-up lost time for this lane group.

21-7. Using the data in Table 21.21, calibrate the passenger-car equivalent for heavy vehicles for the lane group depicted and the heavy-vehicle adjustment factor. Demonstrate that they yield the same results.

21-8. The capacity of a signalized intersection lane group is estimated using the HCM methodology, with standard values as follows:

- $s_o = 1,900$ pc/hg/ln
- $N = 2$ lanes
- $F = 0.75$ (product of all adjustment factors)
- $g/C = 0.60$

If a capacity of 1,900 veh/h for this lane group was measured in the field, what normalized value of s_o should be used to adjust the HCM methodology to yield the correct estimate?

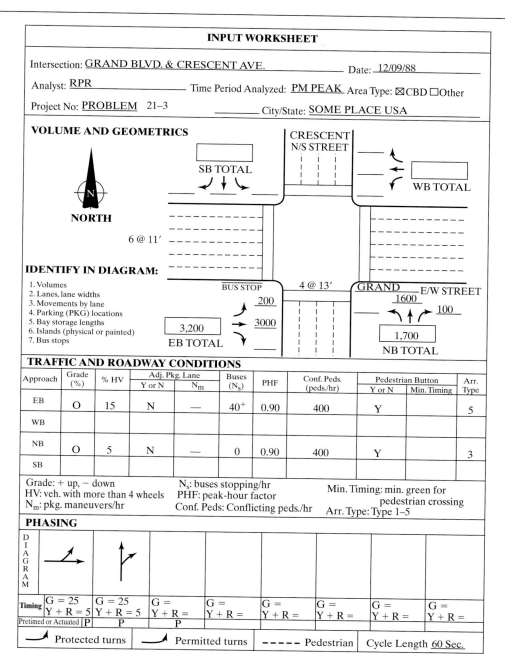

Figure 21.16: Input Worksheet for Problem 21-3

Table 21.21: Data for Problems 21-6 and 21-7

Data for Lane 1, Left Lane

Veh. in Queue	Observed Headways (s)				
	Cycle 1	Cycle 2	Cycle 3	Cycle 4	Cycle 5
1	2.8	2.9	3.0	3.1	2.7
2	2.6	2.6	2.5	3.5H	2.6
3	3.9L	2.3	2.2	2.9	2.5
4	10.2H	2.1	2.0	2.5	2.0
5	8.7	4.0L	1.9	2.2	1.9
6	3.0	9.9L	2.2	2.0	1.9
7	<u>2.9</u>	9.8	2.9H	1.9	3.6HL
8	5.0	3.3	2.6	<u>1.8</u>	9.0
9	7.1	2.8	<u>2.1</u>	7.0	<u>4.0</u>
10	9.0	2.2	4.0	8.0	4.9
11		<u>1.9</u>	5.0		9.0
12		5.5			
13		4.0			
14					
15					

Data for Lane 2, Center Lane

Veh. in Queue	Observed Headways(s)				
	Cycle 1	Cycle 2	Cycle 3	Cycle 4	Cycle 5
1	2.8	2.9	2.9	2.7	2.9
2	2.7	2.5	2.5	2.6	2.3
3	2.3	2.2	2.1	2.3	2.1
4	2.1	2.0	2.0	1.9	2.1
5	2.8H	1.9	1.8	1.9	1.9
6	2.3	1.9	2.0	1.9	2.0
7	2.6H	2.0	2.1	1.8	2.4H
8	2.1	<u>2.1</u>	1.9	2.0	<u>2.5H</u>
9	<u>1.9</u>	4.5	1.8	1.9	6.0
10	5.0	4.4	<u>2.1</u>	<u>2.0</u>	9.0
11			5.6	7.1	
12			3.3		
13					
14					
15					

(Continued)

Table 21.21: Data for Problems 21-6 and 21-7   (*Continued*)

Data for Lane 3, Right Lane

Veh. in Queue	Observed Headways(s)				
	Cycle 1	Cycle 2	Cycle 3	Cycle 4	Cycle 5
1	3.0	2.8	3.1	3.9R	2.8
2	2.5	2.5	2.7	2.8	2.6
3	2.1	2.1	2.8R	2.1	2.1
4	1.9	1.9	2.3	1.9	1.8
5	2.5H	2.0	3.2RH	1.9	1.9
6	2.3	2.1	2.5	1.8	1.9
7	2.4R	2.5R	2.3	<u>1.7</u>	2.1
8	<u>2.2</u>	2.1	2.0	3.7	1.8
9	4.4	<u>1.9</u>	1.8	5.0	<u>1.9</u>
10	6.0	3.5	<u>2.6R</u>		4.7
11		4.0	7.0		
12		5.0			
13		2.9			
14					
15					

Notes: H = heavy vehicle;

L = left turn;

R = right turn

Underline = last vehicle in standing queue.

CHAPTER
22

Applications of Signalized Intersection Analysis

In Chapter 21, the signalized intersection analysis methodology of the 4th Edition of the *Highway Capacity Manual* [1] was presented, discussed, and illustrated in significant detail. It was noted that few, if any, analyses conducted using this methodology are done by hand due to the overwhelming complexity of the analytic models that underlie the procedures.

In this chapter, more complicated problems will be explored using the methodology, but through the application of software. No attempt to address these problems by hand is made.

22.1 Software Packages

At this writing (Spring 2002), there are two primary software packages that address the full range of methodologies and computations in the HCM 2000. These two are:

- Highway Capacity Software (HCS)—maintained and sold through the McTrans Center of the University of Florida at Gainesville
- HiCAP 2000—maintained and sold through Catalina Engineering, Inc., of Tucson, AZ.

Both include a module for running signalized intersection analyses in addition to other highway capacity and level of service analysis methodologies. Neither of these packages (at this writing) contains software for the implementation of Appendix B of Chapter 16 of the HCM, which specifies a model for estimating the average signal timing of an actuated signal based on detector locations and controller settings. The average timing, which is required as an input to run the methodology, must be either measured in the field or estimated using some other procedure at the present time. It is expected that both packages will incorporate a solution as soon as one is available.

There are also a number of specialty software programs that address specific portions of the HCM 2000. Many of these focus on signal analysis and augment the analysis model with programs to optimize signal timing. One such program is SIG/Cinema (or HCM/Cinema). It runs the HCM signalized intersection methodology, essentially conducting a level-of-service analysis of current or future operations. It is also capable of optimizing signal timing based on specified inputs and objectives. Because Polytechnic University is a codeveloper of this software and because it yields interesting graphic summary outputs, it has been selected for use in solving problems in this chapter and for the display of output. It should be noted that SIG/Cinema (or HCM/Cinema) marry an HCM 2000 analysis of a signalized intersection to a Netsim simulation of the intersection, which yields additional useful information concerning queuing, emissions, and fuel consumption.

One of the emerging discussions is the handling of various software packages that claim to faithfully replicate the procedures of the *Highway Capacity Manual*. It should be noted that the Highway Capacity and Quality of Service Committee of the Transportation Research Board (the official creator of the HCM) *does not* review software nor make any statements concerning the degree to which it faithfully replicates the HCM. It is not unusual to find two different packages yielding slightly different results to the same problem. Substantial differences are less frequent, but they occur nevertheless. Differences in software package results can stem from a number of causes. The most common is that the "rounding off" rules and procedures may not be uniform among packages; this yields small differences but not major ones. Differences in interpretation of some aspects of the HCM can lead to bigger discrepancies, and simple (or complex) programming errors can always exist. Given the time-line for updating and producing new generations of software packages, it is unlikely that *any* package will ever reach a truly "bug-free" status before it gives way to a new generation of software. Nevertheless, users of any package should always pay close attention to updates and know how to acquire them.

It is also important that users not simply blindly accept any result from a software package. The user should have a basic feel for the range of reasonable results and should recognize situations in which computational answers fall outside such a "reasonable range." Such occurrences are a strong indication that an error in the software has been identified. When this occurs, direct communication with the software developer is recommended.

22.2   A Sample Problem

Figure 22.1 illustrates a signalized intersection in an urban area. A two-phase signal is in place, even though some of the turning volumes are significant. All current conditions at the intersection are specified.

22.2.1   Base Case: Existing Conditions

The first case to be examined is the existing operation. When existing volumes and physical conditions are entered into HCM-Cinema, the results summarized in Figure 22.2 are obtained.

The results indicate that the WB and SB approaches are critical. The v/c (X) ratios for these lane groups are high: 0.97 for the WB approach, and 0.87 for the SB approach. The WB approach has significantly poorer operations, with the higher v/c ratio and a delay of 52.3 s/veh vs. 27.3 s/veh on the SB approach. This suggests that even for the existing conditions, the signal is somewhat mistimed for these peak conditions. Some green time should be shifted from the N–S phase to the E–W phase to provide more balanced operation. Nevertheless, the existing operation is not unreasonable, producing an intersection critical v/c ratio of 0.91 and an average delay of 32.9 s/veh, which is level of service C.

The permitted left turns may be examined more closely by viewing the saturation flow rate worksheet, (not shown here). It reveals that the left-turn adjustment factors (f_{LT}) for the four approaches are 0.627 (EB), 0.729 (WB), 0.680 (NB), and 0.602 (SB). Thus, while the intersection operates acceptably at present, it can be seen that the opposed left turns have a significant impact, particularly on the critical SB approach, where almost 40% of the lane group's capacity is lost solely because of the impact of opposed left turns. While not required at present, these values would normally cause the traffic engineer to at least consider the possibility of left-turn lanes and phases in the future, should demand grow.

22.2.2   New Scenario: Additional Traffic Due to Development

A small shopping center is opened just south of the intersection. During the peak hour, this shopping center is expected to generate an additional 300 veh/h in the NB direction, with 10% right turns and 10% left turns. During the same hour, 150 SB through vehicles are expected as

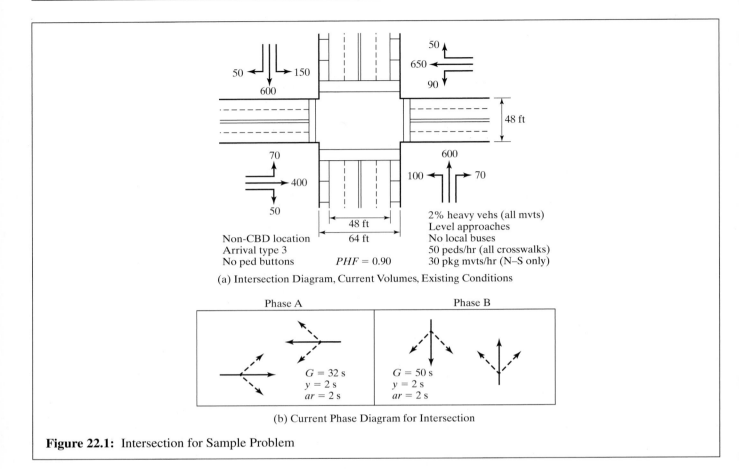

(a) Intersection Diagram, Current Volumes, Existing Conditions

Non-CBD location
Arrival type 3
No ped buttons $PHF = 0.90$

2% heavy vehs (all mvts)
Level approaches
No local buses
50 peds/hr (all crosswalks)
30 pkg mvts/hr (N–S only)

Phase A

$G = 32$ s
$y = 2$ s
$ar = 2$ s

Phase B

$G = 50$ s
$y = 2$ s
$ar = 2$ s

(b) Current Phase Diagram for Intersection

Figure 22.1: Intersection for Sample Problem

well as 20 EB right-turning vehicles and 12 WB left-turning vehicles.

The initial estimate of impact will assume no geometric or signalization changes in the intersection. Thus, the next analysis will simply increment the demand volumes as appropriate, adding the following:

- EB RT   +20 veh/h
- WB LT   +12 veh/h
- SB TH   +150 veh/h
- NB LT   +0.10 * 300 = 30 veh/h
- NB TH   +0.80 * 300 = 240 veh/h
- NB RT   +0.10 * 300 = 30 veh/h

These are added to the demand volumes, and the results shown in Figure 22.3 are obtained.

The new demand volumes during the peak hour clearly cause the intersection to fail. The critical v/c ratio (X_c) rises to 1.11, and the intersection delay rises to 84.4 s/veh, level of service F. The WB and NB approaches are critical; the NB approach exhibits no delays over 100 s/veh. A quick review of left-turn adjustment factors (f_{LT}) reveals that they range from a low of 0.536 for the SB approach to a high of 0.693 for the WB approach. The conclusion that the opposed left turns are a significant problem, especially on the N–S artery, is not too difficult to reach.

The following improvements are suggested for an initial trial:

1. Remove parking on the NB and SB approaches.

2. Install 12-ft left-turn lanes on the NB and SB approaches.

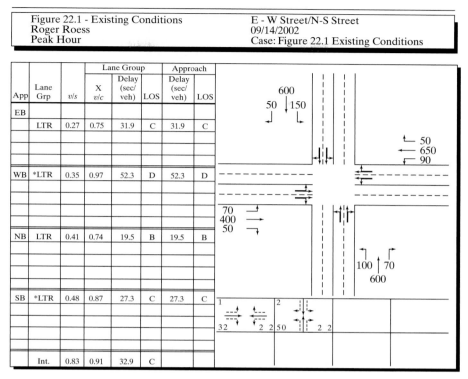

Figure 22.2: HCM-Cinema Results for Existing Conditions (Used with permission of KLD & Associates and Polytechnic University.)

3. The 64-ft width of the N–S street means that through/right-turn lanes would be increased to a width of 13 ft in this scenario.

4. Initiate an exclusive LT phase for the N–S street.

For an initial signal timing, the cycle length of 90 s will be retained. A 10-s LT phase will be provided for the N–S street, while increasing the E–W green to 33 s (the WB approach had a v/c ratio of 1.04 with a 32-s green). The N–S through phase must be *reduced*, therefore, by 10 s (to accommodate the LT phase), 1 s (to accommodate the increased E–W green), and by 4 s (to accommodate an additional set of lost times). Thus, the N–S through phase becomes $50-15 = 35$ s. It is pointed out that this is a starting point for the signal timing. It is unlikely that this scenario will be completely satisfactory. Based on the results of such an analysis, however, a more optimal signalization can be deduced. The analysis of this scenario is shown in Figure 22.4.

The first trial solution did not work out too badly. The intersection parameters—a critical v/c ratio (X_c) of 0.87 and an intersection delay of 37.7 s/veh (LOS D)—are acceptable. Some imbalances, however, should be corrected. The WB lane group (still critical) has a v/c ratio of 1.00 and high delay at 57.5 s/veh. Better balance might be provided by shifting some green time from the N–S through lane groups to the E–W lane groups. The N–S through lane groups, despite having their green time reduced from 50 s to 35 s, work quite well. This is another indication that the primary difficulty involved the opposed left turns from these lane groups.

22.2.3   Adjusting the Signal Timing

It is possible to reallocate the green time among the three phases such that all v/c ratios are equivalent to X_c,

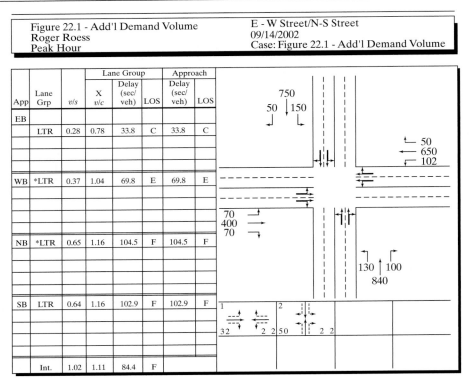

Figure 22.3: HCM-Cinema Results for Additional Demand Scenario (Used with permission of KLD & Associates and Polytechnic University.)

or 0.87. The equation for allocating green time on the basis of v/c ratios is:

$$g_i = (v/s)_i * \frac{C}{X_i} \qquad (22\text{-}1)$$

For our case, the cycle length, C, is 90 s, all X_i values are set at 0.87, and the v/s ratios for the critical lane groups are taken from Figure 22.4: 0.37 WB, 0.30 NB TH, and 0.09 WB LT. Then:

$$g_{E-W} = 0.37 * \frac{90}{0.87} = 38.3 \text{ s}$$

$$g_{N-S,TH} = 0.30 * \frac{90}{0.87} = 31.0 \text{ s}$$

$$g_{N-S,LT} = 0.09 * \frac{90}{0.87} = 9.3 \text{ s}$$

To check, the sum of $38.3 + 31.0 + 9.3 + 12.0 = 90.6$ s, which is 0.6 seconds higher than it should be. This is because we are using v/s ratios from Figure 22.4 that have been rounded to two decimal places. To adjust, we will take 0.1 s from the left-turn phase, 0.2 s from the N–S TH phase, and 0.3 s from the E–W phase. The final signalization is:

$$g_{E-W} = 38.0 \text{ s}$$

$$g_{N-S,TH} = 30.8 \text{ s}$$

$$g_{N-S,LT} = 9.2 \text{ s}$$

Figure 22.5 shows the results of retiming the signal in this manner.

The results are a distinct improvement. Overall delay is reduced to 34.4 s/veh (LOS C), which is close to the delay in the existing condition. The v/c ratios of the critical lane groups are far better balanced, and the delay to the

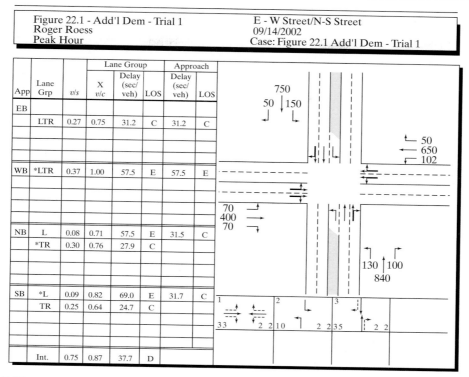

Figure 22.4: HCM-Cinema Results for Additional Demand Scenario—Trial 1 (Used with permission of KLD & Associates and Polytechnic University.)

WB lane group is much reduced. The cost of this, however, is increased delay to N–S left-turning and through/right vehicles. In fact, the SB left-turning movement has a delay of 82.5 s/veh, which is level-of-service F. It is quite common for vehicles in minor phases to have higher delays than others, but this is a bit extreme. In the next trial, the LT phase will be increased to 11.0 s, removing 0.9 s from each of the other phases.

$$g_{E-W} = 37.1 \text{ s}$$

$$g_{N-S, TH} = 29.9 \text{ s}$$

$$g_{N-S, LT} = 11.0 \text{ s}$$

The full results of the second retiming are shown in Figure 22.6. The results indicate a slight increase in overall intersection delay to 34.7 s/veh, still level of service C. The extreme delays in the LT lane groups, however, have been significantly reduced without significant increases elsewhere. The second retiming scheme is, therefore, preferable, and provides effective remediation of the negative impacts of the new development.

It is, of course, possible to fine-tune the green splits further, if so desired. There is no guarantee that such fine-tuning would result in a more favorable result. In fact, it is virtually certain that small changes, given a fixed cycle length and the three-phase plan proposed, will result in only small incremental changes in delay and v/c ratios. Investigating changes in either the cycle length and/or the phase plan, however, might provide additional benefits.

22.2.4   Investigating the Cycle Length

The absolute minimum cycle length that can be employed is given by the following expression:

$$C_{min} = \frac{L}{1 - \sum_i (v/s)_{ci}} \qquad (22\text{-}2)$$

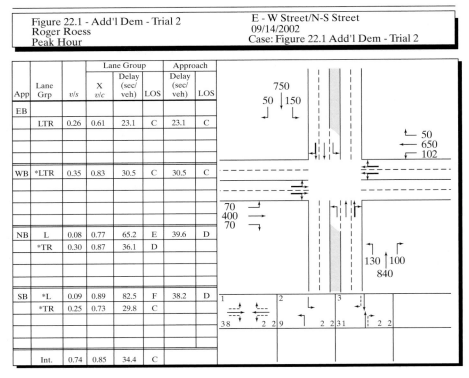

Figure 22.5: HCM-Cinema Results for Additional Demand Scenario—Trial 2 (Used with permission of KLD & Associates and Polytechnic University.)

In our case, the sum of the lost times per cycle, L, is 12 s. The sum of the critical v/s ratios has been between 0.74 and 0.75 for each of the three-phase scenarios tried. Using 0.75, we obtain:

$$C_{min} = \frac{12}{1 - 0.75} = 48.0s$$

For a pretimed signal, this suggests a minimum cycle length of 50 s. The three-phase signal timing can be re-timed using various cycle lengths in increments of 10 s from 50 s to the practical maximum of 120 s. The allocation of green time (i.e., the proportion of effective green time going to each phase) will be maintained from the final signal timing of Figure 22.6. In that signal timing, there were $90 - 12 = 78$ seconds of effective green time. By proportion, these phases should receive:

$$g_{EW} = 37.1/78.0 = 0.476$$

$$g_{NS,TH} = 29.9/78.0 = 0.383$$

$$g_{NS,LT} = 11.0/78.0 = 0.141$$

of the available effective green time. HCM-Cinema runs were conducted for each cycle length under this scenario. The results are shown in the spreadsheet and accompanying plot of Figure 22.7.

As indicated in Figure 22.7, the range of delays is not huge, but it is significant. The worst delay occurs at

Cycle Length C (s)	X_c	Intersection Delay d (s)	LOS
50	0.95	34.9	C
60	0.91	31.6	C
70	0.88	31.7	C
80	0.86	32.9	C
90	0.85	34.7	C
100	0.85	36.7	D
110	0.84	38.9	D
120	0.84	41.2	D

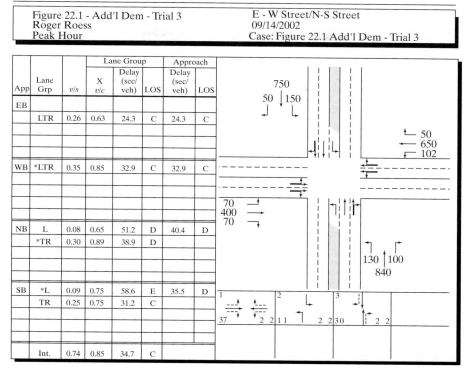

App	Lane Grp	v/s	X v/c	Delay (sec/veh)	LOS	Approach Delay (sec/veh)	LOS
EB							
	LTR	0.26	0.63	24.3	C	24.3	C
WB	*LTR	0.35	0.85	32.9	C	32.9	C
NB	L	0.08	0.65	51.2	D	40.4	D
	*TR	0.30	0.89	38.9	D		
SB	*L	0.09	0.75	58.6	E	35.5	D
	TR	0.25	0.75	31.2	C		
	Int.	0.74	0.85	34.7	C		

Figure 22.1 - Add'l Dem - Trial 3
Roger Roess
Peak Hour

E - W Street/N-S Street
09/14/2002
Case: Figure 22.1 Add'l Dem - Trial 3

Figure 22.6: HCM-Cinema Results for Additional Demand Scenario—Trial 3 (Used with permission of KLD & Associates and Polytechnic University.)

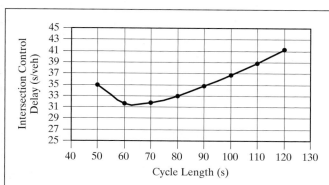

Figure 22.7: Impact of Cycle Length on Delay and X_c

the longest cycle length, 41.2 s/veh, while the best occurs at C = 60 s, when delay is 31.6 s/veh. It is also clear that delay is more or less at a minimum with a cycle length of anywhere between 60 and 70 s. As the cycle length is increased beyond what is necessary to avoid individual cycle failures, delay increases. More

vehicles are forced to wait for the green when there is unused demand on other phases. There is no spike in delay for the shorter cycle lengths, because we began this examination by establishing a minimum cycle length for stable operations. Had cycle lengths less than the minimum of 48 s been attempted, delay would sharply increase, and v/c ratios would exceed 1.00 in the critical lane groups.

Given that there are no constraints on cycle length (such as a system cycle length for coordination), a cycle length of 60 s would be chosen to minimize delays. The green times would conform to the split established. In a 60-s cycle of three phases, using standard lost times, there are 48 s of effective green. Then:

$$g_{EW} = 0.476 * 48 = 22.8 \text{ s}$$

$$g_{NS,LT} = 0.141 * 48 = 6.8 \text{ s}$$

$$g_{NS,TH} = 0.383 * 48 = 18.4 \text{ s}$$

This would be the recommended signal timing that would provide remediation for the negative impacts of the newly generated traffic from the shopping center. Parking would be removed from the vicinity of the intersection, and left-turn lanes would be added to the NB and SB approaches.

22.2.5 Another Option: Protected-Plus-Permitted Phasing

So far, the analysis of alternatives has focused on a simple three-phase signal timing with an exclusive LT phase for the N–S artery. Another option to be considered is allowing N–S left turns to be made on a permitted basis during the N–S through phase. This would provide for compound, or protected-plus-permitted phasing on the N–S street. An analysis of a 60-s cycle

with the splits previously established but with protected-plus-permitted phasing is shown in Figure 22.8.

The allowance of permitted left turns has a substantial impact on the delay to N–S left-turning vehicles, reducing these values to 17.0 s/veh SB and 14.7 s/veh NB. These values were both over 50 s/veh with fully protected phasing. There is virtually no impact on X_c, which remains 0.91.

This points out an essential discrepancy in the HCM 2000 model for protected-plus-permitted phasing. The most important factor in analyzing protected-plus-permitted phasing is the demand split (i.e., the number of vehicles making permitted turns vs. protected turns). In capacity analysis, the HCM assumes that the full capacity of the first phase (in this case the protected phase) must be consumed before demand is allocated to the second phase. In this case, since all the

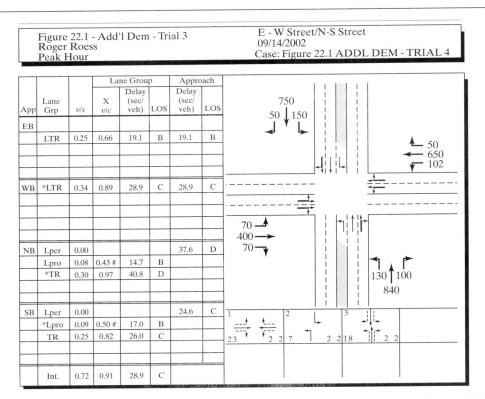

Figure 22.8: HCM-Cinema Results for Additional Demand Scenario—Trial 4 (Used with permission of KLD & Associates and Polytechnic University.)

left-turners "fit" within the protected portion of the phase, none were assigned to the permitted portion of the phase. The delay model, however, allocates vehicles between the permitted and protected portions of the phase. This is why our result shows no impact on *v/c* ratio, but a significant impact on delay. Both models may be a bit extreme. Obviously, if a permitted phase is included, some vehicles will execute their turns during this portion, especially if the opposing flow does not block them. This will have positive benefits on delay to left-turning vehicles. The absolute accuracy of the delay model for compound phasing, however, does not have a substantial data base, and the magnitude of the delay savings indicated herein may be somewhat overstated.

22.2.6   Other Options?

Different phasing options could be attempted. In this case, however, there is no strong argument for splitting the protected left-turn phases on the N–S street, as left-turn demand is relatively equal in the two directions (150 veh/h SB; 130 veh/h NB). There is also no strong argument for adding a left-turn phase for the E–W street, as they seem to operate acceptably without one, and inclusion of such lanes would require taking additional right-of-way.

Thus, it does not appear to be desirable to further complicate the phasing to provide what would be at best small improvements in delay.

22.3   Additional Sensitivities

Consider a very simple problem involving one single-lane approach lane group at a signalized intersection. For simplicity, it is assumed that all vehicles are traveling straight through the intersection and that there are no adjustments that must be applied to ideal saturation flow rate. To investigate some basic sensitivities, cycle lengths ranging from 30 s to 120 s (in 10-s increments) will be tested. It will further be assumed that effective green time is evenly split (i.e., this lane group gets 50% of effective green time) and that the total lost time per cycle

is 8 s, assuming a two-phase signal. An arrival type of 3 is assumed, and there are no initial queues. Demand flow rates of 400 veh/h, 600 veh/h, 800 veh/h, and 1,000 veh/h will be investigated.

Computations for this case are greatly simplified. The capacity of the lane group is given by:

$$c = s_o N F \left(\frac{g}{C} \right) \qquad (22\text{-}3)$$

where c = capacity of the lane group, veh/h
s_o = ideal saturation flow rate, 1,900 pc/hg/ln
N = number of lanes in the lane group, one lane
F = product of all applicable adjustment factors, 1.00
g/C = green-to-cycle length ratio

Consider the capacity of the lane group for a cycle length of 30 s. The cycle length contains 8.0 s of lost time; thus, the available effective green time in the cycle is 30 − 8 or 22 s. The subject lane group has a green time that is one-half this total, or 11 s. The g/C ratio is, therefore, 11/30 = 0.367, and the capacity of the lane group is:

$$c = 1,900 * 1 * 1 * 0.367 = 697 \text{ veh/h}$$

for the conditions described. The g/C ratio is different for each cycle length, as the lost time per cycle remains constant and one-half of the available effective green is allocated to the subject lane group.

Using the delay equations, the control delay can be computed for each of the cycle lengths and each of the demand levels specified. Note that in using the delay formulas, the progression factor (*PF*) is 1.00, the analysis time period is 0.25 h, the adjustment factor for type of controller (*k*) is 0.50 (pretimed control assumed), and the adjustment factor for upstream filtering (*I*) is 1.00 (for an isolated intersection).

The results of these computations are shown in Table 22.1. Three graphic illustrations of the data appear (Figures 22.9, 22.10, and 22.11) to illustrate some of the critical relationships that follow from the HCM 2000 algorithms for capacity and control delay.

Table 22.1: Data for Sensitivity Analyses

Cycle Length (s)	Effective Green Time (s)	Demand (veh/h)	Capacity (veh/h)	Evaluation Criteria			
				v/c Ratio	d_1 (s/veh)	d_2 (s/veh)	d (s/veh)
30	11	400	697	0.574	7.6	3.4	11.0
40	16	400	760	0.526	9.1	2.6	11.7
50	21	400	798	0.501	10.7	2.2	12.9
60	26	400	823	0.486	12.2	2.0	14.2
70	31	400	841	0.475	13.8	1.9	15.7
80	36	400	855	0.468	15.3	1.8	17.2
90	41	400	866	0.462	16.9	1.8	18.7
100	46	400	874	0.458	18.5	1.7	20.2
110	51	400	881	0.454	20.0	1.7	21.7
120	56	400	887	0.451	21.6	1.7	23.3
30	11	600	697	0.861	8.8	13.2	22.0
40	16	600	760	0.789	10.5	8.2	18.7
50	21	600	798	0.752	12.3	6.5	18.8
60	26	600	823	0.729	14.1	5.6	19.7
70	31	600	841	0.713	15.9	5.1	21.0
80	36	600	855	0.702	17.7	4.8	22.5
90	41	600	866	0.693	19.5	4.5	24.0
100	46	600	874	0.686	21.3	4.4	25.7
110	51	600	881	0.681	23.1	4.2	27.4
120	56	600	887	0.677	24.9	4.1	29.1
30	11	800	697	1.148	10.4	82.9	93.3
40	16	800	760	1.053	12.4	47.4	59.8
50	21	800	798	1.003	14.5	32.5	47.0
60	26	800	823	0.972	16.6	25.2	41.8
70	31	800	841	0.951	18.8	21.1	39.9
80	36	800	855	0.936	20.9	18.6	39.5
90	41	800	866	0.924	23.0	16.9	40.0
100	46	800	874	0.915	25.2	15.8	40.9
110	51	800	881	0.908	27.3	14.9	42.2
120	56	800	887	0.902	29.5	14.2	43.7
30	11	1,000	697	1.435	12.7	204.1	216.8
40	16	1,000	760	1.316	15.2	151.4	166.6
50	21	1,000	798	1.253	17.8	124.2	141.9
60	26	1,000	823	1.215	20.3	107.7	128.0
70	31	1,000	841	1.188	22.9	96.6	119.6
80	36	1,000	855	1.170	25.5	88.8	114.3
90	41	1,000	866	1.155	28.2	82.9	111.1
100	46	1,000	874	1.144	30.8	78.4	109.2
110	51	1,000	881	1.135	33.4	74.8	108.2
120	56	1,000	887	1.128	36.0	71.9	107.9

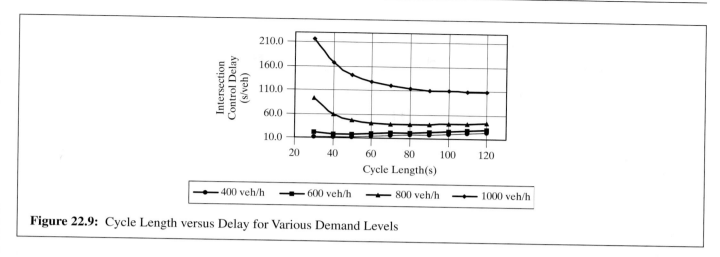

Figure 22.9: Cycle Length versus Delay for Various Demand Levels

22.3.1   Cycle Length Versus Delay

Figure 22.9 shows the relationship between cycle length and delay for the four demand levels tested. In each demand case, there is a cycle length that will minimize control delay for the lane group. For a demand of 400 veh/h, the minimum delay occurs at the shortest cycle length, 30 s. At this light demand, there is considerable unused green time, which inflates delay.

With demand at only 400 veh/h, even the 30-s cycle has too much unused green time to minimize delay. However, since 30 s is the minimum practical cycle length, no smaller cycle lengths are tested.

When the demand is increased to 600 veh/h, the minimum delay occurs at cycle lengths between 40 and 50 s. There is very little change in delay for cycle lengths in this range. At 800 veh/h, delay is minimal and relatively flat between cycle lengths of 70 to 90 s. The rise in delay for longer cycle lengths is relatively small.

At a demand of 1000 veh/h, however, the minimum delay cycle length is 120 s. That is because there is no cycle length that will accommodate this demand with an allocation of 50% of effective green time. All cycle lengths result in oversaturation (i.e., v/c ratios in excess of 1.00). As the cycle lengths get shorter, the impact on increasing delay becomes more severe, as queues continue to grow. It should be noted

that these cases *do not* include the impact of initial queues. These computations reflect the *first* 15-min interval in which demand is 1,000 veh. There will be a residual queue left at the end of this period, which becomes an initial queue for the next. Delay will grow rapidly with successive 15-min time periods until the demand flow drops below the capacity of the lane group.

For any given situation, there will be a cycle length that produces minimum delay for the intersection or for any given lane group. This assumes that the demand flow is heavy enough to warrant signalization and that there is a reasonable cycle length that can accommodate the demand. In general, Figure 22.9 shows two key characteristics:

- When the cycle length is too short, resulting in v/c ratios greater than 1.00, delay increases rapidly.

- When the cycle length is too long, there is too much unused green time built into the cycle, and delay will increase gradually.

In fact, the "bottom" of the curve is relatively flat. A general rule is that any cycle length in the range of 0.75 to 1.50 times the absolute minimum delay cycle length will have little impact on delay.

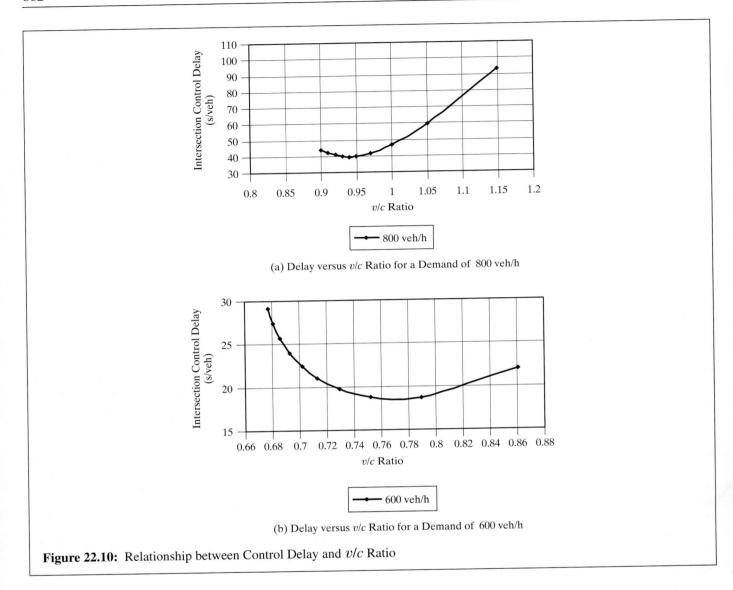

(a) Delay versus *v/c* Ratio for a Demand of 800 veh/h

(b) Delay versus *v/c* Ratio for a Demand of 600 veh/h

Figure 22.10: Relationship between Control Delay and v/c Ratio

22.3.2 Delay versus *v/c* Ratio

The influence of v/c ratio on delay is somewhat complex. That is because the v/c ratio is the *result* of a number of other conditions. It is clearly affected by demand (v), but is also impacted by any condition that affects capacity (c). Capacity is affected by number of variables, including:

- Saturation flow rate
- Number of lanes in the lane group
- Green time
- Cycle length

Of course, the saturation flow rate is itself influenced by a large number of variables, including the base saturation flow rate and 11 adjustment factors, each accounting for one or more prevailing conditions in the lane group.

Thus, it is impossible to examine the relationship between v/c ratio and delay while holding other variables constant. Since capacity varies, for example, with cycle length and green splits, it is impossible to divorce

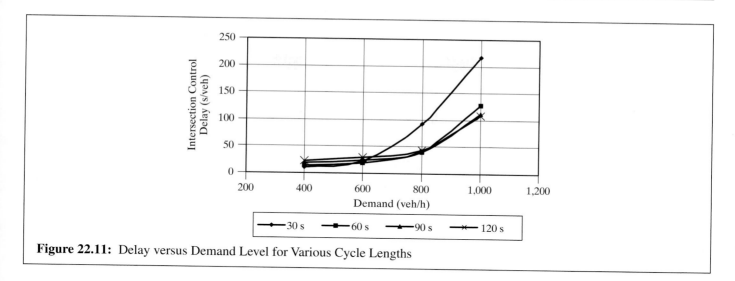

Figure 22.11: Delay versus Demand Level for Various Cycle Lengths

varying v/c ratios from changes in the signal timing. The equation for uniform delay, for example, depends upon cycle length, the g/C ratio, and the v/c ratio. Those three variables, however, are themselves interrelated, and one cannot be varied without varying the others.

Figure 22.10 shows a plot of delay versus v/c ratio for two illustrative demand levels, using the data from Table 22.1.

In Figure 22.10 (a), the demand level leads to v/c ratios in excess of 1.00 at shorter cycle lengths. Obviously, delay increases rapidly when $v/c > 1.00$. As in previous examples, this reflects only the first period during which $v/c > 1.00$, and does not include the cumulative affect of initial queues that would exist in subsequent analysis periods. The minimum delay for this case occurs when the v/c ratio is a relatively high 0.94. With high demand, the inclusion of even small amounts of unused green time (reflected in lower v/c ratios) causes an increase in delay.

In Figure 22-10 (b) demand does not approach a v/c ratio of 1.00. Here, the effects of very low v/c ratios lead to the highest delay levels. Again, the implication is that excessive amounts of unused green time in the cycle will cause additional delay. Minimum delay in this case occurs with a v/c ratio in the range of 0.76–0.78, with little impact anywhere in the range of 0.75–0.80.

Perusing Table 22.1, when demand is as low as 400 veh/h, delay continually decreases as v/c ratio increases, an indication that the minimum delay occurs at some higher

v/c ratio than those shown in the table. Similarly, when demand is 1,000 veh/h, all cases have $v/c > 1.00$; minimum delay would obviously require a v/c ratio below 1.00.

It is hard to draw generalized conclusions. In a way, the variation of delay with v/c ratio mirrors the relationship with cycle length. Cycle length (given the constant allocation of green) really determines capacity in this example, so the similarity of these trends is impossible to avoid.

Minimum delay for any given case will occur at a reasonably high v/c ratio. Low v/c ratios imply lots of unused green time, which breeds delay. Higher v/c ratios approach the range where individual cycle failures can occur (even in a stable analysis period), which also causes delay to increase. Depending upon the intensity of demand on a given lane group, minimum delay will generally occur with v/c ratios in the range of 0.70 to 0.95; as indicated in previous examples, a target v/c ratio in the range of 0.80–0.95 is reasonable for most demand levels.

22.3.3   Demand Versus Delay

For any given signal timing, delay will rise with increasing demand. This is illustrated in Figure 22.11.

Again, it is impossible to detach these trends from those previously discussed. For any given cycle length, as demand increases the v/c ratio increases. Interestingly, the longer cycle lengths have higher delay when demand is low than shorter cycle lengths. This reflects the fact that they have more unused green time than the shorter cycle lengths. Obviously, as demand increases the unused green time reduces, and the shorter cycle lengths move to higher delay.

22.3.4 Summary

All of these sensitivities revolve around a delicate balance of providing sufficient cycle lengths and green times (which means capacity) to prevent individual cycle failures within an analysis period as well as the failure of the entire analysis period. On the other hand, provision of too much capacity implies excessive amounts of unused green time (i.e., vehicles waiting for service in some lane groups, while there is no demand for the lane group having the green). The traffic engineer, in timing the signal, must find the appropriate or optimal condition.

This analysis assumed that arrival type 3 (random arrivals) exists for this lane group. Clearly, providing a better (or worse) progression with other signals would have a far larger impact on the delay on this lane group than any of the factors investigated here. That is why it is appropriate to have a common system cycle length, required for progression, even though such a cycle length is excessive for most of the individual signalized intersections in the progression.

22.4 Closing Comments

Any model of signalized intersection operation must be extremely complex because of the many variables that can affect operational outcomes. Computer tools help the engineer do a more detailed analysis of all existing or proposed conditions than can be done with simpler manual methods, such as the signal-timing approach of Chapter 18.

The tools must be understood, however, and the engineer must be able to recognize results that are clearly counterintuitive. No software package is perfect; none has been fully validated across the entire range of potential conditions. As valuable as these tools are, glitches and errors will always exist; therefore, it is important that the engineer identify them when they occur.

When such errors are found, the engineer should notify the software developer. Most software developers welcome such information and will respond by fixing such problems in updated releases of their software products.

Problem

22-1. Consider the intersection of Grand Blvd. and Crescent Avenue shown below. Both are busy one-way streets in an urban CBD area.

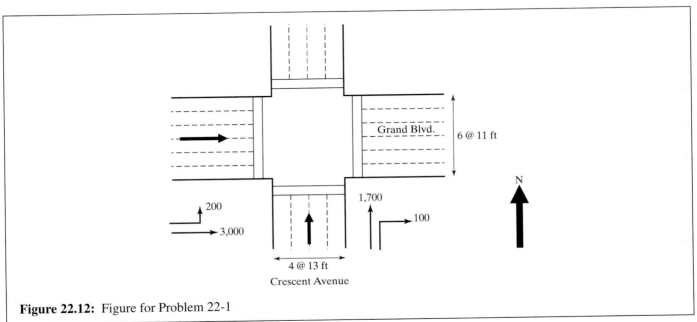

Figure 22.12: Figure for Problem 22-1

A simple two-phase signal is to be provided. The following conditions may be assumed to exist:

$y = 2.0$ s/phase

$ar = 2.0$ s/phase

$\ell_1 = 2.0$ s/phase

$e = 2.0$ s/phase

Level grades on all approaches

15% HV EB; 5% HV NB

20 buses/h EB; 0 buses/h NB

No curb parking on either street

$PHF = 0.90$

400 conflicting pedestrians in each crosswalk

Pedestrian push buttons are provided

Arrival Type = 5 EB; 3 NB

(a) Propose a timing plan for the signalized intersection.

(b) For the proposed timing plan, determine the control delay, v/c ratio, and level of service for each lane group.

(c) Vary the cycle length in increments of 10 s from 30 s to 120 s. Allocate green times according to two rules: (1) to keep the critical lane group v/c ratios equal and (2) to keep the 50% of the effective green assigned to each lane group.

(d) Plot the resulting intersection control delay versus cycle length for each of these trials.

(e) Comment on the results.

CHAPTER
23

Analysis of Unsignalized Intersections

Since 1985, the *Highway Capacity Manual* [*1*] has included a chapter on the analysis of unsignalized intersections. The original chapter was based on a European model [*2*] and enabled a general analysis of situations in which a minor street entering a major street was controlled by YIELD or STOP signs. These intersections could include two approaches (two one-way streets, one controlled), three approaches (T-intersection with the stem leg controlled), or four approaches (minor street controlled); they were classified as "two-way STOP-controlled intersections" (TWSC), even though in some cases only one leg was controlled.

The methodology, based on gap acceptance theory, has since been improved significantly. A major research effort sponsored by the National Cooperative Highway Research Program [*3*] resulted in additional modeling, and material has been added on all-way STOP-controlled intersections and roundabouts. While these aspects have been added since 1985, the method no longer addresses the analysis of YIELD-controlled intersections.

23.1 Analysis of Two-Way STOP-Controlled Intersections

The model for TWSC intersections is based on gap acceptance theory, and a relatively rigid view of the "priority" of various vehicular and pedestrian movements in the intersection. Priority of movements is important, as gaps in the major street traffic stream are sought by a number of different movements. Most gaps, however, may be used by only one vehicle. Thus, where more than one vehicle is waiting for a gap, the first arriving gap is used by the vehicle (or pedestrian) in the movement having the highest priority.

Figure 23.1 illustrates the priority of movements at a typical four-leg and a typical T-intersection. In a four-leg intersection, the highest priority (Priority 1) movements include the through and right-turn movements on the major street, and the pedestrian movements crossing the minor street. These movements have the right of way over all minor-street movements at a STOP sign. Priority 2

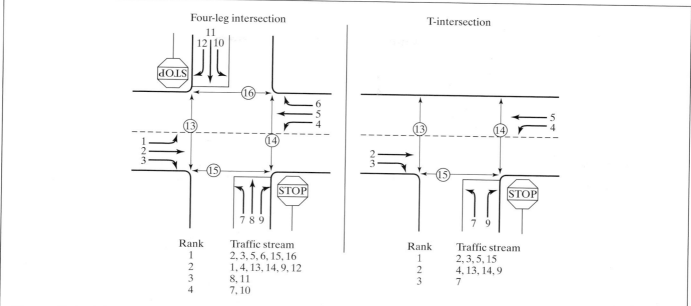

Figure 23.1: Priority of Movements at a TWSC Intersection (Used with permission of Transportation Research Board, National Research Council, *Highway Capacity Manual*, 4th Edition, Washington DC, 2000, Exhibit 17-3, pg. 17-4.)

movements include left turns from the major street, right turns from the minor street, and pedestrians crossing the major street. Left turns from the major street have first access to gaps in the opposing vehicular traffic stream; right turns from the minor street have first access to merge into gaps in the right-most approaching major-street lane. Pedestrians crossing the major street also have the right of way over vehicles seeking the same gaps. The through movements from the minor street are Priority 3, while left turns from the minor street are Priority 4. As shown in Figure 23.1, the priorities are simplified for a T-intersection, where many of the movements do not exist.

Figure 23.2 shows a basic flow chart of the procedure for analysis of TWSC intersections.

Note that the input data includes pedestrian volumes by movement as well as information on nearby signals (on the major street) and heavy-vehicle percentages. Hourly volumes may be used in mixed veh/h, or 15-min rates of flow can be entered directly. In the former case, the volumes are adjusted to reflect 15-min rates of flow by dividing by the peak-hour factor (*PHF*).

Critical gap and follow-up times are combined with conflicting traffic flows to determine a potential capacity, which is then modified to reflect a number of prevailing conditions. Modified movement or lane capacities are then used to estimate queue lengths and delays. Level-of-service criteria based upon control delay are shown in Table 23.1. These and other computational steps are described in the sections that follow.

23.1.1. Determining Conflicting Volume

Each movement seeking gaps does so through a different set of conflicting traffic movements, as illustrated in Figure 23.3.

- Major-street left turns seek gaps through the opposing through movement, the opposing right-turn movement, and pedestrians crossing the far side of the minor street.

- Minor-street right turns seek to merge into the right-most lane of the major street, which contains through and right-turning vehicles. Each

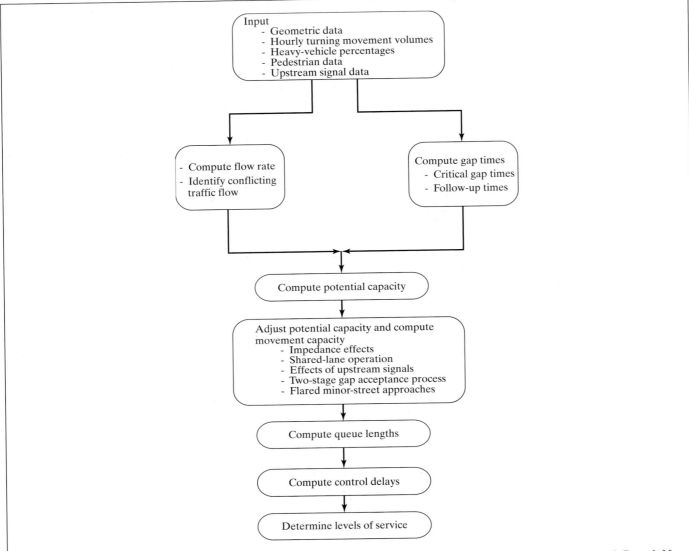

Figure 23.2: Flow Chart for Analysis of TWSC Intersections (Used with permission of Transportation Research Board, National Research Council, *Highway Capacity Manual*, 4th Edition, Washington DC, 2000, Exhibit 17-1, pg. 17-2.)

right turn from the minor street must also cross the two pedestrian paths shown.

- Through movements from the minor street must cross all major street vehicular and pedestrian flows.

- Minor street left turns must deal not only with all major-street traffic flows but with two pedestrian

flows and the opposing minor-street through and right-turn movements.

The "conflicting volume" (v_{cx}) for movement "*x*" is found by following the equations in Figure 23.3. It is important to note the following:

- It is critical that movement numbers be clearly identified as part of the input information, as these

Table 23.1: Level-of-Service Criteria for TWSC Intersections

Level of Service	Average Control Delay (s/veh)
A	0–10
B	>10–15
C	>15–25
D	>25–35
E	>35–50
F	>50

(Used with permission of Transportation Research Board, National Research Council, *Highway Capacity Manual*, 4th Edition, Washington DC, 2000, Exhibit 17-2 pg. 17-2.)

are critical to computing the appropriate conflicting volume values.

- The footnotes are very important. Some right-turning movements included in conflicting volume totals may be discounted if they are separated by islands or channelized right-turn lanes. Approaching right turns from the major street do not actually "conflict" with many movements, but a portion of them are included in "conflicting volume" totals, as not all drivers use turn signals, and minor-street drivers may not be able to tell that such drivers intend to turn right.

- Two-stage gap acceptance occurs where the major street has a median wide enough to store one or more vehicles. In such cases, drivers treat crossing the directional roadways as separate operations, and the amount of median storage affects overall operation. Two-stage gap acceptance is discussed in a later section. For roadways without such a median, the total conflicting traffic is the sum of the conflicting volumes shown for both stages in Figure 23.3.

It is important to note that conflicting traffic volumes are computed as the sum of vehicles plus pedestrians that are deemed to be in conflict with the subject movement. All computations should be in terms of peak 15-min rates of flow.

23.1.2 Critical Gaps and Follow-up Times

The "critical gap" (t_{cx}) for movement "x" is defined as the minimum average acceptable gap that allows intersection entry for one minor-street (or major-street left-turn) vehicle. The term "average acceptable" means that the average driver would accept (or choose to utilize) a gap of this size. The gap is measured as the clear time in the traffic stream(s) defined by all of the conflicting movements. Thus, the model assumes that all gaps shorter than t_{cx} are rejected (or unused), while all gaps equal to or larger than t_{cx} would be accepted (or used).

The "follow-up time" (t_{fx}) for movement "x" is the minimum average acceptable time for a second queued minor-street vehicle to use a gap large enough to admit two or more vehicles. This measure is similar to the saturation flow rate at a signalized intersection.

Table 23-2 shows base (or unadjusted) values of the critical gap and follow-up time for various movements.

Base critical gaps and follow-up times must be adjusted to account for a number of conditions, including heavy-vehicle presence, grade, and the existence of two-stage gap acceptance. Adjusted values are computed as follows:

$$t_{cx} = t_{cb} + t_{cHV}P_{HV} + t_{cG}G - t_{cT} - t_{3LT}$$

$$t_{fx} = t_{fb} + t_{fHV}P_{HV} \qquad (23\text{-}1)$$

where: t_{cx} = critical gap for movement x, s

t_{cb} = base critical gap from Table 23-2, s

t_{cHV} = critical gap adjustment factor for heavy vehicles, s

P_{HV} = proportion of heavy vehicles

t_{cG} = critical gap adjustment factor for grade, s

G = grade, decimal or percent/100

t_{cT} = critical gap adjustment factor for two-stage gap acceptance, s

t_{3LT} = critical gap adjustment factor for intersection geometry, s

t_{fx} = follow-up time for movement x, s

t_{fb} = base follow up time from Table 23-2, s

t_{fHV} = follow-up time adjustment factor for heavy vehicles, s

Adjustment factors are summarized in Table 23.3.

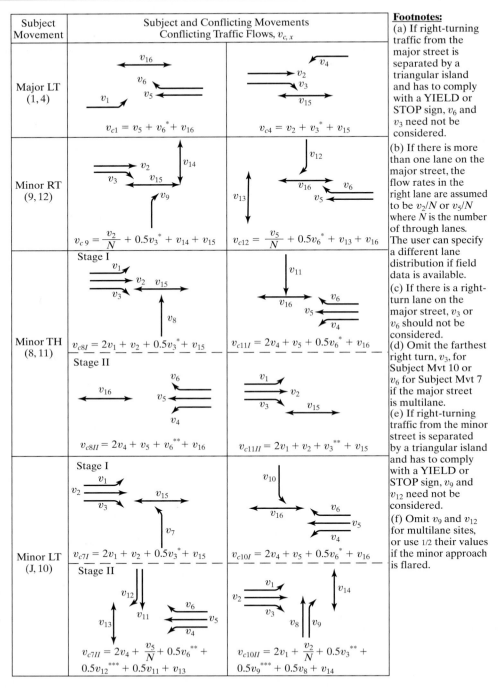

Figure 23.3: Computation of Conflicting Volumes, v_{cx} (Used with permission of Transportation Research Board, National Research Council, *Highway Capacity Manual*, 4th Edition, Washington DC, 2000, Exhibit 17-2, pg. 17-2.)

Table 23.2: Base Critical Gap and Follow-Up Times

Vehicle Movement	Base Critical Gap, t_{cb} (s)		Base Follow-Up Times, t_{fb} (s)
	Two-Lane Major Street	Four-Lane Major Street	
LT from major street	4.1	4.1	2.2
RT from minor street	6.2	6.9	3.3
TH from minor street	6.5	6.5	4.0
LT from minor street	7.1	7.5	3.5

(Used with permission of Transportation Research Board, National Research Council, *Highway Capacity Manual;* 4th Edition, Washington DC, 2000, Exhibit 17-5, pg. 17-7.)

Table 23.3: Adjustments to Base Critical Gap and Follow-Up Times

Adjustment Factor	Values (s)	
t_{cHV}	1.0	Two-lane major streets
	2.0	Four-lane major streets
t_{cG}	0.1	Movements 9 and 12
	0.2	Movements 7, 8, 10, and 11
	1.0	Otherwise
t_{cT}	1.0	First or second stage of two-stage process
	0.0	For one-stage process
T_{3LT}	0.7	Minor-street LT at T-intersection
	0.0	Otherwise
t_{fHV}	0.9	Two-lane major streets.
	1.0	Four-lane major streets

23.1.3 Determining Potential Capacity

Once the conflicting volume, critical gap, and follow-up time are known for a given movement, its potential capacity can be estimated using gap acceptance models. The concept of "potential capacity" assumes that *all* available gaps are used by the subject movement (i.e., that there are no higher priority vehicular or pedestrian movements waiting to use some of the gaps). It also assumes that each movement operates out of an exclusive lane. After "potential capacity" is estimated, it must be modified to reflect impedance from higher-priority movements and the sharing of lanes, as well as other factors. The potential capacity of a movement is computed as:

$$c_{px} = v_{cx}\left[\frac{e^{-(v_{cx}\,t_{cx}/3,600)}}{1 - e^{-(v_{cx}\,t_{fx}/3,600)}}\right] \quad (23\text{-}2)$$

where: c_{px} = potential capacity of movement x, veh/h

v_{cx} = conflicting flow for movement x, conflicts/h

t_{cx} = critical gap for movement x, s

t_{fx} = follow-up time for movement x, s

23.1.4 Accounting for Impedance Effects—Movement Capacity

Potential capacities must first be adjusted to reflect the impedance effects of higher-priority movements that may utilize some of the gaps sought by lower-priority movements. This impedance may come from both pedestrian and vehicular sources.

Priority 2 vehicular movements (LTs from major street and RTs from minor street) are not impeded by

Table 23.4: Tabulation of Impeding Flows

Subject Vehicular Flow	Impeding Vehicular Flows	Impeding Pedestrian Flows
1 (major LT)	None	16
4 (major LT)	None	15
7 (minor LT)	1,4,11,12	15, 13
8 (minor TH)	1, 4	15, 16
9 (minor RT)	None	15, 14
10 (minor LT)	1,4,8,9	16, 14
11 (minor TH)	1, 4	15, 16
12 (minor RT)	None	16, 13

any other vehicular flow, as they represent the highest priority movements seeking gaps. They are impeded, however, by Rank 1 pedestrian movements. Priority 3 vehicular movements are impeded by Priority 2 vehicular movements and Priority 1 and 2 pedestrian movements seeking to use the same gaps. Priority 4 vehicular movements are impeded by Priority 2 and 3 vehicular movements, and Priority 1 and 2 pedestrian movements using the same gaps. Table 23.4 lists the impeding flows for each subject movement in a four-leg intersection. For a T-intersection, these impedances are simplified, and there are no Priority 4 movements.

In general, the movement capacity is found by multiplying the potential capacity by an adjustment factor. The adjustment factor is derived as the product of the probability that each impeding movement will be blocking a subject vehicle. For vehicular impedances, this computation requires the movement capacity of the impeding movements. Thus, it is critical that computations begin with the highest-order movements and proceed in order of decreasing priority. Movement capacity is found as:

$$c_{mx} = c_{px} \prod_{i,j} p_{vi}\, p_{pj} \qquad (23\text{-}3)$$

where: c_{mx} = movement capacity, movement x, veh/h

c_{px} = potential capacity, movement x, veh/h

p_{vi} = probability that impeding vehicular movement "i" is not blocking the subject flow; (also referred to as the "vehicular impedance factor" for movement "i")

p_{pj} = probability that impeding pedestrian movement "j" is not blocking the subject flow; (also referred to as the "pedestrian impedance factor for movement "j")

The probability that impeding vehicular movement "i" is not blocking the subject movement is computed as:

$$p_{vi} = 1 - \frac{v_i}{c_{mi}} \qquad (23\text{-}4)$$

where: v_i = demand flow for impeding movement "i," veh/h

c_{mi} = movement capacity for impeding movement "i," veh/h

Thus, if the v/c ratio for impeding movement "i" is, for example, 0.35, it is $1 - 0.35 = 0.65$ likely that the subject movement is not impeded by movement "i."

For Rank 4 vehicular flows (LTs from the minor street at a four-leg intersections), there must be a modification to vehicular impedances, as two may overlap. Such flows are impeded by both major-street LTs and by the opposing minor street TH movement. However, the presence of major-street LTs may affect the impedance caused by the opposing minor street through movement. In this case, the product of the three overlapping impedances is replaced by a modified value, as follows:

$$p'' = p_{v1}{}^* p_{v4}{}^* p_{vTH}$$

$$p' = 0.65p'' - \left(\frac{p''}{p'' + 3}\right) + 0.6\sqrt{p''} \qquad (23\text{-}5)$$

where: p'' = unadjusted product of three impedance factors

p' = adjusted product of three impedance factors

p_{v1} = impedance factor for vehicular movement 1, a major-street LT

p_{v4} = impedance factor for vehicular movement 4, a major-street LT

p_{vTH} = impedance factor for the opposing minor-street TH movement (movement 8 for movement 10; movement 11 for movement 7)

The adjusted product is used in the movement capacity determination instead of the product of the three individual impedances. Impedance factors for right turns from the minor street and pedestrian movements are still included in the estimate.

Another modification occurs when major street left turns do not operate out of an exclusive lane, which is the assumed operation in computing p_{v1} and p_{v4}. Where these left turns operate out of a shared major-street lane, these impedance factors are modified to reflect this:

$$p_{v1/4}^* = 1 - \left[\frac{1 - p_{v1/4}}{1 - \left(\frac{v_{mTH}}{s_{mTH}} + \frac{v_{mRT}}{s_{mRT}} \right)} \right] \quad (23\text{-}6)$$

where: $p_{v1/4}^*$ = modified impedance factor for movement 1 or 4

$p_{v1/4}$ = unmodified impedance factor for movement 1 or 4

v_{mTH} = major-street through flow rate in direction of impeding movement, veh/h

v_{mRT} = major-street right-turn flow rate in direction of impeding movement, veh/h

s_{mTH} = saturation flow rate for the major-street through movement, veh/h (1,500–1,700 veh/h may be used as a default value).

s_{mRT} = saturation flow rate for the major-street right-turning traffic, veh/h (1,500–1,700 veh/h can be used as a default value)

Pedestrian impedance factors are computed as:

$$p_{pj} = 1 - \left[\frac{v_j \left(\frac{w}{S_p} \right)}{3{,}600} \right] \quad (23\text{-}7)$$

where: p_{pj} = pedestrian impedance factor for impeding pedestrian movement "j"

v_j = pedestrian flow rate, impeding movement "j," peds/h

w = lane width, ft

S_p = pedestrian walking speed, ft/s

The equation is based on the total aggregate time all crossing pedestrians would block a single exit lane for the subject vehicular movement. It is only a single lane that is considered, as vehicles can avoid pedestrians blocking one lane by maneuvering to another.

23.1.5 Determining Shared-Lane Capacity

Movement capacities still represent an assumption that each minor-street movement operates out of an exclusive lane. Where two or three movements share a lane, its combined capacity is given as:

$$c_{SH} = \frac{\sum_y v_y}{\sum_y \left(\frac{v_y}{c_{my}} \right)} \quad (23\text{-}8)$$

where: c_{SH} = shared lane capacity, veh/h

v_y = flow rate, movement "y," sharing lane with other minor-street flow(s)

c_{my} = movement capacity of movement "y," sharing lane with other minor-street flows

Basically, what this equation means is that if three movements sharing a lane, having v/c ratios of 0.25, 0.30, and 0.10, respectively, then the proportion of capacity consumed by the three movements in one lane is their sum, or 0.65. The capacity is then estimated as the total flow rate in the lane divided by the proportion of capacity it represents.

23.1.6 Adjusting for Upstream Signals and Platoon Flow

The algorithms for gap acceptance used in this methodology all assume that major-street flow is random. If there are signalized intersections 0.25 miles or closer to the subject STOP-controlled approach, this is not true, and platoon flow must be considered. The 0.25 miles is a more or less arbitrary boundary. Platoon effects can exist at much longer distances between signals, perhaps even as long as 2.0 miles.

As illustrated in Figure 23.4, three different flows may enter at the signalized intersection, creating platoons at the location of the STOP-controlled intersection. If there is a signalized intersection within 0.25 miles on both sides of the STOP-controlled subject location, then the potential for six different platoons exists.

A driver waiting to make a maneuver at a STOP-controlled intersection may face one of four different situations concerning approaching major-street platooned flows:

- No platoon in either direction is crossing the STOP-controlled approach (Regime 1).

- A platoon from the left is crossing the STOP-controlled approach, but no platoon from the right is crossing the location (Regime 2).

- A platoon from the right is crossing the STOP-controlled approach, but no platoon from the left is crossing the location (Regime 3).

- A platoon in both directions is crossing the STOP-controlled approach (Regime 4).

All of these conditions present different conflicting flow rates to the minor-street driver, and capacities must be adjusted to reflect the probability of each of these conditions existing.

The model for determining this is exceedingly complex, and requires detailed knowledge of the upstream signals and demands. It is quite cumbersome to make these computations by hand, and use of software (such as the HiCAP 2000 or HCS packages) is required. For modeling details of this feature, please consult the HCM directly [1].

At this writing (April 2003), the HCQS Committee has determined that this adjustment is significantly understated in the HCM 2000. Errata are expected to be issued in 2004.

23.1.7 Two-Stage Gap Acceptance

Figure 23.5 illustrates operations where "two-stage" gap acceptance exists. This occurs when the median of the major street is wide enough to store one or more vehicles. Drivers essentially behave as if they have two separate STOP-controlled maneuvers. In these situations, queuing conditions and the storage capability of the median have a significant impact on total capacity and operations.

Where the situation depicted in Figure 23.5 exists, both the minor-street through and left-turn movements must be treated as two-stage operations. Right turns from the minor street are not affected. For each affected movement, movement capacities are computed for Stage 1 (c_{Ix}), Stage 2 (c_{IIx}), and as if a one-stage operation were in effect (c_{mx}). The total capacity of the two-stage process is then found as:

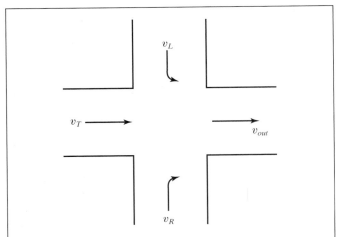

Figure 23.4: Potential Platoons Approaching a STOP-Controlled Intersection (Used with permission of Transportation Research Board, National Research Council, *Highway Capacity Manual*, 4th Edition, Washington DC, 2000, Exhibit 17-11, pg. 17-15.)

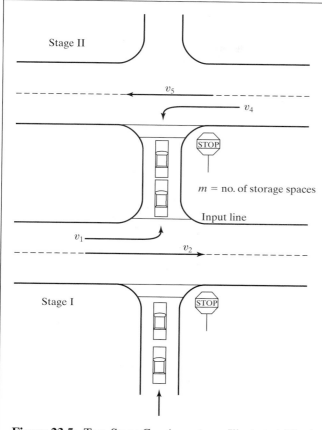

Figure 23.5: Two-Stage Gap Acceptance Illustrated (Used with permission of Transportation Research Board, National Research Council, *Highway Capacity Manual*, Washington DC, 2000, Exhibit 17-17, pg. 17-20.)

$$a = 1 - 0.32e^{-1.3\sqrt{m}} \quad (m > 0)$$

$$y = \frac{c_{Ix} - c_{mx}}{c_{IIx} - v_L - c_{mx}}$$

$$c_{Tx} = \left[\frac{a}{y^{m+1} - 1}\right][y(y^m - 1)$$

$$(c_{IIx} - v_L)(y - 1)c_{mx}] \quad (y \neq 1)$$

$$c_{Tx} = \left[\frac{a}{m + 1}\right][m(c_{IIx} - v_L) + c_{mx}] \quad (y = 1)$$

$$(23\text{-}9)$$

where: c_{Tx} = capacity of two-stage movement x, veh/h

c_{Ix} = capacity of stage I of movement x, veh/h

c_{IIx} = capacity of stage II of movement x, veh/h

c_{mx} = capacity of two-stage movement x, computed as if it were a one-stage movement, veh/h

v_L = flow rate of major street left turns entering the median storage area for the subject movement (either v_1 or v_4), veh/h

a, y = intermediate variables

Like the model for upstream signal effects, this model is quite cumbersome to implement manually. Software (such as the HiCAP 2000 and HCS packages) is most often used. For additional modeling details, consult the HCM directly [1].

23.1.8 Analysis of Flared Approaches

Figure 23.6 illustrates the operation of a STOP-controlled intersection approach with a flared approach. The flared approach allows for some efficiencies, and the approach actually operates somewhere between the extremes of a fully shared lane and a separate lane for the right-turn movement.

First, the average queue length for each movement sharing the right lane is estimated as if it had a separate lane:

$$Q_{sep,i} = \frac{d_{sep,i}\, v_i}{3600} \quad (23\text{-}10)$$

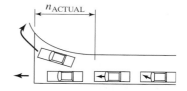

Figure 23.6: Illustration of a Flared Approach (Used with permission of Transportation Research Board, National Research Council, *Highway Capacity Manual*, Washington DC, 2000, Exhibit 17-18, pg. 17-21.)

where: $Q_{sep,i}$ = average queue length for movement "i" (assuming a separate lane), vehs

$d_{sep,i}$ = control delay per vehicle (s) for movement "i" (assuming a separate lane) (See next section for delay computations.)

v_i = demand flow rate for movement "i," veh/h

Then, the maximum queue size among these movements is established. This is actually the length of the storage area required for the movements to operate as if there were separate lanes:

$$n_{max} = \max_i [Int(Q_{sep,i} + 1)] \qquad (23\text{-}11)$$

where: n_{max} = storage length needed for operation as separate lane, vehs

Int = integer function; round to next highest integer of function value

max = maximum function; select highest value among movements "i"

Then, the capacity of the flared approach can be computed as:

$$c_{act} = \left(\sum_i c_{sep,i} - c_{SH} \right) \frac{n}{n_{max}} + c_{SH}$$

$$(n \leq n_{max})$$

$$c_{act} = \sum_i c_{sep,i} \qquad (n > n_{max}) \qquad (23\text{-}12)$$

where: c_{act} = capacity of the flared approach, veh/h

c_{SH} = shared-lane capacity assuming that all movements in the right lane and the flare used a single lane, veh/h

$c_{sep,i}$ = capacity of movement "i," assuming it used a separate lane, veh/h

n_{max} = storage capacity of flare needed to operate as a separate lane, vehs

n = actual storage capacity of the flare, vehs

Once again, this model is cumbersome to implement by hand, and software (such as the HiCAP 2000 or HCS) is most often used. Consult the HCM directly for additional modeling details [1].

23.1.9 Determining Control Delay

The control delay per vehicle for a movement or movements in a separate lane is given by:

$$d_x = \frac{3,600}{c_{mx}} + 900T \left[\left(\frac{v_x}{c_{mx}} - 1 \right) + \sqrt{\left(\frac{v_x}{c_{mx}} - 1 \right)^2 + \frac{\left(\frac{3,600}{c_{mx}} \right)\left(\frac{v_x}{c_{mx}} \right)}{450T}} \right] + 5$$

$$(23\text{-}13)$$

where: d_x = average control delay per vehicle for movement x, s/veh

c_{mx} = capacity of movement or shared-lane x, veh/h

T = analysis period, h (15 min = 0.25 h)

v_x = demand flow rate, movement or shared-lane x, veh/h

Where movements share a lane, a single value of delay for the lane is computed. In the preceding equation, c_{SH} replaces c_{mx}, and v_{SH} replaces v_x.

Delay may also be aggregated using flow rate–weighted averages for an approach or for the entire intersection. In averaging delay for the entire intersection, the delay to major-street vehicles of Rank 1 movements is assumed to be "0." Assigning an intersection level of service based on average intersection delay is, however, quite misleading because of this.

Under unusual circumstances, some delay to major-street Rank 1 vehicles can occur. HCM 2000 provides an algorithm for this. As with other aspects of this methodology, it is complex and is most often implemented through use of appropriate software.

The measure "control delay" is the level-of-service criterion used for both unsignalized and signalized intersection analysis. It is called "control delay," as it represents delay caused by the control measure—in this case, a STOP sign. In terms of traditional delay measures, it is closest to "time in queue delay" plus delay due acceleration and deceleration.

23.1.10 Estimating Queue Length

The HCM 2000 also provides an equation to estimate the 95th percentile queue length for a movement in a separate lane. The equation is similar to the delay equation:

$$Q_{95x} = \frac{3,600}{c_{mx}} + 900T\left[\left(\frac{V_x}{c_{mx}} - 1\right) + \right.$$
$$\left. \sqrt{\left(\frac{V_x}{c_{mx}} - 1\right)^2 + \frac{\left(\frac{3,600}{c_{mx}}\right)\left(\frac{V_x}{c_{mx}}\right)}{150T}}\right]\left(\frac{c_{mx}}{3,600}\right)$$

$$(23-14)$$

where: Q_{95x} = 95th percentile queue length, movement x, vehs

All other variables are as previously defined.

Again, where a shared lane exists, v and c for the shared lane replace movement values.

23.1.11 Sample Problem in TWSC Intersection Analysis

The following illustrative problem is to be analyzed using the HCM 2000 procedure for unsignalized intersections (TWSC). The procedure will result in estimation of appropriate v/c ratios and delays. Delay is used to establish the prevailing level of service that is to be expected.

Note that for simplicity, a T-intersection is used and that there are no flared approaches, two-stage processes, or nearby signalized intersections involved. Thus, all computations can be illustrated clearly without resorting to software. The stem of the T is STOP-controlled.

Step 1:   Find Conflicting Flow Rates

Note from Figure 23-7 that:

- Major-street left turn (Rank 2, v_4) conflicts with vehicle movements 2 and 3 and pedestrian movement 15.
- Minor-street right turn (Rank 2, v_9) conflicts with vehicle movements 2 and 3 and pedestrian mvts 14 and 15.
- Minor-street left turn (Rank 3, v_7) conflicts with vehicle movements 2, 3, 4, and 5 and pedestrian movements 13 and 15.

Using Figure 23.3:

$$v_{c4} = v_2 + v_3 + v_{15} = 200 + 30 + 30$$
$$= 260 \text{ conflicts/hr}$$
$$v_{c9} = (v_2/N) + 0.5v_3 + v_{14} + v_{15}$$
$$= (200/1) + 0.5(30) + 10 + 30$$
$$= 255 \text{ conflicts/hr}$$
$$v_{c7} = 2v_4 + (v_5/N) + v_{13} + v_2$$
$$+ 0.5v_3 + v_{15} = 2(20) + (400/1)$$
$$+ 15 + 200 + 0.5(30) + 30$$
$$= 700 \text{ conflicts/h}$$

Note that in the computation of v_{c7} for a T-intersection, the values of v_1, v_6, v_{11}, and v_{12} are "0," causing these terms to disappear from the computation.

Step 2:   Find Critical Gaps and Follow-Up Times for Each Movement

Critical gaps are as follows:

$$t_c = t_{cb} + t_{cHV}P_{HV} + t_{cG}G - t_{cT} - t_{3LT}$$

Base critical gaps are found in Table 23.2; adjustments are found in Table 23.3.

- t_{cb} = 4.1 s (movement 4); 6.2 s (movement 9); 7.1 s (movement 7)
- t_{cHV} = 1.0 for all movements (two-lane major street)
- P_{HV} = 0.10 for all movements
- t_{cG} = 0.1 for movement 9; 0.2 for movement 7; 1.0 for movement 4

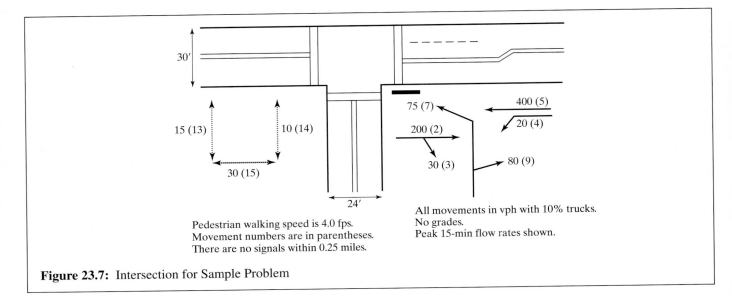

All movements in vph with 10% trucks.
No grades.
Peak 15-min flow rates shown.

Pedestrian walking speed is 4.0 fps.
Movement numbers are in parentheses.
There are no signals within 0.25 miles.

Figure 23.7: Intersection for Sample Problem

- $G = 0.0$ for all movements
- $t_{cT} = 0.0$ for all movements (only one-stage gap acceptance present)
- $t_{3LT} = 0.7$ for movement 7; 0.0 for movements 4 and 9

$$t_{c4} = 4.1 + 1.0(0.10) + 1.0(0.0) - 0.0 \\ - 0.0 = 4.2 \text{ s}$$

$$t_{c9} = 6.2 + 1.0(0.10) + 0.1(0.0) - 0.0 \\ - 0.0 = 6.3 \text{ s}$$

$$t_{c7} = 7.1 + 1.0(0.10) + 0.2(0.0) - 0.0 \\ - 0.7 = 6.5 \text{ s}$$

Follow-up times are as follows:

$$t_f = t_{fb} + t_{fHV} P_{HV}$$

Base follow-up times are found in Table 23.2; adjustments are found in Table 23.3.

- $t_{fb} = 2.2$ s (movement 4); 3.3 s (movement 9); 3.5 s (movement 7)
- $t_{fHV} = 0.9$ for all movements
- $P_{HV} = 0.10$ for all movements

$$t_{f4} = 2.2 + 0.90(0.10) = 2.29 \text{ s}$$
$$t_{f9} = 3.3 + 0.90(0.10) = 3.39 \text{ s}$$
$$t_{c9} = 3.5 + 0.90(0.10) = 3.59 \text{ s}$$

Step 3:   Compute Potential Capacities

Potential capacities are computed from the conflicting volumes, critical gaps, and follow-up times as follows:

$$c_{px} = v_{cx} \frac{e^{-(v_{cx} t_{cx}/3,600)}}{1 - e^{-(v_{cx} t_{fx}/3,600)}}$$

This computation is carried out in Table 23.5.

Step 4: Determine Impedance Effects and Movement Capacities

Vehicular impedance probabilities are computed as:

$$p_{vi} = 1 - \left(\frac{v_i}{c_{mi}}\right)$$

Pedestrian impedance probabilities are computed as:

$$p_{pi} = 1 - \frac{v_j\left(\dfrac{w}{S_p}\right)}{3,600}$$

Movements 4 and 9 are Rank 2 movements. They are not impeded by any vehicular flows. Movement 4, is, however, impeded by pedestrian movement 15,

Table 23.5: Potential Capacity Results for Sample Problem

Movement	v_{cx} (vph)	t_c (sec)	t_f (sec)	c_{px} (vph)
4	260	4.2	2.29	1,259
9	255	6.3	3.39	765
7	700	6.5	3.59	394

and movement 9 is impeded by pedestrian movements 14 and 15. Thus, these pedestrian impedance factors must be known to compute the movement capacities of vehicular movements 4 and 9:

$$p_{p14} = 1 - \frac{(10)(10)/4}{3,600} = 1 - 0.007 = 0.993$$

$$p_{p15} = 1 - \frac{(30)(12)/4}{3,600} = 1 - 0.025 = 0.975$$

Then:

$$c_{m4} = 1,259(0.975) = 1,228 \text{ vph}$$

$$c_{m9} = 765(0.975)(0.993) = 741 \text{ vph}$$

Movement 7 is a Rank 3 movement at a T-intersection. It is impeded by pedestrian movements 13 and 15 and by vehicular movement 4, which may use the same gaps as movement 7 but is of higher rank:

$$p_{p13} = 1 - \frac{(15)(15)/4}{3,600} = 1 - 0.016 = 0.984$$

$$p_{v4} = 1 - \frac{20}{1,228} = 0.984$$

Then:

$$c_{m7} = 394(0.984)(0.984)(0.975) = 372 \text{ vph}$$

Step 5:   Shared Lane Capacities

Movement 4 has its own turning lane. Movements 7 and 9, however, share a lane. The capacity of this lane is found as:

$$c_{SH} = \frac{75 + 80}{\left(\frac{75}{372}\right) + \left(\frac{80}{741}\right)}$$

$$= \frac{155}{0.202 + 0.108} = 500 \text{ vph}$$

Step 6: Adjustments for Upstream Signals, Flared Lanes, and Two-Stage Gap Acceptance

None of these conditions exist, so no such adjustments are required.

Step 7: Delay Computations and Levels of Service

Delay is computed for the major street left turn (movement 4) and for the shared lane on the minor street (serving movements 7 and 9). The equation for delay is given by:

$$d_x = \frac{3,600}{c_{mx}} + 900T \left[\frac{v_x}{c_{mx}} - 1 \right.$$
$$\left. + \sqrt{\left(\frac{v_x}{c_{mx}} - 1\right)^2 + \frac{\left(\frac{3,600}{c_{mx}}\right)\left(\frac{v_x}{c_{mx}}\right)}{450T}} \right] + 5$$

For movement 4, $v = 20$ veh/h and $c = 1,228$ veh/h; for the shared lane (movements 7 and 9), $v = 155$ veh/h and $c = 500$ veh/h. The value of T is taken as 0.25 hr or 15 min for both computations. The computations are implemented in the spreadsheet shown in Table 23.6. This indicates that the major-street left turn operates at level of service A, while the minor-street operates at level of service C (just over the boundary for B). An average delay could be computed for the intersection as a whole. To do this, the 630 veh/h in other movements are assumed to have "0" delay.

$$d_I = \frac{(20 \times 7.98) + (155 \times 15.40) + (630 \times 0)}{(20 + 155 + 630)}$$

$$= 3.16 \text{ s/veh}$$

Table 23.6: Delay Results for Sample Problem

Movement	v (vph)	c (vph)	T (hr)	d (s/veh)
4	20	1228	0.25	7.98
7,9	155	500	0.25	15.40

This is level of service A for the intersection as a whole. The intersection is operating at acceptable levels, and no changes in control, design, or operation are recommended.

23.2 Analysis of Roundabouts

HCM 2000 presents a simple algorithm for determining the capacity of a roundabout approach. While hardly a complete methodology, it at least begins to address a type of intersection that is increasing in popularity in the United States, particularly as part of traffic-calming programs.

The algorithm applies to a single-lane roundabout (one circulating lane). It should be noted that roundabouts

have no additional control devices, and that circulating traffic has the right of way at all times. Where other operating regimes are present, the configuration is called a "traffic circle," and the model presented in the HCM 2000 does not apply.

Figure 23.8 shows the general configuration of a roundabout approach with respect to circulating traffic for which the model applies.

The capacity of a roundabout approach is estimated as:

$$c_a = \frac{v_c\, e^{-(v_c\, t_c/3600)}}{1 - e^{-(v_c\, t_f/3600)}} \qquad (23\text{-}15)$$

where: c_a = capacity of the roundabout approach, veh/h

v_c = conflicting circulating traffic, veh/h

t_c = critical gap for roundabout entry, s

t_f = follow-up time for roundabout entry, s

Table 23.7 gives upper and lower bound values for critical gap and follow-up time at roundabouts. A range of values is given, as this is a general estimate covering a wide range of specific design conditions. Studies in the

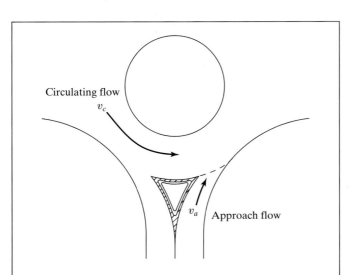

Figure 23.8: Flow on a Roundabout (Used with permission of Transportation Research Board, National Research Council, *Highway Capacity Manual*, 4th Edition, Washington DC, 2000, Exhibit 17-36, pg. 17-46.)

Table 23.7: Upper and Lower Bounds for Critical Gap and Follow-Up Time at Roundabouts

	Critical Gap, s	Follow-Up Time, s
Upper Bound	4.1	2.6
Lower Bound	4.6	3.1

(Used with permission of Transportation Research Board, National Research Council, *Highway Capacity Manual*, 4th Edition, Washington DC, 2000, Exhibit 17-37, pg. 17–46.)

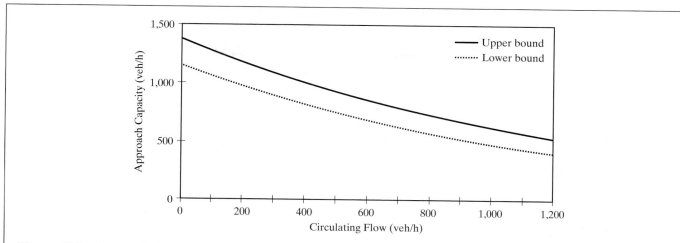

Figure 23.9: Approach Capacity at Roundabouts (Used with permission of Transportation Research Board, National Research Council, *Highway Capacity Manual*, 4th Edition, Washington DC, 2000, Exhibit 17-38, pg. 17–47.)

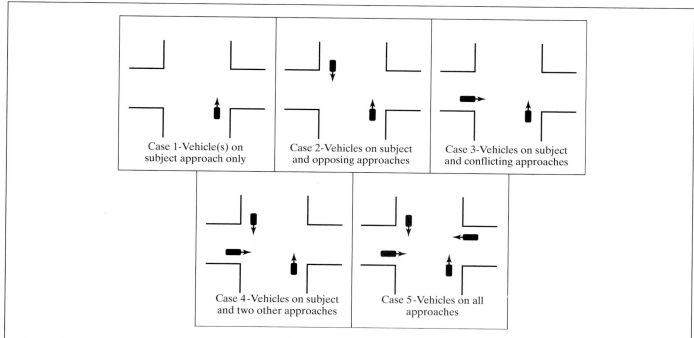

Figure 23.10: Potential Situations at AWSC Intersections (Used with permission of Transportation Research Board, National Research Council, *Highway Capacity Manual*, 4th Edition, Washington DC, 2000, Exhibit 10-25, pg. 10-30.)

United States have not been sufficient to correlate observed differences in critical gap and follow-up time to specific geometric or traffic factors.

In an analysis of a roundabout, it is critical that all movements be converted to the appropriate circulating flows, v_c, passing each approach. U-turn movements, where they are prevalent, must also be included in this determination.

Figure 23.9 depicts the relationship between circulating flow and approach capacity using the model and the criteria of Table 23.7.

23.3 Analysis of All-Way STOP-Controlled Intersections (AWSC)

The HCM 2000 contains a detailed procedure for the determination of capacity and level of service at all-way STOP-controlled intersections. The model is extremely complex and requires multiple (sometimes hundreds) of iterations to come to a conclusion. For this reason, the model is only implemented through the use of software.

Figure 23.10 illustrates the basic premise of the model. The figure depicts the five basic traffic configurations that can occur at a typical four-leg, all-way STOP-controlled intersection. Based on input volumes, the model evaluates the probability that each of these will occur and then assesses the impact of each occurrence on the capacity and delay for each approach. As each approach affects operations on the other, resulting v/c ratios for each approach affect the probabilities for each case, which results in revised capacities and v/c ratios. This creates the need for a multiply iterative procedure.

Students interested in the details of this complex model are urged to consult the HCM 2000 directly.

References

1. *Highway Capacity Manual*, 4th Edition, Transportation Research Board, National Research Council, Washington DC, 2000.

2. Siegloch, W., *Die Leistungsermittlung an Knotenpunkten ohne Lichtsignalsteurung*, Schriftenreihe Strassenbau und Strassenverkehrstechnik, Heft 154, Bonn, Germany, 1973.

3. Kyte, M, et al., *Capacity and Level of Service at Unsignalized Intersections, Vol. 1—Two-Way STOP-Controlled Intersections*, NCHRP Web Document 5, National Cooperative Highway Research Program, Washington DC, 1997.

Problems

All problems refer to the intersection shown in Figure 23.11.

23-1. Determine conflicting volumes for movements 1, 7, 8, and 9.

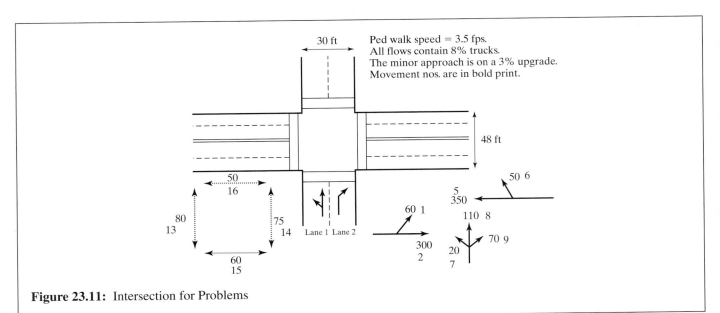

Ped walk speed = 3.5 fps.
All flows contain 8% trucks.
The minor approach is on a 3% upgrade.
Movement nos. are in bold print.

Figure 23.11: Intersection for Problems

23-2. Determine the potential capacities for movements 1, 7, 8, and 9.

23-3. Determine the movement capacities for movements 1, 7, 8, and 9.

23-4. Determine the shared-lane capacity for movements 7 and 8.

23-5. Determine the control delay and level of service for individual movements and for the intersection as a whole, as appropriate. Comment on the results, making recommendations for any improvements you believe may be justified by the results of the analysis.

CHAPTER

24

Signal Coordination for Arterials and Networks

24.1 Basic Principles of Signal Coordination

In situations where signals are close enough together so that vehicles arrive at the downstream intersection in platoons, it is necessary to coordinate their green times so that vehicles may move efficiently through the *set* of signals. It serves no purpose to have drivers held at one signal watching wasted green at a downstream signal, only to arrive there just as the signal turns red.

In some cases, two signals are so closely spaced that they should be considered to be one signal. In other cases, the signals are so far apart that they may be considered as isolated intersections. However, vehicles released from a signal often maintain their grouping for well over 1,000 feet. Common practice is to coordinate signals less than one mile apart on major streets and highways.

24.1.1 A Key Requirement: Common Cycle Length

In coordinated systems, all signals must have the same cycle length. This is necessary to ensure that the beginning

of green occurs at the same time relative to the green at the upstream and downstream intersections. There are some exceptions, where a critical intersection has such a high volume that it may require a double cycle length, for instance, but this is done rarely and only when no other solution is feasible.

24.1.2 The Time-Space Diagram and Ideal Offsets

The time-space diagram is a plot of signal indications as a function of time for two or more signals. The diagram is scaled with respect to distance, so that one may easily plot vehicle positions as a function of time. Figure 24.1 is a time-space diagram for two intersections. Standard conventions are used in the figure: a green signal indication is shown by a blank line ($\rule{2em}{0.8em}$), yellow by a shaded line ($\rule{2em}{0.8em}$) and red by a solid line ($\rule{2em}{0.8em}$). In many cases, such diagrams show only effective green and effective red, as shown in Figure 24.1. This figure illustrates the path (trajectory) that a vehicle takes as time passes. At $t = t_1$, the first signal turns green. After some lag, the vehicle starts and moves

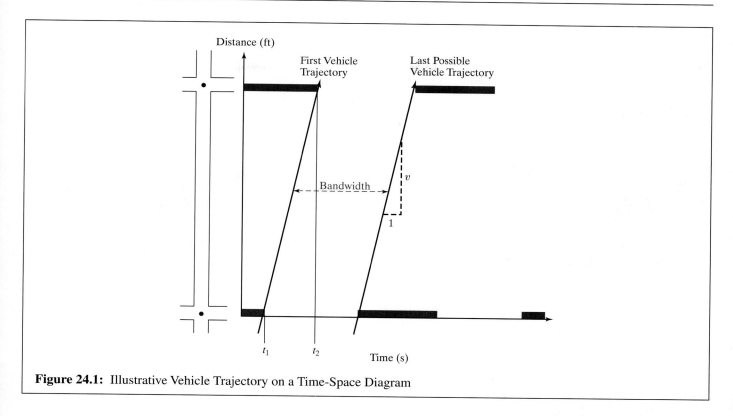

Figure 24.1: Illustrative Vehicle Trajectory on a Time-Space Diagram

down the street. It reaches the second intersection at some time $t = t_2$. Depending on the indication of that signal, it either continues or stops.

The difference between the two green initiation times (i.e., the difference between the time when the upstream intersection turns green and the downstream intersection turns green), is referred to as the *signal offset*, or simply the *offset*. In Figure 24.1, the offset is defined as t_2 minus t_1. Offset is usually expressed as a positive number between zero and the cycle length. This definition is used throughout this and other chapters in this text.

There are other definitions of offset used in practice. For instance, offset is sometimes defined relative to one reference upstream signal, and sometimes it is defined relative to a standard zero. Some signal hardware uses "offset" defined in terms of red initiation, rather than green; other hardware uses the end of green as the reference point. Some hardware uses offset in seconds;

other hardware uses offset as a percentage of the cycle length.

The "ideal offset" is defined as exactly the offset such that, as the first vehicle of a platoon just arrives at the downstream signal, the downstream signal turns green. It is usually assumed that the platoon was moving as it went through the upstream intersection. If so, the ideal offset is given by:

$$t_{ideal} = L/S \qquad (24\text{-}1)$$

where: t_{ideal} = ideal offset, s

L = distance between signalized intersections, ft

S = average speed, ft/s

If the vehicle were stopped and had to accelerate after some initial start-up delay, the ideal offset could be represented by Equation 24-1 plus the start-up time at the first intersection (which would usually add 2 to 4 seconds). In general, the start-up time would be included only at the *first* of a series of signals to be coordinated, and often not at all. Usually, this will reflect the ideal offset desired for maximum bandwidth, minimum delay, and minimum stops. Even if the vehicle is stopped at the first intersection, it will be moving through most of the system.

Figure 24.1 also illustrates the concept of *bandwidth*. Bandwidth is the amount of green time that can be used by a continuously moving platoon of vehicles through a series of intersections. In Figure 24.1, the bandwidth is the entire green time at both intersections, because several key conditions exist:

- The green times at both intersections are the same.
- The ideal offset is illustrated.
- There are only two intersections.

In most cases, the bandwidth will be less, perhaps significantly so, than the full green time.

Figure 24.2 illustrates the effect of offset on stops and delay for a platoon of vehicles leaving one intersection and passing through another. In this example, a 25-s offset is ideal, as it produces the minimum delay and the minimum number of stops. The effect of allowing a poor offset to exist is clearly indicated: delay can climb to 30 s per vehicle and the stops to 10 per cycle. Note that the penalty for deviating from the ideal offset is usually not equal in positive and negative deviations. An offset of $(25 + 10) = 35$ s causes much more harm than an offset of $(25 - 10) = 15$ s, although both are 10 s from the ideal offset. Figure 24.2 is illustrative, as each situation would have similar but different characteristics.

24.2 Signal Progression on One-Way Streets

Signal progression on a one-way street is relatively simple. For the purpose of this section, it will be assumed that a cycle length has been chosen and that the green allocation at each signal has been previously determined.

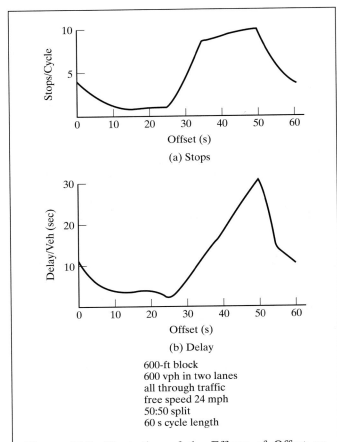

(a) Stops

(b) Delay

600-ft block
600 vph in two lanes
all through traffic
free speed 24 mph
50:50 split
60 s cycle length

Figure 24.2: Illustration of the Effects of Offset on Stops and Delay

24.2.1 Determining Ideal Offsets

Consider the one-way arterial shown in Figure 24.3, with the link lengths indicated. Assuming no vehicles are queued at the signals, the ideal offsets can be determined if the platoon speed is known. For the purpose of illustration, a desired platoon speed of 60 ft/s will be used. The cycle length is 60 s, and the effective green time at each intersection is 50% of the cycle length, or 30 s. Ideal offsets are computed using Equation 24-1, and are listed in Table 24.1.

Note that neither the cycle length nor the splits enter into the computation of ideal offsets. In order to

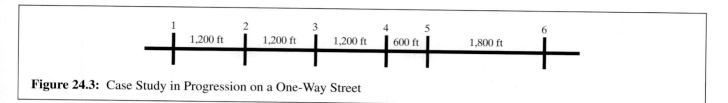

Figure 24.3: Case Study in Progression on a One-Way Street

Table 24.1: Ideal Offsets for Case Study

Signal	Relative to Signal	Ideal Offset
6	5	1,800/60 = 30 s
5	4	600/60 = 10 s
4	3	1,200/60 = 20 s
3	2	1,200/60 = 20 s
2	1	1,200/60 = 20 s

see the pattern that results, the time-space diagram should be constructed according to the following rules:

1. The vertical should be scaled so as to accommodate the dimensions of the arterial, and the horizontal so as to accommodate at least three to four cycle lengths.

2. The beginning intersection (Number 1, in this case) should be scaled first, usually with main street green (MSG) initiation at $t = 0$, followed by periods of green and red (yellow may be shown for precision). See Point 1 in Figure 24.4.

3. The main street green (or other offset position, if MSG is not used) of the next downstream signal should be located next, relative to $t = 0$

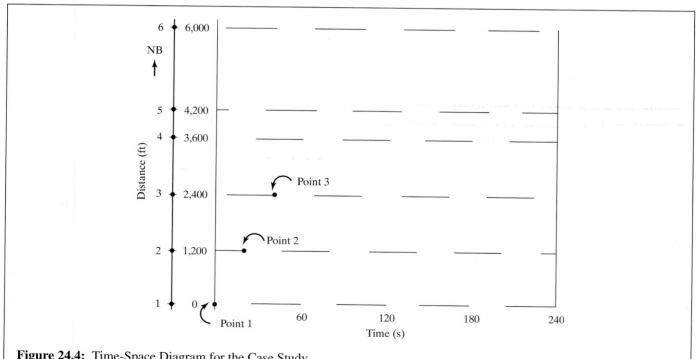

Figure 24.4: Time-Space Diagram for the Case Study

and at the proper distance from the first intersection. With this point located (Point 2 in Figure 24.4), fill in the periods of effective green and red for this signal.

4. Repeat the procedure for all other intersections, working one at a time. Thus, for Signal 3, the offset is located at point 3, 20 s later than Point 2, and so on.

Figure 24.4 has some interesting features that can be explored with the aid of Figure 24.5.

First, if a vehicle (or platoon) were to travel at 60 fps, it would arrive at each of the signals just as they turn green; this is indicated by the solid trajectory lines in Figure 24.5. The solid trajectory line also represents the speed of the "green wave" visible to a stationary observer at Signal 1, looking downstream. The signals turn green in order, corresponding to the planned speed of the platoon, and give the visual effect of a wave of green opening before the driver. Third, note that there is a "window" of green in Figure 24.5, with its end indicated

by the dotted trajectory line; this is also the trajectory of the *last* vehicle that could travel through the progression without stopping at 60 ft/s. This "window" is the bandwidth, as defined earlier. Again, in this case it equals the green time because all signals have the same green time and have ideal offsets.

24.2.2   Potential Problems

Consider what would happen if the actual speed of vehicle platoons in the case study was 50 ft/s, instead of the 60 ft/s anticipated. The green wave would still progress at 60 ft/s, but the platoon arrivals would lag behind it. The effect of this on bandwidth is enormous, as shown in Figure 24.6. Only a small window now exists for a

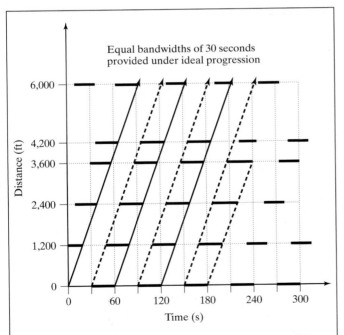

Figure 24.5: Vehicle Trajectory and "Green Wave" in a Progressed Movement

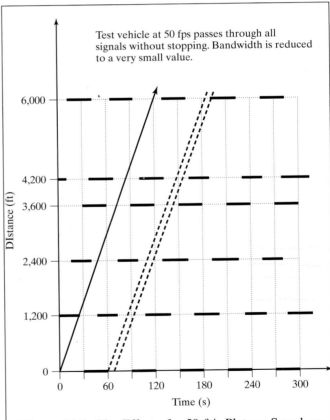

Figure 24.6: The Effect of a 50-ft/s Platoon Speed on Progression

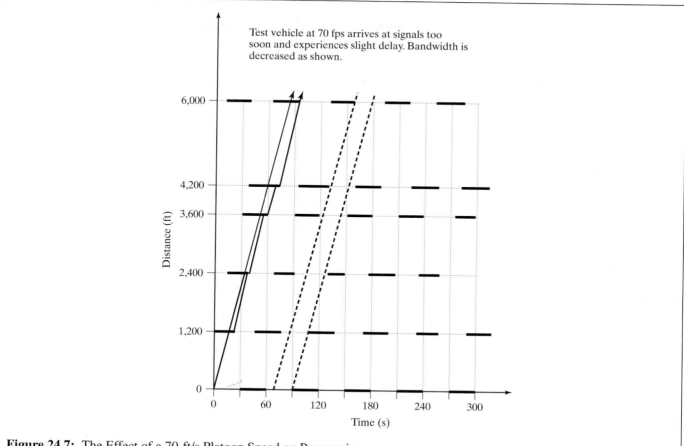

Figure 24.7: The Effect of a 70-ft/s Platoon Speed on Progression

platoon of vehicles to continuously flow through all six signals without stopping.

Figure 24.7 shows the effect of the vehicle traveling faster than anticipated, 70 ft/s in this illustration. In this case, the vehicles arrive a little too early and are delayed; some stops will have to be made to allow the "green wave" to catch up to the platoon.

In this case, the effect on bandwidth is not as severe as in Figure 24.6. In this case, the bandwidth impact of *underestimating* the platoon speed (60 ft/s instead of 70 ft/s) is not as severe as the consequences of *overestimating* the platoon speed (60 ft/s instead of 50 ft/s).

24.3 Bandwidth Concepts

Bandwidth is defined as the time difference between the first vehicle that can pass through the entire system without stopping and the last vehicle that can pass through without stopping, measured in seconds.

The bandwidth concept is very popular in traffic engineering practice because the windows of green are easy visual images for both working professionals and public presentations. The most significant shortcoming of designing offset plans to maximize bandwidths is that internal queues are often overlooked in the bandwidth approach. There are computer-based maximum bandwidth solutions that go beyond the historical formulations, such as PASSER [2].

24.3.1 Bandwidth Efficiency

The efficiency of a bandwidth is defined as the ratio of the bandwidth to the cycle length, expressed as a percentage:

$$EFF_{BW} = \left(\frac{BW}{C}\right) * 100 \qquad (24\text{-}2)$$

where: EFF_{BW} = bandwidth efficiency, %

BW = bandwidth, s

C = cycle length, s

A bandwidth efficiency of 40% to 55% is considered good. The bandwidth is limited by the minimum green in the direction of interest.

Figure 24.8 illustrates the bandwidths for one signal-timing plan. The northbound efficiency can be estimated as $(17/60) * 100\% = 28.3\%$. The southbound bandwidth is obviously terrible—there is no bandwidth through the defined system. The northbound efficiency is only 28.3%. This system is badly in need of retiming, at least on the basis of the bandwidth objective. Just looking at the time-space diagram, one might imagine sliding the pattern at Signal 4 to the right and the pattern at Signal 1 to the left, allowing some coordination for the southbound vehicles.

24.3.2 Bandwidth Capacity

In terms of vehicles that can be put through the system of Figure 24.8 without stopping, the northbound bandwidth can carry $17/2.0 = 8.5$ vehicles per lane per cycle in a nonstop path through the defined system, assuming that the saturation headway is 2.0 s/veh. Thus the northbound direction can handle 8.5 veh/cycle * 1 cycle/60 s * 3,600 s/hr = 510 veh/h/ln very efficiently if they are organized into eight-vehicle platoons when they travel through this system.

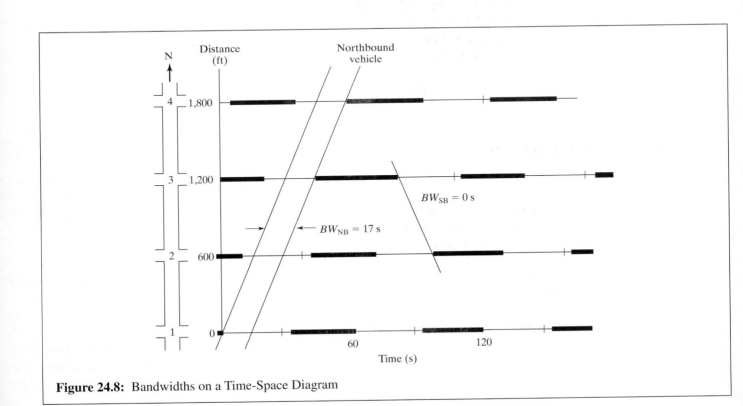

Figure 24.8: Bandwidths on a Time-Space Diagram

If the per lane demand volume is less than 510 vphpl and if the flows are well organized (and if there is no internal queue development), the system will operate well in the northbound direction, even though better timing plans might be obtained.

In general terms, the number of vehicles that can pass through a defined series of signals without stopping is called the *bandwidth capacity*. The illustrated computation can be described by the following equation:

$$c_{BW} = \frac{3,600 * BW * L}{C * h} \qquad (24\text{-}3)$$

where: c_{BW} = bandwidth capacity, veh/h

BW = bandwidth, s

L = number of through lanes in the indicated direction

C = cycle length, s

h = saturation headway, s

Equation 24-3 does not contain any factors to account for nonuniform lane utilization and is intended only to indicate some limit beyond which the offset plan will degrade, certainly resulting in stopping and internal queuing. It should also be noted that bandwidth capacity is *not* the same as lane group capacity. Where the bandwidth is less than the full green time, there is additional lane group capacity outside of the bandwidth.

24.4   The Effect of Queued Vehicles at Signals

To this point, it has been assumed that there is no queue standing at the downstream intersection when the platoon from the upstream signal arrives. This is generally not a reasonable assumption. Vehicles that enter the traffic stream between platoons will progress to the downstream signal, which will often be "red." They form a queue that partially blocks the progress of the arriving platoon. These vehicles may include stragglers from the last platoon, vehicles that turned into the block from unsignalized intersections or driveways, or vehicles that came out of parking lots or parking spots. The ideal offset must be adjusted to allow for these vehicles, so as to

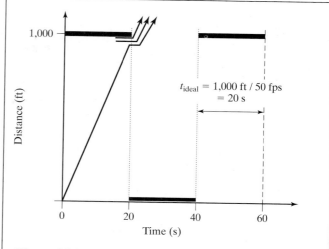

Figure 24.9: The Effect of Queued Vehicles at a Signal

avoid unnecessary stops. The situation without such an adjustment is depicted in Figure 24.9, where it can be seen that the arriving platoon is delayed behind the queued vehicles as the queued vehicles begin to accelerate through the intersection.

To adjust for the queued vehicles, the ideal offset is adjusted as follows:

$$t_{adj} = \frac{L}{S} - (Qh + \ell_1) \qquad (24\text{-}4)$$

where: t_{adj} = adjusted ideal offset, s

L = distance between signals, ft

S = average speed, ft/s

Q = number of vehicles queued per lane, veh

h = discharge headway of queued vehicles, s/veh

ℓ_1 = start-up lost time, s

The lost time is counted only at the first downstream intersection, at most. If the vehicle(s) from the preceding intersection were themselves stationary, their start-up causes a shift that automatically takes care of the start-up at subsequent intersections.

Offsets can be adjusted to allow for queue clearance before the arrival of a platoon from the upstream intersection. Figure 24.10 shows the situation for use of the modified ideal offset equation.

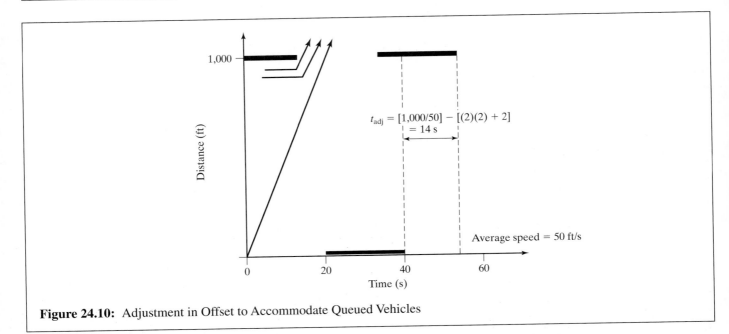

Figure 24.10: Adjustment in Offset to Accommodate Queued Vehicles

Figure 24.11 shows the time-space diagram for the case study of Figure 24.5, given queues of two vehicles per lane in all links. Note that the arriving vehicle platoon has smooth flow, and the lead vehicle has 60 ft/s travel speed. The visual image of the "green wave," however, is much faster, due to the need to clear the queues in advance of the arriving platoon.

The "green wave," or the progression speed as it is more properly called, is traveling at varying speeds as it moves down the arterial. The "green wave" will appear to move ahead of the platoon, clearing queued vehicles in advance of it. The progression speed can be computed for each link as shown in Table 24.2.

It should be noted, however, that the bandwidth and, therefore, the bandwidth capacity is now much smaller. Thus, by clearing out the queue in advance of

the platoon, more of the green time is used by queued vehicles and less is available to the moving platoon.

The preceding discussion assumes that the queue is known at each signal. In fact, this is not an easy number to know. However, if we know that there is a queue and know its approximate size, the link offset can be set better than by pretending that no queue exists.

Consider the sources of the queued vehicles:

- Vehicles turning in from upstream side streets during their green (which is main-street red).
- Vehicles leaving parking garages or spaces.
- Stragglers from previous platoons.

There can be great cycle-to-cycle variation in the actual queue size, although the average queue size may

Table 24.2: Progression Speeds in Figure 24.11

Link	Link Offset (s)	Speed of Progression (ft/s)
Signal 1 → 2	$(1{,}200/60) - (4 + 2) = 14$	$1{,}200/14 = 85.7$
Signal 2 → 3	$(1{,}200/60) - (4) = 16$	$1{,}200/16 = 75$
Signal 3 → 4	$(1{,}200/60) - (4) = 16$	$1{,}200/16 = 75$
Signal 4 → 5	$(600/60) - (4) = 6$	$600/6 = 100$
Signal 5 → 6	$(1{,}800/60) - (4) = 26$	$1{,}800/26 = 69.2$
Total Offset = 78 sec		

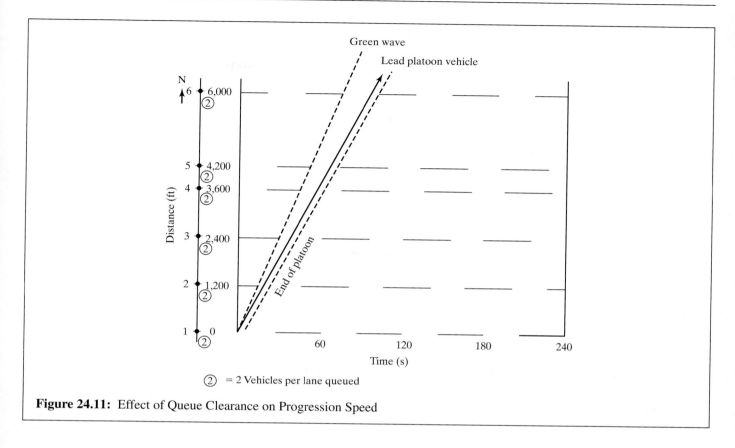

Figure 24.11: Effect of Queue Clearance on Progression Speed

be estimated. Even at that, queue estimation is a difficult and expensive task. Even the act of adjusting the offsets can influence the queue size. For instance, the arrival pattern of the vehicles from the side streets may be altered. Queue estimation is therefore a significant task in practical terms.

24.5 Signal Progression for Two-Way Streets and Networks

The task of progressing traffic on a one-way street is relatively straightforward. To highlight the essence of the problem on a two-way street, assume that the arterial shown in Figure 24.5 is a two-way street rather than a one-way street. Figure 24.12 shows the trajectory of a *southbound* vehicle on this arterial. The vehicle is just fortunate enough not to be stopped until Signal 2, but is then stopped again for Signal 1, for a total of two stops

and 40 s of delay. There is no bandwidth, meaning that it is not possible to have a vehicle platoon pass along the arterial nonstop.

Of course, if the offsets or the travel times had been different, it might have been possible to have a southbound bandwidth through all six signals.

24.5.1 Offsets on a Two-Way Street

Note that if any offset were changed in Figure 24.12 to accommodate the southbound vehicles, then the northbound bandwidth would suffer. For instance, if the offset at Signal 2 were decreased by 20 s, then the pattern at that signal would shift to the left by 20 s, resulting in a "window" of green of only 10 s on the northbound, rather than the 30 s in the original display (Figure 24.5).

The fact that the offsets on a two-way street are interrelated presents one of the most fundamental problems of signal optimization. Note that inspection of a typical

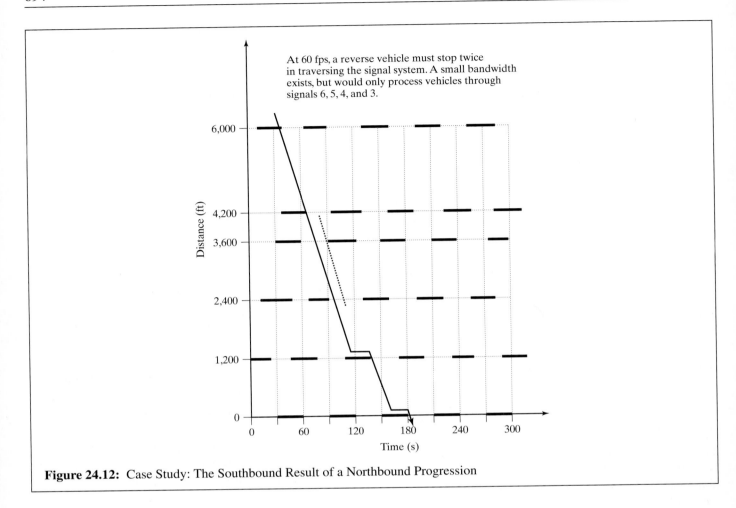

At 60 fps, a reverse vehicle must stop twice in traversing the signal system. A small bandwidth exists, but would only process vehicles through signals 6, 5, 4, and 3.

Figure 24.12: Case Study: The Southbound Result of a Northbound Progression

time-space diagram yields the obvious conclusion that the offsets in two directions add to one cycle length (Figure 24.13a). However, for longer blocks, the offsets might add to two (or more) cycle lengths (Figure 24.13b).

Figure 24.13 illustrates both actual offsets and travel times, which are not necessarily the same. While the engineer might desire the ideal offset to be the same as the travel times, this is not always the case. Once the offset is specified in one direction, it is automatically set in the other. The general expression for the two offsets in a link on a two-way street can be written as:

$$t_{1i} + t_{2i} = nC \qquad (24\text{-}5)$$

where: t_{1i} = offset in direction 1 (link i), s

t_{2i} = offset in direction 2 (link i), s

n = integer value

C = cycle length, s

To have $n = 1$ (Figure 24.13a), $t_{1i} \leq C$; to have $n = 2$, $C < t_{1i} \leq 2C$.

Any actual offset can be expressed as the desired "ideal" offset, plus an "error" or "discrepancy" term:

$$t_{actual(i,j)} = t_{ideal(i,j)} + e_{ij} \qquad (24\text{-}6)$$

where j represents the direction and i represents the link. In a number of signal optimization programs that are used for two-way arterials, the objective is to minimize some function of the discrepancies between the actual and ideal offsets.

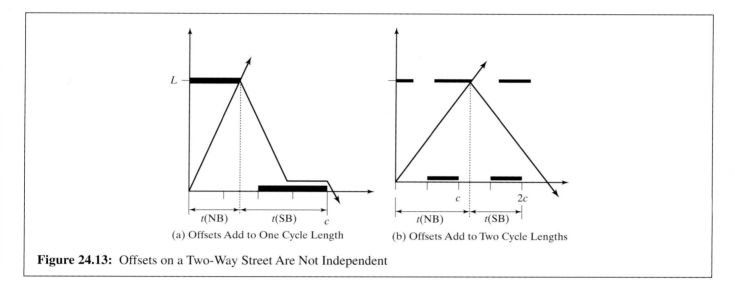

(a) Offsets Add to One Cycle Length (b) Offsets Add to Two Cycle Lengths

Figure 24.13: Offsets on a Two-Way Street Are Not Independent

24.5.2 Network Closure

The relative difficulty of finding progressions on a two-way street, as compared with on a one-way street, might lead one to conclude that the best approach is to establish a system of one-way streets to avoid the problem. A one-way street system has a number of advantages, not the least of which is elimination of left turns against opposing traffic. One-way streets simplify network signalization, but they do not eliminate closure problems, and they carry other practical disadvantages. See Chapter 27 for a complete discussion of one way streets.

Figure 24.14 illustrates network closure requirements. In any set of four signals, offsets may be set on three legs in one direction. Setting three offsets, however, fixes the timing of all four signals. Thus, setting three offsets fixes the fourth.

Figure 24.15 extends this to a grid of one-way streets, in which all of the north-south streets are independently specified. The specification of one east-west street then "locks in" all other east-west offsets. Note that the key feature is that an open tree of one-way links can be completely independently set and that it is the closing or "closure" of the open tree that presents constraints on some of the links.

To develop the constraint equation, refer to Figure 24.14 and walk through the following steps, keying to the green in all steps:

Step 1. Begin at Intersection 1 and consider the green initiation to be time $t = 0$.

Step 2. Move to Intersection 2, noting that the link offset t_1 specifies the time of green initiation at this intersection, relative to its upstream neighbor. Thus, green starts at Intersection 2 facing northbound at $t = 0 + t_1$.

Step 3. Recognizing that the westbound vehicles get released after the N–S green is finished, green begins at Intersection 2 facing west at

$$t = 0 + t_1 + g_{NS,2}$$

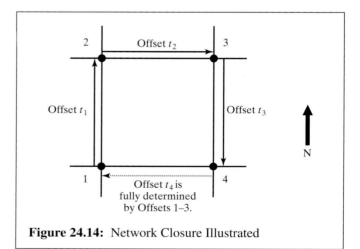

Figure 24.14: Network Closure Illustrated

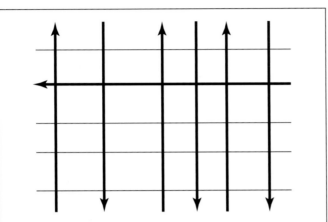

Figure 24.15: Impact of Closure on a Rectangular Street Grid

Step 4. Moving to intersection 3, the link offset specifies the time of green initiation at Intersection 3 relative to Intersection 2. Thus the green begins at Intersection 3, facing west at

$$t = 0 + t_1 + g_{NS,2} + t_2$$

Step 5. Similar to Step 3, the green begins at Intersection 3, but facing south, after the E–W green is finished at time:

$$t = 0 + t_1 + g_{NS,2} + t_2 + g_{EW,3}$$

Step 6. Moving to Intersection 4, the green begins in the southbound direction after the offset t_3 is added:

$$t = 0 + t_1 + g_{NS,2} + t_2 + g_{EW,3} + t_3$$

Step 7. Turning at Intersection 4, it is the NS green that is added to be at the start of green facing east:

$$t = 0 + t_1 + g_{NS,2} + t_2 + g_{EW,3} + t_3 + g_{NS,4}$$

Step 8. Moving to Intersection 1, it is t_4 that is relevant to be at the start of green facing east:

$$t = 0 + t_1 + g_{NS,2} + t_2 + g_{EW,3} + t_3 + g_{NS,4} + t_4$$

Step 9. Turning at Intersection 1, green will begin in the north direction after the EW green finishes:

$$t = 0 + t_1 + g_{NS,2} + t_2 + g_{EW,3} + t_3 + g_{NS,4} + t_4 + g_{EW,1}$$

This will bring us back to where we started. Thus, this is either $t = 0$ or a multiple of the cycle length.

The following relationship results:

$$nC = 0 + t_1 + g_{NS,2} + t_2 + g_{EW,3} + t_3 + g_{NS,4} + t_4 + g_{EW,1} \qquad (24\text{-}7)$$

where the only caution is that the g values should include the change and clearance intervals.

Note that Equation 24-7 is a more general form of Equation 24-5, for the two-way arterial is a special case of a network. The interrelationships stated in Equation 24-7 are constraints on freely setting all offsets. In these equations one can trade off between green allocations and offsets. To get a better offset in Link 4, one can adjust the splits as well as the other offsets.

While it is sometimes necessary to consider networks in their entirety, it is common traffic engineering practice to decompose networks into non-interlocking arterials whenever possible. Figure 24.16 illustrates this process.

Decomposition works well where a clear center of activity can be identified and where few vehicles are expected to pass through the center without stopping (or starting) at or near the center. As the discontinuity in all progressions lies in and directly around the identified center, large volumes passing through can create significant problems in such a scheme.

In summary, if offsets are set in one direction on a two-way street, then the reverse direction is fixed. In a network, you can set any "open tree" of links, but links that close the tree already have their offsets specified.

The reader is advised to check the literature for the optimization programs in current use. TRANSYT [3] is used extensively for arterials and networks and PASSER II [2] for certain arterials. CORSIM [4], VISSIM [5],

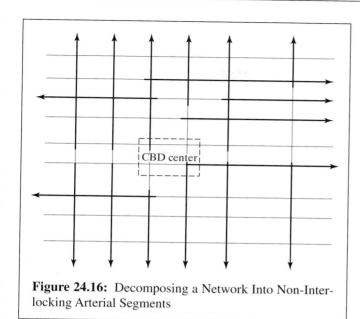

Figure 24.16: Decomposing a Network Into Non-Interlocking Arterial Segments

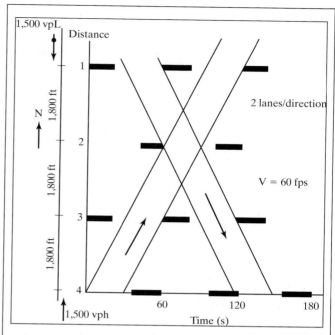

Figure 24.17: Case Study: Four Intersections with Good Two-Way Progression

and PARAMICS [6], to name a few, are used for network simulation.

24.5.3 Finding Compromise Solutions

The engineer usually wishes to design for maximum bandwidth in one direction, subject to some relation between bandwidths in the two directions. Sometimes, one direction is completely ignored. Much more commonly, the bandwidths in the two directions are designed to be in the same ratio as the flows in the two directions.

There are computer programs that do the computations for maximum bandwidth that are commonly used by traffic engineers as mentioned earlier. Thus, it is not worthwhile to present an elaborate manual technique herein. However, to get a feel for the basic technique and trade-offs, a small "by-hand" example will be shown.

Refer to Figure 24.17, which shows four signals and decent progression in both directions. For purposes of illustration, assume it is given that a signal with 50:50 split must be located midway between Intersections 2 and 3. Figure 24.18 shows the possible effect of inserting the new signal into the system. It would appear that there is no way to include this signal without destroying one or the other bandwidth, or cutting both in half.

To solve this problem, the engineer must move the offsets around until a more satisfactory timing plan develops. A change in cycle length may even be required.

Note that the northbound vehicle takes $3,600/60 = 60$ s to travel from Intersection 4 to Intersection 2, or—given $C = 60$ s—one cycle length. If the cycle length had been $C = 120$ s, the vehicle would have arrived at Intersection 2 at $C/2$, or one-half the cycle length. If we try the 120-s cycle length, a solution presents itself.

Figure 24.19 shows one solution to the problem, for $C = 120$ s, which has a 40-s bandwidth in both directions for an efficiency of 33%. The 40-s bandwidth can handle $(40/2.0) = 20$ vehicles per lane per cycle. Thus, if the demand volume is greater than $3,600(40)(2)/(2.0 \times 120) = 1,200$ veh/h, it will not be possible to process the vehicles nonstop through the system.

As indicated in the original information (see Figure 24.17), the northbound demand is 1,500 veh/h. Thus there will be some difficulty in the form of excess

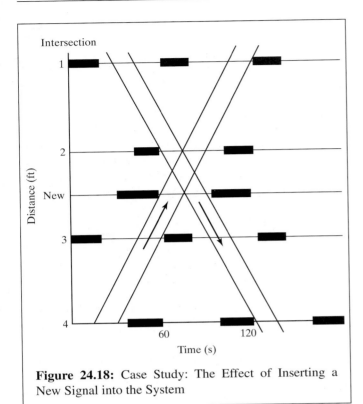

Figure 24.18: Case Study: The Effect of Inserting a New Signal into the System

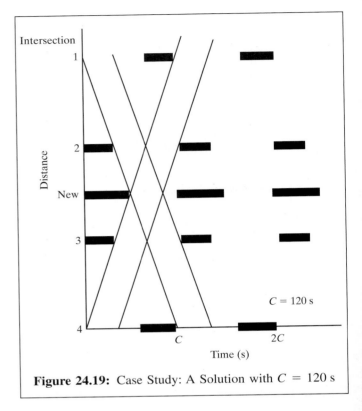

Figure 24.19: Case Study: A Solution with $C = 120$ s

vehicles in the platoon. They can enter the system but cannot pass Signal 2 nonstop. They will be "chopped off" the end of the platoon and will be queued vehicles in the next cycle. They will be released in the early part of the cycle and arrive at Signal 1 at the beginning of red. Figure 24.20 illustrates this, showing that these vehicles then disturb the next northbound through platoon.

Note that Figure 24.20 illustrates the limitation of the bandwidth approach when internal queuing arises, disrupting the bandwidth. The figure also shows the southbound platoon pattern, suggesting that the demand of exactly 1,200 veh/h might give rise to minor problems of the same sort at Signals 3 and 4.

If one were to continue a trial-and-error attempt at a good solution, it should be noted that:

- If the green initiation at Intersection 1 comes earlier in order to help the main northbound platoon avoid the queued vehicles, the southbound

platoon is released sooner and gets stopped or disrupted at Intersection 2.

- Likewise, shifting the green at Intersection 2 cannot help the northbound progression without harming the southbound progression.

- Shifting the green at Intersection 3 cannot help the southbound progression without harming the northbound progression.

- Some green can be taken from the side street and given to the main street.

- It is also possible that the engineer may decide to give the northbound platoon a more favorable bandwidth because of its larger demand volume.

This illustration showed insights that can be gained by simple inspection of a time-space diagram, using the concepts of bandwidth, efficiency, and an upper bound on demand volume that can be handled nonstop.

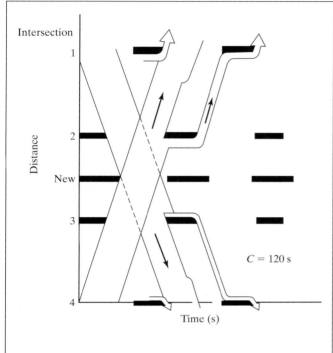

Figure 24.20: Case Study: Effect on Platoons with Demand Volume 1,500 Veh/h

24.6 Common Types of Progression

24.6.1 Progression Terminology

The sole purpose of this section is to introduce some common terminology:

- Simple progression
- Forward progression
- Flexible progression
- Reverse progression

Simple progression is the name given to the progression in which all signals are set so that a vehicle released from the first intersection will arrive at all downstream intersections just as the signals at those intersections initiate green. That is, each offset is the ideal offset, set by Equation 24-4 with zero queue. Of necessity, simple progressions are effective only on one-way streets or on two-way streets on which the reverse flow is small or neglected.

Because the simple progression results in a green wave that advances with the vehicles, it is often called a *forward progression*, taking its name from the visual image of the advance of the green down the street.

It may happen that the simple progression is revised two or more times in a day so as to conform to the direction of the major flow or to the flow level (since the desired platoon speed can vary with traffic demand). In this case, the scheme may be referred to as a *flexible progression*.

Under certain circumstances, the internal queues are sufficiently large that the ideal offset is negative; that is, the downstream signal must turn green before the upstream signal to allow sufficient time for the queue to start moving before the arrival of the platoon. Figure 24.21 has link lengths of 600 ft, platoon speeds of 60 ft/s, and internal queues averaging seven vehicles per lane at each intersection. The visual image of such a pattern is of the green marching upstream toward the drivers in the platoon. Thus it is referred to as a *reverse progression*. Figure 24.21 also illustrates one of the unfortunate realities of so

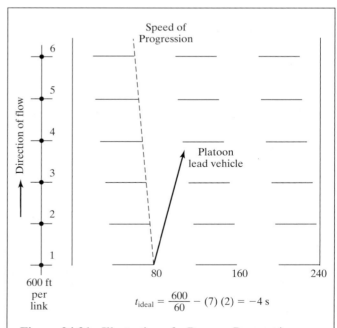

$$t_{ideal} = \frac{600}{60} - (7)(2) = -4 \text{ s}$$

Figure 24.21: Illustration of a Reverse Progression

many internal queued vehicles: the platoon's lead vehicle only gets to Signal 4 before encountering a red indication. As the platoon passes Signal 3, there are only 12 seconds of green to accommodate it, resulting in all vehicles beyond the sixth (i.e., 12/2 = 6) being cut off at Signal 3.

In the next several sections, common progression systems that can work extremely effectively on two-way arterials and streets are presented. As will be seen, these systems rely on having uniform block lengths and an appropriate relationship among block length, progression speed, and cycle length. Since achieving one of these progressions has major benefits, the traffic engineer may wish to set the system cycle length based on progression requirements, introducing design improvements at intersections where the system cycle length would not provide sufficient capacity. Rather than increase the system cycle length to accommodate the needs of a single intersection, redesign of the intersection should be attempted to provide additional capacity at the desired system cycle length.

24.6.2   The Alternating Progression

For certain uniform block lengths and all intersections with a 50:50 split of effective green time, it is possible to select a feasible cycle length such that:

$$\frac{C}{2} = \frac{L}{S} \qquad (24\text{-}8)$$

where: C = cycle length, s

L = block length, ft

S = platoon speed, ft/s.

In this situation, the progression of Figure 24.22 can be obtained. There is no limit to the number of signals that may be included in the progression.

The name for this pattern is derived from the "alternate" appearance of the signal displays: as the observer at Signal 1 looks downstream, the signals alternate—red, green, red, green, and so forth.

The key to Equation 24-8 is that the ideal offset in either direction (with zero internal queues) is L/S. That is, the travel time to each platoon is exactly one-half the

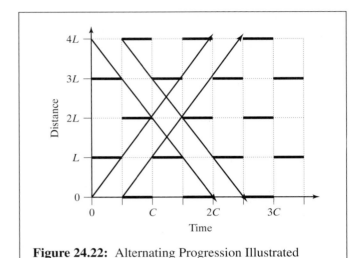

Figure 24.22: Alternating Progression Illustrated

cycle length, so that the two travel times add up to the cycle length.

The efficiency of an alternate system is 50% in each direction, because all of the green is used in each direction. The bandwidth capacity for an alternating progression is found using Equation 24-3 and noting that the bandwidth, BW, is equal to one-half the cycle length, C. If a saturation headway of 2.0 s/veh is assumed, then:

$$C_{BW} = \frac{3{,}600 * BW * L}{h * C} = \frac{3{,}600 * 0.5C * L}{2.0 * C} = 900L$$

where all terms are as previously defined. This is an approximation based on the assumed saturation headway of 2.0 s/veh. The actual saturation headway may be determined more accurately using the *Highway Capacity Manual's* procedure for intersection analysis.

Note that if the splits are not 50:50 at some signals, then (1) if they favor the main street, they simply represent excess green, suited for accommodating miscellaneous vehicles, and (2) if they favor the side street, they reduce the bandwidths.

As a practical matter, note the range of the block lengths for which alternating patterns might occur. Using Equation 24-8, appropriate block lengths are computed for platoon speeds of 30 and 50 mi/h (i.e., 44.0 and 73.3 ft/s), and cycle lengths of 60 and 90 s. The results are shown in Table 24.3. These results are illustrative; other combinations are clearly possible as well. All of these signal spacings imply a high-type arterial, often in a suburban setting.

24.6.3 The Double-Alternating Progression

For certain uniform block lengths with 50:50 splits, it is not possible to satisfy Equation 24-8, but it is possible to select a feasible cycle length such that:

$$\frac{C}{4} = \frac{L}{S} \qquad (24\text{-}9)$$

In this situation, the progression illustrated in Figure 24.23 can be obtained.

The key is that the ideal offset in either direction (with zero internal queues) over *two* blocks is one-half of a cycle length, so that two such travel times (one in each direction) add up to a cycle length. There is no limit to the number of signals that can be involved in this system, just as there was no limit with the alternate system.

The name of the pattern is derived from the "double alternate" appearance of the signal displays—that is, as the observer at Signal 1 looks downstream, the signals alternate in pairs: green, green, red, red, green, green, red, red, and so forth.

The efficiency of the double alternate signal system is 25% in each direction, because only half of the green is used in each direction. The upper limit on the bandwidth capacity may be approximated by assuming a 2.0 s/veh saturation headway and noting that the BW is one-quarter of C.

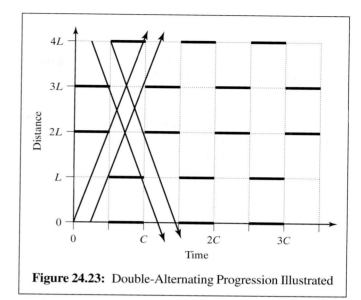

Figure 24.23: Double-Alternating Progression Illustrated

As with the alternate system, if the splits are not 50:50 at some signals, then (1) if they favor the main street, they simply represent excess green, suited for accommodating miscellaneous vehicles, and (2) if they favor the side street, they reduce the bandwidths.

$$C_{BW} = \frac{3{,}600 * BW * L}{h * C} = \frac{3{,}600 * 0.25C * L}{2.0 * C} = 450L$$

Table 24.4 shows some illustrative combinations of cycle length, platoon speed, and block lengths for which a double-alternating progression would be appropriate. Other combinations are, of course, possible as well. Some of these signal spacings represent a high-type arterial. With the shorter cycle lengths, however, some urban facilities could also have the necessary block lengths.

Table 24.3: Some Illustrative Combinations for Alternating Progression

Cycle Length	Platoon Speed	Matching Block Length
60 s	44.0 fps	1,320 ft
60 s	73.3 fps	2,199 ft
90 s	44.0 fps	1,980 ft
90 s	73.3 fps	3,299 ft

Table 24.4: Some Illustrative Combinations for Double-Alternating Progression

Cycle Length	Platoon Speed	Matching Block Length
60 s	44.0 fps	660 ft
60 s	73.3 fps	1,100 ft
90 s	44.0 fps	990 ft
90 s	73.3 fps	1,649 ft

24.6.4 The Simultaneous Progression

For very closely spaced signals or for rather high vehicle speeds, it may be best to have all the signals turn green at the same time. This is called a simultaneous system, since all the signals turn green simultaneously. Figure 24.24 illustrates a simultaneous progression.

The efficiency of a simultaneous system depends on the number of signals involved. For N signals:

$$EFF(\%) = \left[\frac{1}{2} - \frac{(N-1)*L}{S*C} \right] * 100 \quad (24\text{-}10)$$

For four signals with $L = 400$ ft, $C = 80$ s, and $S = 45$ ft/s, the efficiency is 16.7%. For the same number of signals with $L = 200$ ft, it is 33.3%.

Simultaneous systems are advantageous only under a limited number of special circumstances. The foremost of these special circumstances is very short block lengths. The simultaneous system has an additional advantage, however, that is not at all clear from a bandwidth analysis: under very heavy flow conditions, it forestalls breakdown and spillback. This is so because (1) it allows for vehicle clearance time at the downstream intersection where queues inevitably exist during heavy flow, and (2) it cuts platoons off in a way that

generally prevents blockage of intersections. This works to the advantage of cross traffic. Specific plans for controlling spillback under heavy traffic conditions are discussed in Section 24.7.

24.6.5 Insights Regarding the Importance of Signal Spacing and Cycle Length

It is now clear that:

- All progressions have their roots in the desire for ideal offsets.
- For certain combinations of cycle length, block length, and platoon speed, some very satisfactory two-way progressions can be implemented.
- Other progressions can be designed to suit individual cases, using the concept of ideal offset and queue clearance, trial-and-error bandwidth-based approaches, or computer-based algorithms.

A logical first step in approaching a system is simply to ride the system and inspect it. As you sit at one signal, do you see the downstream signal green but with no vehicles being processed? Do you arrive at signals that have standing queues but were not timed to get them moving before your platoon arrived? Do you arrive on the red at some signals? Is the flow in the other direction significant, or is the traffic really a one-way pattern, even if the streets are two way?

It is very useful to sketch out how much of the system can be thought of as an "open tree" of one-way links. This can be done with a local map and an appreciation of the traffic flow patterns. Referring to Figure 24.16, a distinction should be made among:

- Streets that are one way
- Streets that can be treated as one way, due to the actual or desired flow patterns
- Streets that must be treated as two way
- Larger grids in which streets (one way and two way) interact because they form unavoidable "closed trees" and are each important in that they

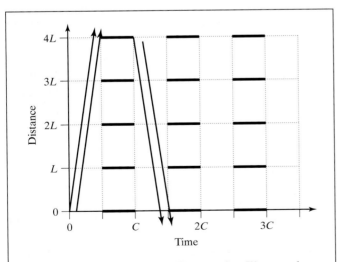

Figure 24.24: Simultaneous Progression Illustrated

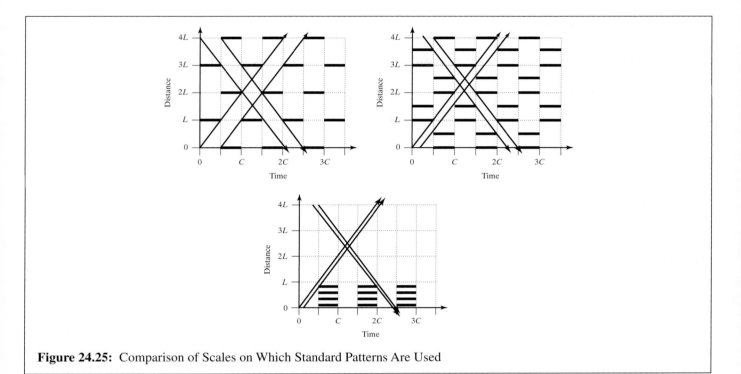

Figure 24.25: Comparison of Scales on Which Standard Patterns Are Used

cannot be ignored for the sake of establishing a "master grid" that is an open tree

- Smaller grids in which issue is not coordination but rather local land access and circulation, so that they can be treated differently (Downtown grids may well fall into the latter category, at least in some cases.)

The next most important issue is the cycle length dictated by the signal spacing and platoon speed. Attention must focus on the combination of cycle length, block length, and platoon speed, as shown earlier in this chapter.

Figure 24.25 shows the three progressions of the preceding sections—alternating, double alternating, and simultaneous—on the same scale. The basic "message" is that as the average signal spacing decreases, the type of progression best suited to the task changes.

Figure 24.26 illustrates a hypothetical arterial that comes from a low-density suburban environment with a larger signal spacing, into the outlying area of a city, and finally passes through one of the city's CBDs. As

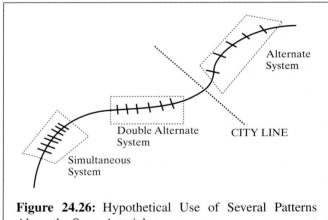

Figure 24.26: Hypothetical Use of Several Patterns Along the Same Arterial

the arterial changes, the progression used may also be changed, to suit the dimensions.[1]

Note that the basic lesson here is that a system can sometimes be best handled by breaking it up into several smaller systems. This can be done with good effect on even smaller systems, such as ten consecutive signals, of which a contiguous six are spaced uniformly and the other four also uniformly, but at different block lengths. Note that to the extent that block lengths do not exist perfectly uniformly, these plans can serve as a basis from which adaptations can be made. Note also that the suitability of the cycle length has been significant. It is often amazing how often the cycle length is poorly set for system purposes.

24.7 Coordination of Signals for Oversaturated Networks

It is well recognized that the oversaturated traffic environment is fundamentally different from the undersaturated environment. In undersaturated networks, capacity is adequate and queue lengths are generally well contained within individual approaches. On the other hand, the oversaturated environment is characterized by an excess of demand relative to capacity, unstable queues that tend to expand over time with the potential of physically blocking intersections (blockage, spillback), thus slowing queue discharge rates and, in effect, reducing capacity when it is most needed. Control policies for oversaturated networks thus focus on maintaining and fully exploiting capacity to maximize productivity (vehicle throughput) of the system by controlling the inherent instability of queue growth.

24.7.1 System Objectives for Oversaturated Conditions

When networks are congested, the explicit objectives change from minimize delay and stops to:

- *Maximize system throughput.* This is the primary objective. It is achieved by (a) avoiding queue spillback, which blocks intersections and wastes green time; (b) avoiding starvation, the tardy arrival of traffic at the stop-bar which wastes green time; and (c) managing queue formation to yield the highest service rate across the stop-bar.
- *Fully utilize storage capacity.* This objective seeks to confine congested conditions to a limited area by managing queue formation in the context of a "feed forward" system.
- *Provide equitable service.* Allocate service to cross-street traffic and to left-turners so that all travelers are serviced adequately and the imperative of traffic safety is observed.

Because intersection blockage can so degrade the network, its removal must be the prime objective of the traffic engineer. The overall approach can be stated in a logical set of steps:

- Address the root causes of congestion—first, foremost, and continually.
- Update the signalization, for poor signalization is frequently the cause of what looks like an incurable problem.
- If the problem persists, use novel signalization to minimize the impact and spatial extent of the extreme congestion.
- Provide more space by use of turn bays and parking restrictions.
- Consider both prohibitions and enforcement realistically—do they represent futile effort that will only transfer the problem?
- Take other available steps, such as right-turn-on-red, recognizing that the benefits will generally not be as significant as either signalization or more space.
- Develop site-specific evaluations where there are conflicting goals, such as providing local parking versus moving traffic, when the decision is ambiguous.

---

[1]Of course, if the flow is highly directional—as may well be from the suburbs in the morning—then these suggestions are superseded by the simple expedient of treating the streets as one-way streets and imposing a simple forward progression, with queue clearance if needed.

This list was constructed in Reference [7] with some allowance for ease of implementation: it is generally easier to change signalization than to remove urban parking; it is generally easier to treat spot locations than entire arterials; and so forth.

24.7.2   Signal Remedies

It is difficult to overstate how often the basic problem is poor signalization. Once the signalization is improved through reasonably short cycle lengths, proper offsets (including queue clearance), and proper splits, then many problems disappear. Sometimes, of course, there is just too much traffic. At such times *equity offsets*, to aid cross flow, and *different splits*, to manage the spread of congestion, may be appropriate if other options cannot be called upon. These options may be used as distinct treatments or as part of a *metering plan*.

Metering Plans

Three forms of metering can be applied within a congested traffic environment, characterized by demand exceeding supply (i.e., *v/c* deficiencies):

1. Internal
2. External
3. Release

Internal metering refers to the use of control strategies within a congested network so as to influence the distribution of vehicles arriving at or departing from a critical location. The vehicles involved are stored on links defined to be part of the congested system under control, so as to eliminate or significantly limit the occurrence of either upstream or downstream intersection blockage.

Note that the metering concept does not explicitly minimize delay and stops, but rather manages the queue formation in a manner that maximizes the productivity of the congested system.

Figure 24.27 and 24.28 show situations in which internal metering might be used: (1) controlling the volume being discharged at intersections upstream of a critical intersection (*CI*), thus creating a "moving storage" situation on the upstream links; (2) limiting the turn-in flow from cross-streets, thus preserving the arterial for its through flow; and (3) metering in the face of a backup from "outside."

External metering refers to the control of the major access points to the defined system, so that inflow rates into the system are limited if the system is already too congested (or in danger of becoming so).

External metering is convenient conceptually, because the storage problem belongs to "somebody else," outside the system. However, there may be limits to how much metering can be done without creating major

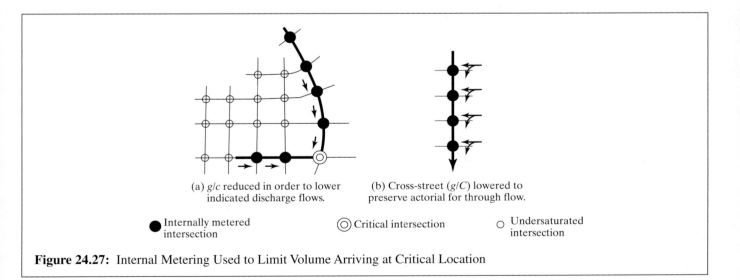

(a) *g/c* reduced in order to lower indicated discharge flows.

(b) Cross-street (*g/C*) lowered to preserve actorial for through flow.

● Internally metered intersection   ◎ Critical intersection   ○ Undersaturated intersection

Figure 24.27: Internal Metering Used to Limit Volume Arriving at Critical Location

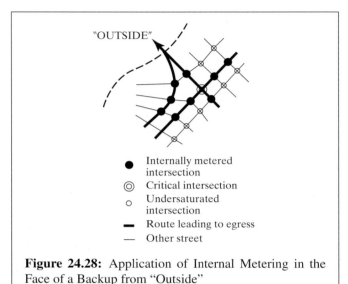

Figure 24.28: Application of Internal Metering in the Face of a Backup from "Outside"

problems in the "other" areas. Figure 24.29 shows a network with metering at the access points.

As a practical matter, there must be a limited number of major access points (e.g., river crossings, a downtown surrounded by water on three sides, a system that receives traffic from a limited number of radial arterials, and so forth). Without effective control of access, the control points can potentially be bypassed by drivers selecting alternative routes.

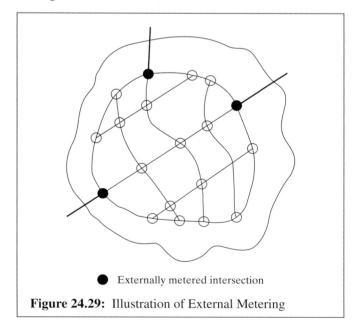

Figure 24.29: Illustration of External Metering

Release metering refers to cases in which vehicles are stored in such locations as parking garages and lots, from which their release can (at least in principle) be controlled. The fact that they are stored "off street" also frees the traffic engineer of the need to worry about their storage and their spillback potential.

Release metering can be used at shopping centers, megacenters, and other concentrations. While there are practical problems with public (and property-owner) acceptance, this could even be—and has been—a developer strategy to lower discharge rates so that adverse impacts are avoided. Such strategies are of particular interest when the associated roadway system is distributing traffic to egress routes or along heavily congested arterials.

Shorter Cycle Lengths

Chapter 17 demonstrated that increasing the cycle length does not substantially increase the capacity of the intersection. In fact, as the cycle length increases, so do the stored queue lengths and the length of discharged platoons. The likelihood of intersection blockage increases, with substantial adverse impacts on system capacity. This is particularly acute when short link lengths are involved.

Note that a critical flow of v_i nominally discharges $v_i C/3{,}600$ vehicles in a cycle. If each vehicle requires D ft of storage space, the length of the downstream link would have to be:

$$L \geq \left(\frac{v_i C}{3{,}600} \right) \times D \qquad (24\text{-}11)$$

where L is the available downstream space in feet. This may be set by the link length or by some lower value, perhaps 150 ft less than the true length (to keep the queue away from the discharging intersection or to allow for turn-ins).

Equation 24-11 may be rearranged as:

$$C \leq \left(\frac{L}{D} \right)\left(\frac{3{,}600}{v_i} \right) \qquad (24\text{-}12)$$

Note that v_i in this case is the discharge volume per downstream lane, which may differ from the demand

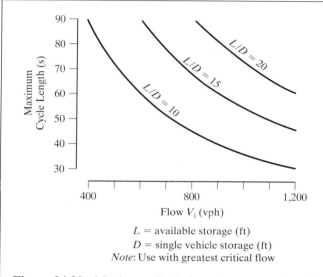

L = available storage (ft)
D = single vehicle storage (ft)
Note: Use with greatest critical flow

Figure 24.30: Maximum Cycle Length as a Function of Block Spacing

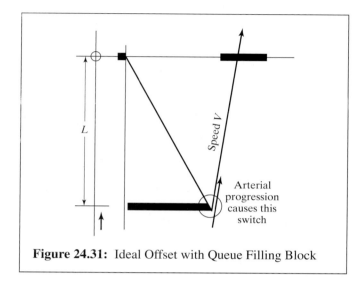

Figure 24.31: Ideal Offset with Queue Filling Block

volume, particularly at the fringes of the "system" being considered. Refer to Figure 24.30 for an illustration of this relationship. Note that only rather high flows (≥ 800 veh/h/ln) and short blocks will create very severe limits on the cycle length. However, these are just the situations of most interest for conditions of extreme congestion.

Equity Offsets

Offsets on an arterial are usually set to move vehicles smoothly along the arterial, as is logical. If no queues exist on the arterial, the ideal offset is L/S, where L is the signal spacing in feet, and S is the vehicle speed in feet per second. If a queue of Q vehicles exist, the ideal offset t_{ideal} is:

$$t_{\text{ideal}} = \frac{L}{S} - Q * h \qquad (24\text{-}13)$$

where h is the discharge headway of the queue, in seconds. Clearly, as Q increases, t_{ideal} decreases, going from a "forward" progression to a "simultaneous" progression to a "reverse" progression.

Unfortunately, as the queue length approaches the block length, such progressions lose meaning, for it is quite unlikely that both the queue *and* the arriving vehicles will be passed at the downstream intersection. Thus the arrivals will be stopped in any case.

At the same time, the cross-street traffic at the upstream intersection is probably poorly served because of intersection blockages. Figure 24.31 illustrates the time at which t_{ideal} would normally cause the upstream intersection to switch to green (relative to the downstream intersection).

Consider the following case, illustrated in Figure 24.32: Allow the congested arterial to have its green at the upstream intersection until its vehicles just begin to move; then switch the signal, so that these vehicles flush out the intersection, but no new vehicles continue to enter.

At the same time, this gives the cross-street traffic an opportunity to pass through a clear intersection. This concept, defined as *equity offset*, can be translated into the equation:

$$t_{equity} = g_i C - \frac{L}{S_{\text{acc}}} \qquad (24\text{-}14)$$

where g_i is the *upstream* main street (i.e., the congested intersection) green fraction, and S_{acc} is the speed of the "acceleration wave" shown in Figure 24.33.

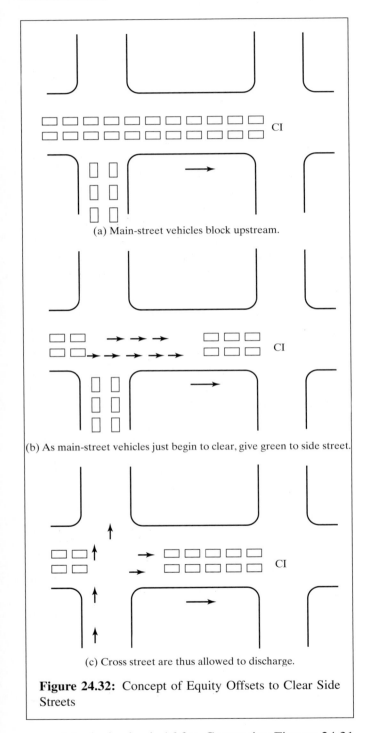

(a) Main-street vehicles block upstream.

(b) As main-street vehicles just begin to clear, give green to side street.

(c) Cross street are thus allowed to discharge.

Figure 24.32: Concept of Equity Offsets to Clear Side Streets

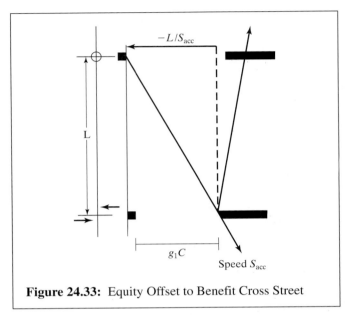

Figure 24.33: Equity Offset to Benefit Cross Street

would have caused it to switch to green in this particular case. This is not surprising, for the purpose is different—equity offsets are intended to be fair (i.e., equitable) to cross-street traffic.

Simulation tests using a microscopic simulation model have shown the value of using equity offsets: congestion does not spread as fast as otherwise and may not infect the cross streets at all.

Figure 24.34(a) shows a test network used to test the equity-offset concept. Link 2 is upstream of the critical intersection (*CI*). For the demands and signal splits shown, it is likely to accumulate vehicles, with spillback into its upstream intersection likely. If this occurs, the discharge from Link 1 will be blocked, and its queue will grow. In the extreme, congestion will spread.

The equity offset is computed as

$$t_{equity} = (0.60)(60) - \frac{600}{16} = -1.5 \text{ s}$$

using Equation 24-14. (At 25 ft per vehicle and a platoon speed of 50 ft/s, Equation 24-4 would have yielded $t_{ideal} = (600/50) - (24)(2) = -36$ s for progressed movement. Of course, progressed movement is a silly objective when 24 vehicles are queued for 30 s of green.)

A typical value is 16 fps. Comparing Figures 24.31 and 24.33, it is clear that equity offset causes the upstream signal to go red just when "normal" offsets

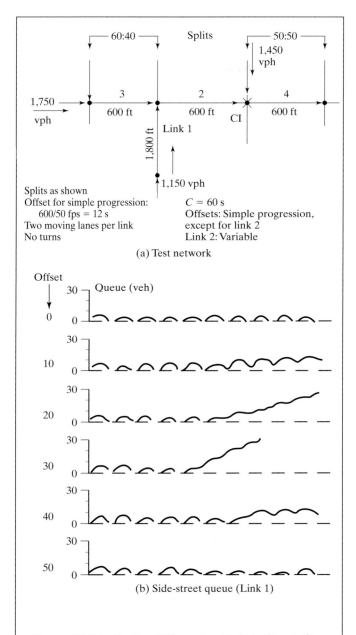

Splits as shown
Offset for simple progression:
600/50 fps = 12 s
Two moving lanes per link
No turns

$C = 60$ s
Offsets: Simple progression,
except for link 2
Link 2: Variable

(a) Test network

(b) Side-street queue (Link 1)

Figure 24.34: Equity Offsets Avoid Side-Street Congestion, Despite Spillbacks

Figure 24.34(b) shows the side-street queue (i.e., the Link 1 queue) as a function of the main-street offset. Note that an offset of -36 s is the same as an offset of $+24$ s when $C = 60$ s, due to the periodic pattern of the offsets. Figure 24.34(b) shows the best result for allowing the side street to clear when the equity offset (offset $= -1.5$ s) is in effect, and, in this case, the worst results when the queue-adjusted "ideal offset" (offset $= 24$ s) would have been in effect.

The foregoing discussion assumes that the cross-street traffic does not turn into space opened on the congested arterial. If a significant number of cross-street vehicles do turn into the arterial, a modification in the offset is appropriate to assure that the upstream traffic on the congested arterial also has its fair share.

The equity offset concept has been used to keep side-street flows moving when an arterial backs up from a critical intersection (CI). It may also be used to keep an arterial functioning when the cross streets back up across the arterial from their critical intersections.

Imbalanced Split

For congested flow, the standard rule of allocating the available green in proportion to the relative demands could be used, but it does not address an important problem. Consider the illustration of Figure 24.35. If the prime concern is to avoid impacting Route 347 and First Avenue (but with little concern for the minor streets in between, if any), it is not reasonable to use a 50:50 split.

Considering that the relative storage available is 750 ft in one direction and 3,000 ft in the other and we wish neither to be adversely affected, the impact could be delayed for the longest time by causing the excess-vehicle queue to grow in proportion to the available storage. The two critical-lane discharge flows f_i would have to be set such that:

$$\frac{d_1 - f_1}{d_2 - f_2} = \frac{L_1}{L_2} \tag{24-15}$$

and:

$$f_1 + f_2 = CAP \tag{24-16}$$

where d_i are the demands (veh/h/ln), L_i is the storage, and CAP is the sum of the critical-lane flows (i.e., the capacity figure).

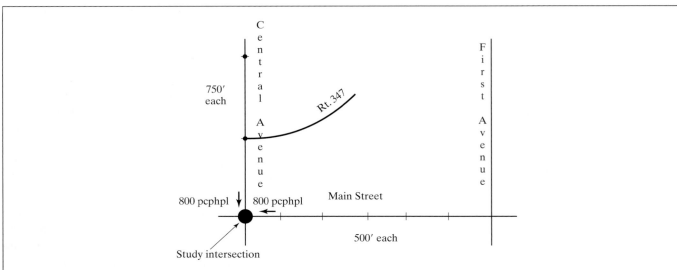

Figure 24.35: An Illustration of Split Determination (Used with permission of Transportation Research Board, "Traffic Determination in Oversaturated Street Networks," *NCHRP Report 194*, pg. 123, Washington DC.)

For the illustrative problem, using $CAP = 1,550$ veh/h/ln, the previous equations result in $f_1 = 790$ veh/h/ln and $f_2 = 760$ veh/h/ln, where direction 1 is the shorter distance. This is a 51:49 split.

Note that in the extreme, if only one direction has a cross route that should not be impacted, much of the green could be given to that direction (other than some minimum for other phases) in order to achieve that end.

24.8 Computer-Controlled Traffic Systems

Computers are now used to control traffic signals along arterials and in networks in most modern cities throughout the world. This section acquaints the reader with the basic issues and concepts involved in computer control of surface street traffic.

With the current emphasis on Intelligent Transportation Systems (ITS), computer control of systems is now classified as Advanced Transportation Management Systems (ATMS) and the control centers themselves as Transportation Management Centers (TMCs). However, the existence of such systems predates the term "ITS" by a few decades.

24.8.1   System Characteristics

Figure 24.36 shows the most basic concept: the computer sends out commands that control the signals along one or more arterials. There is no "feedback" of information from detectors in the field, and the traffic signal plans are not "responsive" to actual traffic conditions.

How are the plans for such a system developed? They are typically generated "off line" based on earlier traffic data. They are not developed in "real time" by a high-speed computer based on the latest traffic data. The major advantages of this "limited" system include:

1. *Ability to update signals from a central location.* As a practical matter, before computerized systems, signals were not updated in many areas simply because sufficient staff does not exist to do the task. Sending one or two people along an entire arterial to re-time the signals individually at each intersection is a time-consuming task. The ability to do this from a central location opened the potential for getting a necessary task done conveniently.

2. *Ability to have multiple plans and special plans.* In many localities a three-dial controller

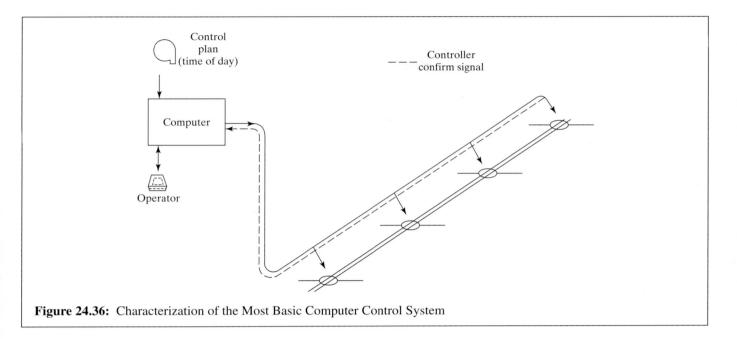

Figure 24.36: Characterization of the Most Basic Computer Control System

is quite sufficient: if traffic is generally regular, three basic plans (AM peak, PM peak, off peak) can meet most needs. The computer opens the possibility to have an N-dial controller, with special plans stored for Saturday, Sunday, severe rain, games at the local stadium, and so forth. With appropriate plans stored for each such event, the plans can be "called up" by time of day or by operator intervention. The plans can be said to be stored in a "library" and called up as needed, most often by time of day.

3. *Information on equipment failures.* The early systems simply took control of electromechanical controllers, driving the cam-shaft from the central computer and receiving a confirmation signal. Failure to receive this confirmation signal indicated trouble, perhaps the controller "hanging up" temporarily. The information provided by the control computer allowed such failures to be detected and repair crews dispatched.

4. *Performance data on contractor or service personnel.* With a failure detected and notification made, the system can log the arrival of the crew

and/or the time at which the intersection is returned to active service.

These advantages alone can greatly enhance the service provided to the public.

24.8.2 Collection and Use of Data

Figure 24.37 shows the preceding system, but with the refinement that detectors in the field are feeding information back to the central location. The refinement makes use of the computer's ability to receive great amounts of information and to process it. However, the information is not being used in an "on-line" setting; it still does not influence the current plan selection.

In any case, this enhancement is still useful, for it provides the basic data for doing the off-line development of up-to-date plans, for identifying seasonal and other adjustments, and for observing general growth in the traffic level and/or changes in the pattern. Typically, the computer is being used as the tool for collection of permanent or long-term count data.

Computer control systems can also make use of traffic data to aid in plan selection. This may be done in one of three principle ways:

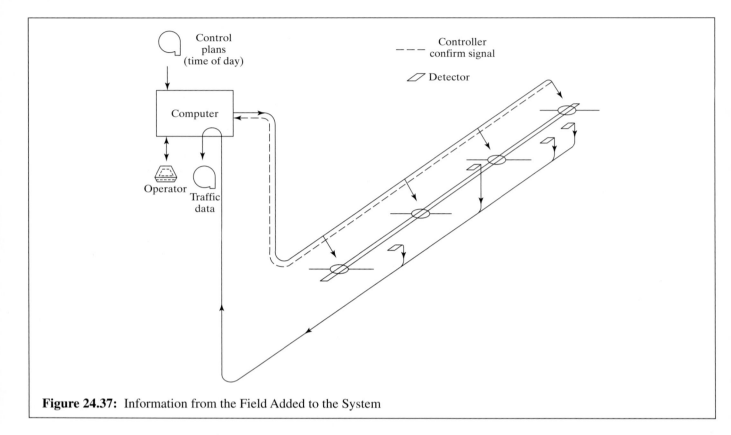

Figure 24.37: Information from the Field Added to the System

1. *Use library; monitor deviations from expected pattern.* This concept uses a time-of-day approach, looking up in a library both the expected traffic pattern and the preselected plan matched to the pattern. The actual traffic pattern (as estimated from the field data) can be compared to the expected, and if a deviation occurs, the computer can then look through its library for a closer "match" and use the appropriate plan. The check can be done before putting the plan into effect, or there can be periodic checking to evaluate whether the current plan should be superseded.

2. *Use library; match plan to pattern.* This is a variation on the first concept, with the observed pattern being matched to the most appropriate prestored pattern and the corresponding plan being used. The feature missing is the history of what is *expected* to be happening, based on time of day.

3. *Develop plan on line.* This concept depends on the ability to do the necessary computations within a deadline either as a "background" task or on a companion computer dedicated to such computations. This approach is possible due to the increased computing power and memory that now far exceeds that available even a few years ago, at radically lower costs.

There are limitations to how "responsive" a system of traffic signals can be. This is because the implementation of a new system cycle length and system of offsets cannot be implemented instantaneously. Rather, the transition may take some time, as offsets are "rolled" forward from key signal locations. During the transition, offset patterns are disrupted, and large queues may begin to form. In general, it is not feasible to implement many different coordination patterns within a short period of time.

24.8.3   An Overview of Modern Systems

Traffic control with the use of modern digital computers is now commonplace. Toronto was one of the pioneers [8] in the early 1960s, as were Glasgow [9], Munich [10], San Jose [8], and Wichita Falls [8]. By the late 1970s, FHWA had identified some 200 such systems in the United States [11], and the pace has continued.

In the 1970s, a testbed was established in Washington DC, and three generations of Urban Traffic Control Systems (UTCS) were undertaken [12]. There generations were, respectively, targeted for control with (1) off-line computation, updates every 15 minutes, based on traffic conditions; (2) on-line computation, updates every 10–15 minutes; (3) on-line computation, updates every 3–6 minutes. The third generation was tested but not implemented in the testbed.

It is now common to have imaging systems that create multiple "virtual" detectors by use of video and other cameras [e.g., 13] and algorithms that depend upon intelligent scanning of the pavement as well as objects in motion.

These systems open the door to direct observation of queues, classification of vehicles, and many more observation points for speed and flow, including lane-by-lane observation. A new control policy can reasonably anticipate these and other variables being observable, at least at key locations. At the same time, there is a large installed base of existing systems, most of which depend upon a limited number of conventional loop detectors. It is unlikely that these systems can easily add a new generation of traffic detectors and integrate them into the system. It is therefore important that a new control policy *not* be predicated upon the use of such state-of-the-art detectors, but rather that it can add value even to existing systems.

The Toronto System

The Toronto computer control system stands alone in its scale, pioneering effort, and success [8]. Consider the following:

- It was the first computer control system, installed in 1963 and subject to all the risks of a pioneering effort.

- There were 864 intersections involved.

- The capital outlay was recovered within six months, in terms of user benefits.

- The Metropolitan Department of Roads and Traffic invested energy and professional staff in a continual upgrading and evaluation of improved control policies.

- In cooperation with other agencies within Canada, the Metropolitan Department of Roads and Traffic undertook a major effort in 1974–1976 to "study, develop, test, and evaluate existing and new traffic signal control strategies, including both off-line signal optimization techniques and real-time computer traffic responsive control concepts."

The last effort cited resulted in a series of three very substantial reports on the state of the art [14], the evaluation of off-line strategies [15], and the development and evaluation of a real-time computerized traffic control strategy [16].

The evaluation of off-line optimization programs in the 1974–1976 work used three computer programs judged to be the best available at the time: (1) SIGOP, (2) TRANSYT, and (3) a combination method. These are described in References [17], [18], and [19], respectively.

The SIGOP program used an offset optimization algorithm to minimize the discrepancy between the actual offsets and a set of ideal offsets (calculated by the program or specified by the user). The TRANSYT program optimizes splits and offsets by minimizing a "network performance index" that is the sum of weighted link delays and stops, employing a traffic flow model to generate dispersion. The combination method determined a set of signal offsets to optimize a combination of total delay and stops, given a set of splits.

After an extensive evaluation project, the report [15] concluded:

> It is doubtful whether the floating car speed and delay survey is an appropriate technique for accurately measuring on-street levels of performance ... the field sampling rate was necessarily limited. ...

The use of system rates and travel times led to somewhat inconclusive results, primarily due to the fact that aggregated system data tends to conceal fluctuations of volume and travel time on individual links. . . .

The link performance evaluation technique seemed to have produced the most conclusive results.

These quotes are extracted because of the powerful lessons contained in them, drawn from one of the most successful computer control projects in existence: (1) it was not feasible to get enough travel-time runs to identify differences, if they existed, and (2) the system-level aggregated statistics hid some variations simply by averaging them out.

In a related paper [20], it is stressed that:

Whichever optimization program is chosen to design urban network signal settings, it is imperative that the user have a thorough understanding of the selected program and also a comprehensive knowledge of the signal system. In addition, a commitment must be made to carefully review the program output to ascertain its validity. While these off-line programs can be utilized as engineering aids in network signal setting design, they should not be used as replacements for engineering judgment and expertise.

The three programs used all performed comparably to the base condition. However, it was noted that TRANSYT had greater successes elsewhere and that:

The superiority of TRANSYT was not evident from the results in Toronto. This is perhaps due to the fact that the program was used without prior calibration of some of the program parameters for local conditions (such as the smoothing factor used in the platoon dispersion model) . . . There is sufficient reason to expect that it will perform much better if the program is calibrated.

This last quote is singled out because of the stress on checking the validity of the "default values" or parameters in such models. Given the common use of such tools as TRANSYT, it is almost certain that most users do not do this systematically.

Note that although the Toronto test program did not calibrate TRANSYT to the specific locale, the results were comparable to the well-refined existing signal plan. Thus, while other users may not get the full benefit without local calibration, they will still get rather good results—if their experience is as fortunate as that in Toronto.

The Toronto work also involved the development and evaluation of an on-line optimization program known as RTOP (real-time optimization program), which is of the "second-generation" type. (This term will be defined in the following section; see Table 24.5 for the characteristics of each "generation" of control.) RTOP used a 15-min control period. The prediction algorithm uses information from three sources: historic trends, current trends, and a directional relationship (the last one is based on volume trend on a "representative" link). Further detail is contained in Reference [16].

In the two areas in which RTOP was tested (one central, one suburban), it surpassed the existing signal plans in the central area but was much less effective in the suburban area. The factors judged to be contributory were (1) the traffic fluctuations were sharper in the suburban area, making it harder to follow a trend, given the 15-min period used; (2) the detector density was less in the suburban area, providing less accurate volume estimates; and (3) the larger number of signals in the suburban area led to greater computation times, sometimes exceeding the available time and thus causing the last policy to be "held over" for one more period.

The Urban Traffic Control System

The Urban Traffic Control System (UTCS), another landmark project [12] was an FHWA demonstration and research project in traffic control systems. The stated objectives were to:

- Develop and test, in the real world, new computer-based control strategies that would improve traffic flow.

- Document system planning, design, installation, operation, and maintenance to assist traffic engineers with installing their own systems.

- Simulate modernization of traffic control equipment.

Table 24.5: Comparison of Key Features: Three Generations of Control

Feature	First Generation	Second Generation	Third Generation
Optimization	Off-line	On-line	On-line
Frequency of Update	15 minutes	5 minutes	3–6 minutes
No. of Timing Patterns	Up to 40 (7 used)	Unlimited	Unlimited
Traffic Prediction	No	Yes	Yes
Critical Intersection Control	Adjusts split	Adjusts split and offset	Adjusts split, offset, and cycle
Hierarchies of Control	Pattern selection	Pattern computation	Congested and medium flow
Fixed Cycle Length	Within each section	Within variable groups of intersections	No fixed cycle length

(From "The Urban Traffic Control System in Washington, DC," US Department of Transportation, Federal Highway Administration, 1974.)

In 1972 the system was initiated, with 114 intersections under control. This was extended by 1974 to 200 intersections. The plan was to demonstrate and evaluate three "generations" of control, each more sophisticated than its predecessor. The three generations are as follows:

- *First generation*—uses a library of prestored timing plans, each developed with off-line optimization programs. The plan selected can be based on time of day (TOD), measured traffic pattern (traffic responsive, or TRSP), or operator specification. The update period is 15 minutes. First generation allows critical intersection control (CIC) and has a bus priority system (BPS). For further information on the BPS, see References [21 and 22]. As is necessary, first generation has a signal transition algorithm.

- *Second generation*—uses timing plans computed in real time, based on forecasts of traffic conditions, using detector observations input into a prediction algorithm.

- *Third generation*—was conceived as a highly responsive control, with a much shorter control period than second generation and without the restriction of a cycle-based system. Third generation included a queue management control at critical intersections (CIC/QMC). The third generation was not implemented due in large part to

the detector requirements, which exceeded the level present in the testbed.

Table 24.5 summarizes the principal features of the three generations of control. This terminology—"first generation" and so forth—has become pervasive in the discussion of traffic control systems.

One of the very interesting products of the overall UTSC project is the microscopic traffic simulation program known as TRAF-NETSIM (named "UTCS-1" in its original form). This simulation program is now very commonly used, incorporated into the well-known program CORSIM.

By 1984, some 40 cities had, or were planning to have, UTCS-based systems [8].

Other Projects Throughout the World

There have also been other significant projects undertaken that will be discussed in this section.

Great Britain More recent developments implemented in Glasgow and Coventry have concentrated on adaptive control. The method known as SCOOT (split, cycle, and offset optimization technique) is designed to minimize congestion [23]. The basic philosophy of this method is adjustment of signal timing in small, frequent intervals. SCOOT uses the signal-optimization logic previously developed for the TRANSYT signal-timing program [24]. Detector data is stored in the computer in terms of "cyclic flow profiles" [25]—that is, histograms of traffic-flow variation over a signal cycle. These "profiles" are used to develop a timing pattern that achieves the optimum degree of coordination. Detectors are located at the upstream end of key links; the measured occupancy of these detectors indicates queue lengths that may become critical. This information is then used to adjust signal timing so as to reduce the likelihood of the queue's blocking the upstream junctions. The TRRL report on the SCOOT program concludes:

> At its present stage of development, the traffic responsive SCOOT method of signal coordination is likely to achieve savings in delay which average about 12% compared to control by a high standard of up-to-date fixed time plans..... SCOOT is likely to be most effective where traffic demands are heavy and approach the maximum capacity of the junctions where the demands are variable and unpredictable and where the distances between junctions are short. SCOOT is likely to give further benefits compared to fixed time plans which, as is often the case, are out of date. Although the evidence is limited, it would appear that the delay reductions achieved by SCOOT are likely to double from 12 to over 20% when the fixed time plans are from 3 to 5 years old.

Australia   Australia shares, with the United States and England, the lead in theoretical and conceptual development for network computer traffic control under congested conditions. For instance, A.J. Miller [26] has developed some of the basic approaches for both intersection and network control. Although originally developed for undersaturated conditions, the method of approach has been found useful in the development of control strategies for the congested regime. The most significant development in the implementation of computer control signal systems in Australia has been in Sydney.

The Sydney Coordinated Adaptive Traffic (SCAT) system is a distributed-intelligence, three-level, hierarchical system using microprocessors and minicomputers [27]. The system is capable of real-time adjustment of cycle, split, and offset in response to detected variations in demand and capacity. It is designed to calibrate itself.

For control purposes, the total system is divided into a large number of comparatively small subsystems varying from one to ten intersections. This system configuration is in software. As far as possible, the subsystems are chosen to be traffic entities, and for many traffic conditions they will run without relation to each other. As traffic conditions demand, the subsystems "marry" with adjacent subsystems to form a number of larger systems or one large system. This "marriage" of subsystems is calculated in much the same way as are the interrelationships between intersections within a subsystem. Thus, there is a hierarchy of control as distinct from a hardware hierarchy.

Recent improvements in the control algorithms are designed to lead to improved real-time evaluation of both changes in offset patterns and the marriage of adjacent subsystems. The effects of possible changes in both intra- and inter-subsystems' offset patterns are evaluated using actual volumes and bandwidth parameters based on progression at free speeds [28].

If the degree of saturation exceeds a preset level (i.e., congested flow), cycle length is increased, with the additional time assigned to the phase showing saturation. At the same time, coordination will be forced into operation if congestion due to critical queues is imminent. There is no direct critical-queue detection; the system philosophy depends on detectors near the stop line. Critical queues are inferred from the detection of excessive decreases in flow at upstream intersections.

Spain   Two-level hierarchical, fully adaptive computerized traffic control systems have been implemented in Barcelona [29] and Madrid [30]. At the higher level, cycles are selected, subareas defined, and coordination sequences established. At the lower level, within each subarea, offsets and splits are computed. Maximum bandwidth techniques [31] are used for undersaturated conditions. For saturated (congested) traffic flow, delays and stops are optimized by an algorithm that assumes equal distribution of traffic over the entire green part of the signal cycle for every intersection. The algorithm selects optimum offsets for every pair of intersections and the associated split.

A check is made on resulting queue length and on the speed of backward propagation of the queue to determine possible conflicts (blockage). If such is the case, the optimal offset is modified iteratively, until an offset is found that results in the longest permissible queue. Signal splits are optimized through a minimum-delay criterion and adjusted to reflect route and network effects. Green time may thus be reduced to diminish the probability of excessive queues at downstream intersections or when excessive queues are noted by occupancy detectors.

Japan The world's largest computer-controlled signal system—ultimate size, 8,000 intersections—is installed in Tokyo [*32*]. Large-scale systems are installed in other Japanese cities: the Osaka and Nagoya systems control well in excess of 1,000 intersections [*33*]. These systems use a number of different control criteria ranging from minimizing stops under light traffic conditions to maximizing capacity when traffic demand is heavy. The control mode is assigned on the basis of the degree of congestion from detector volume and occupancy data. Splits are selected on the basis of volume ratios and modified so as to balance queue lengths on all approaches. Offsets are adjusted in response to the derivative of delay with respect to offset. The computer-controlled signal system is a part of an integrated TSM (Transportation Systems Management) system that also includes advisory routing information using radio, telephone, and changeable message signs.

Germany The major effort at computerized network signal control in Germany is the PBIL system installed in Aachen [*34*]. The PBIL system operates by minimizing delay, which is defined as the difference between actual travel time and expected travel time under free-flow conditions. When critical queue lengths on the approaches to critical intersections are exceeded, the delay for the critical approach input into the optimizing algorithm is artificially increased and the critical intersection and the immediately adjacent related one (star network) are taken out of the system. The control algorithm then determines splits and offsets so as to minimize delay for this network.

France Although a great deal of theoretical development work on congestion control has been done in France (e.g., *35*), only one network computer-controlled signal system has been installed. This system, in Bordeaux, uses 1-GC control but is noteworthy for the extensive amount of traffic metering included [*36*].

24.8.4 Adaptive Signal Control

Over the last decade or so, FHWA has focused on real-time traffic adaptive control systems (RT-TRACS) for arterials and networks, all of which seek to satisfy an objective function usually expressed in terms of vehicle delay and stops.

SCOOT and SCATS, discussed earlier, are examples of adaptive signal control. PB Faradyne developed OPAC (optimized policies for adaptive control), an on-line signal timing algorithm. While all of these systems have relief of congestion as a primary objective, existing RT-TRACS are not designed to expressly manage oversaturated traffic environments. Much of the current research, therefore, is now focusing on oversaturated conditions.

Real-time policies can adjust green time every cycle, yet they try to maintain full productivity without adjusting offsets, thereby creating a signal transition problem. The only time it is necessary to change the offsets is when important changes in the traffic environment take place, such as (1) volume changes that cause undersaturated approaches to become saturated (or vice versa); (2) incidents occur, (3) travel patterns change considerably, or (4) changes are necessary to provide equitable service. For any of these changes, the system would determine a new base signal-timing plan. The green times of this new policy can then be adjusted, as frequently as required, until the next update becomes necessary.

24.9 Closing Comments

This chapter has introduced the basic considerations and concepts of signal coordination for undersaturated flows on one-way and two-way arterials and in networks. This chapter also addressed oversaturated conditions on surface streets. The problem of congestion and over saturation is widespread and is not often approached in any consistent manner. Definite measures can be taken, but preventive action addressing the root causes must be given a high priority. Among the possible measures, those relating to signalization generally can have the greatest impact. The non-signal remedies are in no way to be minimized, particularly those that provide space either for direct productivity increases or for removing impediments to the principal flow.

References

1. *Highway Capacity Manual*, Transportation Research Board, Washington DC, 2000.

2. *PASSER II-90 Microcomputer User's Guide*, distributed by the McTrans Center, Gainsville, Fl, 1991.

3. *TRANSYT-7F Users Guide, Methodology for Optimizing Signal Timing* (MOST) Volume 4, USDOT, Federal Highway Administration, Washington DC, 1991.

4. *TSIS User's Guide*, Version 4.0, *CORSIM User Guide*, Federal Highway Administration, Washington, DC, 1996.

5. VISSIM, PTV Vision Software Suite, Innovative Transportation Concepts, Inc., Seattle, WA, 2001.

6. Paramics Microscopic Simulation Software, Version 4.0, Quadstone, Limited, 2000.

7. "Traffic Control in Oversaturated Street Networks," *National Cooperative Highway Research Program*, (NCHRP) *Report 194*, Transportation Research Board, National Research Council, Washington DC, 1985.

8. Stockfish, C.R., *Selecting Digital Computer Signal Systems*, USDOT, FHWA Report FHWA-RD-72-20, Washington DC,1972.

9. Halroyd, J. and Hillier, J.A., *Area Traffic Control in Glasgow*, Traffic Engineering and Control, September, 1969.

10. Bolke, W., *Munich's Traffic Control Centre*, Traffic Engineering and Control, September, 1967.

11. CCSAG Newsletter, Institute of Transportation Engineers, Vol. 2, No. 4, December 1979.

12. *The Urban Traffic Control System in Washington*, DC., USDOT, FHWA, September 1974.

13. Autoscope System, Image Sensing Systems, Inc., St Paul MN.

14. *Improved Operation of Urban Transportation Systems*: Volume 1, *Traffic Signal Control Strategies, A State of the Art*, Metropolitan Toronto Department of Roads and Traffic, Toronto, Canada, March 1974.

15. *Improved Operation of Urban Transportation Systems*: Volume 2, *The Evaluation of Off-Line Area Traffic Control Strategies*, Metropolitan Toronto Department of Roads and Traffic, Toronto, Canada, November 1975.

16. *Improved Operation of Urban Transportation Systems*: Volume 3, *The Development and Evaluation of a Real-Time Computerized Traffic Control Strategy*, Metropolitan Toronto Department of Roads and Traffic, Toronto, Canada, November 1976.

17. *SIGOP: Traffic Signal Optimization Program, User's Manual*, NTIS PB-182-835, U.S. Bureau of Public Roads, 1968.

18. Robertson, D.I., "TRANSYT: A Traffic Network Study Tool," Report LR 253, Road Research Laboratory, Crowthorne, Berkshire, U.K., 1969.

19. Huddart, K. and Turner, E., "Traffic Signal Progressions—GLC Combination Method," *Traffic Engineering and Control*, Vol. 11, No. 7, November 1969.

20. Rach, L., *et al.*, "An Evaluation of Off-Line Traffic Signal Optimization Techniques," prepared for the 54th Annual TRB Meeting, Washington DC, 1975.

21. Raus, J., "Urban Traffic Control/Bus Priority System (UTCS/BPS): A Status Report," *Public Roads*, Vol. 38, No. 4. March 1975.

22. MacGowan, J. and Fullerton, I.J., "Development and Testing of Advanced Control Stategies in the Urban Traffic Control System," *Public Roads*, Vol. 43, No. 3, December 1979.

23. Hunt, P.B., *et al.*, "SCOOT—A Traffic Responsive Method of Coordinating Signals," *TRRL Report 1014*, Road Research Laboratory, Crowthorne, Berkshire, England, 1981.

24. Robertson, D.I., "TRANSYT: A Traffic Network Study Tool," *Report LR 253*, Road Research Laboratory, Crowthorne, Berkshire, U.K., 1969.

25. Robertson, D.I., "Cyclic Flow Profiles," *Traffic Engineering and Control*, Vol. 15, No. 4, June 1974.

26. Miller, A.J., "A Computer Control System for Traffic Networks," *Proceedings of the Second International Symposium on the Theory of Traffic Flow*, London, U.K., 1963, OECD, Paris, France, 1965.

27. Sins, A.G. and Dobinson, K.W., "S.C.A.T.—The Sydney Coordinated Adaptive Traffic System—Philosophy and Benefits," in *Proceedings of the International Symposium on Traffic Control Systems*, Vol. 2B, University of California, Berkeley, CA, 1979.

28. Luk, J.Y.K. and Sims, A.G., "Selection of Offsets for Sub-Area Linkage in SCATS," *Australian Road Research*, Vol. 12, No. 2, June 1982.

29. Garcia Roman, J., Lopez Montejano, A., and Sanchez Hernandez, D., "New Area Traffic Control System in the City of Barcelona," in *Traffic Control and Transportation Systems*, American Elsevier Publishing, New York, NY, 1974.

30. Guehrer, H.H., "Area Traffic Control—Madrid," *Proceedings of the First International Symposium on Traffic Control*, IFAC/IFIP, Versailles, France, 1970.

31. Little, J.D., Martin, B.V., and Morgan, J.T., *Synchronizing Traffic Signals for Maximal Bandwidth*, Department of Civil Engineering, Massachusetts Institute of Technology, Cambridge, MA, 1969.

32. Inose, H., Okamoto, H., and Yumoto, N., "A Multi-Computer Urban Traffic Control and Surveillance System in Tokyo," in *Traffic Control and Transportation Systems*, American Elsevier Publishing, New York, NY, 1974.

33. Hasegawa, T., "Traffic Control Systems in Japan," in *Research Directions in Computer Control of Urban Traffic Systems*, ASCE, New York, NY, 1979.

34. Albrecht, H. and Phillips, P., "Ein Programmsystem zur verkehrsabhaengingen Signalsteuerung nach dem Verfahren der Signalprogrammbildung (PBIL)," *Strassenbau und Strassenverkehrstechnic*, Heft 240, Bundesminister für Verkehr., Bonn, Germany, 1977.

35. Ministere des Transports, *Traitement de la Saturation: Approache Theorique et Applications Practiques*, Center d'Etudes des Transports Urbains, Bagneux, France, 1982.

36. Morrish, D.W., "Area Traffic Control in Bordeaux: A Contrast with British Practice," *Traffic Engineering and Control*, Vol. 21, No. 8/9, August 1980.

Problems

24-1. Two signals are spaced at 1,000 ft on an urban arterial. It is desired to establish the offset between these two signals, considering only the primary flow in one direction. The desired progression speed is 40 mph. The cycle length is 60 s. Saturation headway may be taken as 2.0 s/veh and the start-up lost time as 2.0 s.

(a) What is the ideal offset between the two intersections, assuming that vehicles arriving at the upstream intersection are already in a progression (i.e., a moving platoon), at the initiation of the green?

(b) What is the ideal offset between the two intersections, assuming that the upstream signal is the first in the progression (i.e., vehicles are starting from a standing queue)?

(c) What is the ideal offset, assuming that an average queue of three vehicles per lane is expected at the downstream intersection at the initiation of the green? Assume the base conditions of part (a).

(d) Consider the offset of part (a). What is the resulting offset in the opposite (off-peak) direction? What impact will this have on traffic traveling in the opposite direction?

(e) Consider the offset of part (a). If the progression speed were improperly estimated and the actual desired speed of drivers was 45 mph, what impact would this have on the primary direction progression?

24-2. Consider the time-space diagram in Figure 24.38 for this problem. For the signals shown:

(a) What is the NB progression speed?

(b) What is the NB bandwidth and bandwidth capacity? Assume a saturation headway of 2.0 secs/veh.

(c) What is the bandwidth in the SB direction for the same desired speed as the NB progression speed? What is the SB bandwidth capacity for this situation?

(d) A new development introduces a major driveway which must be signalized between Intersections 2 and 3. It requires 15 s of green out of the 60-s system cycle length. Assuming that you had complete flexibility as to the exact location of the new driveway, where would you place it? Why?

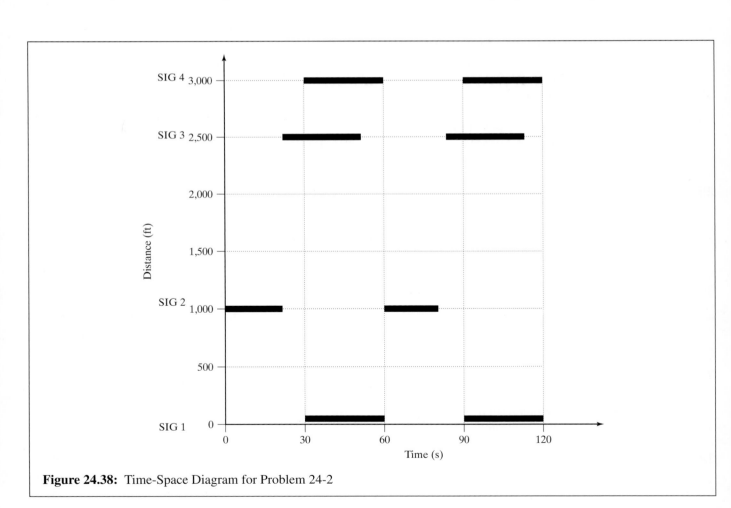

Figure 24.38: Time-Space Diagram for Problem 24-2

24-3. A downtown grid has equal block lengths of 750 ft along its primary arterial. It is desired to provide for a progression speed of 30 mph, providing equal service to traffic in both directions along the arterial.

(a) Would you suggest an alternating or a double-alternating progression scheme? Why?

(b) Assuming your answer of (a), what cycle length would you suggest? Why?

24-4. Refer to Figure 24.39. Trace the lead NB vehicle through the system. Do the same for the lead SB vehicle. Use a platoon speed of 50 ft/s. Estimate the number of stops and the seconds of delay for each of these vehicles.

24-5. Refer to Figure 24.39. Find the NB and the SB bandwidths (in seconds). Determine the efficiency of the system in each direction and the bandwidth capacity. There are three lanes in each direction. The progression speed is 50 ft/s.

24-6. (a) If vehicles are traveling at 60 fps on a suburban road and the signals are 2,400 ft apart, what cycle length would you recommend? What offset would you recommend?

(b) If an unsignalized intersection is to be inserted at 600 ft from one of the signalized intersections, what would you recommend? (See Figure 24.40.)

24-7. You have two intersections 3,000 ft apart and have achieved some success with a 50:50 split, 60-s cycle length, and simultaneous system.

(a) Draw a time-space diagram and analyze the reason for your success.

(b) A developer who owns the property fronting on the first 2,000 ft of the subject distance plans a major employment center. She plans a major driveway and asks your advice on its location. What is your recommendation, and why?

24-8. (a) Consider four intersections, spaced by 500 ft. The platoon speed is 40 ft/s. Recommend a set of offsets for the eastbound direction, considering only the eastbound traffic.

(b) If there are queues of three vehicles at each of the intersections, recommend a different set of offsets (if appropriate).

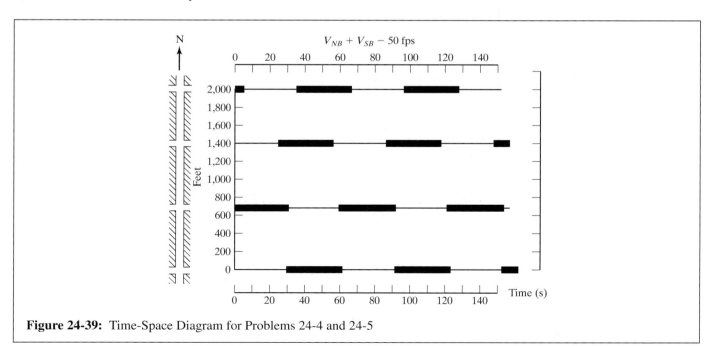

Figure 24-39: Time-Space Diagram for Problems 24-4 and 24-5

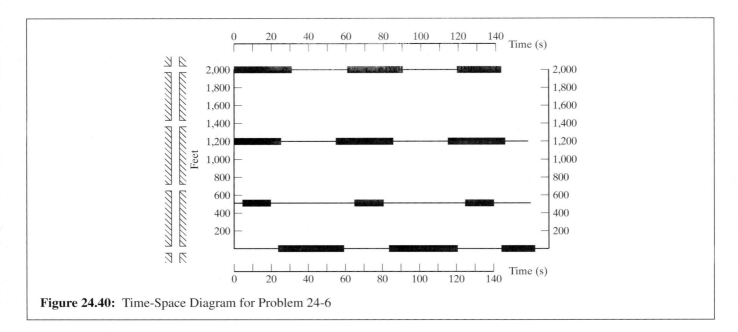

Figure 24.40: Time-Space Diagram for Problem 24-6

24-9. (a) Construct a time-space diagram for the following information and estimate the northbound bandwidth and efficiency for platoons going at 50 ft/s:

Data for Problem 24-9

Signal No.	Offset (s)	Cycle length	Split (MSG first)
6	16	60	50:50
5	16	60	60:40
4	28	60	60:40
3	28	60	60:40
2	24	60	50:50
1		60	60:40

All of the offsets are relative to the preceding signal. All signals are two-phase. There are two lanes in each direction. All block lengths are 1,200 feet.

(b) Estimate the number of platooned vehicles that can be handled nonstop northbound and southbound.

24-10. For the situation in Problem 24-3, design a better timing plan (if possible), under two different assumptions:

(a) Only the northbound flow is important.

(b) The two directions are equally important.

24-11. Find the offset for a link of 1,500 ft, no standing queue at the downstream signal, and a platoon traveling at 40 ft/s. Re-solve: is there is a standing queue of eight vehicles per lane.

24-12. Develop an arterial progression for the situation shown in Figure 24.41. Use a desired platoon speed of 40 ft/s. For simplicity, the volumes shown are already corrected for turns and *PHF*.

24-13. Throughout this chapter, the emphasis was on platoons of vehicles moving through the system, with no desire to stop. However, buses travel slower than most passenger cars and must stop. This problem addresses the timing of signals solely for the bus traffic.

(a) For the situation shown in Figure 24.42, time the signals for the eastbound bus. Draw a time-space diagram of the solution.

(b) Now consider the westbound bus. Locate the westbound bus stops approximately

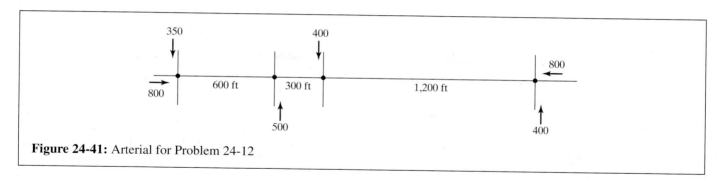

Figure 24-41: Arterial for Problem 24-12

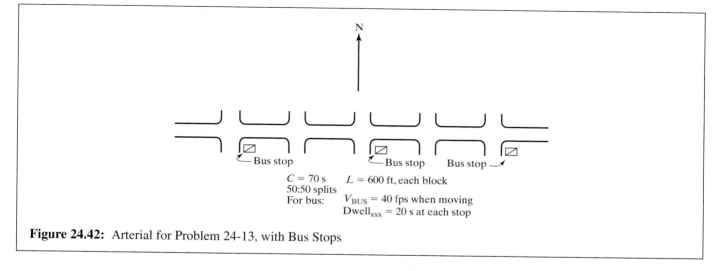

Figure 24.42: Arterial for Problem 24-13, with Bus Stops

every two blocks and adjust the offsets to make the best possible path for the westbound bus, without adversely affecting the eastbound bus. Draw the revised time-space diagram.

(c) Show the trajectories of the eastbound and westbound lead passenger cars going at 60 fps.

24-14. Refer to Figure 24.43. Second street is southbound with offsets of + 15 s between successive signals. Third street is northbound with offsets of + 10 s between successive signals. Avenue A is eastbound, with a + 20-s offset of the signal at Second Street and Avenue A relative to the signal at Third Street and Avenue A. Given this information, find the offsets along Avenues B through J. The directions alternate, and all splits are 60:40,

with the 60 on the main streets (Second and Third Streets).

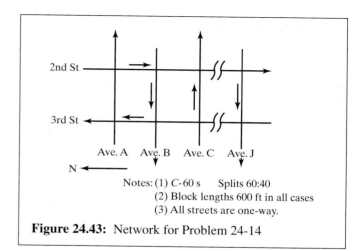

Figure 24.43: Network for Problem 24-14

24-15. Given three intersections, spaced 600 ft apart, each with $C = 60$ s and 50:50 split, find an offset pattern that equalizes the bandwidth in the two directions. [*Hint*: Set the first and the third relative to each other, and then do the best you can with the second intersection. This is a good way to start.]

24-16. A major development is proposed abutting a suburban arterial as shown in Figure 24.44. The arterial is 60 ft wide, with an additional 5 ft for shoulders on each side, and no parking. There is moderate development along the arterial now. Platoons of vehicles travel at 60 ft/s in each direction. The center lane shown in the figure is for turns only. The proposed development is on the north side, with a major driveway to be added at 900 ft along the arterial, requiring a signal. Evaluate the impact of this development in detail. Be specific, and illustrate your points and recommendations.

24-17. Refer to Figure 24.45. Find the unknown offset X. The cycle length is 80 s. The splits are 50:50.

24-18. Given an arterial with 20 consecutive signals, spaced at 1,500 ft with vehicles moving at 50 ft/s, which coordination scheme is the best: simultaneous, alternate, or double alternate? What cycle length should be used?

24-19. Given the information for the indicated arterial, in Figure 24.46, with $C = 70$ s and 50:50 splits:

(a) Plot the time-space diagram.

(b) Find the two bandwidths. Show them graphically, and find the numeric values. If they do not exist, say so.

(c) An intersection is to be placed midway between Intersections 3 and 4, with $C=70$ s and 50:50 split. Recommend an appropriate offset.

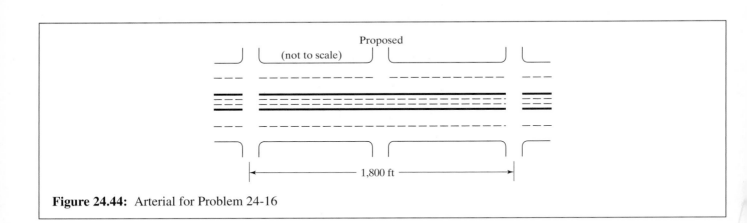

Figure 24.44: Arterial for Problem 24-16

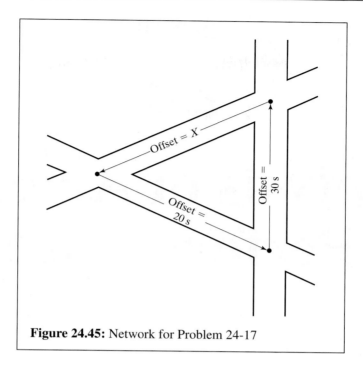

Figure 24.45: Network for Problem 24-17

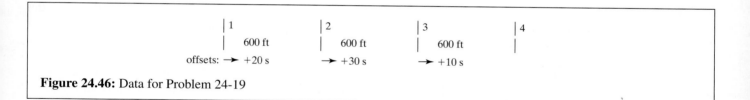

Figure 24.46: Data for Problem 24-19

Analysis of Arterial Performance

This chapter looks at the performance of an arterial as an overall facility. The primary function of an arterial is moving through vehicles; thus, the arterial level of service in the *Highway Capacity Manual* [1] is based on the average travel speed of the through vehicles on the arterial under consideration. This is the basic measure of performance. There is no numeric computation of "arterial capacity," such as is defined at intersections or for freeway segments.

The average travel speed on an arterial section may be determined from:

$$ART\ SPD = \frac{3,600L}{\left(\sum_i RT_i * L_i \right) + \left(\sum_i d_i \right)} \quad (25\text{-}1)$$

where:

$ART\ SPD$ = arterial average travel speed, mi/h
L = length of the arterial section under study, mi
RT_i = total running time per mile on segment *i* of the arterial, s
L_i = length of segment *i* of the arterial, mi

d_i = control delay at signalized intersection *i* in the lane group that includes the through movement, s

The analysis of arterials involves both *segments* and *sections*, although this particular language is no longer used in the HCM. Figure 25.1 illustrates the definition of an arterial segment. An arterial segment is a one-way segment of arterial that begins just downstream of a signalized intersection and extends through (and includes) the next downstream signal. Thus, an arterial segment contains *one* signalized intersection at its downstream end. An arterial *section* is a series of adjacent segments that are analyzed as a unit.

The average arterial speed, defined by Equation 25-1, includes the control delay at signalized intersections and the running time between signalized intersections. While the delay component ideally includes only through vehicles, the reality is that control delay at intersections is predicted only for lane groups. In cases involving shared lane groups, the lane group that includes through vehicles also includes some turning vehicles. Thus, the delay component may not be completely devoid of delay to some turning vehicles.

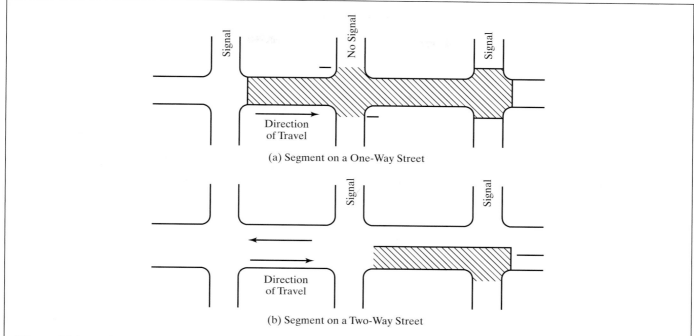

Figure 25.1: Definition of an Arterial Segment (Used with permission of Transportation Research Board, *Highway Capacity Manual*, 4th Edition, Washington DC, Exhibit 15-13, pg. 15-13.)

It should be noted that the most significant value of the HCM procedure is in the methodological approach and the level-of-service definitions. Specific values for average travel speed can be obtained from a number of sources, including local data, validated simulations, or default estimates from the HCM itself. Field data should be considered as the best and most accurate alternative.

The average travel speed is strongly influenced by the number of signals per mile. On a given facility, such factors as inappropriate signal timing, poor progression, and increasing traffic flow can substantially degrade the arterial LOS. Arterials with high signal densities are even more susceptible to these factors.

Unlike most other facility types in the HCM, arterials are divided into four different class types, depending upon the functional and design category that best describes the arterial in question. The arterial levels of service are defined according to the class of arterial, as shown in Table 25.1 on the next page.

25.1 Determining Arterial Class

Arterial class is a categorization of arterials based on signalized intersection spacing, free flow speed, and location. For different arterial classes, drivers have different expectations of performance, and thus the level of service breakpoints vary by class. In the HCM 2000, only arterials and two-way, two-lane rural highways have different classifications. This model is being studied for use with other facility types, such as freeways, for future versions of the HCM.

Consider the arterials depicted in Figure 25.2. Figure 25.2(a) shows a divided arterial with left-turn bays, longer block lengths, and separation from the local-access traffic. Figure 25.2(b) shows an arterial with one lane in each direction, much shorter blocks, and a narrower right of way. In this latter case, some would question whether the street is an arterial. However, there are such streets that do carry sufficient through traffic to be so classified.

Table 25.1: Arterial Level-of-Service Defined

Urban Street Class	I	II	III	IV
Range of free-flow speeds (FFS)	55 to 45 mi/h	45 to 35 mi/h	35 to 30 mi/h	35 to 25 mi/h
Typical FFS	50 mi/h	40 mi/h	35 mi/h	30 mi/h
LOS	Average Travel Speed (mi/h)			
A	>42	>35	>30	>25
B	>34–42	>28–35	>24–30	>19–25
C	>27–34	>22–28	>18–24	>13–19
D	>21–27	>17–22	>14–18	>9–13
E	>16–21	>13–17	>10–14	>7–9
F	≤16	≤13	≤10	≤7

(Used with permission of Transportation Research Board, *Highway Capacity Manual*, 4th Edition, Washington DC, Exhibit 15-2, pg. 15-3.)

In the two illustrations shown in Figure 25.2 on the next page, drivers logically have different expectations of the service they will be provided on such streets. The "expectation" is influenced by a set of factors (parking, right-of-way width, intersection spacing, pedestrians, and so on) in such a way that is most obviously reflected in the speeds with which a driver is satisfied for the given street.

There are four arterial classes defined in the HCM, with classes that are generally defined as follows [2]:

- *Class I*. Arterials in nonrural areas with free-flow speeds of at least 45 mi/h and a signal density of less than two signalized intersections per mile—or arterials in rural developed areas.

- *Class II*. Arterials with free-flow speeds of 35 to 45 mph and a signal density of 2 to 4.5 signals per mile.

- *Class III*. Arterials with free-flow speeds of 30 to 35 mph and a signal density of at least 4.5 signals per mile.

- *Class IV*. Arterials in urban downtown areas of population 750,000 or more—or arterials with free-flow speeds of 25 to 35 mph and a signal density of 7 to 12 signals per mile.

Determining arterial class is an important first step in doing an arterial analysis. The HCM 2000 provides guidance based on the arterials functional and design categories. Table 25.2 is used for this purpose.

Consider the following example: An arterial serves primarily through traffic, providing connections to area freeways and major trip generators. It is a multilane, undivided arterial, with no parking permitted. Many, but not all, intersections have separate left-turn lanes. The posted speed limit is 35 mi/h, and there is some pedestrian activity. The density of unsignalized entry points and development could be described as moderate. There are about four signals per mile on this arterial. What class of arterial is this?

The dominance of through traffic and service of major generators defines this as a "principal arterial." The other information conforms most closely to the description of an "intermediate" design category. From Table 25.2, this combination of characteristics is considered to be a Class II arterial.

Note that the selection of arterial class can be made using a measured free-flow speed (FFS), directly from Table 25.1. For an arterial, the FFS is defined as the speed of a vehicle when a green wave is present and where light traffic does not inhibit drivers from freely selecting speeds.

25.2 Basic Performance Concepts

25.2.1 Arterial Speed Concepts

There are three relevant speed measures on an arterial:

1. Free-flow speed
2. Running speed
3. Average travel speed

Table 25.2: Determination of Arterial Class

	(a) Functional and Design Categories Defined	
	Functional Category	
Criterion	**Principal Arterial**	**Minor Arterial**
Mobility function	Very important	Important
Access function	Very minor	Substantial
Points connected	Freeways, important activity centers, major traffic generators	Principal arterials
Predominant trips served	Relatively long trips between major points and through trips entering, leaving, and passing through the city	Trips of moderate length within relatively small geographical areas

	Design Category			
Criterion	**High-Speed**	**Suburban**	**Intermediate**	**Urban**
Driveway/access density	Very low density	Low density	Moderate density	High density
Arterial type	Multilane divided; undivided or two-lane with shoulders	Multilane divided; undivided or two-lane with shoulders	Multilane divided or undivided; one-way, two-lane	Undivided one-way, two-way, two or more lanes
Parking	No	No	Some	Significant
Separate left-turn lanes	Yes	Yes	Usually	Some
Signals/mi	0.5–2	1–5	4–10	6–12
Speed limit	45–55 mi/h	40–45 mi/h	30–40 mi/h	25–35 mi/h
Pedestrian activity	Very little	Little	Some	Usually
Roadside development	Low density	Low to medium density	Medium to moderate density	High density

(b) Arterial Class Defined

	Functional Category	
Design Category	**Principal Arterial**	**Minor Arterial**
High-Speed	I	N/A
Suburban	II	II
Intermediate	II	III or IV
Urban	III or IV	IV

(Used with permission of Transportation Research Board, *Highway Capacity Manual*, 4th Edition, Washington DC, Exhibits 10-3 and 10-4, pg. 10-6.)

(a) Illustration of street that is an obvious arterial, designed as such

(b) Illustration of a street that is an arterial only by virtue of amount of through traffic

Figure 25.2: Illustration of Arterials with Clearly Different Driver Expectations

Free-flow speed is the speed most drivers would choose on the arterial if they had green indications and were not part of a platoon) but had to deal with all other prevailing conditions, such as block spacing, contiguous land use, right-of-way characteristics (profile, medians, etc), pedestrian activity, parking, and other such factors.

Three major items characterize an arterial: environment, traffic, and signals. The free-flow speed is determined by the total "environment." The free-flow speed may vary by time of day. An urban arterial can have different free-flow speeds at 6:00 PM and 3:00 AM. However, in most cases this will not influence the identification of arterial class. It does influence how observable free-flow speed is. Free-flow speed should be measured at just the time when all the factors are present, except for the prevailing traffic levels (which are also present, of necessity) and red indications. Thus, common advice would be to measure the speed of individual vehicles away from platoons at midblock, away from these influences, as an approximation of free-flow speed.

Running speed, or average running speed, is the average speed of vehicles while in motion (i.e., the speed does not include the impact of stopped delay). It might be 5 to 10 mi/h lower than the free-flow speed.

Average travel speed is the actual speed with the additional effect of signals and all other stops and delays added. It is the measure by which level of service is defined.

25.2.2   Determination of Arterial Speed

Examining Equation 25-1, it can be seen that the time spent on an arterial is composed of two factors: control delay at the signalized intersections and running time between signalized intersections.

Control Delay Components at Signalized Intersections

Control delay for the through vehicles is found from the following equations:

$$d = d_1(PF) + d_2 + d_3 \tag{25-2}$$

$$d_1 = \frac{0.5C\left(1 - \dfrac{g}{C}\right)^2}{1 - \left[\min(1, X)\dfrac{g}{C}\right]} \tag{25-3}$$

$$d_2 = 900T\left[(X - 1) + \sqrt{(X - 1)^2 + \frac{8kIX}{cT}}\,\right] \quad (25\text{-}4)$$

$d_3 = $ delay due to pre-existing queue (see Chapter 21).

where: d = control delay (s/veh)

        d_1 = uniform delay (s/veh)

        d_2 = incremental delay (s/veh)

        d_3 = initial queue delay (s/veh)

      PF = progression adjustment factor

        g = effective green time (s)

       C = cycle length (s)

       X = volume to capacity ratio

       T = duration of analysis period (h)

       k = incremental delay adjustment for actuated control

       I = incremental delay adjustment for filtering/metering by upstream signals

       c = capacity of the lane group (veh/h)

All the factors used in the calculation of delay can be found in the signalized intersection chapter of this book (Chapter 21) except for the upstream filtering adjustment factor, "I." This factor accounts for the effects of filtered arrivals from the upstream signal. If the intersection is isolated, then a value of 1.0 is used. On an arterial, however, vehicles arrive in platoons, filtered from the upstream intersection, thereby decreasing the variance in the number of vehicles that arrive per cycle at the downstream intersection. Table 25.3 shows the HCM recommended values for "I" for non-isolated intersections.

Mid-Block Delay Components

In special cases, there may be mid-block delays due to pedestrian crosswalks or other mid-block locations where vehicles must regularly stop. Such delays may be added as a third term in the denominator of Equation 25-1.

Estimating Running Time on Arterial Segments

The HCM method for estimating running time is shown in Table 25.4 on the next page. It can be seen that segment length and free flow speed are the factors that influence running time in this methodology. There is no accounting for the influence of traffic volume on running time.

Consider the following situation: A Class I arterial segment has a length of 0.30 miles. What is the expected running time for the segment?

From Table 25.4, using the default *FFS* of 50 mi/h, the running time per mile is shown to be 95 s/mi. Since the link is 0.30 miles long, the expected running time for this link is 95 * 0.30 = 28.5 s.

It seems logical, however, that the volume of vehicles on the arterial would also influence the running time. The Florida DOT has sponsored research into running times and speeds [3] and has found this to be true. FDOT is currently using the values based on this research, as shown in Table 25.5 on the next page, in its Quality/LOS handbook and accompanying software for predicting running time. The Highway Capacity and Quality of Service Committee is looking into replacing its table with Florida's version.

Note that the Florida model uses only free-flow speed to categorize the arterial and tabulates running

Table 25.3: Recommended "I" Values for Lane Groups with Upstream Signals

	Degree of Saturation at Upstream Intersection, X_u						
	0.40	0.50	0.60	0.70	0.80	0.90	≥ 1.0
I	0.922	0.858	0.769	0.650	0.500	0.314	0.090

Note: $I = 1.0 - 0.91 X_u^{2.68}$ and $X_u \leq 1.0$.
(Used with permission of Transportation Research Board, *Highway Capacity Manual*, 4th Edition, Washington DC, Exhibit 15-7, pg. 15-8.)

Table 25.4: Segment Running Time per Mile (HCM 2000)

Urban Street Class	I			II			III		IV		
FFS (mi/h)	55[a]	50[a]	45[a]	45[a]	40[a]	35[a]	35[a]	30[a]	35[a]	30[a]	25[a]
Average Segment Length (mi)	Running Time per Mile (s/mi)										
0.05	b	b	b	b	b	b	–	–	–	227	265
0.10	b	b	b	b	b	b	145	155	165	180	220
0.15	b	b	b	b	b	b	135	141	140	150	180
0.20	b	b	b	109	115	125	128	134	130	140	165
0.25	97	100	104	104	110	119	120	127	122	132	153
0.30	92	95	99	99	102	110	d	d	d	d	d
0.40	82	86	94	94	96	105	d	d	d	d	d
0.50	73	78	88	88	93	103	d	d	d	d	d
1.00	65[c]	72[c]	80[c]	80[c]	90[c]	103[c]	d	d	d	d	d

Notes:

(a) It is best to have an estimate of *FFS*. If there is none, use the table above, assuming the following default values:

For Class	FFS (mi/h)
I	50
II	40
III	35
IV	30

(b) If a Class I or II urban street has a segment length less than 0.20 mi, (a) reevaluate the class and (b) if it remains a distinct segment, use the values for 0.20 mi.

(c) For long segment lengths on Class I or II urban streets (1 mi or longer), *FFS* may be used to compute running time per mile. These times are shown in the entries for a 1.0-mi segment.

(d) Likewise, Class III or IV urban streets with segment lengths greater than 0.25 mi should first be reevaluated (i.e., the classification should be confirmed). If necessary, the values above 0.25 mi can be extrapolated.

Although this table does not show it, segment running time depends on traffic flow rates; however, the dependence of intersection delay on traffic flow rate is greater and dominates in the computation of travel speed.

(Used with permission of Transportation Research Board, *Highway Capacity Manual*, 4th Edition, Washington DC, Exhibit 15-3, pg. 15-4.)

speed instead of running time. In the short problem previously posed (Type I arterial—50 mi/h *FFS*—0.30 mile segment), the result would be a range of values depending upon the intensity of demand. First, the link length of 0.30 miles converts to 1.00/0.30 = 3.33 signals/mile. Using the "3 signals/mile" category, the expected running speed would range from 46.8 mi/h (at 200 veh/h/ln) to 41.4 mi/h (at 1,000 veh/h/ln), interpolating between a *FFS* of 55 mi/h and 45 mi/h.

The corresponding running times over a link of 0.30 miles would be between:

$$RT = \frac{0.30*3600}{46.8} = 23.1 \text{ s}$$

$$RT = \frac{0.30*3600}{41.4} = 26.1 \text{ s}$$

Both of these values are lower than the 28.5 s suggested by the HCM 2000 methodology.

Table 25.5: Segment Running Speeds in FDOT Q/LOS Handbook (mi/h)

Signal Spacing	Demand (veh/mi/ln)	Free Flow Speed			
		55 mi/h	45 mi/h	35 mi/h	25 mi/h
1 signal per mile	200	55	45	35	25
	400	53.1	43.3	33.1	24.1
	600	51.4	41.8	31.6	23.4
	800	49.3	40.4	30.4	22.7
	1000	47.7	39.8	29.4	22.1
3 signals per mile	200	50.6	42.9	34.1	
	400	49.6	41.5	32.6	
	600	48.4	40.3	31.4	
	800	46.8	39.1	30.4	
	1000	44.7	38.1	29.4	
5 signals per mile	200	46.7	40.6	33.1	
	400	46.1	39.8	32.2	
	600	44.9	38.8	31.2	
	800	43.4	37.5	30.4	
	1000	41.4	36	29.4	
7 signals per mile	200	43.5	38.5	32.2	
	400	43.2	38.1	31.6	
	600	42.1	37.2	30.9	
	800	40.6	35.6	29.8	
	1000	38.6	33.8	28.6	
9 signals per mile	200	41	36.4	31.4	
	400	40.6	36.3	31	
	600	39.6	35.5	30.4	
	800	37.9	34	29.3	
	1000	35.8	31.7	28.1	

(Used with permission of Florida Department of Transportation.)

Example 25-1:

A Class II, 4-lane urban arterial has a free flow speed of 45 mph. It consists of three identical segments of 1,950 feet, the cycle length is 120 seconds, g/C ratios are 0.42, arrival type is 4, with pretimed controllers at all intersections. The demand volume is 887 through vehicles. The *PHF* is 0.925, and the adjusted saturation flow rate at each intersection is 1,900 veh/h/ln. Determine the average travel speed for each segment for the section as a whole.

Solution:

Step 1: Find the delay at each intersection, using Equations 25-2, 3, and 4:

$$d = d_1 \, PF + d_2 + d_3$$

$$d_1 = \frac{0.5C\left(1 - \dfrac{g}{C}\right)^2}{1 - \left[\min(1, X)\dfrac{g}{C}\right]}$$

$$d_2 = 900T\left[(X - 1) + \sqrt{(X - 1)^2 + \frac{8kIX}{cT}}\right]$$

where: $g/C = 0.42$ (given)

$T = 0.25$ h (assumed analysis period)

$C = 120$ s (given)

$$v/c = X = (887/0.925)/(1,900*0.42*2)$$
$$= 0.60$$

$$k = 0.50, \text{ for pretimed signals}$$
$$I = 0.769 \text{ (Table 25.3)}$$
$$c = 1900*2*0.42 = 1,596 \text{ veh/h}$$
$$PF = 0.867 \text{ (See Chapter 21; Table 21.8}$$
$$AT = 4; g/C = 0.42)$$
$$d_3 = 0.0 \text{ s/veh (no preexisting queues assumed)}$$

Note that the v/c ratio for the first intersection of the arterial will be dependent on the segment being analyzed. If before the first intersection in the segment there is no upstream signal, then the I factor should be set to 1.0 for random arrivals. In this case, it is assumed that there is a similar intersection upstream, since its arrival type is 4, and 0.60 is assumed for all the upstream v/c ratios. Then:

$$d_1 = \frac{0.5*120*(1-0.42)^2}{1-[0.60*0.42]} = 27.0 \text{ s/veh}$$

$$d_2 = 900*0.25\left[(0.60-1)\right.$$
$$+ \left.\sqrt{(0.60-1)^2 + \frac{8*0.50*0.769*0.60}{1596*0.25}}\right]$$
$$= 1.29 \text{ s/veh}$$
$$d = 27.0*0.867 + 1.29 + 0.00 = 24.7 \text{ s/veh}$$

This is the expected delay at *each* of the three signalized intersections in the section described.

Step 2: Determine Running Time, RT, using Table 25.4

For a Class II arterial with a free-flow speed of 45 mph and average segment length of 1,950/5,280 = 0.37 mi, the running time per mile is 95.5 s/mi. For a link length of 1,950/5,280 = 0.37 miles, the running time per segment is 95.5 * 0.37 = 35.3 s.

25.3 Sensitivities

25.3.1 The Impact of Signal Spacing on Arterial Performance

Figure 25.3 shows the impact of signal spacing (segment length) on arterial speed (for a constant value of delay

Step 3: Find Arterial Speed on Each Segment

The arterial speed on each segment is:

$$ART\ SPD = \frac{3,600*0.37}{35.3 + 24.7} = 22.2 \text{ mi/h}$$

Because the three segments are identical, this is also the arterial speed for the section as a whole:

$$ART\ SPD = \frac{3,600*(3*0.37)}{(3*35.3)+(3*24.7)} = 22.2 \text{ mi/h}$$

Step 4: Determine LOS, using Figure 25.1

The level-of-service for a Class II arterial with an arterial speed of 22.2 mi/h is LOS C. This is the LOS for each segment, as well as for the section as a whole, as the arterial speeds are uniform throughout.

Using Table 25.5 to find running time would have resulted in an arterial speed of 22.4 mph, with LOS = C. In general, the Florida research [3] found that the HCM was too conservative in predicting running times.

Using Table 25.5:

$$\frac{1}{0.37 \text{ mile}} = 2.7 \text{ signals/mile}$$

Using the "3 signals per mile" category

$$d = \frac{887}{0.925} = 959 \text{ veh/h}$$

3 signals per mile

Demand	FFS: 45 mi/h
800	39.1
959	$38.1 + \frac{1000 - 959}{1000 - 800} \times (39.1 - 38.1) = 38.3$
1000	38.1

$$RT = \frac{0.37 \times 3,600}{38.3} = 34.8 \text{ s}$$

$$ART\ SPD = \frac{3,600 \times 0.37}{34.8 + 24.7} = 22.4 \text{ mi/h}$$

per intersection). As the signal spacing decreases—that is, the number of signals per mile increases—the arterial performance is degraded. As the number of signals increases, additional control delay is added to the arterial travel time. Further, as the link length decreases, running time per mile increases, also adding to the arterial travel time.

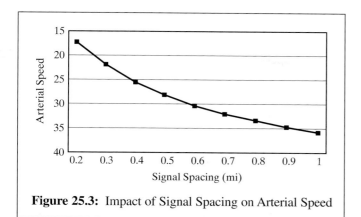

Figure 25.3: Impact of Signal Spacing on Arterial Speed

This is a critical characteristic of arterials. Even if each intersection operates at a high level of service (LOS A, for example), if there are many of them per mile, the arterial level of service can be poor, due to the high signal density.

This can also be seen by rewriting the basic equation for arterial speed, Equation 25-1, in terms of the number of signals on the arterial, as shown later in Equation 25-5. The more signals per mile, the more total delay and running time on the arterial. This translates to a slower average travel speed.

25.3.2 The Impact of Progression Quality on Arterial Speed

Figure 25.4 shows the effect of the quality of progression on arterial speed for a condition in which all other

variables are held constant. As quality of progression degrades, the delay at signalized intersections increases significantly. There is a dramatic improvement when an exceptional progression plan is installed.

Intersection level of service can vary by two levels in some cases, due solely to changes in the quality of the progression. This is not atypical and can be extended to the arterial level of service.

25.3.3 Impact of Cycle Length on Arterial Speed

Figure 25.5 shows the effect of decreasing the cycle length on the arterial speed of a hypothetical arterial, with all other variables held constant.

As seen in Figure 25.5, the impact of cycle length on arterial speed is limited. Cycle length has an impact on intersection delay, both directly and because the v/c ratio will increase as the cycle length gets smaller, given a fixed demand flow rate and green splits.

In a given system, the change of cycle length might have a relatively small effect on arterial speed. That is not to say that choosing a cycle length is unimportant. On the contrary, choosing of the *system cycle length* is very important. It is the interaction of cycle length, block length, and platoon speed on which one should focus. In Chapter 24, there is a detailed discussion on the importance of choosing the correct system cycle length.

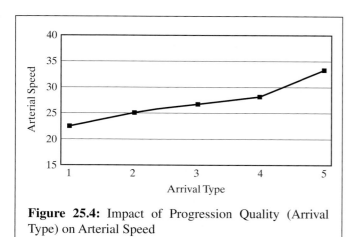

Figure 25.4: Impact of Progression Quality (Arrival Type) on Arterial Speed

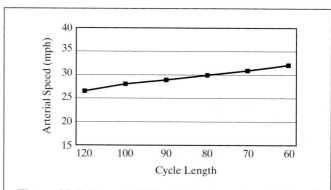

Figure 25.5: Impact of Cycle Length on Arterial Speed

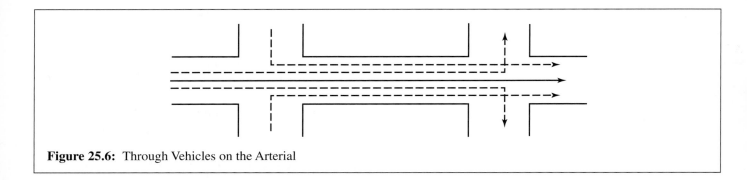

Figure 25.6: Through Vehicles on the Arterial

25.4 Through Vehicles on the Arterial

The HCM defines arterial performance in terms of through vehicles. This is appropriate, given that the primary function of an arterial is to move through vehicles. The true through vehicles, (i.e., the only vehicles that should be included in an arterial analysis), are those vehicles that neither enter nor leave the link, as illustrated in Figure 25.6.

Vehicles that have just entered an arterial (i.e., those that turn in from the upstream intersection), expect a different (and poorer) treatment in the first link, just as they expect a different (and poorer) treatment in the last link as they seek to turn off the arterial. Thus the measure should reflect only the "true through" portion of their trip.

The differences are most interesting. As shown in Figure 25.7, the average travel speed of the through vehicles is 4–5 mi/h higher than the average travel speed of all vehicles on any given link [4].

As noted previously, however, the HCM is forced to compromise this principle, as it is not possible to estimate the delay to through vehicles only at intersections having shared lane groups.

25.5 Arterial vs. Intersection LOS

The arterial level of service reflects solely how good the trip is for through vehicles, over the defined segment. Intersections and arterials serve two distinct purposes and have quality-of-flow measures suited to their function.

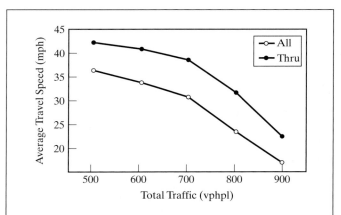

Figure 25.7: Speed of Through Vehicles versus All Vehicles on an Arterial Link (From W. McShane, *et al.*, "Insights into Access Management Details Using TRAF-NETSIM," presented at the Second National Conference on Access Management, 1996.)

The intersection's function is to move vehicles among competing flows past a point. The arterial's function is to move vehicles along a longitudinal distance.

To gain insight, Equation 25-1 can be rewritten in terms of intersections per mile as follows, assuming a constant link length and intersections with similar control delay:

$$ART\ SPD = \frac{3,600}{RT_{mi} + N*d_i} \qquad (25\text{-}5)$$

where: RT_{mi} = running time per mile, s/veh

d_i = control delay per intersection, s/veh

N = number of intersections per mile

An intersection with a stopped delay of 6.4 s per vehicle is certainly level of service A in terms of its function. However, for a Class III arterial with a free-flow speed of 35 mph and 10 intersections per mile, the arterial travel speed is 3,600/[145 + (10)(6.4)] = 17.2 mi/h. This could not be rated A under any circumstances.

The engineer or planner must now present an analysis that shows 10 consecutive intersections at LOS A, resulting in an overall trip (in this example) with LOS D.

This can be explained only in terms of the function of the two traffic facilities, as presented above. This is testimony to the importance of signal spacing: it can and does affect the quality of the arterial trip.

25.6   Design Implications

The preceding discussion leads to an interesting implication: even great intersection performance may not be good enough to obtain good arterial performance, if there are too many signals per mile. Stated in another way, for even reasonable numbers of signals per mile, it is true that good arterial performance requires outstanding intersection performance in many cases.

Consider an existing arterial with observed running speed of 40 mph and five signals per mile, determined to be a Class I arterial. Local officials desire the arterial to let drivers average between 30 and 35 mi/h. Using Equation 25-5 and noting that a running speed of 40 mph implies a running time of 3,600/40 = 90 s per mile:

$$35 \text{ mi/h} = \frac{3,600}{90 + (5)(delay)}$$

Each intersection must have (on average) 2.57 s of control delay per vehicle for the lane group containing the through traffic. For an average travel speed of 30 mi/h, the answer would have been 6.0 s/veh; for 32.5 mi/h, 4.15 s/veh.

There are only two ways to assure such performance in most circumstances: (1) take green time from the side street to give to the main street, with consequent delay and inconvenience to the side-street vehicles, and (2) design the arterial to achieve the target intersection delay (for the arterial through vehicles) by some combination of signal coordination, number of lanes, and/or other factors.

The simple fact is that what would have been looked upon as *over*-design, based only on the operation of an intersection, has become necessary to achieve good arterial operation.

25.7   Summary

This chapter has discussed the methodology for analyzing the performance of an arterial as defined in the *Highway Capacity Manual*. The authors recognize that in the real world, arterials are often analyzed by using alternative simulation software programs, such as COR-SIM and SYNCHRO and many others.

Additionally, this chapter looked at the analysis of an arterial only from the point of view of the automobile. In Chapter 26, the planning and design of an arterial is discussed from a multimodal perspective.

References

1. *Highway Capacity Manual*, Transportation Research Board, Washington DC, 2000.

2. *Florida Quality of Service and Level of Service (QLOS) Handbook*, Florida Department of Transportation, 2000.

3. Prassas, E., "Improved Running Times for HCM Table 11-4 and Related Observations on Average Travel Speed," *Transportation Research Record 1678*, Washington DC, 1999, pgs. 9–17.

4. McShane, W., *et al.*, *Insights into Access Management Details Using TRAF-NETSIM*, presented at The Second National Conference on Access Management, August 1996.

Problems

25-1. An arterial has a free flow speed of 45 mph and an observed running time of 75 s per mile. Each of its intersections has 4.5 s of control delay per vehicle for the through-lane group. For eight signals per mile, compute the arterial average travel speed and the arterial level of service. Do the same for 12 signals per mile.

25-2. An arterial has the same traffic demand, phasing, and splits on its northbound and southbound directions. Yet field studies show markedly different average travel speeds (and thus levels of service) in the two directions. Specify the most obvious reason(s) for this.

25-3. A 2.4-mi divided multilane urban street is a Class II arterial and has eight signalized intersections at 0.30-mi spacing. The arterial has the following characteristics:

Field-measured $FFS = 39$ mph, cycle length is 70s, g/c ratio is 0.60 at every cycle, lane group capacity is 1,850 vph, arrival type is 5 for all segments (pretimed signals). The v/c ratios at each intersection are as follows: 0.59, 0.61, 0.61, 0.57, 0.59, 0.60, 0.60, and 0.58.

For a one-hour analysis period, find the LOS for each segment and for the entire length of the arterial.

25-4. Re-solve Problem 25-3 if the arrival type at each intersection is 4 instead of 5.

25-5. A field study has shown the information in Table 25.6 along a northbound one-way arterial:

The special mid-block delay is a pedestrian crossing. The arterial has been found to be Class I. Evaluate the arterial level of service, by segment and overall, and the intersection levels of service at each intersection. Plot the speed profile along the arterial, and show the overall travel speed on the same plot. The length of each link is 0.2 mi.

25-6. Consider the eastbound direction of a Class II arterial, with 28 mph running speed and the following information at each signal for the lane group containing the through traffic:

$$C = 70 \text{ sec}$$
$$g/C = 0.60$$
$$X = 0.80$$
$$c = 1950$$

Table 25.6: Data for Problem 25-5

Block	Running Time (s)	Intersection Approach Delay (s/v)	Mid-Block Special Delays (s/v)
1	18.0	5.4	–
2	21.5	7.9	–
3	19.3	10.1	–
4	23.0	10.1	15.0
5	22.0	10.1	–
6	19.7	8.3	–
7	17.4	6.0	–

The arterial segments are 0.15 mi each. The signals are pretimed.

Investigate the effect of the quality of progression by considering all possibilities and tabulating the arterial segment average travel speed and LOS that result, as well as the intersection stopped delay and LOS.

25-7. A number of test car runs in each direction on a two-way arterials yields the information in Table 25.7:

The travel times are cumulative and include all delay time. All segments are 0.20 mi long. The observed free-flow speed is 48 mph.

Determine the level of service in each segment and overall for each direction.

25-8. For the following arterial section shown in Figure 25.8, find the arterial speed and level of service for each segment and for the section as a whole. A review of existing conditions in Table 25.8 suggests that the arterial could be reasonably classified as a Class II or a Class III arterial. What impact does this choice have on the final results?

Table 25.7: Data for Problem 25-7

Northbound		
Segment	**Cumulative Avg. Travel Time (s)**	**Average Control Delay (sec/veh)**
NB Start — →	0.0	
1	23.0	4.1
2	47.2	6.2
3	82.4	13.2
4	122.9	17.3
5	161.8	16.1
6	196.3	12.7
7	226.4	9.3
8	255.1	8.2
9	281.8	6.7
10	307.5	5.9

Southbound		
Segment	**Cumulative Avg. Travel Time (s)**	**Average Control Delay (sec/veh)**
SB Start — →	0.0	
10	37.0	14.6
9	72.1	13.7
8	111.5	17.1
7	149.9	16.3
6	194.2	20.2
5	235.3	17.8
4	276.9	19.7
3	320.9	21.7
2	366.3	22.8
1	414.0	24.2

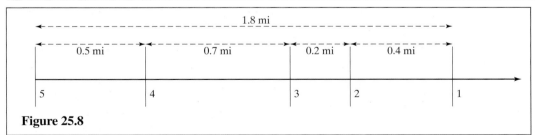

Figure 25.8

Table 25.8: Data for Problem 25-8

Int	Delay on EB TH Lane Group to Intersection
1	15.4 s/veh
2	31.2 s/veh
3	12.8 s/veh
4	13.5 s/veh

CHAPTER
26

Arterial Planning and Design

Arterials are used by many varied modes of travel. At a planning level, it is important to evaluate the impacts of a project on the quality of service provided to users of all modes. On an arterial, the usual modes recognized are automobile, transit, pedestrian, and bicycle.

26.1 Arterial Planning Issues and Approaches

Highway capacity and quality of service can be viewed to exist at three levels: operational, conceptual, and generalized.

At an operational level, very detailed analyses are made, often using simulation packages to meet those needs. Particularly for arterials, where the interaction between signalized intersections is important as vehicles move in platoons down the arterial, programs such as CORSIM [1] and SYNCHRO [2] are needed to evaluate the arterial's operational quality.

Generalized planning is applicable for broad applications such as statewide analyses, initial problem identification, and future-year analyses. Generalized planning

is applicable when the desire is for a quick, "in the ball-park" estimate of LOS; it makes extensive use of default values. Service volume tables are generally used for these estimates of LOS. Florida DOT's Quality/LOS Handbook [3] publishes service volume tables that are presently the most frequently used service volume tables in the United States.

Conceptual planning (preliminary engineering) is done when there is a need to obtain a good estimate of a facility's LOS without doing detailed, comprehensive operational analyses. Examples of conceptual planning include:

- Trying to reach a decision on an initial design concept and scope for a facility
- Conducting alternative analyses (e.g., a four-lane versus a six-lane arterial)

In the next section, LOS planning methodologies are provided for each of the modes commonly using the arterial. When an analysis is performed, the methodologies should be solved simultaneously, giving LOS results for each mode.

26.2 Multimodal Performance Assessment

26.2.1   Bicycle Level-of-Service

For bicycle LOS, one of the best methodologies is the model developed by Sprinkle Consulting, Inc. (SCI). It is currently being used across the country, with excellent results. The operational model can be found in Reference [4]. Herein we present the planning model, which is a simplified version of the complete model. Bicycle LOS is based on the following equation:

$$BLOS = 0.507 \ln\left(V_{15}/N\right)$$
$$+ 0.199 * SPt * (1 + 10.38 * P_T)^2$$
$$+ 7.066 * \left(1/PR\right)^2 - 0.005 * W_e^2 + 0.76 \quad (26\text{-}1)$$

where $BLOS$ = bicycle LOS points

V_{15} = peak 15-minute automobile volume, vehs

N = number of through lanes

SPt = Speed factor = $1.1199 \ln(S_R - 20)$ + 0.8103

S_R = Average running speed, found in running speed table of Chapter 25 (but if unavailable, may use posted speed limit as surrogate)

P_T = percentage of heavy vehicles in the traffic stream

PR = FHWA's pavement condition rating; if the pavement condition is undesirable then $PR = 4.5$; if the pavement condition is typical, then $PR = 3.5$; and if the pavement condition is desirable, then $PR = 2.5$

W_e = Average effective width of outside lane, where W_e is found as follows:

$ADT = (V_{15} * 4)/(K * D)$: Then if $ADT > 4,000$, $W_e = W_t$; and if $ADT \leq 4,000$, then $W_e = W_t * (2 - 0.00025 * ADT)$, where W_t is the width of the outside

lane. Then, if there is a bike lane add another 10 feet to W_e.

It can be seen that the important factors in bicycle LOS are average effective width of outside lane, vehicle volume, vehicle speed, percent heavy vehicles, and pavement condition. The equation gives a point score on which LOS is based, as shown in Table 26.1

26.2.2   Pedestrian Level-of-Service

To determine pedestrian level-of-service, there are two very different models commonly being used. Both have advantages and disadvantages. The *Highway Capacity Manual* (HCM) model is based on pedestrian volume and calculating pedestrian density to determine LOS. In large cities with very high volumes of pedestrians, this model is most useful. The details of the HCM model is not included here, and the reader is referred to Chapter 19 of the HCM 2000 [5].

In many other areas, where the pedestrian volume is low to moderate, the pedestrian environment determines the quality of service and thus the LOS score for pedestrians. The SCI model [6] presented below is not dependent upon the volume of pedestrians using the facility; rather, LOS is completely dependent upon environmental factors.

The pedestrian LOS is based on the following equation:

$$PLOS =$$
$$-1.2276 * \ln\left(W_{ol} + W_l + 0.2 * OSP + f_{sw} * W_s * f_b\right)$$
$$+ 0.0091 * \left(V_{15}/\text{Int}N\right) + 0.0004 * S_R^2 + 6.0468$$
$$(26\text{-}2)$$

Table 26.1: Bicycle and Pedestrian Level of Service Categories

Level of Service	Score
A	≤ 1.5
B	>1.5 and ≤ 2.5
C	>2.5 and ≤ 3.5
D	>3.5 and ≤ 4.5
E	>4.5 and ≤ 5.5
F	>5.5

where $PLOS$ = pedestrian LOS score

W_{ol} = width of the outside lane, ft

W_l = width of shoulder or bike lane, ft

OSP = percent of segment with on-streett parking

f_{sw} = sidewalk presence coefficient $(= 6 - 0.3Ws)$

W_s = width of sidewalk

f_b = Buffer area barrier coefficient (5.37 for trees spaced 20 ft on center, otherwise 1.00)

V_{15} = peak 15-min vehicle volume, vehs

S_R = average running speed of vehicles, mi/h

The equation for pedestrian LOS gives a point score that is then used to determine the pedestrian LOS grade. Table 26.1 gives the LOS letter grades dependent upon the point score.

The existence of a sidewalk is the most important factor in pedestrian LOS. The other factors in pedestrian LOS are similar to those for bicycles—namely, the effective width of the outside lane, vehicle volume, and vehicle speed.

One of the disadvantages of these methodologies is that the point system that is used to determine LOS is not intuitive—that is, it does not refer to any one measure of effectiveness that the transportation professional, the traveler, or elected officials can readily grasp.

26.2.3 Bus Level-of-Service

For an analysis of bus LOS, the *Transit Capacity and Quality of Service Manual* (TCQSM) [7] covers a complete range of conditions, with full operational analyses.

For a planning analysis, however, the most important factor determining quality and level-of-service for riders is the waiting time for a bus to arrive [3]. FDOT has incorporated this idea into its planning handbook, with the approval and help of the authors of the TCQSM. Table 26.2 shows the service frequency LOS thresholds, based on the number of adjusted buses per hour providing service. Readers might need to look at additional factors that are important in their specific communities—for example, load factor—and are referred to the TCQSM for LOS definitions based on other factors besides waiting time.

To find the adjusted service frequency of buses per hour, the actual number of buses arriving per hour is adjusted by a pedestrian adjustment factor (based on the quality of service to pedestrians on the roadway), by a connection adjustment factor, and by a crossing

Table 26.2: Bus Level-of-Service Criteria

LOS	Adjusted Service Frequency (buses per hour)	Headway	Comments
A	>6.0	<10	Passengers don't need schedules.
B	4.01 to 6.0	10 to 14	Frequent service; passengers consult schedules.
C	3.0 to 4.0	15 to 20	Maximum desirable time to wait if bus is missed.
D	2.0 to 3.0	21 to 30	Service unattractive to choice riders.
E	1.0 to 2.0	31 to 60	Service available during hour.
F	<1.0	>60	Service unattractive to all riders.

adjustment factor, as follows:

$$\text{Adjusted buses} = \text{BusesPerHour} * PedAdj \\ * ConnectAdj * CrossAdj \quad (26\text{-}3)$$

where:

$PedAdj$ = the pedestrian adjustment factor, found in Table 26.3

$ConnectAdj$ = the connection adjustment factor (= 1 if no sidewalk or protective roadway barrier exists, and = 0.9 if one does exist)

$CrossAdj$ = the crossing adjustment factor, found in Table 26.4.

Table 26.3: Pedestrian LOS Adjustment Factors Used to Calculate Bus LOS

Pedestrian LOS	Adjustment Factor
A	1.15
B	1.10
C	1.05
D	1.00
E	0.80
F	0.55

26.3 Preserving the Function of an Arterial

It is essential to learn from prior experience. Perhaps the most important lesson is that a state or regional arterial system can be put in place, only to be degraded over time by localized growth in communities along its route. To preserve the arterial function in the face of such growth, it is necessary to think in terms of:

- *Separating local frictions* from arterial traffic, where "local frictions" include parking, double parking, land access, deliveries, and perhaps local bus service
- *Providing alternate means* in the initial plan to meet all of these future needs, lest the pressure to "accommodate" them in the future become unbearable
- *Allowing for efficient platooned movement* by proper signal spacing on two-way arterials and by use of one-way pairs
- *Removing built-in frictions* such as left turns across competing flows, by use of one-way facilities and

Table 26.4: Roadway Crossing Adjustment Factors

	Conditions That Must Be Met:			
Arterial Class	Median	Number of Mid-Block Through Lanes	Automobile LOS	Crossing Adjustment Factor
I	All situations	2	A or B	
II	All situations	2	A, B, or C	
III	All situations	≤4	A or B	1.05
IV	All situations	≤4	All levels of service	
I	None or Nonrestrictive	≥4	B, C, D, E, or F	
	Restrictive	≥8	All levels	
II	None or Nonrestrictive	≥4	C, D, E, or F	0.80
	Restrictive	≥8	All levels	
III	None or Nonrestrictive	≥4	D, E, or F	
	Restrictive	≥8	All levels	
All cases not included in conditions for factor 1.05 and 0.80				1.00

by design features such as turn lanes in the median (to allow EB lefts to become WB rights, for instance)

This advance planning may influence not only the basic layout of various routes, but also the right of way needed for full development of some facilities (particularly two-way arterials). Refer to Figure 26.1, which shows an innovative and a standard treatment of a shopping center driveway.

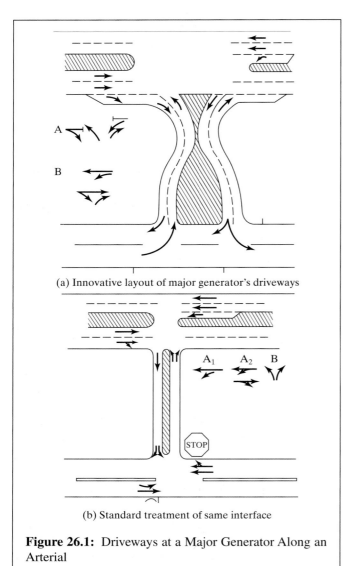

(a) Innovative layout of major generator's driveways

(b) Standard treatment of same interface

Figure 26.1: Driveways at a Major Generator Along an Arterial

Historically, arterials have served "tidal flow." The AM peak period is inbound, and the PM is outbound. Generally, traffic engineers and others think of arterials in these terms, and impacts are assessed (sometimes implicitly) in these terms. In fact, the directional split has been changing over the years. The development of distributed centers in a region ("megacenters," concentrations of 0.5 million square feet or more) can accentuate this, leading to true bi-directional peaks. For two-way arterials, a bi-directional peak means that two-way progressions are essential. This, in turn, means that signal spacing becomes critical and must be taken into account: The standard approach of treating a two-way arterial as a de facto one-way arterial (with tidal flow) will no longer suffice. Requirements in impact assessments have not generally caught up with this trend. However, it is an essential element of future planning.

26.3.1 Design Treatments

In order to accomplish the planning objective, good functional design is essential. Chapter 25 addressed signal-spacing requirements, emphasizing that poor signal spacing will assure that future generations of traffic engineers will be kept busy coping with an inherent problem.

To assure that local activity such as land access, parking, and short trips does not disrupt the arterial function, some arterials now have an "inner" and an "outer" roadway, just as some freeways do; refer to Figure 26.2. This construction (with "service roads") also ensures that the arterial function is defined for the future as well.

Discussions have appeared in the literature [8] for many years concerning "future arterial design," in which limited-access arterials are built as intermediate facilities between freeways and the surface street system. These facilities would include design features to minimize left turns, pedestrian conflicts, and even the number of signals. Such arterials would have the advantage of providing a "transition" from the freeway to the local streets and vice versa, in terms of free-flow speed and quality of flow.

Discussions of future arterial design also focus on one-way pairs in both urban and suburban settings.

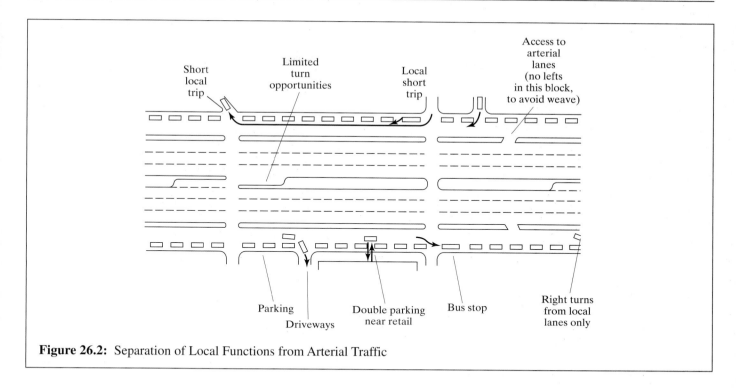

Figure 26.2: Separation of Local Functions from Arterial Traffic

Chapter 19 also presented a number of concepts for limiting the negative impact of left turns on arterial operation, including jug-handles, continuous flow intersections, and various U-turn treatments, as well as unique intersection channelizations that allow the timing of signals in two directions to be decoupled. Reference [9] has additional information.

26.3.2 Reallocation of Arterial Space

Even when some desirable features are not included in the initial design, rehabilitation projects offer another opportunity to greatly enhance the facility for future decades. This is important particularly because much of the road system will routinely be rehabilitated during the careers of those reading this text. In rehabilitation work, much can be done, particularly within rights of way typical of suburban arterials. However, urban arterials and even suburban arterials dominated by strip development often lack the right of way with which to introduce radical changes. Further, there are improvements that can—and sometimes must—be introduced prior to major rehabilitation.

In addition to the obvious remedy of updating the signalization, the next most viable remedy is often the reallocation of the street space. This may take a number of forms, including the following:

- Changing parking regulations to allow curb use by trucks or other vehicles that might otherwise double park
- Removing parking, to provide a moving lane or to provide right-turn bays
- Reducing lane widths, to provide additional lane(s), left-turn bays, or two-way turn lanes

In addition, lane usage may be varied by time of day (i.e., reverse some lane directions) in order to increase capacity in the peak direction.

26.3.3 Other Aspects of Operation

Because arterial management is an overall process and perspective, it is also necessary to look at the individual components, decisions, and elements in the context of the impact on the overall operation.

On the intersection level, the operations aspects that generally deserve the most attention for this purpose are:

- Turn bays (left and right)—their existence and length
- RTOR
- Turn prohibitions
- Arrival patterns of platoons
- Signalization

For example, right-turn-on-red (RTOR) allows some localized delay reductions. However, it might also contribute to an imbalance between the numbers of upstream arterial and turn-in vehicles entering a link, or it might inhibit discharge from a generator (e.g., a parking lot) within a link.

On the arterial level, the operations aspects that generally deserve the most attention for this purpose are:

- Signal spacing
- Parking
- Transit
- delivery activity
- land access function
- special-use lanes

Special-use lanes include HOV lanes (high-occupancy vehicles), reversible lanes, and two-way turn lanes.

Table 26.5 shows the several classes of transportation system management (TSM) actions that might be taken in an overall approach to facility management. In this context, TSM is defined as follows [10]:

TSM is a planning and operating process designed to conserve resources and energy and to improve the quality of urban life. All existing transportation facilities are viewed as elements of a single system; the objective is to organize these elements into one efficient, productive, and integrated transportation system.

The same reference further states that:

TSM ... is an important component of a comprehensive transportation plan. It should be considered before embarking on a capital-intensive set of options.... It should respond to current and future needs.

There is a distinct and important emphasis on (relatively) low-cost improvement and management techniques prior to new construction.

26.4 Access Management

Access management is one of the emerging themes in traffic engineering, with several states creating the explicit legislation and administrative infrastructure needed. A number of the points addressed in the previous section touch on access management issues. This section summarizes some of the operational aspects related to access management.

The primary operation measures that may be taken in access management are:

- Achieving proper signal spacing
- Minimizing conflicts by proper median treatements
  - (a) Two-way left-turn lanes (TWLTL)
  - (b) Restricted or raised medians
- Minimizing frictions by controlling driveway number, placement, and design
- Separating and/or directing flows by use of back streets, side-street access, and lanes divided from the through lanes for local service (frontage roads, in effect)

Much has been said about the first item, and it will not be addressed further in this section other than to note that it is perhaps the primary measure. The several operational measures just listed do have overlaps. For instance, installing a raised median does determine what driveway movements are feasible. Likewise, separating or directing flows to other access points is in effect influencing the number—or existence—of driveways. The same is true of concentrating flows into fewer driveways.

Some comments are in order. Two-way left-turn lanes (TWLTLs) are very popular in some jurisdictions

Table 26.5: Classes of TSM Actions

Class	Strategy Group	Actions
Traffic Management Aimed at improving vehicle movements by increasing the capacity and safety of the existing facilities and systems	Traffic Operations	Intersection and roadway widening One-way streets Turn-lane installation Turning-movement and land-use restrictions New freeway lane using shoulders
	Traffic Control	Local intersection signal improvement Arterial signal system Area signal system Freeway diversion and advisory signing Freeway surveillance and control

	Roadway Assignment	Exclusive bus lane—arterial
		  Take-a-lane
		  Add-a-lane
		Bus-only street
		Contraflow bus lane
		Reversible lane systems
		Freeway HOV bypass
		Exclusive HOV lane—freeway
		  Take-a-lane
		  Add-a-lane

	Pedestrian and Bicycle	Widen sidewalks Pedestrian grade separation Bikeways Bike storage Pedestrian control barriers
Transit Management Designed to increase ridership by providing expanded and more efficient public transportation	Transit Operations	Bus route and schedule modifications Express bus service Bus traffic signal preemption Bus terminals
	Simplified Fare Collection	Marketing program Maintenance improvements
	Transit Management	Vehicle fleet improvements Operations monitoring program
	Inter-Modal Coordination	Park and ride facilities Transfer improvements
Demand Management Oriented toward reducing trips or number of vehicles by encouraging other types of transportation services	Paratransit	Carpool matching programs Vanpool programs Taxi/group riding programs Dial-a-ride

(Continued)

Table 26.5: Classes of TSM Actions (*Continued*)

Class	Strategy Group	Actions
	Work Schedule	Jitney service Elderly and handicapped service Staggered work hours and flex-time Four-day week
Restraint Measures Aimed at discouraging vehicle use mostly through restrictive controls	Parking Management	Curb parking restrictions Residential parking control Off-street parking restrictions HOV preferential parking Parking-rate changes
	Restricted Areas	Area licensing Auto-restricted zones Pedestrian malls Residential traffic control
	Commercial Vehicle	On-street loading zones Off-street loading zones Peak-hour on-street loading prohibition Truck route system
	Pricing	Peak-hour tolls Low-occupancy vehicle tolls Gasoline tax Peak/off-peak transit fares Elderly and handicapped fares Reduced transit fares

(Used with permission of Institute of Transportation Engineers, from ITE Technical Committee 6Y-19. "Planning Urban Arterial and Freeway Systems: Proposed Recommended Practice" *ITE Journal*, Copyright © 1981 Institute of Transportation Engineers.)

because they (1) remove turning traffic from through lanes, (2) do not require radical changes in access to existing land uses, and (3) allow some movements such as lefts from a driveway to be made in two stages—one move to the TWLTL, followed by another merging into the traffic—so that two independent and smaller gaps are needed, not one combined and larger gap.

At the same time, some of the literature has focused on the reality that raised medians have lower mid-block accident experiences than TWLTLs. This is somewhat obvious, if only because the number of conflicts is reduced when the median is raised; refer to Figure 26.3. However, it does not tell the complete story, for in some cases the raised median is not a viable option. Reference

[11] reports that at mid-block locations, the accident rate per million vehicle miles of traffic is:

- 2.43 accidents/MVM for undivided
- 1.66 accidents/MVM for TWLTL
- 1.09 accidents/MVM for restrictive medians

Reference [12] also addresses such differences, including the statistical significance of its findings.

Reference [13] is a 1996 TRB Circular on driveway and street intersection spacing and is especially relevant. It includes a discussion of practices in various states and other jurisdictions. It also includes a discussion of minimum driveway spacings to avoid facing the

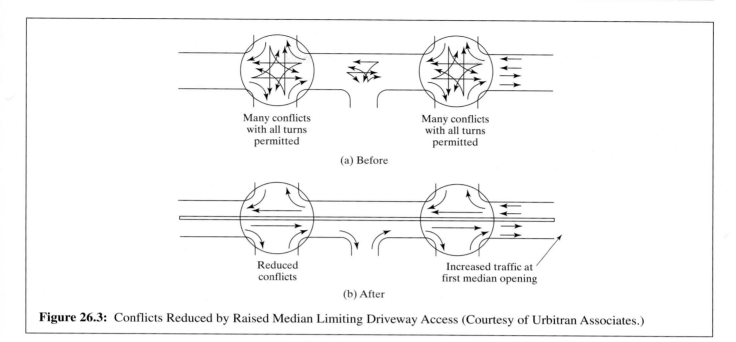

Figure 26.3: Conflicts Reduced by Raised Median Limiting Driveway Access (Courtesy of Urbitran Associates.)

driver with too many overlapping decisions. Refer to Figure 26.4 for one illustration of the problem.

Reference [14] uses simulation to gain insight into the effects of driveways, including eliminating left turns at the driveway and having decel/accel lanes associated with the driveway. Figure 26.5 shows the left-turns-eliminated results on the westbound through traffic for the case indicated, with the initial case being 30% lefts and 70% rights and the second case being the same volume but 100% right turns (as assured by a raised median

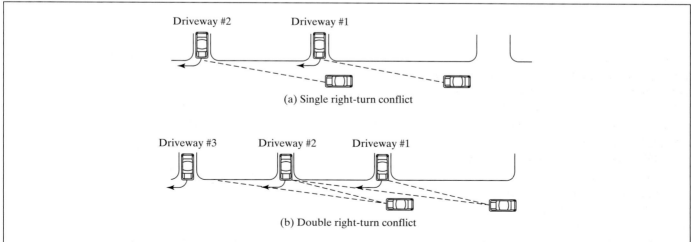

Figure 26.4: Right-Turn Conflict Overlap (Used with permission of Transportation Research Board, *Driveway and Street Intersection Spacing,* Circular 456, Washington DC, 1996.)

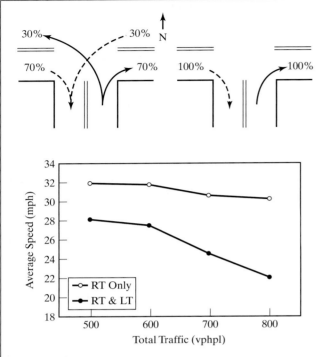

Figure 26.5: The Effect of Eliminating Left Turns at Driveways (From W. McShane, "Insights Into Access Management Details Using TRAF-NET-SIM," presented at the 2nd National Conference on Access Management, 1996.)

or signing, for instance). There was no median storage area in this case, and the results are quite plausible, due to frictions. There was no significant effect on the eastbound through traffic, and it is not shown.

The same reference found that the presence of deceleration lanes into the driveway had a significant (2 mph) benefit for the eastbound through traffic, but that an acceleration lane had little benefit for eastbound through traffic (although it may have benefited the departing driveway traffic itself). Even when there was a +2 mi/h benefit for the through traffic, it must be noted that this recovered only one-third of a −6 mph detriment due to the existence of the driveway.

The question of redirecting flows to back streets, side streets, and/or restricting local access to frontage roads seems straightforward. If it can be done, the number and intensity of effect of driveways on principal arterials can be minimized. Two of the related issues are

(1) do the alternative path and access exist? (2) Are the intersections properly designed to accept the additional load from the "minor" streets, which now include this redirected traffic?

26.4.1 Goods Activity on Arterials

Trucks and other goods-related vehicles have a significant impact on arterial operation, both because of their passenger-car equivalent (pce) effect and because of their pickup and delivery activity.

A number of cities have well-designed back-street or "alley" systems for goods delivery, zoning requirements on a number of off-street loading bays, and/or curb-use regulations that accommodate truck needs to a large degree. However, in many areas, the basic mode of operation is that truck double-parking is endemic, with trucks parking within some 100 feet of the land use they are serving. If the location of the land use is such that the truck tends to block a moving lane at either the entry or discharge end of a link, the impact on the arterial traffic can be very significant. Indeed, it is equivalent to the short-term loss of one lane, with associated turbulence as vehicles move around the obstacle. For v/c ratios already close to 1.00, this can be a severe impact.

A number of observations are in order:

- There is a growing body of knowledge on trip generation rates of various land uses (retail, office, residential) for truck activity, with the rate often expressed in terms of trips per 100 sq ft of space for a given land use. Within retail, the rate varies significantly by business (e.g., fast food, shoe store, department store).

- There are associated characteristic time-of-day distributions and duration-of-stay distributions for each land use.

- Combining the rates and temporal distributions with the spatial distribution of land uses and the proclivity of truckers to be in very close proximity to the land use being visited (for security of cargo even more than for convenience), a detailed estimate of truck activity can be obtained.

- Given this estimate of truck activity, a positive program of mitigation (off-street bays, curb-space allocation, and similar measures) can be planned, or the adverse impact can be estimated according to double-parking and lane blockages.

To illustrate just some of these considerations, refer to Figures 26.6 and 26.7, which show (respectively) the temporal distribution of truck trips by time of day and purpose in a 1972 study, and the temporal distribution of trucks as a percentage of the traffic stream in a 1985 study. The profile is (fortunately) rather characteristic: truck activity peaks outside of the general commuting peak hours.

Remedies based on "making the truck traffic go away" are often counterproductive for many reasons. Consider some examples:

- Chasing double-parked trucks away simply puts them in motion, circulating throughout the network and driving up the total VMT as they circle the block.

- Levying of fines is an added cost of doing business but is often inexpensive compared to downtown storage space (in some downtown areas, package

delivery trucks can be stationary for much of the day, with the truck being used as a base of operations).

- Restricting numbers of trucks may not be viable. Some industries simply do not have sufficient on-premises storage to accommodate even one day's inventory of raw materials, for historic reasons in the development of the industry in the particular city.

- Restricting delivery hours to "off hours" may not be viable for the same reason, plus the reality that the business must then be staffed for off-hours acceptance of delivery.

Truck policies that are too restrictive affect the economic viability of the industry within the city or region, thus attaining the ultimate in counterproductivity for a locality—loss of jobs, tax base, and economic activity.

Against this must be balanced traffic operations and efficiency, leading one to a clear conclusion regarding the need for effective advance planning of well-estimated truck needs.

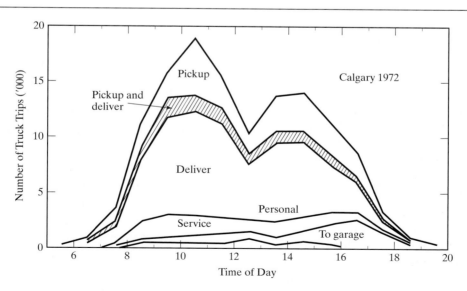

Figure 26.6: Truck Trips by Time of Day, Calgary, 1972 (Used with permission of Prentice Hall, Homburger, W., *Transportation and Traffic Engineering Handbook*, 2nd Edition, Washington DC, 1982.)

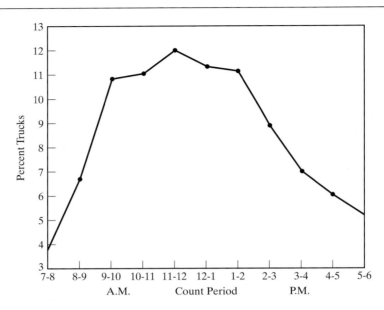

Figure 26.7: Percent of Trucks in the Traffic Stream, Austin TX, 1985 (Used with permission of Institute of Transportation Engineers, Gerard, D., "Truck Operations on Arterial Streets," *Strategies to Alleviate Traffic Congestion*, Washington DC, 1988.)

26.5 Signal Policies

The following are four arterial management issues related to signalization:

- Transitions from one plan to another
- Coordinating multiphase signals
- Multiple and sub-multiple cycle lengths
- Diamond interchanges

26.5.1 Transitions from One Plan to Another

Very little appears in the literature on finding a "best" or optimum way of moving from one plan to another in an orderly and efficient way. Nonetheless, it is an important problem; some engineers feel that the transition from one plan to another during peak loads is more disruptive than having the wrong plan in operation.

The fundamental problem is to get from "Plan A" to "Plan B" without allowing:

- Red displays so short as to leave pedestrians stranded in front of moving traffic
- Such short green displays that drivers get confused and have rear-end accidents as one stops but the other does not
- Excessive queues to build up during excessively long greens
- Some approaches to be "starved" for vehicles due to long red displays upstream, thus wasting their own green

Further, based on lessons learned in one demonstration system, Reference [*16*] reports that:

No more than two signal cycles should be used to change an offset, and the offset should be changed by lengthening the cycle during transition if the new offset falls within 0 and 70% of the cycle length. The cycle should be shortened to reach the new offset if the offset falls within the last 30% (70–100%) of the cycle.

The most basic transition algorithm is the "extended main-street green" used in conventional hardware. At each signal, the old plan is kept in force until main-street green (MSG) is about to end, at which time MSG is extended until the time at which the new plan calls for its termination. Clearly, some phase durations will be rather long. However, the policy is simple, safe, and easily implemented.

Reference [*17*] reported on the test of six transition algorithms (some of which were boundary cases and not field-implementable algorithms) because of the logical importance of effective transitions in computer-controlled systems, which update their plans frequently. Over the range of situations simulated (volume increasing, decreasing, constant), the "extended main-street green" algorithm was no worse than any of the special designs.

26.5.2 Coordinating Multiphase Signals

Multiphase signals (more than two phases) are sometimes required by local policy, dictated by safety considerations (for instance, when lefts must cross a very wide opposite direction), or needed to reduce the pce of the left-turners in the face of an opposing flow.

When multiphase signals exist along a two-way arterial, they actually introduce another "degree of freedom" in attempts to get good progressions in two directions. Consider the time-space diagram of Figure 26.8, with the northbound progression set first, as shown. The usual challenge is to find a "best" southbound progression that does not disrupt this northbound success; the usual constraint is that the southbound and northbound greens must occur at the same time, so that perhaps only part of the "window" can be used for southbound platoon movement. This is illustrated in Figure 26.8.

However, there is some new flexibility. Refer to Figure 26.9, which shows that the SB through green can be "moved around" by the simple expedient of locating the protected left turns in various places. In this particular example, that gives the SB through a flexibility of ±10 seconds relative to the (fixed) NB through-green initiation.

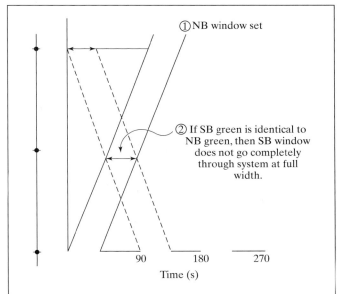

Figure 26.8: Illustrative Problem for Multiphase Signal Coordination

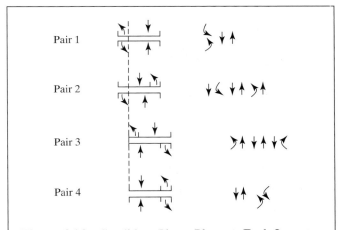

Figure 26.9: Candidate Phase Plans at Each Intersection of a Multiphase Arterial

Observe that the SB window can be made wide by "pushing" the SB through at any intersection to an extreme in most cases. Further, the direction of the "push" generally alternates SB down the arterial (first one extreme and then the other) except for fortuitous spacings.

Thus the actual selected settings will frequently alternate between "Pair 2" and "Pair 3" in Figure 26.9.

Crowley [18] and Messer el al. [19] have both addressed the optimization of signal progressions along two-way multiphase arterials. Messer used his work as the basis for the PASSER program, which implemented this policy.

26.5.3   Multiple and Sub-Multiple Cycle Lengths

Coordination systems operate on the principle of moving platoons of vehicles efficiently through a number of signals. In order to do this, a common cycle length is almost always assumed.

Chapter 24 emphasized that the selection of cycle length is a system consideration. In addition, (1) delay is rather insensitive to cycle-length variation over a range of cycle lengths; (2) capacity does not increase significantly with increased cycle length; and (3) real net capacity is likely to decrease if large platoons are encouraged, because of storage and spillback problems.

Nonetheless, there are situations in which multiple or sub-multiple cycle lengths can work to advantage or when other combinations are necessitated. As a matter of definition, if the system is at $C = 60$ s and one intersection is put at $C = 120$ s, it is a "multiple" of the system cycle length. If on the other hand, the system is at $C = 120$ s and one intersection is put at $C = 60$ s, it is a "sub-multiple" of the system cycle length.

Other combinations are, of course, possible, but the pattern they induce will repeat only after the lowest period (a "supercycle length" so to speak) has passed. For a 60-s and a 90-s pair of cycle lengths, the common period is 180 s. For a 60-s and a 75-s cycle length, the common period is 300 s.

Figure 26.10 shows one of the effects of having a multiple cycle length in the system: The platoon discharged from the greater cycle length (and thus greater phase duration) moves into the downstream link, to be processed in two parts; with very good offset, this could be no problem, for much of the platoon is kept moving. However, for poor offsets, the entire platoon could become a queue, if only for a short time.

A number of situations might give rise to use of multiple or sub-multiple cycle lengths:

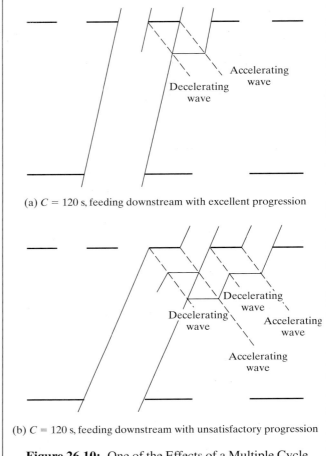

(a) $C = 120$ s, feeding downstream with excellent progression

(b) $C = 120$ s, feeding downstream with unsatisfactory progression

Figure 26.10: One of the Effects of a Multiple Cycle Length in the System

- If there is a very closely spaced diamond interchange somewhere along an urban arterial that has $C = 120$ s (or even $C = 90$ s), then $C = 60$ s might well benefit the diamond operation by keeping platoons small and avoiding internal storage problems and even spillback.

- At a very wide intersection of two major arterials, at least one of which has a $C = 60$ s, the added lost time per phase (due to longer clearance intervals associated with the wide intersection) may present a problem, and it may be best to reduce the number of cycles per hour at this one intersection by increasing the cycle length to $C = 120$ s, if storage permits.

- If turns dictate that multiphasing is required and if the geometry of the overall arterial indicates that the system cycle length should not be changed, then it may be best to allow this one intersection to have a different (greater) cycle length.
- At the intersection of two major arterials, each with its own system cycle length dictated by its own geometry, it may be necessary to accept a different cycle length at the common intersection.

Other examples could be constructed, but these serve for illustration and are representative.

In considering the use of multiple or sub-multiple cycle lengths, attention must be paid to upstream and downstream storage, relative g/C ratios, and the length of the common period (the least common denominator). Some savings can be achieved in particular cases (e.g., reference [20]), but usually at the expense of a markedly more complicated analysis.

26.5.4   The Diamond Interchange

Figure 26.11 shows a sketch of a typical or "conventional" diamond interchange at the juncture of an arterial and a freeway. Such diamond interchanges are relatively inexpensive, do not need signalization for typical initial volumes, and do not consume much space. They are well suited to locations where an interchange is needed for service, but the volumes are modest.

Unfortunately, volumes grow at such locations, due to local development and/or the simple expansion of the urban area. By the time this growth is a problem, the contiguous land has often been developed, and its cost is prohibitive (not to mention the practical and political problems connected with acquiring such land). Thus the option of a total redesign is often not open.

Historically, much of the literature has been concerned with signal optimization at diamond interchanges (see, e.g., References [21] and [22]). The reason is simple: with the space unavailable, signalization is one of the few hopes for coping with the problem.

The inherent problems of a conventional diamond interchange under heavy volumes are considerable; as illustrated in Figure 26.11(b), there are numerous conflicting movements. Further, the bridge over the highway is relatively short (typically 300 ft) and must be used for storage of left-turners during certain phases. Last, volumes from the freeway tend to back up onto the freeway, substantially degrading its performance. All of

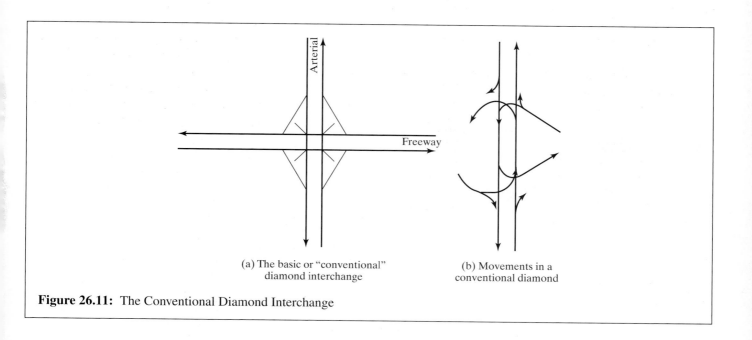

(a) The basic or "conventional" diamond interchange

(b) Movements in a conventional diamond

Figure 26.11: The Conventional Diamond Interchange

this is exacerbated by periodic intersection blockages as queues develop due to the left turns from the arterial onto the freeway.

The true problem is that the initial design did not anticipate the traffic growth, and the initial economic analysis—if any—did not include "life cycle" costs reflecting traffic growth, delay costs, travel-time costs, accident costs, and land and construction capital costs [*23*].

One design alternative that can be considered is the split diamond illustrated in Figure 26.12(a).

- The storage leading off the freeway is greater, for it includes the EW space between the NB and SB sections of the "split" arterial.

- There are no left-turn flows crossing or competing for green with opposing flows. (Refer to Figure 26.12(b).)

- All signals can therefore be two phase, simplifying the operation. (Refer to Figure 26.12(c).)

- The EW storage space and the two-phase operation allow for narrower bridge widths.

- If the split of the arterial is sufficiently wide, drivers may not realize that they are driving through a "diamond" configuration, for the two arterial directions could be considerable distances from each other, and the enclosed land could be fully developed.

Some signal optimization is in order, for the "closure" equations for optimum progressions must be tailored to the geometry and the traffic patterns. There are other diamond configurations and variations on each. However, the conventional and split serve to introduce the principal issues in this rather familiar—and troublesome—configuration.

One of the inescapable lessons of any consideration of the diamond interchange is that the true solution is much more than an optimum signalization. The literature is filled with better signalization schemes, because the given condition in so many localities has always been, "We have a problem, but cannot provide more space: What can we do?" The answer is then, "Patch and cope, using signalization."

For the reader, however, the true solution—and lesson—is to avoid such outcomes, by initial design, if

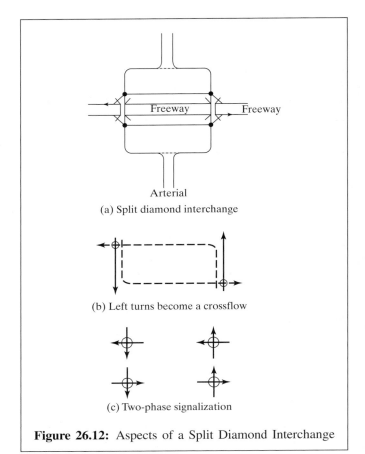

(a) Split diamond interchange

(b) Left turns become a crossflow

(c) Two-phase signalization

Figure 26.12: Aspects of a Split Diamond Interchange

at all possible. Cost-effective alternatives based on life-cycle costs are one method of enhancing the chance of "building in" a better solution.

26.6 Summary

This chapter has covered many different areas of arterial planning and has discussed the role of underlying design principles. It also introduced some information on analyzing arterials as multimodal facilities. This is an important trend in traffic engineering, which is being emphasized by all levels of DOTs. There has been a substantial amount of research on multimodal quality of flow indices, much of it initiated by the Florida Department of Transportation (FDOT). Likewise, there has been considerable work done on the individual modes—vehicular, fixed-route transit, bicycle, and pedestrian. At the same time, this same work has raised fundamental issues on how to establish comparability across the modes, for the comprehension of users, elected officials, and transportation professionals. Should we construct a set of levels of service that are comparable in meaning (excellent to unacceptable, typically) for (a) any interested users to observe and consider in their mode choice or (b) only those users who are inclined (or able) to select the specific mode? The essence of the multimodal approach is to provide realistic choices to travelers. Even employing user perceptions, we must firm up the concept: Which users decide for which modes, or do all users decide for the set of modes? There are still many questions to be answered in regard to multimodal analyses, which will be discussed in the traffic engineering community over the next few years.

One-way street systems and transit operations also affect and are part of arterial operations. These may also occur, however, on local streets and are therefore treated in Chapter 27.

References

1. Florida Department of Transportation, *Quality/Level of Service Handbook*, Tallahassee, Florida, 2001

2. *CORSIM User's Guide*, Federal Highway Administration, 1991.

3. SYNCHRO Plus SimTraffic 5.0, Trafficware Corporation, California, 2001.

4. Landis, B., *et al.*, "Real-time Human Perceptions: Toward a Bicycle Level of Service, *Transportation Research Record 1578*, Transportation Research Board, Washington DC, 1997

5. *Highway Capacity Manual*, Transportation Research Board, National Research Council, Washington DC, 2000.

6. Landis, B., *et al.*, "Modeling the Roadside Walking Environment: A Pedestrian Level of Service" *Transportation Research Record 1773*, Transportation Research Board, Washington DC, 2001

7. *Transit Capacity and Quality of Service Manual.*

8. Levinson, H.S., "Operational Measures—Future," report presented at the 32nd Annual Meeting, Institute of Transportation Engineers, Denver CO, August 1962, published in the *ITE Proceedings*, 1962.

9. Kramer, R.P., "New Combinations of Old Techniques to Rejuvenate Jammed Suburban Arterials," *Strategies to Alleviate Traffic Congestion, Proceedings of the 1987 National Conference*, Institute of Transportation Engineers, Washington DC, 1988.

10. "Planning Urban Arterial and Freeway Systems: Proposed Recommended Practice," Institute of Transportation Engineers Technical Committee 6Y-19, *ITE Journal*, April 1985.

11. Long, G.D., *et al.*, "Safety Impacts of Selected Median and Access Design Features," report to Florida Department of Transportation, Transportation Research Center, University of Florida, Gainsville, FL, 1995.

12. Bowman, B.L. and Vecellio, R.L., "The Effect of Urban/Suburban Median Types on Both Vehicular and Pedestrian Safety," paper presented at the 73rd Annual Meeting of the Transportation Research Board, January 1994.

13. "Driveway and Street Intersection Spacing," *Circular 456*, Transportation Research Board, National Research Council, Washington DC, March 1996.

14. McShane, W., *et al.*, *Insights into Access Management Details Using TRAF-NETSIM*, presented at the second annual conference on Access Management, August 1996.

15. Homburger, W.S., *et al.* (editors), *Transportation and Traffic Engineering Handbook*, 2nd Edition, Prentice Hall, Englewood Cliffs, NJ, 1982.

16. Bissell, H.H., and Cima, B.T., "Dallas Freeway Corridor Study," *Public Roads*, Vol. 45, No. 3, 1982.

17. Ross, P., "An Evaluation of Network Signal Timing Transition Algorithms," *Transportation Engineering*, September 1977.

18. Crowley, K.W., "Arterial Signal Control," Ph.D. Dissertation, Polytechnic Institute, Brooklyn, NY, 1972.

19. Messer, C.J., *et al.*, "*A Variable-Sequence Multiphase Progression Optimization Program*," *Highway Research Record 445*, Transportation Research Board, National Research Council, Washington DC, 1973.

20. Kreer, J.B., "When Mixed Cycle Length Signal Timing Reduces Delay," *Traffic Engineering*, March 1977.

21. Messer, C.J., *et al.*, "Optimization of Pretimed Signalized Diamond Interchanges Using Passer III," *Transportation Research Report 644*, Transportation Research Board, National Research Council, Washington DC, 1977.

22. Messer, C.J., *et al.*, "A Real-Time Frontage Road Progression Analysis and Control Strategy," *Transportation Research Report 503*, Transportation Research Board, National Research Council, Washington DC, 1974.

23. Oh, Y.T., "The Effectiveness for the Selections of Various Diamond Interchange Designs," Ph.D. Dissertation, Polytechnic University, Brooklyn, NY, 1988.

24. "A Recommended Practice for Proper Location of Bus Stops," Institute of Transportation Engineers, Washington DC, 1967.

25. *Evaluation of UTCS/BPS Control Strategies*, U.S. Department of Transportation, Federal Highway Administration, Washington DC, March 1975.

26. Tarnoff, P.J., "The Result of Urban Public Traffic Control Research: An Interim Report," *Traffic Engineering*, Vol. 45, No. 4, April, 1975.

27. Allen, P.V., "*Georgia County Implements Traffic Signal Priority Control Program*" *Public Works*, Vol. 132, No. 12, 2001.

28. Skehan, S., "Traffic Signal Priority For Metro Rapid Buses: The Los Angeles Experience," ITE 2001 Annual Meeting, Compendium of Papers, 2001.

Problems

26-1. Find the level of service for bicycles on a six-lane arterial (three per direction), with 12-ft lanes, 2,400 veh/h demand volume, 6% trucks, and typical pavement conditions. The running speed is 42 mi/h. Find the LOS for the case where a bike lane exists and for the case where a bike lane does not exist.

26-2. What is the level of service for a bus service that is running three buses per hour? The pedestrian level of service on the street is B, the street has a sidewalk. It is a Class-II arterial with two through lanes and automobile level of service of B.

26-3. What is the level of service for pedestrians on an urban street with a sidewalk and the following conditions: a two-lane arterial, with the outside lane being 13 ft wide. The demand volume is 1,800 vph, 4% trucks, 5-ft bike lane, 45 mph running speed, no on-street parking, and a 6-ft sidewalk.

26-4. Re-solve Problem 26-3 if there was no sidewalk.

CHAPTER
27

Traffic Planning and Operations for Urban Street Networks

Various chapters of this text have addressed the design, control, and operation of major traffic facilities and systems. For the most part, however, the focus has been on the movement of through traffic on facilities that are intended to cater to that component of traffic: freeways and expressways, rural and suburban highways, and arterials. The largest mileage of traffic facilities, however, consists of local streets, providing primarily access service in a variety of settings from densely populated urban areas to remote rural areas. This chapter focuses on some of the unique problems involved in the design, control, and operation of local street networks.

27.1 Goals and Objectives

Local street networks provide vital access service to abutting lands. Access is required by both goods and people and must be carefully managed to strike a balance between providing all of the needed service without creating too much congestion.

"Access" service comes in many forms, including:

- On-street curb parking
- Access to off-street parking facilities

- Curb loading zones for commercial vehicles delivering goods
- Access to off-street loading docks for commercial vehicles delivering goods
- Curb space for bus transit stops and taxi stands

Inclusion of appropriate space and locations for these and similar types of facilities must be incorporated into a cohesive plan for a local street network.

As noted, local street networks serve a wide variety of land-use patterns. However, in rural or other sparsely populated areas, traffic engineering problems are minimal, due to low demand. Provision of adequate street widths and access to driveways is generally all that is required for local streets to function well in such areas. Note, of course, that all of the normal safety criteria (such as sight distances) for design must be met for local streets, as for major highways.

The worst traffic engineering problems arise in urban networks, where even local streets can have significant demand levels. Different networks in different situations have, however, different problems. The local streets of Manhattan, for example, are not comparable to the local streets of a residential suburban community. Further, in many areas, local street systems fall into one of three general categories, depending upon the predominant land use(s) being served. The three primary categories of local streets are (1) residential, (2) industrial, and (3) commercial.

The primary objectives of local street system design, management, control, and operation are:

1. The design of the network should reinforce its function and purpose.

2. Adequate allocation of curb space should be provided among the competing uses (e.g. parking, loading and unloading, transit stops and taxi stands).

3. Safety of motorists and pedestrians must be assured.

4. Blockage of local streets to through movement must be controlled and minimized.

5. Appropriate access to and through the local network must be provided for emergency vehicles.

All of these speak to logical concerns and focus on a common sense principle: local streets must be designed, managed, controlled, and operated in a manner that enhances and preserves their intended function.

The sections that follow discuss particular aspects of traffic engineering that affect local street networks.

27.2   Functional Issues

In a perfect world, all traffic on local street networks is either accessing a particular land use at a particular location or circulating while seeking a parking place or a location. There should be little "through" flow (i.e., motorists traveling significant distances along local streets—that is, more than two or three blocks). Large amounts of circulating traffic generally indicate a shortage of parking supply. Excessive amounts of through traffic on local streets generally reflect congestion on nearby arterials, with motorists seeking alternative routes.

The typical urban or suburban local street is 40 ft wide. This accommodates two 8-ft parking lanes (or space for loading zones, transit stops, etc.) and two 12-ft

travel lanes. The typical cross section is intended to minimize right-of-way requirements while providing space for stopped and moving vehicles. With parking on both sides, the design encourages relatively low-speed operation. When two vehicles must pass (in opposite directions), this is somewhat uncomfortable if there are parked cars on both sides of the street. Where local streets serve commercial and industrial areas, they are often wider to accommodate significant truck traffic.

There are two functional issues that dominate traffic engineering on local street networks:

1. The design of local street networks should reinforce their function and purpose as such.

2. Through traffic should be discouraged through a variety of traffic engineering measures.

These functional issues are addressed using a variety of measures.

27.3   Control of Left-Turn Conflicts

The frequency of left-turning movements is a critical element of urban networks. Most intersections of local streets are either uncontrolled or STOP/YIELD controlled. Intersections of local streets with collectors or arterials may be signalized, but the local-street phase rarely has an *exclusive* left-turn phase or left-turn bay. Because of this, many networks in high-density urban areas experience a great deal of congestion due to opposed left turns at signalized intersections within the network.

At a signalized intersection, left turns can be expedited through the use of left-turn lanes and/or signals. Left turns may be prohibited at key locations. For the most thorny network left-turn problems, a one-way local street system may be installed. The latter is discussed in the following section.

Table 27.1 summarizes all left-turn control options, with key advantages and disadvantages noted.

It should be noted that every left turn that is prohibited causes a minimum diversion of an "around-the-block" movement involving three right turns. The benefits of the prohibition must be weighed against the impact of these additional movements.

Table 27.1: Left-Turn Alternatives for Signalized Street Systems

Option	Key Advantages	Key Disadvantages
Two-Phase Signalization	Minimizes lost times; efficient where there are few LTs.	Left turns are opposed; Congestion may appear with many LTs.
Two-Phase Signalization with Left-Turn Lane(s)	Removes waiting LTs from through lanes.	Does not address the limits of opposing flow on LT operations.
Multiphase Signalization with Protected LT Phase and Left-Turn Lane(s)	Provides unopposed LTs; Reduces congestion caused by LTs.	Requires LT lane(s); Increases cycle length; may increase total delay.
Multiphase Signalization with Compound LT Phase and Left-Turn Lane(s)	Minimized cycle length needed to provide for protected LTs; Reduces delay to LTs by allowing permitted LTs throughout the green phase.	Complex phasing is difficult to convey to motorists; may confuse motorists.
Prohibition of Left Turns	Avoids LT problems.	Diverts LTs to other locations where problems may arise; causes driver inconvenience.
One-Way Streets	Removes opposing flow for all LTs.	Requires compatible system geometry; may increase average trip lengths and VMT.

27.4 One-Way Street Systems

One-way street systems represent the ultimate solution to elimination of left-turn conflicts at intersections and the congestion that they may cause. It is hard to imagine, for example, the Manhattan street system operating as two-way streets, yet this was the case until the early 1960s, when NYC traffic engineer Henry Barnes introduced the one-way street concept. For high-density street networks with many signalized intersections, one-way streets are attractive, because:

- Providing signal progression is a relatively simple task with no special geometric constraints and no multiphase signals.
- There are no opposing flows to create operational problems for left-turning vehicles.
- There are related benefits for safety and capacity in many cases.

Safety and capacity benefits should not be overlooked. The elimination of opposed left turns removes

a major source of intersection accidents. A one-way street may allow more efficient lane striping and additional capacity. A 50-ft street width would normally be striped for four lanes, two in each direction, as a two-way street. As a one-way street, it is possible to stripe for five lanes.

One study [1] reported that a conversion of a two-way street system to one-way operation resulted in:

- 37% reduction in average trip time
- 60% reduction in the number of stops
- 38% reduction in accidents

While these results are certainly illustrative, every case has unique features, and these percentage improvements will not be automatically achieved in every case.

A successful one-way street program, however, depends upon having paired "couplets" within close proximity. A pair of streets with similar cross-sections and numbers of lanes must be available so that capacity and travel times in two directions can be balanced. It would also be difficult to operate a couplet where one street is a "main street," with many large trip generators and the paired street is not.

Business owners often object to installing one-way street systems, fearing lost business due to changing circulation patterns. This would clearly be an issue in the pairing of a main street with a back street even if their capacities were similar. Some businesses will be affected more than others. A gas station, for example, that is located on a far corner of two one-way streets, through which turns are no longer made is a typical example of this. Nevertheless, most studies show that one-way street systems in commercial areas generally result in enhanced economic activity.

The most noticeable negative impact of one-way street systems is that trip lengths are increased as illustrated in Figure 27.1.

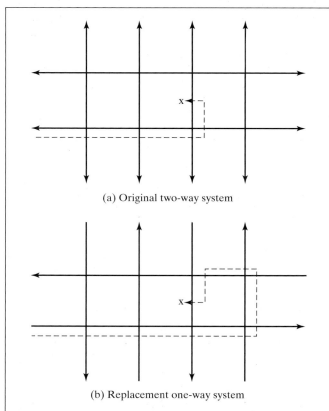

(a) Original two-way system

(b) Replacement one-way system

Figure 27.1: Increased Trip Lengths on a One-Way Street System Illustrated

Another impact of one-way street systems affects transit routing. Bus routes that formerly operated on one street under two-way operation must now use two streets, one block apart, under one-way operation. This is more confusing to users, and lengthens some pedestrian trips from bus stops to the desired destination. It becomes a particular problem when a major generator or generators are located on one of the two streets, such as a subway station or a major department store. In this case, a substantial pedestrian flow may be generated as transit users access the system.

While there are clearly some disadvantages associated with one-way street systems, where widespread network congestion exists and where reasonably paired couplets exist, it is often a simple way of accomplishing many traffic efficiencies. Table 27.2 provides a summary of the advantages and disadvantages of one-way street systems.

27.5 Special-Use Lanes

Special-use lanes on urban streets can be used to improve circulation and flow under a variety of circumstances. The most common forms for use on urban street network are:

- Exclusive transit or HOV lanes
- Reversible lanes
- Two-way left-turn lanes

Exclusive transit lanes are provided where bus volumes are high, and where congestion due to regular traffic produces unreasonable delays to transit vehicles. While a single transit lane can be provided along the right curb of a local street, where sufficient lanes are available, it is recommended that two lanes be provided, as illustrated in Figure 27.2. As such lanes are provided when bus volumes are relatively high, a second bus lane is a virtual requirement to allow buses to pass each other, particularly where there are multiple routes using the facility that do not have the same bus stop locations. The provision of the second lane allows buses to "go around" each other as needed, without encroaching on mixed traffic lanes.

Table 27.2: Advantages and Disadvantages of One-Way Street Systems

Advantages

General Benefits:
Improved ability to coordinate traffic signals
Removal of opposed left turns
Related quality of flow benefits such as increased average speed and decreased delays
Better quality of flow for bus transit; lower transit operating costs
Left-turn lanes not needed
More opportunity to maneuver around double-parked or slow-moving vehicles
Ability to maintain curb parking longer than otherwise possible (due to capacity benefits)

Capacity Benefits:
Reduced left-turn pces
Fewer signal phases (at signalized intersections)
Reduced delay
Better utilization of street width

Safety Benefits:
Intersection LT conflicts removed
Midblock LT conflicts removed
Improved driver field of vision

Disadvantages

Increased trip lengths for some/most/all vehicles, pedestrians, and transit routes
Some businesses negatively affected
Signal coordinated in grid still poses closure problem
Transit route directions now separated by at least one block
For transit routes, a 50% reduction in right-hand lanes; may create bus stop capacity problem
Concern of businesses about potential negative impacts
Fewer turning opportunities
Additional signing needed to designate "one-way" designations, turn prohibitions, and restricted
  entry, as required by MUTCD

Such lanes expedite transit flow and act as an encouragement for travelers to use public transportation as opposed to driving. There two notable disadvantages of such lanes:

- Right-turning vehicles from mixed lanes must be allowed to use the bus lanes at intersections where right turns are permitted.

- Enforcement is difficult; no parking, standing, or stopping can be permitted in the bus lanes.

Transit-only lanes are often heavily criticized by taxi operators. In some cases, disputes are avoided by allowing all high-occupancy vehicles (defined as two or more, or three or more persons per vehicle) to use the lanes. Taxis, however, may cause congestion within the lane by making curbside pickups or drop-offs at undesignated locations. Occupancy requirements are also difficult to enforce on urban streets.

Reversible lanes are used where highly directional traffic distributions exist and there are sufficient lanes on facilities to allow this. Generally, a minimum of five lanes is required for effective reversible lane operation. In such a plan, one or more lanes serve traffic in different directions, according to the time of day. Reversible lanes are best controlled using overhead signals with green and

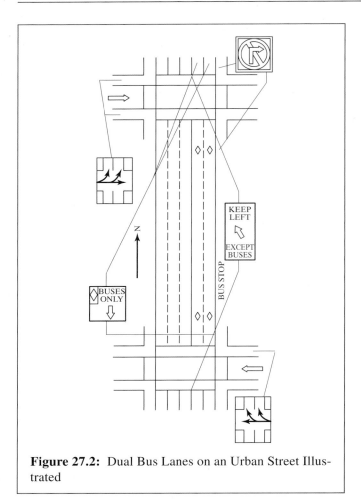

Figure 27.2: Dual Bus Lanes on an Urban Street Illustrated

red Xs to signify which direction is permitted. Such signals should be supplemented by clear regulatory signing, indicating the hours of operation in each direction. Reversible lanes cannot be used where intersections have exclusive LT lanes. The benefit of reversible lanes is that capacity in each direction can be varied in response to demand fluctuations. The most serious disadvantage is the cost to control such lanes, along with the confusion to drivers.

The two-way left-turn lane is provided on two-way urban streets where (1) there is an odd number of lanes, such as 3, 5, or 7; and (2) where mid-block left turns are a significant cause of congestion. When employed, such a lane allows vehicles making a mid-block left turn into or from a driveway to use the center lane as a refuge. In effect, vehicles waiting to make a left turn no longer block a through lane of traffic.

27.6 Managing the Curb

On most urban street networks, curb space allocation is a most contentious and important issue. Curb space is needed for parking, truck loading and unloading, transit stops, taxi stands, and miscellaneous other uses. Failure to provide for effective regulation and enforcement can lead to massive problems, generally involving illegal use of the curb and/or double parking.

Parking is a most complex issue that is dealt with in greater detail in Chapter 11. The need for and usage of curb parking varies with land use and with the supply of off-street parking. In general, curb parking is permitted on most local streets. For visibility around corners, it is customary to prohibit parking within 20 ft of the STOP line. Parking meters can be used to regulate parking duration and turnover. In residential areas, long-term parking is generally permitted and parking meter programs are installed only as a revenue measure. In commercial areas, turnover is encouraged through shorter parking duration limits. Most meters in commercial areas are set for 30-minute to 2-hour durations, depending upon the type of commercial enterprises that are located on adjacent properties. Longer durations would be provided for a furniture store than for a supermarket, for example. For commercial purposes, most parkers wish to be within 400 to 600 ft of their destinations. More detail on acceptable walking distances is found in Chapter 11.

The amount of curb space that must be devoted to transit bus stops depends upon many factors, including bus frequency, dwell times, the type of bus stop (on-line, off-line), signal timing, and others. The *Highway Capacity Manual* [2] defines an algorithm for estimating the capacity of a bus stop:

$$B_S = N_{EB}\left[\frac{3,600 * (g/c)}{t_c + (g/c)t_d + (Z_a c_v t_d)}\right] \quad (27\text{-}1)$$

where: B_S = maximum number of buses per hour that can use the bus stop

   N_{EB} = number of effective bus berths at the bus stop

   g/C = effective green to cycle-length ratio for street on which bus arrives; 1.00 for unsignalized locations

t_c = clearance time between successive buses, s

t_d = average dwell time, s

Z_a = one-tail normal variate corresponding to probability that queues will form behind the bus stop

c_v = coefficient of variation of dwell times

The number of effective bus berths (N_{EB}) is not the same as the physical number of berths provided. As additional curb bus berths are added, the efficiency with which each may be used goes down. The number of effective bus berths is 1.00 for a single-berth bus stop, 1.85 for a two-berth bus stop, and 2.45 for a three-berth bus stop. There is rarely a need for more than two berths on most local streets, but there are exceptions in dense urban areas. The length of a bus berth varies according to the size of buses being used and the type of stop. Specific design guidelines for bus stops are given in Reference [3].

Bus clearance time (t_c) is the time it takes for a bus to start up and travel its own length. In the case of off-line stops, reentry delay must be added to allow the bus to rejoin the right-most travel lane. Where signal queue clearance delays bus reentry, alternative procedures are used. Reference [4] recommends that a total clearance time of between 10 s and 15 s be used. A reasonable guideline is to use $t_c = 10$ for on-line bus stops and 15 s for off-line bus stops.

Dwell time (t_d) is the amount of time a bus spends stopped to serve boarding and alighting passengers. It is best measured in the field, as it varies with the number of passengers served (per bus, at the particular bus stop in question), fare collection systems, and type of doors used. In the absence of field data, recommended default values [5] are 60 s for CBD stops, transit center, major transfer, and major park-and-ride stops; 30 s for major outlying stops; and 15 s for typical outlying stops.

The coefficient of variability in bus dwell times (c_v) depends upon a number of factors as well. If measured, it is defined as the standard deviation in dwell times divided by the average dwell time. A default value of 0.60 is used in the absence of field data.

The value of Z_a depends upon the bus stop failure rate that is considered to be tolerable. "Failure" occurs when a queue forms at a bus stop because more buses than effective berths are present. Recommended values

[6] are 1.96 (2.5% failure rate) to a low of 1.44 (7.5% failure rate) for outlying stops, with 1.04 (15% failure rate) acceptable for CBD bus stops.

Equation 27-1 can be manipulated to determine the number of effective berths needed at any given bus stop:

$$N_{EB} = B_S \left[\frac{t_c + (g/C)*t_d + (Z_a c_v t_d)}{3{,}600(g/C)} \right] \quad (27\text{-}2)$$

where all variables are as previously defined. Table 27.3 gives the number of effective berths required for various bus volumes, assuming a t_c of 10 s for on-line stops and 15 s for off-line stops; dwell times (t_d) of 30 s and 15 s; a Z_a of 1.96 (corresponding to a 2.5% queue failure rate); and a coefficient of variability (c_v) of 0.6.

For an on-line, other outlying bus stop at an unsignalized location, a bus volume of 25 buses per hour would require 1.48 effective bus berths (Table 27.3). This would require two physical bus berths to be provided at this bus stop location.

Truck-loading facilities are best located off-street. Of course, for most residential areas, this is not possible, nor is the frequency of truck loading sufficient to warrant specified zones. In commercial areas, loading zones may be required to avoid severe double-parking blockage. Depending upon the length of the block and the number of deliveries expected, one or two truck loading zones might be set aside. Their length should accommodate the largest expected truck with appropriate clearance for entry and exit movements. For security and convenience, most trucks will not unload more than a few car lengths from the final destination. For this reason, provision of truck loading zones must be carefully considered and vigorously enforced if they are to be effective.

27.7 Traffic Calming

Traffic calming techniques have been under development since the 1970s, with much of the early work done in Europe. With the 1991 Intermodal Surface Transportation Efficiency Act (ISTEA) and its successor act in 1998, the Transportation Equity Act for the 21st Century (TEA-21), new attitudes toward traffic and its management began to develop in the United States.

Table 27.3: Typical Numbers of Bus Berths Required

g/C	Type of Stop	Bus Volume Per Hour					
		10	25	40	55	60	75
0.20	On-Line, Major Outlying	0.71	1.78	2.85	3.92	4.27	5.34
	On-Line, Other Outlying	0.43	1.06	1.70	2.34	2.55	3.19
	Off-Line, Major Outlying	0.78	1.95	3.13	4.30	4.69	5.86
	Off-Line, Other Outlying	0.50	1.24	1.98	2.72	2.97	3.71
0.30	On-Line, Major Outlying	0.75	1.88	3.02	4.15	4.52	5.65
	On-Line, Other Outlying	0.45	1.12	1.79	2.46	2.68	3.35
	Off-Line, Major Outlying	0.82	2.06	3.29	4.53	4.94	6.18
	Off-Line, Other Outlying	0.52	1.29	2.06	2.84	3.10	3.87
0.40	On-Line, Major Outlying	0.80	1.99	3.18	4.38	4.77	5.97
	On-Line, Other Outlying	0.47	1.17	1.87	2.57	2.80	3.50
	Off-Line, Major Outlying	0.87	2.16	3.46	4.76	5.19	6.49
	Off-Line, Other Outlying	0.54	1.34	2.15	2.95	3.22	4.03
0.50	On-Line, Major Outlying	0.84	2.09	3.35	4.60	5.02	6.28
	On-Line, Other Outlying	0.49	1.22	1.95	2.68	2.93	3.66
	Off-Line, Major Outlying	0.91	2.27	3.63	4.99	5.44	6.80
	Off-Line, Other Outlying	0.56	1.39	2.23	3.07	3.35	4.18
0.60	On-Line, Major Outlying	0.88	2.20	3.52	4.83	5.27	6.59
	On-Line, Other Outlying	0.51	1.27	2.04	2.80	3.05	3.82
	Off-Line, Major Outlying	0.95	2.37	3.79	5.22	5.69	7.11
	Off-Line, Other Outlying	0.58	1.45	2.31	3.18	3.47	4.34
0.7	On-Line, Major Outlying	0.92	2.30	3.68	5.06	5.52	6.90
	On-Line, Other Outlying	0.53	1.32	2.12	2.91	3.18	3.97
	Off-Line, Major Outlying	0.99	2.48	3.96	5.45	5.94	7.43
	Off-Line, Other Outlying	0.60	1.50	2.40	3.30	3.60	4.49
1.00 No Signal	On-Line, Major Outlying	1.05	2.61	4.18	5.75	6.27	7.84
	On-Line, Other Outlying	0.59	1.48	2.37	3.26	3.55	4.44
	Off-Line, Major Outlying	1.12	2.79	4.46	6.13	6.69	8.36
	Off-Line, Other Outlying	0.66	1.65	2.65	3.64	3.97	4.96

In its most simple terms, traffic calming is about preserving the function of local streets. While the initial focus of traffic calming activities was on residential local street systems, broadened interest has resulted in applications to other street networks as well. References [6] and [7] represent excellent treatments of the state of the art in traffic calming through 2000.

As has been noted previously, local streets are intended to provide primarily land-access service. When traffic begins to use local streets for through movements, this leads to volume and speed levels that are incompatible with this primary function. Traffic calming is a set of traffic engineering measures and devices that are intended to address these problems. The specific goals of traffic calming are to:

1. Reduce traffic volumes on local streets.

2. Reduce traffic speeds on local streets.

3. Reduce truck and other commercial traffic on local streets.

4. Reduce accidents on local streets.

5. Reduce negative environmental impacts of traffic, such as air and noise pollution and vibrations.

6. Provide a safer and more inviting environment for pedestrians and children.

Most of the objectives of traffic calming are best achieved through design. In fact, most traffic calming techniques are virtual retrofits to a local street system that was improperly designed in the first place. Figure 27.3 shows how a residential local-street design can accomplish most of these goals and how a standard grid design thwarts them.

In the design of Figure 27.3 (a), through movements and high speeds are discouraged by the basic street design. Only the designated collectors (shown in heavy black) have direct access to the arterial system. The local streets involve curved alignments or cul-de-sacs, and do not run continuously in any given direction for more than a few blocks.

An open-grid network does none of this. Because local streets run basically parallel to arterials, they become attractive alternative routes during times of congestion. The straight layout does not discourage high speeds; frequent use of STOP signs is often substituted

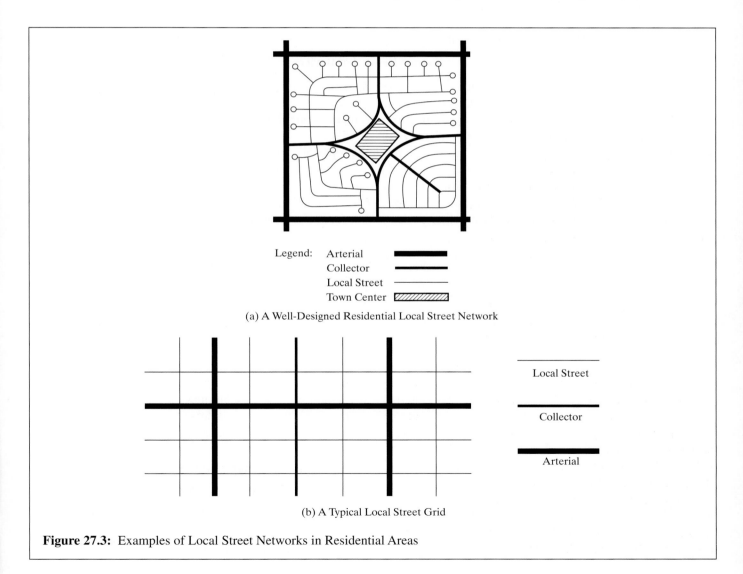

Legend: Arterial
Collector
Local Street
Town Center

(a) A Well-Designed Residential Local Street Network

Local Street

Collector

Arterial

(b) A Typical Local Street Grid

Figure 27.3: Examples of Local Street Networks in Residential Areas

to reduce speeds; this use, however, is technically prohibited by the *Manual on Uniform Traffic Control Devices* [8].

27.7.1 Traffic Calming Approaches

Over the past 20 years, traffic engineering experience has established a number of practices that have been successful in achieving the goals of traffic calming. These measures and devices are generally divided into categories of volume reduction and speed reduction and are summarized in Table 27.4.

Traffic calming devices also contribute to better pedestrian and bicycle environments and reduce accident occurrence and severity. A number of illustrations of the use of these devices follow.

In both of the examples in Figure 27.4, through movement on the local street is physically barred. In (a), two dead-end streets are created with a mid-block barrier. The barrier can be landscaped and may also provide for pedestrian refuge and crossings. In (b), a cul-de-sac is created by blocking the intersection access of the local street. This also creates the opportunity for creative landscaping, and provides for uninterrupted pedestrian movement on one side of the intersection.

Table 27.4: Summary of Traffic Calming Devices

Volume Reduction Devices	
Full Closures	Creation of cul-de-sacs and dead ends by blocking all movements at a designated point on a local street.
Partial Closures	Closing a street to movement in one direction; closing a street to designated movements only.
Semi-Diverters	Devices used to implement partial closures.
Diagonal Diverters	Complete diagonal blockage of an intersection, forcing all traffic to turn.
Median Barriers	Prohibiting designated turning movements and side-street crossing movement; also achieves a narrowing of travel lanes.
Forced Turn Islands	Channelizing islands that force some movements to turn.
Speed Reduction Devices	
Speed Humps	Gentle speed bumps that require drivers to slow; designs must permit normal operation of emergency vehicles.
Speed Tables	A raised segment of the roadway with a flat profile for some distance; must allow normal operation of emergency vehicles.
Raised Crosswalks	Combining a speed table with a pedestrian crosswalk, or a speed table encompassing an entire intersection area.
Textured-Pavement Intersections	Use of textured materials for pavement, creating a rough riding sensation.
Neighborhood Traffic Circles	Small traffic circles fitted within the normal intersection right-of-way.
Roundabouts	Intersection designed to operate as a rotary with flared and channelized approach paths.
Chicanes	Use of devices to create a serpentine or twisting roadway.
Chokers	Devices used to narrow the traveled way at designated locations.
Neckdowns	Use of chokers to narrow intersection approaches at and in the intersection area.
Realigned Intersections	Use of devices to realign the intersection, effectively prohibiting designated movements.

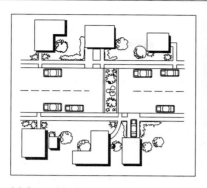
(a) Street blockage creating two dead
ends.

(b) Creation of cul-de-sac by blocking street access
at intersection.

Figure 27.4: Examples of Full Street Closures (Used with permission of the Federal Highway Administration and the Institute of Transportation Engineers, Ewing, R., *Traffic Calming: State of the Art*, Washington DC, August 1999, pg. 24. Photo courtesy of City of Portland Department of Transportation.)

Figure 27.5 illustrates diagonal diverters. These devices require all vehicles to turn as they pass through the intersection and provide for pedestrian refuge and connectivity. Diagonal diverters can be used at many intersections in a grid street network, essentially providing for the type of non-straight streets illustrated in the design of Figure 27.3 (a). They correct the most difficult aspect of grid networks: they are straight and encourage through movements and higher than desirable speeds.

Figure 27.6 illustrates neighborhood traffic circles and roundabouts (or rotaries). Both require drivers to slow down before traversing the intersection, even when intending to go straight through the intersection. Traffic circles and roundabouts have some fundamental differences. While traffic circles have a primary use as a traffic calming measure, roundabouts can handle relatively high traffic volumes when properly designed. A traffic circle usually (but not always) fits within the existing intersection area, while a roundabout involves enlarging the right of way when standard intersections are replaced.

At a traffic circle, normal right-of-way rules prevail, unless specific alternative regulations are posted. In

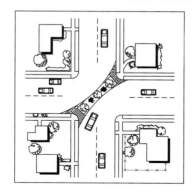

Figure 27.5: Semi-Diverters Blocking Through Movement at an Intersection (Used with permission of the Federal Highway Administration and the Institute of Transportation Engineers, Ewing, R., *Traffic Calming: State of the Art*, Washington DC, August 1999, pg. 27. Photo courtesy of City of Portland Department of Transportation.)

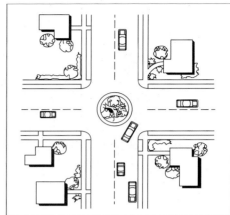

(a) Neighborhood Traffic Circle Illustrated

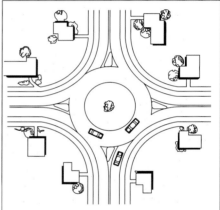

(b) Roundabouts

Figure 27.6: Traffic Circles and Roundabouts (Used with permission of the Federal Highway Administration and the Institute of Transportation Engineers, Ewing, R., *Traffic Calming: State of the Art*, Washington DC, August 1999, pgs. 47 and 48. Photos courtesy of City of Portland Department of Transportation.)

roundabouts, traffic within the rotary always has the right of way over entering vehicles.

In terms of usage, traffic circles are generally intended to reduce speeds and perhaps discourage some through traffic from using the local street. Depending upon the design, the circle can be nicely planted, but is normally not used for pedestrian refuge, as pedestrians continue to cross in the normal pattern in marked or unmarked crosswalks.

The use of chicanes to both narrow a traveled way and to create a serpentine driving path in a straight right-

of-way is illustrated in Figure 27.7. As with most other traffic calming devices, they are intended primarily to reduce speed, but they also discourage through movement and can reduce volumes as well. Continuous use of chicanes can create longitudinal areas in which enhanced environments for pedestrians, bicycles, and residents can be created.

The principal traffic-calming device to reduce speeds is the traffic hump. These devices are made in one of four profiles (sinusoidal, circular, parabolic, and flat-topped) and force drivers to slow down to avoid an

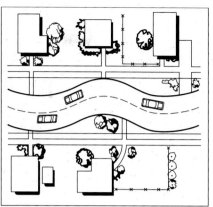

Figure 27.7: Use of Chicanes Illustrated (Used with permission of the Federal Highway Administration and the Institute of Transportation Engineers, Ewing, R., *Traffic Calming: State of the Art*, Washington DC, August 1999, pg. 49. Photo courtesy of City of Portland Department of Transportation.)

unpleasant sensation while going over the hump. While the device is intended to make drivers uncomfortable, they are not designed to damage a vehicle or throw it out of control if the driver crosses it without slowing down. Advance warning signs are always posted. The specific designs of these devices and their location should always be discussed with police, fire, ambulance and rescue services, as their safe operation at elevated speeds must be assured.

Figure 27.8 illustrates a variety of traffic calming devices not discussed in detail herein. All contribute to the general objectives of traffic calming and in the creation of an improved environment for residents, pedestrians, children, and users other than motorists.

It is important that a comprehensive plan for traffic calming be developed for a defined network, as opposed to the sequential placement of individual devices and treatments. Extensive public hearings and involvement of all interested groups is a must if a successful plan is to be developed. As a minimum, such groups must include residents; owners of commercial establishments; customers; police, fire, and emergency service personnel; and local officials.

The goals and objectives of each plan should be carefully crafted and clearly stated. It is also valuable to recognize that overall plans should be implemented on a trial basis, so that impacts can be observed. Changes to

the plan are almost always necessary, as unanticipated impacts are almost always present. The plan, while reducing volume and speed, making streets safer, and improving the environment for pedestrians, residents, and others, must still serve its basic transportation function. It is not feasible to take a major multilane arterial and close it, unless alternative facilities are going to be provided to handle the diverted traffic. While traffic calming devices and approaches are intended to reinforce the function and purpose of local-street systems, they will be far less successful if employed to *change* the function and purpose of a network, such as from an arterial to a local street.

Figure 27.9 provides an illustration of a traffic-calming plan for an urban local-street network serving a primarily residential neighborhood with local stores and merchants present. Note that while many different traffic calming treatments are incorporated into the overall plan, not all street links are directly involved. Nevertheless, there are only one or two paths left where a motorist could drive straight through the area on a local street. These were doubtless carefully chosen and probably represent collector-type facilities where local stores are located. More devices, or course, could be added. However, it is always advisable to use the minimum treatments necessary to achieve the specific objectives of the project.

(a) Sinusoidal Speed Bump

(b) Partial Closure of Street

(c) Crosswalk Combined with Narrowing
with Speed Hump

(d) Raised Pedestrian Crosswalk Combined

(e) Partial Closing at T-Intersection

(f) Partial Closing to Through Movement

Figure 27.8: Miscellaneous Traffic Calming Devices illustrated (Photos courtesy of City of Portland Department of Transportation.)

27.7.2   Impacts and Effectiveness of Traffic Calming Measures

The three most tangible objectives of traffic calming are reductions in speed, volumes, and accidents on local-street networks. Reference [7] contains detailed summaries of results of hundreds of traffic calming projects along with the reductions achieved in these three critical areas. Table 27.5 summarizes some of these results.

The student is cautioned that the data of Table 27.5 is illustrative. Note that some of the sample sizes are relatively small and that all of the average results reflect relatively large standard deviations. It is clear

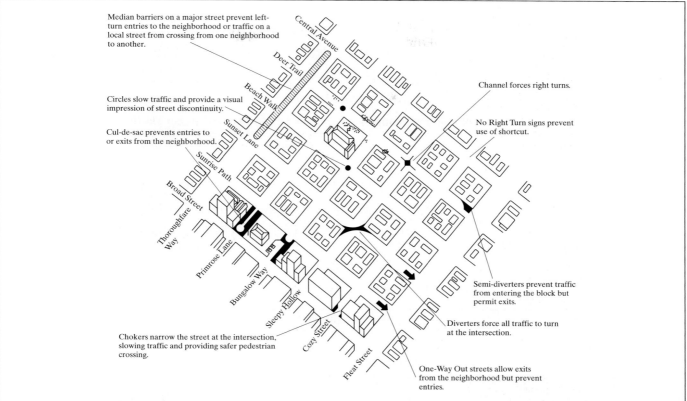

Figure 27.9: Illustration of Traffic Calming Devices Applied to a Neighborhood Grid (Used with permission of Federal Highway Administration, Smith, D.T., *State of the Art: Residential Traffic Management*, FHWA Research Report RD-80-092, December 1980.)

from the data that speed humps and speed tables are devices that are effective in all three areas, reducing speeds, volumes, and accident occurrence. Volume reductions are largest for devices that partially or completely block through vehicles, which is an expected result.

The safety impacts of traffic circles are also impressive, although most of the data represents one city—Seattle. It is also logical, as most local-street accidents occur at intersections. Traffic circles substantially eliminate intersection conflicts, as illustrated in Figure 27.10.

Safety benefits of 43 international studies are even more impressive [9]. Summary results are shown in Figure 27.11. Traffic circles and chicanes each resulted in average reductions in accident frequency of 82%. The

large safety benefit of chicanes is somewhat surprising and appears to relate to the increased driver vigilance and the lower speeds required to negotiate the serpentine path created by the devices. Another interesting result is that physical traffic calming measures produce greater accident reductions than any regulatory measures such as speed limit reductions and STOP signs.

Table 27.6 summarizes a generalized overview of the effectiveness of various traffic calming measures with respect to a number of objectives and operational outcomes. The results come from a comprehensive study conducted in Phoenix AZ.

As noted previously, traffic calming is best built into the design of facilities and networks. However, as the nature of traffic and the surrounding lands served

Table 27.5: Reported Impacts of Traffic Calming Measures

Speed Impacts

Traffic Calming Measure	Average Impact on 85th Percentile Speed		
	Number of Samples	Δ Speed (mi/h)	Percent Decline (%)
12-ft Speed Humps	179	−7.60	22%
14-ft Speed Humps	15	−7.70	23%
22-ft Speed Tables	58	−6.60	18%
Longer Speed Tables	10	−3.20	9%
Raised Intersections	3	−0.30	1%
Traffic Circles	45	−3.90	11%
Narrowings	7	−2.60	4%
One-Lane Slow Points	5	−4.80	14%
Half Closures	16	−6.00	19%
Diagonal Diverters	7	−1.40	4%

Volume Impacts

Traffic Calming Measure	Average Impact on Volumes		
	Number of Samples	Δ Volume (veh/day)	Percent Decline (%)
12-ft Speed Humps	143	−355	18%
14-ft Speed Humps	15	−529	22%
22-ft Speed Tables	46	−415	12%
Traffic Circles	49	−293	5%
Narrowings	11	−263	10%
One-Lane Slow Points	5	−392	20%
Full Closures	19	−691	44%
Half Closures	53	−1611	42%
Diagonal Diverters	27	−501	35%
Other Volume Controls	10	−1167	31%

Safety and Accident Impacts

Traffic Calming Measure	Average Impact on Accident Occurrence		
	Number of Samples	Δ Accidents (Acc.Yr.)	Percent Decline (%)
12-ft Speed Humps	50	−0.33	13%
14-ft Speed Humps	5	−1.73	40%
22 ft Speed Tables	8	−3.05	45%
Traffic Circles	130	−1.55	71%

(Compiled from Reference [7], Tables 5.1, 5.2, and 5.7.)

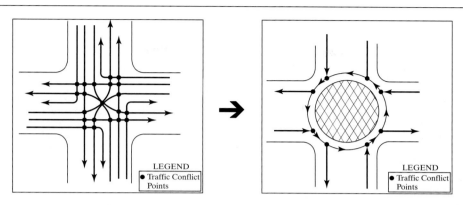

Figure 27.10: Potential Conflicts Reduced by Traffic Circles (Used with permission of Institute of Transportation Engineers, Stein, H., et al., "Portland's Successful Experience with Traffic Circles," *Compendium of Technical Papers*, Washington DC, 1992, pgs. 39–44.)

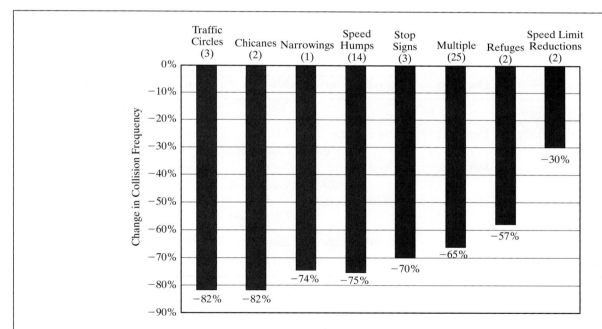

Figure 27.11: Average Reduction in Collisions from International Studies (Used with permission of Insurance Corporation of British Columbia, Geddes, E., *et al.*, *Safety Benefits of Traffic Calming*, Columbia, Vancouver BC, Canada, 1996, pg. 38.)

changes, it is often necessary to reinforce the intended purpose of local street networks by using retrofit measures that separate them from through traffic and other unwanted elements. The state of the art is expanding rapidly, as is the understanding of the impacts of implementing such measures on a broad scale.

Table 27.6: Generalized Assessment of Traffic Calming Measures

Device	Traffic Volume Reduction	Speed Reduction	Noise and Pollution	Safety	Access Restrict.	Emergency Vehicle Access	Maintenance Problems	Level of Violations	Cost
Speed Bumps	Possible	Limited	Increase Noise	No Problem	None	Minor Problem	None	N/A	Low
STOP Signs	Unlikely	None	Increase	Unclear	None	No Problem	None	Potentially High	Low
NO LT/RT Signs	Yes	None	Decrease	Improved	No Turns	No Problem	Vandalism	Potentially High	Low
One-Way Street	Yes	None	Decrease	Improved	One Direction	One Direction	None	Low	Low
Chokers	Unlikely	Minor	No Change	Improved for Peds	None	No Problem	Trucks Hit Curbs	N/A	Mod.
Traffic Circles	Possible	Likely	No Change	Unclear	None	Some Constraint	Vandalism	Low	Mod.
Median Barriers	Yes	None	Decrease	Improved	RT Only	Minor Constraint	None	Low	Mod.
Forced Turn Channel	Yes	Possible	Decrease	Improved	Some	Minor Constraint	Vandalism	Potentially High	Mod.
Semi-Diverter	Yes	Likely	Decrease	Improved	One Direction	Minor Constraint	Vandalism	Potentially High	Mod.
Diagonal Diverter	Yes	Likely	Decrease	Improved	Thru Traffic	Some Constraint	Vandalism	Low	Mod.
Cul-de-Sac	Yes	Likely	Decrease	Improved	Total	Some Constraint	Vandalism	Low	High

(Used with permission of Federal Highway Administration and Institute of Transportation Engineers, Ewing, R., *Traffic Calming: State of the Practice*, Washington DC, Aug 1999, Table 3-4, pg. 20.)

27.8 Closing Comments

This chapter has attempted to provide a general overview of a number of important traffic engineering and management strategies that affect the operation of local street networks. This is not intended to be an exhaustive coverage of all of the topics presented, and the reader is encouraged to consult the literature for more extensive treatments of these topics.

References

1.   Karmeier, D., "Traffic Regulations," *Traffic and Transportation Engineering Handbook*, 2nd Edition, Institute of Transportation Engineers, Washington DC, 1982.

2.   *Highway Capacity Manual*, 4th Edition, Transportation Research Board, National Research Council, Washington DC, 2000.

3. Fitzpatrick, K., Hall, K., Perdinson, D., and Nowlin, L., *Guidelines for the Location of Bus Stops*, *Transit Cooperative Research Program Report 19*, Transportation Research Board, National Research Council, Washington DC, 1996.

4. St. Jacques, K. and Levinson, H., "Operational Analysis of Bus Lanes on Arterials," *Transit Cooperative Research Program Report 26*, Transportation Research Board, National Research Council, Washington DC, 1997.

5. Levinson, H., *INET Transit Travel Times Analysis*, Urban Mass Transportation Administration, U.S. Department of Transportation, Washington DC, April 1982.

6. O'Brien, A. and Brindle, R., "Traffic Calming Applications," *Traffic Engineering Handbook*, 5th Edition, Institute of Transportation Engineers, Washington DC, 1999.

7. Ewing, R., *Traffic Calming: State of the Practice*, Federal Highway Administration and the Institute of Transportation Engineers, Washington DC, August 1999.

8. *Manual on Uniform Traffic Control Devices*, Millennium Edition, Federal Highway Administration, U.S. Department of Transportation, Washington DC, 2000.

9. Geddes, E., *Safety Benefits of Traffic Calming*, Insurance Corporation of British Columbia, Columbia, Vancouver BC, Canada, 1996.

Problems

27-1. The street network shown in Figure 27.12 serves a small downtown street network. The main street (shown as the thick black line) has six lanes, three in each direction. Most but not all of the major generators are located on this street. Two bus routes use this street as well. All other local streets shown have standard 40-ft street widths with metered parking on both sides. There is no parking on the main street. Intersection congestion has

created significant delays throughout the network. A one-way street system and other measures are under consideration. Is a one-way street system an appropriate solution? Why or why not? If it is, propose a specific plan. Propose a control plan for the network that would alleviate some of the existing problems. State any assumptions made.

The spacing between North-South streets is 600 ft. The spacing between East-West streets is 700 ft.

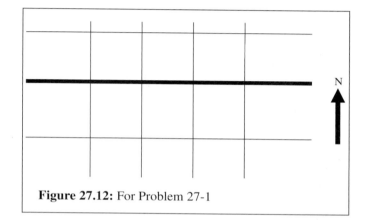

Figure 27.12: For Problem 27-1

27-2. A local street with off-line bus stops (created by a lane of curb parking) services two bus routes, one with 10 buses/hr, the other with 15 buses/hr. If the average dwell time of a bus is 20 seconds, how long would each bus stop have to be to prevent queuing problems? Is this reasonable? If not, are there any potential remedies? Block lengths are 500 ft.

27-3. Class Project

Establish working groups of from three to four people. Each group will select a local street network exhibiting problems that might be addressed by traffic calming measures. The group will survey the network and analyze what specific traffic problems need to addressed. The objectives to be achieved should be clearly stated. A plan of action should be proposed, with each group making a public presentation to the class. If possible, involve local officials or groups in the project, and invite them to the presentations.

Index